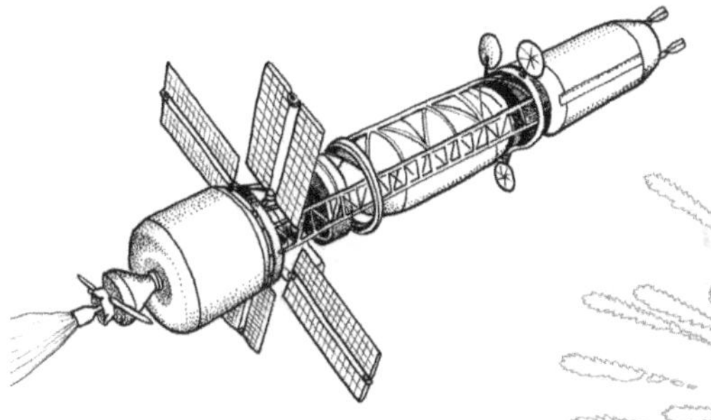

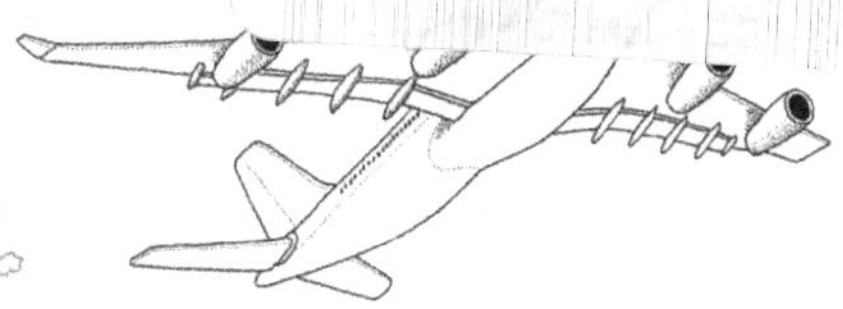

The CHEMICAL ELEMENTS

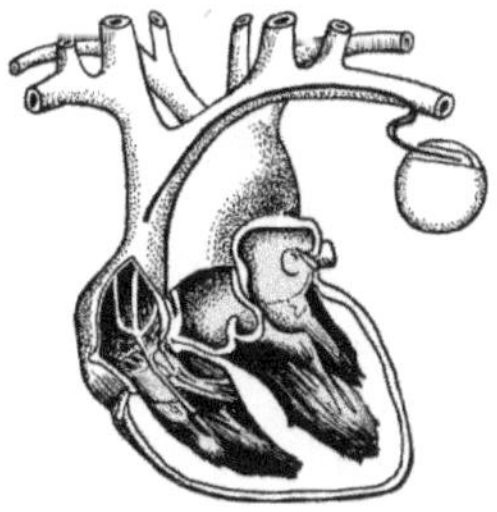

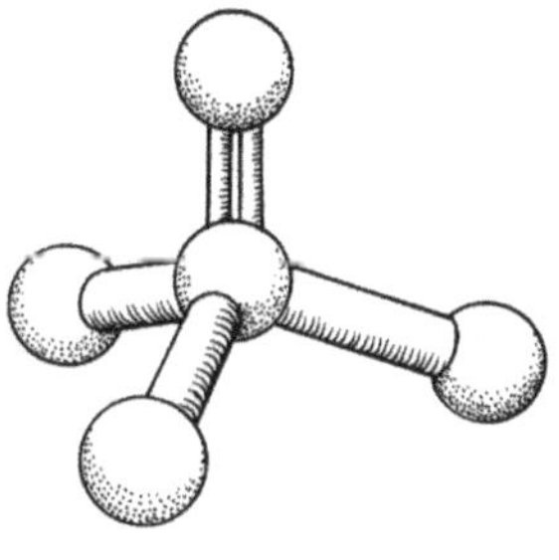

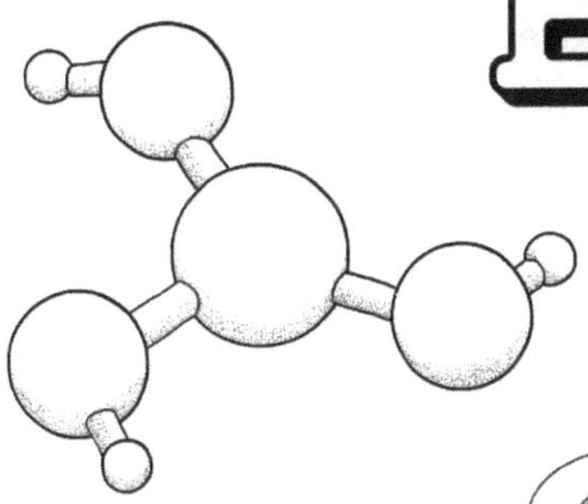

Coloring and Activity Book

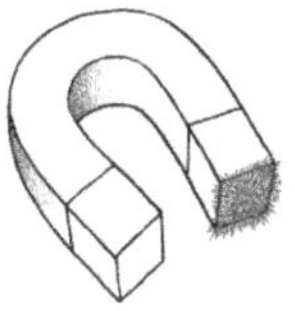

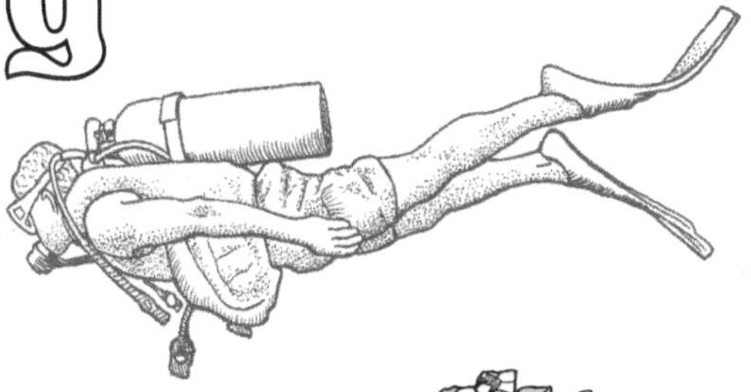

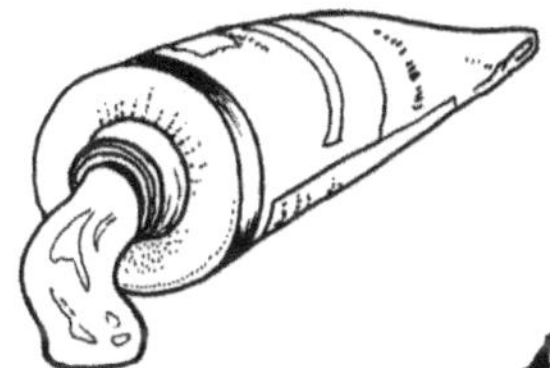

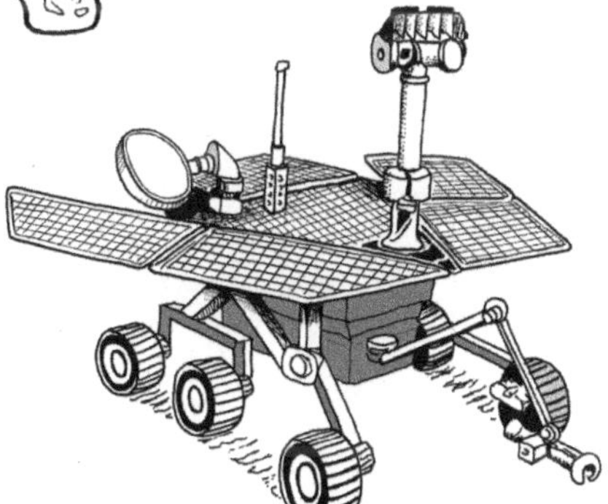

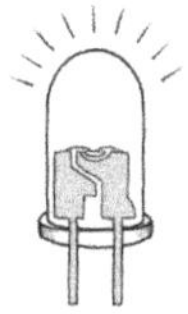

by Ellen Johnston McHenry

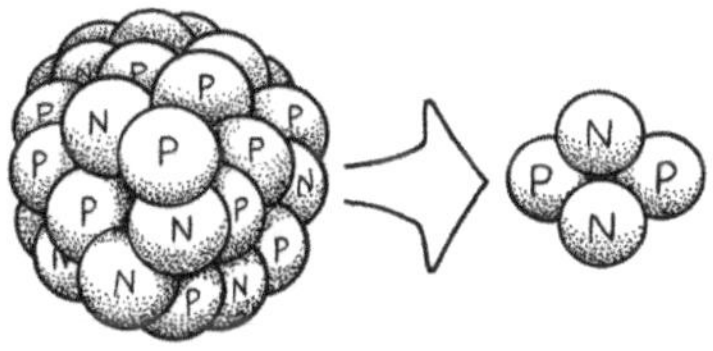

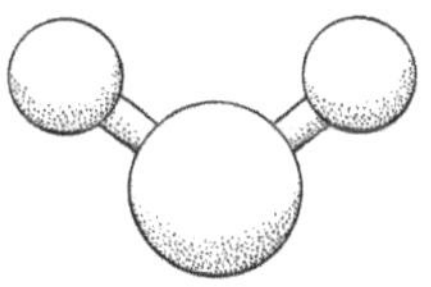

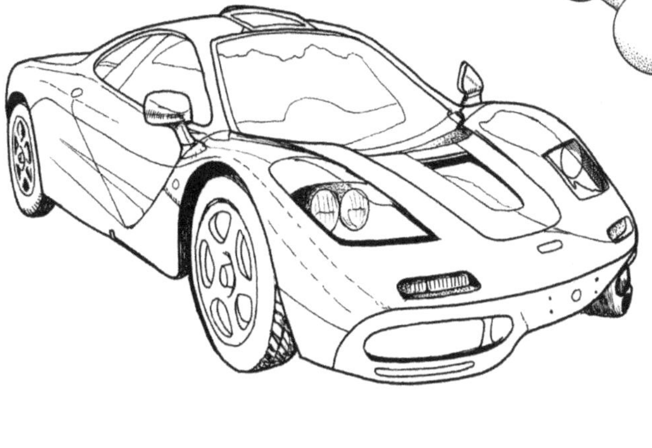

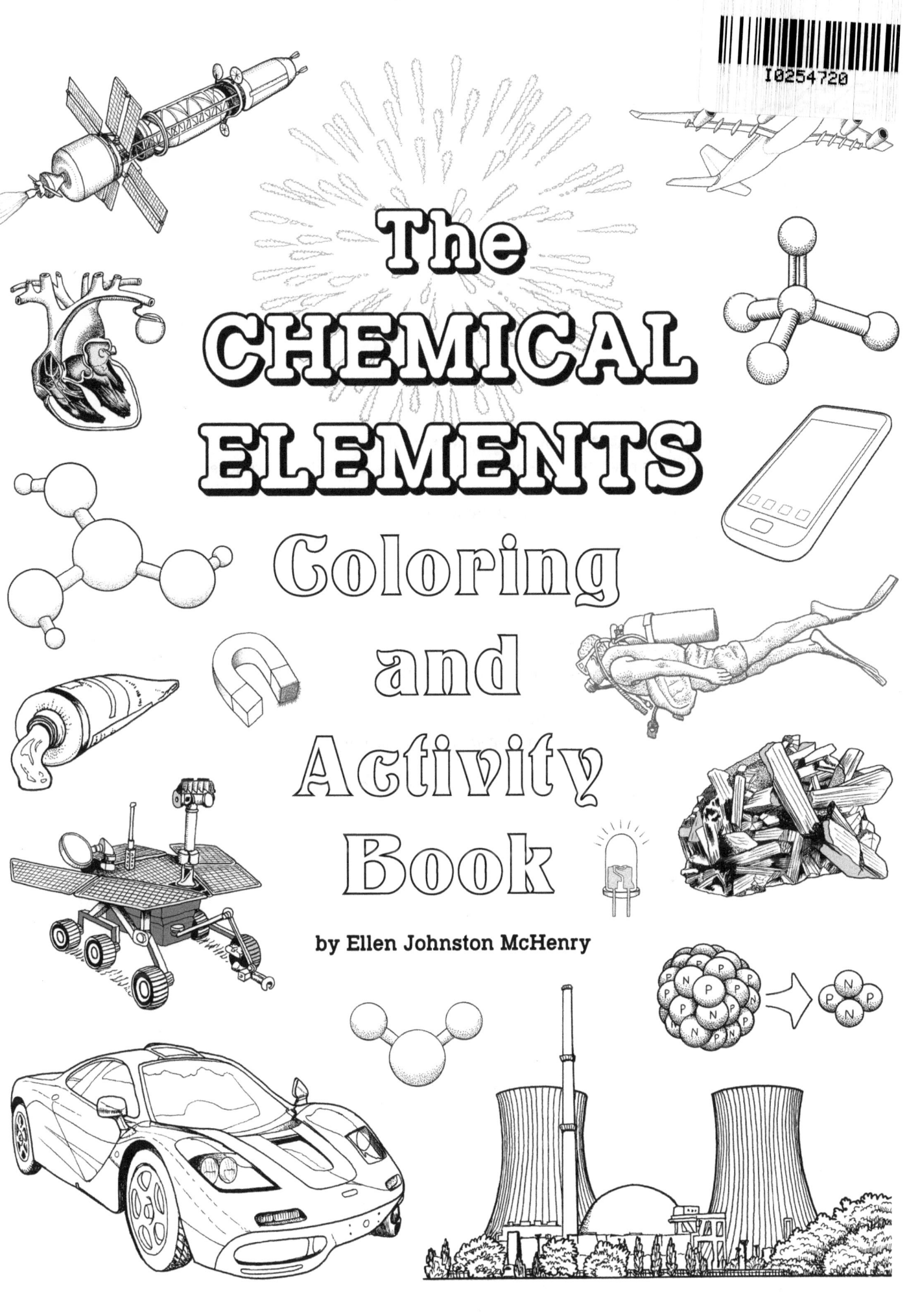

ISBN: 978-1-7374763-0-6
Published by Ellen McHenry's Basement Workshop, Pennsylvania, USA

Find other books by this author by searching your favorite online book distributor or by visiting www.ellenjmchenry.com.

Retailers may order this book through IngramContent.com.

4_6_26

What is an element?

An element is a substance that is made entirely from one type of atom. For example, the element oxygen consists of all atoms that are identified as oxygen atoms. All atoms of a given element have the same number of protons in their nucleus. It is the number of protons that gives atoms their identity. Oxygen atoms always have 8 protons.

What is an atom?

The atom is the most basic unit of matter that has its own identity. Atoms are made of little pieces called **electrons, protons, and neutrons**. Each element has a unique number of protons. If you change the number of protons, the atom becomes a different element. In the past few decades, scientists have made new elements by adding more protons to atoms that are already very large.

What do atoms look like?

We don't really know, because atoms are too small to see under a microscope. The best images we have are made using electron microscopes. Regular microscopes work by bending light rays to make things look larger. Electron microscopes use a beam of electrons and a special detector that can sense what the electrons do as they pass through a sample. The data gathered by the detector goes into a computer that turns it into a picture we can see on a screen. Images of atoms always look blurry and they can't show us the protons, neutrons or electrons.

If you search for pictures of atoms, almost every image you will find will be a diagram, not a photo. Sometimes the diagram will look a bit like the solar system, with rings going around a blob in the center. Other diagrams will look like balloons tied together, or like a series of spherical shells nested inside each other. None of these diagrams accurately depicts what an atom really looks like. Each type of diagram represents some of the facts we've learned about atoms. No single drawing can give us a complete picture of an atom.

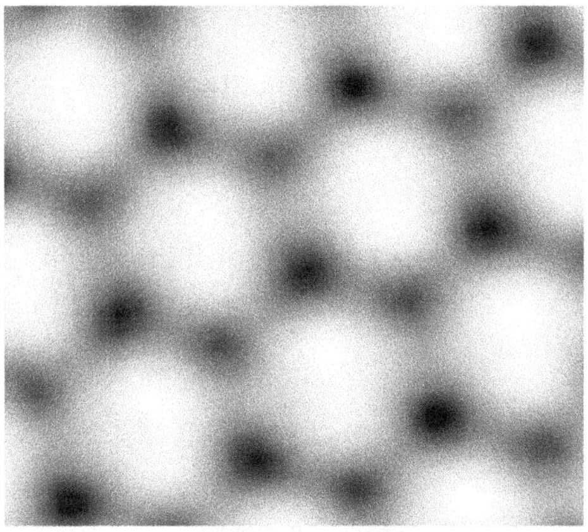

The fuzzy black areas are carbon atoms. We can't zoom in any closer. Carbon atoms often attach to each other in this hexagonal pattern.

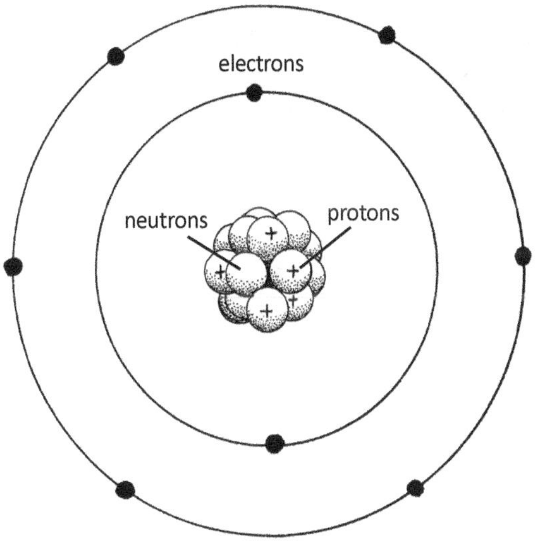
electrons
neutrons
protons

The "solar system" model shows the nucleus in the center (made of protons and neutrons) surrounded by electrons traveling in "orbits." This type of diagram is great for showing how many electrons occupy each energy level (or "shell"), but it does not show the correct shape of the orbitals, and it makes the particles too large.

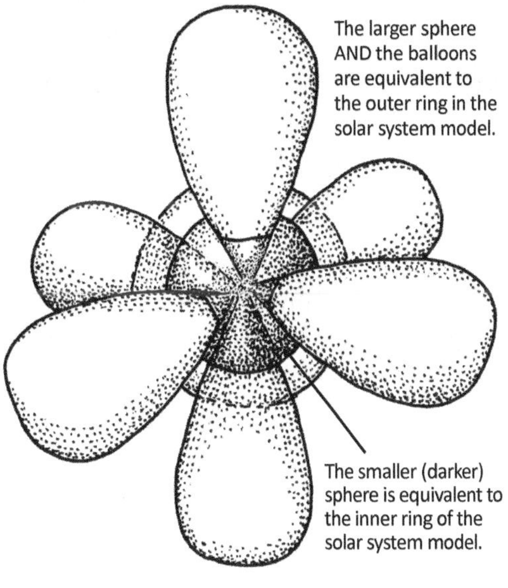
The larger sphere AND the balloons are equivalent to the outer ring in the solar system model.
The smaller (darker) sphere is equivalent to the inner ring of the solar system model.

This diagram shows that electron orbitals look more like balloons than rings. Each "balloon" shows the area where one electron is most likely to be found.
One of the downsides to this type of diagram is that it doesn't tell us anything about the nucleus.

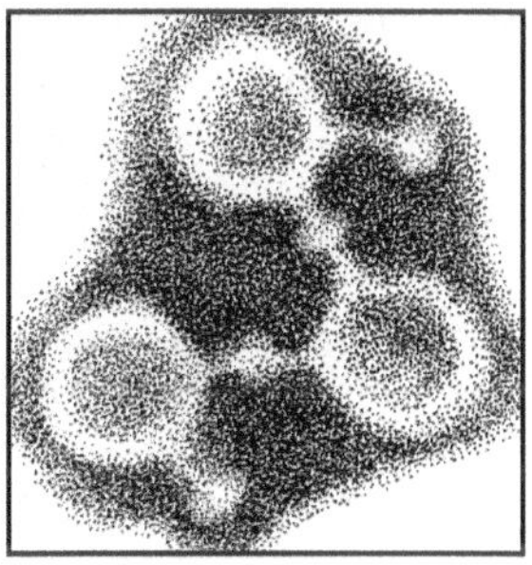

One of the most recent advances in atomic imaging is "non-contact atomic force spectroscopy." Imagine someone reading Braille, then substitute silver instead of paper, a molecule instead of a bump, and an oxygen atom instead of a finger. The oxygen atom's interaction with the molecule is interpreted by a computer which makes the data into an image. The hexagons are made of six carbon atoms. The "balls" are also carbons.

If we can't see atoms, how do we know their structure?

It took decades of experiments by many scientists to figure out the structure of an atom. Logic and math played a large role in the process. The spacing between the nucleus and the electrons was revealed by Ernest Rutherford's famous gold foil experiment in 1908. Most of the alpha particle "bullets" were going right through the gold foil, as if it wasn't even there, suggesting that atoms are made of mostly empty space. When an alpha particle bounced back, this was interpreted as a hit to the nucleus, and the number of hits was used to calculate the approximate size of the nucleus. J. J. Thomson discovered the electron and determined that it carried a negative electrical charge. James Chadwick's experiment revealed that the nucleus contained not only positively charged particles, but neutral ones, as well. Niels Bohr looked at the results of many experiments being done in the early 1900s and figured out that they suggested that the electrons were not just flying around the nucleus randomly, but were arranged into distinct layers, or shells, which resulted in the solar system model of the atom. Further experiments, combined with the laws of probability, would tell us that electrons occupy distinct areas (orbitals) and we can't know for sure where an electron is located inside that area at any given time. Balloon shapes were then used to try to depict this idea.

What is atomic mass?

The word "mass" is a more technical and precise word for "weight." The difference between the two words becomes easy to understand if you think of weighing yourself here on earth and then using the same scale to weigh yourself on the moon. The moon's gravity is much less than the earth's gravity, so the scale on the moon will give you a smaller number, saying that you weigh less. Your mass has not changed, though.

The mass of an atom is the sum of all its pieces: electrons, protons, and neutron. Electrons are so tiny, however, that they don't contribute much to the total mass, so they are usually ignored in the calculation. To calculate the mass of an atom, you simply **add the number of protons and the number of neutrons**.

Why aren't all atomic masses whole numbers?

You can't cut protons or neutrons into pieces. The number of protons or neutrons is always a whole number (the regular numbers we use to count). So how is it possible, then, to have an atomic mass that is not a whole number?

The atomic mass listed for each element is an **average**. Thousands and thousands of atoms were "weighed" and then the weights were averaged. Sometimes an atom can have an extra neutron, or be missing a neutron. When these unusual atoms are included in the data being averaged, it will make the answer come out to be a decimal number. For example, 99% of all carbon atoms have 6 neutrons, but about 1% have 7 neutrons, and about .01% have 8 neutrons (these are known as Carbon-14). If you average these numbers, you come out with an atomic mass of 12.01.

What is an isotope?

Since it is the number of protons that defines the identity of an element, the number of neutrons can vary. For example, most carbon atoms have 6 protons and 6 neutrons, but a few of them have 7 neutrons, and a very few of them have 8. These variations are called isotopes. "Iso" means "equal," and "tope" means "place." Isotopes are all "equal" in regards to their "place" as members of their element.

Isotopes can be written in various ways: Carbon-14, C-14, ^{14}C

What is the role of neutrons?

"Like charges" repel, so all of the positive protons in the nucleus are pushing against each other. However, there is another force at work in the nucleus, called the "strong force," which acts only at short range but can keep protons or neutrons locked together. The presence of neutrons allows the protons to be spaced just far enough apart to give the strong force the advantage over the force of repulsion.

What is an alpha particle?

An alpha particle is 2 protons and 2 neutrons stuck together. An alpha particle is identical to the nucleus of a helium atom. Alpha particles are often ejected from the nuclei of unstable atoms. If the alpha particle can pick up 2 electrons, it will become a helium atom. (The presence of helium inside mineral crystals suggests that there are some unstable radioactive atoms in the mineral.) Alpha particles are often used in research.

What is radioactivity?

Each type of atom has an ideal ratio of protons to neutrons in the nucleus. For smaller atoms, this ratio is close to 50/50. Larger atoms require more neutrons to help hold the protons together. Eventually, however, nuclei get so large that no matter what the ratio of protons to neutrons is, the nucleus is likely to fall apart. Sometimes the nucleus simply spits out a neutron. More often, the nucleus will eject an alpha particle (2 protons and 2 neutrons). Another possibility is that a neutron will turn into a proton and an electron, and the electron will be ejected as a "beta particle." Or, a proton can turn into a neutron and a positron (a positively charged electron) will be ejected. Sometimes the process of falling apart will cause the atom to emit a burst of very high energy called gamma radiation.

How do atoms stick together?

If two atoms happen to bump into each other, their outer rings of electrons will touch. That outer ring acts like a bumper and protects the nucleus and all the inner rings from coming into contact with anything outside the atom. The number and arrangement of electrons in an atom's outer ring will determine how it interacts with other atoms. For smaller atoms, a general rule holds true: the atom wants 8 electrons in its outer shell. If it has more or less than 8, the atom is likely to borrow or give away some of its electrons to get closer to the ideal 8. ("8 is great!") For larger atoms, it gets more complicated, but the general idea still holds true, that the atom will interact with other atoms in ways that provide it with a better arrangement of electrons in its outer shells.

Some atoms are very lucky and they have their outer shells full. These are very content atoms, the ones we call "noble" or "inert," meaning that they do not react with other atoms under normal conditions. We find these atoms in the last column of the Periodic Table.

The atoms that we classify as non-metals (most notably carbon, nitrogen, oxygen, sulfur, and phosphorus) will bond to other atoms in a way that we call "covalent." "Co" means "with" so this type of bonding involves sharing of electrons with another atom. The classic example of covalent bonding is the water molecule. In water, we see an oxygen atom sharing electrons with two hydrogen atoms. The hydrogen has only one electron in its tiny outer shell (which is so small it holds only two total) and would like to either give away the one it has or find an extra to make a pair. Oxygen has 6 electrons in its outer shell, so it would like to gain 2 more to make a total of 8.

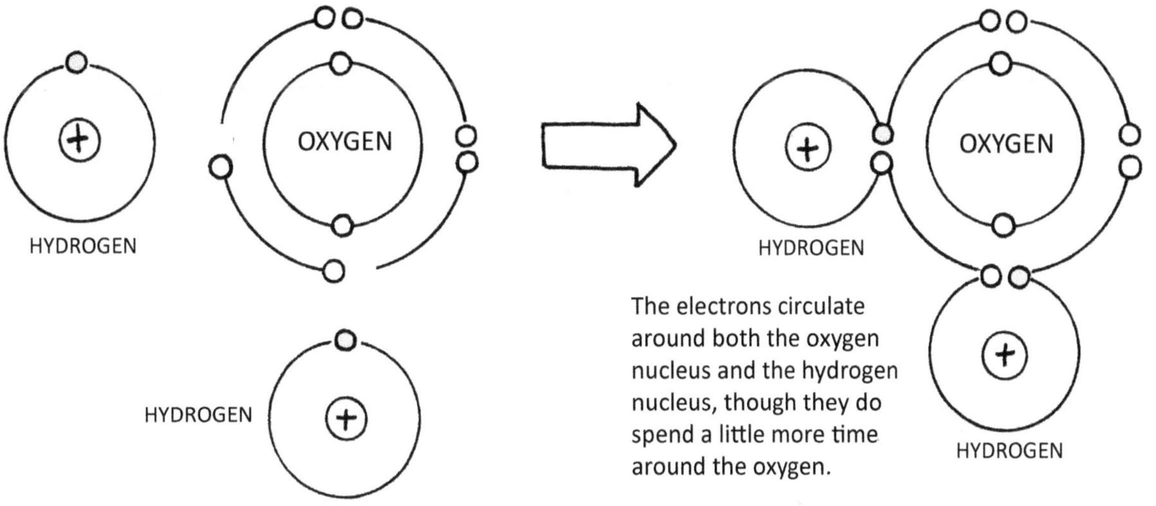

The electrons circulate around both the oxygen nucleus and the hydrogen nucleus, though they do spend a little more time around the oxygen.

Another way that atoms can interact is called "ionic" bonding. An ion is an atom that has gained or lost electrons, and therefore carries a negative or positive electrical charge. We find this happening to atoms which have only 1 or 2 electrons in their outer shell, or atoms that have close to 8, with 6 or 7 electrons in their outer ring. The atoms with 1 or 2 think it is better to have 0 in the outer shell, so they will give away those electrons any chance they get. The atoms with 6 or 7 are desperate to get a few more to make the perfect 8, so they will take electrons from other atoms. The classic example of ionic bonding is NaCl, sodium chloride (table salt).

The sodium atom gives away its one electron, and chlorine gladly accepts it. This improves the situation in their outer shells, but at the same time it upsets the internal balance between their electrons and protons. In their pure form, all atoms have an equal number of negative electrons and positive protons. This makes the atom electrically neutral. Before sodium gives away its outer electron it has 11 protons and 11 electrons. After that outer electron is gone, sodium will have 11 protons and 10 electrons, giving it a charge of +1. In chlorine, the opposite happens. When it gains an electron, it will then have 17 protons and 18 electrons, giving it a charge of -1.

We now have two ions, one with a +1 charge and one with a -1 charge. We know that opposite charges attract, so these ions are attracted to each other and therefore stay next to each other.

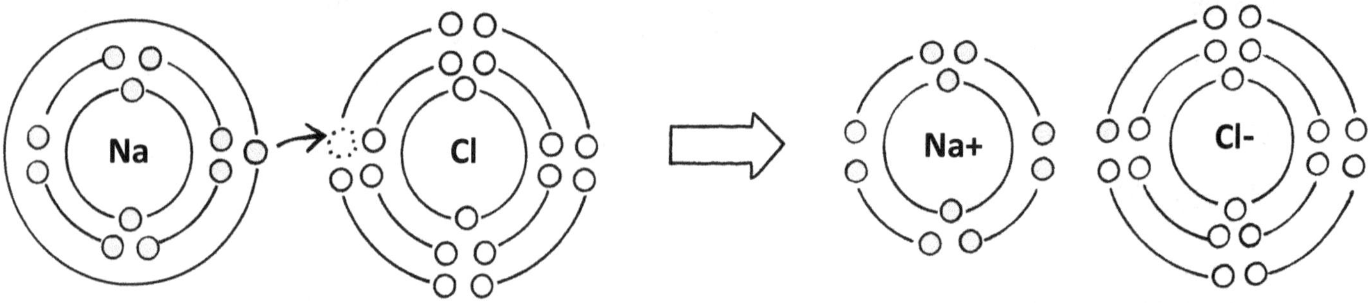

A third type of bonding is called "metallic" bonding, and, as the name suggests, we find this happening in metals, particularly those in the middle section of Periodic Table, such as iron, silver, gold, copper, zinc, nickel and platinum. The electron configuration of these atoms is such that they are able to let their outer electrons float around. It's like a big group share, with all the electrons belonging to everyone at the same time. Because the electrons can move about, these metals can be used to make electrical wires. Electricity is made of moving electrons and can easily move right through all these shared electrons.

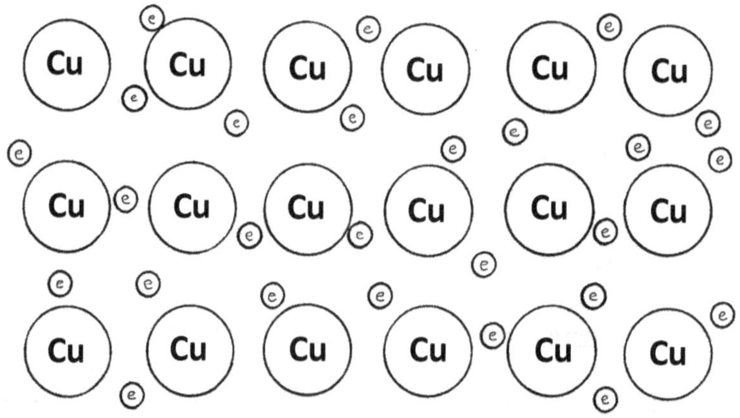

How do we draw bonds between atoms?

Covalent bonds are usually represented by a line, or stick, going between two atoms. The atoms are drawn as balls so neither the nucleus nor the electron orbitals are shown. Two lines are used when the atoms are doing a "double share" of electrons, making a double bond).

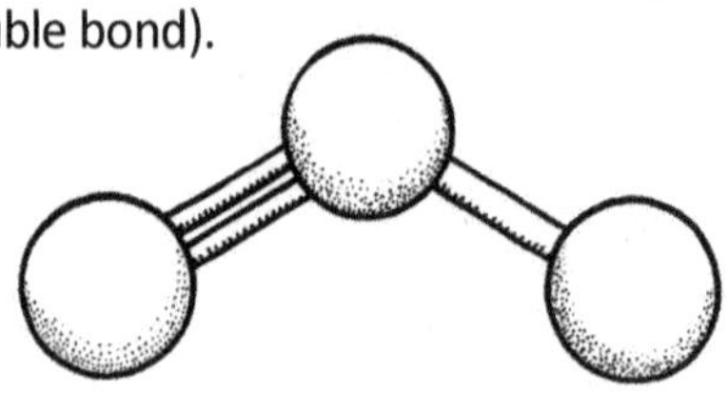

Ionic bonds are sometimes also shown as sticks, especially if the atoms are in a crystal, but in smaller molecules you often see nothing but empty space between the balls.

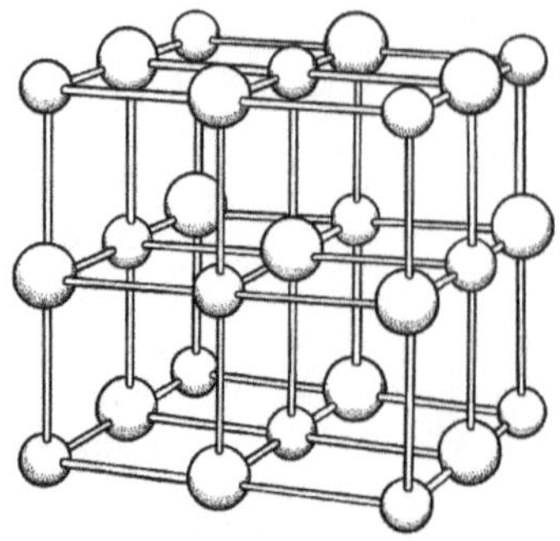

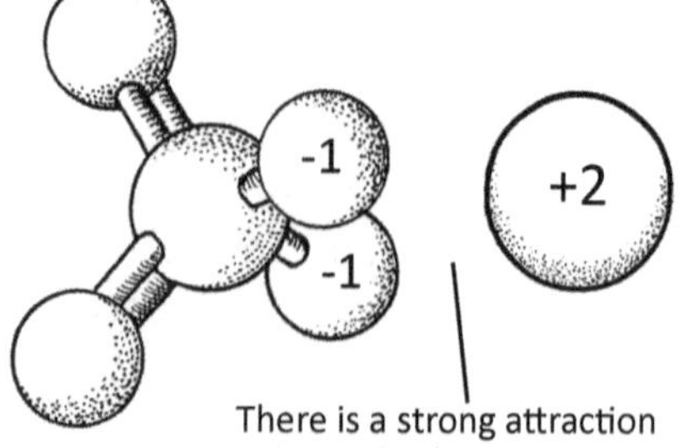

There is a strong attraction going on in this empty space.

How were the elements discovered?

Ancient peoples discovered that when you heated certain rocks to a very high temperature, a liquid would come out of them. When this liquid cooled, it would turn into a hard metal that could be shaped into knives, tools, or jewelry. The picture shows ancient Egyptians smelting copper from mineral rocks called malachite and azurite. The men at the sides are using foot pumps to blow air into the kiln to make it extra hot. A "normal" fire does not get hot enough to melt most mineral ore rocks. Blowing lots of extra oxygen into a fire allows the temperature to get much hotter.

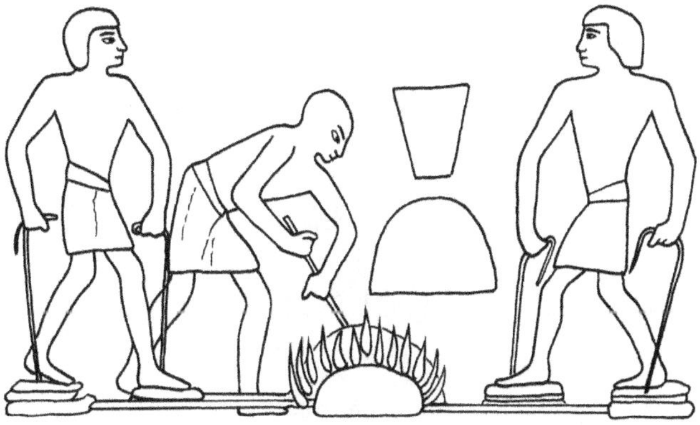

The ancients discovered and used seven metals that we know today to be elements: copper, gold, silver, tin, lead, iron and mercury. Of course, they had no idea that these metals were actually pure elements. It would be many centuries until people figured out that only some substances are pure elements and everything else is a mixture of these elements.

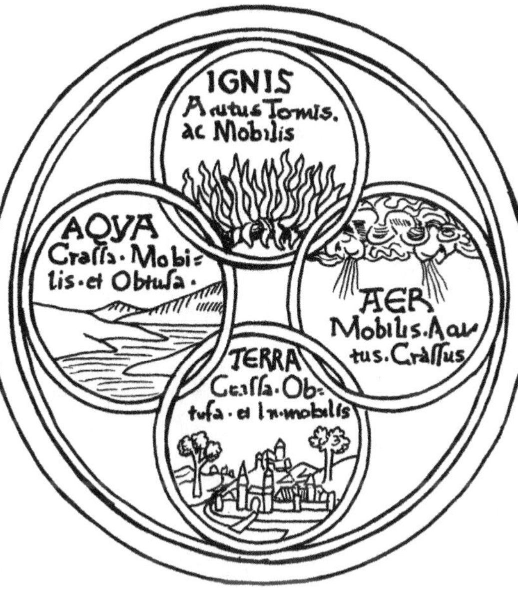

Ancient peoples correctly guessed that there were fundamental "elements" that combined in different ways to make the wide variety of things they saw in the world. However, many wrong ideas about elements existed for a long time. Some of the earliest wrong ideas came from the Greeks, who theorized that all matter is made of just four elements: **water, earth, fire and air.** This idea seems ludicrous to us today, but before we pass judgment on the ancient Greeks, we should bear in mind that real, true facts about chemistry can also seem very strange. For example, how can two gases, hydrogen and oxygen, combine to make a liquid (water)? In the process of combustion (burning), carbon from solid objects can end up in the air as a gas. Real chemistry can sound pretty strange too!

Shown here is a woodcut from the Middle Ages, depicting the four basic elements. This belief was widely held around the world for well over a thousand years.

One of the first chemists to investigate the nature of elements was **Jabir ibn Hayyan**, who lived in Persia (now Iran) in the 700s. He studied mathematics, astronomy, geography, medicine, philosophy, physics and chemistry. As far as we know, he was the first person in history to do "modern" scientific experiments. He designed pieces of equipment that could boil and separate mixtures, catch steam, and collect particles. He began working with substances that we now call “acids,” and found one that could dissolve gold. He was also the first person to begin to classify rocks and minerals according to their chemistry. He discovered that some substances would turn to steam when they were heated, such as mercury and sulfur. Others, such as gold, silver, and copper would turn into a liquid when heated. There were also substances, such as ordinary stones, that would neither turn into steam nor melt, and could only be crushed into a powder. Hayyan had discovered a key to figuring out what matter was made of.

Hayyan studied the writings of the Greeks, and from them he accepted the idea that water, fire, air, and earth were essential elements.

This stamp from Syria commemorates the work of Hayyan.

However, his experiments suggested to him that there were more than just four elements. He believed that mercury and sulfur were also elements. No matter what you did to them, they remained the same. They could be combined with other things, but they could not be broken apart.

The Arabic alchemists that came after ibn Hayyan added "salt" to their list of elements. This was a step backward, but they had no way of knowing this. The nature of salt would not be revealed until the 1800s. They also made another mistake—they thought that it might be possible to turn one element into another. They still recognized that some substances were basic elements, but thought that perhaps this did not rule out turning one element into another.

In the early days of the Renaissance in Europe (the 1200s), Europeans began reading these ancient Arabic alchemy texts. They were especially fascinated by the idea that one metal might be turned into another. Perhaps they could find a way to turn copper into gold!

While European alchemists were hard at work trying to turn things into gold, they made some interesting discoveries. Like ibn Hayyan, they began noticing that some substances seemed very resistant to change, as if they *couldn't* change. The list of basic elements began to grow. The list included the substances that the ancients had already discovered—**gold, silver, copper, mercury, lead, sulfur, iron, and tin**, and some new ones: **antimony, arsenic and bismuth**. Bismuth was usually discovered by miners who were looking for lead or silver. It would be found sandwiched between lead and silver. Since bismuth was more shiny than lead, but not as shiny as silver, this led the miners to believe that lead would eventually become silver, and bismuth was an intermediate stage, halfway between lead and silver. When the miners found a layer of bismuth they would say, "Oh no, we came too soon!"

The elemental nature of bismuth was first noticed by a German scientist named **Georgius Agricola**, in the early 1500s. Agricola spent a large part of his adult life studying mines and the stuff that came out of them. He wrote one of the first books on mining techniques. He believed that bismuth was a completely different substance from either lead or silver. He also began to realize the relationship between rocks, minerals and elements. Rocks were made of minerals and minerals were made of elements. Elements were not made of anything; they were the ultimate basic ingredients. Figuring out the nature of elements would prove to be quite a challenge.

This print is from Agricola's book De Re Metallica, *published in 1556. Miners are digging up rocks that contain metals.*

The next great leap in understanding the elements came in Germany in 1669. An alchemist by the name of **Hennig Brandt** was trying to make gold. By this time, alchemists had tried just about every rock and mineral on the planet, so Hennig decided to boil something that was yellowish-gold but wasn't a rock or mineral. He collected hundreds of gallons of urine and boiled it down until it became a thick paste. When he began heating this paste as hot as he could get it, something amazing happened—it began to glow! It glowed with a bright, white light. Hennig had discovered the element **phosphorus**. (The word phosphorus means "bearer of light.")

The discovery of phosphorus caused a major shift in the thinking of scientists all over the world. Brandt had proved that organic things were made of elements, too. Phosphorus could be found in both living and non-living things.

In the early 1700s, the word "alchemy" was replaced by the word "chemistry." Chemists in the 1700s were no longer interested in trying to turn things into gold. They understood that gold was an element. Other metallic elements on their list were copper, nickel, iron, silver, tin, lead, zinc, and bismuth.

During the 1800s, chemists were eager to discover new elements, so they tested every mineral sample they could get. They improved their equipment and invented new tools and techniques. One of the most important new technologies of the 1800s was electricity.

Humphry Davy was one of the leaders in the use of electricity to discover new elements. In the early 1800s, electricity was produced by voltaic piles—stacks of metal discs separated by pads soaked in salt water. Even with these primitive batteries, Davy was able to produce a strong enough electrical current to pull elements out of a solution. He discovered **sodium, magnesium, calcium, potassium, strontium, and barium**.

Another key to discovering new elements was the invention of refrigeration. **Sir William Ramsay** was able to chill a sample of air to such low temperatures that the gases in the air turned to liquid. As the temperature was slowly allowed to rise, the liquid gases would go back into their gaseous form and Ramsay would collect each gas. This is the way that the noble gases were discovered (**neon, argon, krypton, xenon, radon**).

Humphry Davy

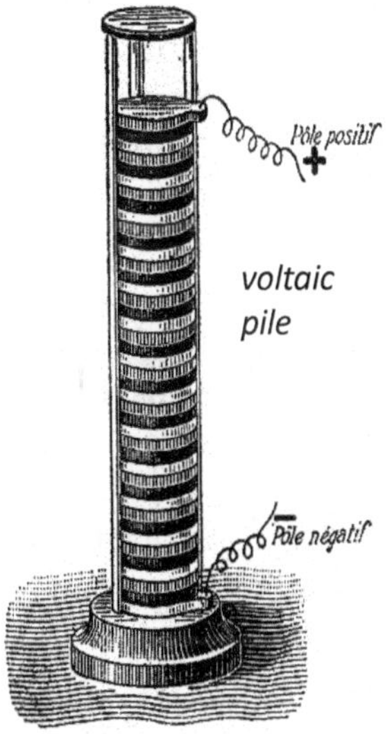

voltaic pile

Sir William Ramsay

early refrigeration equipment

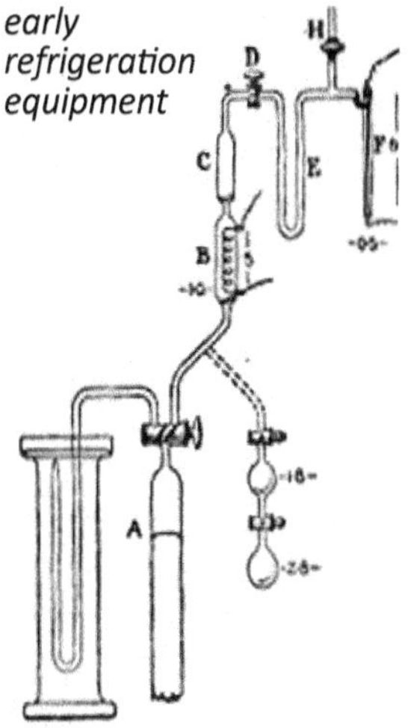

Fig. XIV.3.

The mineralogists also continued their work with rocks, and were able to discover some rare earth elements and more metals. The list of elements grew rapidly during this century. By the late 1800s, there were 63 known elements.

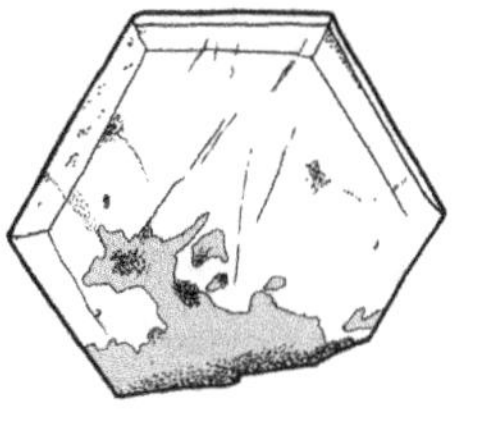

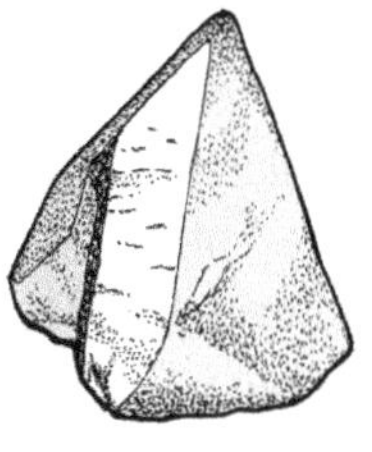

In the late 1800s, chemists were starting to make charts of all the known elements. Several scientists began working on ways to classify the elements, but only one is well remembered today: **Dmitri Mendeleev**. *(men-del-LAY-ev)* Mendeleev's stroke of genius was to realize that there were quite a few elements that had not yet been discovered. He made a rectangular chart and left blank spaces in places where he believed there was an element missing. He turned out to be absolutely right in every case. Soon, those missing elements were discovered just as Mendeleev predicted. Today we call this chart the Periodic Table. There are many more elements known today than in Mendeleev's day, but the basic structure of the table has remained the same.

Reihen	Gruppo I. — R^2O	Gruppo II. — RO	Gruppo III. — R^2O^3	Gruppo IV. RH^4 RO^2	Gruppo V. RH^3 R^2O^5	Gruppo VI. RH^2 RO^3	Gruppo VII. RH R^2O^7	Gruppo VIII. — RO^4
1	H=1							
2	Li=7	Be=9,4	B=11	C=12	N=14	O=16	F=19	
3	Na=23	Mg=24	Al=27,3	Si=28	P=31	S=32	Cl=35,5	
4	K=39	Ca=40	—=44	Ti=48	V=51	Cr=52	Mn=55	Fe=56, Co=59, Ni=59, Cu=63.
5	(Cu=63)	Zn=65	—=68	—=72	As=75	Se=78	Br=80	
6	Rb=85	Sr=87	?Yt=88	Zr=90	Nb=94	Mo=96	—=100	Ru=104, Rh=104, Pd=106, Ag=108.
7	(Ag=108)	Cd=112	In=113	Sn=118	Sb=122	Te=125	J=127	
8	Cs=133	Ba=137	?Di=138	?Ce=140	—	—	—	— — — —
9	(—)	—	—	—	—	—	—	
10	—	—	?Er=178	?La=180	Ta=182	W=184	—	Os=195, Ir=197, Pt=198, Au=199.
11	(Au=199)	Hg=200	Tl=204	Pb=207	Bi=208	—	—	
12	—	—	—	Th=231	—	U=240	—	— — — —

In 1898, **Marie Curie** discovered some of the last naturally-occurring elements. She spent many months boiling pitchblende, a mineral ore that was known to contain uranium. Radioactivity in uranium had just been discovered by Henri Becquerel and she wanted to find out more about it. After all the uranium had been removed, the ore was still emitting radiation. She managed to extract two more radioactive elements from the ore: **polonium and radium**.

Many other scientists then joined the race to discover radioactive elements, and actinium, protactinium and francium were discovered. **Ernest Rutherford** proposed that radioactive atoms "decay" because their nucleus is large and unstable. He had observed a form of radiation he called "alpha particles" coming from radioactive elements. He figured out that these alpha particles contained protons, and theorized that when an atom lost or gained protons it would change into a different element. In 1919 he did an experiment where he shot alpha particles at nitrogen atoms and turned them into oxygen.

This began a new era in the search for elements. If you can't find them, why not just make them yourself?

Making elements isn't easy. It takes huge machines that fill entire buildings just to make a few atoms of a new element. The first machine capable of manufacturing a new atom was the cyclotron built by **Ernest Lawrence** and his team at the Berkeley Lab in California. The first man-made elements produced in the cyclotron were **neptunium**, in 1940, and **plutonium**, in 1941. In the next few years, the work at the lab focused on learning how to use radioactive atoms to make weapons that could be used during World War 2. During this research, more new elements were made: **americium, curium, berkelium and californium**. After the war, research continued. New elements were found in the air after the explosion of an atomic bomb. In the late 1900s, improvements were made in the equipment, and particle accelerators and the new cyclotrons were able to make super heavy elements that only existed for a few seconds at a time. The researchers did not want to stop making elements until they had finished the bottom row on the Periodic Table. Element 118 was finally made in 2002 in Dubna, Russia.

The cyclotron in the Berkeley Radiation Lab (later called Berkeley National Lab) in 1940.

What is the Periodic Table?

The Periodic Table is the complete list of all elements that exist in the universe. Anything you can think of is made of these elements.

Why is the table "periodic"?

It is "periodic" because periodically (every now and then) certain patterns are repeated. The first person to notice these repeating patterns was the inventor of the table, Dmitri Mendeleev *(mend-dell-AY-ev)*

The columns are called groups, and the rows are called periods. The elements in each group (column) have similar chemical properties. The elements in the last column on the right all have full outer shells and will not form molecules (under standard conditions). The elements in the first column, group 1, are extremely reactive and will burn if placed in water. The elements in group 11 are the most well-known precious metals: copper, silver and gold. The elements in group 17 are called the halogens; they are all toxic and smell bad but will form non-toxic, useful salts.

The periods (rows) give you information about the atom's electron structure. When a new row starts, this means that the electron arrangement has begun a new round of its pattern of orbital shapes. The pattern always starts with spheres, then goes to balloon shapes, then to more complicated balloon shapes. (The letters s, p, d and f are used to represent these shapes.)

Group → ↓ Period	1	2	3	4	5	6	7	8	9	10	11	12	13	14	15	16	17	18
1	1 H																	2 He
2	3 Li	4 Be											5 B	6 C	7 N	8 O	9 F	10 Ne
3	11 Na	12 Mg											13 Al	14 Si	15 P	16 S	17 Cl	18 Ar
4	19 K	20 Ca	21 Sc	22 Ti	23 V	24 Cr	25 Mn	26 Fe	27 Co	28 Ni	29 Cu	30 Zn	31 Ga	32 Ge	33 As	34 Se	35 Br	36 Kr
5	37 Rb	38 Sr	39 Y	40 Zr	41 Nb	42 Mo	43 Tc	44 Ru	45 Rh	46 Pd	47 Ag	48 Cd	49 In	50 Sn	51 Sb	52 Te	53 I	54 Xe
6	55 Cs	56 Ba		72 Hf	73 Ta	74 W	75 Re	76 Os	77 Ir	78 Pt	79 Au	80 Hg	81 Tl	82 Pb	83 Bi	84 Po	85 At	86 Rn
7	87 Fr	88 Ra		104 Rf	105 Db	106 Sg	107 Bh	108 Hs	109 Mt	110 Ds	111 Rg	112 Cn	113 Nh	114 Fl	115 Mc	116 Lv	117 Ts	118 Og

Lanthanides	57 La	58 Ce	59 Pr	60 Nd	61 Pm	62 Sm	63 Eu	64 Gd	65 Tb	66 Dy	67 Ho	68 Er	69 Tm	70 Yb	71 Lu
Actinides	89 Ac	90 Th	91 Pa	92 U	93 Np	94 Pu	95 Am	96 Cm	97 Bk	98 Cf	99 Es	100 Fm	101 Md	102 No	103 Lr

Why are there two rows underneath the table?

The lanthanide and actinide rows are put at the bottom so that the table will be less wide. On the next page you can see what the table looks like when those rows are placed where they ought to be. When the table is very wide, it does not fit onto a page very well. The letters end up being so small you can hardly see them. Making the format closer to square allows the letters to be larger and therefore easier to see.

Is this a bit harder to read?

1 H																															2 He
3 Li	4 Be																									5 B	6 C	7 N	8 O	9 F	10 Ne
11 Na	12 Mg																									13 Al	14 Si	15 P	16 S	17 Cl	18 Ar
19 K	20 Ca															21 Sc	22 Ti	23 V	24 Cr	25 Mn	26 Fe	27 Co	28 Ni	29 Cu	30 Zn	31 Ga	32 Ge	33 As	34 Se	35 Br	36 Kr
37 Rb	38 Sr															39 Y	40 Zr	41 Nb	42 Mo	43 Tc	44 Ru	45 Rh	46 Pd	47 Ag	48 Cd	49 In	50 Sn	51 Sb	52 Te	53 I	54 Xe
55 Cs	56 Ba	57 La	58 Ce	59 Pr	60 Nd	61 Pm	62 Sm	63 Eu	64 Gd	65 Tb	66 Dy	67 Ho	68 Er	69 Tm	70 Yb	71 Lu	72 Hf	73 Ta	74 W	75 Re	76 Os	77 Ir	78 Pt	79 Au	80 Hg	81 Tl	82 Pb	83 Bi	84 Po	85 At	86 Rn
87 Fr	88 Ra	89 Ac	90 Th	91 Pa	92 U	93 Np	94 Pu	95 Am	96 Cm	97 Bk	98 Cf	99 Es	100 Fm	101 Md	102 No	103 Lr	104 Rf	105 Db	106 Sg	107 Bh	108 Hs	109 Mt	110 Ds	111 Rg	112 Cn	113 Nh	114 Fl	115 Mc	116 Lv	117 Ts	118 Og

What are the "families" of elements?

The families are sets of elements that are similar in important ways. Their similarities often correlate to which group or period they are in, but there are some exceptions.

Alkali metals: Group 1 elements (first column on left)
Alkali earth metals: Group 2 elements (second column on left)
Transition metals: Elements 21 to 30, 39 to 48 and 71 to 80 (also called the "d" block)
"True" metals: Al, Ga, In, Tl, Sn, Pb, Bi
Semi-metals, or metalloids: B, Si, Ge, As, Sb, Te, Po (some lists include At)
Non-metals: C, N, O, P, S, Se (and also H, though it is not close to these on the table)
Halogens: F, Cl, Br, I (some lists include At)
Noble gases: He, Ne, Ar, Kr, Xe, Rn
Lanthanides (often referred to as the rare earths): 57 to 70
Actinides: 89 to 102
Super-heavy elements: 102 to 118

NOTE: These families are shown on the back cover of this book. There isn't any significance to the colors. Any color can be used for any family.

Which natural elements are most abundant?

The most common elements are generally the ones in the top rows. Both living and non-living things in the natural world are made mostly of the elements H, C, N, O, F, Na, Mg, Mn, Al, Si, P, S, Cl, K, Ca, and Fe.

This graph shows the abundance of elements found in rocks and minerals. Notice that the line goes up and down in a zig-zag pattern. Atoms that have an even number of protons are more abundant than those that have an odd number.

Wikipedia maintains a list of the elements in order of their abundance.

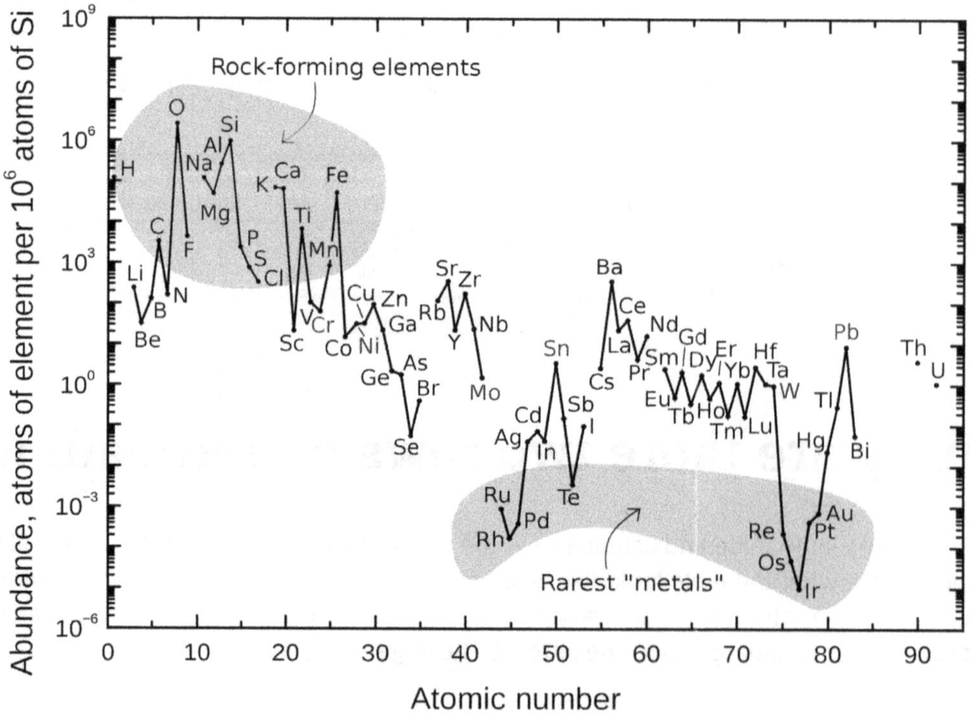

Periodic Table of the Elements

Atomic Number → 1 | Symbol ← H | Name → Hydrogen | Atomic Weight ← 1.0

1	2	3	4	5	6	7	8	9	10	11	12	13	14	15	16	17	18
1 **H** Hydrogen 1.0																	2 **He** Helium 4.0
3 **Li** Lithium 6.9	4 **Be** Beryllium 9.01											5 **B** Boron 10.8	6 **C** Carbon 12.01	7 **N** Nitrogen 14.0	8 **O** Oxygen 15.9	9 **F** Fluorine 18.9	10 **Ne** Neon 20.2
11 **Na** Sodium 22.9	12 **Mg** Magnesium 24.3											13 **Al** Aluminium 26.9	14 **Si** Silicon 28.0	15 **P** Phosphorus 30.9	16 **S** Sulfur 32.0	17 **Cl** Chlorine 35.4	18 **Ar** Argon 39.9
19 **K** Potassium 39.1	20 **Ca** Calcium 40.0	21 **Sc** Scandium 44.9	22 **Ti** Titanium 47.8	23 **V** Vanadium 50.9	24 **Cr** Chromium 51.9	25 **Mn** Manganese 54.9	26 **Fe** Iron 55.8	27 **Co** Cobalt 58.9	28 **Ni** Nickel 58.7	29 **Cu** Copper 63.5	30 **Zn** Zinc 65.4	31 **Ga** Gallium 69.7	32 **Ge** Germanium 72.6	33 **As** Arsenic 74.9	34 **Se** Selenium 78.9	35 **Br** Bromine 79.9	36 **Kr** Krypton 83.8
37 **Rb** Rubidium 85.4	38 **Sr** Strontium 87.6	39 **Y** Yttrium 88.9	40 **Zr** Zirconium 91.2	41 **Nb** Niobium 92.9	42 **Mo** Molybdenum 95.9	43 **Tc** Technetium (98)	44 **Ru** Ruthenium 101	45 **Rh** Rhodium 102.9	46 **Pd** Palladium 106.4	47 **Ag** Silver 107.8	48 **Cd** Cadmium 112.4	49 **In** Indium 114.8	50 **Sn** Tin 118.7	51 **Sb** Antimony 121.7	52 **Te** Tellurium 127.6	53 **I** Iodine 126.9	54 **Xe** Xenon 131.3
55 **Cs** Caesium 132.9	56 **Ba** Barium 137.3	57 **La** Lanthanum 138.9	72 **Hf** Hafnium 178.4	73 **Ta** Tantalum 180.9	74 **W** Tungsten 183.8	75 **Re** Rhenium 186.2	76 **Os** Osmium 190.2	77 **Ir** Iridium 192.2	78 **Pt** Platinum 195.0	79 **Au** Gold 196.9	80 **Hg** Mercury 200.6	81 **Tl** Thallium 204.3	82 **Pb** Lead 207.2	83 **Bi** Bismuth 208.9	84 **Po** Polonium (209)	85 **At** Astatine (210)	86 **Rn** Radon (222)
87 **Fr** Francium (223)	88 **Ra** Radium (226)	89 **Ac** Actinium (227)	104 **Rf** Rutherfordium (267)	105 **Db** Dubnium (268)	106 **Sg** Seaborgium (269)	107 **Bh** Bohrium (270)	108 **Hs** Hassium (269)	109 **Mt** Meitnerium (278)	110 **Ds** Darmstadtium (281)	111 **Rg** Roentgenium (282)	112 **Cn** Copernicium (285)	113 **Nh** Nihonium (286)	114 **Fl** Flerovium (289)	115 **Mc** Moscovium (289)	116 **Lv** Livermorium (293)	117 **Ts** Tennessine (294)	118 **Og** Oganesson (294)

NOTE: The masses of super-heavy elements, especially those beyond Rf, are difficult to determine, and chemists don't all agree on a single correct number. Therefore, the numbers listed here could be slightly different from other tables.

58 **Ce** Cerium 140.1	59 **Pr** Praseodymium 140.9	60 **Nd** Neodymium 144.2	61 **Pm** Promethium (145)	62 **Sm** Samarium 150.3	63 **Eu** Europium 151.9	64 **Gd** Gadolinium 157.25	65 **Tb** Terbium 158.9	66 **Dy** Dysprosium 162.5	67 **Ho** Holmium 164.9	68 **Er** Erbium 167.25	69 **Tm** Thulium 168.9	70 **Yb** Ytterbium 173	71 **Lu** Lutetium 174.9
90 **Th** Thorium (232)	91 **Pa** Protactinium (231)	92 **U** Uranium (238)	93 **Np** Neptunium (237)	94 **Pu** Plutonium (244)	95 **Am** Americium (243)	96 **Cm** Curium (247)	97 **Bk** Berkelium (247)	98 **Cf** Californium (251)	99 **Es** Einsteinium (252)	100 **Fm** Fermium (257)	101 **Md** Mendelevium (258)	102 **No** Nobelium (259)	103 **Lr** Lawrencium (266)

Color the families of elements. The Periodic Table on the back cover can help you see where the families are, but you can choose your own colors if you'd like.

Alphabetical list

Actinium	Ac	89
Aluminum	Al	13
Americium	Am	95
Antimony	Sb	51
Argon	Ar	18
Arsenic	As	33
Astatine	At	85
Barium	Ba	56
Berkelium	Bk	97
Beryllium	Be	4
Bismuth	Bi	83
Bohrium	Bh	107
Boron	B	5
Bromine	Br	35
Cadmium	Cd	49
Calcium	Ca	20
Californium	Cf	98
Carbon	C	6
Cerium	Ce	58
Cesium	Cs	55
Chlorine	Cl	17
Chromium	Cr	24
Cobalt	Co	27
Copernicium	Cn	112
Copper	Cu	25
Curium	Cm	96
Darmstadtium	Ds	110
Dubnium	Db	105
Dysprosium	Dy	66
Einsteinium	Es	99
Erbium	Er	68
Europium	Eu	63
Fermium	Fm	100
Flerovium	Fl	114
Fluorine	F	9
Francium	Fr	87
Gadolinium	Gd	64
Gallium	Ga	31
Germanium	Ge	32
Gold	Au	79

Hafnium	Hf	72
Hassium	Hs	108
Helium	He	2
Holmium	Ho	67
Hydrogen	H	1
Indium	In	49
Iodine	I	53
Iridium	Ir	77
Iron	Fe	26
Krypton	Kr	36
Lanthanum	La	57
Lawrencium	Lr	103
Lead	Pb	82
Lithium	Li	3
Livermorium	Lv	116
Lutetium	Lu	71
Magnesium	Mg	12
Manganese	Mn	25
Meitnerium	Mt	109
Mendelevium	Md	101
Mercury	Hg	80
Molybdenum	Mo	42
Moscovium	Mc	115
Neodymium	Nd	60
Neon	Ne	10
Neptunium	Np	93
Nickel	Ni	28
Nihonium	Nh	113
Niobium	Nb	41
Nitrogen	N	7
Nobelium	No	102
Oganesson	Og	118
Osmium	Os	76
Oxygen	O	8
Palladium	Pd	46
Phosphorus	P	15
Platinum	Pt	78
Plutonium	Pu	94
Potassium	K	19
Praseodymium	Pr	59

Promethium	Pm	61
Protactinium	Pa	91
Radium	Ra	88
Radon	Rn	386
Rhenium	Re	75
Rhodium	Rh	45
Roentgenium	Rg	111
Rubidium	Rb	37
Ruthenium	Ru	44
Rutherfordium	Rf	104
Samarium	Sm	62
Scandium	Sc	21
Seaborgium	Sg	106
Selenium	Se	34
Silicon	Si	14
Silver	Ag	47
Sodium	Na	11
Strontium	Sr	38
Sulfur	S	16
Tantalum	Ta	73
Technetium	Tc	43
Tellurium	Te	52
Tennessine	Ts	117
Terbium	Tb	65
Thallium	Tl	81
Thorium	Th	90
Thulium	Tm	69
Tin	Sn	50
Titanium	Ti	22
Tungsten	W	74
Uranium	U	92
Vanadium	V	23
Xenon	Xe	54
Ytterbium	Yb	70
Yttrium	Y	39
Zinc	Zn	30
Zirconium	Zr	40

COLORING PAGES

H

1 proton
1 electron
Atomic mass: 1.0

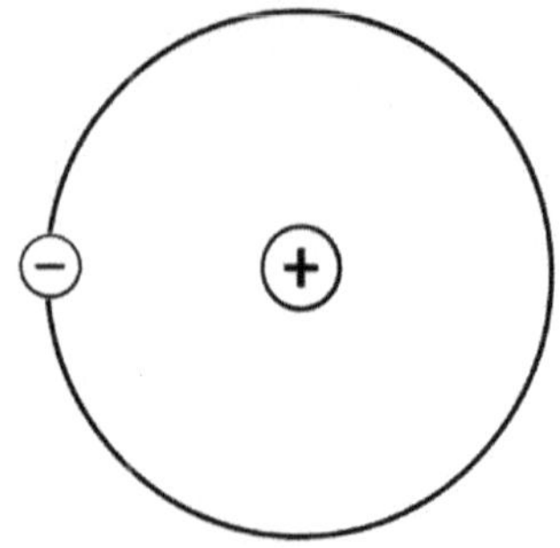

Hydrogen

From Greek words "hydro" (water) and "genes" (make)

Hydrogen is the smallest and lightest of all the elements. It is made of just one proton and one electron. Most of the time it doesn't have any neutrons. On rare occasions when hydrogen does gain a neutron, we call it "heavy hydrogen." Adding a neutron doesn't change its identity; it will still be hydrogen because it has one proton.

Hydrogen is a gas. If you put hydrogen into balloons, they will float. But don't try this because hydrogen is very flammable (catches fire easily). There was a terrible accident in 1937 in New Jersey, USA, when a hydrogen-filled blimp caught fire. That was the last time anyone put hydrogen into a blimp or balloon. The flammability of hydrogen can be put to good use, however, by using it as fuel in rocket engines. On a smaller scale, welders use tanks of hydrogen as a source of intense heat for joining steel parts.

Stars, including our sun, are made primarily of burning hydrogen gas. The extreme heat causes the hydrogen atoms to bump into each other and sometimes they combine to form larger elements such as helium, lithium or sodium.

Hydrogen bonds to many other elements and is found in thousands of molecules. The reason it likes to bond to other atoms is because its one electron is "lonely" and would like to be part of a pair. There are many atoms who would be very happy to have hydrogen come over and share its electron with them. Atoms that frequently bond with hydrogen include oxygen, carbon, nitrogen, and chlorine. When carbon atoms join together to make very long chains, hydrogen atoms will attach themselves to any free place they can find along the chain. This type of molecule (a chain of carbon atoms with hydrogens attached) is called a "hydrocarbon." Hydrocarbon molecules include methane (natural gas), octane (liquid gasoline), vegetable oils, animal fats, wax, and many types of plastics.

You can assign your own colors to the atoms, but here is what a professional scientific illustrator would be most likely to use:

White: Hydrogen (blank)
Red: Oxygen (O)
Black: Carbon (C)
Green: Chlorine (Cl)
Blue: Nitrogen (N)

The hydrogens look pretty big in these models. They are actually much smaller than these other atoms, but it looks nicer if the balls are close to the same size.

H_2 hydrogen gas

H_2O Water

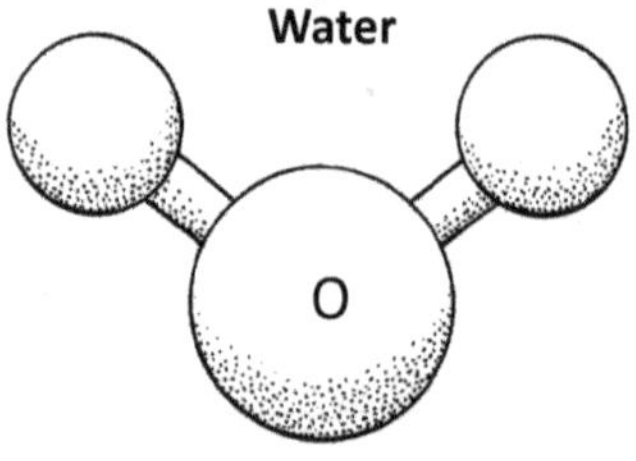

NH_3 Ammonia

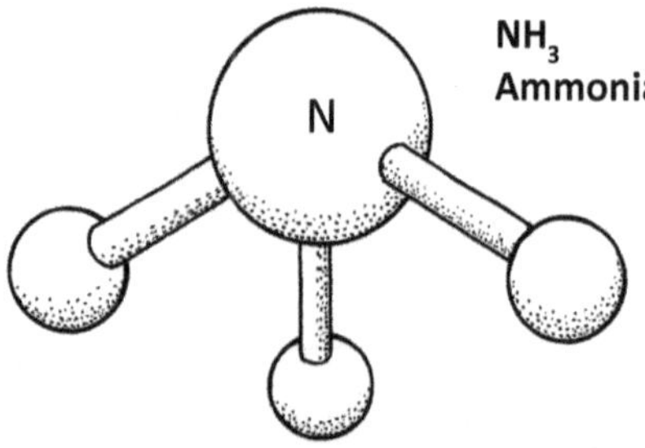

CH_4 Methane (natural gas)

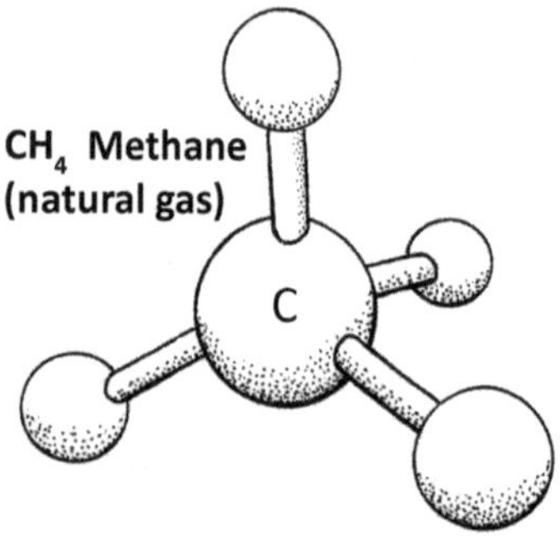

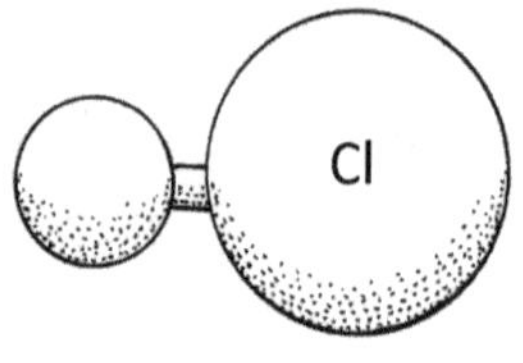

HCl Hydrochloric acid

Hydrogen peroxide H_2O_2

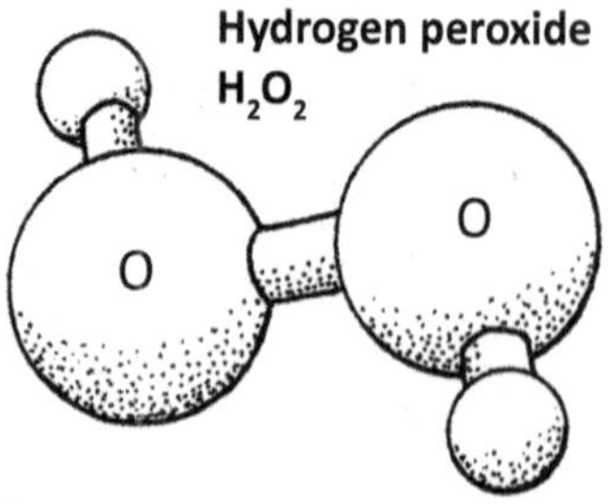

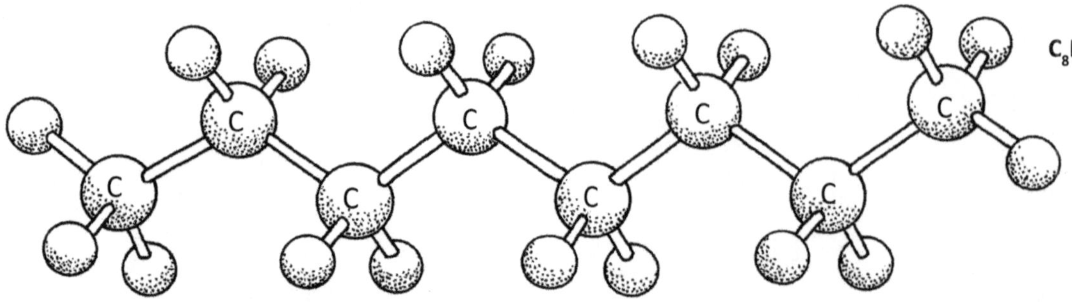

C_8H_{18} Octane

Octane is liquid gasoline (petrol). Hydrocarbon chains can grow to be very long. Plastics are made of chains that contain thousands of carbon atoms.

1 H

Hydrogen

Stars, including our sun, use hydrogen as fuel.

Water is H_2O.

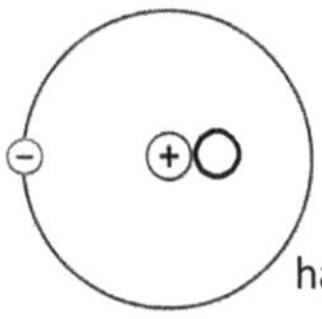

Deuterium, or "heavy hydrogen," has a neutron.

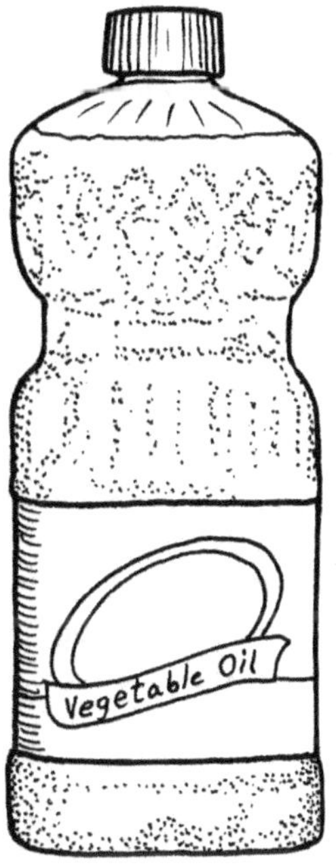

Hydrogenated oils are made of chains of carbon atoms with hydrogen atoms attached.

Gasoline (petrol) is made of carbon and hydrogen.

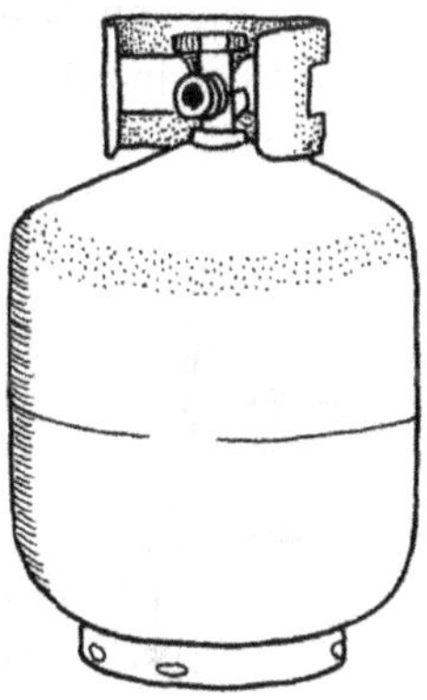

Natural gas is methane, CH_4.

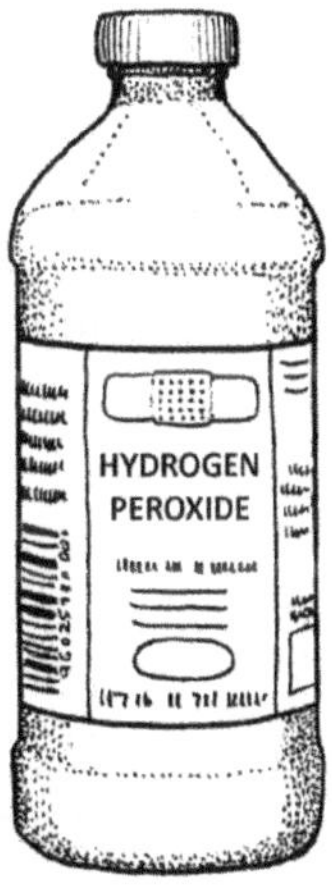

Hydrogen peroxide, H_2O_2 is used to clean wounds.

Some cleaning products contain ammonia, NH_3.

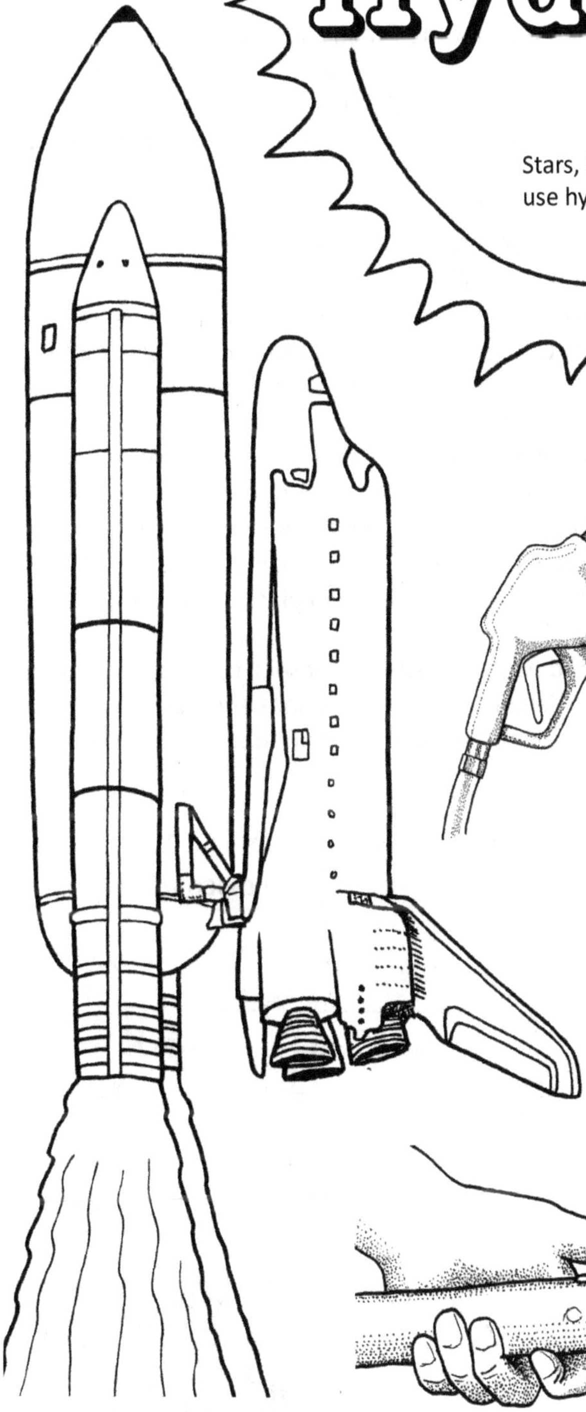

Liquid hydrogen is often used as rocket fuel (in combination with liquid oxygen).

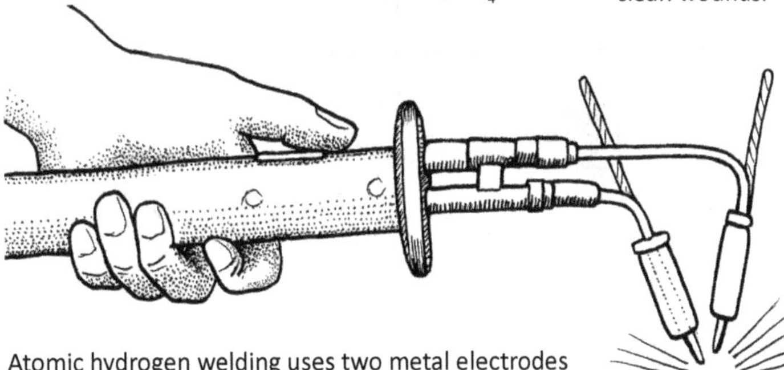

Atomic hydrogen welding uses two metal electrodes with a stream of hydrogen gas between them.

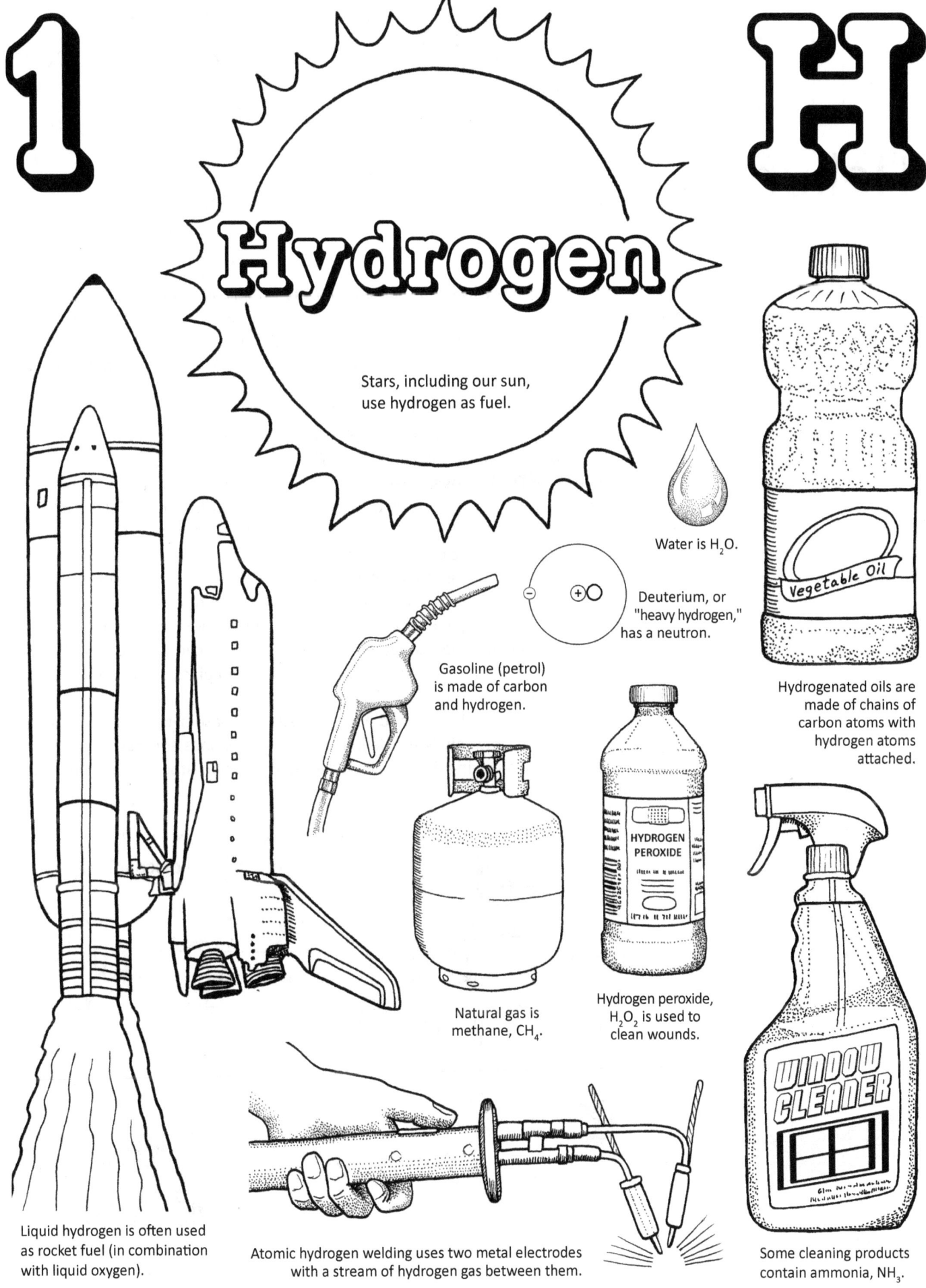

He

Helium

From the Greek word for sun: "helios"

2 protons
2 neutrons
2 electrons

Atomic mass: 4.0

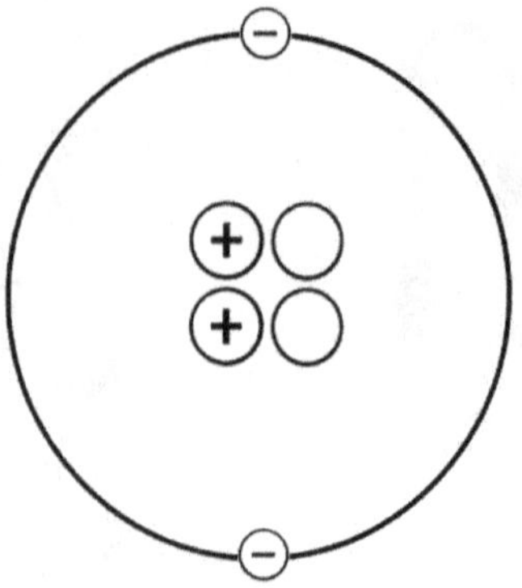

Helium was first discovered in the sun, which is why it was named after the Greek god of the sun, Helios. Scientists in the 1860s were beginning to use a new tool, called the spectrometer, to look at light produced by various things, including the elements as they were burned. They noticed that each burning element seemed to give off a unique light pattern, almost like a fingerprint, by which it could be identified. When they saw a new light pattern as they looked at the sun, they knew it must be a new element. In 1868, Norman Lockyer announced the discovery of a new element that he had named "helium." Then, in 1895, William Ramsay discovered helium in a sample of rock that contained the element uranium. Helium was not just in the sun, but on earth as well! It was later found that helium is produced as uranium atoms break apart, or "decay."

The element helium is a very light gas. Unlike hydrogen, helium is not flammable. Put a spark to helium and nothing happens. This makes it very safe to put in blimps, party balloons, and weather balloons. Helium is so nonreactive that it can be put into rocket engines that are filled with hydrogen. It is also used as a "shield gas" in arc welding, surrounding and insulating the dangerously hot arc of electricity.

Another place the safety of helium comes in handy is in air tanks used by scuba divers. The air around us is mostly nitrogen, with some oxygen mixed in. If divers take normal air down with them, the nitrogen can do something harmful. If the divers come up too quickly, the nitrogen can bubble into their blood, much like bubbles appear when you open a carbonated beverage. Bubbles in your blood is not good! This dangerous condition is called "the bends." (Divers hurt so much they bend over with the pain.) However, if helium is used in place of nitrogen, divers can come back up without having to worry about getting "the bends."

Helium has other technological uses. A mixture of helium and neon is used in red lasers, the kind that are used to read bar codes at check outs in stores. Extremely cold liquid helium is used in machines and devices that need extremely powerful magnets, such as MRI machines in hospitals, and the particle accelerators used by physicists to do experiments with electrons, protons, and neutrons.

Helium atoms don't bond to other atoms. They float around by themselves.

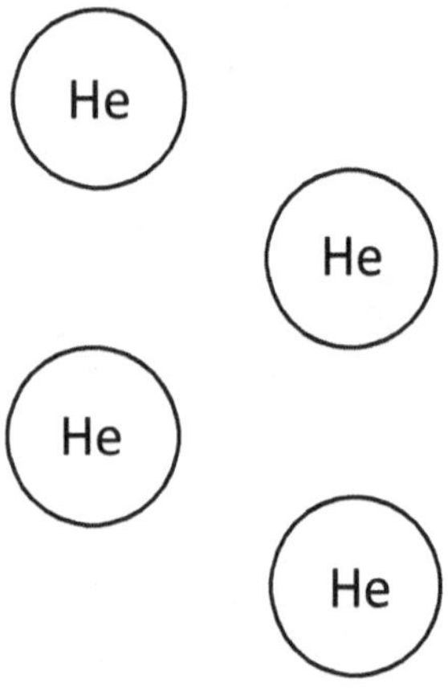

This is a spectrometer from the 1800s, similar to the one used to discover helium.

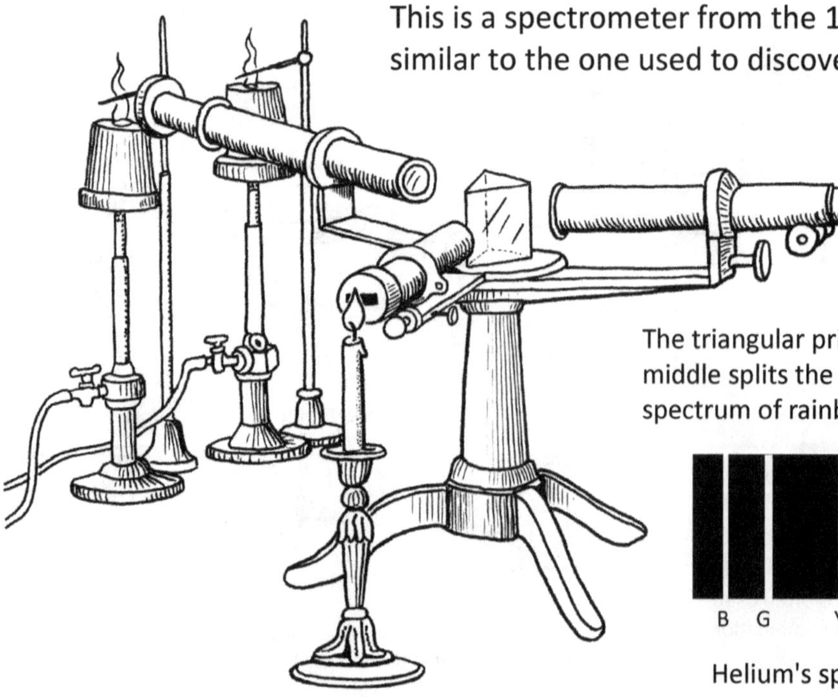

The triangular prism in the middle splits the light into a spectrum of rainbow stripes.

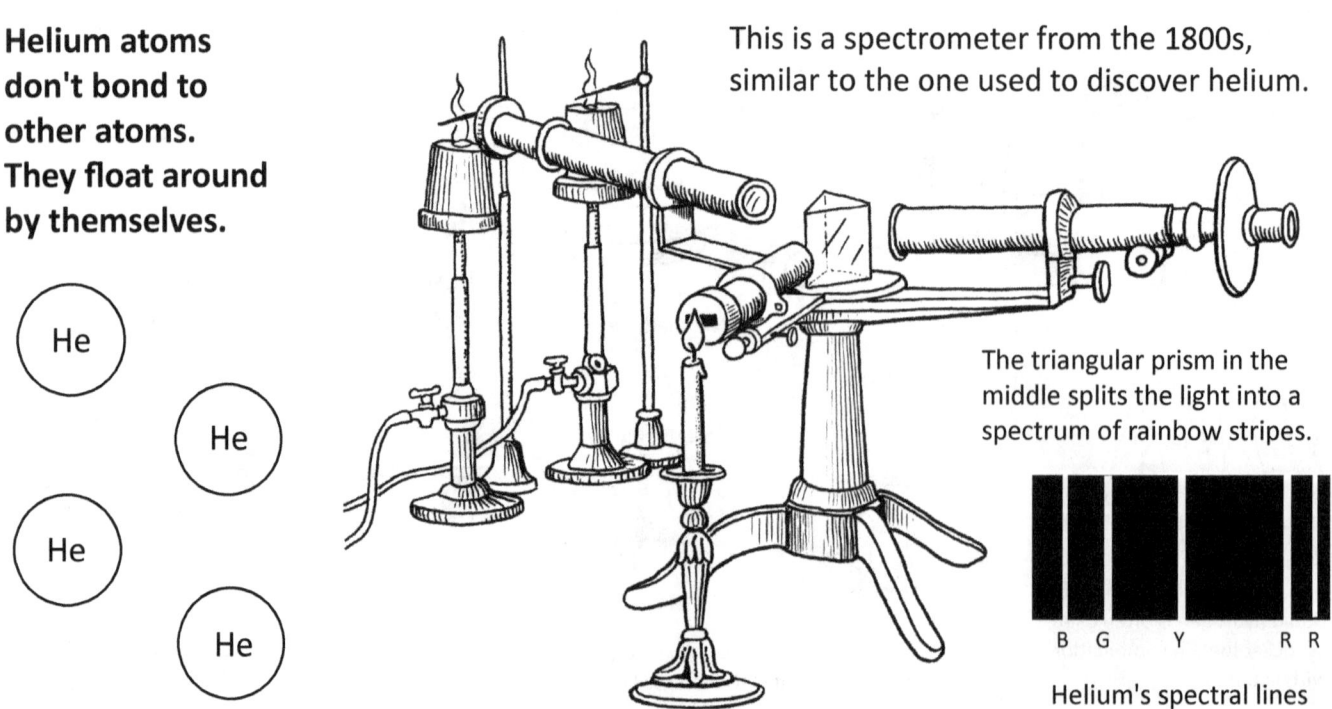

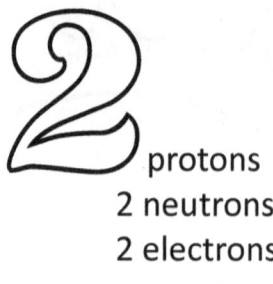

Helium's spectral lines

2 He Helium

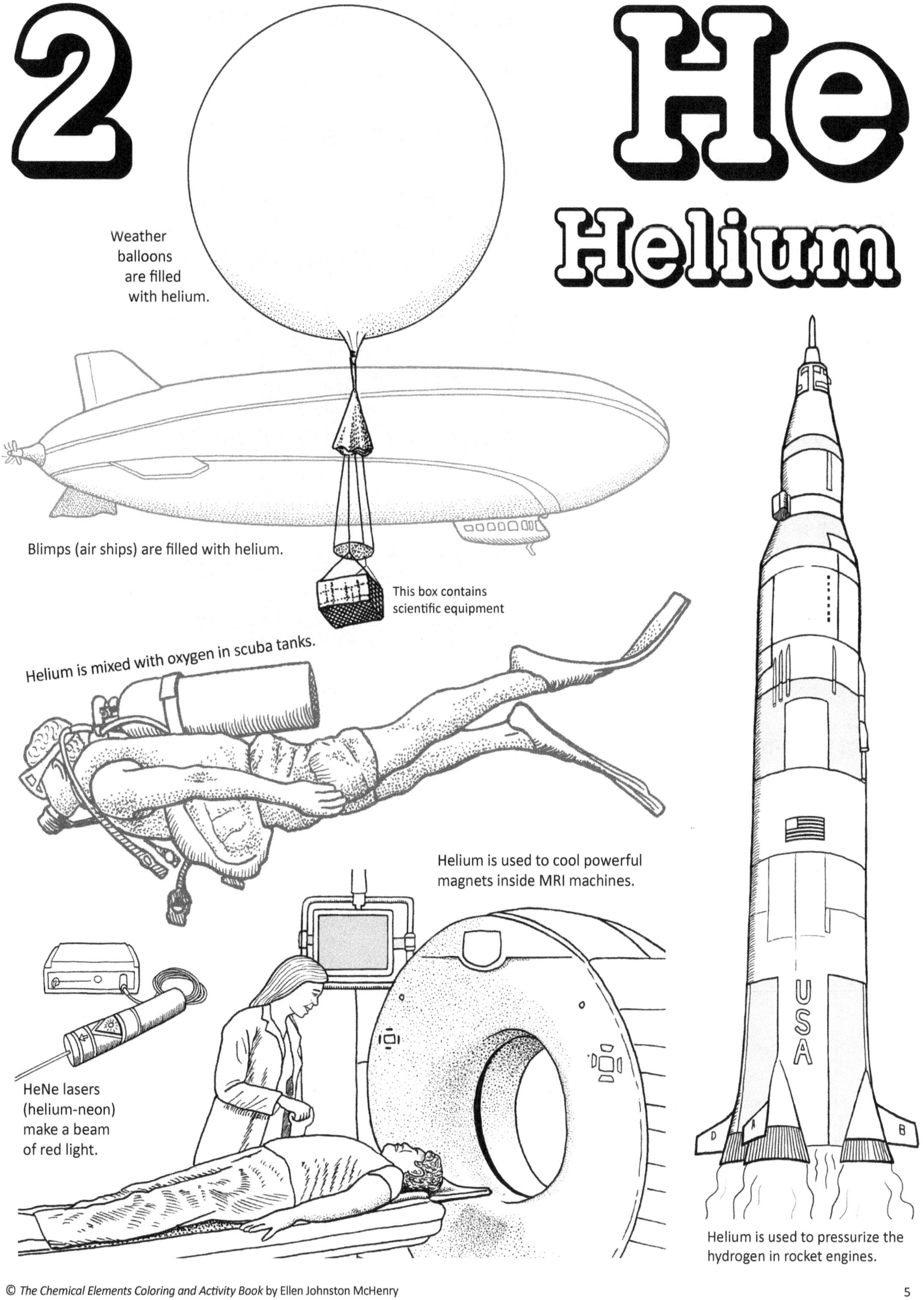

Weather balloons are filled with helium.
Blimps (air ships) are filled with helium.
This box contains scientific equipment
Helium is mixed with oxygen in scuba tanks.
Helium is used to cool powerful magnets inside MRI machines.
HeNe lasers (helium-neon) make a beam of red light.
USA
Helium is used to pressurize the hydrogen in rocket engines.

Li

3

protons
4 neutrons
3 electrons

Atomic mass: 6.94

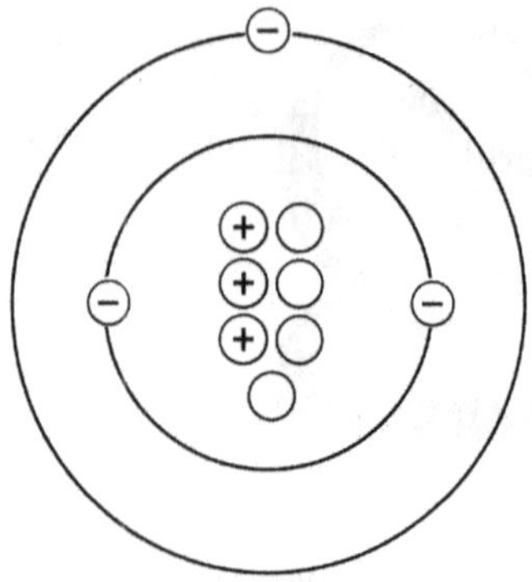

Lithium

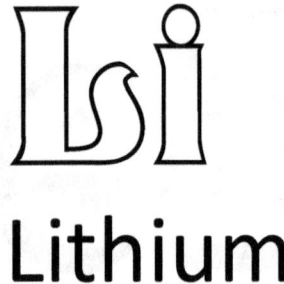

From the Greek word for stone: "lithos"

Lithium is most well-known for its use in long-life batteries, but it's also used in lubricants, fuels, metal alloys, glass making, and even medicines. Some of its useful qualities come from its electron configuration. Do you see that one lonely electron in the outer ring? It's very "unhappy" because it doesn't have a partner to pair up with. The two electrons in the inner ring are paired up, and are therefore very content. The unpaired electron in the outer ring is so "unhappy" that it would rather go off and be part of another atom than stay where it is. Some atoms, like fluorine, chlorine, and bromine (members of the halogen family) are desperate to grab an extra electron that doesn't belong to them, so if they run into a lithium atom, it's a perfect match. Molecules like LiF, LiCl and LiBr are relatively easy to make in a lab. LiF, lithium fluoride, takes the form of a clear crystal which can be used in optical lenses and in radiation detectors. LiCl, lithium chloride, is a white powder that is used in fireworks and emergency flares because it produces a bright reddish-pink flame. LiBr, lithium bromide, can be used to trap moisture in air conditioning systems.

Lithium atoms will bond to small groups of atoms, such as the carbonate ion, CO_3^{2-}. Lithium carbonate, Li_2CO_3, is used by the ceramics industry to make glazes and tile adhesives, by the metal industry to process aluminum, by the glass industry to make ovenware, by the pharmaceutical industry to make medicines, and by the battery industry to make long-life lithium ion batteries. Lithium bonds well to the hydroxide ion, OH^-, to form LiOH, a compound that can remove carbon dioxide from the air that circulates inside an airplane. Lithium will also bond to metals such as aluminum, copper and manganese, making lightweight alloys (metal mixtures) that are used to make airplanes.

Lithium atoms are never found alone in nature. To get a pure sample of lithium, a strong electrical current must be used. Pure lithium looks like a silvery metal and is so light it will float on water. It will also react with the water, trying to get rid of that lonely electron, and this will cause it to look like it is burning on top of the water.

When Li bonds to F, Cl, or Br, it forms a crystal shape:

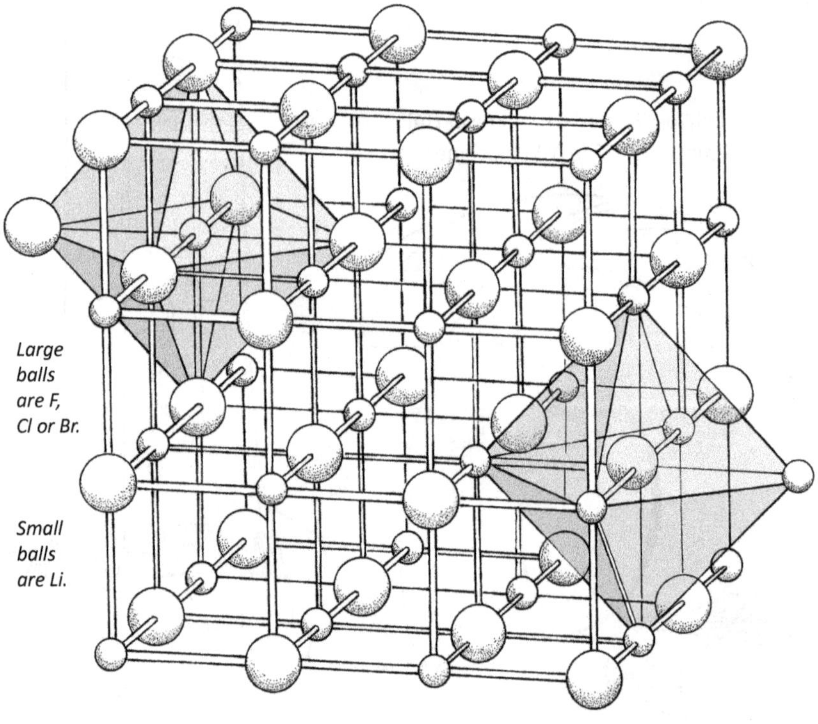

Large balls are F, Cl or Br.

Small balls are Li.

When two Li atoms connect to a CO_3, they don't bond in the way that C and O do. (The sticks represent bonds.) Instead, the Li atoms are held in place by electrical attraction. The positively charged Li atoms (ions) are attracted to the negatively charged oxygen atoms in the CO_3.

This Li_2CO_3 molecule will join with others just like it to form a crystal-like structure.

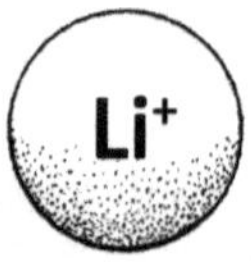

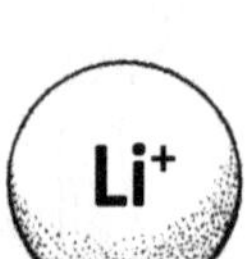

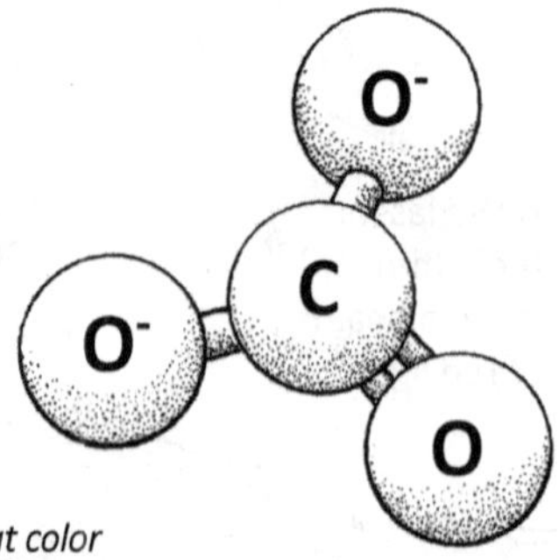

Red: Oxygen (O)
Black: Carbon (C)
You can decide what color to make lithium. A professional artist would probably use purple or pink.

3 Li

Lithium

Lithium makes red sparks when it burns so it is used in fireworks.

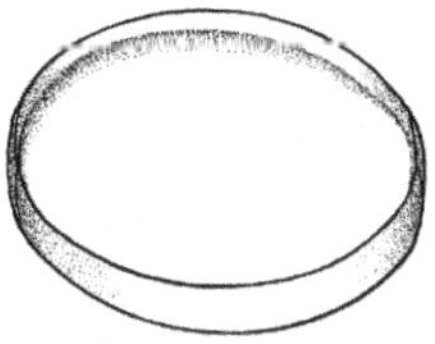

Li is used to make special lenses for high-tech optics.

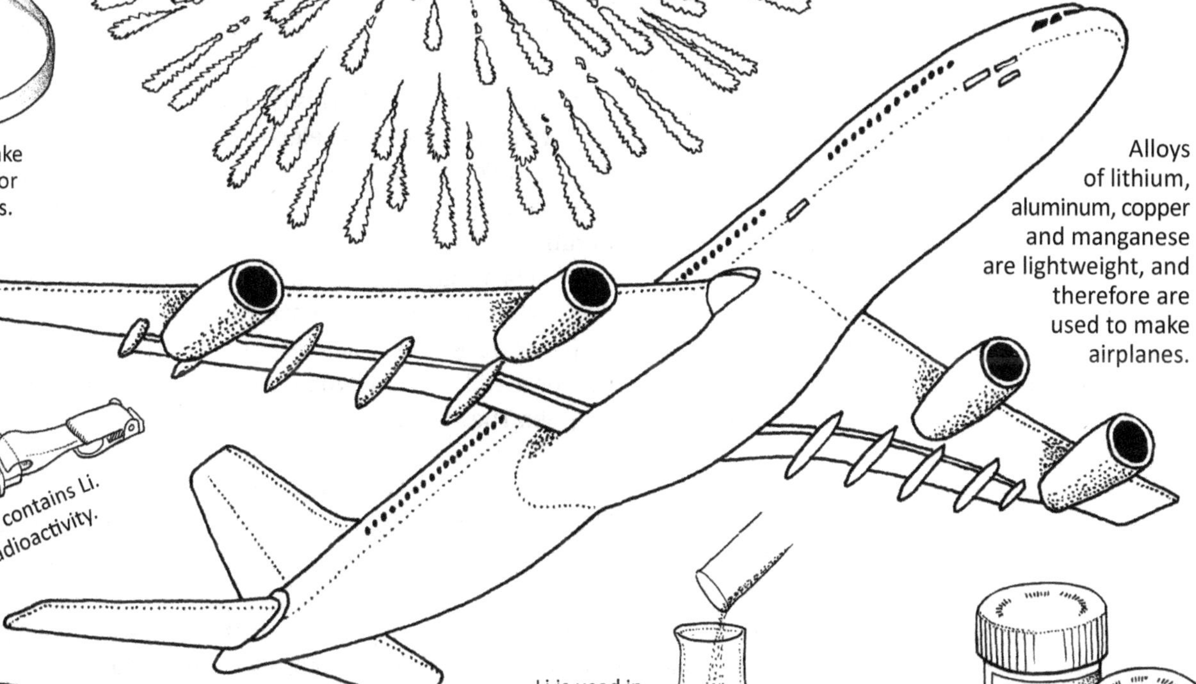

Alloys of lithium, aluminum, copper and manganese are lightweight, and therefore are used to make airplanes.

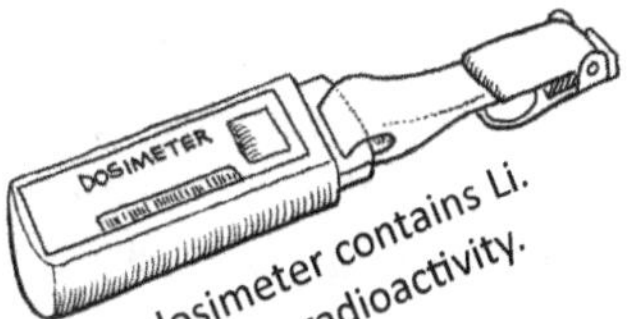

This dosimeter contains Li. It measures radioactivity.

This lubricant contains lithium.

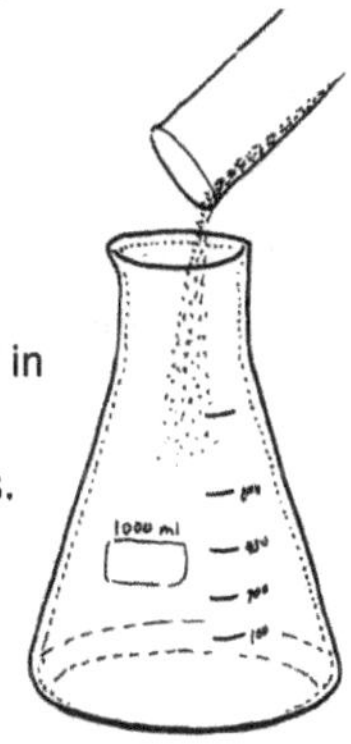

Li is used in chemical reactions.

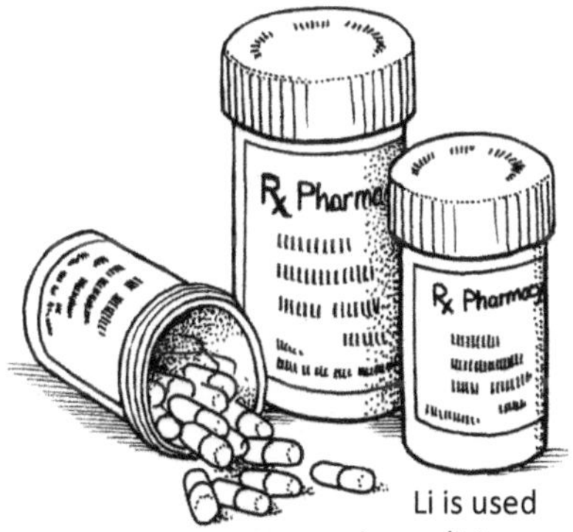

Li is used in medicines.

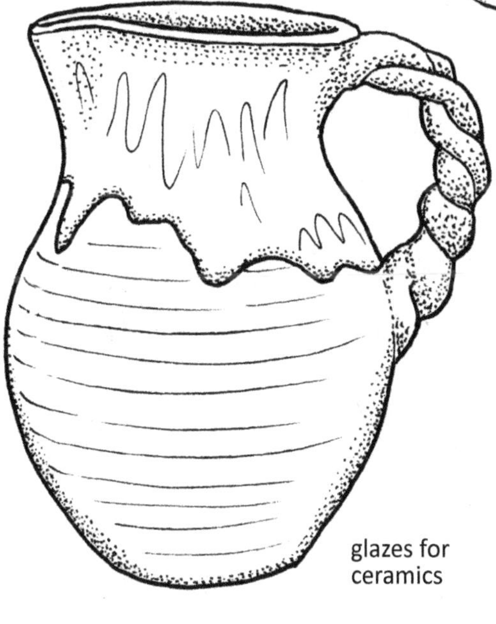

glazes for ceramics

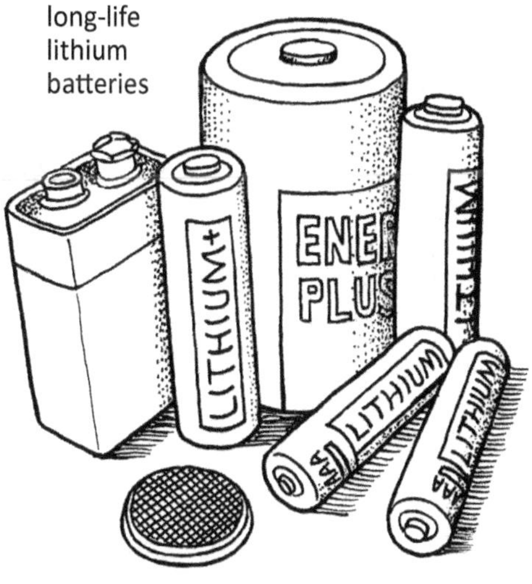

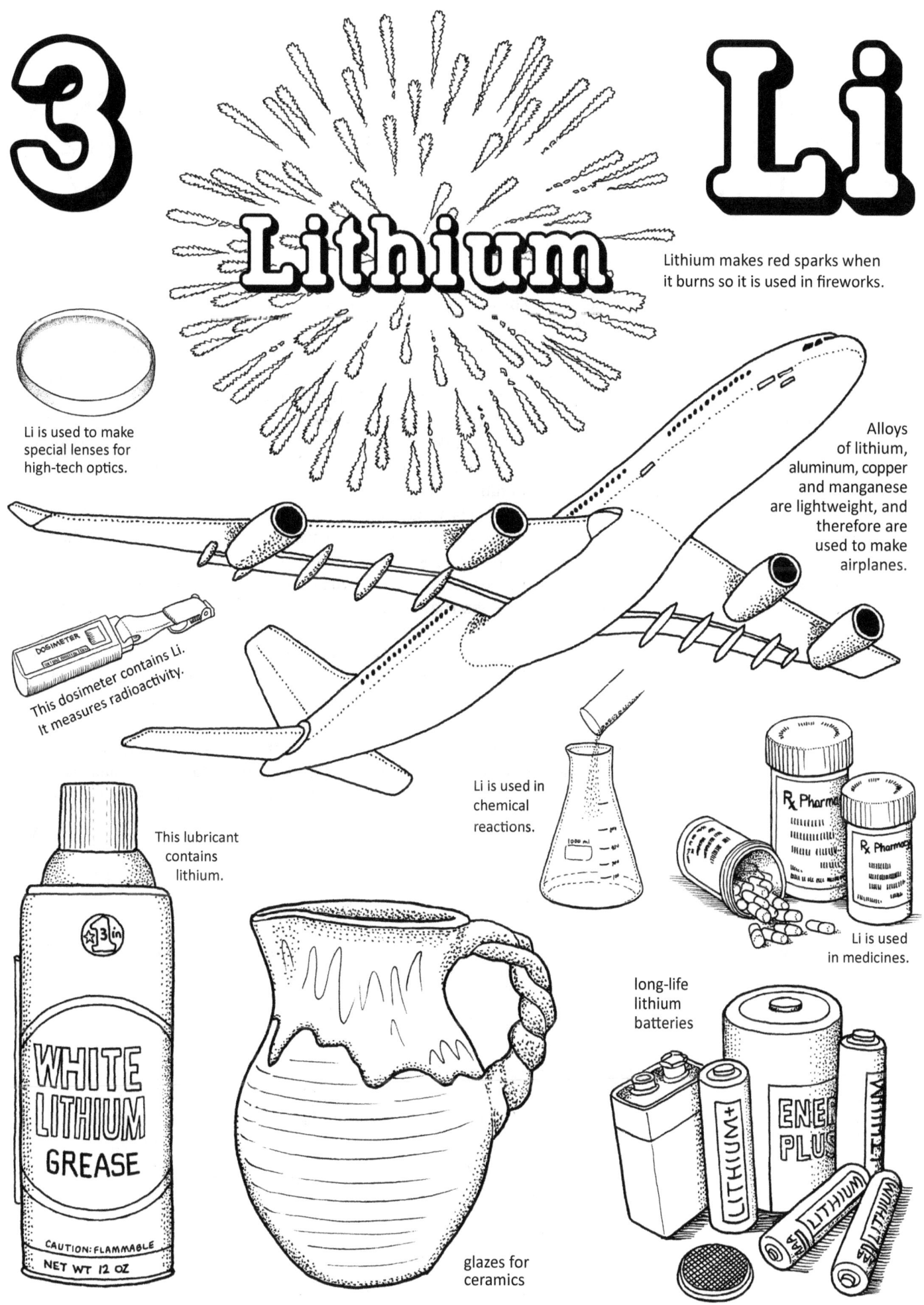

long-life lithium batteries

Be

4

protons
5 neutrons
4 electrons

Atomic mass: 9.01

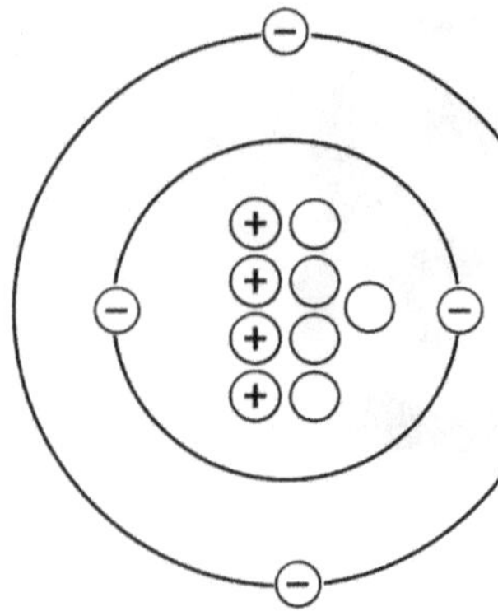

Beryllium

From the mineral "beryl"

Beryllium's name comes from the mineral beryl. Beryl is made of beryllium, aluminum, silicon and oxygen, with this chemical formula: $Be_3Al_2(SiO_3)_6$. When beryl is made into a gemstone, we call it an emerald. Beryllium was first extracted from beryl in 1828 by two people working independently, one in France and one in Germany.

Beryllium is the smallest and lightest member of the alkali earth family (the second column from the left on the Periodic Table). This means that it has two electrons in its outer shell. This is better than just one, but beryllium would prefer to have 8 electrons in its outer shell, so it will easily give up its electrons to another atom or group of atoms. Oxygen makes a natural pairing, since it is looking for two electrons to complete its shell. BeO, beryllium oxide, is used to make parts for rocket engines, as a protective coating on telescope mirrors, as semiconductors in radios, and for ceramic parts in microwave devices, vacuum tubes, and lasers.

Pure beryllium can also be very useful, due to the fact that x-rays will go right through very small atoms. If you want to put a "window" in a vacuum tube, you need a substance that is both strong (won't cave in when the pressure drops inside the tube) and yet will let x-rays pass through. Beryllium is perfect for this.

When a little bit of beryllium is added to another metal, such as copper or aluminum, it makes it stronger. Beryllium bronze is made of 2% beryllium and 98% copper. The strength of beryllium bronze makes it an excellent choice for the manufacturing of parts such as heavy duty springs, which must maintain their shape even under a lot of stress. Beryllium bronze is special in another way, too. It won't create a spark if it strikes another metal, even steel. There are some places where sparks can be very dangerous, and you don't want to take the risk that your tool will start a fire or cause an explosion. Beryllium bronze tools are used on oil rigs, in coal mines, in satellite manufacturing, and by people who repair MRI machines.

Beryllium's claim to fame is that it was used to discover neutrons. In 1932, James Chadwick shot alpha particles (nuclei of helium atoms) at a piece of beryllium, and unknown particles (neutrons) were produced. Beryllium can be used as a source of neutrons for lab experiments, particle accelerators, nuclear power plants, and in atomic bombs.

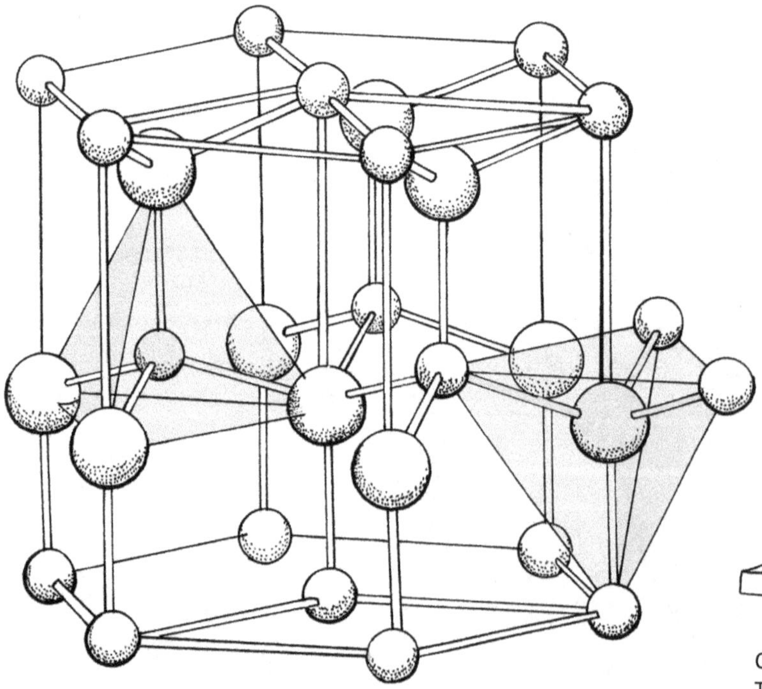

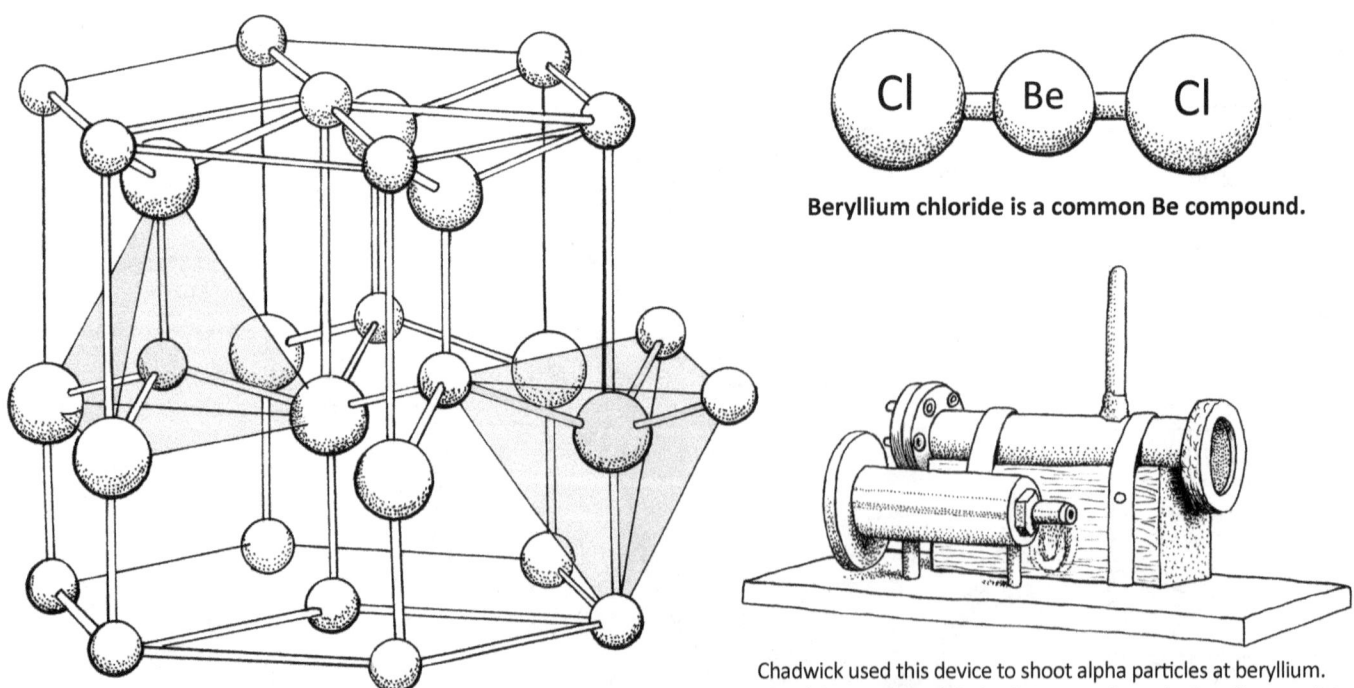

Beryllium oxide (BeO) will make a crystal lattice shape.

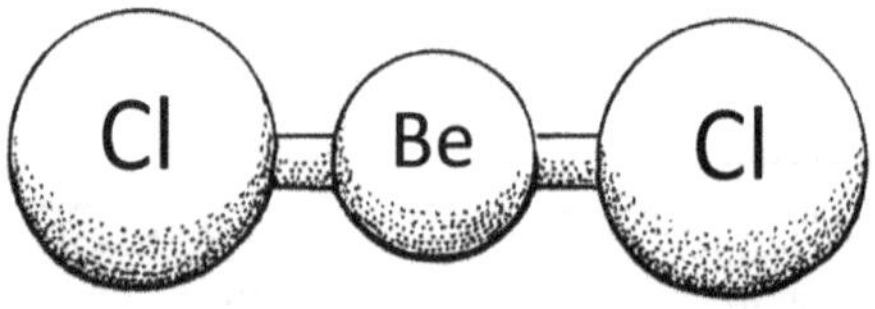

Beryllium chloride is a common Be compound.

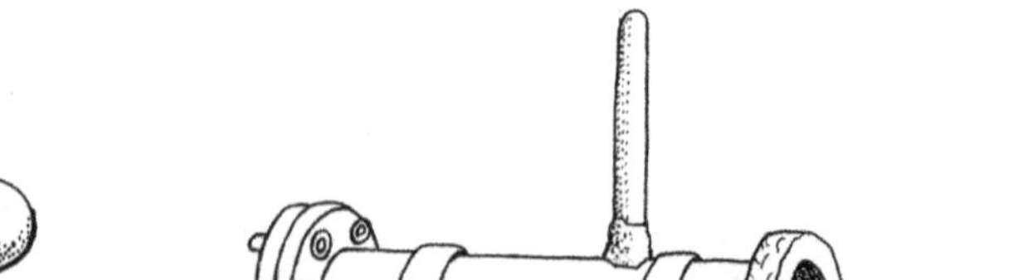

Chadwick used this device to shoot alpha particles at beryllium. The alpha particles dislodged neutrons from the beryllium nuclei.

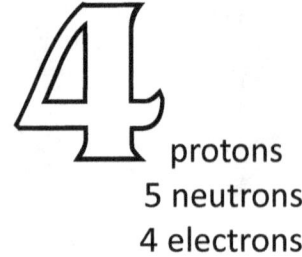

4 Be

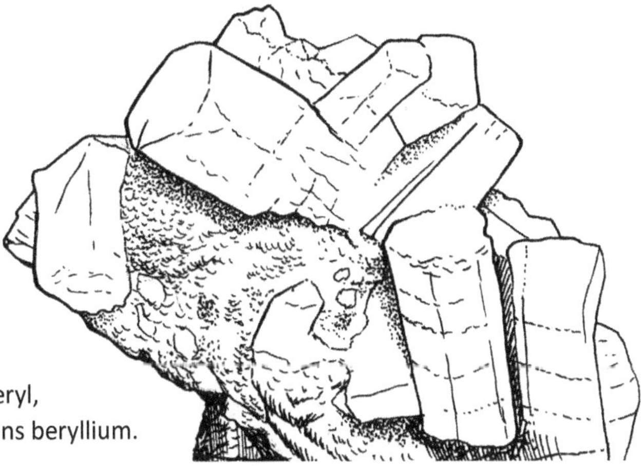

The green mineral beryl, $Be_3Al_2(SiO_3)_6$, contains beryllium.

Beryllium

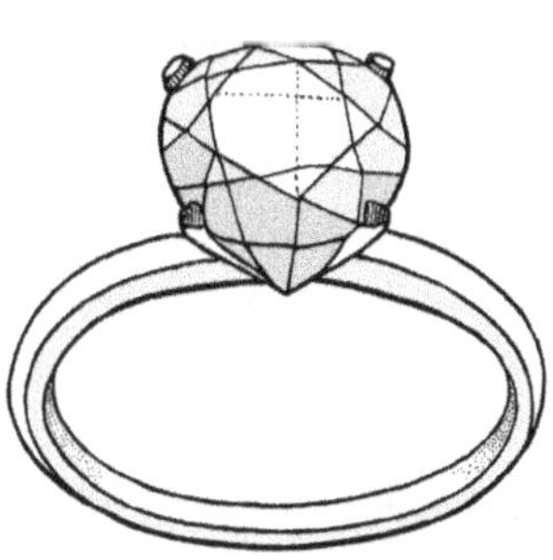

Emeralds are made of beryl.

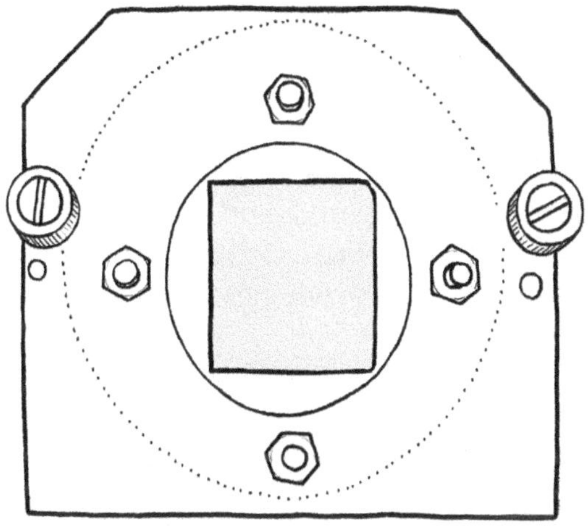

"windows" for x-ray chambers

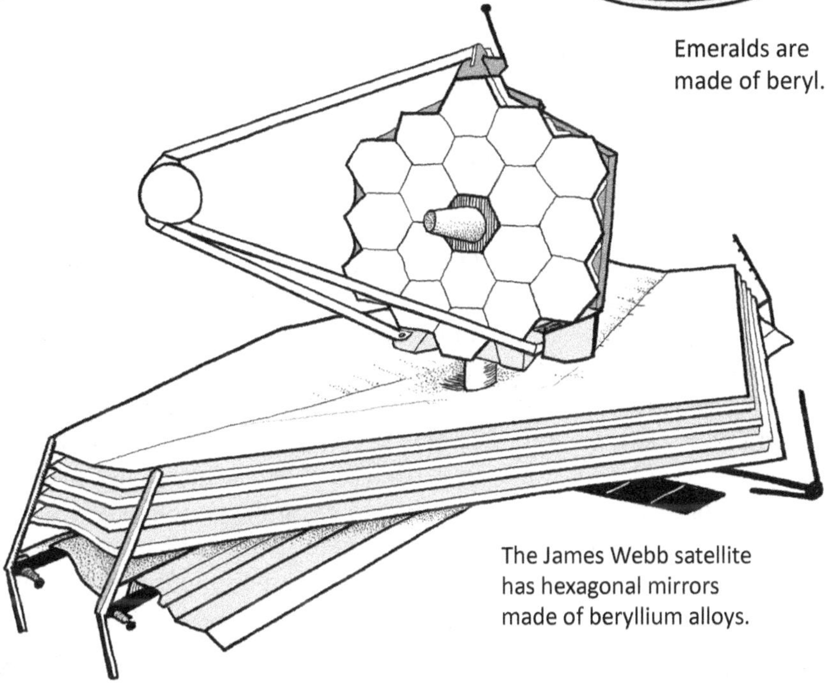

The James Webb satellite has hexagonal mirrors made of beryllium alloys.

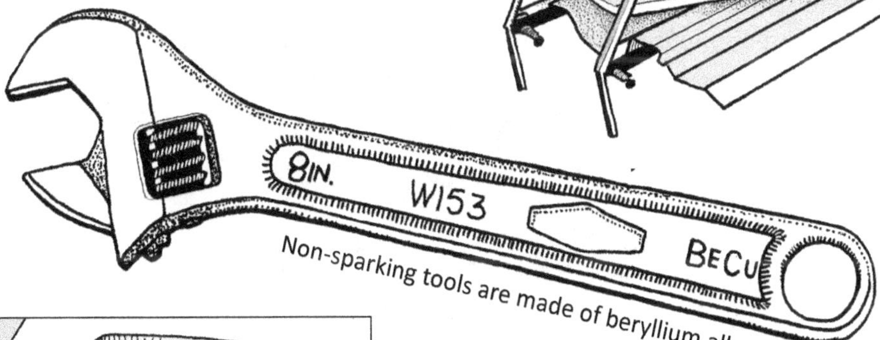

Non-sparking tools are made of beryllium alloys.

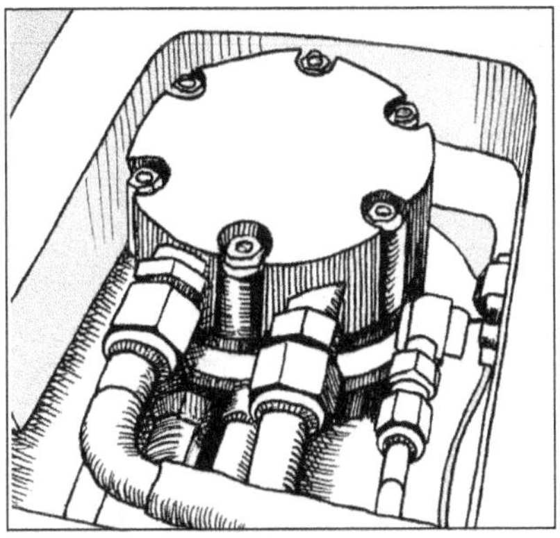

Beryllium inside this machine is used to produce a source of neutrons.

heavy duty springs

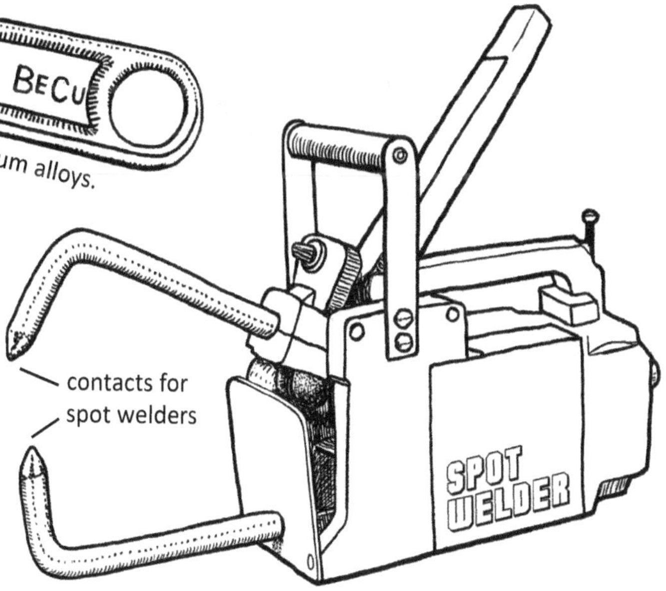

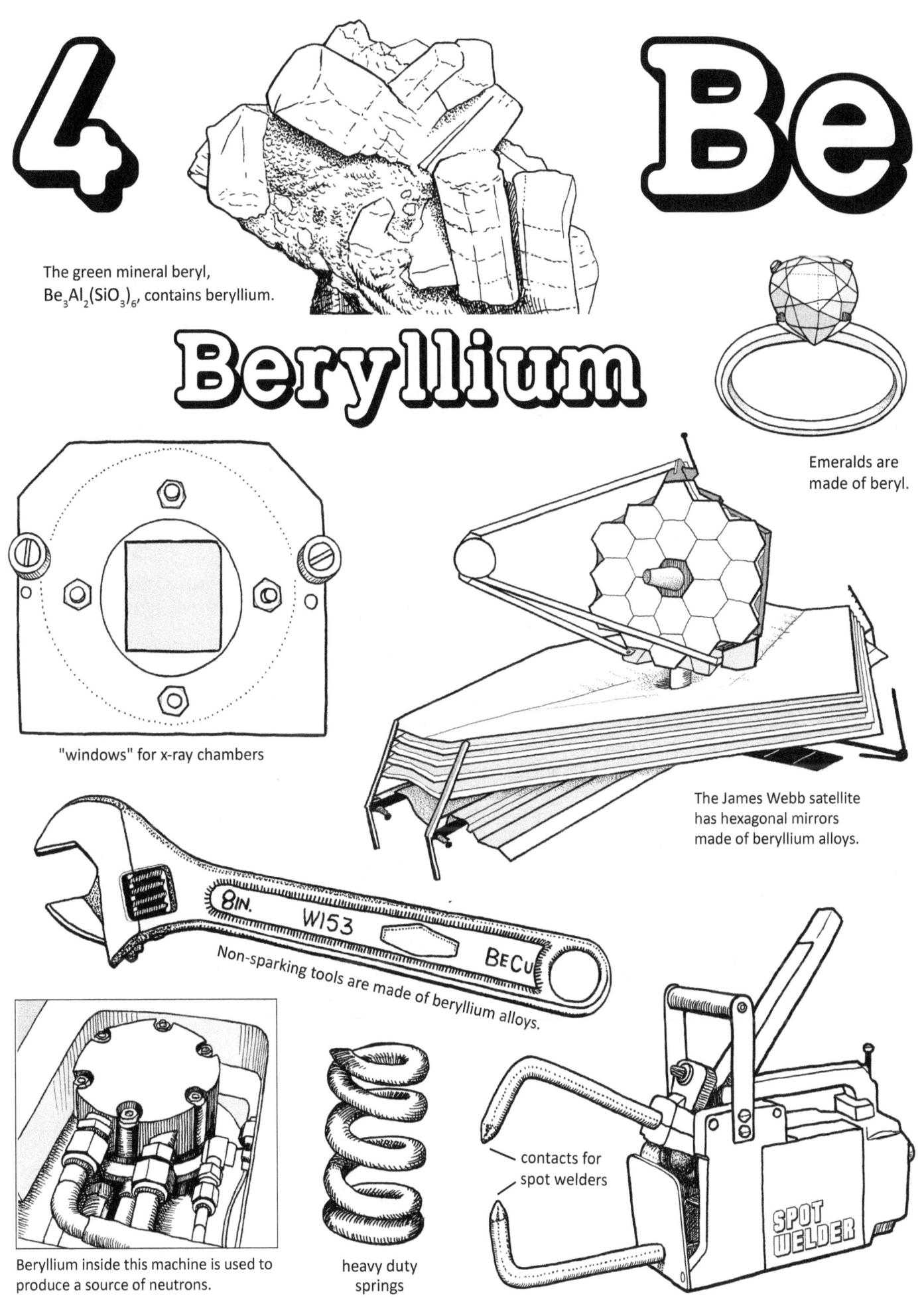

contacts for spot welders

B 5

Boron

From the mineral "borax"

5 protons
5 or 6 neutrons
5 electrons

Atomic mass: 10.81

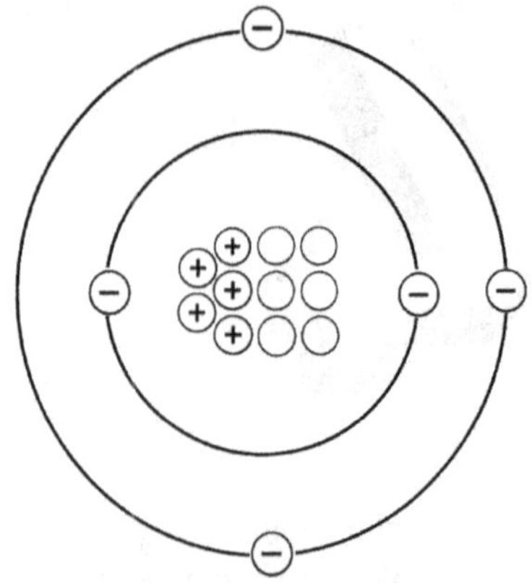

Boron is happy with either 5 or 6 neutrons. It is shown here with 6 neutrons because 80% of all boron atoms have 6. However, if it loses a neutron, it's no big deal. In many atoms, losing a neutron IS a big deal, and this will make the nucleus unstable. (Unstable nuclei tend to fall apart and spit out dangerous particles that can cause damage to plants and animals.) Boron atoms with 5 neutrons can safely add one more. The atoms with 5 neutrons are useful in nuclear power plants that use radioactive (unstable) elements that emit neutrons when they fall apart. Rods containing boron atoms are placed in areas where dangerous free neutrons need to be safely absorbed.

The fact that boron can have either 5 or 6 neutrons explains why its mass is listed as 10.81. The mass is the total number of protons and neutrons in the nucleus. Since boron can have either 5 or 6 neutrons, we must look at as many boron atoms as we can, and then calculate the average. The average turns out to be 10.81, so this is listed as the official atomic mass. But you'll never find a boron atom with 10.81 things in its nucleus! It will always be 10 or 11.

Boron is added to glass to make it less likely to shatter at high temperatures. This "borosilicate" glass is ideal for both kitchens and science labs. (The glassware called Pyrex® is borosilicate glass.) Tiny borosilicate glass beads can be added to paint that is used to put lines on roads. The glass beads in the paint will reflect shining headlights at night. Boron is added to glass that will be spun into the very thin fibers that make fiberglass insulation.

A very useful property of boron is that it won't burn (meaning combustion in the presence of oxygen). Boron compounds, such as zinc borate, can be sprayed onto fabric or wood to make them fire resistant. Boron's presence in fiberglass increases its resistance to fire as well as making the fibers stronger. When boron is used in fireworks, the atoms don't "burn" but they do heat up, showing a bright green color.

Boron is usually extracted from the mineral "borax" ($Na_2B_4O_7$ - $10H_2O$). Borax is used to make laundry washing powder. (Kids might know this powder as the stuff you combine with white glue to make an oozy substance known as "slime" or "goop.") The cleaning power of borax is useful in medicine, too. Borax can be turned into boric acid, H_3BO_3, and put into germ-fighting eye washes. Borax is poisonous to insects and is often used in ant and roach traps.

Boric acid H_3BO_3

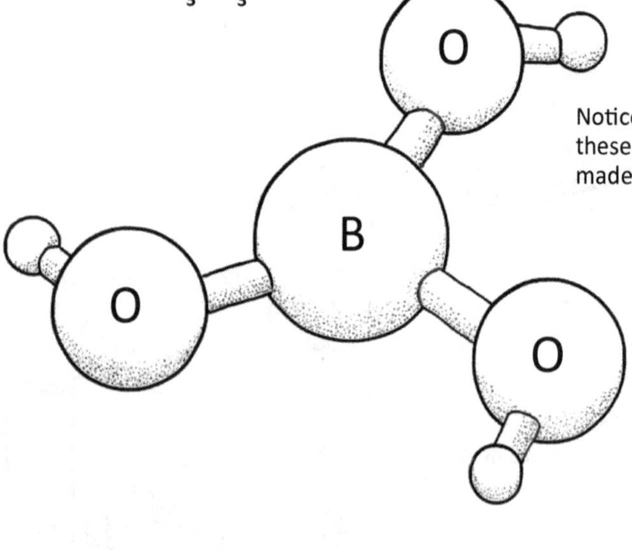

Notice that both of these structures are made of B, O, and H.

You can assign colors as you wish, but scientific illustrators usually make oxygen red and hydrogen white. (The unmarked atoms are H.)

Tetraborate $B_4O_5(OH)_4$

"Tetra" means "4."

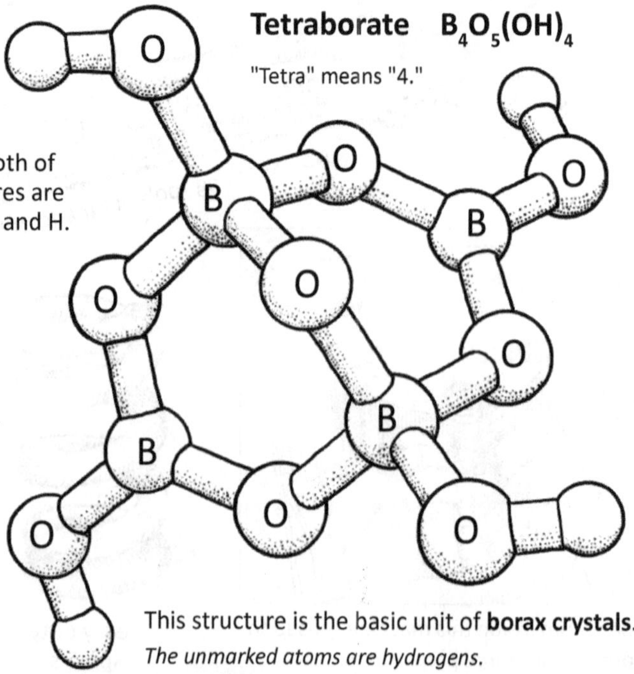

This structure is the basic unit of **borax crystals**.
The unmarked atoms are hydrogens.

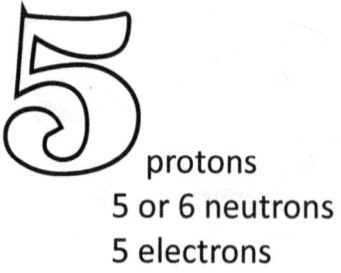

5

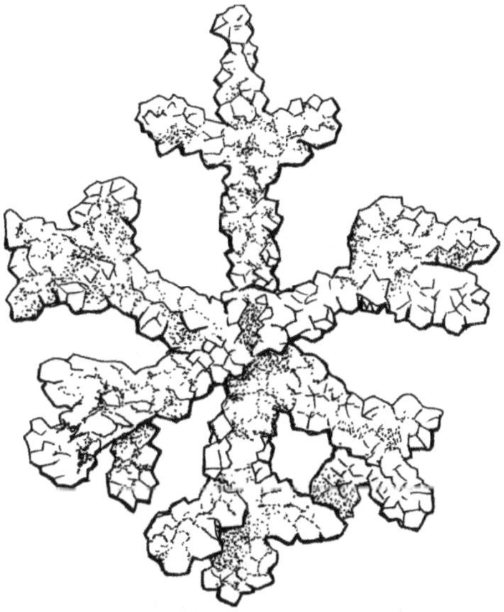

It's fun to watch borax crystals grow on top of shapes you make.

B

Ulexite is a boron mineral with interesting optical properties.

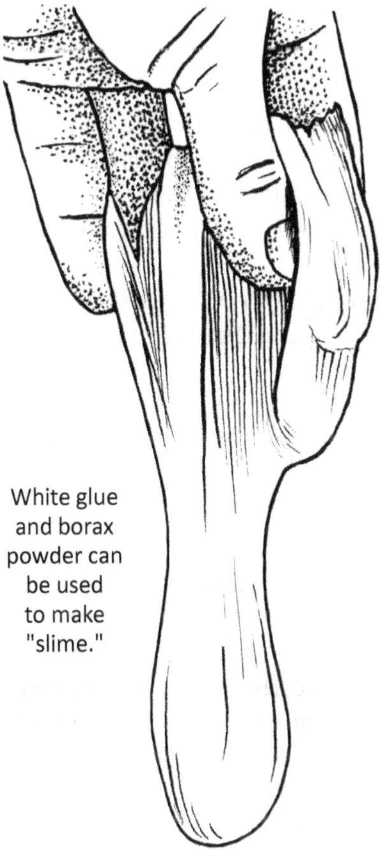

White glue and borax powder can be used to make "slime."

Boron

borosilicate glass

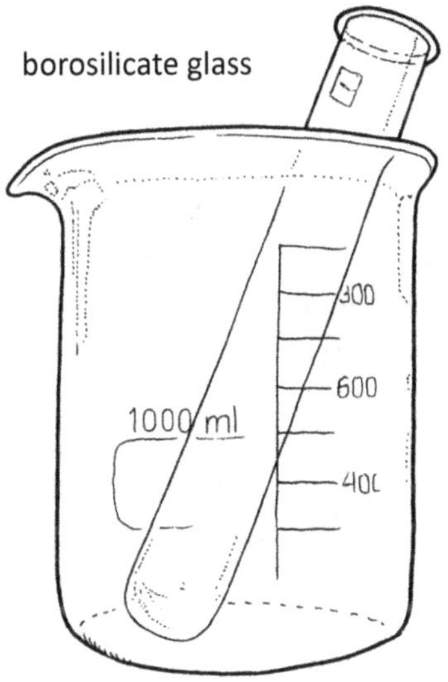

Borax washing powder

fiberglass insulation

antiseptic eye wash

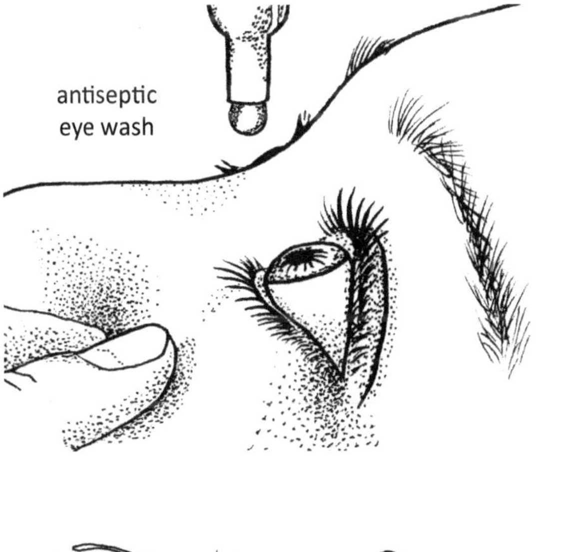

ant poison

C 6

Carbon

6 protons
6 neutrons
6 electrons

Atomic mass: 12.01

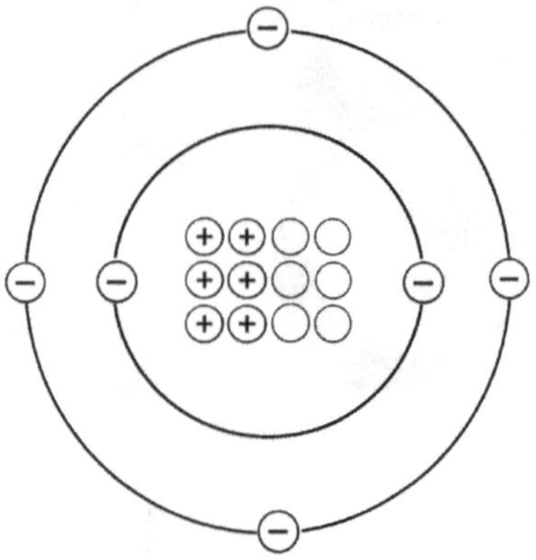

From the Latin word for charcoal: "carbo"

Carbon is the most flexible and "friendly" atom on the Periodic Table. It will bond with many other elements, although its favorites are hydrogen and oxygen. If there are no other atoms around to bond with, carbon will bond to itself, forming pure-carbon substances such as diamonds, graphite and coal. That's right, coal and diamonds are made of the same stuff! The most fascinating pure-carbon structure is the buckyball, a hollow sphere of 60 carbon atoms arranged in the same pattern as a soccer ball (hexagons surrounded by pentagons).

Carbon is found in the air around us as carbon dioxide, CO_2. Vehicles can put both CO_2 and CO (carbon monoxide) into the air as by-products of combustion. CO is very dangerous and many people have CO detectors in their homes if they have a furnace that burns natural gas, CH_4.

Carbon can bond to three oxygen atoms and make the carbonate ion, CO_3^{2-}. If a calcium atom sticks to carbonate, we get calcium carbonate, $CaCO_3$. Calcium carbonate is the main ingredient in the mineral calcite and in the rock known as limestone. Sea shells are a biological form of calcium carbonate.

Hydrocarbon molecules are made of just carbon and hydrogen atoms and can be small (CH_4, natural gas), medium-sized (C_8H_{18}, octane, liquid gasoline) or so long we can't even count the carbon atoms (plastics and rubbers). Carbon and hydrogen atoms can also form a ring known as benzene. The benzene ring, or an adaptation of it, is at the heart of thousands of molecules, including polystyrene plastic, Styrofoam®, food preservatives, cholesterol, natural almond flavor, spot removers, moth balls, paints, and medicines.

Many biological molecules have carbon at their core. Proteins, fats and sugars are all carbon-based substances. DNA, the extremely long ladder-shaped molecule that is like a library of information for living cells, has carbon atoms at key points in its structure. Carbon is also at center of many other molecules essential to life, including enzymes.

Limestone
$CaCO_3$

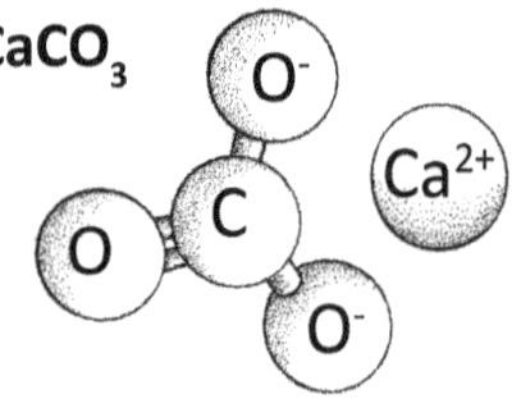

Methane
CH_4

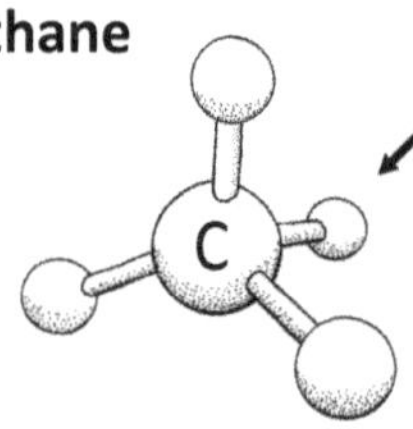

Octane C_8H_{18}

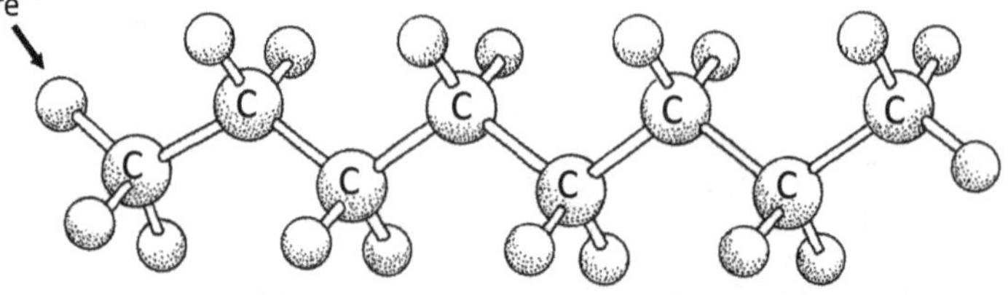

Benzene C_6H_6

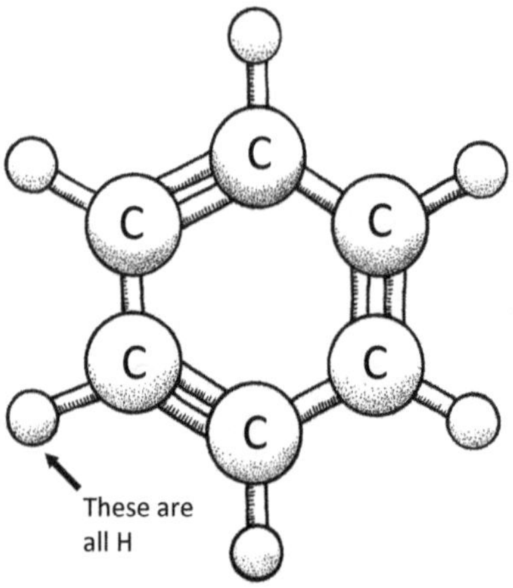

Diamond lattice (pure C)

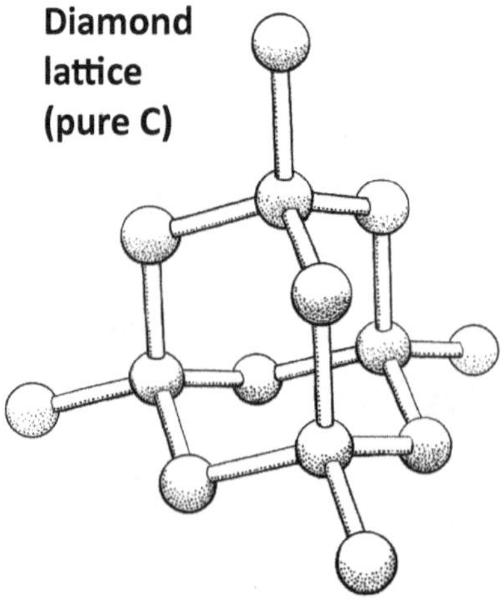

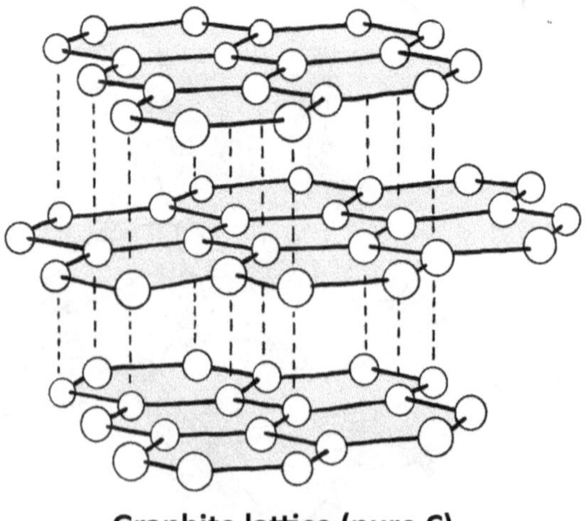

Graphite lattice (pure C)

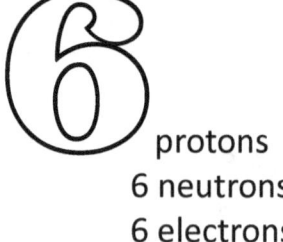

6 Carbon C

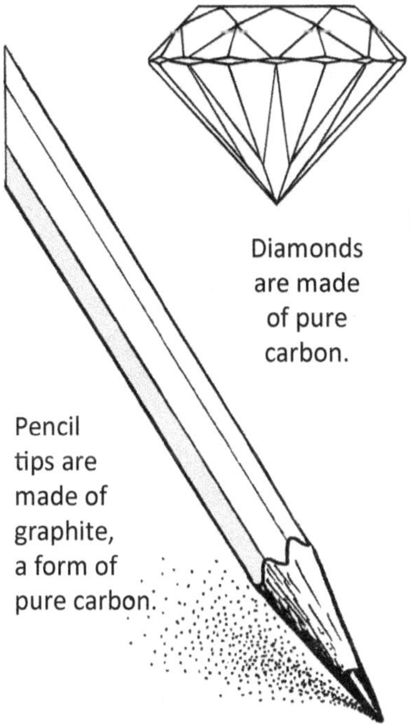

Diamonds are made of pure carbon.

Pencil tips are made of graphite, a form of pure carbon.

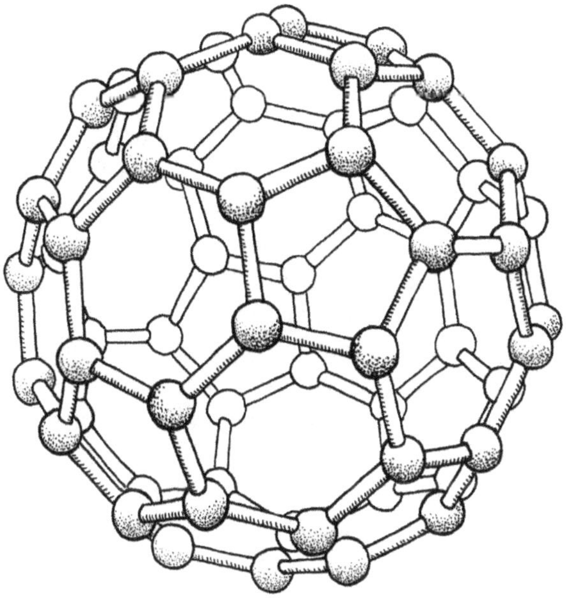

"Buckyballs" are made of 60 carbon atoms.

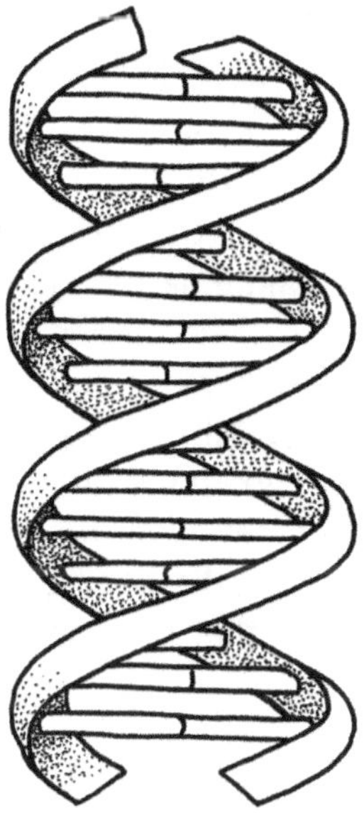

Carbon atoms determine the structure of DNA.

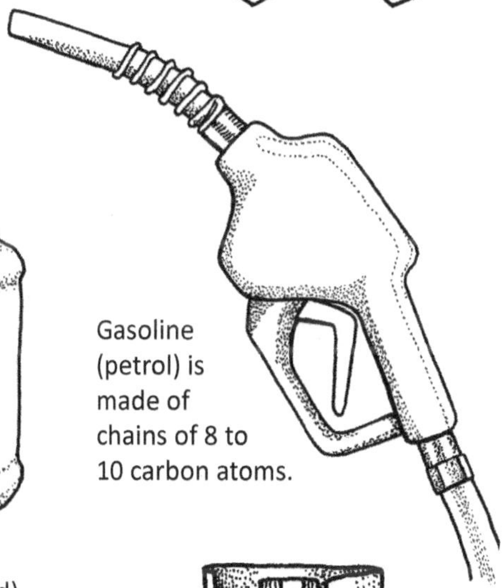

Gasoline (petrol) is made of chains of 8 to 10 carbon atoms.

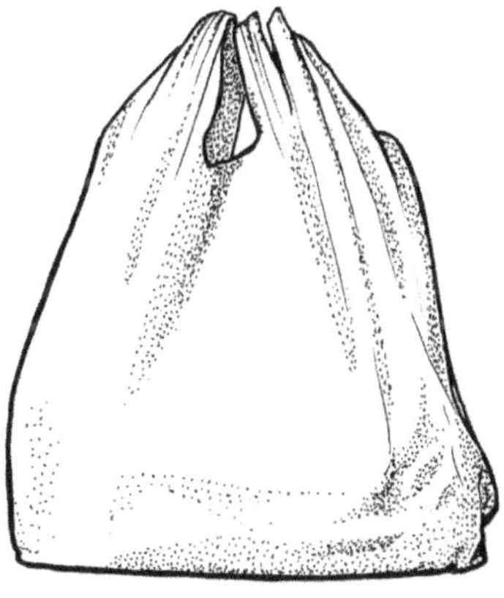

All forms of plastic are made of long chains of carbon atoms (with hydrogens attached).

Coal is also a form of carbon. Miners in the 19th century used mules to pull heavy coal cars out of mines.

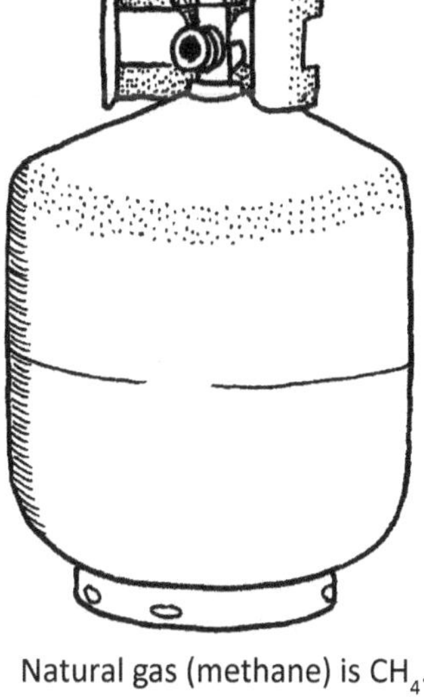

Natural gas (methane) is CH_4.

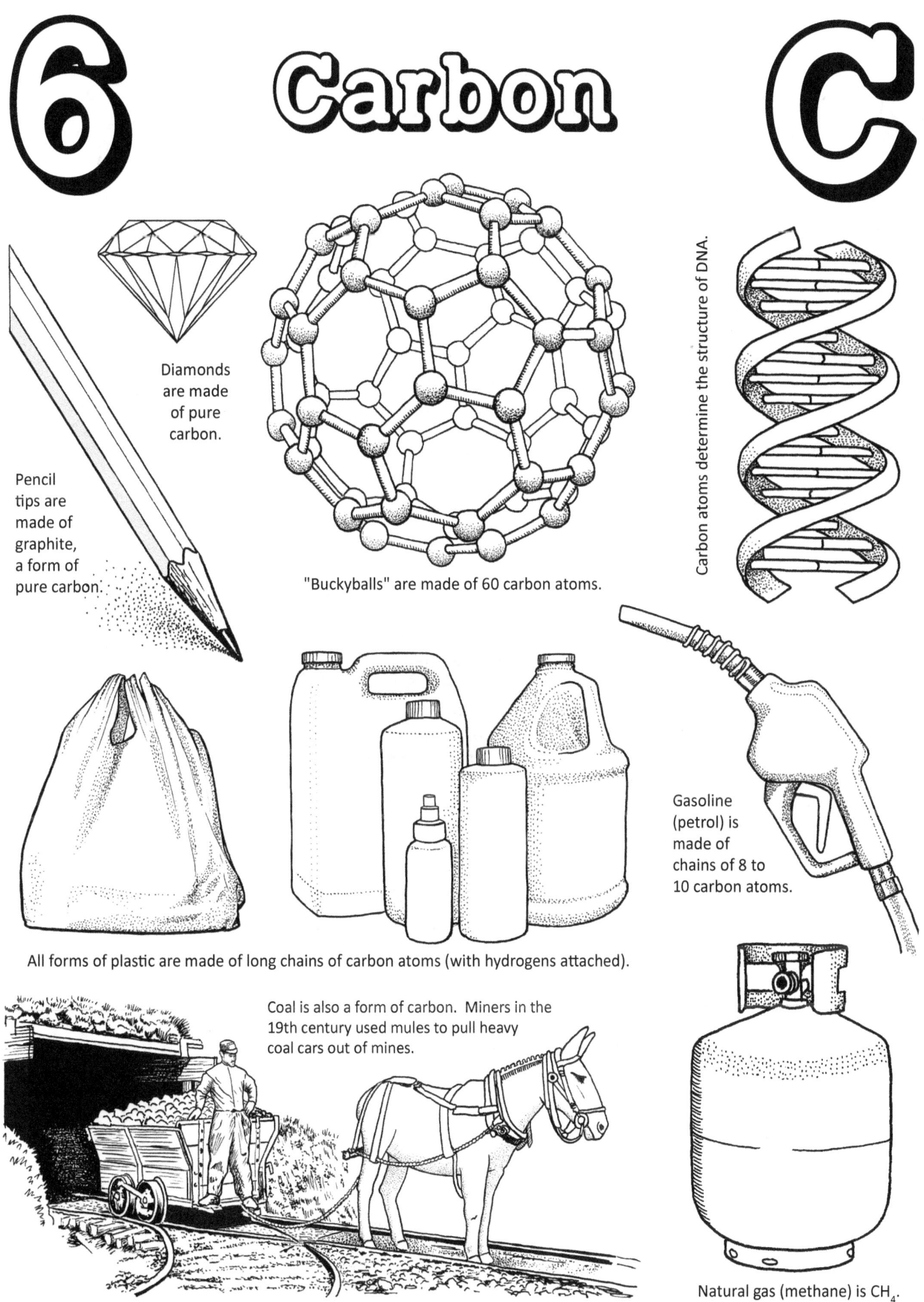

N

7

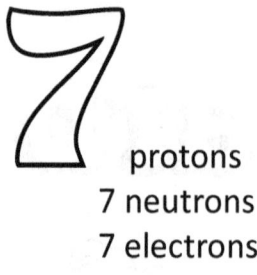

protons
7 neutrons
7 electrons

Atomic mass: 14.0

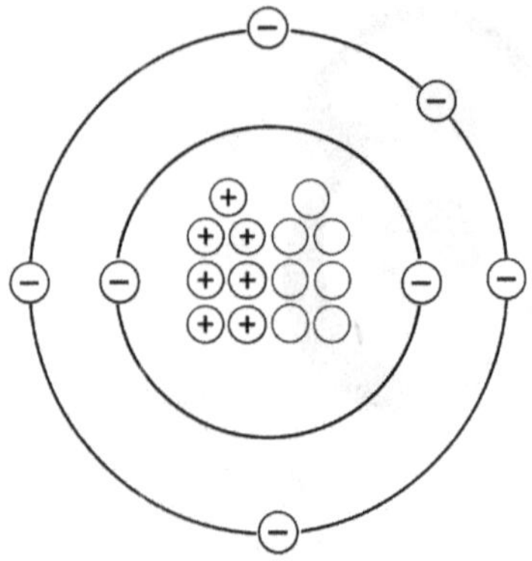

Nitrogen

From Greek words "nitron" (saltpeter) and "genes" (make)

Nitrogen is named after one of the compounds in which it is found, potassium nitrate (KNO_3), which was known in the ancient world as "nitron" and was named "saltpeter" by the Europeans. ("Peter" means "rock.") Since ancient times, this mineral has been used to preserve meats, and nitrates are still used today to keep packaged meats from turning brown. Eventually, it was discovered that saltpeter could be made into gunpowder if charcoal and sulfur were added it to. Gunpowder isn't just for guns; large amounts of gunpowder are also used to make fireworks. Nitrogen is found in other explosive chemicals, such as TNT (trinitrotoluene), nitroglycerin (dynamite), and sodium azide (NaN_3) which is found in air bags in cars. The explosive nature of all these chemicals is due to the fact that they allow nitrogen gas, N_2, to form very quickly.

The air around us is mostly nitrogen in its stable form, N_2. When two nitrogen atoms are bound to each other, they form one of the most stable molecules in nature, unwilling to react with anything around it. N_2 is so stable that it can be used as a fireproof shield in high temperature welding. Pure nitrogen gas can also protect and preserve fruits such as apples, keeping them fresh (in cold storage) for up to two years.

N_2O is nitrous oxide, often called "laughing gas." It has a slightly sweet smell and is used by doctors and dentists as a mild anesthetic for minor surgery or dental procedures. It's easy to confuse N_2O with NO_2, but NO_2 is nitrogen dioxide, a reddish-brown gas that is a common form of air pollution.

NH_3 is ammonia, a gas with a pungent odor that can sting your nose. It's that strong smell that comes from wet diapers or cat litter boxes that have been sitting for a while. NH_3 is used in many industrial processes, including the manufacturing of fertilizers that can put nitrogen into the soil. Plants need nitrogen but can't get it from the air.

In biology, we find nitrogen in more complex molecules, such as amino acids, which are strung together to form proteins. Proteins are the building blocks from which cells and tissues are made.

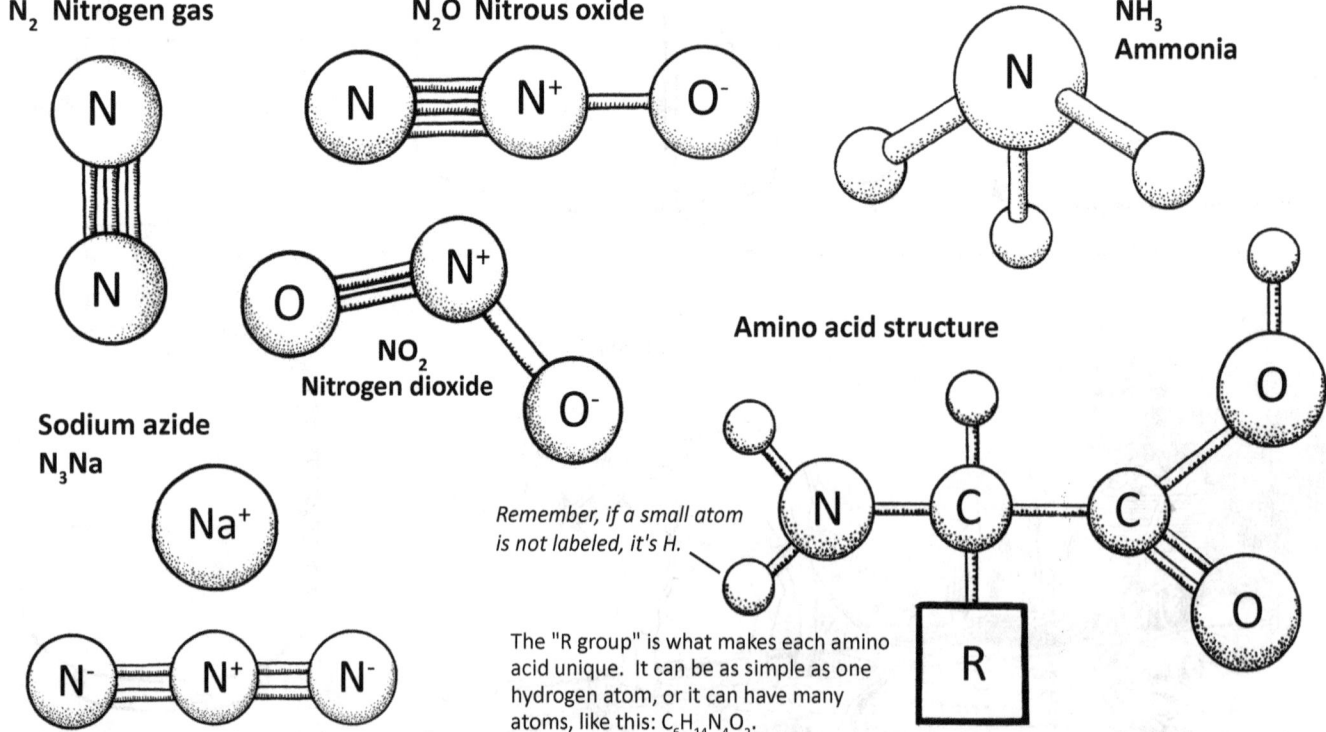

N_2 Nitrogen gas

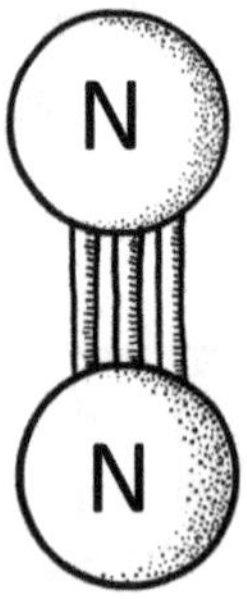

N_2O Nitrous oxide

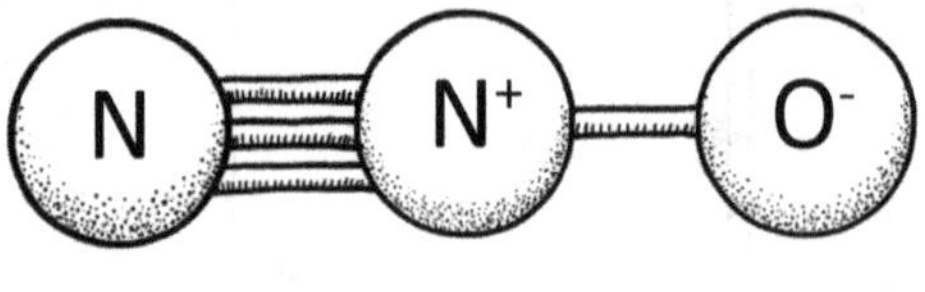

NH_3 Ammonia

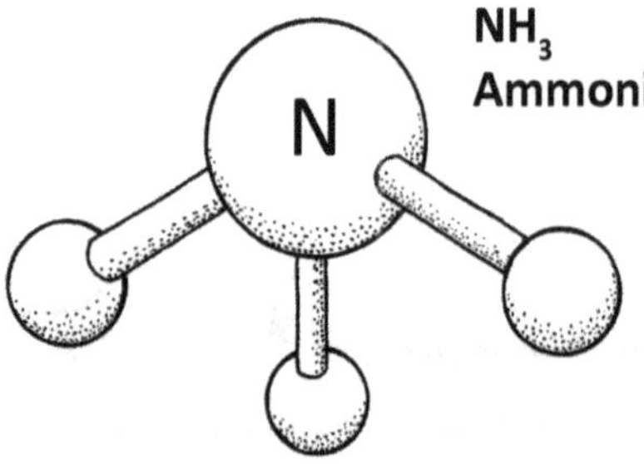

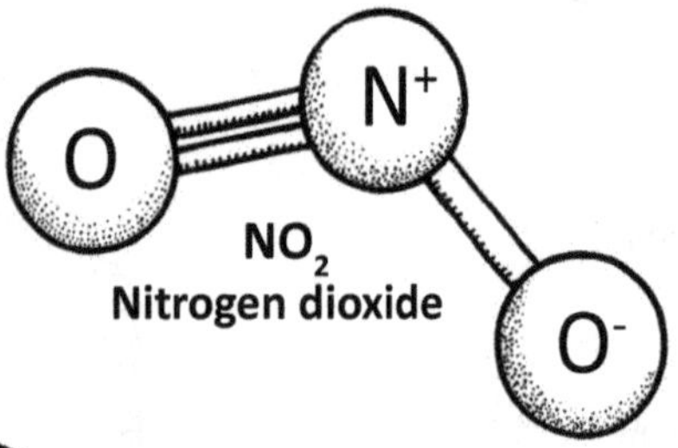

NO_2 Nitrogen dioxide

Sodium azide N_3Na

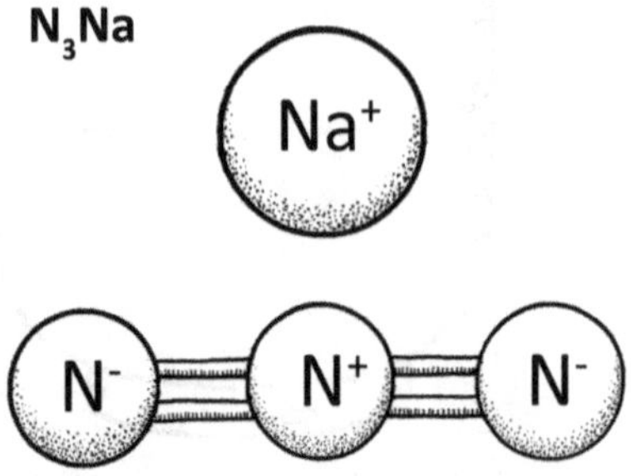

Amino acid structure

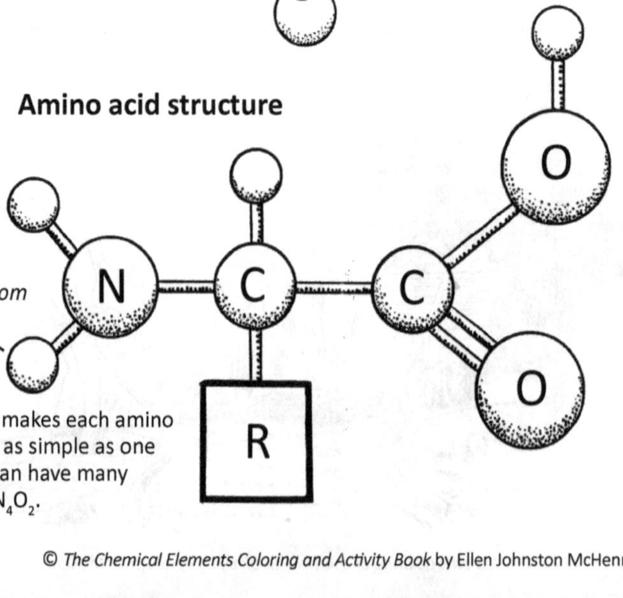

Remember, if a small atom is not labeled, it's H.

The "R group" is what makes each amino acid unique. It can be as simple as one hydrogen atom, or it can have many atoms, like this: $C_6H_{14}N_4O_2$.

7

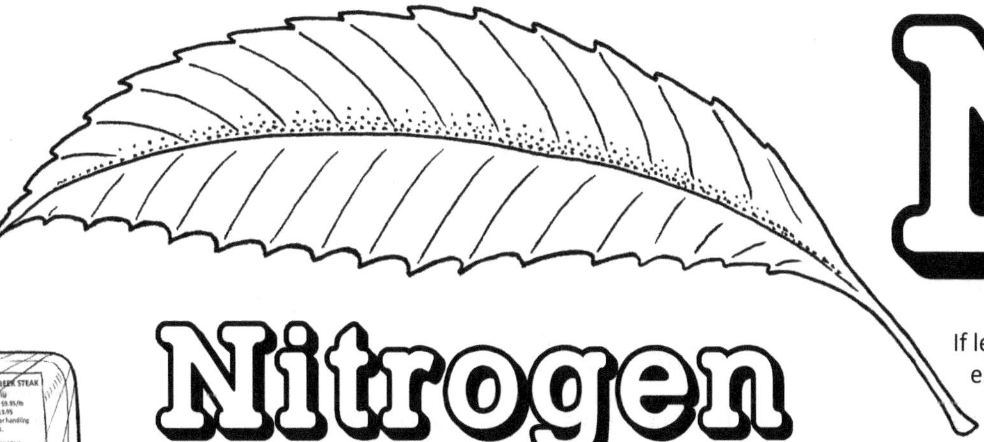

N

Nitrogen

If leaves don't get enough nitrogen they lose their green color.

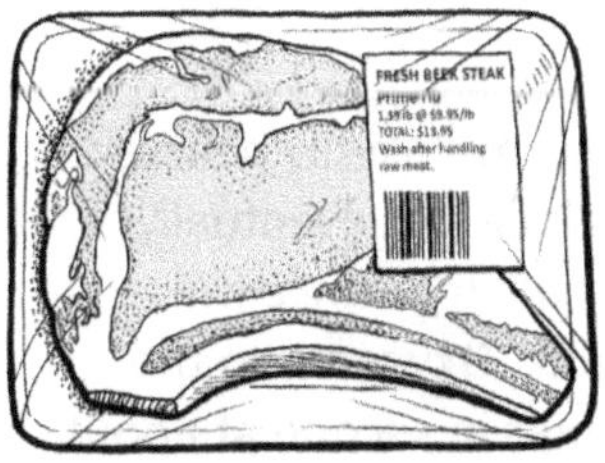

Nitrates and nitrites are used to preserve meats.

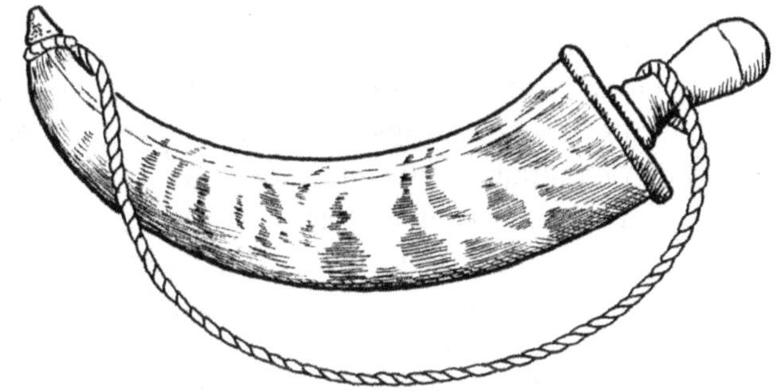

Throughout history, horns have been used to store gunpowder (a nitrogen compound).

Some cleaning products contain ammonia, NH_3.

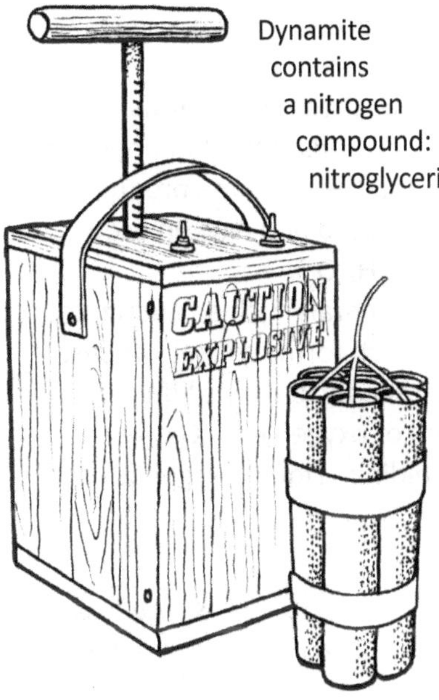

Dynamite contains a nitrogen compound: nitroglycerin.

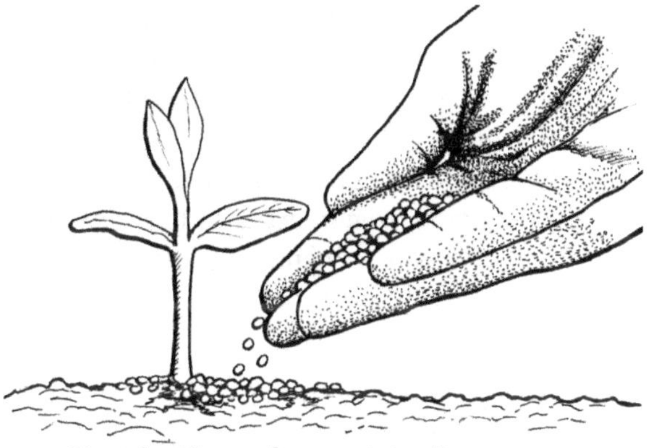

Plant fertilizers often contain nitrogen.

Air bags in vehicles are inflated with nitrogen.

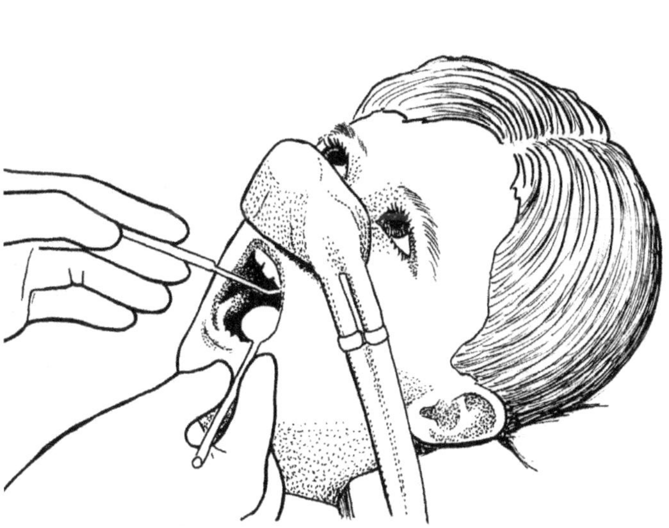

N_2O, nitrous oxide, is used by dentists as a mild anesthetic.

O

8

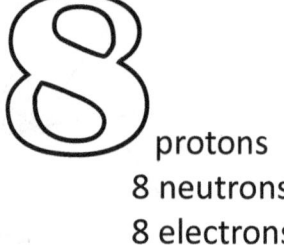

protons
8 neutrons
8 electrons
Atomic mass: 15.9

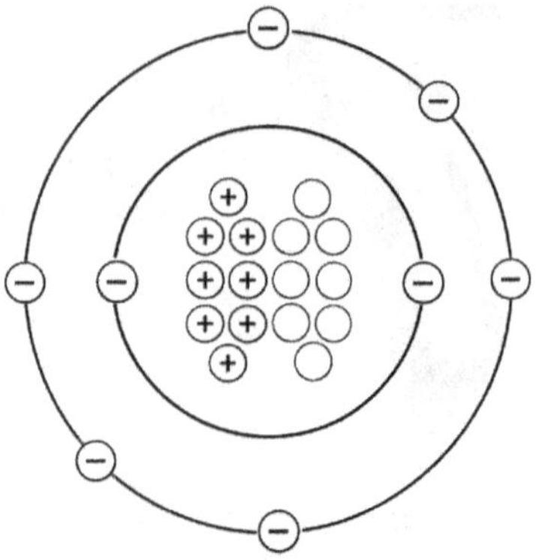

Oxygen

From Greek words "oxy" (sour) and "genes" (make)

Oxygen's name comes from the fact that when it was discovered in the late 1700s, it was mistakenly believed to be the key factor in the formation of acids, which taste sour (lemon juice, for example). This idea was eventually proven wrong, but by that time the name "oxygen" was being used by everyone and it was too late to change it.

Oxygen is the third most abundant element in the universe, after hydrogen and helium. Oxygen is in the air all around us as O_2. Nitrogen, N_2, makes up about 78% of our atmosphere and oxygen comes in second at about 20%. In the upper atmosphere we find three oxygen atoms stuck together to form ozone, O_3. Ozone layers help to protect earth from dangerous ultra-violet radiation produced by the sun.

Oxygen is a very reactive element and will bond to most of the other elements to form compounds whose names usually end in "ate," "ide," or "ite." The most well known oxide compound is water, H_2O. A similar molecule is H_2O_2, hydrogen peroxide, whose usefulness as a disinfectant is due to the fact that the second oxygen atom falls off easily, reverting back to H_2O. A single oxygen atom is very dangerous and will try to steal electrons from any atom or molecule it comes into contact with. We have body cells that use single oxygens as "bullets" to fire at germs.

All forms of animal life need oxygen for cellular respiration, the process by which cells extract energy from sugars and fats. Fish and other sea life use oxygen that is dissolved into the water around them. Plants produce oxygen as a waste product of photosynthesis, so there is a balance between oxygen produced and oxygen used.

Plants use carbon dioxide, CO_2, from the air to make sugar molecules (glucose, $C_6H_{12}O_6$).

Oxygen is found in the crust of the earth bound to the element silicon to form minerals such as quartz, feldspar, mica, and olivine. All these minerals are based on the silicon tetrahedron, SiO_4, which forms crystal lattices.

When oxygen is cooled down to -183° C it becomes a liquid. Oxygen is often transported in its liquid state because it takes up less space. One liter of liquid oxygen expands to become 840 liters of oxygen gas. Oxygen gas is used in medical devices, in steel production, in plastic manufacturing and as fuel in welding. Liquid oxygen is used (along with liquid hydrogen) as rocket fuel.

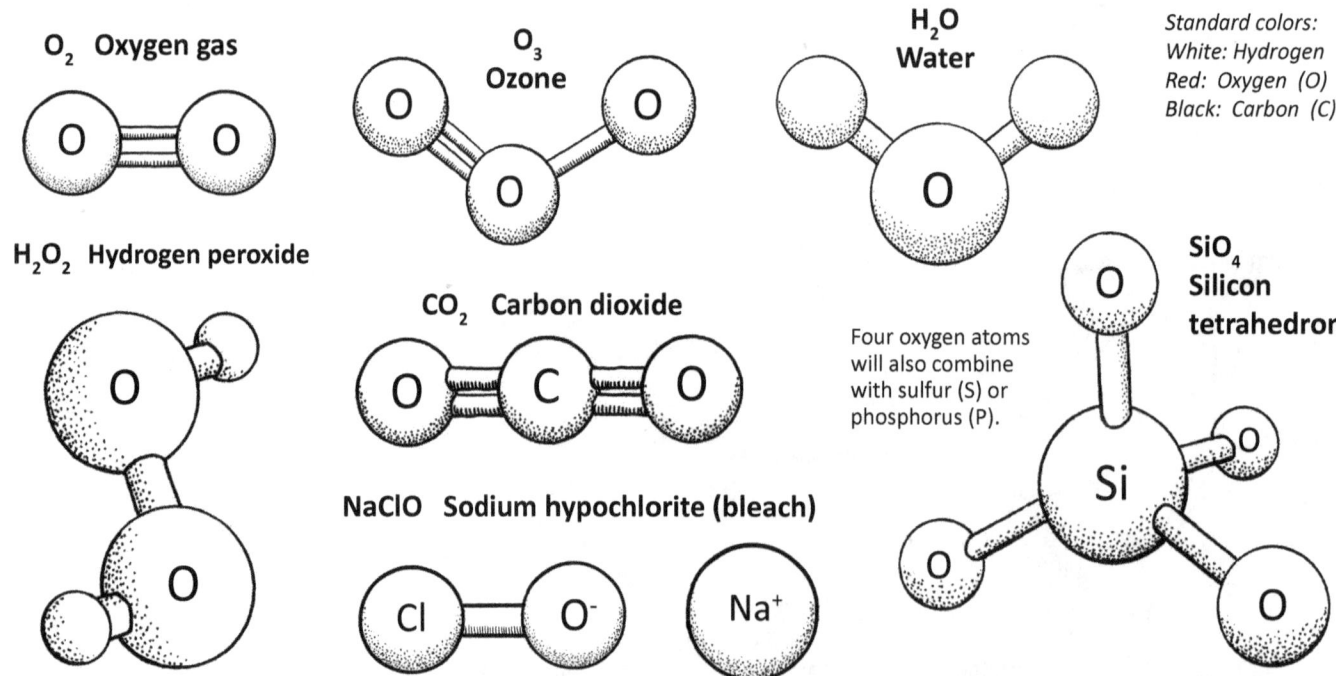

O_2 Oxygen gas

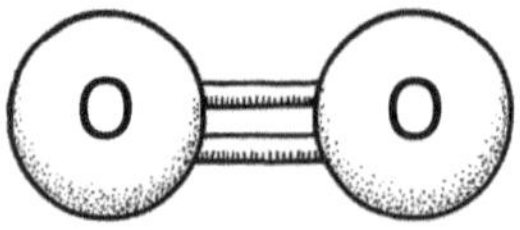

O_3 Ozone

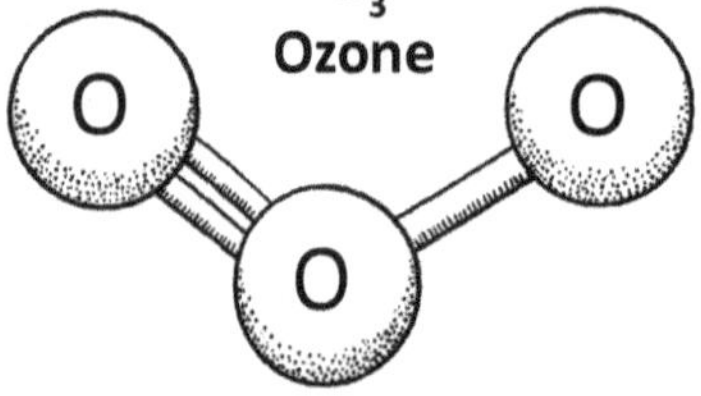

H_2O Water

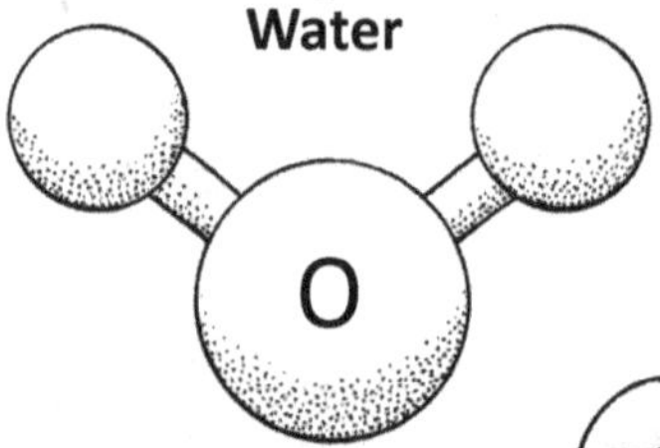

Standard colors:
White: Hydrogen
Red: Oxygen (O)
Black: Carbon (C)

H_2O_2 Hydrogen peroxide

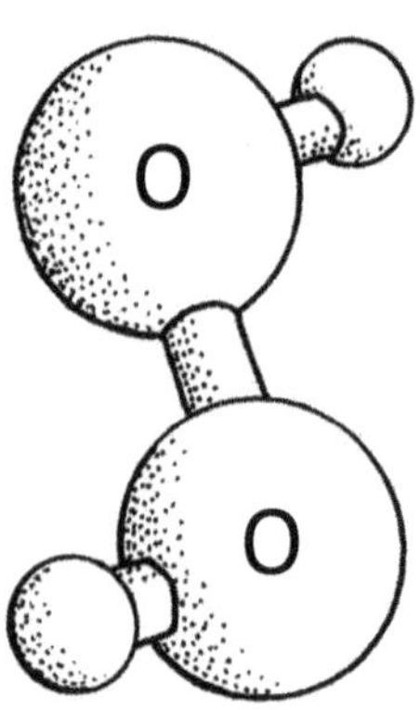

CO_2 Carbon dioxide

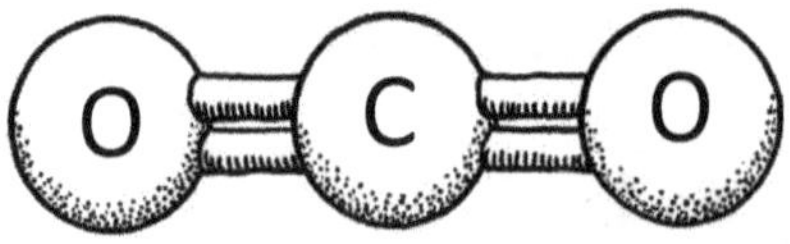

NaClO Sodium hypochlorite (bleach)

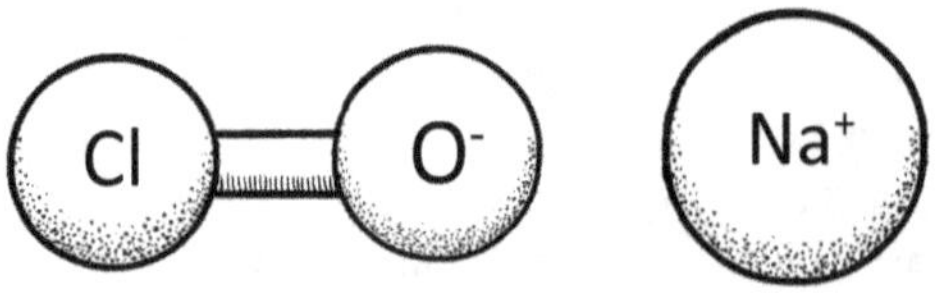

SiO_4 Silicon tetrahedron

Four oxygen atoms will also combine with sulfur (S) or phosphorus (P).

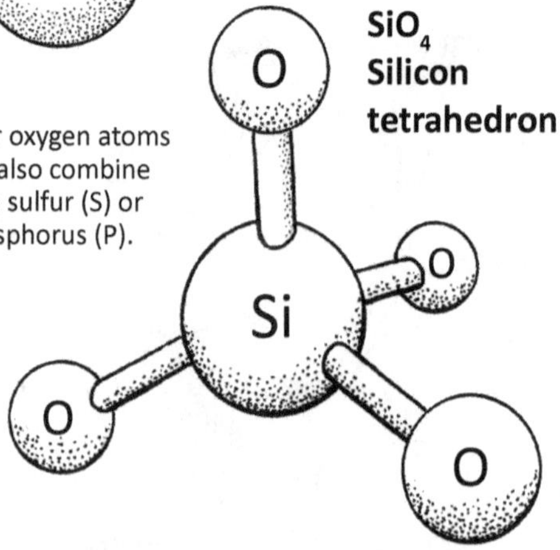

8

O

Oxygen

Combustion produces carbon dioxide.

Oxygen is necessary for combustion.

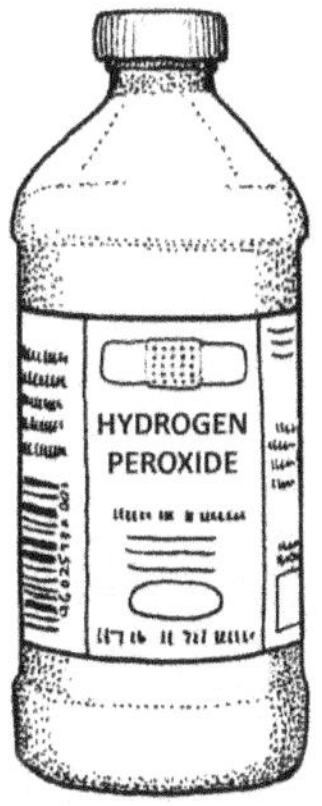

Hydrogen peroxide is used in first aid.

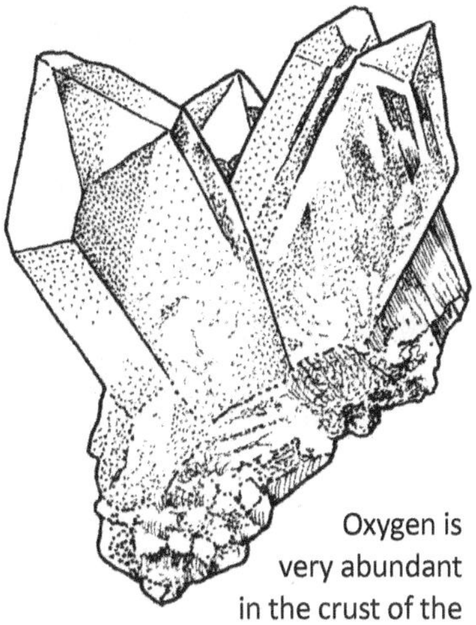

Oxygen is very abundant in the crust of the earth, often as quartz.

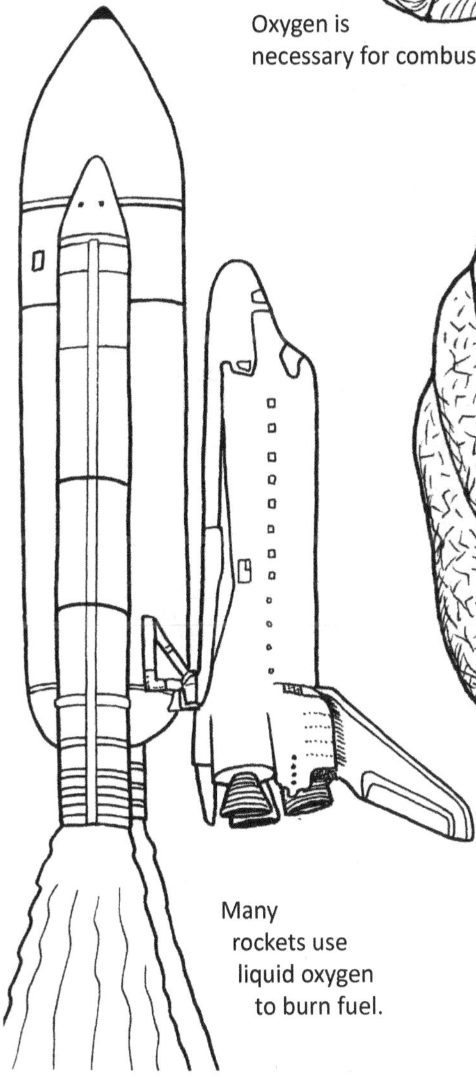

Many rockets use liquid oxygen to burn fuel.

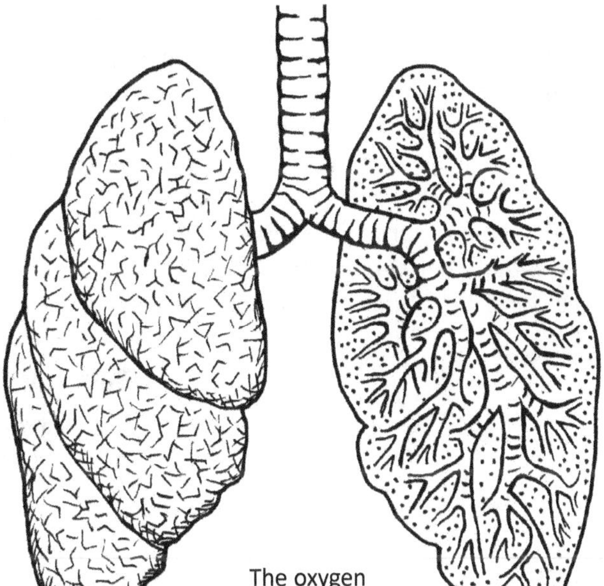

The oxygen we take in with our lungs is used in the process of cellular respiration.

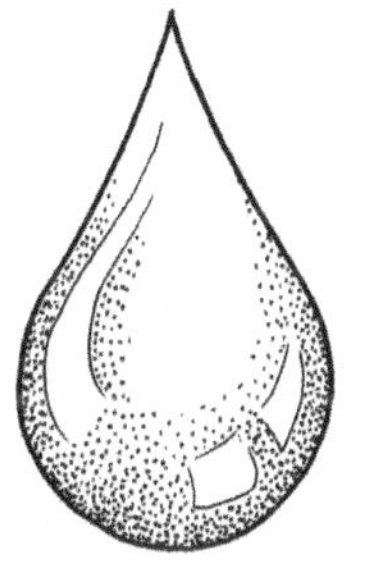

Oxygen and hydrogen make water.

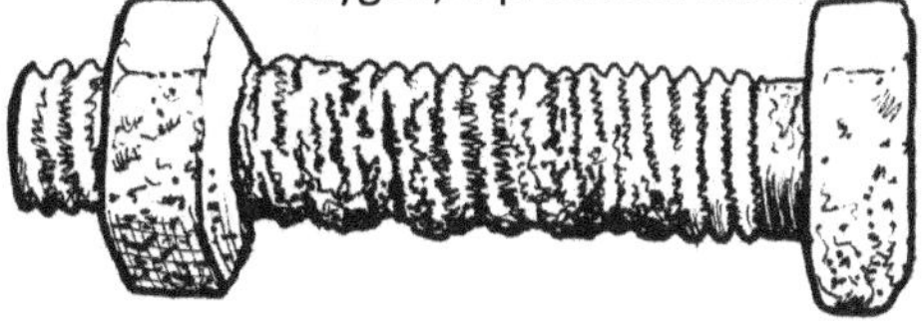

When iron combines with oxygen, it produces rust.

Bleach, NaClO, has an oxygen atom that falls off easily.

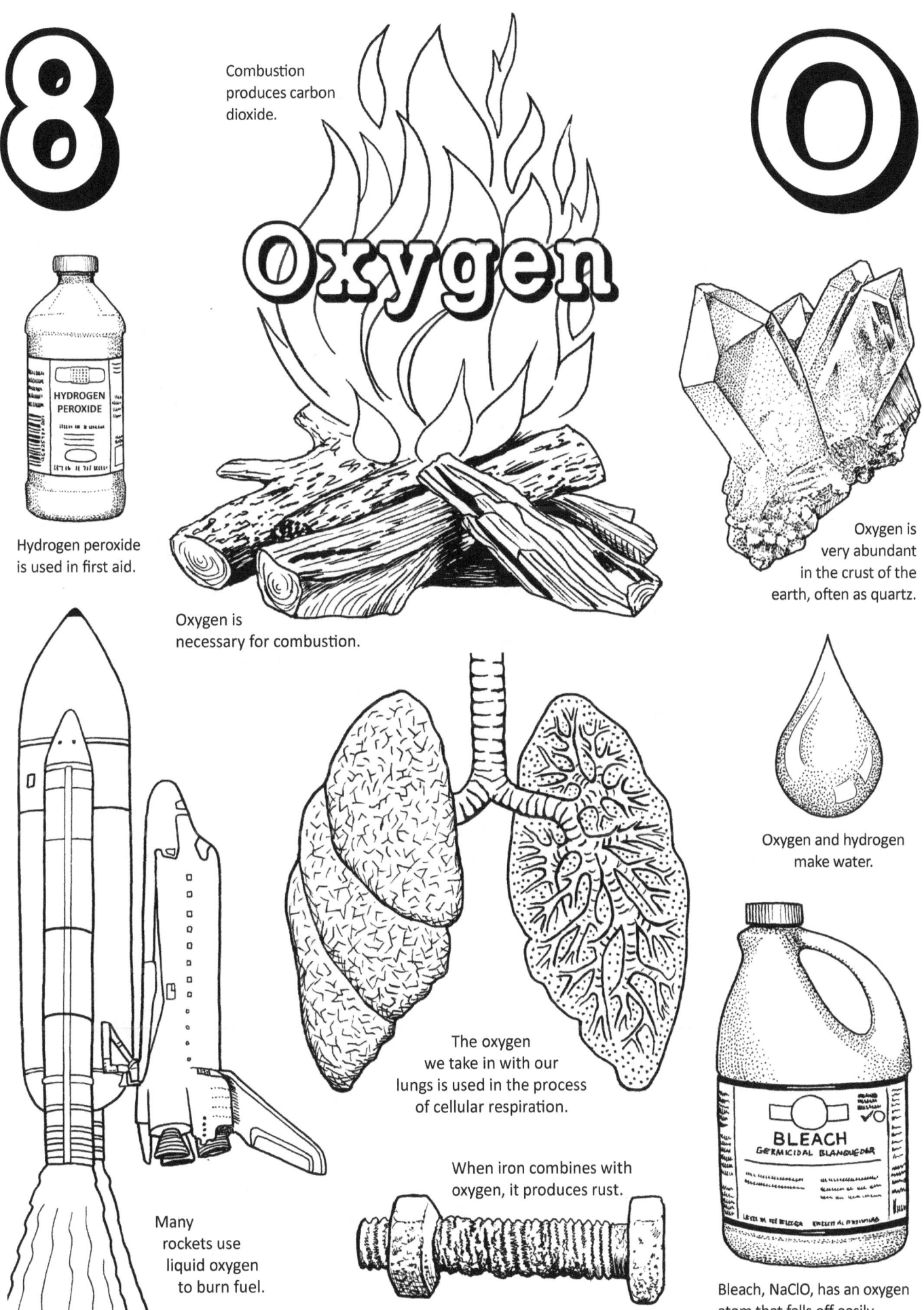

F

9

protons
10 neutrons
9 electrons

Atomic mass: 18.9

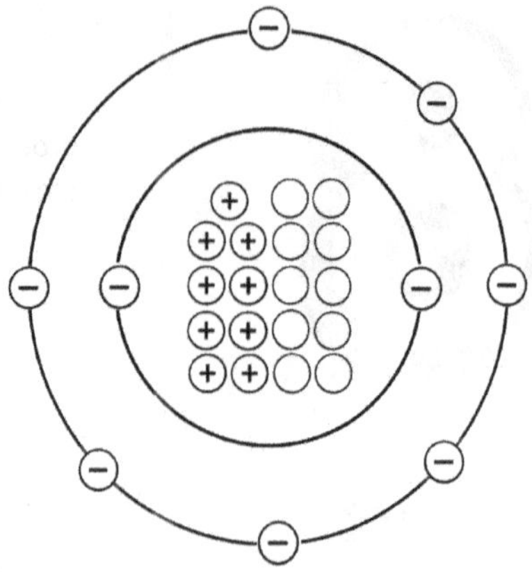

Fluorine

From Latin word "fluere" meaning "to flow"

Fluorine is famous for being the most "electronegative" element on the Periodic Table. This means that it can hold on to other atoms more tightly than any other element can. This is due to its size and its number of electrons. Because fluorine is a fairly small atom, its electrons are very close to the positively charged protons in the nucleus. Opposite charges attract, and fluorine's protons are able to hold the electrons very tightly. (Larger atoms can lose some of their outer electrons.) Because fluorine's outer shell has only 7 electrons, it falls one short of the perfect number: 8. Atoms with one empty place in their outer shell desperately want to steal or borrow an electron to fill that slot. Fluorine is so desperate that it will grab the first available electron it finds, usually an electron belonging to another atom. Thus, fluorine is never found alone in nature. (A single F atom is very dangerous!)

Fluorine is often found in the company of the element calcium, forming calcium fluoride, CaF_2. When found in rocks, CaF_2 is a mineral called fluorite (or fluorspar). Very pure fluorite crystals can be made into camera and telescope lenses. Crystals of lesser value can be crushed into a powder and used as "flux" in metal smelting. The fluorine atoms will grab impurities that metallurgists don't want in the hot, liquid metal. Getting rid of these contaminants makes the hot metal flow more easily. Fluorine's name comes from this ability to make liquid metals flow.

When fluorine grabs a hydrogen atom, hydrofluoric acid, HF, is formed. This acid is very dangerous to work with. It burns flesh and it steals calcium from bones. It is used to etch designs into glass because it is one of the few substances that can dissolve glass. Despite the fact that it is so dangerous, fluorine does play an important role in the body, being one of the minerals that help to make our teeth very strong.

Sulfur atoms can bond to six fluorines, making SF_6, sulfur hexafluoride. Unlike HF, this substance is very safe. SF_6 is a gas that won't react with anything and can be used as insulation. A similar molecule is uranium hexafluoride, UF_6. This molecule is used to "enrich" uranium by collecting U atoms that have an atomic mass of 235.

When fluorine bonds with carbon, it forms C_2F_4, tetra-fluoro-ethylene, better known as Teflon®. Teflon® is very slippery so when pans are coated with it, they become "non-stick." A similar substance is poly-tetra-fluoro-ethylene, better known by the brand name Gore-Tex®. Gore-Tex® fabric is rainproof while still allowing body moisture to escape.

CaF_2 Calcium fluorite (crystal lattice)

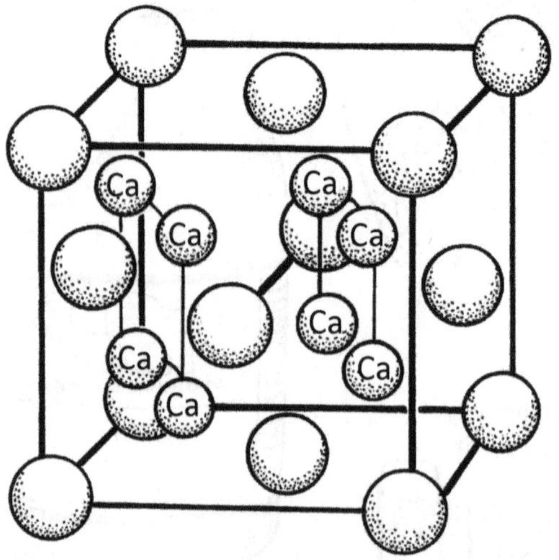

This cube is called a unit cell. We see only part of the crystal, so it seems like the math is wrong. If you looked at the entire crystal, there would be 2 Fs for every Ca.

SF_6 Sulfur hexafluoride

The central atom is sulfur. Artists usually make sulfur yellow. All the other atoms are fluorine.

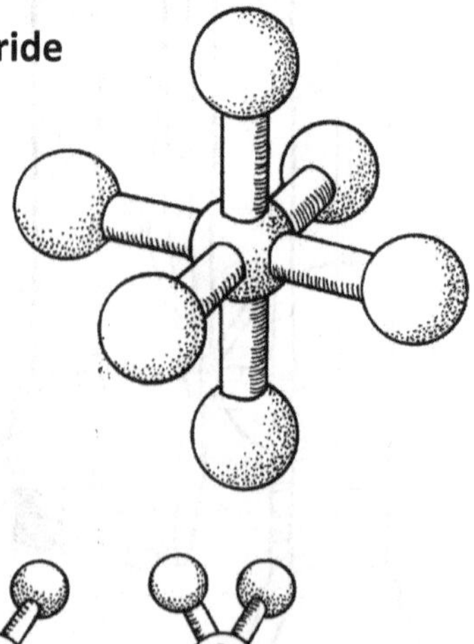

C_2F_4 Teflon®

This molecule is a very long polymer made of thousands of carbon atoms bonded to fluorine atoms.

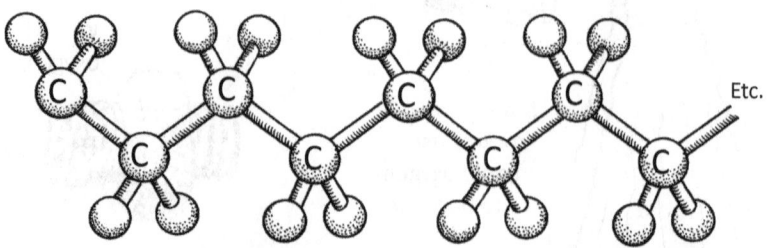

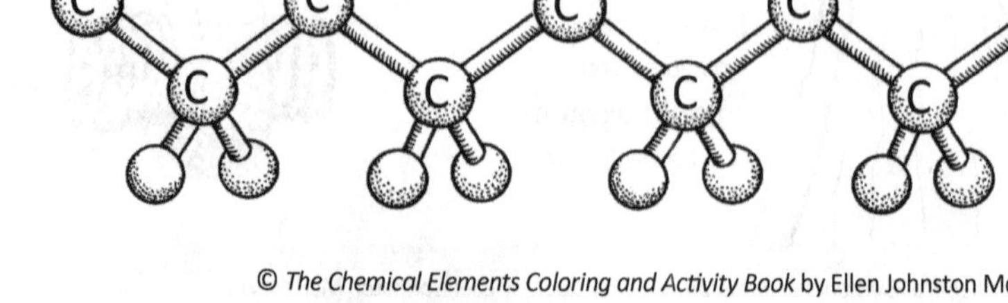

9 F

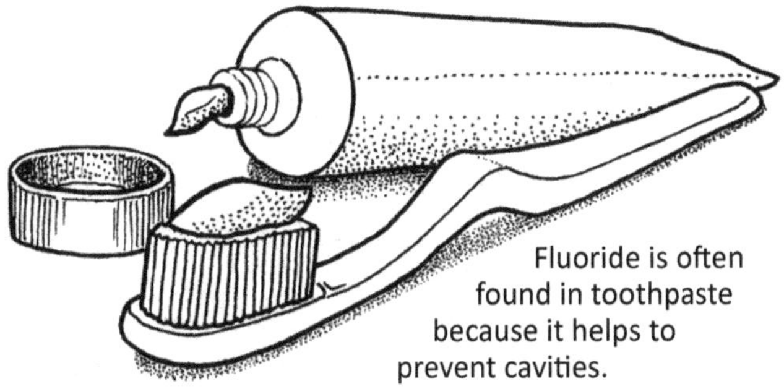

Fluoride is often found in toothpaste because it helps to prevent cavities.

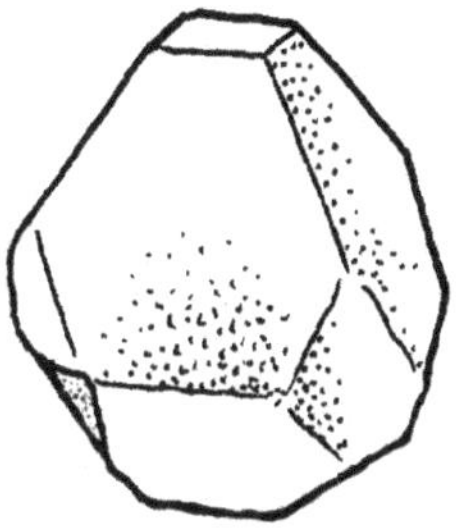

Fluorite crystals are often green or purple.

Fluorine

Gore-Tex® clothes use a modified form of Teflon®.

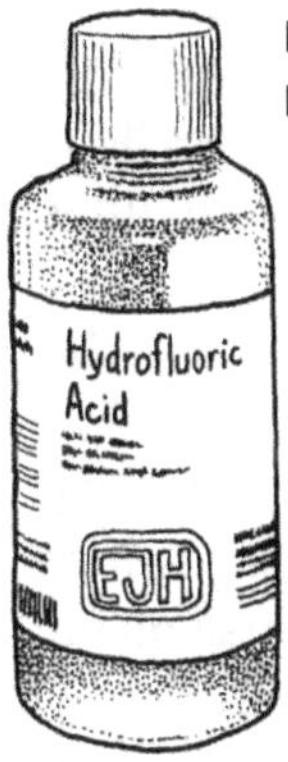

Hydrofluoric acid, HF, is dangerous!

Fluorite is used for high magnification camera lenses.

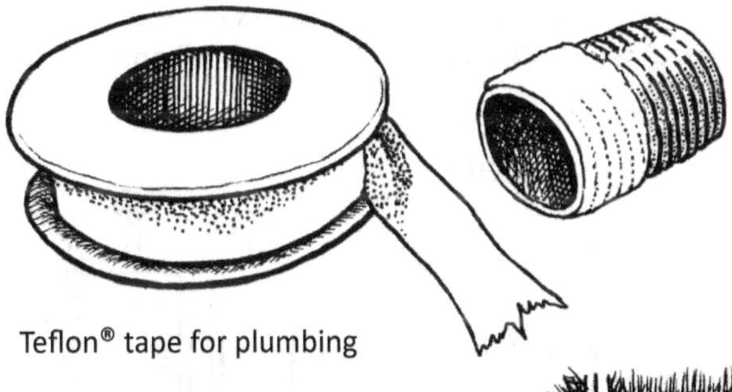

Teflon® tape for plumbing

Teflon® makes cooking pans "non-stick."

Fluorine is used as a "flux" in the smelting of metal. Smelting means heating mineral ores until they become a liquid (often bright yellow in color). Metals can then be collected and cooled.

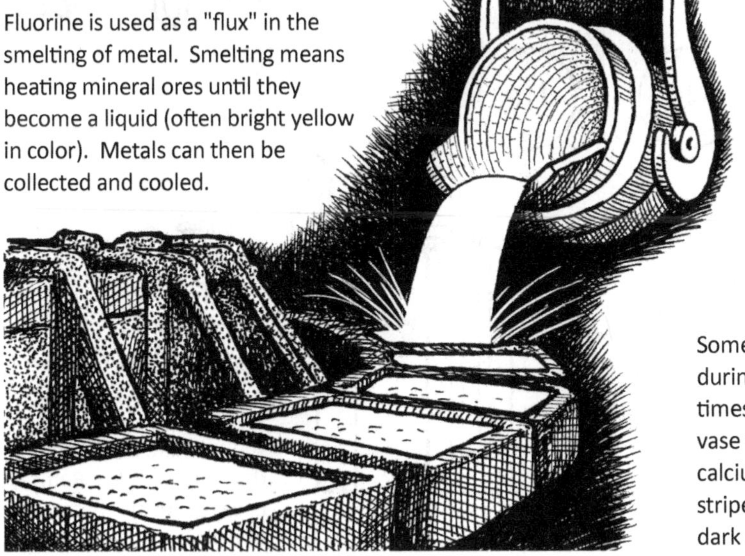

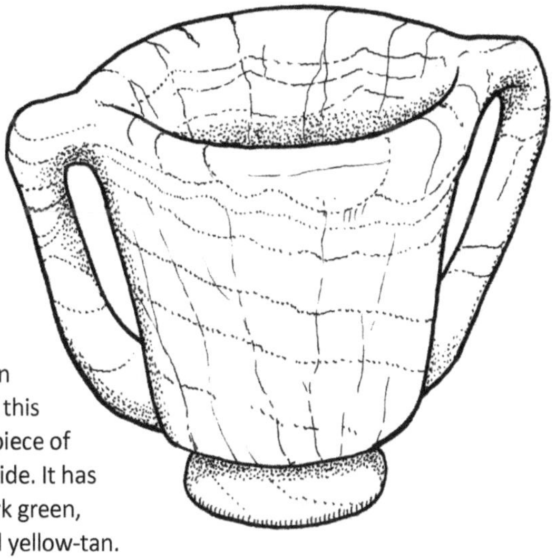

Someone during Roman times carved this vase from a piece of calcium fluoride. It has stripes of dark green, dark red, and yellow-tan.

Ne 10

protons
10 neutrons
10 electrons

Atomic mass: 20.2

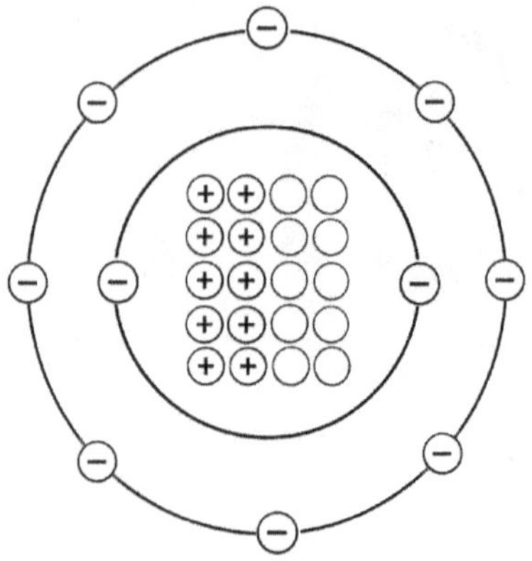

Neon

From the Greek word for "new"

Neon belongs to the family of elements called the noble gases. They are found in the last column on the right side of the Periodic Table. The noble gases are very lucky because their outer electron shells are completely filled. They do not have any empty slots, nor do they have any extra electrons to give away. This is why they will not interact with other atoms. The noble gases are called "inert" because they are so nonreactive. Neon is sometimes called the most inert element on the Periodic Table.

Like most of the noble gases, neon can be safely used in places where there is electricity, such as inside fluorescent light tubes. Neon lights (the ones that actually have neon in them) glow with an orange-red color. Most of the lights that are called "neon" lights are actually filled with other gases and with powdered minerals that produce colors like green, yellow, or blue.

Neon is used in cold-cathode voltage regulator tubes, which look a lot like old-fashioned vacuum tubes. A similar product, called "nixie tubes," are used to make an unusual type of digital clock. Neon can also be used in high-voltage indicator devices, and in structures that absorb lightning strikes. Helium, another noble gas, is used with neon to make helium-neon (HeNe) lasers, which produce a bright red line of light.

Neon is found in the air all around us, but in very small quantities. The way you collect it out of the air is to chill the air to -269° C, the temperature at which all the gases turn to liquid. Then the temperature is turned up very slowly, one degree at a time. At -246° C, liquid neon turns back into a gas and is captured.

Neon was discovered at about the same time as the elements krypton and xenon, in 1898, by Sir William Ramsay and his assistant Morris Travers, using this chilling technique. As they watched the gases appear, some of them were familiar, such as oxygen, nitrogen, helium, and carbon dioxide. Then they found a "new" one, so they named it neon, after the Greek word for new.

Sir William Ramsay in his lab

Neon does not form molecules

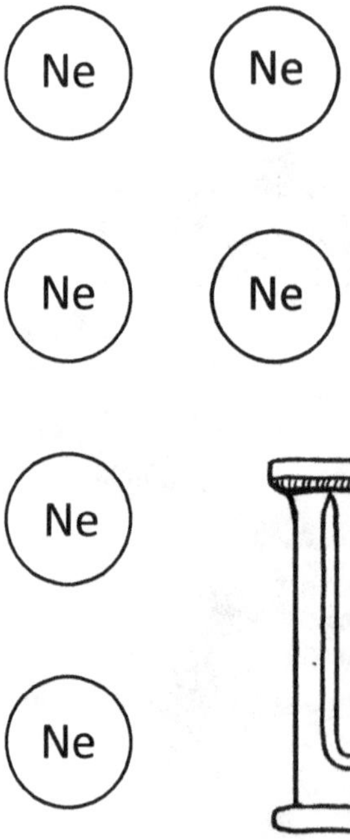

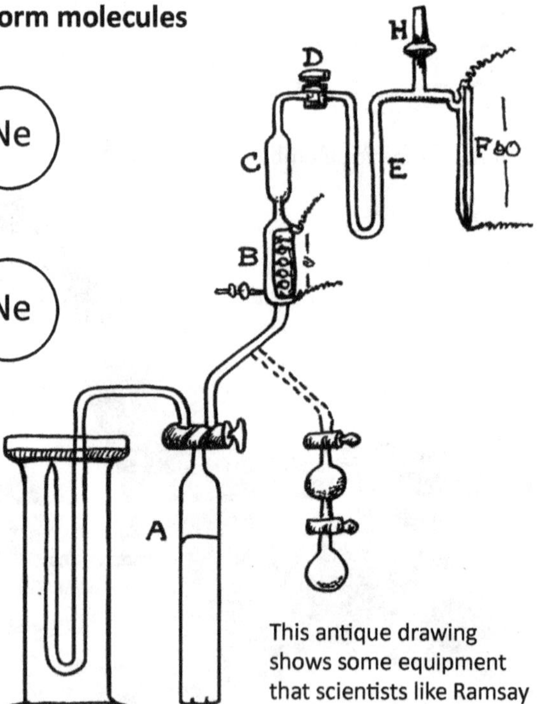

This antique drawing shows some equipment that scientists like Ramsay used to "distill" gases out of the air by chilling them.

10 Neon Ne

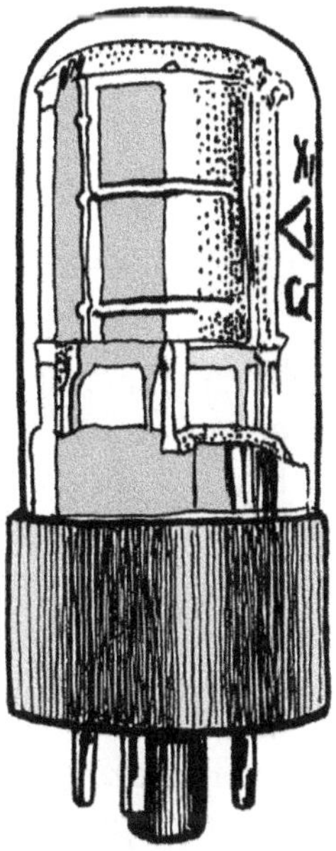

This is a cold-cathode voltage regulator tube. They are used only rarely now, as solid state regulators have mostly replaced them.

Neon signs (that actually have neon in them) glow bright orange-red.

This is a "nixie" tube. It contains tiny neon tubes shaped like numbers. The tubes can be used to make clocks that are purchased by collectors as novelties.

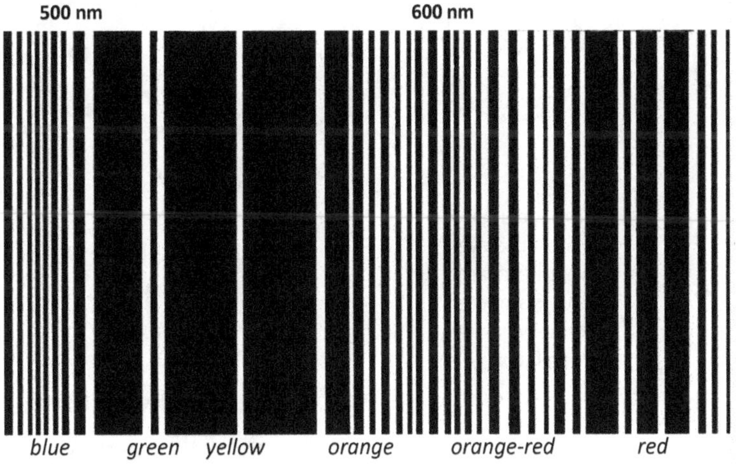

If you look at glowing neon gas through a spectrometer, this is what you will see. The small amount of blue and green light that comes from glowing neon atoms is drowned out by all the orange and red.

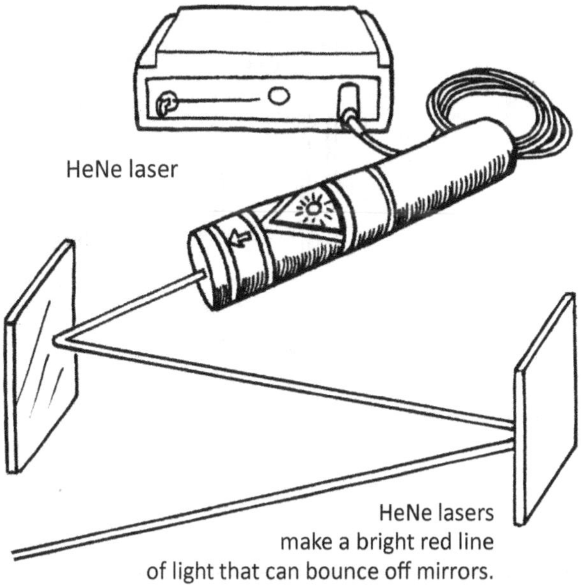

HeNe lasers make a bright red line of light that can bounce off mirrors.

Na 11

Sodium

11 protons
12 neutrons
11 electrons

Atomic mass: 22.9

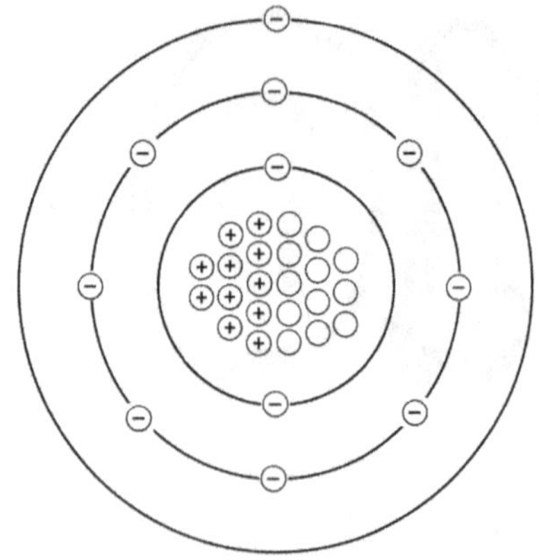

The name comes from the substance in which it was discovered, "caustic soda." The symbol, Na, is from the Latin "natrium," meaning "sodium carbonate."

Sodium was discovered by the famous chemist Sir Humphry Davy, in 1807. He discovered both sodium and potassium in that year, using electricity to pull the atoms out of a solution. The solution he used for sodium was called caustic soda (known to us today as sodium hydroxide, NaOH), and it is from that substance that sodium gets its name. Pure sodium is a very soft, shiny silver metal that quickly turns dark gray if it is exposed to air. Putting sodium into water causes it to burst into bright yellow flames, so chemists keep their samples of sodium in jars of oil, protected from both water and air.

Sodium has only one electron in its outer shell, making it desperate to get rid of that lonely electron, even though getting rid of it will mean upsetting the equal balance of electrons and protons. It prefers having a positive electrical charge to having a lonely electron in an orbit all by itself. After a sodium atom loses that outer electron, it is called an "ion" instead of an atom. An ion is an atom that does not have an equal number of electrons and protons.

Sodium is always found attached to other atoms, and one of its favorites is chlorine, making a molecule of NaCl (sodium chloride). NaCl forms crystals that we know as table salt. Salt has a long history of being useful in many food preparation processes, including preserving meats so they do not spoil.

Sodium lights were used widely in public areas before the invention of LED lights because the bulbs had a very long life. Sodium bulbs give off a very yellow glow, which comes from sodium's spectral lines.

Sodium plays important roles in the body. In the blood, it helps to maintain proper blood pressure. In nerve cells, special pumps transport it across the membrane, allowing the transmission of electrical signals.

NaCl Sodium chloride (table salt) crystal lattice

(The atoms that are not marked are chlorine, Cl.)

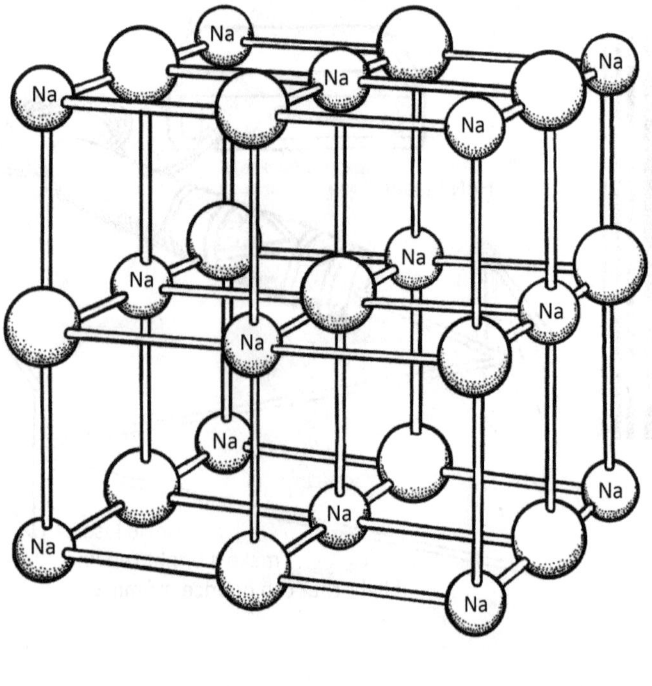
Na

$NaHCO_3$ Sodium bicarbonate
(baking soda)

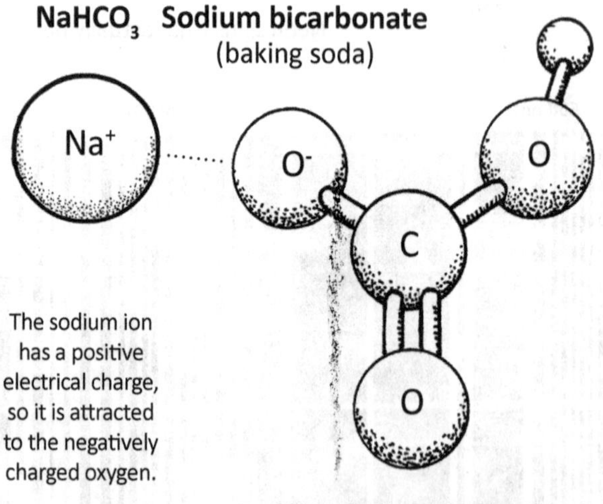
Na^+
O^-
O
C
O
The sodium ion has a positive electrical charge, so it is attracted to the negatively charged oxygen.

NaClO Sodium hypochlorite
(bleach)

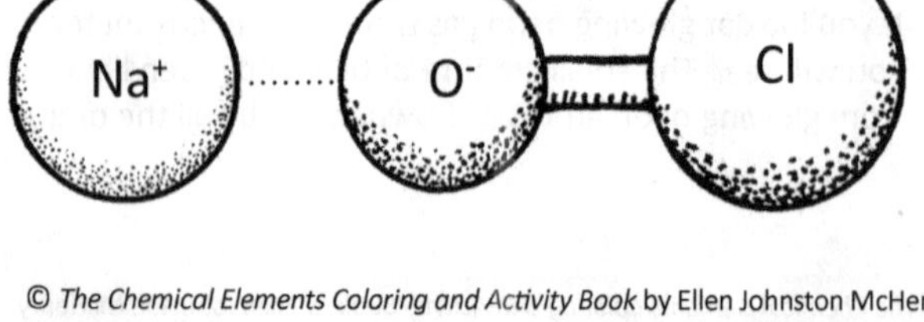
Na^+
O^-
Cl

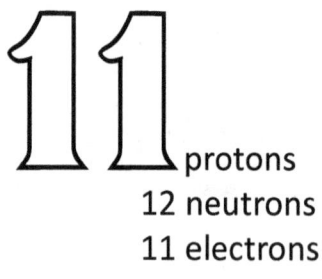

11 Na

Sodium

NaCl crystals are white and have a cubic shape.

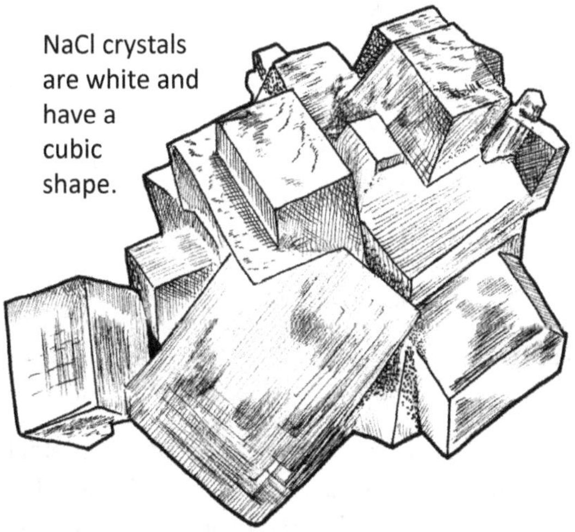

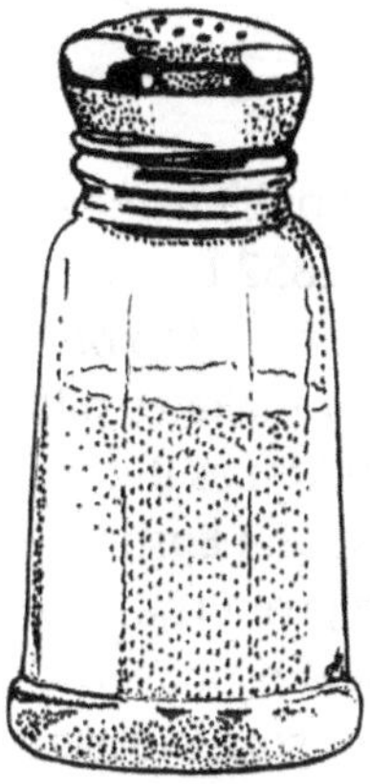

Table salt is NaCl.

Baking soda makes things rise in the oven. Its formula is $NaHCO_3$.

Sodium vapor lights keep the Na atoms at high pressure.

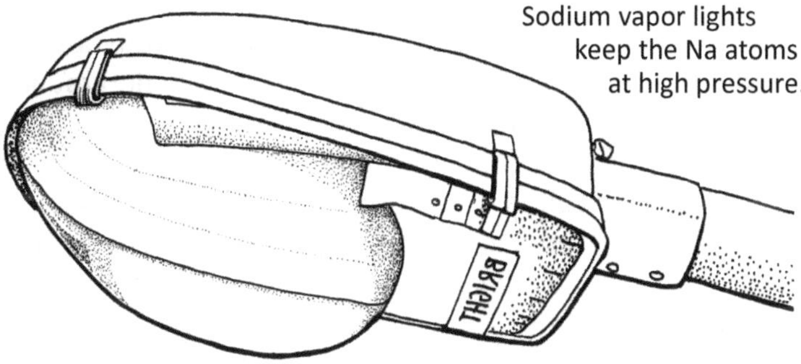

Bleach and Borax both contain sodium.

The sodium-potassium pump is found in cell membranes.

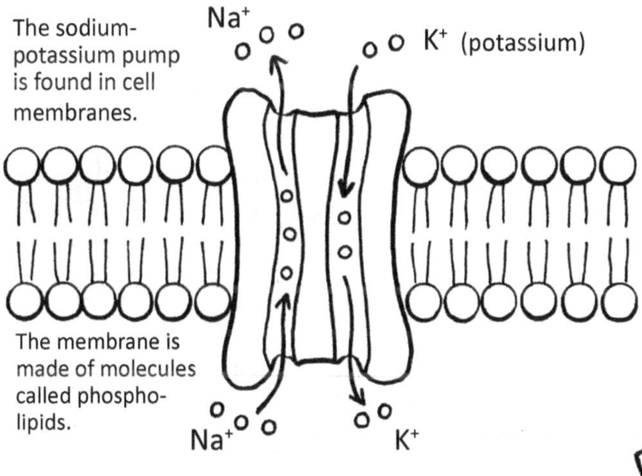

The membrane is made of molecules called phospholipids.

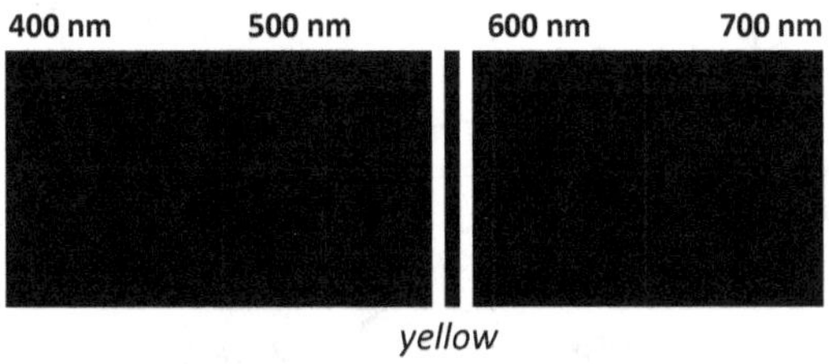

Sodium has a very distinct emission spectrum. Burning sodium produces just two yellow lines.

This is not an air tank. It is a "re-breather." It recycles air, using NaOH and CaO to remove CO_2.

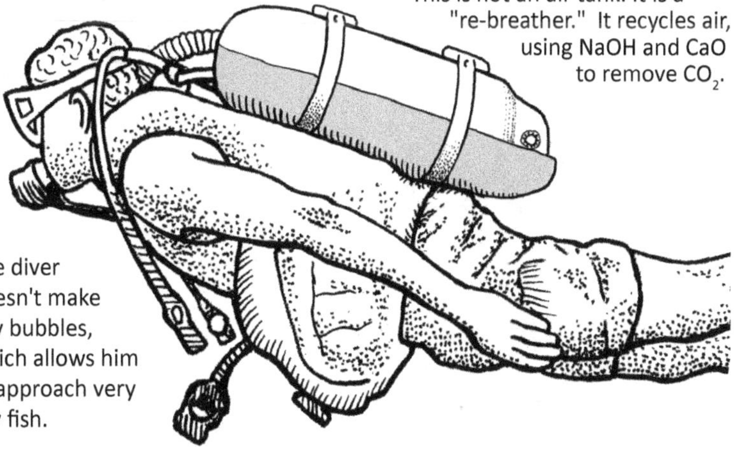

The diver doesn't make any bubbles, which allows him to approach very shy fish.

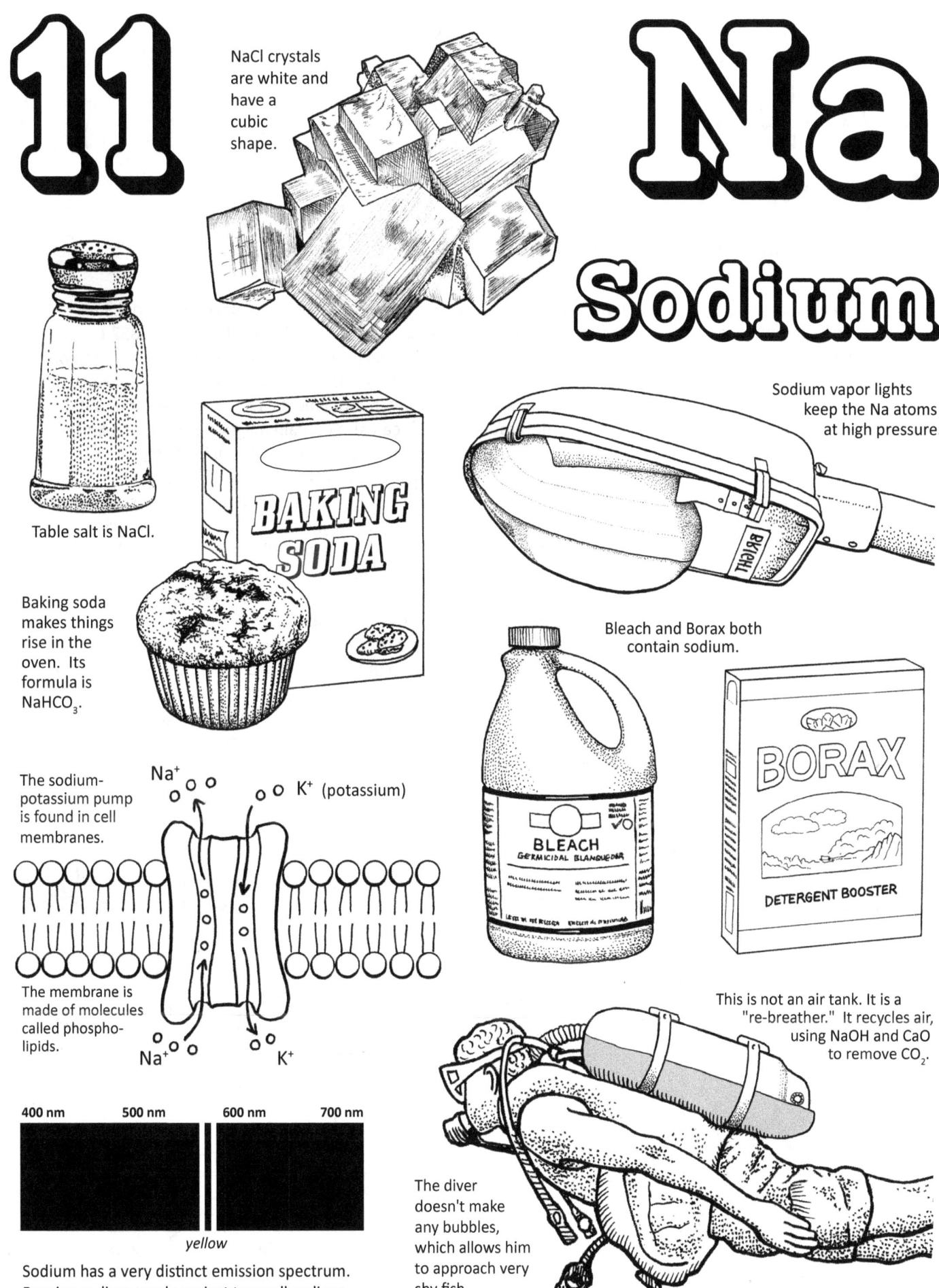

Mg 12

protons
12 neutrons
12 electrons

Atomic mass: 24.3

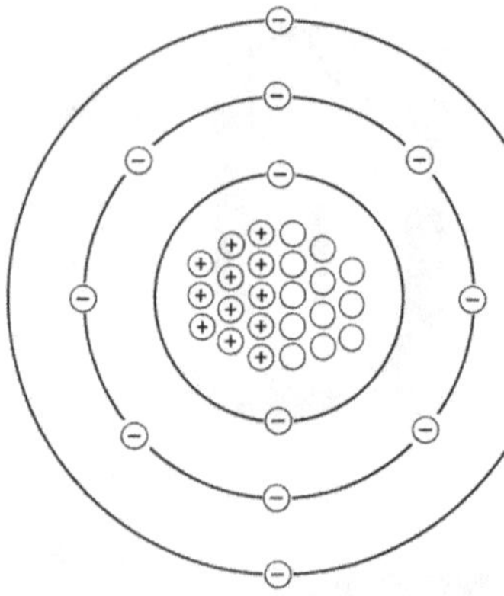

Magnesium

From the area of Greece called Magnesia

Magnesium was discovered and named in 1808 by Sir Humphry Davy. He used electricity to pull magnesium atoms out of a chemical solution. This technique, called electrolysis, was used by Robert Bunsen in 1852 to produce enough magnesium that it could be evaluated for use in many industrial processes. Magnesium was found to be light and strong but melted at low temperatures. It is most useful when combined with aluminum to make an alloy.

Today, magnesium is usually extracted from ocean water, which has almost as much magnesium as it does sodium and chlorine (NaCl, salt). Water from underground sources can also contain magnesium, as John Epsom discovered in the early 1600s. His well water tasted bitter but it turned out have wonderful healing properties, especially for skin. When the bitter water evaporated, it left behind crystals that are known today as Epsom salts, $MgSO_4$. These salts are still commonly used to make soaking baths as a remedy for unhealthy skin or sore muscles. Another health product containing magnesium is "milk of magnesia," $Mg(OH)_2$, which is used as an antacid.

Magnesium is very abundant in the Earth's crust, particularly in a mineral called dolomite, which is very similar to limestone. Limestone is $CaCO_3$, and dolomite is $MgCO_3$. Magnesium can easily replace calcium because of the fact that both of these elements have two electrons in their outer shell. The number of electrons in the outer shell is what gives elements their ability to bond with certain atoms or molecules. Magnesium also likes to bond to oxygen to make MgO, magnesium oxide, a mineral commonly found in rocks.

Metals that contain magnesium are used to make parts for many machines and devices, including airplanes, rockets, cars, sports equipment, and electronic devices.

Magnesium burns with a brilliant white light, which makes it ideal for use in fireworks, sparklers, flares, and tracer bullets. (Tracer bullets produce a flash of light so you can see them as they speed through the air.) Before the age of LEDs, magnesium was used to make flashbulbs for cameras.

MgO Magnesium oxide
MgS Magnesium sulfide

Both MgO and MgS will form lattices.
(The smaller balls represent Mg atoms.)

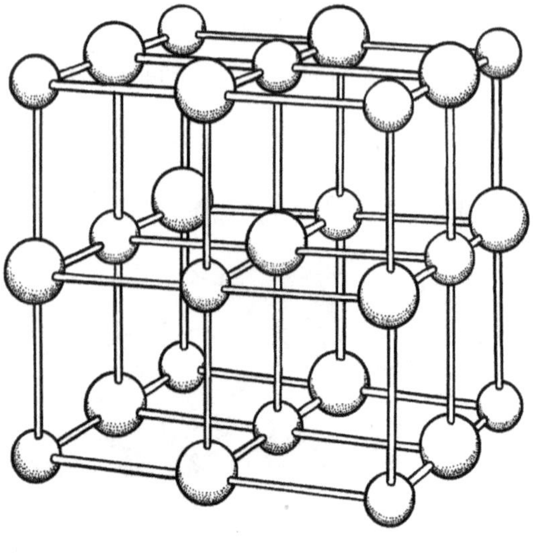

$MgCO_3$ Dolomite

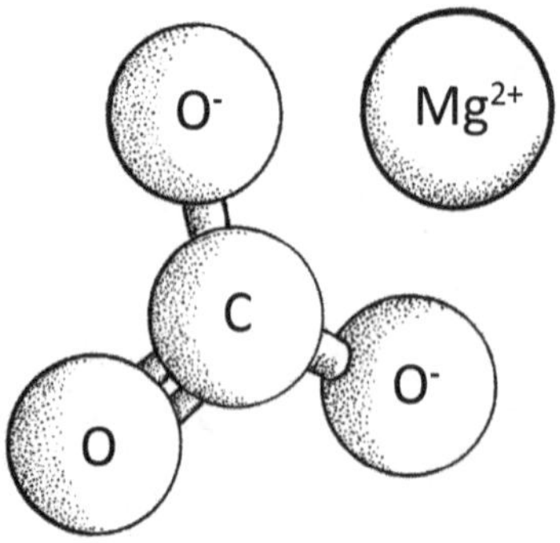

What element is magnesium attracted to in all of these molecules?

$MgSO_4$ Magnesium sulfate
(Epsom salt)

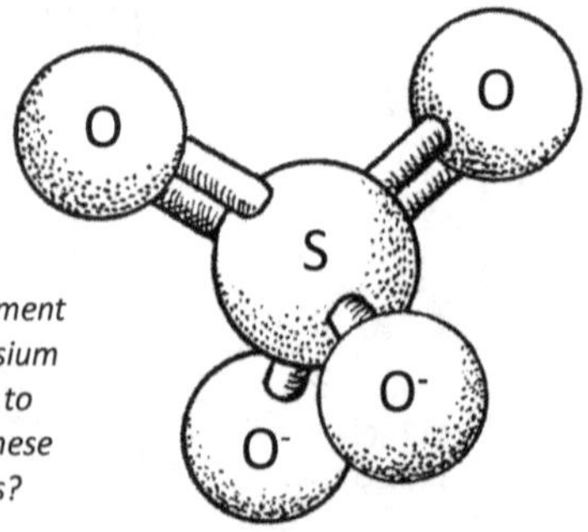

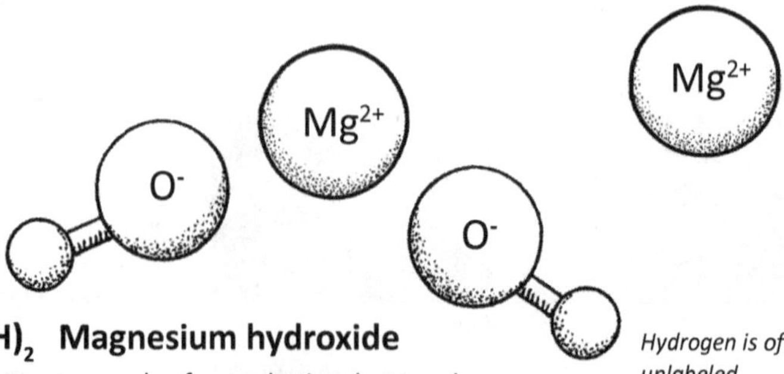

$Mg(OH)_2$ Magnesium hydroxide

This substance can also form a lattice, but much more complicated than MgO or MgS.

Hydrogen is often unlabeled.

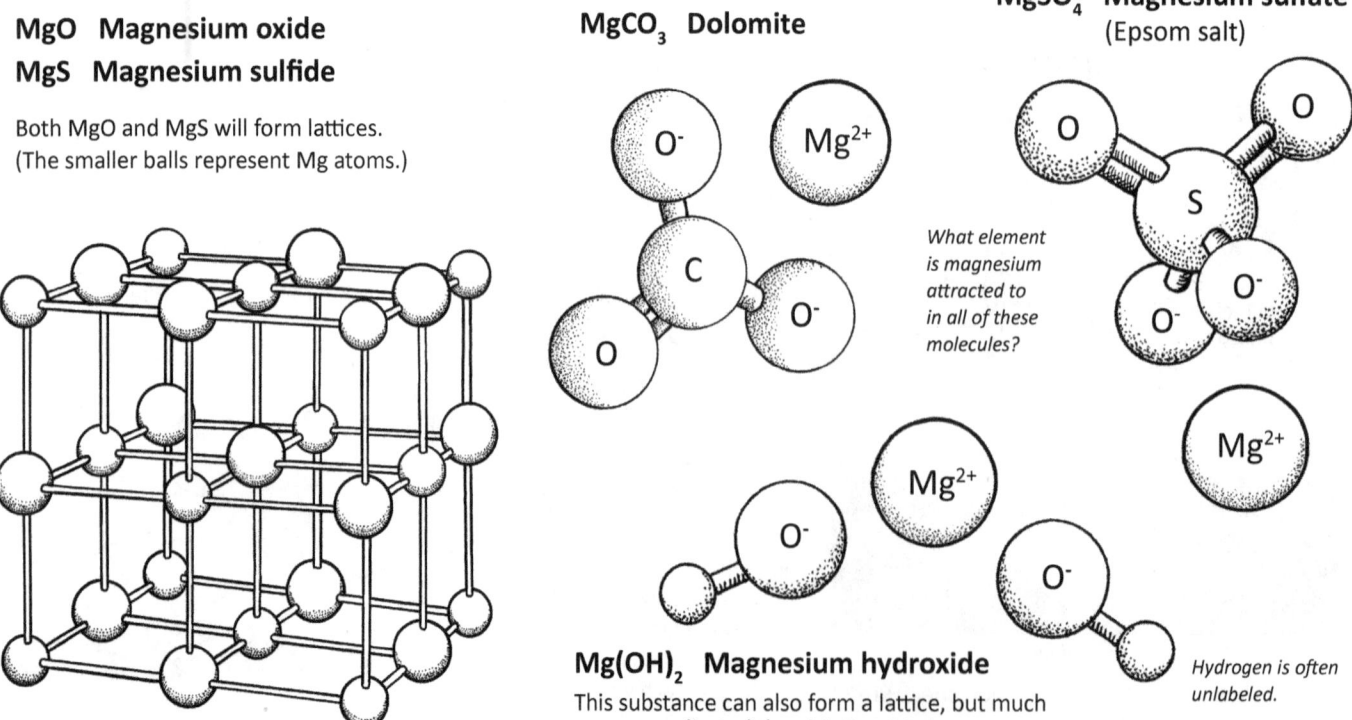

12 Mg

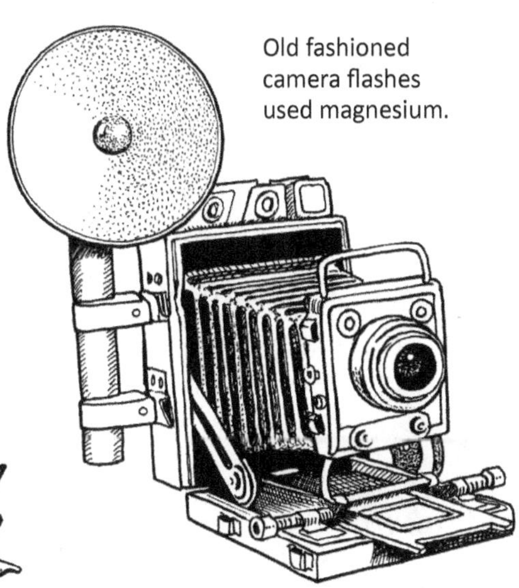

Old fashioned camera flashes used magnesium.

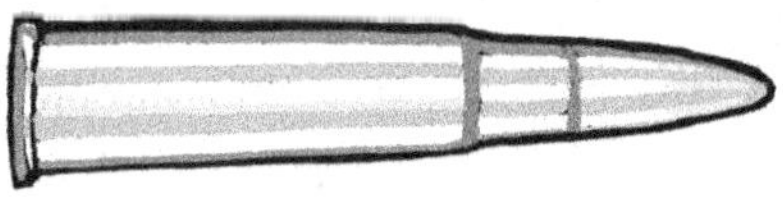

Tracer bullets make a bright flash.

Magnesium

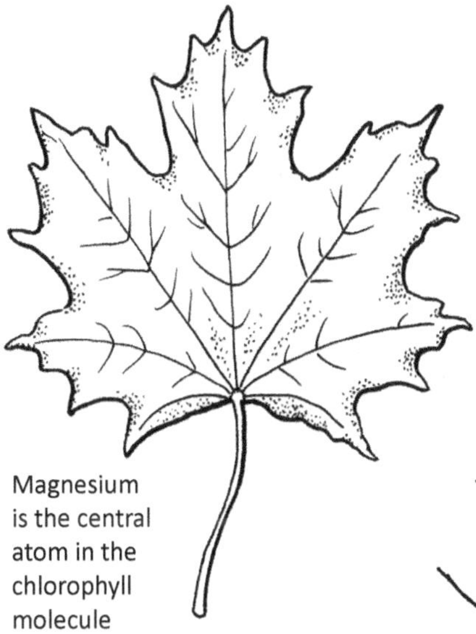

Magnesium is the central atom in the chlorophyll molecule found in leaves.

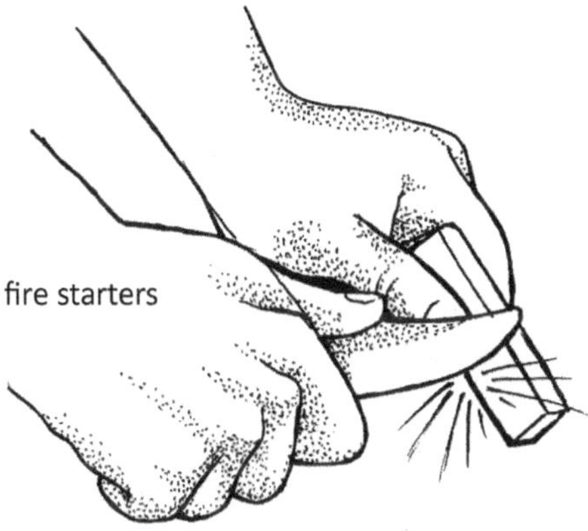

fire starters

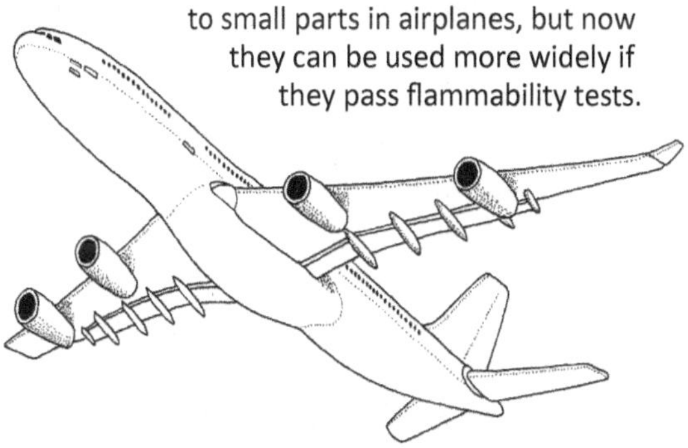

Magnesium alloys used to be restricted to small parts in airplanes, but now they can be used more widely if they pass flammability tests.

Epsom salt is used for soaking baths. It is also great for plants!

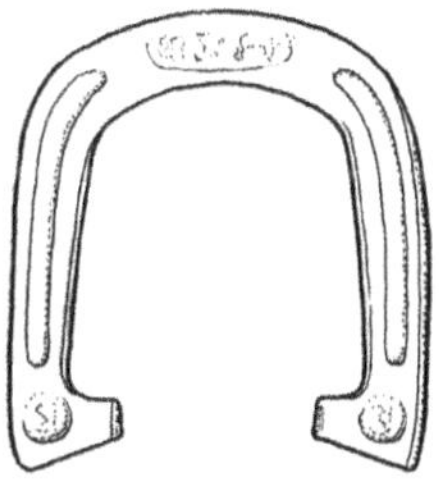

sports equipment, including horseshoes

Some car parts use magnesium alloys.

white sparklers

cell phone cases

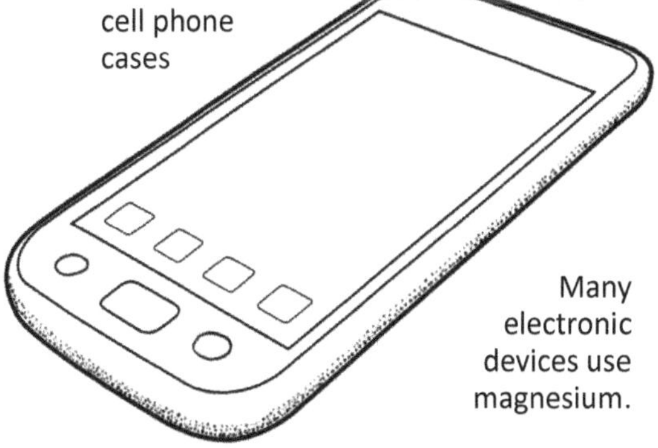

Many electronic devices use magnesium.

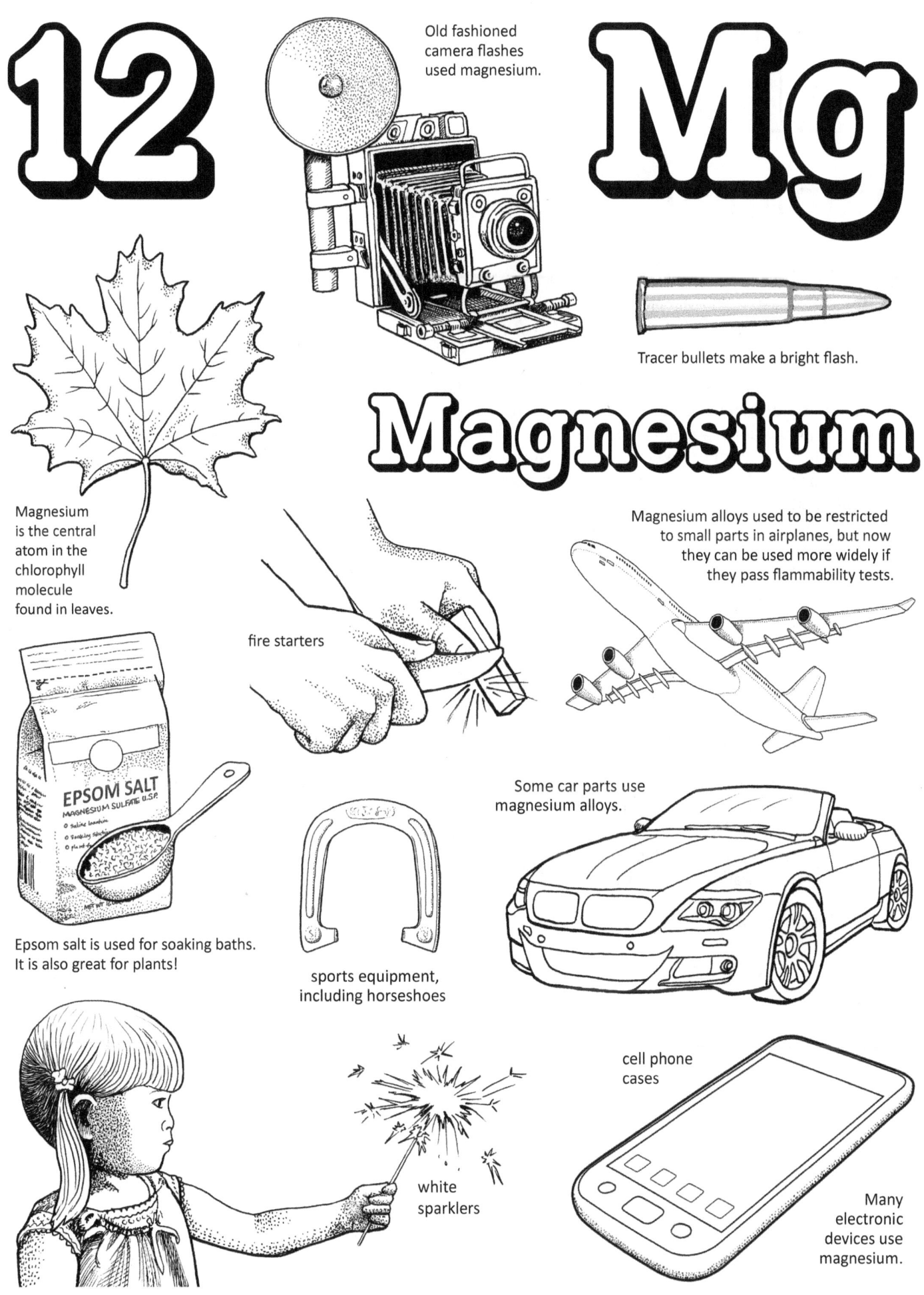

Al 13

protons
14 neutrons
13 electrons

Atomic mass: 26.9

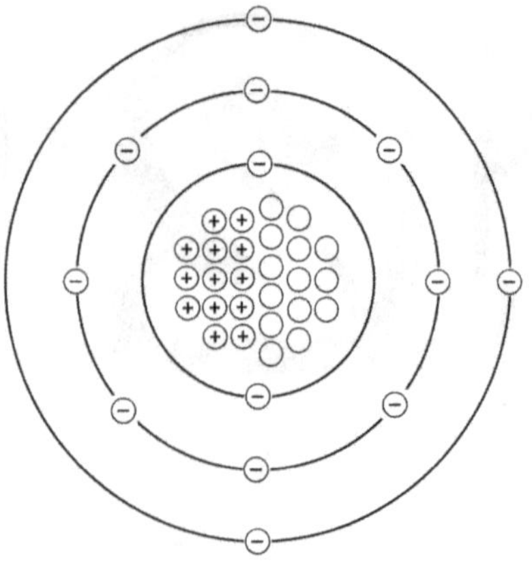

Aluminum
(or Aluminium)

From the chemical compound called "alum"

Alum is a natural mineral compound that has been used since ancient times by doctors (to quickly shrink tissues and to stop bleeding) and in the fabric dyeing industry. In the 1700s, chemists figured out that alum contains either potassium or sodium, and sulfate, SO_4^{2-}, and also an unknown element. In 1754, a German chemist succeeded in making artificial alum by boiling clay (which happened to contain aluminum) with sulfuric acid and potash (wood ashes). In 1824, a Danish chemist managed to pull pure aluminum atoms out of a solution to produce a solid lump of silvery metal that was very lightweight. The metal was not officially named until Humphry Davy began working with it in the 1800s. He chose the name "aluminum," which is the name that is now used in Canada and the U.S. Years later, some scientists in the UK decided that they preferred "aluminium" because it ends in "-ium" like the names of many other elements, and they began using that spelling in their publications.

Today, aluminum is usually extracted from a rock called bauxite *(box-ite)* by grinding the rock into a powder, then making it into a hot liquid solution into which electrodes are placed. The aluminum atoms come out of the solution and stick to one of the electrodes. The pure aluminum will often have small amounts of other metals added to it to make an alloy that is suitable for various industrial processes.

Adding magnesium to aluminum makes it stronger without adding extra weight, so this alloy is widely used in the manufacturing of airplanes, boats, army tanks, and window frames. Sometimes both silicon and magnesium are added to make a three-metal alloy that is strong and very resistant to corrosion—great for making cars and trucks. Copper and zinc are also widely used in alloys because they add strength. The element manganese makes an alloy that is excellent for cooking utensils and beverage cans. Adding nickel and cobalt will make an alloy known as AlNiCo, which is used to make magnets. Aluminum foil and aluminum food trays are made of almost pure aluminum.

Aluminum sulfate, $Al_2(SO_4)_3$, is used in paper manufacturing and as a fertilizer for plants. Aluminum chlorohydrate, $Al_2Cl(OH)_5$, is the active ingredient in many antiperspirants. Aluminum hydroxide, $Al(OH)_3$, is the active ingredient in some brands of antacids.

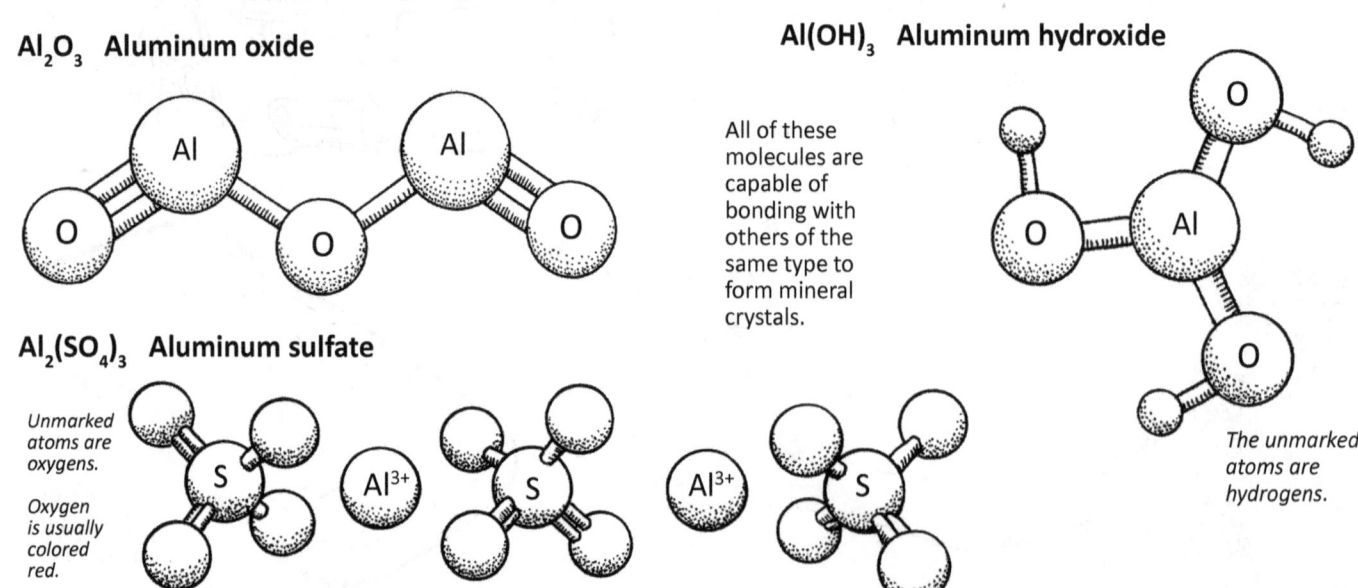

Al_2O_3 Aluminum oxide

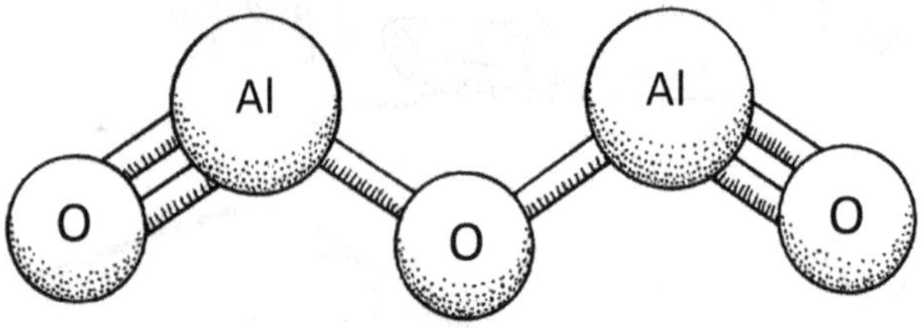

$Al(OH)_3$ Aluminum hydroxide

All of these molecules are capable of bonding with others of the same type to form mineral crystals.

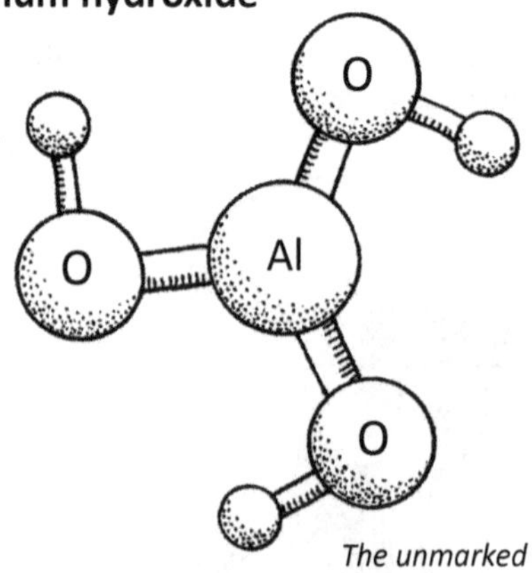

The unmarked atoms are hydrogens.

$Al_2(SO_4)_3$ Aluminum sulfate

Unmarked atoms are oxygens.

Oxygen is usually colored red.

13 Al

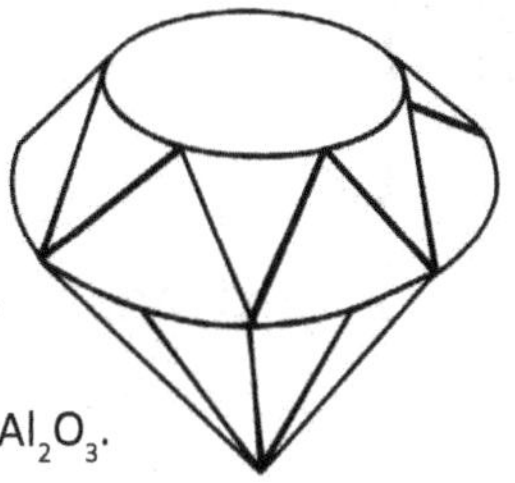

Rubies are a red variation of the mineral corundum, Al_2O_3.

Sapphires are also corundum, but are blue.

Aluminum

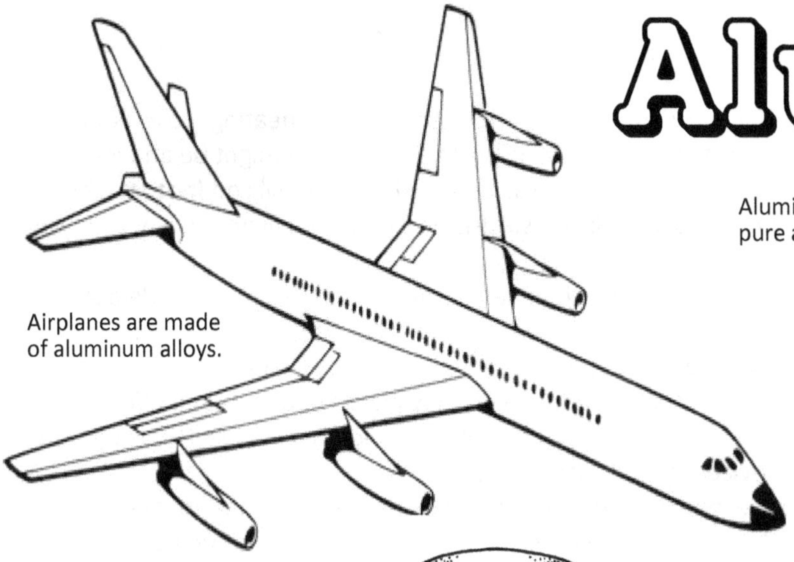

Airplanes are made of aluminum alloys.

Aluminum foil is made of almost pure aluminum.

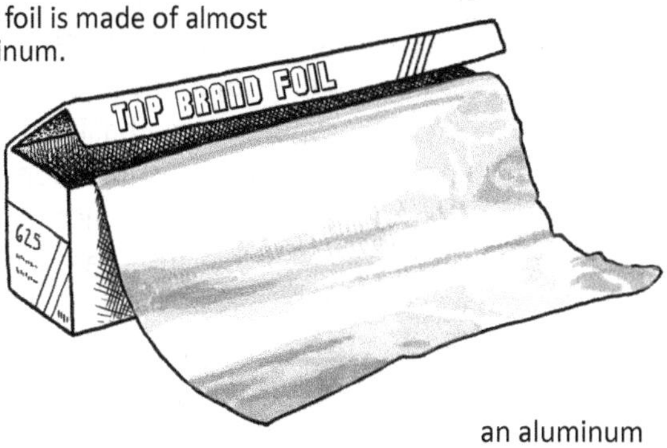

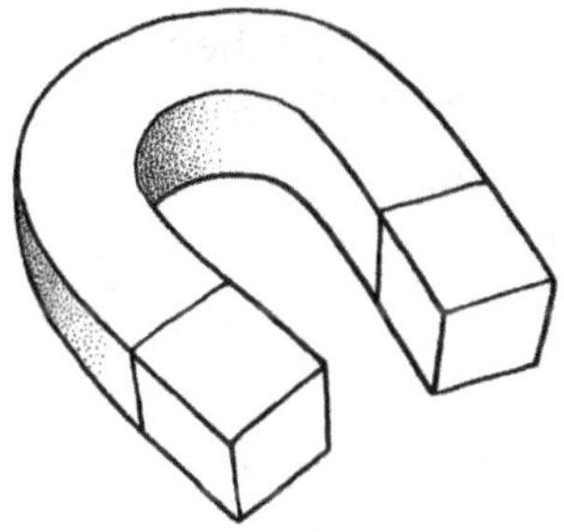

Alnico magnets are made of aluminum, nickel, and cobalt.

Many antiperspirants contain aluminum.

Aluminum is fairly easy to recycle.

an aluminum beverage can

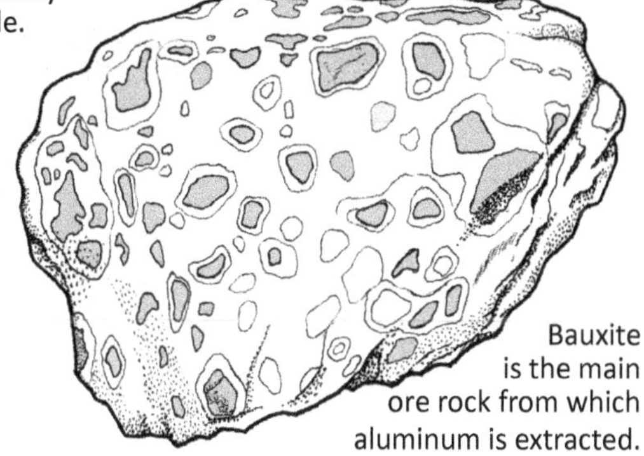

Bauxite is the main ore rock from which aluminum is extracted.

Lightweight bicycle frames are often made of Al.

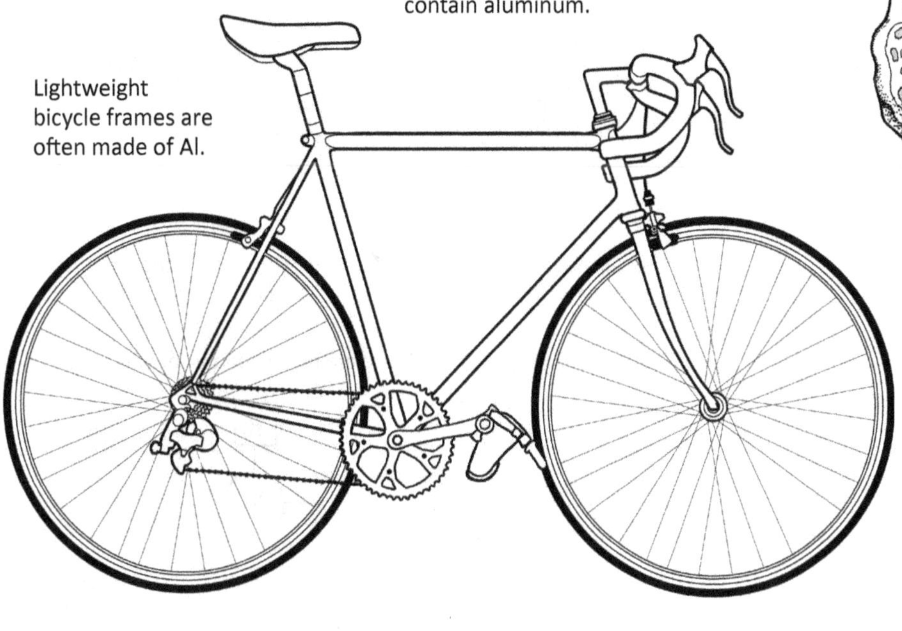

recyclable food pans and trays

Si 14

protons
14 neutrons
14 electrons

Atomic mass: 28.08

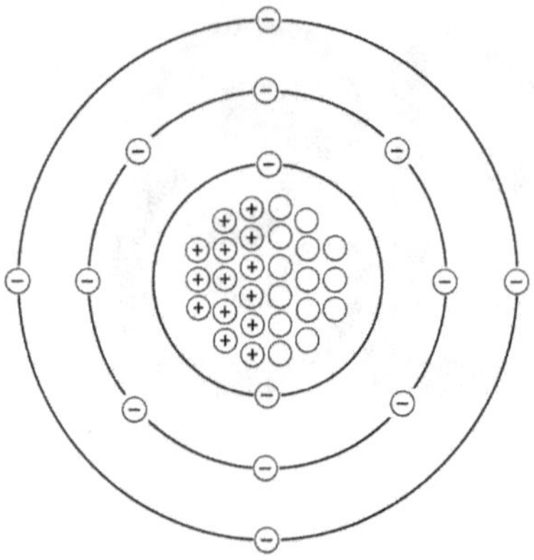

Silicon

From "silex," the Latin word for flint

Silicon was observed for the first time in 1824 by Swedish chemist Jöns Berzelius by heating a compound that contained fluorine, potassium, and silicon. Previous chemists had suspected that silicon might be an element but they were never able to get it separated from the oxygen atoms it was attached to. Pure silicon (not attached to any other atoms) is a dark blue-gray, with a very shiny surface. It is not surprising that pure silicon is shiny because silicon combines with oxygen to make glass, SiO_2.

Silicon and oxygen are the basis for the large family of silicate minerals. Silicate gemstones include agate, amethyst, flint, jasper, carnelian, chalcedony, onyx, and opal. Silicate minerals that are not gemstones include olivine, hornblende, asbestos, mica (biotite and muscovite), and feldspar. Granite is a rock that is made of mixtures of quartz, feldspar and mica. Sand that is light in color is usually made of very tiny pieces of quartz and feldspar.

Quartz is an extremely useful mineral because its crystal structure produces electricity when it is squeezed. Quartz crystals can be used to make clocks, sonar devices, and ultrasound machines.

Pure silicon can be grown into crystals that become very useful when they have small amounts of boron, germanium and arsenic added to them. The crystals are used to make solid-state electronic parts such as micro-chips and transistors, which are found in devices such as computers, tablets, and cell phones.

Another way silicon can combine with oxygen is in long chains, with other small molecules attached. These long chains are called polymers and the substances they make are called silicones. Silicone substances you may be familiar with are Silly Putty®, silicone baking trays, and silicone caulk (used around windows, sinks and tubs).

A few forms of life use silicon to make their shells. Diatoms and radiolarians are beautiful microscopic protozoans with shells made of SiO_2 (glass). One type of sea sponge, the glass sponge, uses silicon to build its skeleton.

SiC Silicon carbide

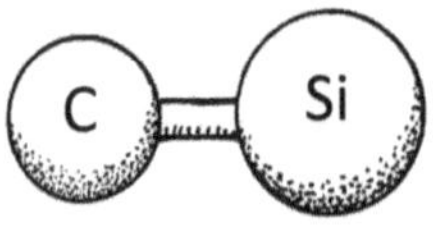

These silicon-carbon units stack together to form the lattice shape shown below.

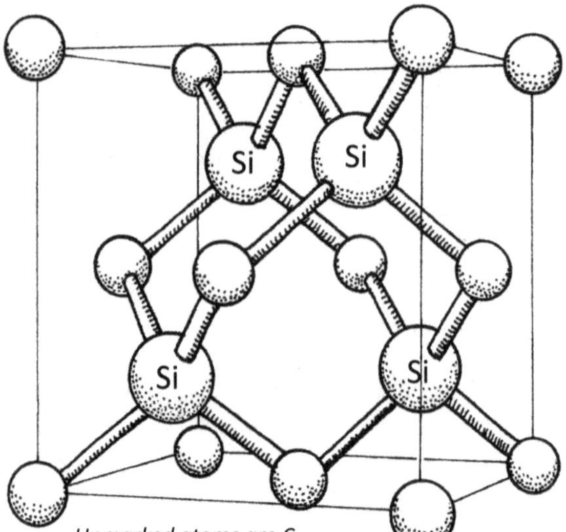

Unmarked atoms are C.

$SiCl_4$ Silicon tetrachloride

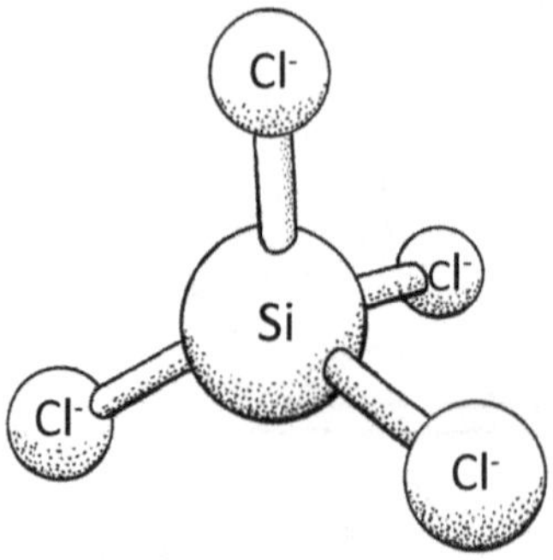

SiO_4 Silicate tetrahedron

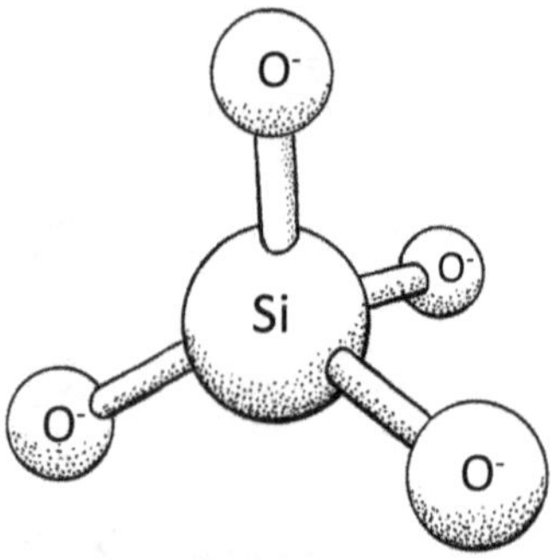

These tetrahedral molecules usually bond with many others identical to themselves to form some kind of lattice shape.

Silicone polymers (There are many options for what R can be.)

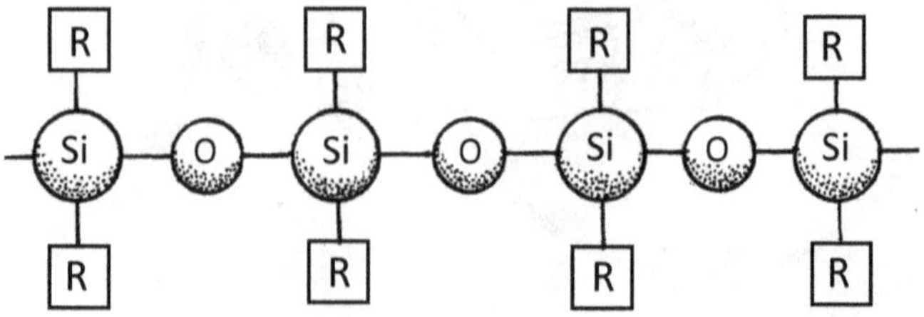

14 Silicon Si

quartz crystals, SiO_2

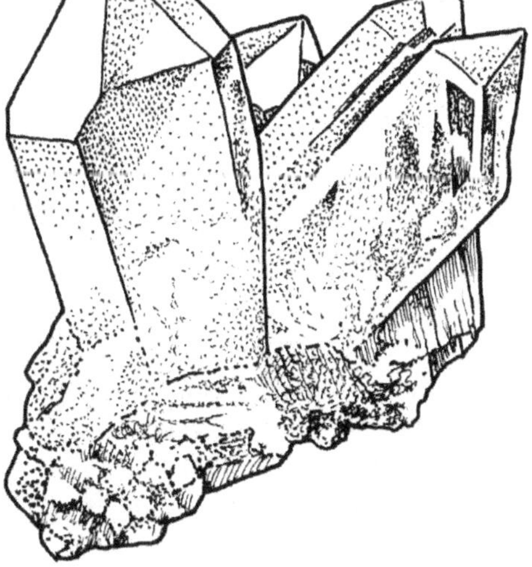

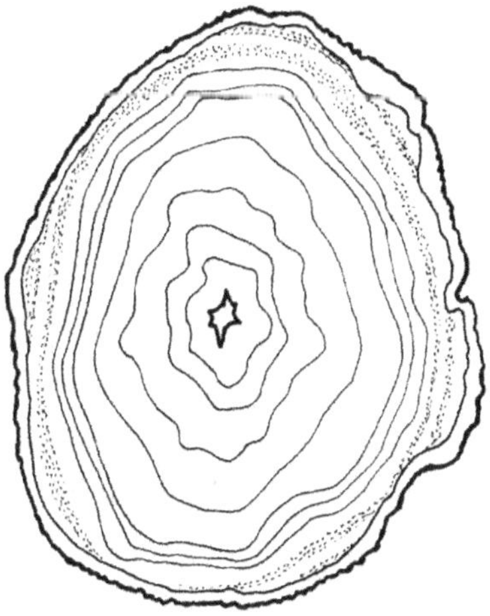

Agates, with their colorful rings, are often sliced and polished.

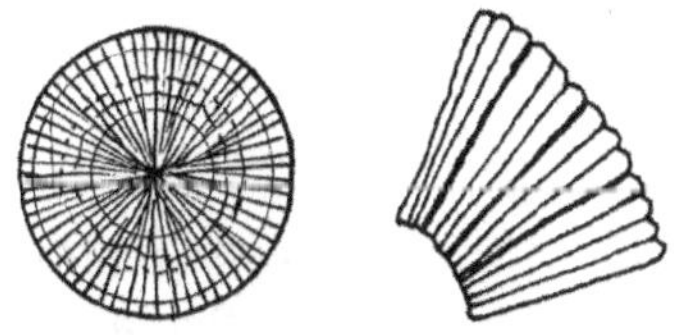

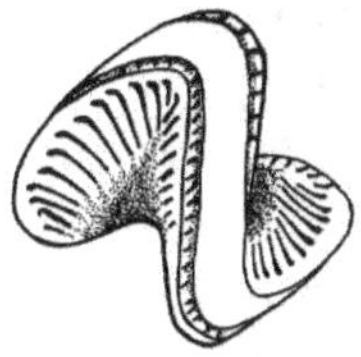

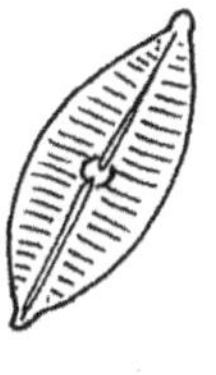

Diatoms are microscopic cells that make "glass" shells using silicon from the water.

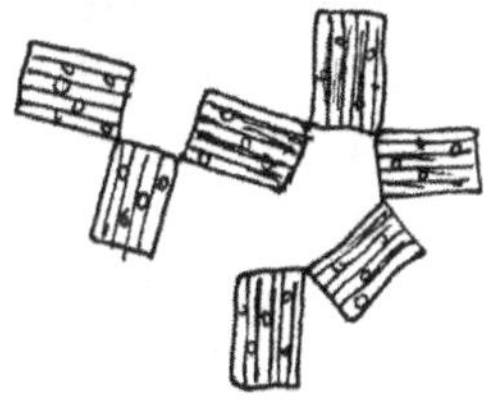

Silicone caulk keeps water out of cracks.

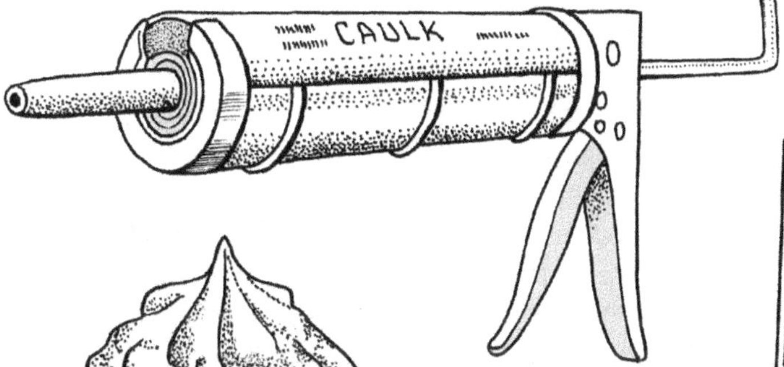

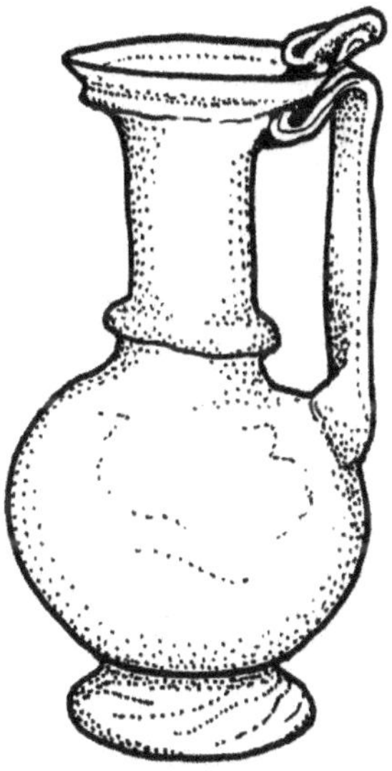

ancient Roman glass vase

Silicon can be made into silicone, which can be molded into flexible baking dishes.

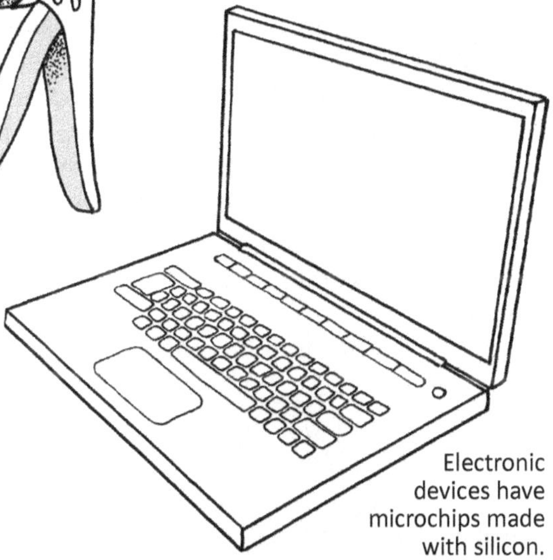

Electronic devices have microchips made with silicon.

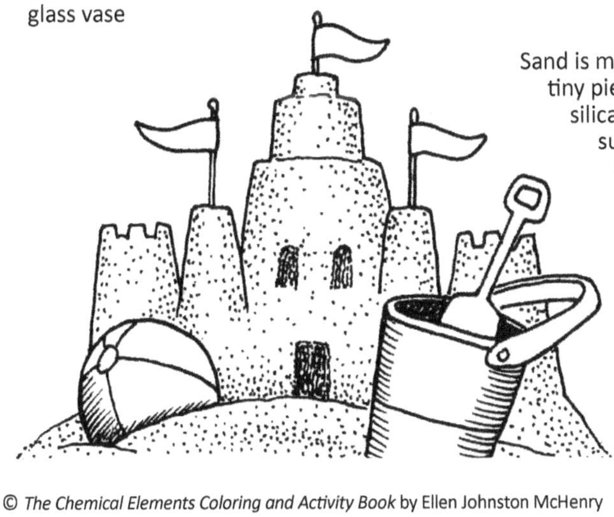

Sand is made of tiny pieces of silicate rock, such as quartz, feldspar, and mica.

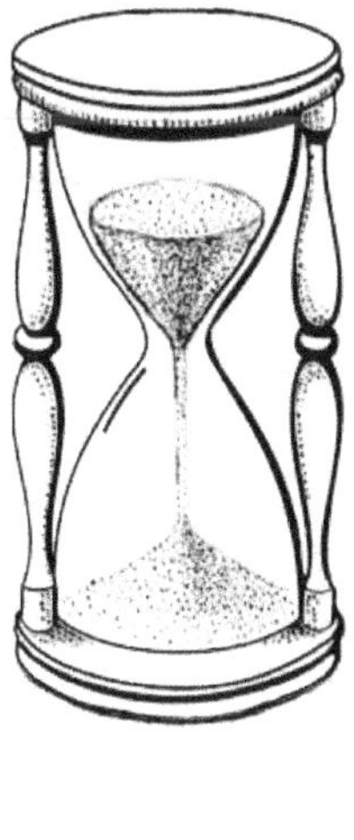

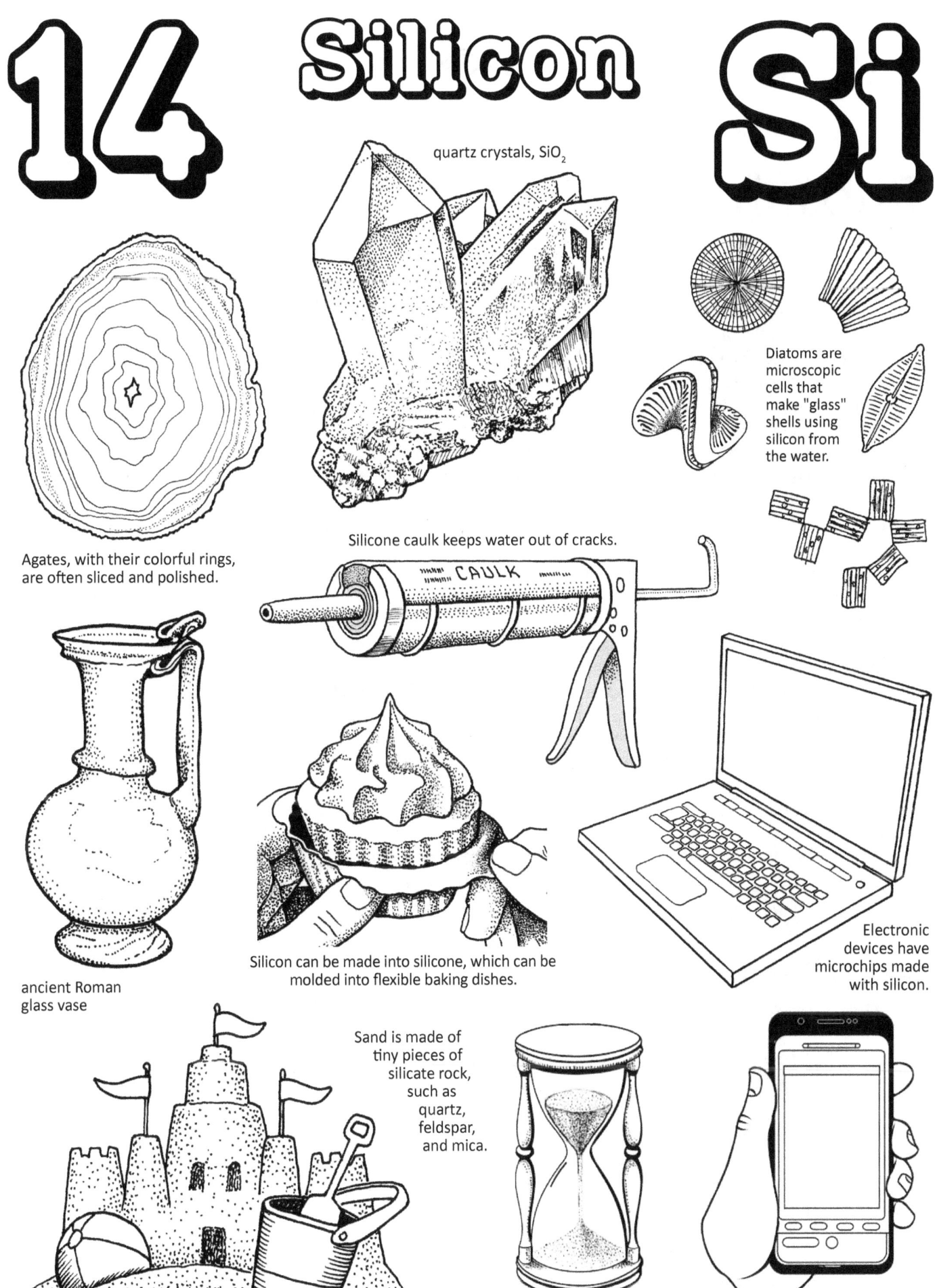

P 15

15 protons
16 neutrons
15 electrons

Atomic mass: 30.97

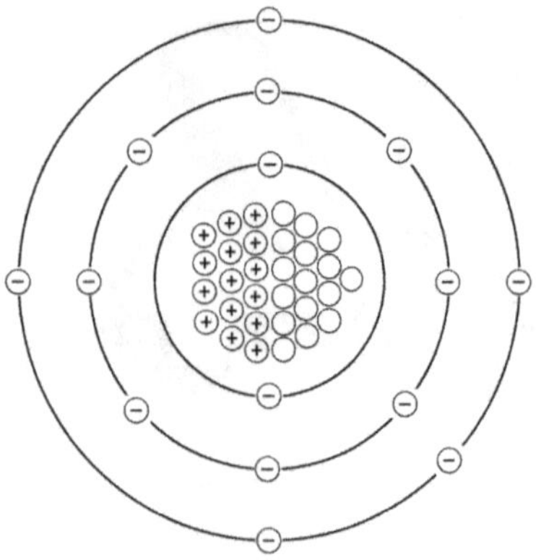

Phosphorus

From Greek words meaning "light-bearer"

The discovery of phosphorus came about quite accidentally. In 1669, a German chemist named Hennig Brand was trying to find a way to make gold. (Chemists at this time didn't know that gold was an element and could not be made.) Since urine was yellow, he thought it might contain very small amounts of gold, so he collected gallons and gallons of urine and starting boiling it down to get the "yellow stuff" out. What he got instead was white stuff that glowed in the dark. He named it *phosphorus mirabilis*, meaning "miraculous bearer of light."

It wasn't until 1769 that another source of phosphorus was discovered. Bones also contained phosphorus, and working with bone ash was certainly less smelly than working with urine. Phosphorus was not recognized as an element until French chemist Antoine Lavoisier began experimenting with it in 1777. In the 1840s, another source of phosphorus was discovered: bird droppings (guano). Vast supplies of bird guano were found on tropical islands and they became an important source of plant fertilizer for European agriculture. Eventually, phosphorus was discovered in rocks in the late 1800s, and rocks continue to be our source of phosphorus today.

Pure phosphorus comes in three colors: white, red, and black. White is the most dangerous to work with and can burst into flames if not kept under water. It is also toxic and had a brief history of being used as a poison. Its flammability gave rise to the invention of the match, as well as the invention of new types of weapons. If white phosphorus is heated to a very high temperature, the phosphorus atoms rearrange their structure and turn red. Red phosphorus is much less dangerous than white, and was much safer for making matches. If heated further, red phosphorus will turn black and will become very safe and stable, but much less useful.

Phosphorus is an essential element to both plants and animals. Phosphate, PO_4^{3-}, is one of the working parts in ATP, the energy molecule for all forms of life. Phosphate is also a structural component of DNA. Phosphoric acid, H_3PO_4 is found in carbonated beverages. Trisodium phosphate, Na_3PO_4, is used in some cleaning products and water softeners. Calcium phosphate, $Ca_3(PO_4)_2$, is used to make baking powder and in the manufacturing of china dishes. Various other compounds that contain phosphorus are used in fluorescent light bulbs.

Na_3PO_4 Trisodium Phosphate

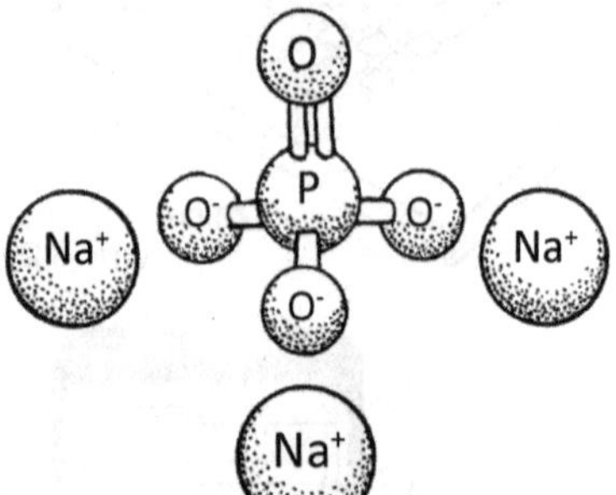

H_3PO_4 Phosphoric acid

Oxygen is usually red.

Unmarked atoms are H.

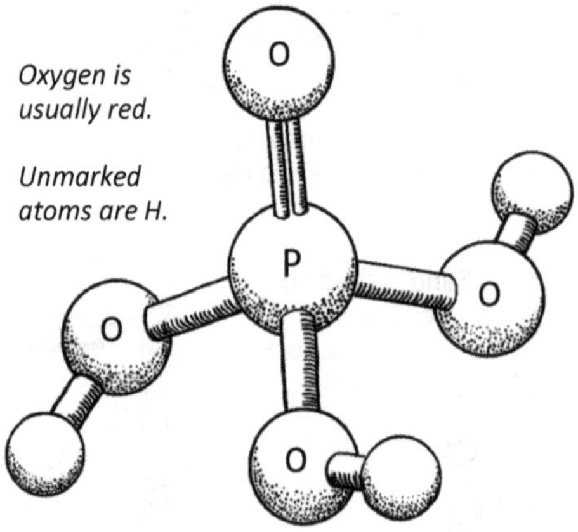

PO_4^{3-} Phosphate

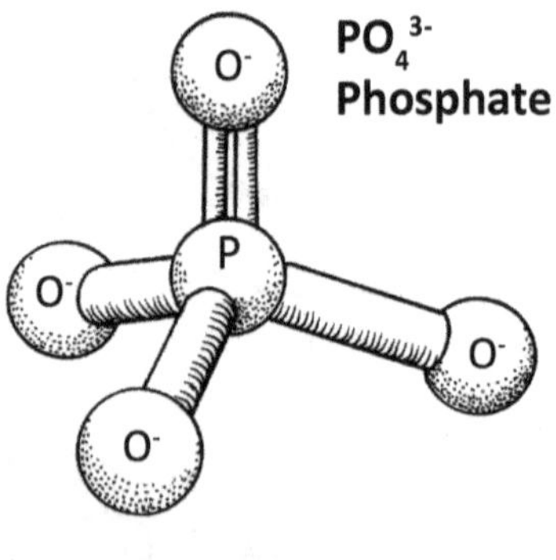

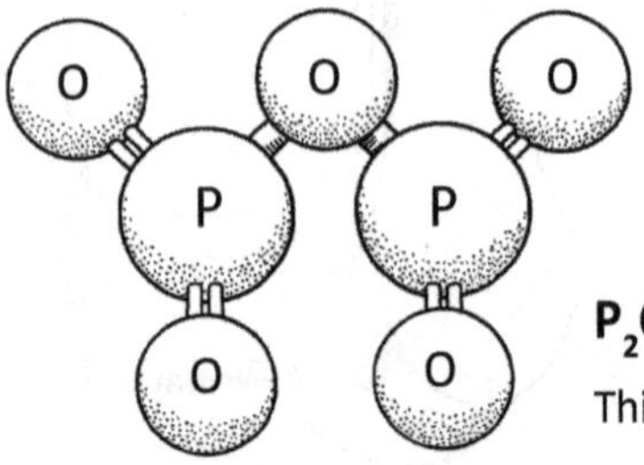

P_2O_5 Phosphorus pentoxide

This compound can form a lattice shape.

$POCl_3$ Phosphoryl chloride

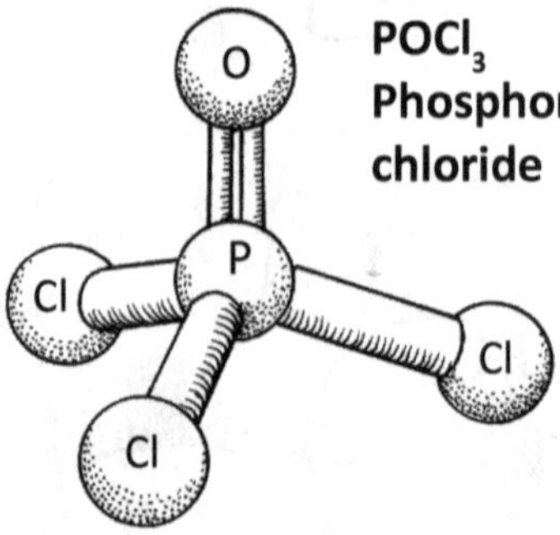

15 P

Matches are made with red phosphorus.

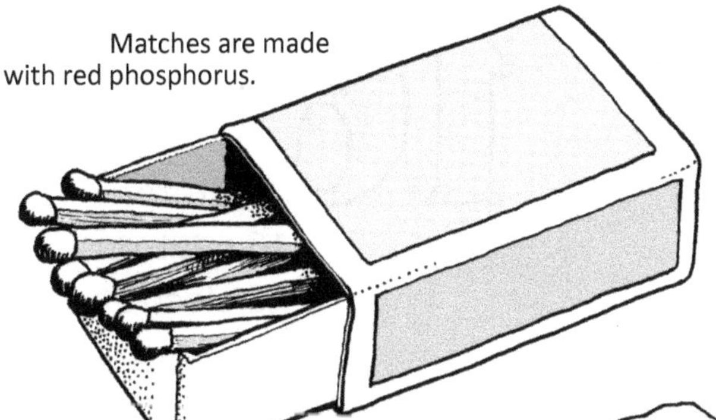

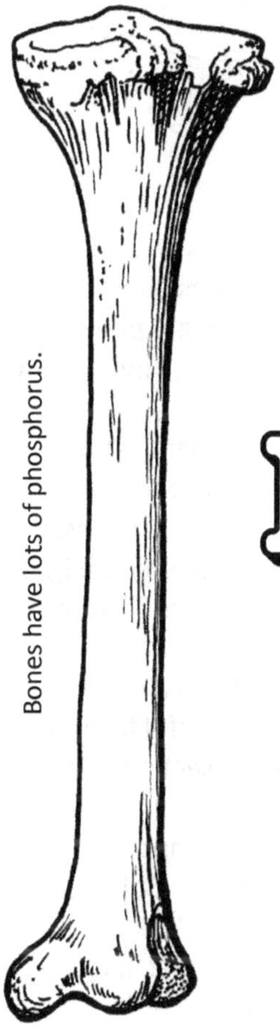

Bones have lots of phosphorus.

Teeth have lots of phosphorus.

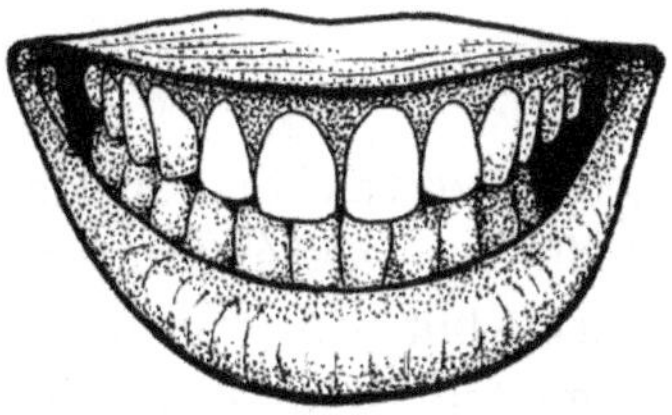

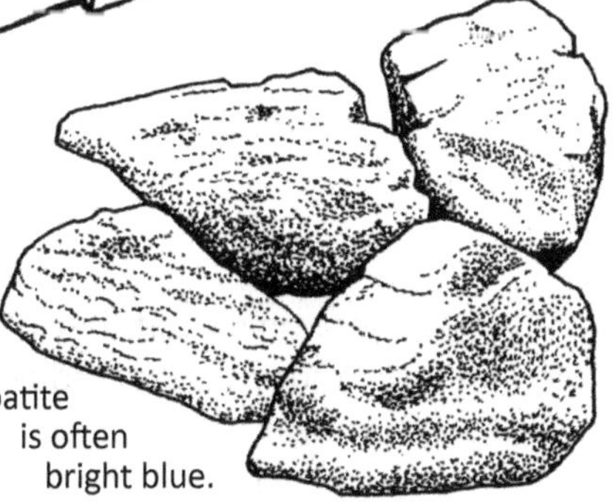

Apatite is often bright blue.

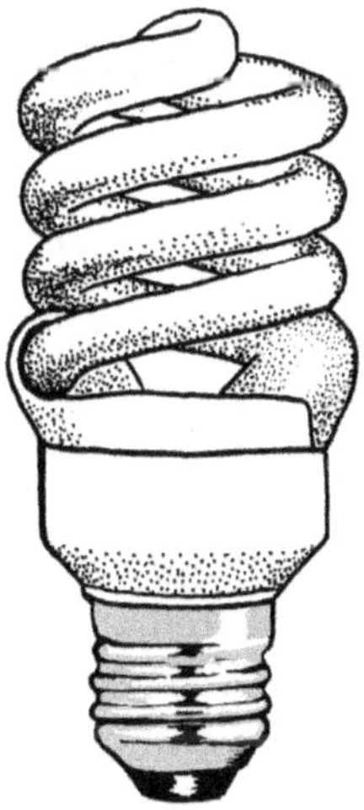

Some fluorescent bulbs contain phosphorus.

Phosphorus

Cleaning products that contain phosphorus are great for cleaning greasy surfaces.

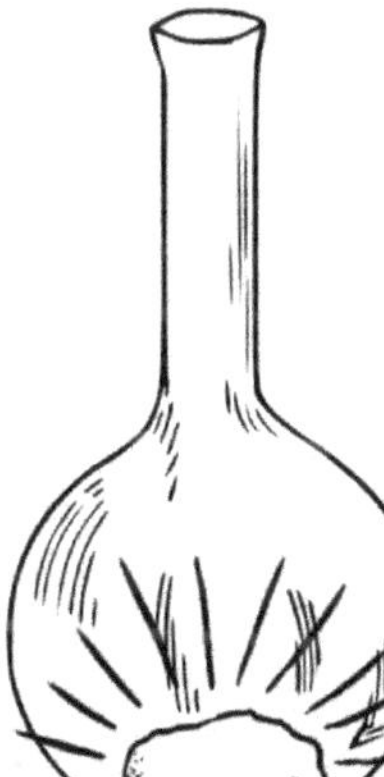

White phosphorus glows bright yellowish green when heated.

Carbonated beverages often contain phosphoric acid.

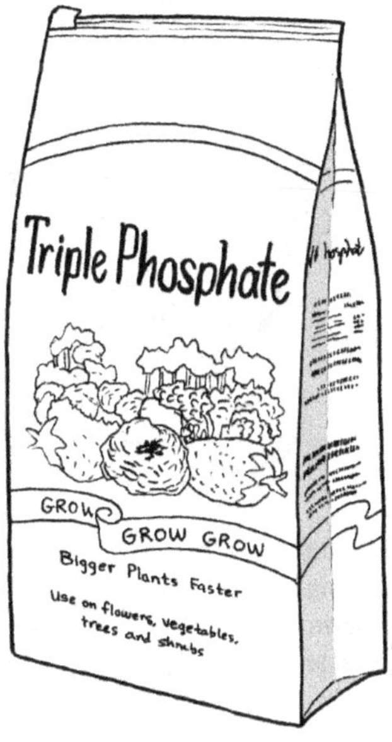

phosphate fertilizer for plants

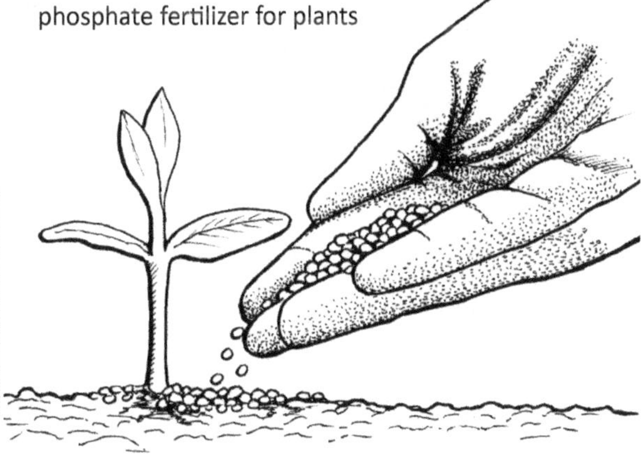

China is made with calcium phosphate.

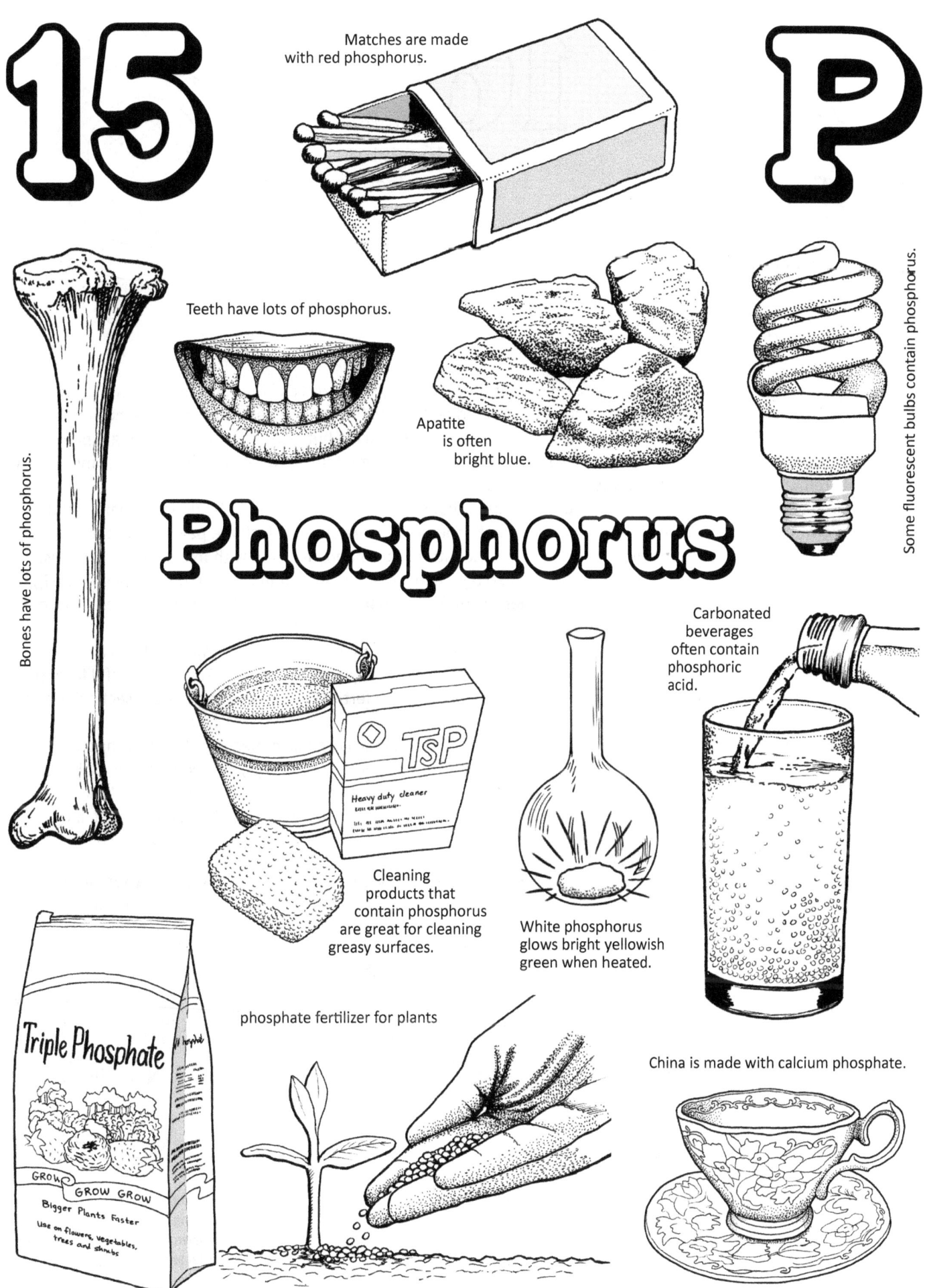

S

16

protons
16 neutrons
16 electrons

Atomic mass: 32.06

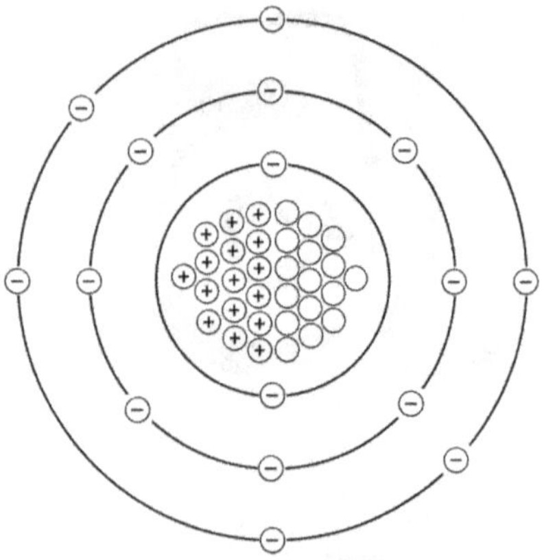

Sulfur
(or Sulphur)

From the Latin "sulphurium"

People knew about sulfur in ancient times, but they did not know it was a chemical element. They used sulfur in some of the same ways we use it today. In the Middle East, sulfur was used as a topical medicine and as an insecticide. In China, they discovered it was not only useful in medicine but also as an ingredient in gunpowder. Sulfur was also known and used in India and Greece. The historical name for it translates as "brimstone," meaning "burning stone," probably because it was often found around volcanoes.

In 1777, French chemist Antoine Lavoisier realized that sulfur was not a compound, but an element. Sulfur is an element that can exist by itself in nature. You can find lumps of pure sulfur, which are pale yellow and smell like a lit match. Though it doesn't need to bond with other atoms, it is happy to bond with many different atoms, becoming a part of lots of different compounds. In geology, sulfur is an ingredient in these minerals: galena (PbS), pyrite (FeS_2), barite ($BaSO_4$), gypsum ($CaSO_4$), sphalerite (ZnS), cinnabar (HgS), and stibnite (Sb_2S_3). Some of these minerals are useful in industry, such as gypsum, which is used to make plasterboard walls for houses.

Most of the sulfur used in industry today comes from petroleum, rather than minerals. Sulfur is a natural by-product when petroleum is refined (made into a usable product like gasoline and oils). The most useful form of sulfur for industry is sulfuric acid, H_2SO_4. Sulfuric acid is used to make fertilizer, lead-acid batteries, insecticides and fungicides, matches, and many other things.

Sulfur is an essential ingredient for all of life and is found in many organic molecules. Sulfur compounds called "thiols" have a strong odor and are found in smelly things like garlic, rotten eggs, and skunk spray. Sulfur is found in three amino acids. The sulfur-containing amino acids form cross links, making tough proteins like keratin, which is found in skin, hair, and feathers. Sulfur cross-linking is also the key to making "vulcanized" rubber, a form of rubber tough enough to be used for vehicle tires.

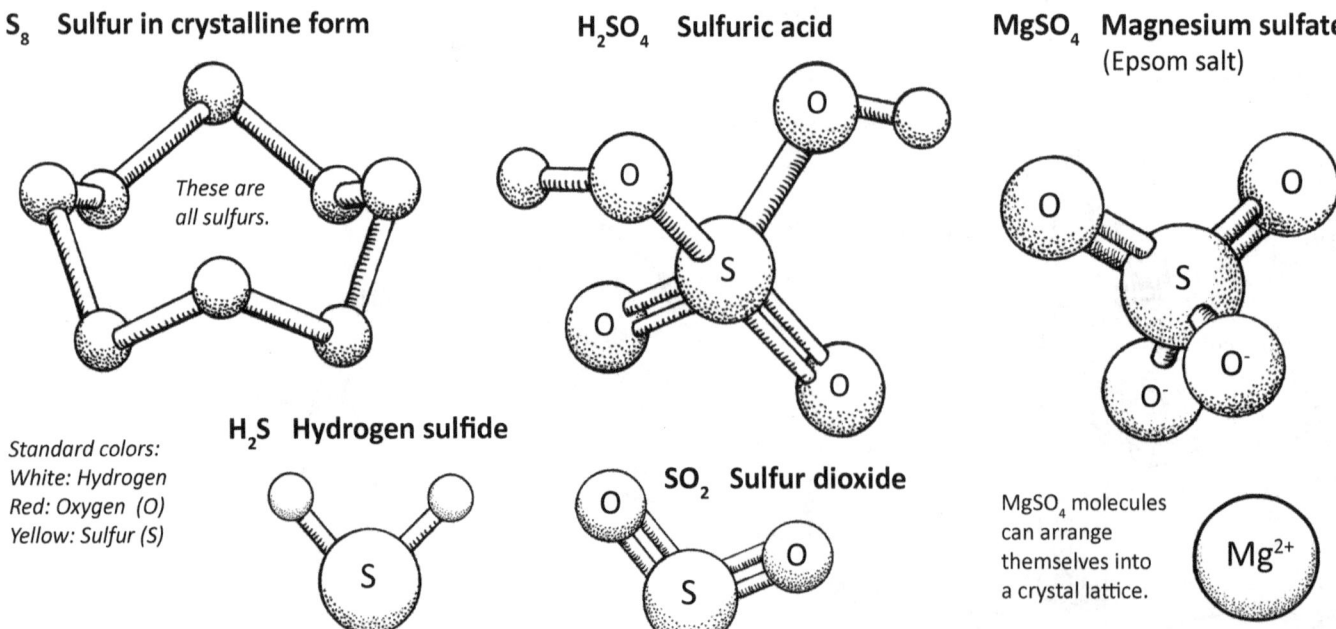

S_8 Sulfur in crystalline form

H_2SO_4 Sulfuric acid

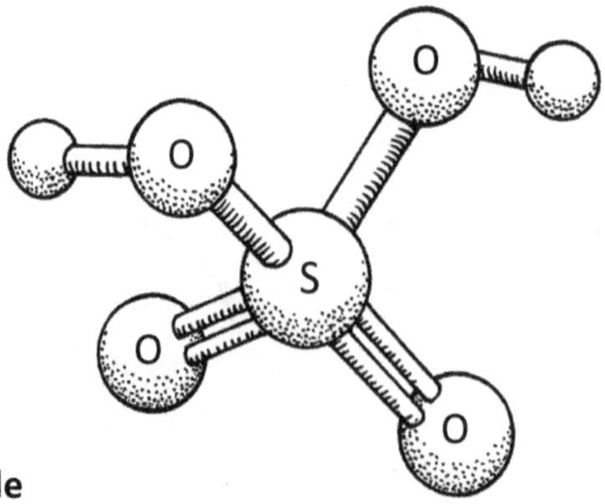

$MgSO_4$ Magnesium sulfate
(Epsom salt)

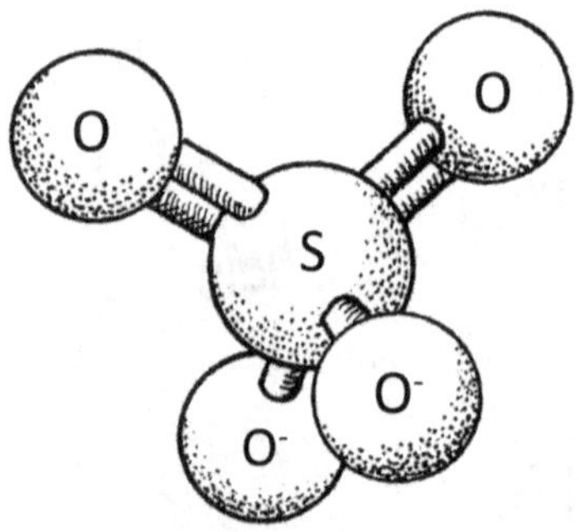

Standard colors:
White: Hydrogen
Red: Oxygen (O)
Yellow: Sulfur (S)

H_2S Hydrogen sulfide

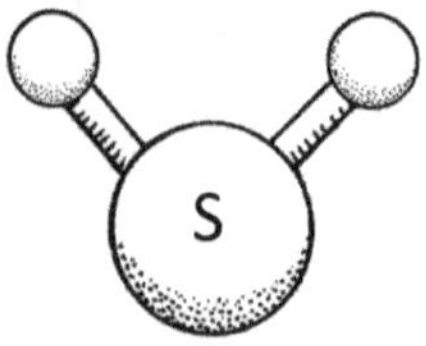

SO_2 Sulfur dioxide

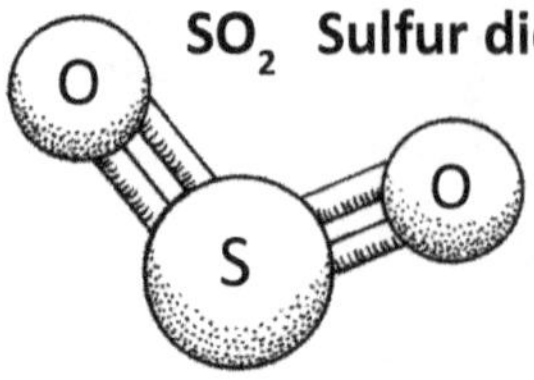

$MgSO_4$ molecules can arrange themselves into a crystal lattice.

16 S

Volcanoes can put SO_2 and H_2S into the air.

Sulfur

Iron pyrite, FeS_2 ("fool's gold") forms shiny gold cubic crystals.

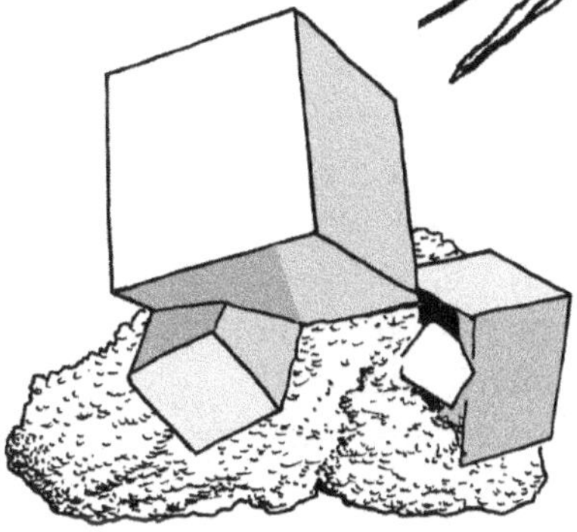

Barite, $BaSO_4$, often forms in this shape, called a desert rose.

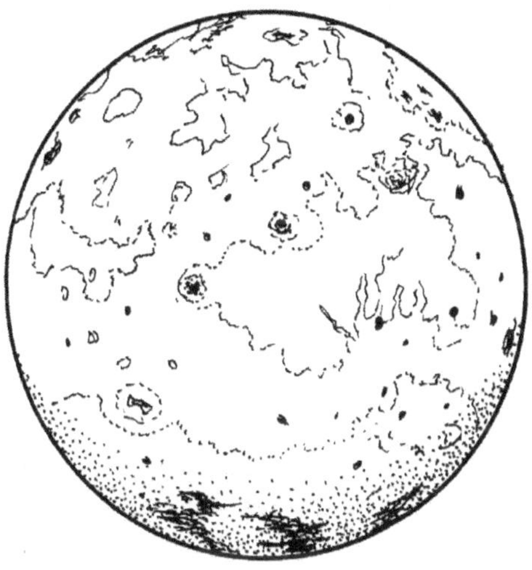

Jupiter's moon, Io, is yellow and orange, due to the large amount of sulfur produced by its many volcanoes.

Sulfur is used to make "vulcanized" rubber for tires.

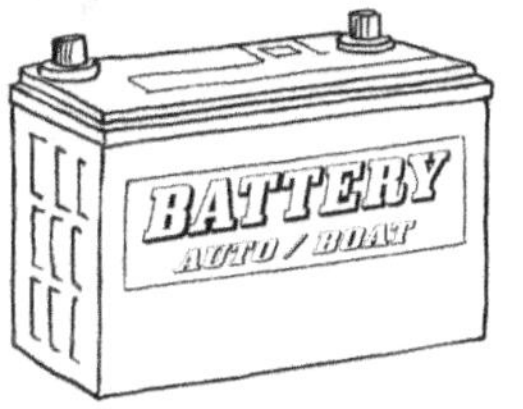

Car batteries use sulfuric acid.

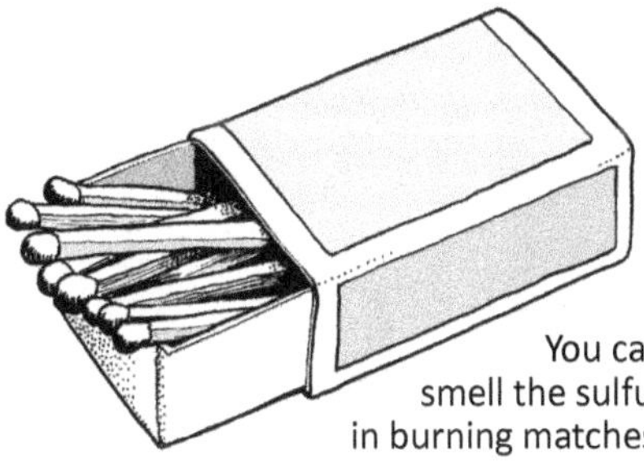

You can smell the sulfur in burning matches.

Epsom salt $MgSO_4$

Sulfur compounds are found in many smelly things, including skunk spray and garlic.

Sulfur dioxide, SO_2, is a common form of air pollution emitted by many factories. The smoke might look light yellow if it contains a lot of sulfur.

Cl 17

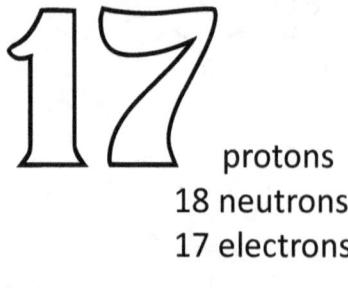

protons
18 neutrons
17 electrons

Chlorine

Atomic mass: 35.45

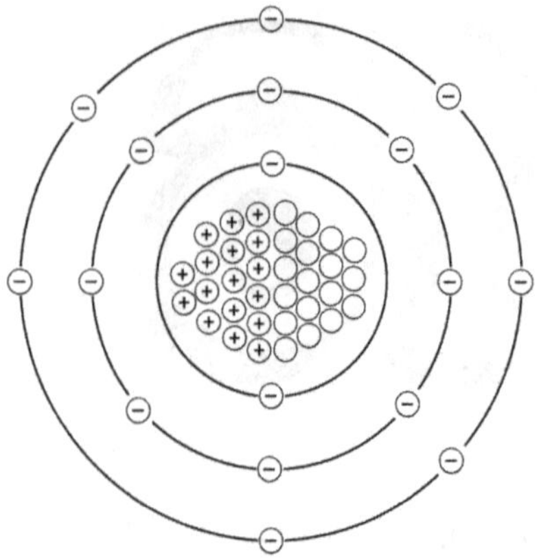

From Greek word "chloros" meaning "light green"

Chlorine is very reactive and is capable of bonding with almost any element on the Periodic Table. It is a member of the halogen family (the "salt makers") along with fluorine, bromine, and iodine. All these elements are very reactive, due to the fact that they have 7 electrons in their outer shells, one short of having the full number, 8. When they bond with elements in the first two columns of the Periodic Table, they form salts. Chlorine will form many salts, including NaCl, KCl, RbCl, $MgCl_2$, $CaCl_2$, and $SrCl_2$.

In its pure form, chlorine is a greenish gas. Several chemists discovered chlorine gas, but did not know what it was. Humphry Davy was the first to realize it was a new element and in 1810 he named it "chlorine."

The most common compound that contains chlorine is NaCl, table salt. Chlorine is one of the essential elements that all living things need, and it plays many roles in the human body. Our stomachs produce HCl, hydrochloric acid, to digest proteins. One of our immune cells uses a chlorine compound to poison and kill germs.

In the early 1800s, chemists discovered chlorine's ability to disinfect, even though germs had not yet been discovered. Chlorine compounds began to be used to clean surfaces and equipment in hospitals. Soon after, they found that it could be added to drinking water to prevent illnesses such as cholera. Today, we still use chlorine-based bleaches, such as sodium hypochlorite, NaClO, to disinfect, and we still add chlorine to drinking water. Public swimming pools are often treated with chlorine to keep them germ-free.

Chlorine is used in thousands of industrial processes, either as a reactant or as part of the final product. It is used in the manufacturing of paper, plastics, medicines, insecticides, textiles, dyes, paints, and solvents. Carbon tetrachloride, CCl_4, is used in dry cleaning. White PVC pipes are made of polyvinyl chloride.

Sadly, chlorine has also been used as a weapon. In World War I, chlorine gas was used in smoke bombs. The gas would go into the lungs and turn into hydrochloric acid, instantly destroying lung tissue.

HCl Hydrogen chloride
(also known as hydrochloric acid)

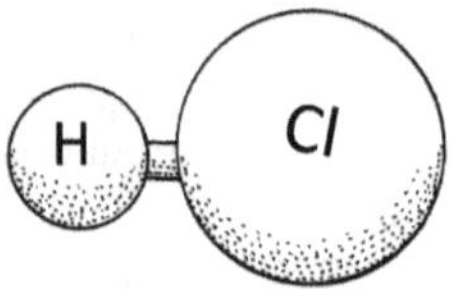

CCl_4 Carbon tetrachloride

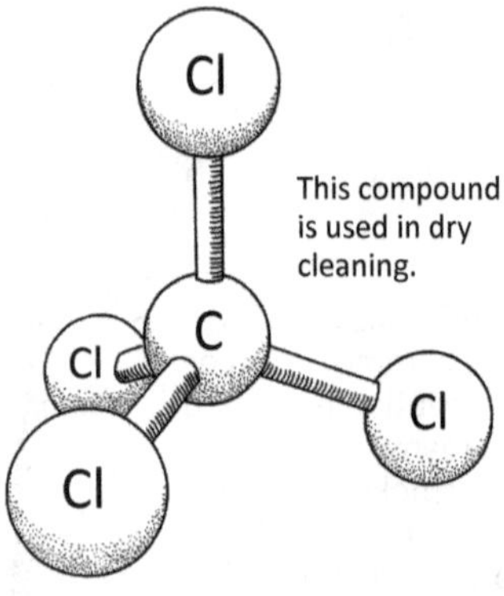

This compound is used in dry cleaning.

$CHCl_3$ Chloroform

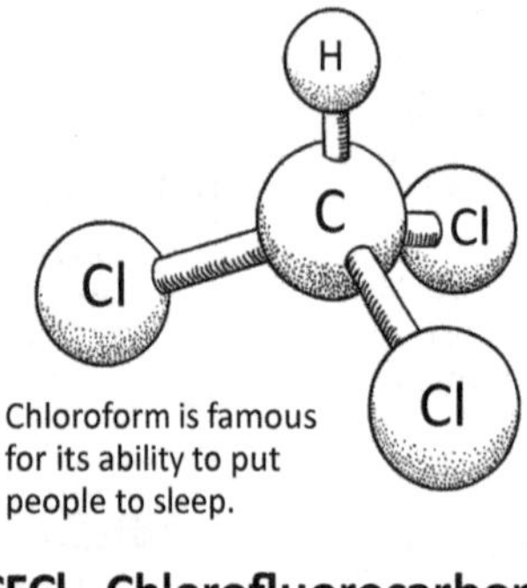

Chloroform is famous for its ability to put people to sleep.

$CFCl_3$ Chlorofluorocarbon

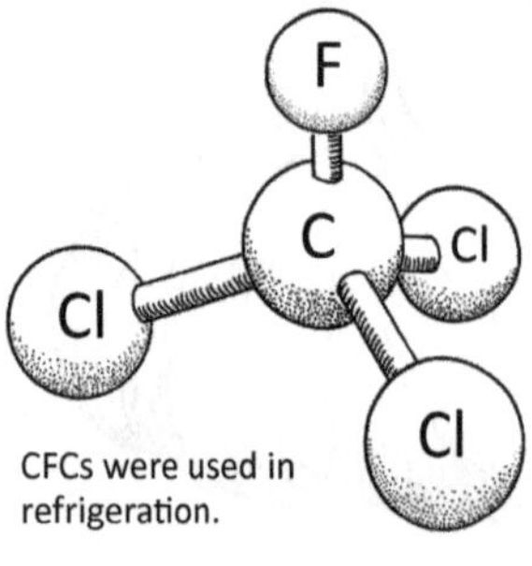

CFCs were used in refrigeration.

NaCl Sodium chloride

Color the larger balls green for chlorine.
Color the smaller balls yellow for sodium.

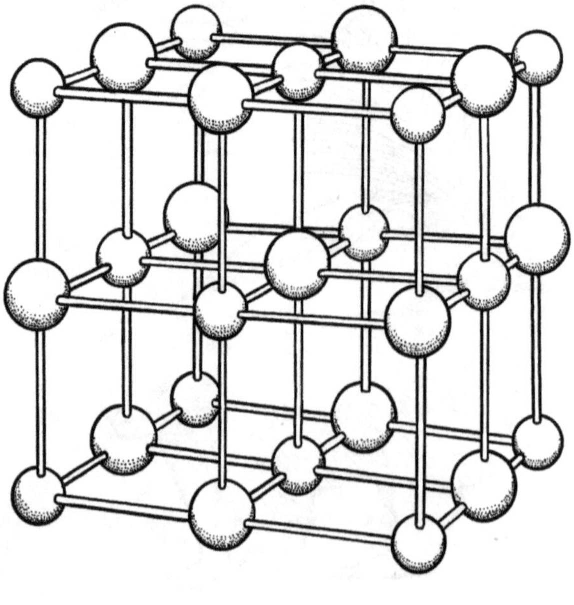

17 Cl

Chlorine

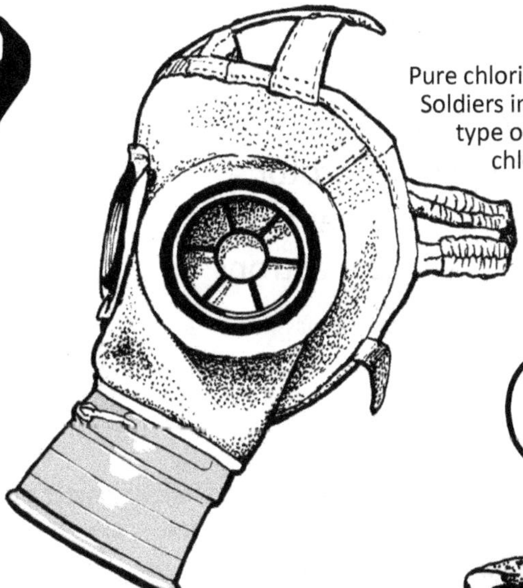

Pure chlorine is a poisonous gas. Soldiers in World War 1 used this type of gas mask during chlorine attacks.

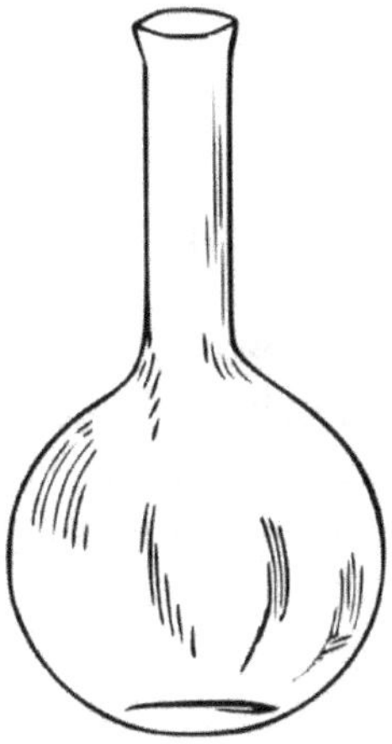

Draw yellowish-green chlorine gas in the flask.

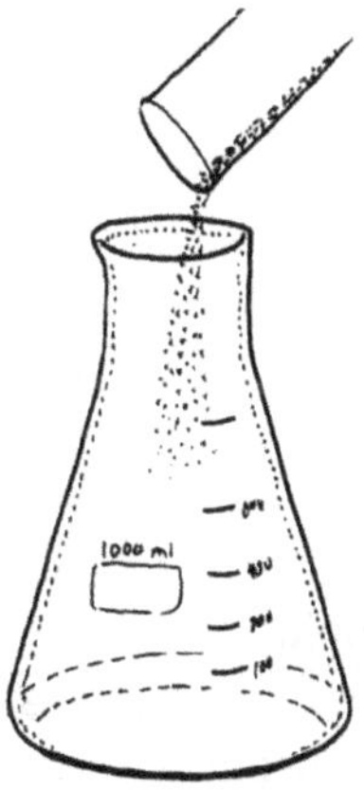

Chlorine is used in many chemical reactions.

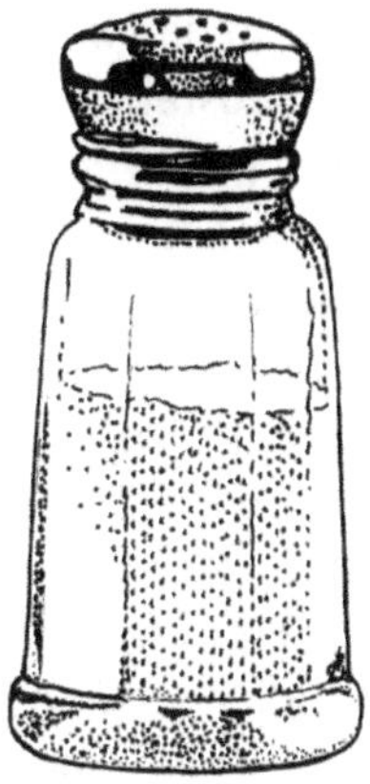

Salt is NaCl.

NaCl makes cubic crystals.

Chlorine bleach can be used to disinfect surfaces, or it can be added to water to kill germs.

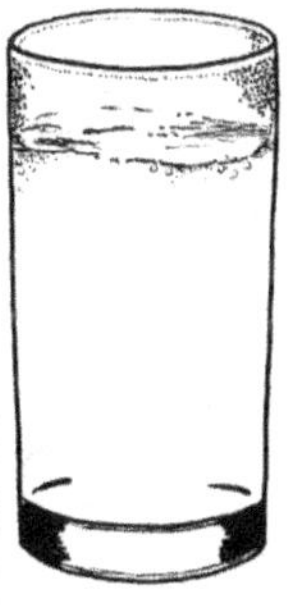

chlorinated drinking water

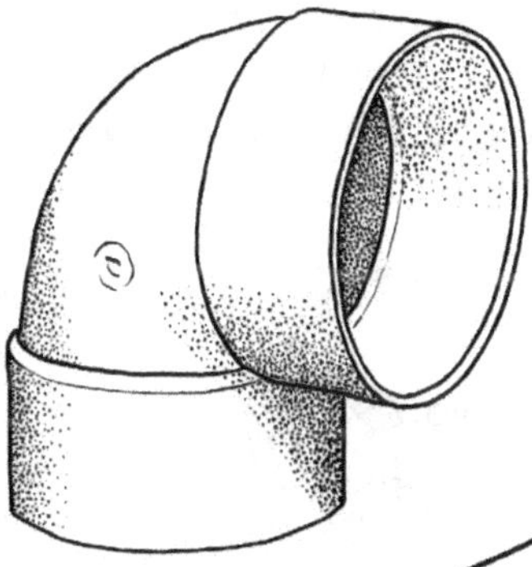

PVC pipes are made of polyvinyl chloride.

Chlorine is used in some paper manufacturing processes.

Chlorofluorocarbons (CFCs) were used in refrigerators and air conditioners until they were banned starting in 1987.

Chlorine is often put into swimming pools.

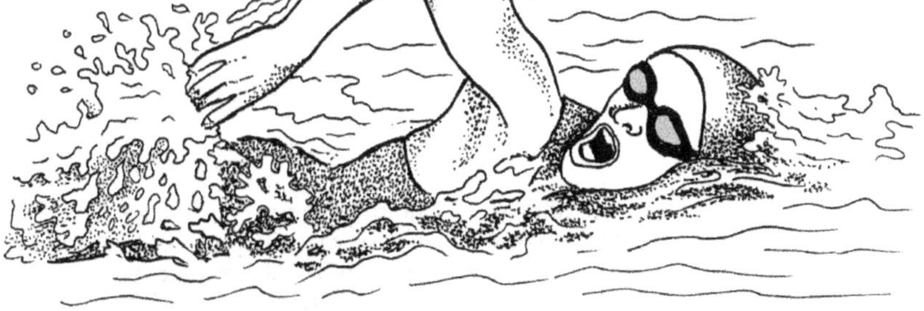

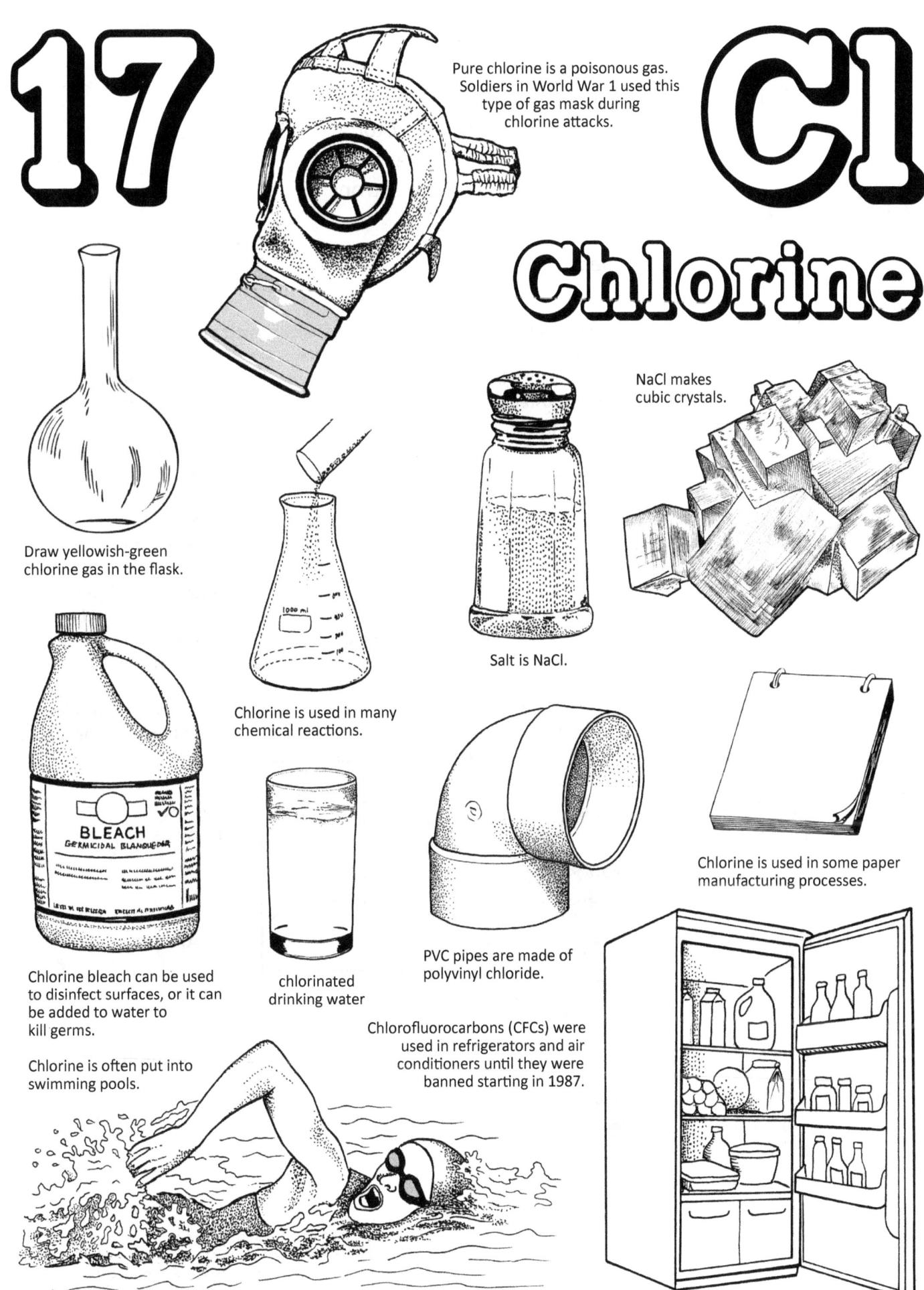

Ar 18

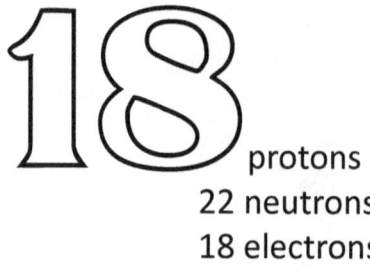

protons
22 neutrons
18 electrons

Atomic mass: 39.9

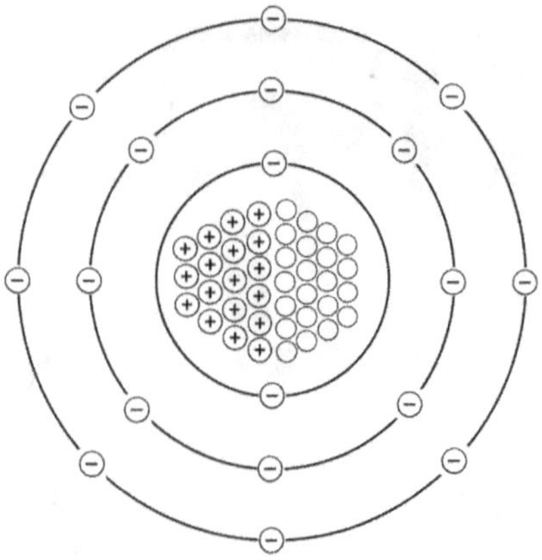

Argon

From the Greek word "argos" meaning "lazy"

Argon is a harmless gas that is found in the air all around us. It makes up almost 1% of our atmosphere, and is twice as abundant as water vapor. It was discovered in 1894 by Sir William Ramsay and Lord Rayleigh. They used a technique suggested by Henry Cavendish, who had investigated gases back in the late 1700s. They exposed normal air to both electricity and a very alkaline substance until all the oxygen, nitrogen, and carbon dioxide were gone. They found that there was still a gas left in the jar. When they tested this gas they found that it would not react with any other element, so they called it "argon," meaning "lazy." Today, argon is produced by simply chilling air until all the gases turn to liquid, then letting the temperature rise slowly and catching each element as it "boils" off and becomes a gas again.

Argon isn't the only lazy gas. It belongs to the family of elements called the noble gases, found in the column all the way to the right on the Periodic Table. These gases don't react with other elements because their outer electron shell is completely full. They don't need to gain or lose any electrons.

Since argon is so nonreactive, it is ideal for use in places where safety around heat is a concern, such as in graphite electric furnaces (used for manufacturing of steel) and in gas metal arc welding. Argon acts like a shield around the intense heat of the weld. It is also perfect for filling all types of light bulbs, both incandescent and fluorescent.

When argon is used in lasers, they emit a blue-green light. These lasers are used in specialized microscopy, in surgery, in some DNA sequencers, for inspecting semiconductor wafers, and in laser light shows.

A lesser-known use for argon is in the poultry industry where it is used to butcher large numbers of chickens very quickly. Argon gas is heavier than air, so it will hover at ground level. Once the birds begin breathing the argon, they fall asleep before being asphyxiated, so they never experience any pain or fear.

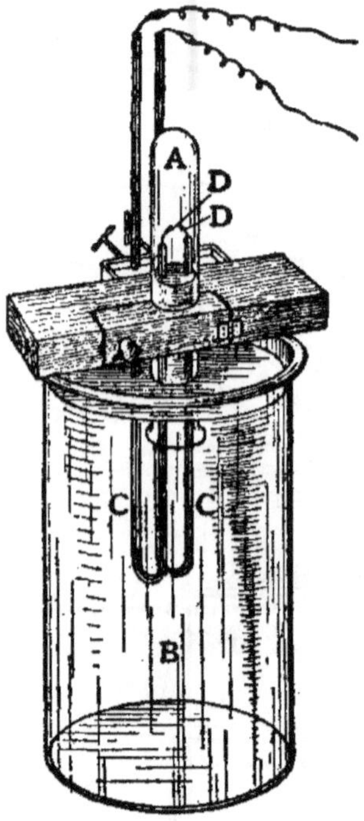

The apparatus that Ramsay and Rayleigh used to isolate argon.

Sir William Ramsay

Lord Rayleigh (John Strutt)

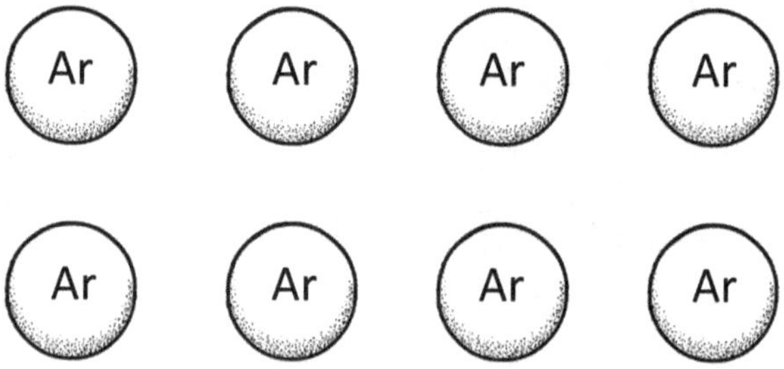

Normally, argon floats around as single atoms.

HArF Argon fluorohydride

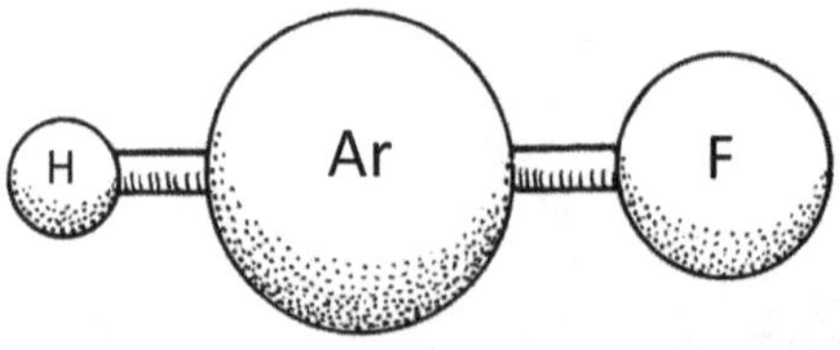

In the year 2000, some Finnish scientists cooled argon down to -265° C, and mixed it with hydrogen fluoride while exposing it to ultraviolet radiation. HArF molecules formed, but they quickly fell apart when the temperature began rising.

18 Argon Ar

Argon is used for humane slaughtering of poultry.

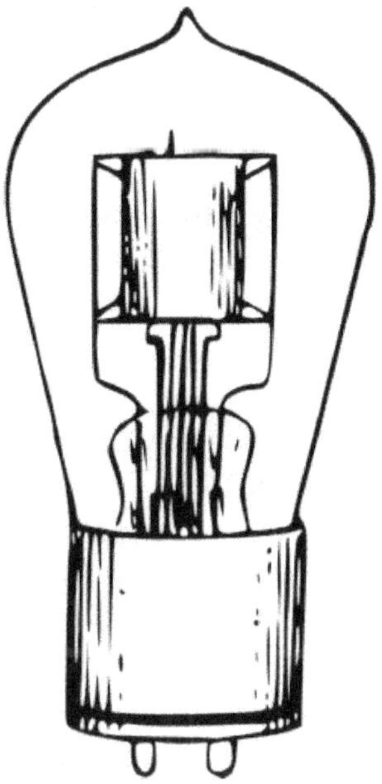

Argon is used to fill all kinds of light bulbs.

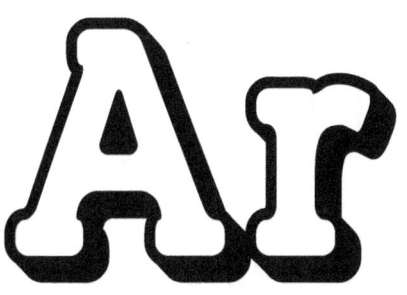

Argon lasers are used for eye surgery. They make a bright blue line of light.

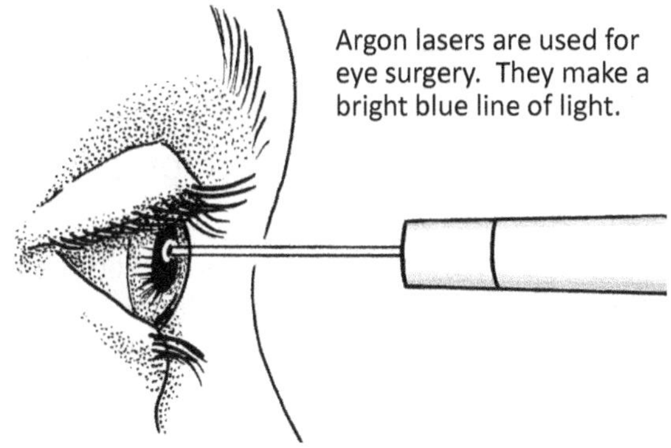

This is a cutaway view of a graphite furnace, used to melt steel. Electricity flows through the graphite rods and produces a bowl of hot, liquid metal.

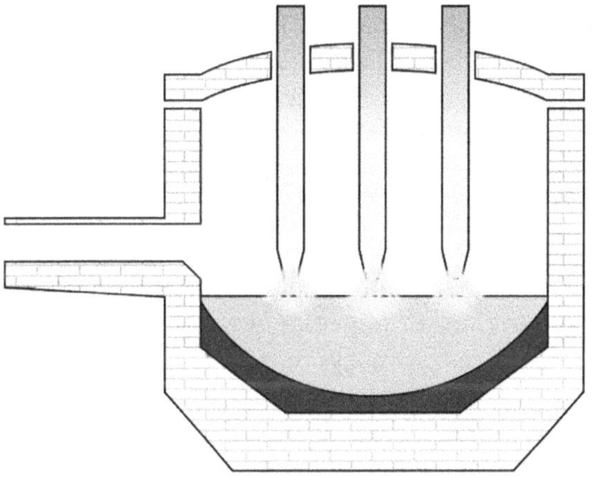

Replacing regular air with argon will keep paint and varnish from drying and oxidizing.

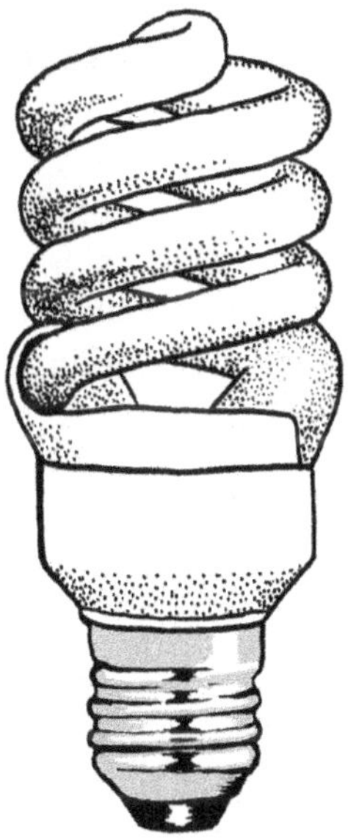

Glove boxes can be filled with an inert gas like argon so that technicians can work with materials that can't be exposed to oxygen or nitrogen.

Gas metal arc welding uses argon to shield the hot welding area from the oxygen in the air.

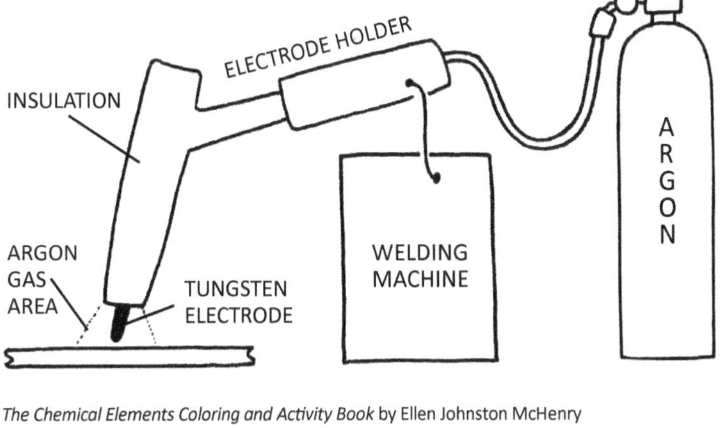

K 19

Potassium

19 protons
20 neutrons
19 electrons

Atomic mass: 39.1

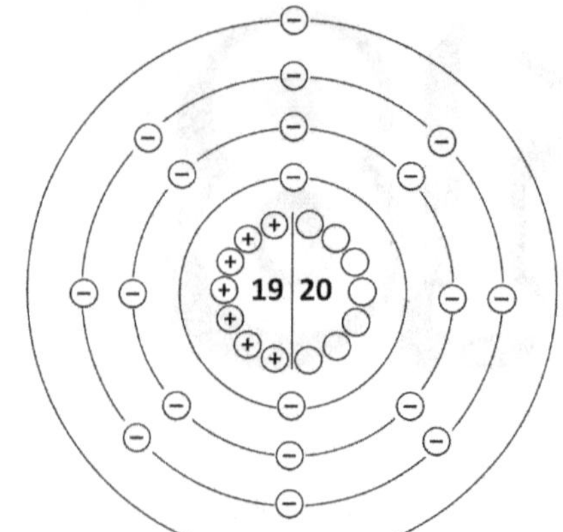

From the word "potash" (ashes from plant leaves)
The letter K is from the Latin "kalium" which came from an Arabic word for potash, "kali

Pure potassium had never been seen before Sir Humphry Davy produced it in 1807 using electricity to draw the potassium atoms out of a solution of caustic potash (KOH). A few months after this, he would use the same technique to isolate pure sodium. Neither of these elements ever occur in their pure state in nature because they are so reactive. Their reactivity is due to the fact that they have one lonely electron in their outer shell, and they are desperate to give it away. Both potassium and sodium in their pure state are soft, shiny silver metals.

In the earth's crust, potassium is found primarily in rocks called feldspar, especially orthoclase feldspar. It also occurs naturally in deposits of saltpeter, KNO_3, and KCl, potassium chloride. Hundreds of years ago, saltpeter was found to be one of the key ingredients needed to make gunpowder, along with sulfur and charcoal.

Potassium is an essential mineral to both animals and plants. (Potash, the source of potassium before the 20th century, was made from the ashes of plants.) In our own bodies, potassium and sodium flow in and out of cells through channels called sodium-potassium pumps. These pumps are especially important in nerve cells, which use them to transmit electrical impulses. Foods high in potassium include potatoes, spinach, avocados, and bananas.

Potassium plays a role in many chemical reactions, both as a reactant and as an end product. Potassium compounds are used in the manufacturing of inks, dyes, stains, soaps, bleach, matches, glass, and tanned leather.

Potassium compounds are used in food preparation. Potassium bisulfate, $KHSO_3$, is a preservative and potassium bromate, $KBrO_3$, is used to strengthen bread dough. Potassium chloride, KCl, is used as a salt (NaCl) substitute.

Potassium chloride, KCl, is a very useful potassium compound. It is used for de-icing sidewalks, as flux in glass manufacturing, as a fertilizer, as a source of radiation in scientific research, in petroleum and natural gas extraction, in heat packs that provide instant heat, and in home water softeners.

KCl Potassium chloride

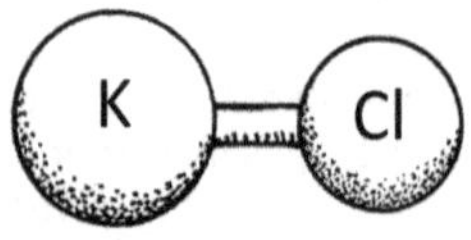

KOH Caustic potash

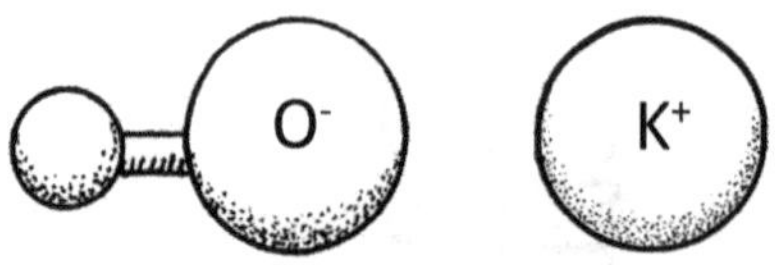

KNO_3 Potassium nitrate (saltpeter, or nitre)

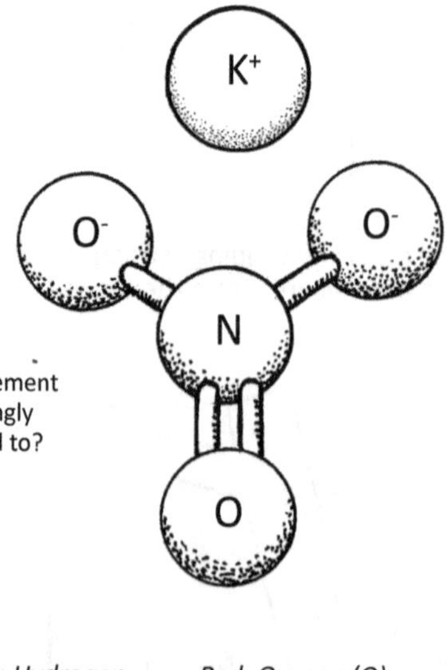

K_2SO_4 Potassium sulfate

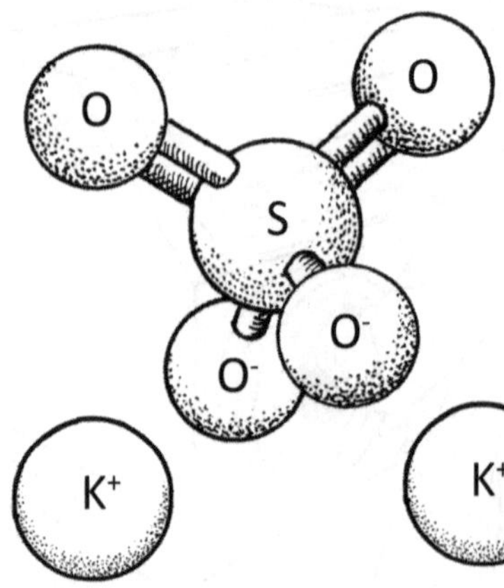

$KBrO_3$ Potassium bromate

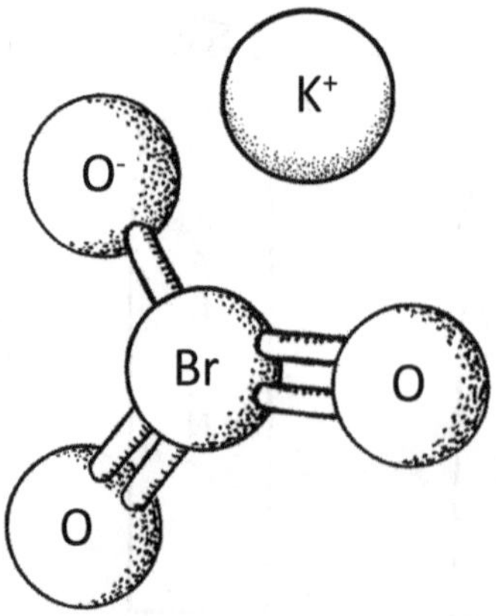

What element is K strongly attracted to?

White: Hydrogen *Red: Oxygen (O)*
Yellow: Sulfur (S) *Purple: Potassium (K)*
Green: Chlorine (Cl) *Orange: Bromine (Br)*

19 K

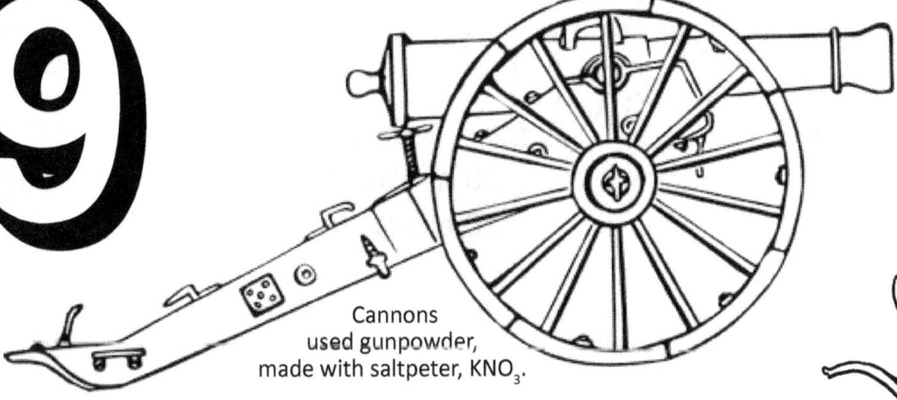

Cannons used gunpowder, made with saltpeter, KNO_3.

Potassium

These foods are high in potassium...

bananas

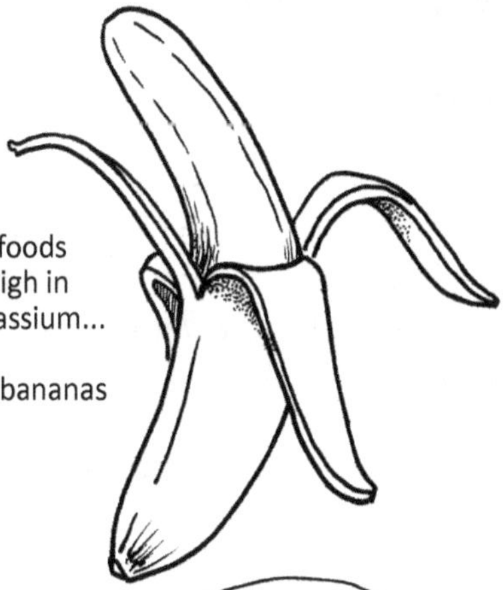

Plants need potassium as one of their essential minerals.

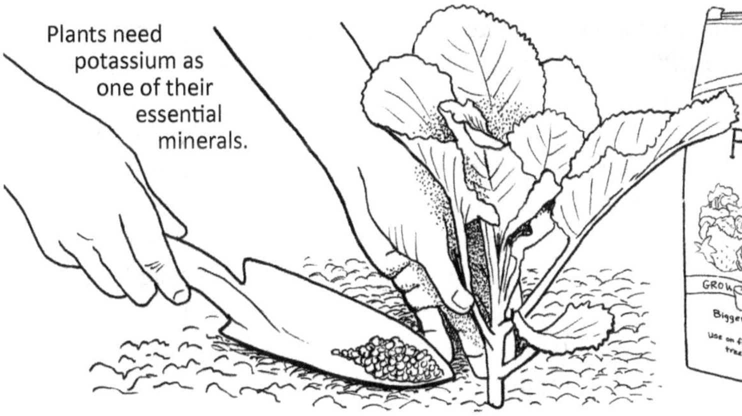

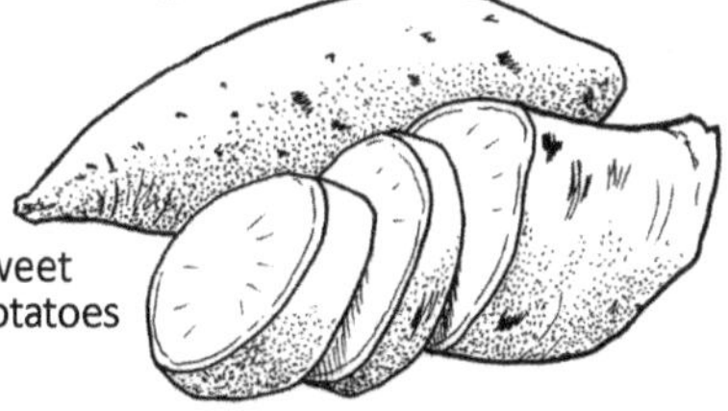

sweet potatoes

avocados

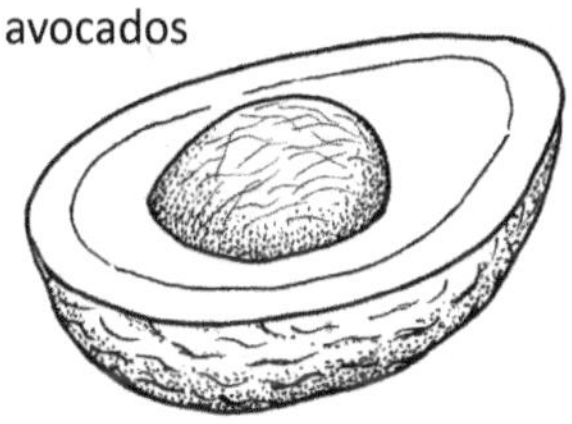

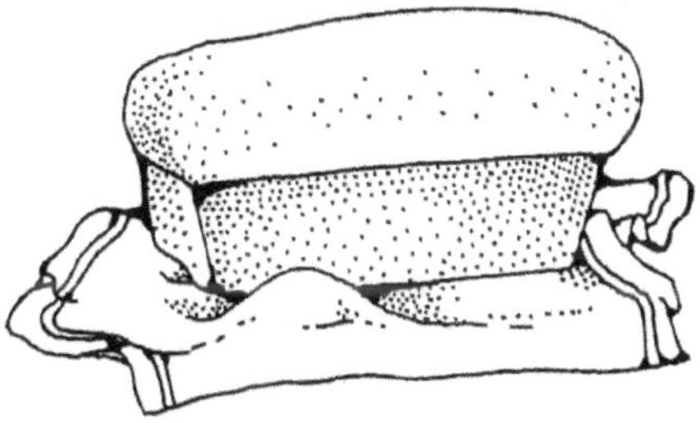

$KBrO_3$ is used in commercial bread baking.

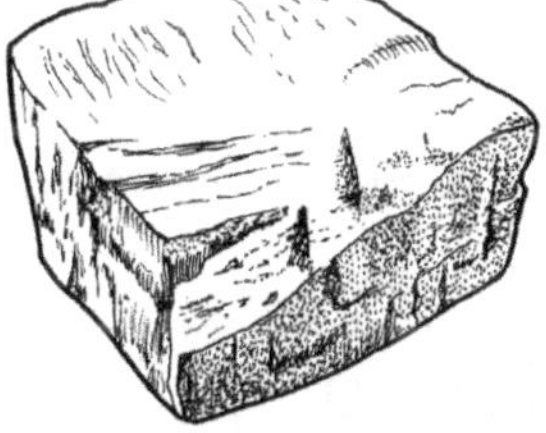

Orthoclase feldspar is often pink or light orange.

Potassium is used in soap making.

Potassium bicarbonate is in some carbonated beverages.

The sodium-potassium pump is found in cell membranes.

The membrane is made of molecules called phospholipids.

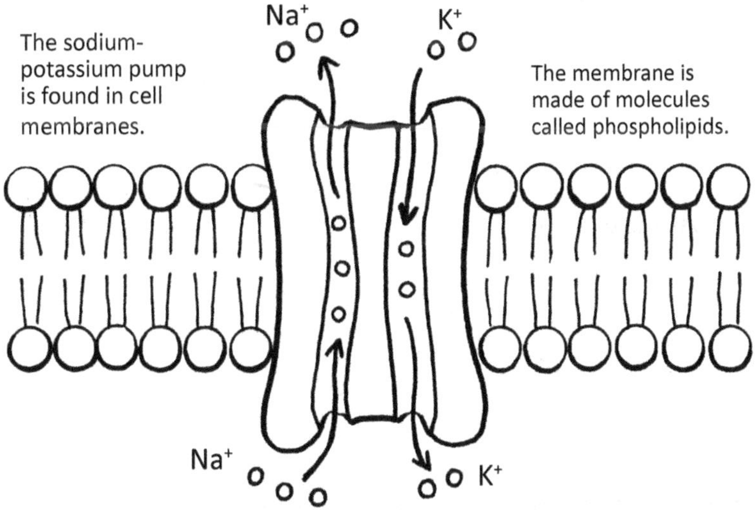

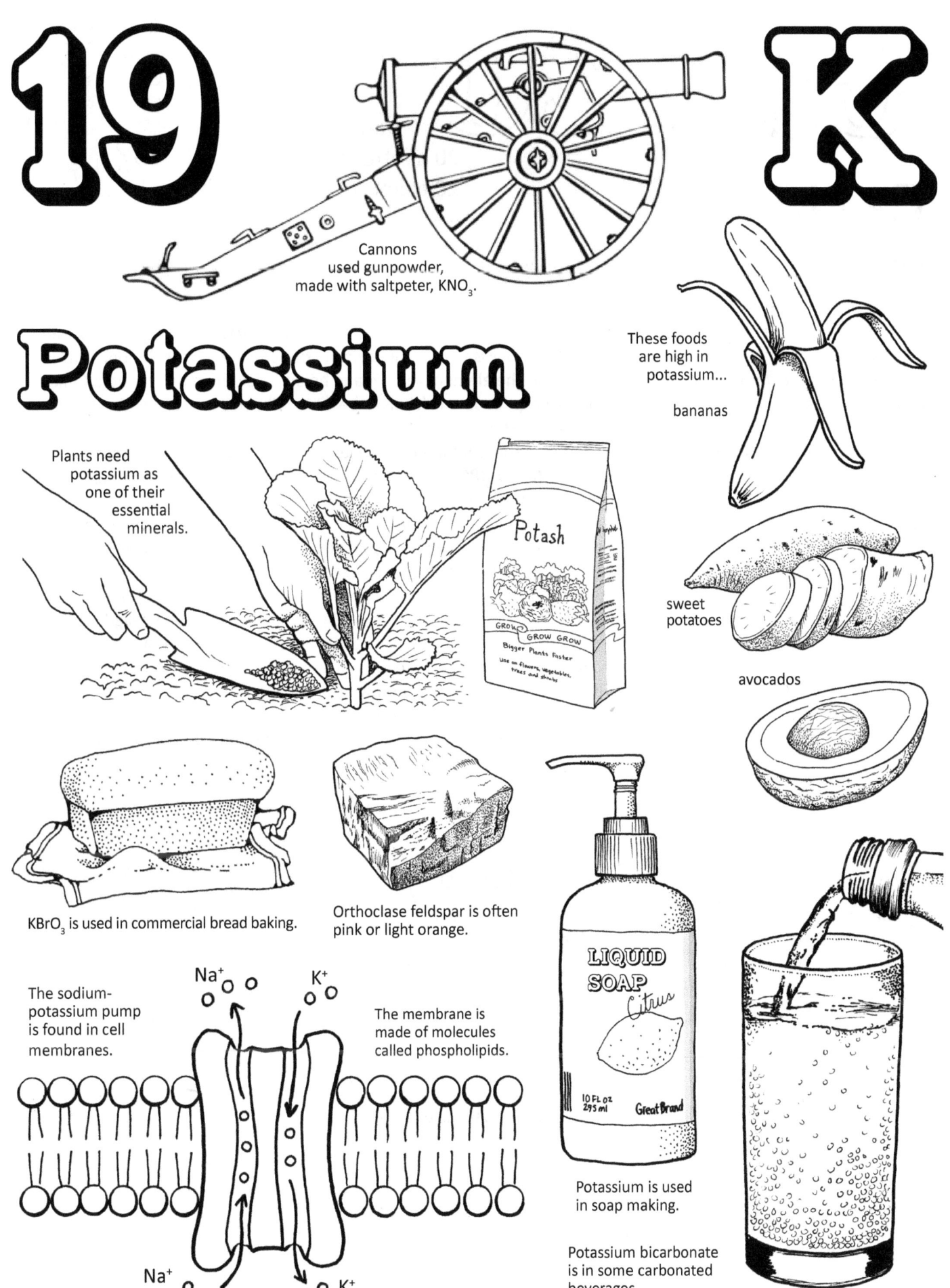

Ca 20

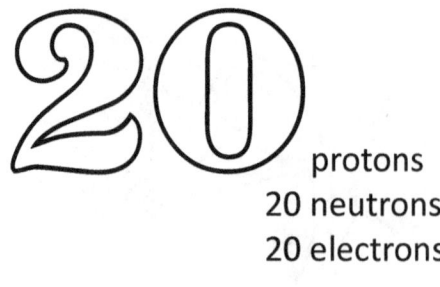

protons
20 neutrons
20 electrons

Atomic mass: 40.08

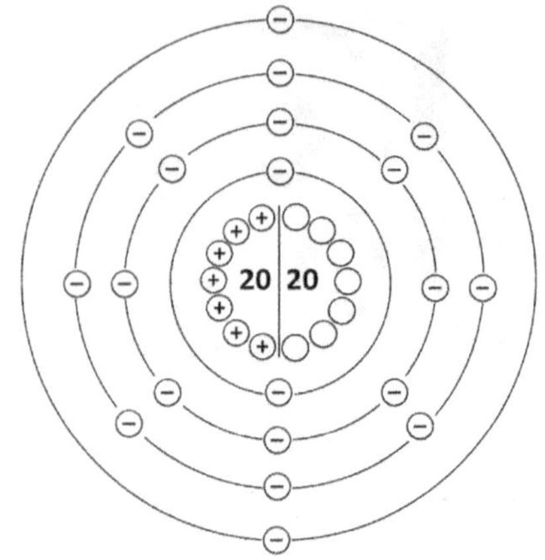

Calcium

From the Greek word "calx," meaning "lime"

Calcium is yet another element discovered by Sir Humphry Davy. Its existence was already suspected, but it was not isolated until 1808, when Davy used electrolysis to pull pure calcium out of a chemical solution. Within the space of a few weeks, Davy had also discovered the elements above and below calcium on the Periodic Table: magnesium, barium and strontium. Most people expect pure calcium to be white, like so many of the calcium compounds they know, but pure calcium is a soft, gray metal.

In the crust of the earth, calcium is found in these minerals: limestone, chalk, aragonite (all of these are $CaCO_3$), gypsum ($CaSO_4$), fluorite (CaF_2), and apatite ($Ca_{10}(PO_4)_6(OH)_2$). When limestone is squeezed and heated it turns into marble, a metamorphic rock that has been widely used for buildings and statues.

Calcium plays many vital roles in the body. Besides being an important building material for bones and teeth, calcium is used in the transmission of signals in the nervous system, in contraction of muscle cells, as cofactors for many enzymes, in protein synthesis, and in the fertilization of an egg cell.

Many mollusks (clams, snails, oysters, etc.) can take calcium out of the sea water and use it to build shells. Natural chalk deposits, such as the White Cliffs of Dover, are made of the shells of microscopic single-celled protozoa called foraminiferans and coccolithophores. England and Denmark have the greatest number of chalk cliffs.

Pure calcium metal is used in steel making, where it binds to oxygen and sulfur. It is added to aluminum alloys to give the metal greater strength. It is used as a "getter" to remove oxygen and nitrogen from tubes of inert gas (such as argon). Calcium hydride, CaH_2 is used as a source of hydrogen.

Calcium compounds are found in many household products such as baking ingredients, drain cleaner, toothpaste, antacids, and medicines. It can also be found in maintenance-free car batteries.

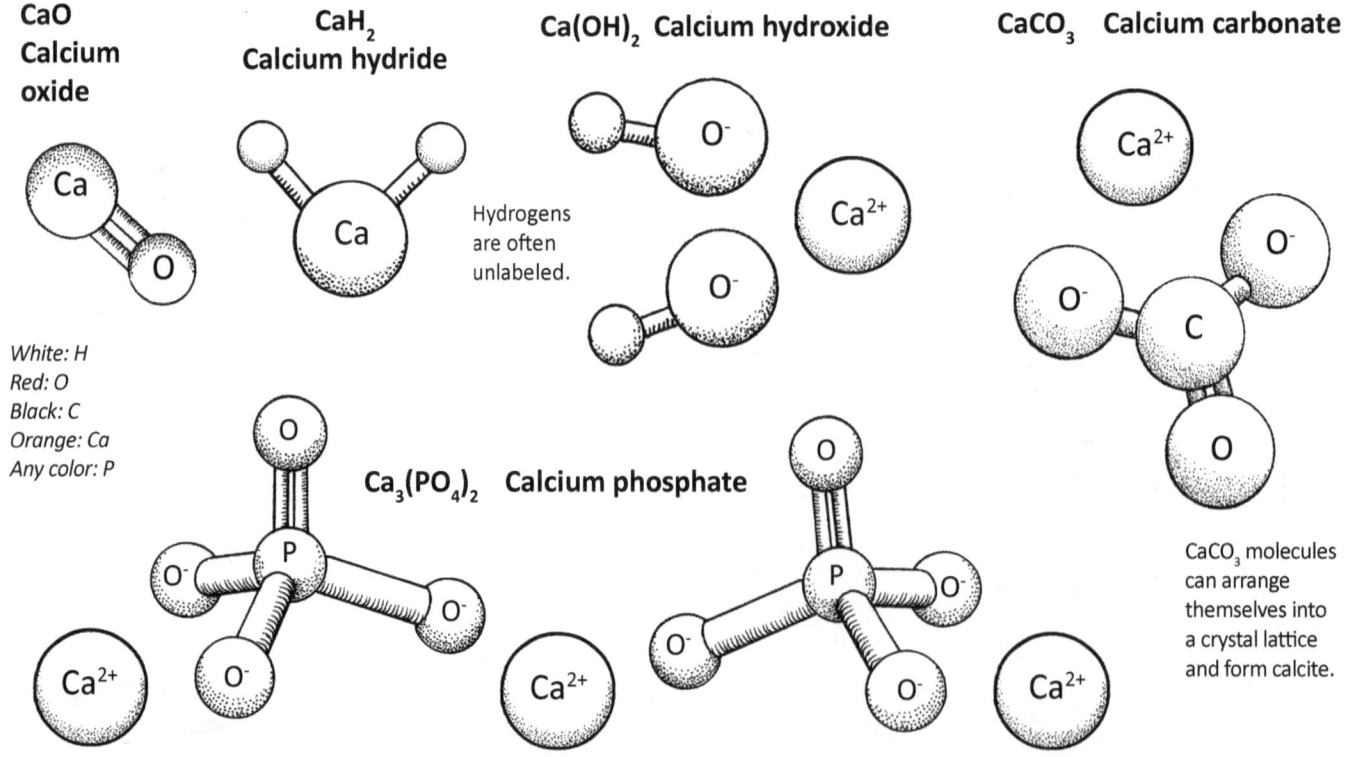

CaO
Calcium oxide

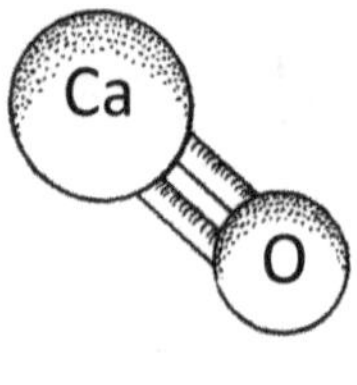

CaH_2
Calcium hydride

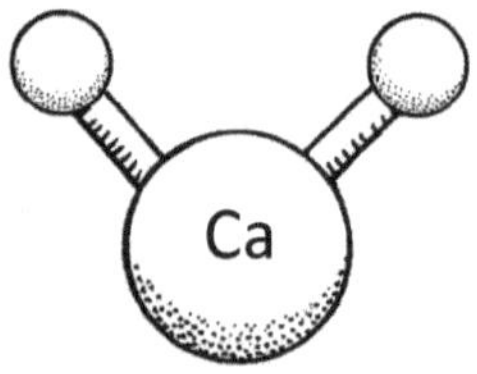

Hydrogens are often unlabeled.

$Ca(OH)_2$ Calcium hydroxide

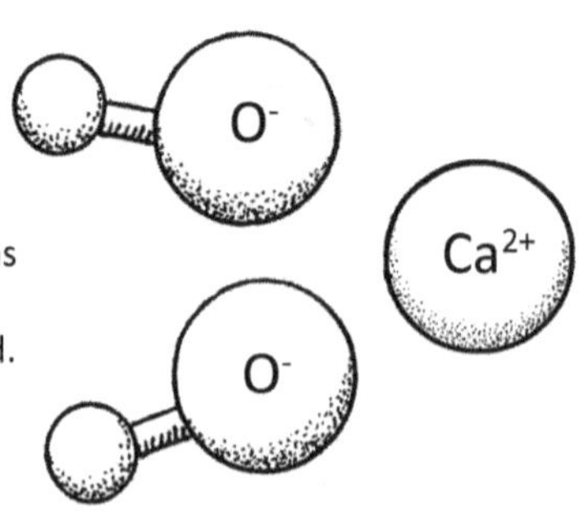

$CaCO_3$ Calcium carbonate

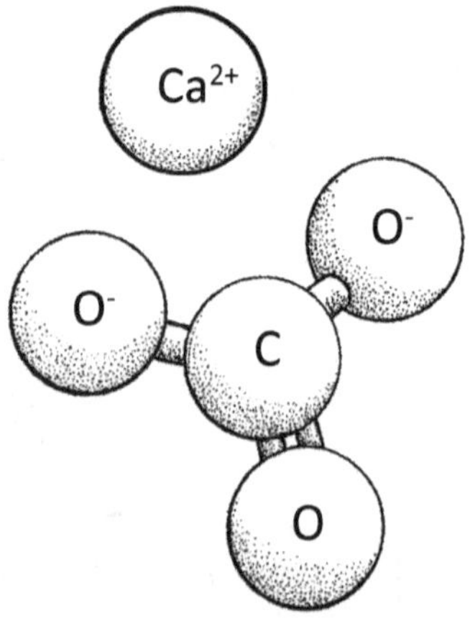

$CaCO_3$ molecules can arrange themselves into a crystal lattice and form calcite.

White: H
Red: O
Black: C
Orange: Ca
Any color: P

$Ca_3(PO_4)_2$ Calcium phosphate

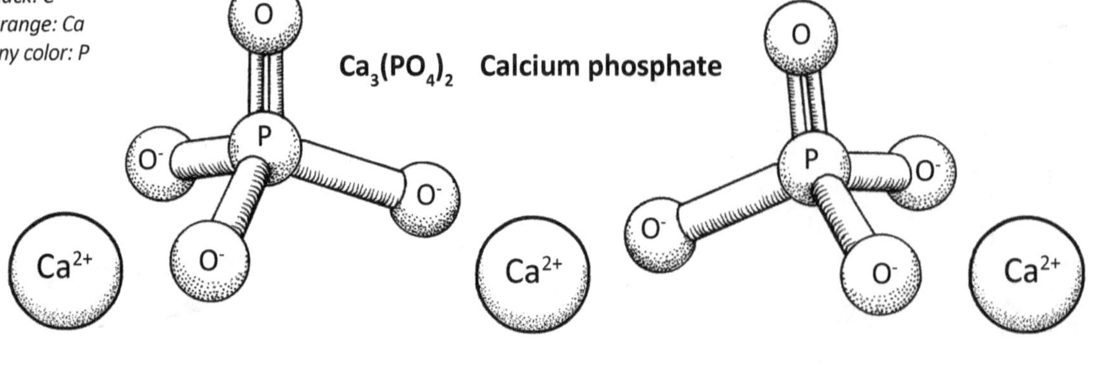

20 Ca

Clams can take calcium from the water.

Calcium

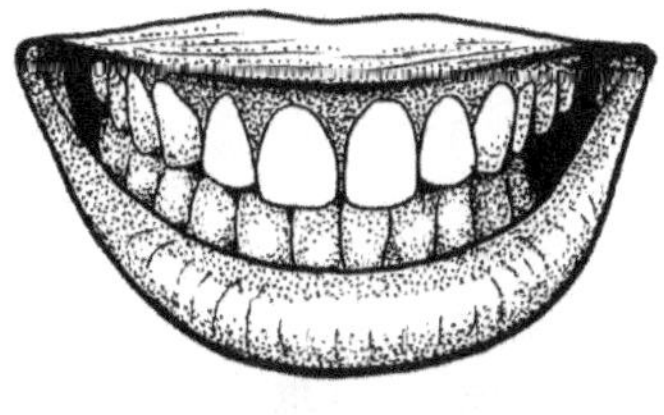

Like bones, teeth need calcium.

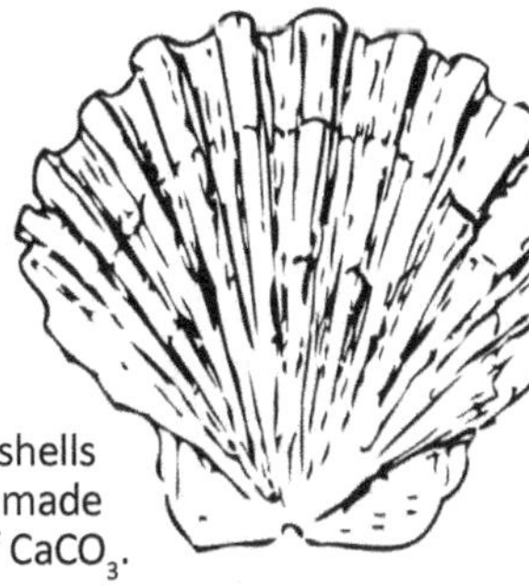

Sea shells are made of $CaCO_3$.

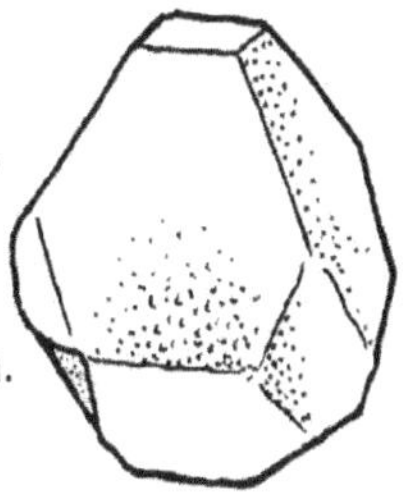

Fluorite, CaF_2, is often purple or green.

This apatite is blue.

The Medici lions are famous marble sculptures in Florence, Italy.

White cliffs are made of chalk.

Broccoli is high in calcium.

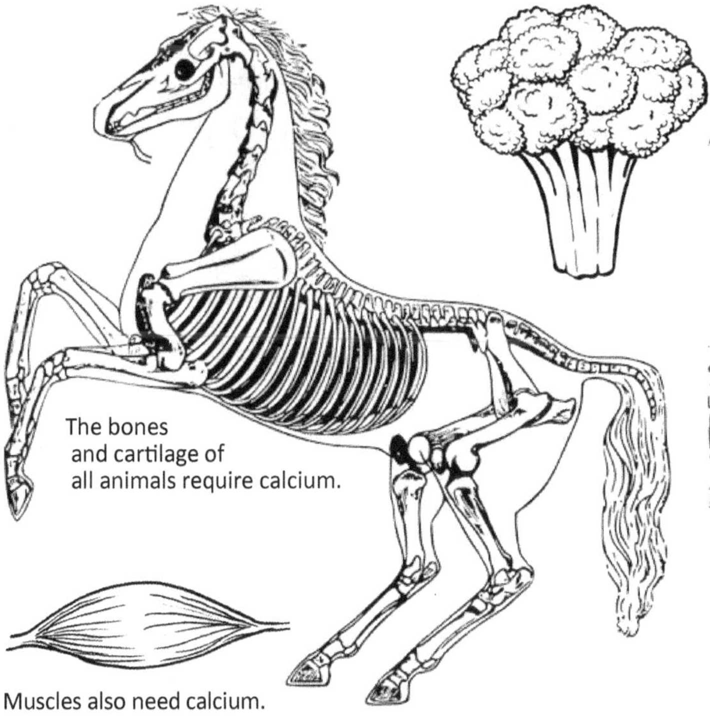

The bones and cartilage of all animals require calcium.

Muscles also need calcium.

Milk from cows or goats is high in calcium and can be used to make cheese, butter, yogurt, and ice cream.

Sc 21

protons
24 neutrons
21 electrons

Atomic mass: 44.96

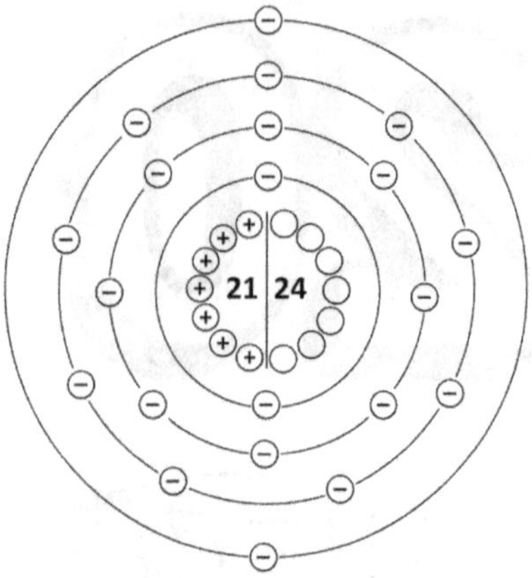

Scandium

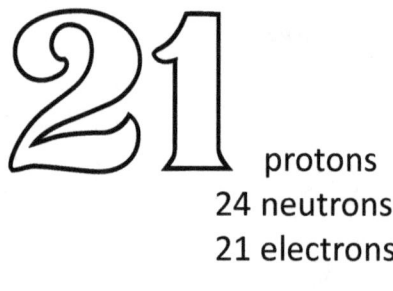

Named after Scandinavia

In the 1860s, when Dmitri Mendeleev *(men-dell-AY-ev)* was drawing his idea for a table to organize all the chemical elements, he suspected that there were elements yet to be discovered. He left a blank space after calcium, and used a temporary name for the missing element: ekaboron. He predicted what the element would look like, what its atomic weight would be, and how it would react with other elements. In 1879, a Swedish scientist named Lars Nilson was working with a mineral called euxenite *(yewks-en-ite)* hoping to find the element ytterbium. Instead, he isolated a new substance that produced an unknown element that he named scandium, after Scandinavia (the area of the world occupied by Sweden and the countries immediately surrounding it). Scandium turned out to be the missing element that Mendeleev had predicted as ekaboron.

Pure scandium is a silvery gray metal. It is the first element in the large group we call the transitional metals, though some chemists prefer to think of it a rare earth metal (elements 58 to 71). Very pure scandium was not produced until 1960.

Scandium is most useful as an alloy, often with aluminum. Scandium-aluminum alloys are used to make airplanes, bicycle frames, baseball bats, and lacrosse sticks.

Scandium is used in high intensity lighting (metal halide bulbs) used in places like outdoor stadiums, where the lights need to be extremely bright. Scandium bulbs produce light that is very close to natural sunlight.

Scandium atoms that have 25 neutrons instead of 24 are radioactive and are used by the oil refining industry as "tracers" to help them track where all the products and byproducts are going.

No one mines for just scandium. Scandium is obtained as a by-product in mining operations that are producing many other elements including thorium, uranium, titanium, and rare earths like europium and gadolinium. The top producers of scandium are China, Russia, Ukraine and the Philippines. There is a limited market for scandium because titanium can often be used instead, giving better results at a lower price.

Scandium hydride

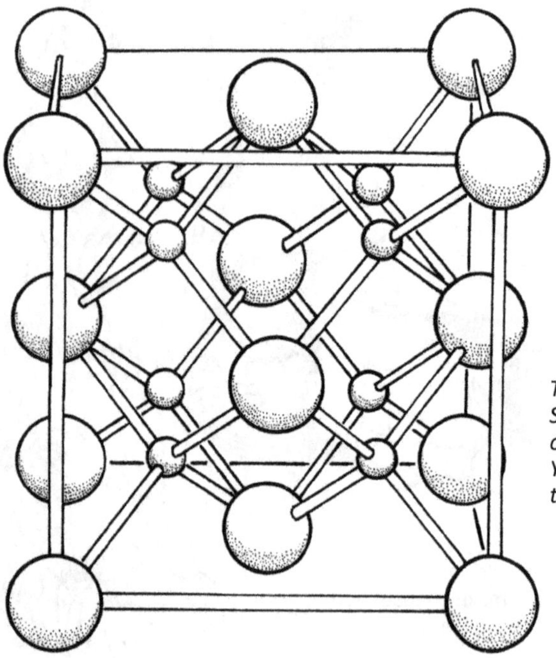

The large balls are Sc and the small ones are H. You can choose the colors.

Scandium hydride is a man-made alloy where hydrogen is added to a sample of pure scandium.

Scandium nitrate
$Sc(NO_3)_3$

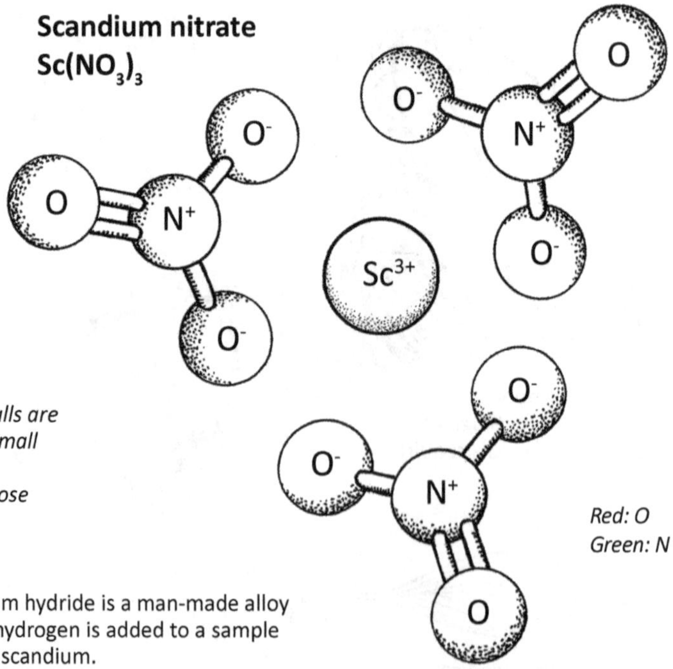

Red: O
Green: N

21 Sc

Scandinavia always includes Sweden, Norway, Denmark and Finland. Some people also like to include Iceland.

Scandium

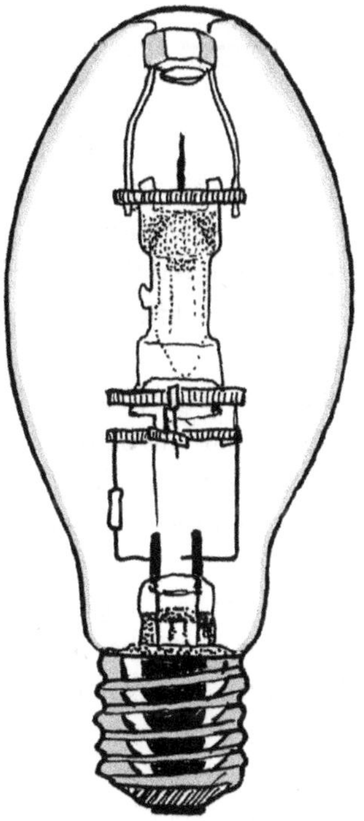

Scandium is used for high intensity bulbs in sports stadiums.

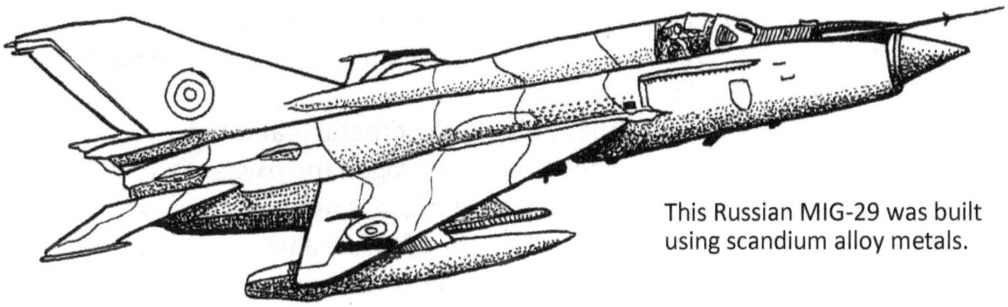

This Russian MIG-29 was built using scandium alloy metals.

Aluminum bats often contain some scandium.

Lacrosse sticks are made of scandium aluminum alloys.

Euxenite is an orange-tan rock that contains yttrium, titanium, tantalum, niobium, and cerium, as well as a little scandium.

Pure scandium is silvery gray.

Frames of expensive bicycles are sometimes made of scandium alloys.

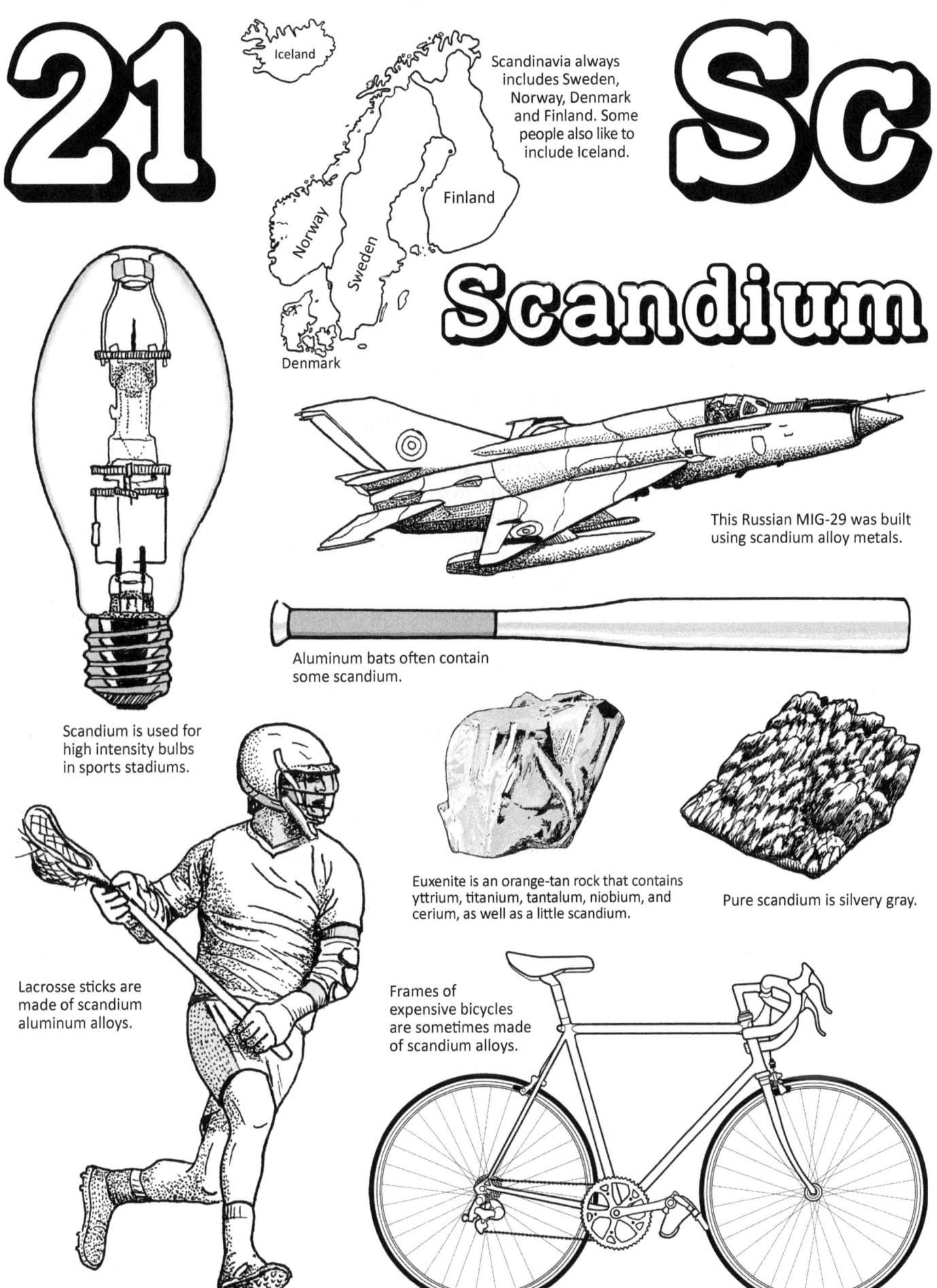

Ti

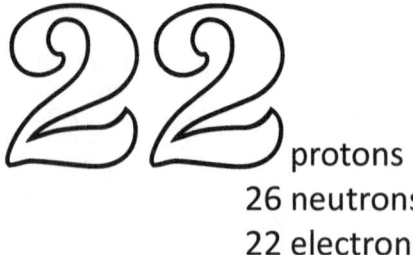

22 protons
26 neutrons
22 electrons

Atomic mass: 47.8

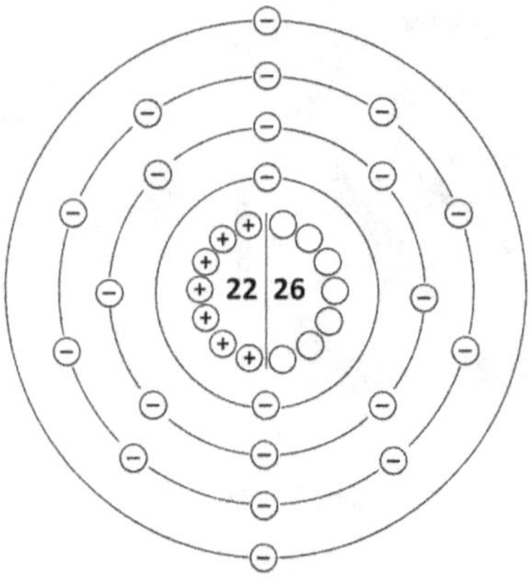

Titanium

Named after the Titans of Greek mythology

Titanium was discovered by two people at about the same time. In 1791, William Gregor discovered a new element in a mineral called ilmenite, in Cornwall, England. In 1795, a Prussian chemist named Martin Klaproth isolated a new element from a mineral called rutile in an area that is now part of Hungary. These new elements turned out to be the same thing, the element we now call titanium. (Klaproth was the one who named it.) We now know that titanium is not rare, but is very abundant in the earth's crust, being found in a number of different minerals, and is even dissolved in sea water.

Titanium in its pure form is a shiny, silver metal that is both light and strong. It has the unusual property of being able to burn in an atmosphere of pure nitrogen (no oxygen) which limits its use in metals that will be directly exposed to flames and high heat. Mostly, titanium is used in alloys, where this feature is not a problem. Titanium is usually mixed with aluminum because it very light, like aluminum, but is much stronger. Other elements are often added, too, such as iron, copper, vanadium, molybdenum, magnesium, or silicon.

Titanium alloys are very resistant to heat and rust. They are perfect for places where corrosion might be a problem, like in boats and planes. The alloys are used for propellers, landing gear, engine parts and exhaust pipes. (Two of the first airplanes to use titanium alloys were the Lockheed A-12 and the SR-17 Blackbird.) You can also find titanium in sports equipment like golf clubs and tennis rackets.

Titanium dioxide, TiO_2, is used to make white pigments for paint. Titanium whites are also used in some types of paper and plastic, and even in toothpaste, since titanium is completely safe and non-toxic.

Because it is strong and non-toxic, titanium has many uses in repairing the human body. Many artificial joints use titanium and it is used for pins that will hold broken bones together. Unlike other metals, titanium is not magnetic, making it safe for people with titanium joints to go into MRI machines for diagnostic imaging.

TiN Titanium nitride

TiN is used as a thin coating on top of steel parts.

Ti: large
N: small

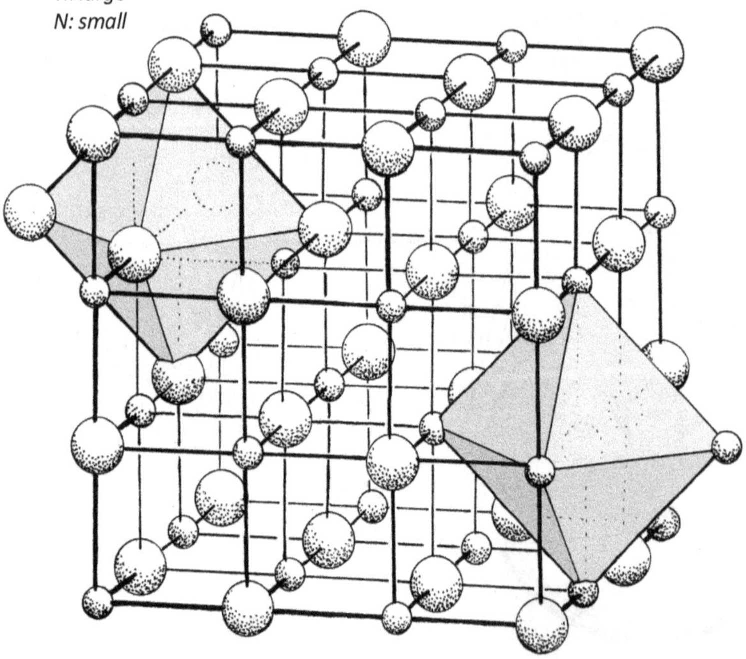

$TiCl_4$
Titanium tetrachloride

$TiCl_4$ is used to produce pure titanium and TiO_2.

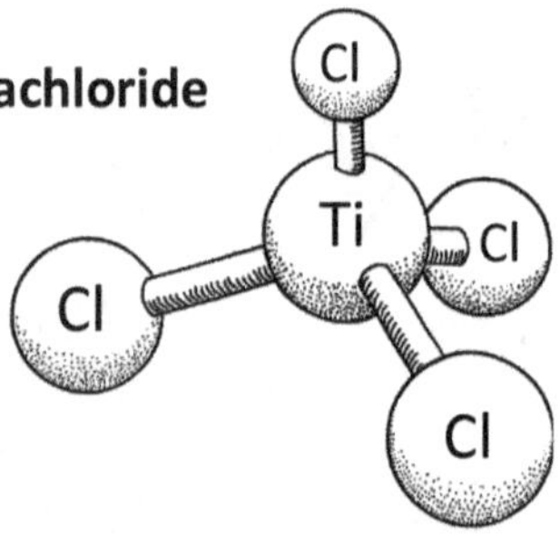

TiO_2 Titanium dioxide

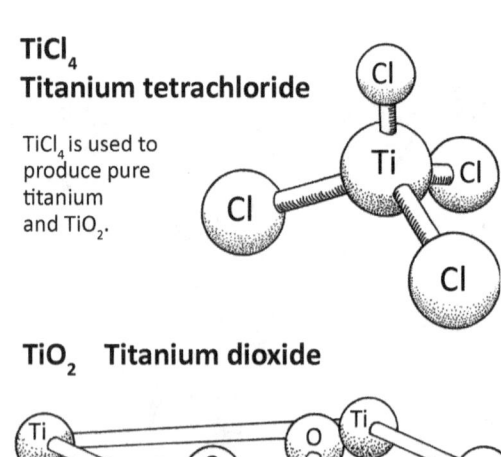

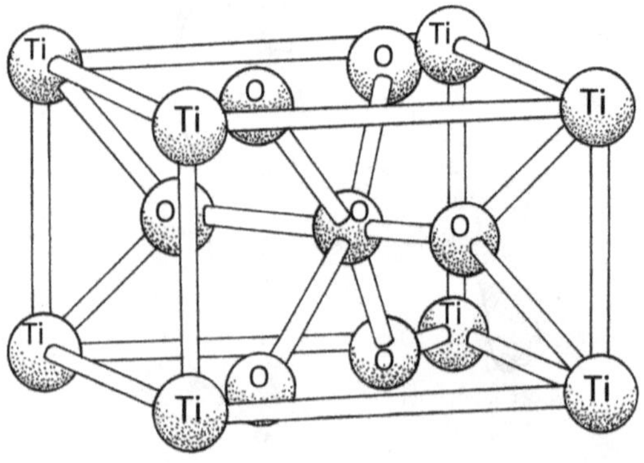

22

Rutile is a shiny black ore of titanium. Its chemical composition is TiO_2, titanium dioxide.

Ti

Some tools are more expensive because they contain titanium.

Titanium alloys are used for both structural and mechanical parts in airplanes.

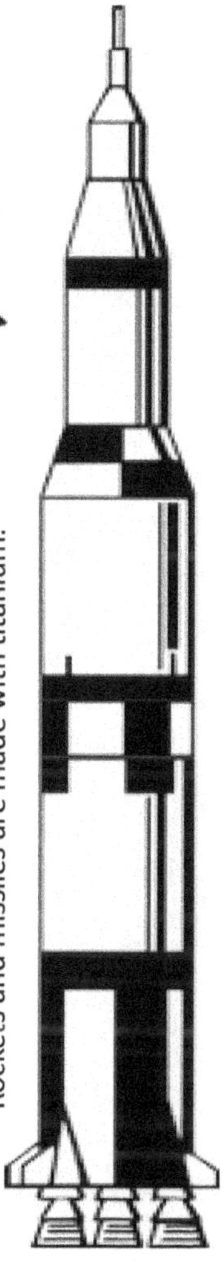

Rockets and missiles are made with titanium.

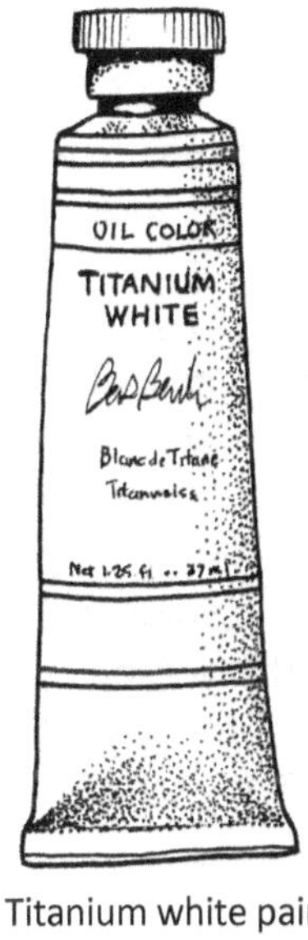

Titanium white paint

Titanium

Titanium horseshoes

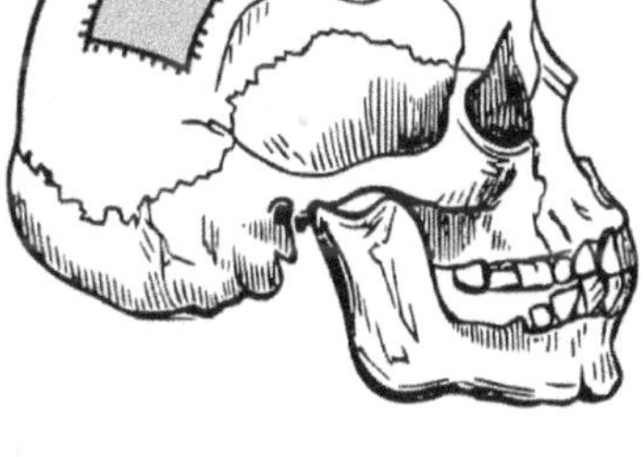

Titanium is perfect for repairing bones.

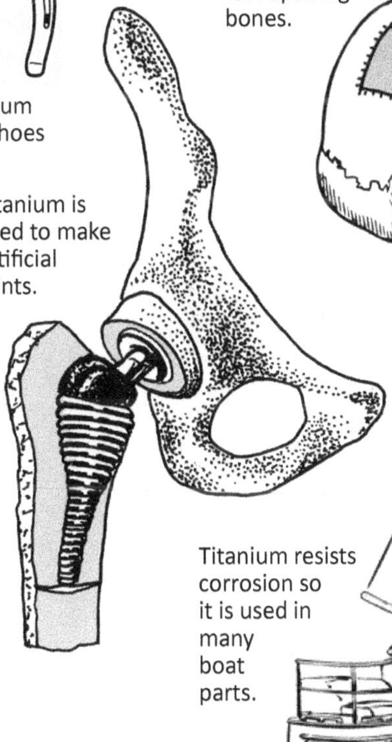

Drill bits can be coated with titanium.

Titanium is used to make artificial joints.

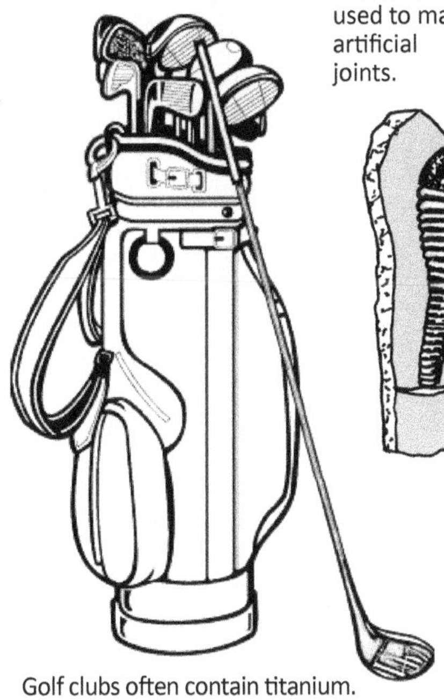

Golf clubs often contain titanium.

Boat propellers are made of titanium alloys.

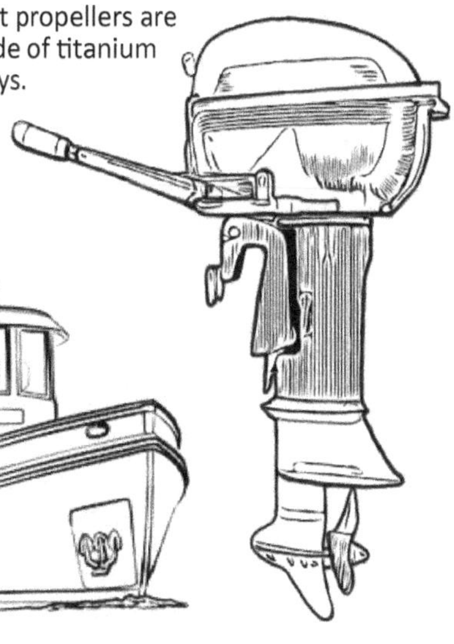

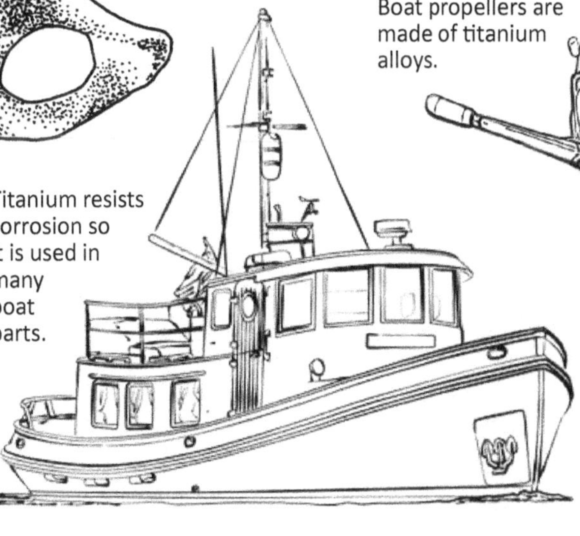

Titanium resists corrosion so it is used in many boat parts.

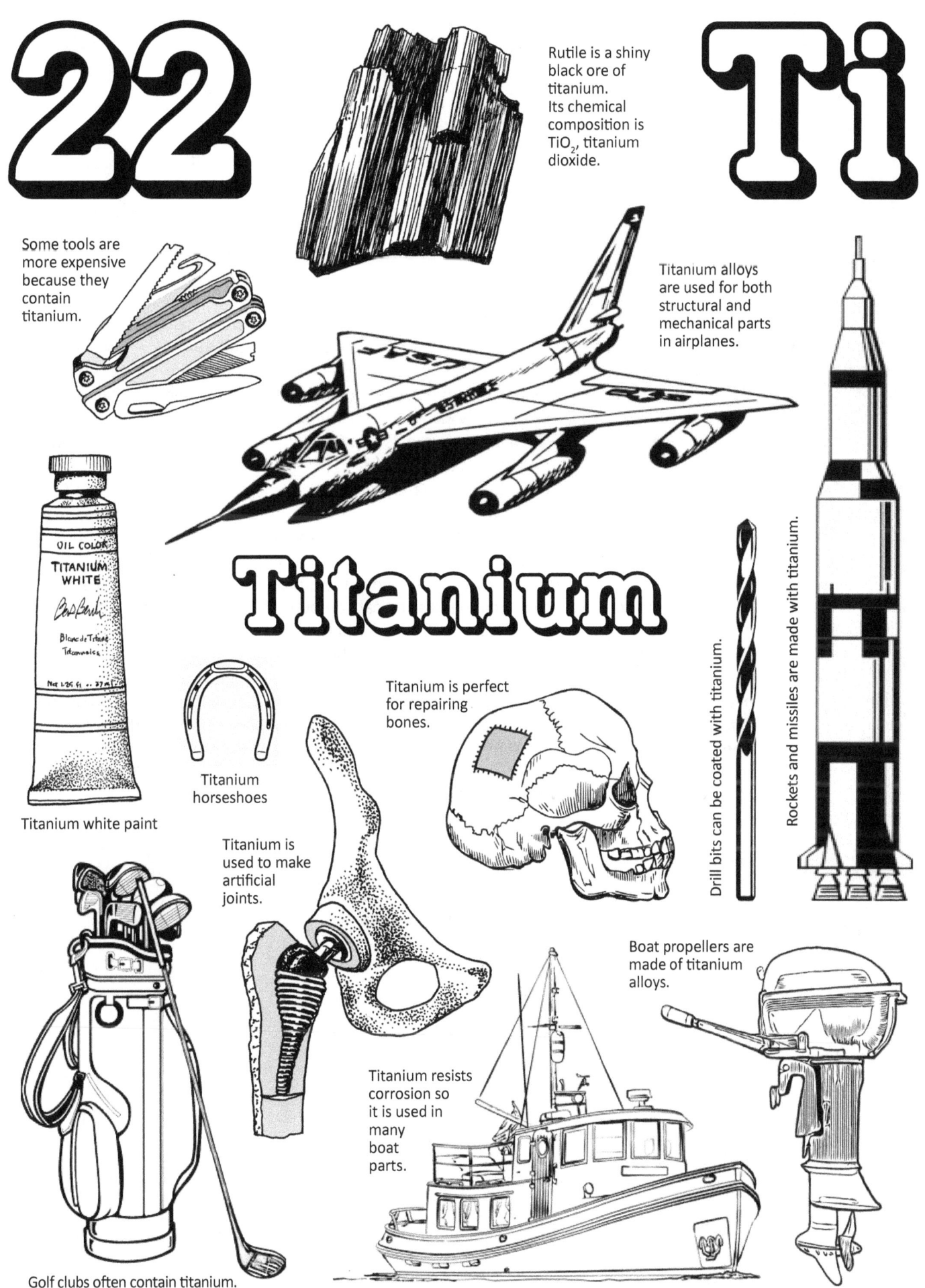

V 23

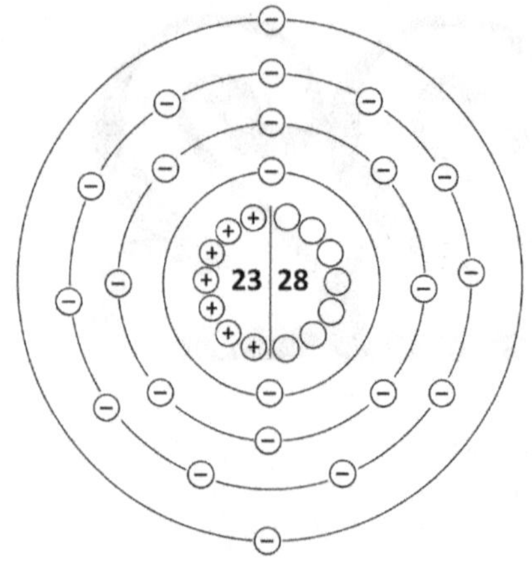

protons
28 neutrons
23 electrons
Atomic mass: 50.9

Vanadium

Named after the Norse goddess Vanadis

In 1801, a scientist in Mexico, Andres Manuel del Rio, discovered a new element in a sample of rock that he called "brown lead." He determined that besides lead, there was a new element in this mineral, an element that would turn red when mixed into many solutions. He wanted to name this element "erythronium," since "erythro" means "red." However, his letter to the Institute in Paris was lost in a shipwreck and his discovery went unrecognized. Thirty years later, a Swedish chemist isolated a new element which turned out to be the same one that del Rio had found. He saw the beautifully colored chemical compounds that could be created with this element, and decided to name it after the Norse goddess of beauty, Vanadis.

The first industrial use of the element vanadium was in the steel alloys used to make the Ford Model T automobile in 1910. During the first half of the 20th century, all vanadium came from a mine in Peru. Then it was discovered that vanadium was often mixed into ores of uranium, especially a rock called carnotite, and as more uranium was mined in the late 20th century, vanadium began to be available through uranium mining.

Vanadium in it pure form is a fairly hard steel-blue metal. When it is added to other metals, it makes the alloys harder and more resistant to corrosion. Steel that has a significant amount of vanadium added to it is called ferrovanadium and is used for machine parts such as axles, gears, crankshafts, and engine parts. High-vanadium alloys are also used for surgical instruments and industrial tools that need to take a lot of wear and tear.

Vanadium is present in most plants and animals, but plays very minor roles in biological processes. Two notable exceptions, however, are mushrooms and tunicates. The *Amanita muscaria* (fly agaric) fungus is a poisonous mushroom with a bright red cap. It contains a very high level of vanadium for unknown reasons. In the ocean, strange creatures called tunicates also collect and store large quantities of vanadium in certain parts of their bodies.

Vanadium dioxide, VO_2, is used in the production of glass coatings Vanadium pentoxide, V_2O_5, is useful in the fabric dye industry where it is used as a mordant (a substance that helps the dye stick to the fabric). Another use for V_2O_5 is in superconducting magnets, where it is mixed with the element gallium.

V_2O_5 Vanadium pentoxide

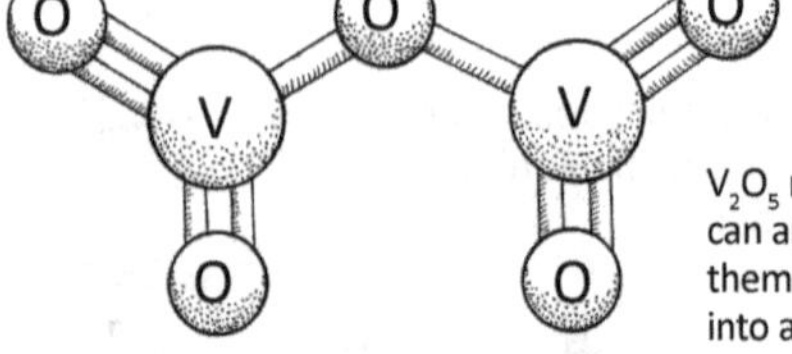

V_2O_5 molecules can arrange themselves into a lattice.

VO_2 Vanadium oxide

Large balls are V
Small balls are O

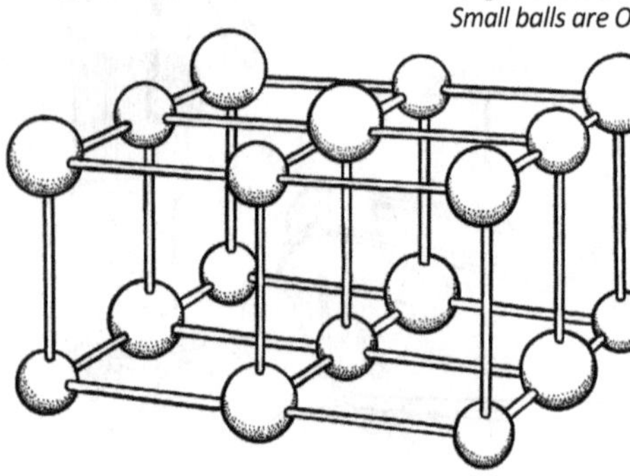

$VO(O_2C_5H_7)_2$ Vanadyl acetylacetonate

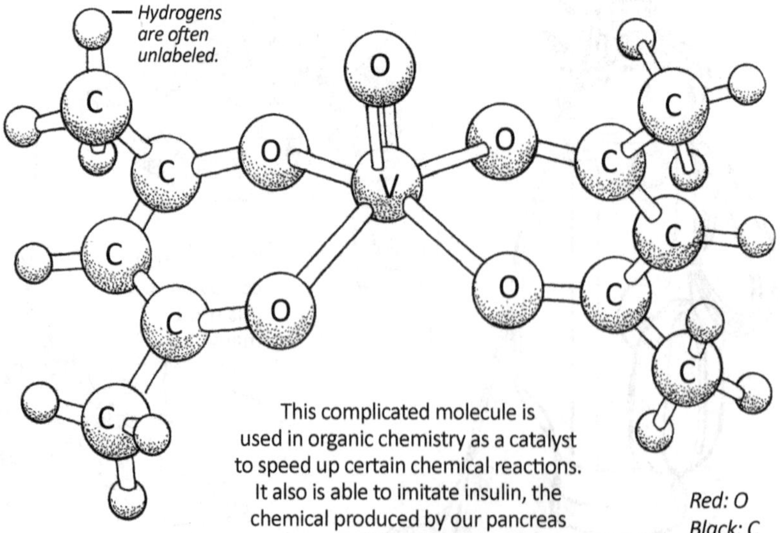

This complicated molecule is used in organic chemistry as a catalyst to speed up certain chemical reactions. It also is able to imitate insulin, the chemical produced by our pancreas to control blood sugar levels.

Red: O
Black: C
White: H

23

V

Vanadanite's formula: $Pb_5(VO_4)_3Cl$

Vanadanite crystals are dark red.

Vanadium

The fly agaric mushroom has a bright red cap with white spots. It stores and uses vanadium.

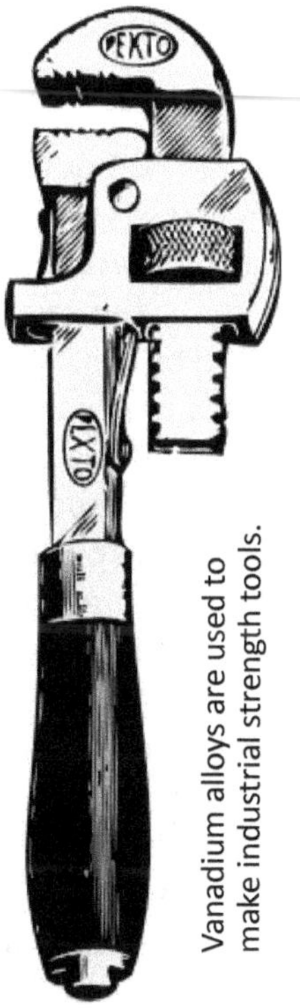

Vanadium alloys are used to make industrial strength tools.

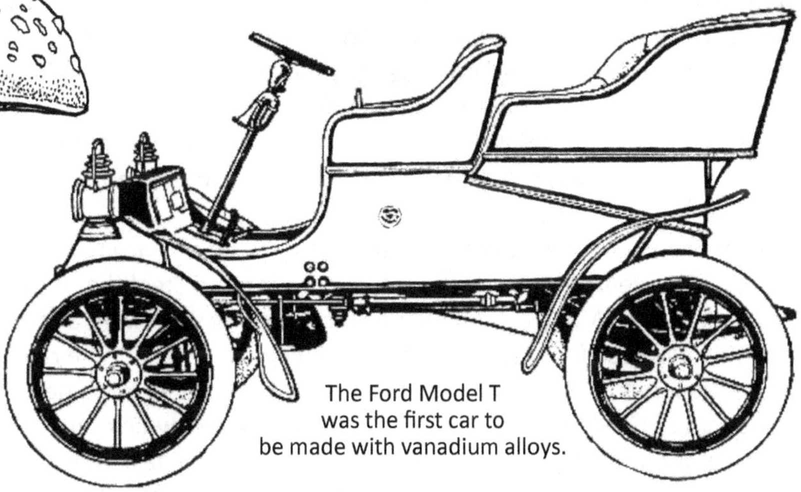

The Ford Model T was the first car to be made with vanadium alloys.

This wrench is labeled "VANADIUM."

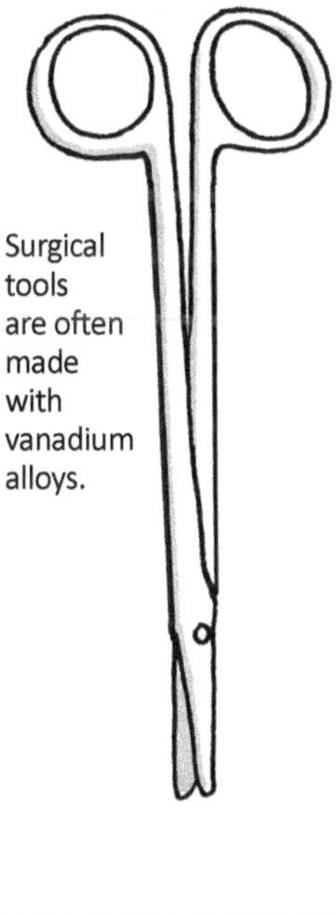

Surgical tools are often made with vanadium alloys.

Sea creatures called tunicates use vanadium. They look like transparent tubes with blue stripes. There is a very dark blue stripe at the top.

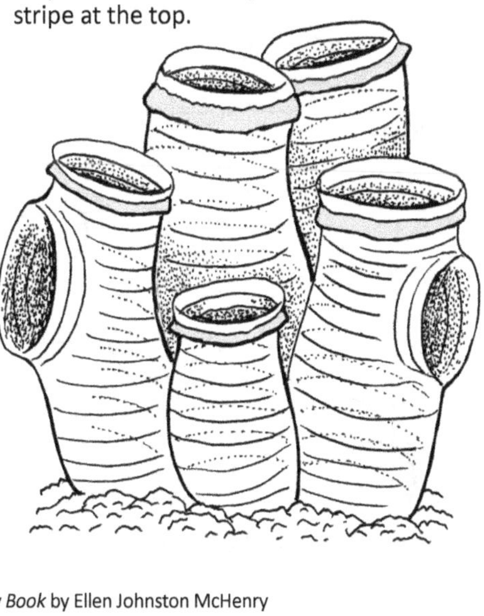

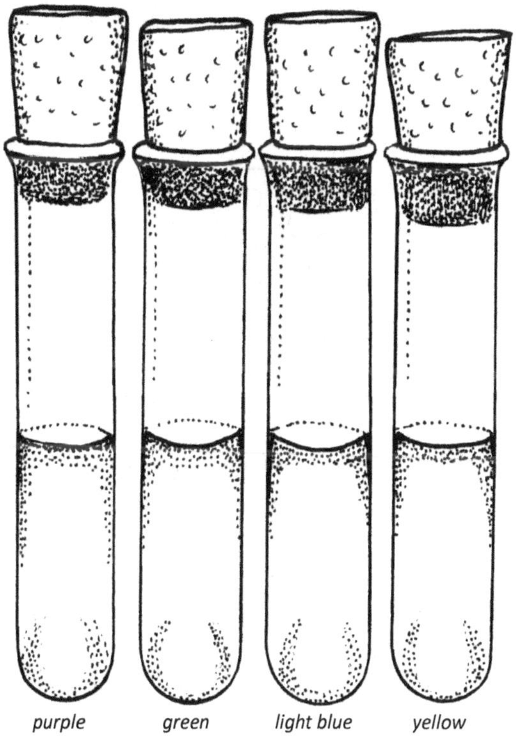

purple *green* *light blue* *yellow*

When vanadium is mixed with solutions that are acidic or basic, many colors can be produced.

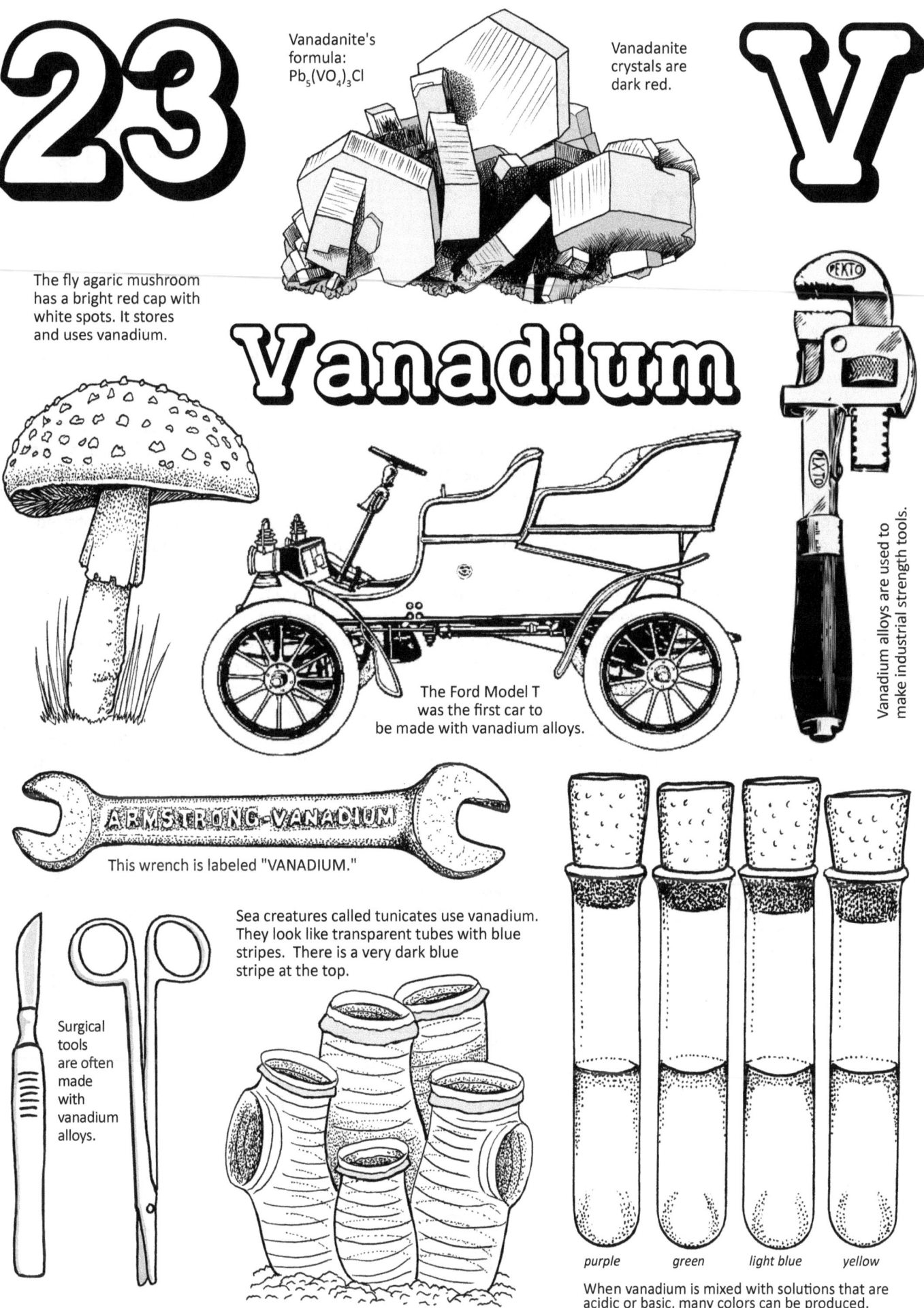

Cr 24 protons

28 neutrons

24 electrons

Atomic mass: 51.9

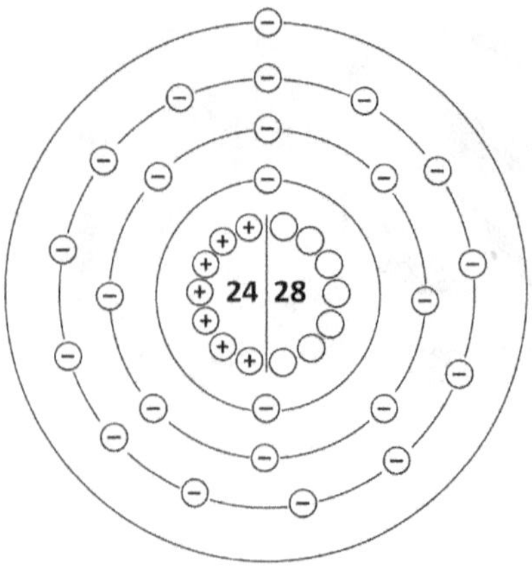

Chromium

"Chroma" is the Greek word for color

Chromium is famous for being shiny, though its name comes from the Greek word for color. Metal parts on cars, trucks, and motorcycles are often coated with a thin layer of chromium to make the metal shine. A process called "electroplating" is used to apply the chromium.

The colorful aspect of chromium is only seen when it is mixed with other elements to form compounds, often oxides. Tiny amounts of chromium in the mineral corundum turn it into bright red rubies. Chromium is also found in a red mineral crystal called crocoite. Chromium yellow was the original "school bus yellow" used in America in the mid 1900s. Chromium yellow isn't used on buses any more, but it is still sold in tubes in art supply stores, and is an important pigment for painters. Chromium can also make green, purple, red and brown pigments. Green chromium oxide is the main ingredient in a paint that is used on military vehicles that need to hide from infrared radar. The green paint reflects light in the same way that green leaves do, so the vehicle will be hard to detect if there is a lot of green foliage around it.

Pure chromium is very hard. The only elements that are harder than chromium are carbon (in diamonds) and boron. Chromium is added to steel to make "stainless steel," a form of steel that is both stronger and more resistant to corrosion than regular steel. Many household products such as silverware and appliances are made with stainless steel. In construction it is used for exterior parts of buildings that need to be rust-proof. Adding nickel to stainless steel turns it into a "superalloy," tough enough for making tanks and jet engines.

Chromium compounds are used in the textile industry as mordants (which cause dyes to be more permanent on fabrics) and in the leather tanning industry to preserve the animal hides. (Chromium bonds to collagen.)

Chromium has a very high melting point, which means it can be used to make molds that will hold very hot liquid metals. It can also be used to make parts for blast furnaces that melt metals.

$PbCrO_4$ Lead chromate (Crocoite)

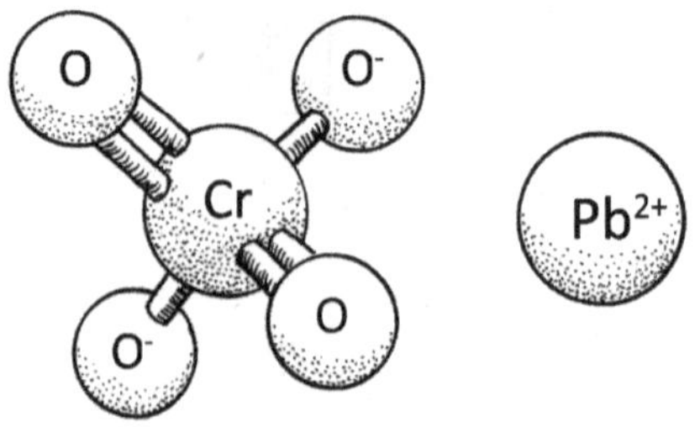

Cr_2O_5 Chromium pentoxide

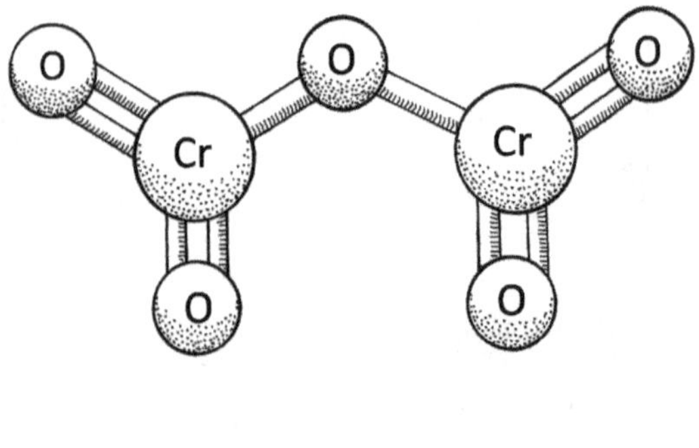

CrO_2 Chromium (IV) oxide

Chromium can bond in more than one way, so Roman numerals are used to keep track of which "oxidation state" chromium is using. You can see how CrO_2 connects to other CrO_2 molecules to make a continuous structure. This compound (in the form of a mineral powder) was used to make magnetic recording tape for cassette players and VCRs.

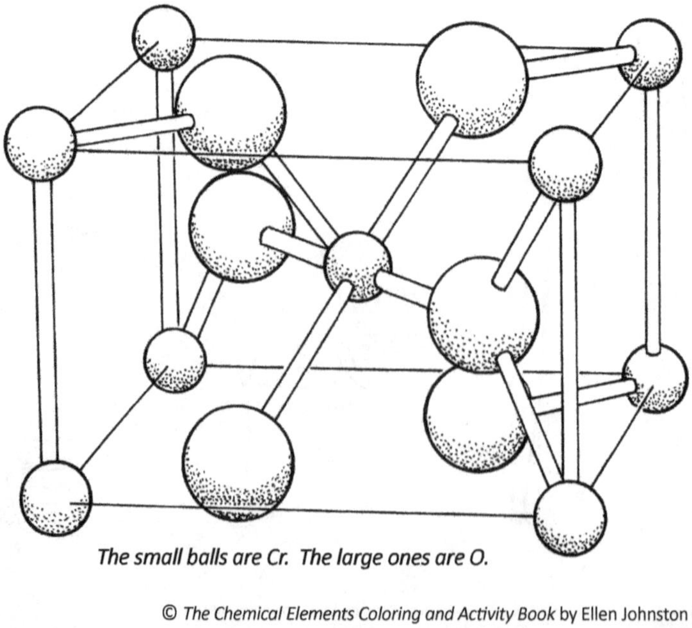

The small balls are Cr. The large ones are O.

24 Cr

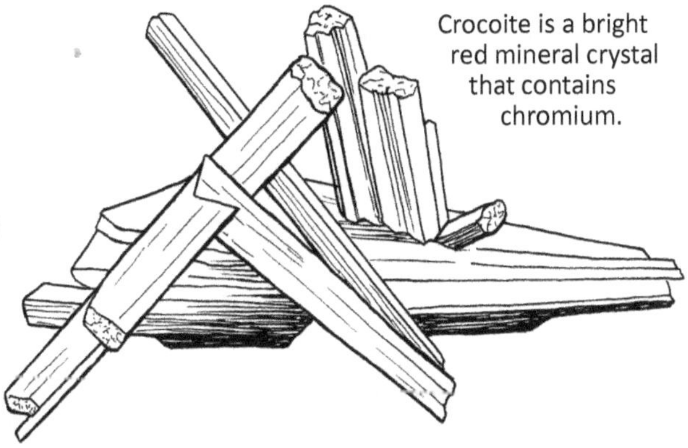

Crocoite is a bright red mineral crystal that contains chromium.

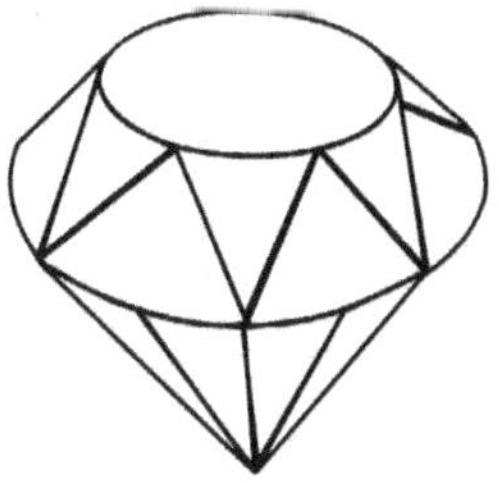

Red rubies are made of the mineral corundum, which has tiny amounts of chromium.

Chromium

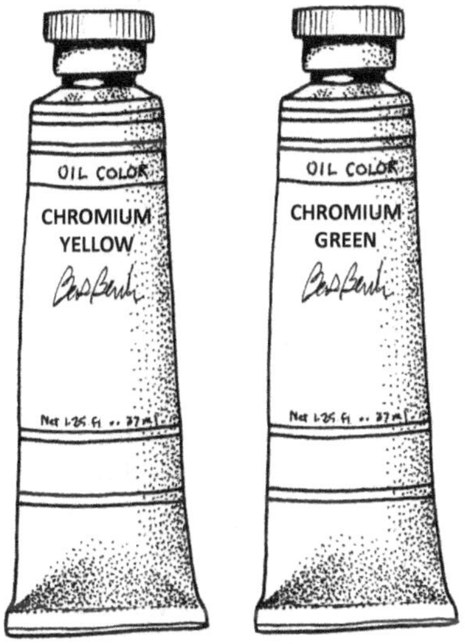

Chromium pigments in paints.

"School bus yellow" was originally made using chromium.

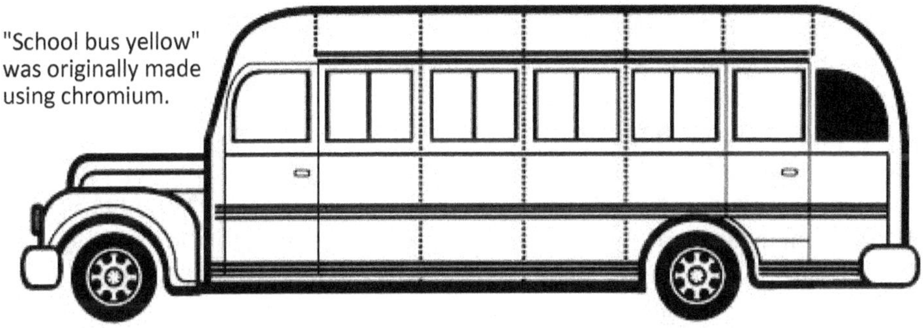

Military vehicles are often painted with a chromium green pigment that hides them from infrared sensors.

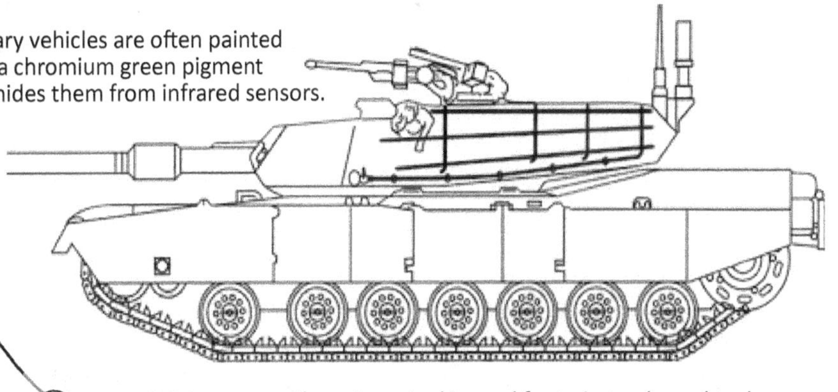

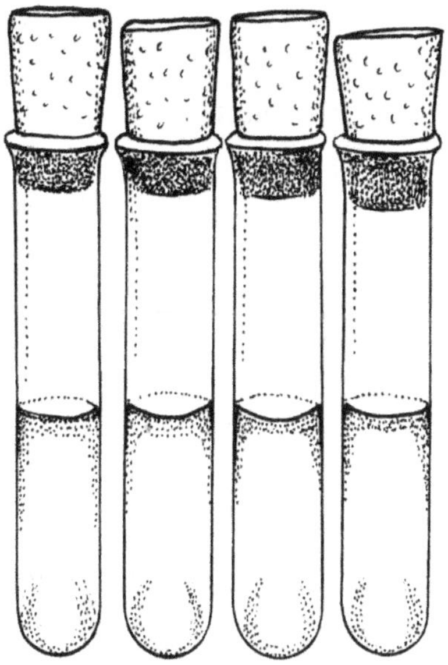

purple green red yellow

Like vanadium, chromium can make many colorful compounds.

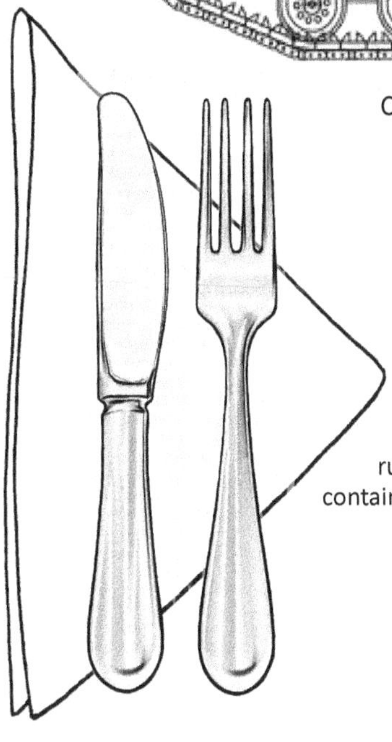

Chromium steel is used for train tracks and tanks.

Stainless steel won't rust because it contains chromium.

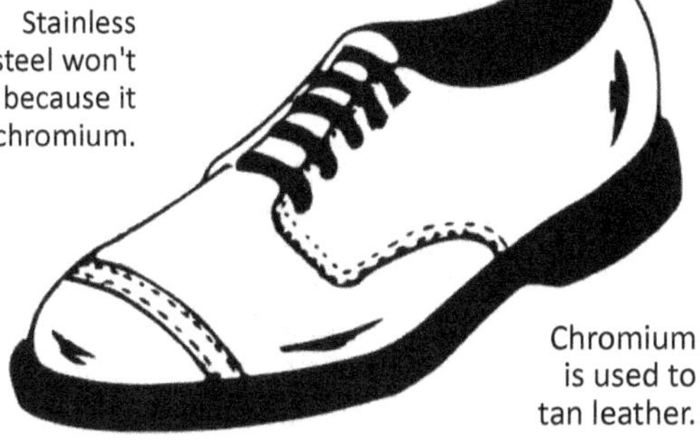

Chromium is used to tan leather.

Mn 25

Manganese

25 protons
30 neutrons
25 electrons
Atomic mass: 54.9

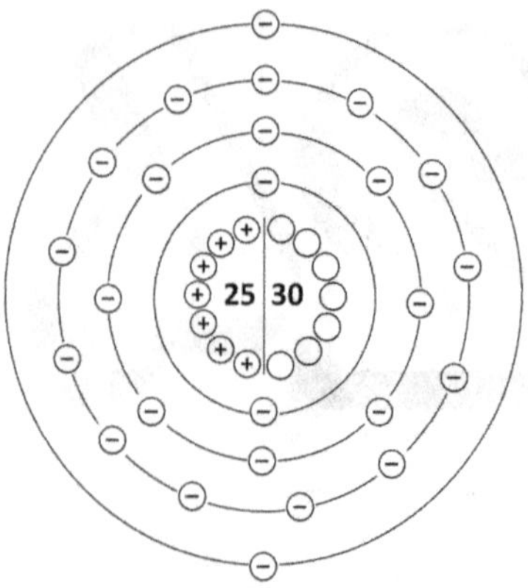

Named after an area of Greece called Magnesia

It is very easy to confuse *manganese* and *magnesium*. Humphry Davy wanted to use a different name for magnesium when he discovered it in 1808, knowing that manganese was already an element (isolated in 1774) and realizing how much confusion it would cause having two elements with similar names. However, other scientists did not share his concern, and now you must learn about both magnesium and manganese.

Manganese is always found mixed with other elements such as iron, copper, nickel, aluminum, and cobalt. In the earth, manganese is most often found in the mineral pyrolusite, MnO_2, and in rhodochrosite crystals, $MnCO_3$. In the ocean, manganese occurs as round balls, called "nodules," lying on the bottom of the ocean. These nodules also contain other metals such as iron, copper, aluminum and cobalt.

Most manganese produced today goes into the production of steel. Manganese can remove unwanted elements from the steel, such as oxygen and sulfur, as well as making the steel easier to shape while it is hot and stronger after it is cold. Railroad tracks are made with steel that contains about 1% manganese. The glass industry also uses manganese to remove unwanted elements from the hot, liquid glass.

Manganese oxide compounds have a very dark color, either purple, brown or black. Brown and black manganese compounds were used by ancient peoples in France to draw pictures on cave walls. Glass makers can use manganese (along with other elements) to make green or brown glass. Potassium permanganate, $KMnO_4$, is a purple mineral with many uses. It can be used to make pink or purple glass, but it is also highly valued in the world of medicine for its ability to kill germs and for its usefulness as a biological stain.

Our bodies need small amounts of manganese in order to make many enzymes work properly. The enzyme *arginase* helps to beak down and recycle protein molecules. Other manganese enzymes protect us against dangerous broken molecules called "free radicals." Good dietary sources of manganese include mussels, brown rice, Lima beans, sweet potatoes, pine nuts, spinach, and pineapple.

Manganese dioxide is used in batteries and as an ingredient in anti-knock additives for gasoline.

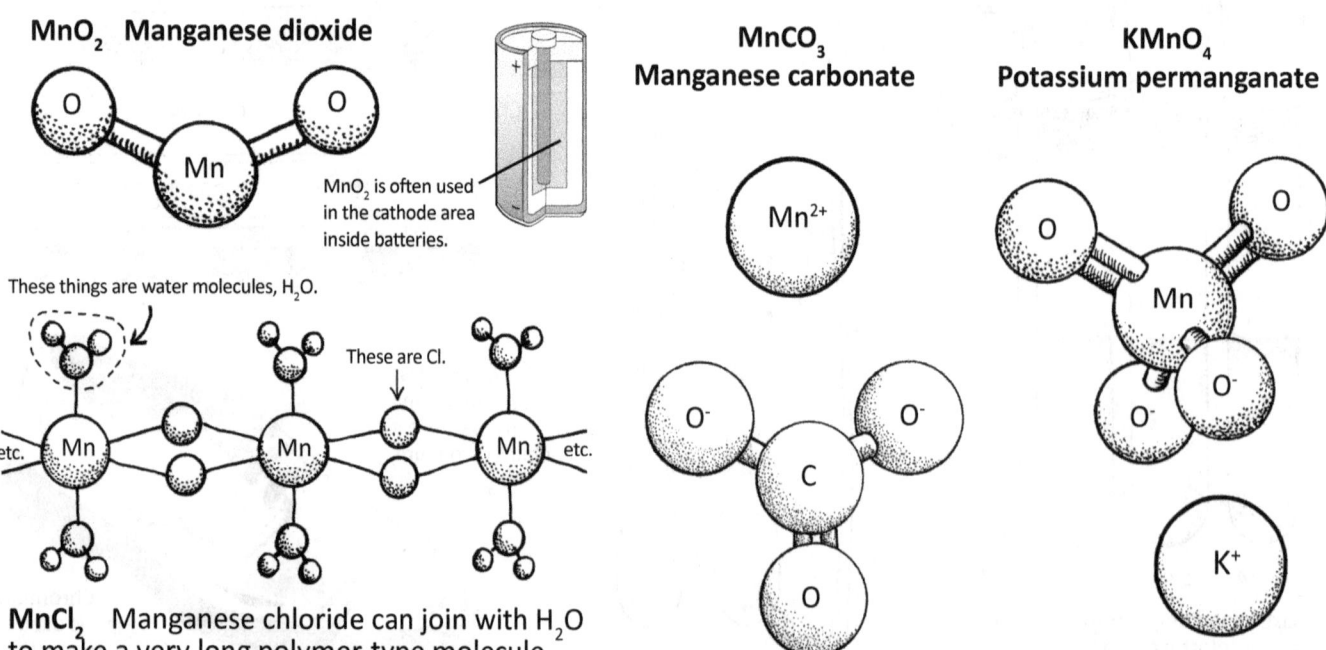

MnO_2 Manganese dioxide

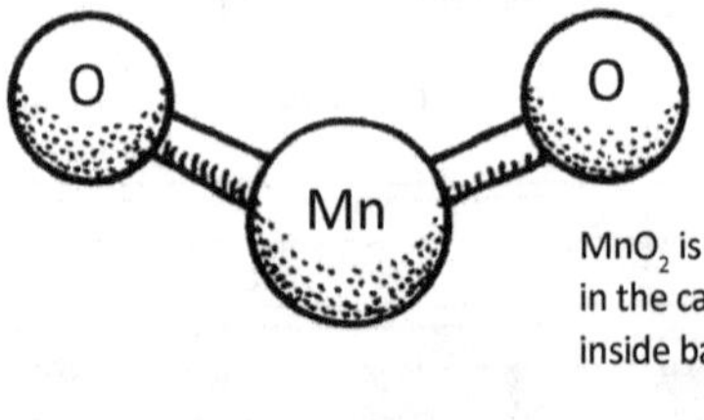

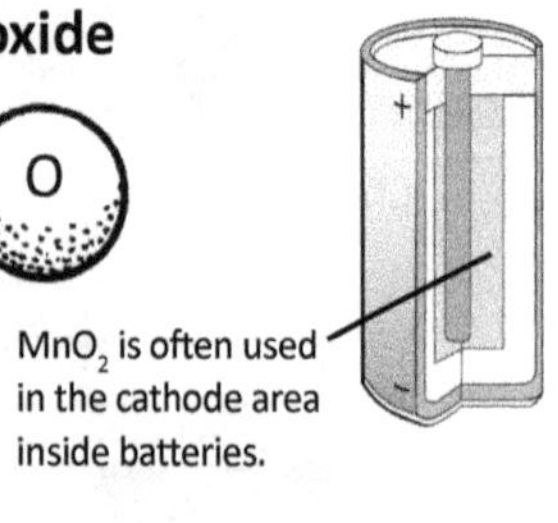

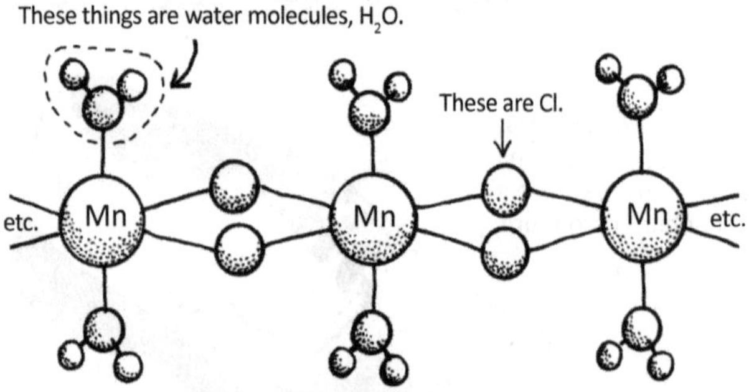

$MnCl_2$ Manganese chloride can join with H_2O to make a very long polymer-type molecule.

$MnCO_3$
Manganese carbonate

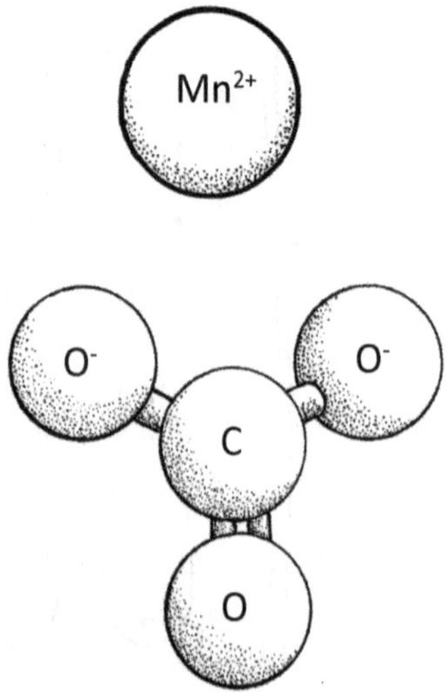

$KMnO_4$
Potassium permanganate

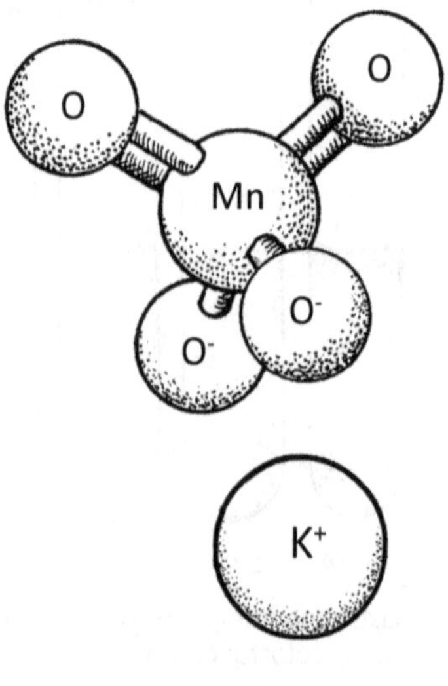

25 Mn

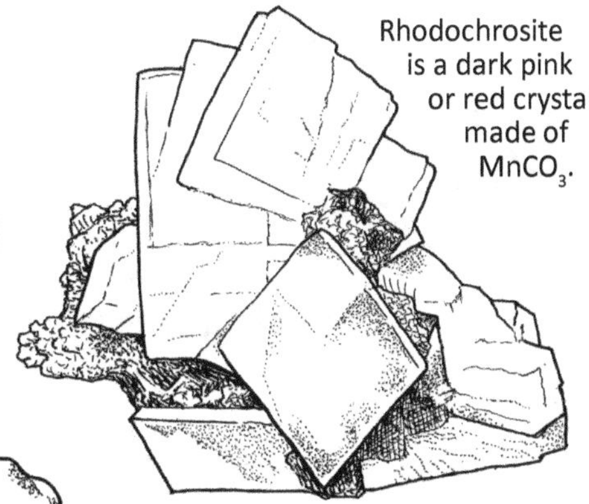

Rhodochrosite is a dark pink or red crystal made of $MnCO_3$.

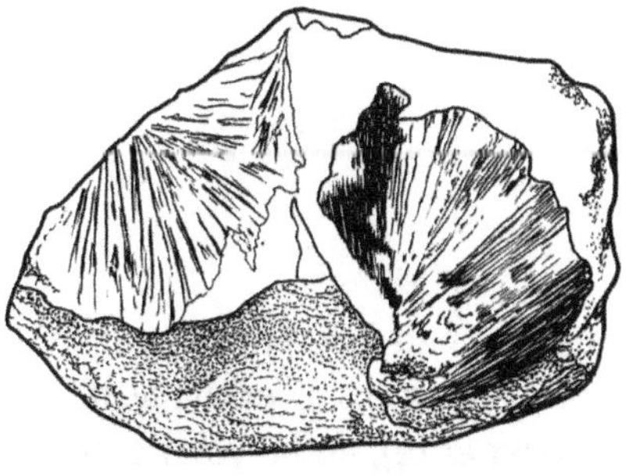

Pyrolusite, MnO_2, is usually dark gray but can have patches of orange-brown.

"Manganese nodules" also contain iron, copper, aluminum, cobalt, and silicon. Millions of them can be found lying on the ocean floors around the world.

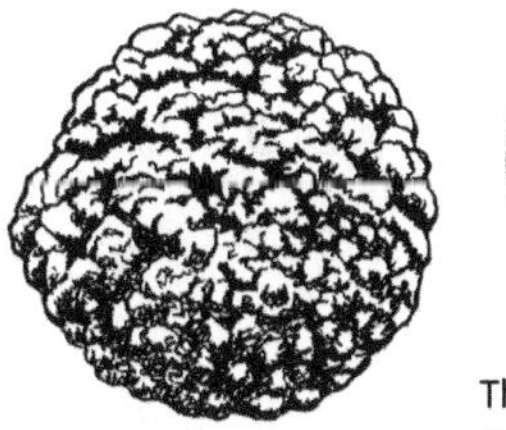

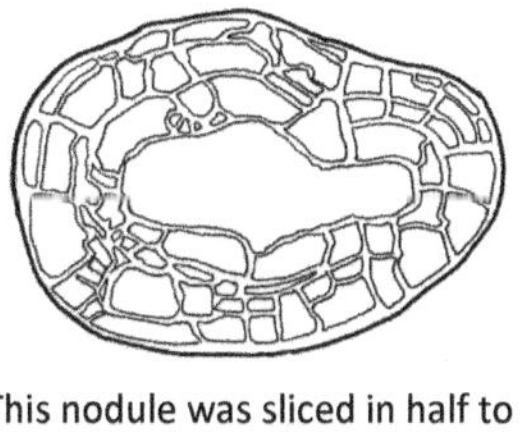

This nodule was sliced in half to reveal the layers on the inside.

Manganese

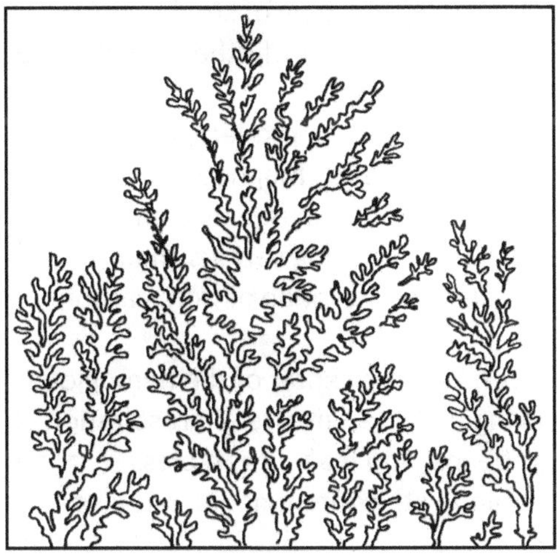

What would you think if you saw this dark green pattern on the side of a rock? These patterns were widely believed to be plant fossils until it was discovered that some manganese compounds can slowly grow into these "dendrite"

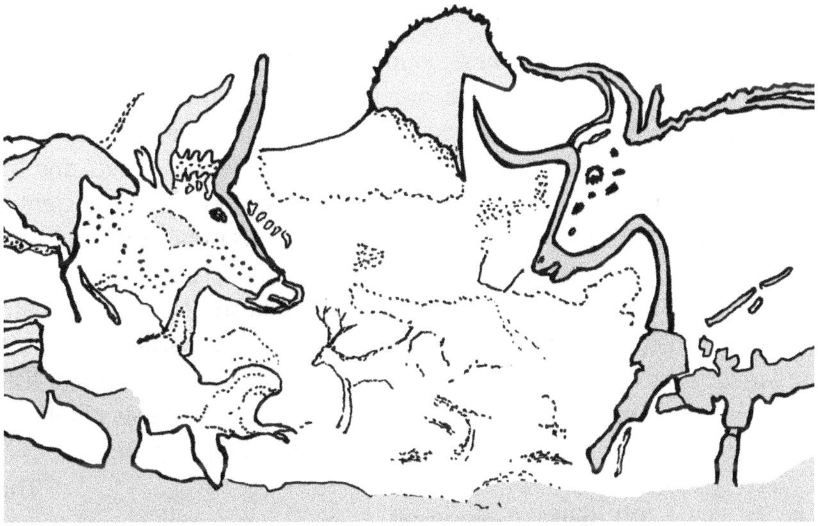

The cave paintings of Lascaux, France, were partly drawn with dark brown and black manganese pigments. The background colors are iron-based reds and tans.

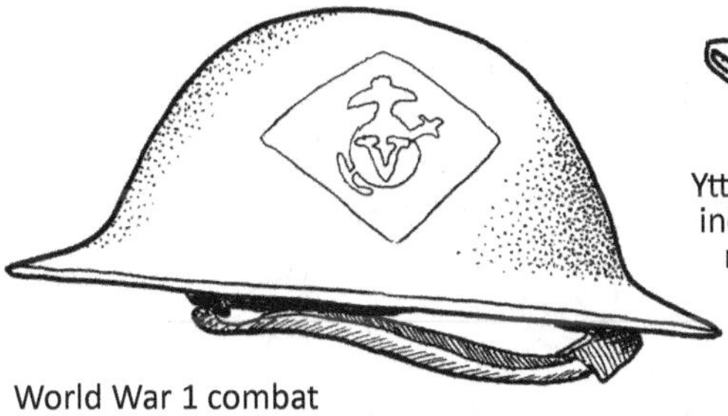

World War 1 combat helmets were made of a manganese alloy steel.

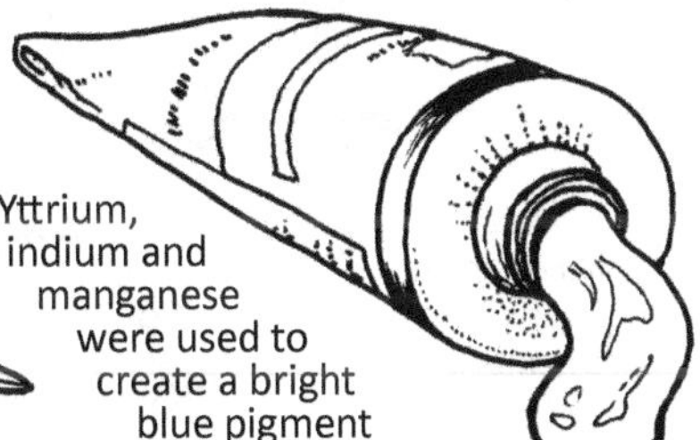

Yttrium, indium and manganese were used to create a bright blue pigment named "YInMn."

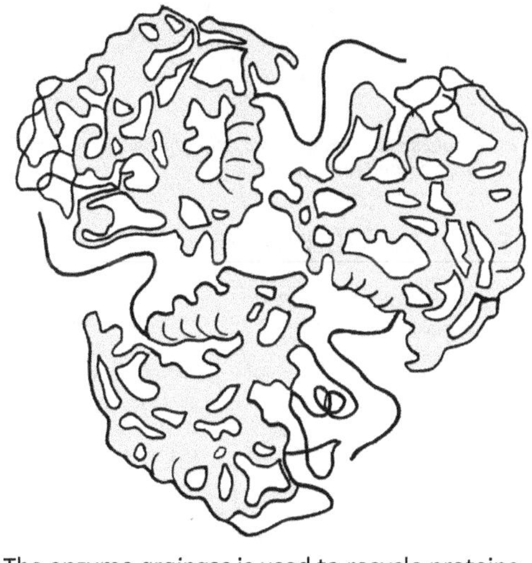

The enzyme *arginase* is used to recycle proteins. Manganese atoms are located near the center. *(Make each of the 3 sub-units a different color.)*

These foods are good dietary sources of manganese.

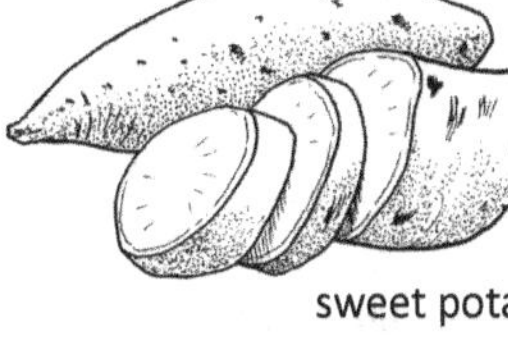

sweet potatoes

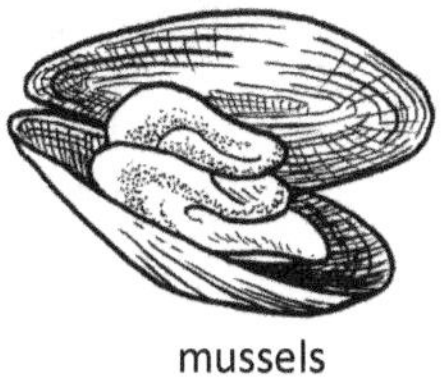

mussels

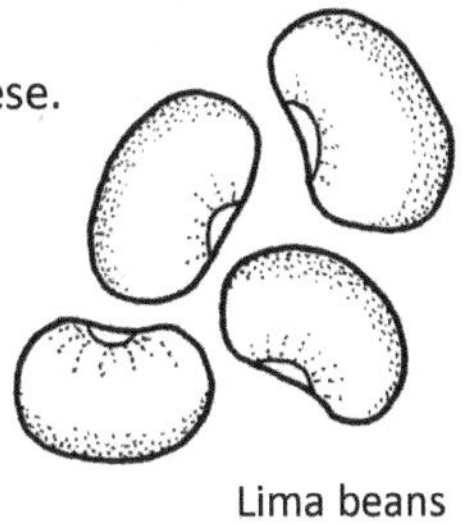

Lima beans

Fe 26

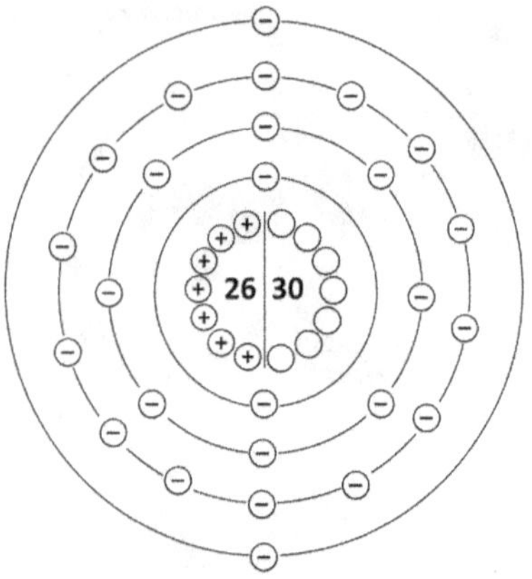

protons
30 neutrons
26 electrons
Atomic mass: 55.8

Iron

From the Old English word "iren"

The symbol, Fe, comes from the Latin word for iron, "ferrum."

The first place iron was discovered thousands of years ago was in meteorites that had fallen to the earth. Ancient peoples learned that the metal in these meteorites could be very useful for making tools. Eventually, it was discovered that iron was present in certain ore minerals (such as hematite and magnetite), and if you heated these ores, you could extract the iron. Pure iron is shiny and silvery gray, but since iron atoms are very reactive, iron quickly combines with oxygen in the air to form a flaky crust that we know as "rust." The next great discovery was that if you added just a little bit of carbon to the iron, you could make steel. "Cast iron" is a type of steel that has more than 2% carbon. Modern steel has not only carbon added to it, but often small amounts of other metals, also, such as magnesium, manganese, aluminum, chromium, vanadium, molybdenum, and nickel.

Iron atoms are magnetic, due to the arrangement of their electrons. Electrons have a property called "spin" and if an atom has many unpaired electrons spinning the same direction, it will be magnetic. Magnetism was first discovered in the mineral magnetite, Fe_3O_4, many centuries ago, and these "lodestones" were used as compasses.

Iron oxide minerals have been used as pigments since ancient times, usually producing various shades of red, brown, and tan, though some rare compounds provide blue or purple. Prussian blue is made of iron, carbon and nitrogen, and was the original blue used in blueprints.

Iron is an essential mineral for both plants and animals. In mammals, iron is at the center of the hemoglobin molecule which transports oxygen in the blood. Iron is also necessary for making enzymes involved in respiration and muscle contraction. Good sources of dietary iron include meat, fish, eggs, beans, and leafy vegetables.

Iron (III) chloride, $FeCl_3$, is used by sewage treatments plants, as it can bind with contaminants. It is also used in the printed circuit board industry to etch copper plates.

Fe_2O_3 Iron (III) oxide (hematite)

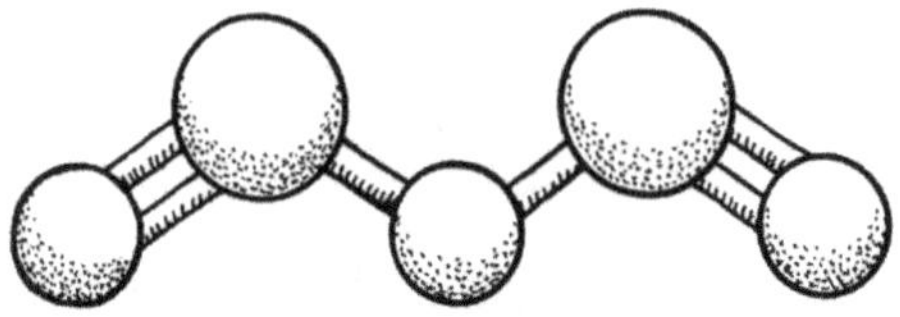

You should be able to figure out what the balls represent. Two are iron and three are oxygen.

$FeCl_3$ Iron (III) chloride

The (III) means iron is making three bonds. The sticks represent the bonds.

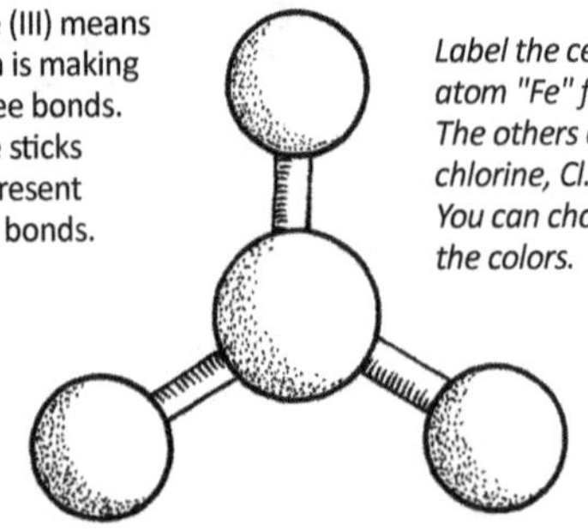

Label the central atom "Fe" for iron. The others are chlorine, Cl. You can choose the colors.

The "heme" molecule

This molecule sits in the middle of a larger protein called **hemoglobin**. The iron atom attracts oxygen.

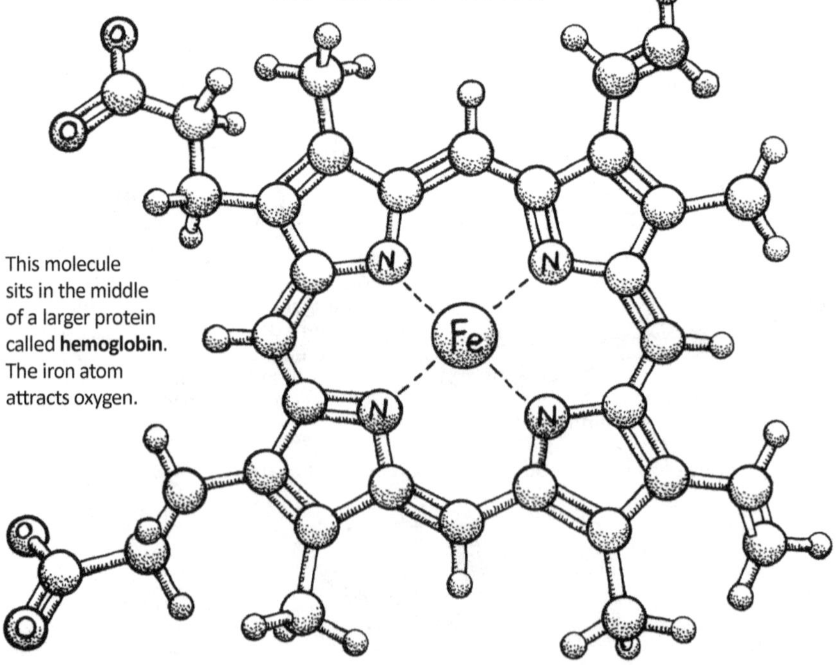

The small unlabeled balls are H, hydrogen. The larger unlabeled balls are C, carbon.

26 Fe

Steel will rust if not protected. The atoms of iron in the steel combine with oxygen from the air.

Iron

This iron bridge in Shropshire, UK, was finished in 1779. (drawing from early 1800s)

Stainless steel uses chromium as well as iron and carbon.

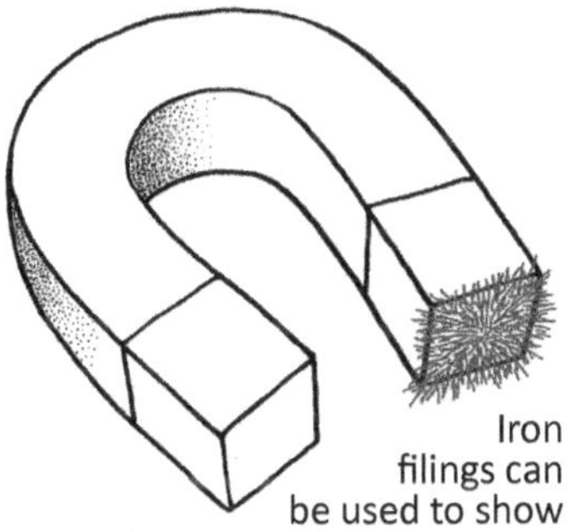

Iron filings can be used to show the fields created by magnets.

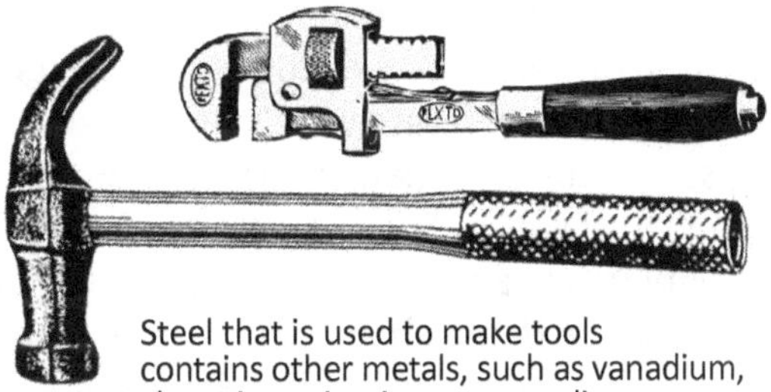

Steel that is used to make tools contains other metals, such as vanadium, chromium, titanium, or scandium.

Cast iron is more than 2% carbon. It is brittle, but makes good pans and pipes.

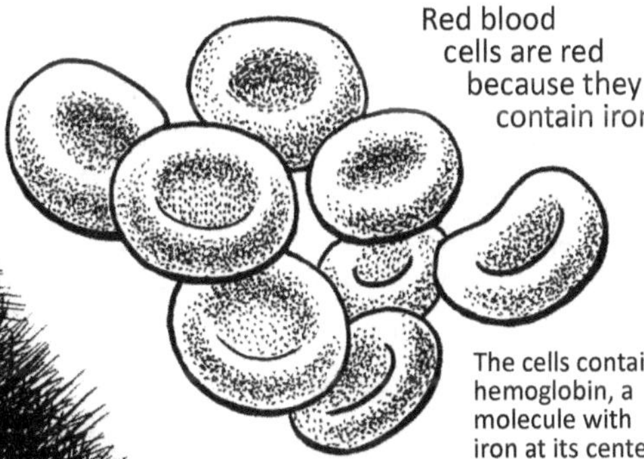

Red blood cells are red because they contain iron.

The cells contain hemoglobin, a molecule with iron at its center.

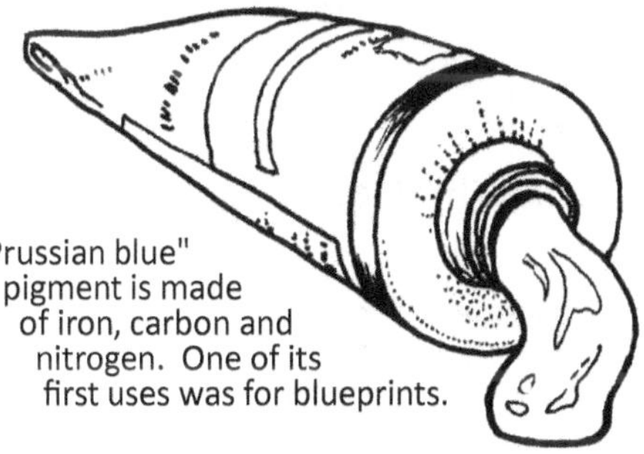

"Prussian blue" pigment is made of iron, carbon and nitrogen. One of its first uses was for blueprints.

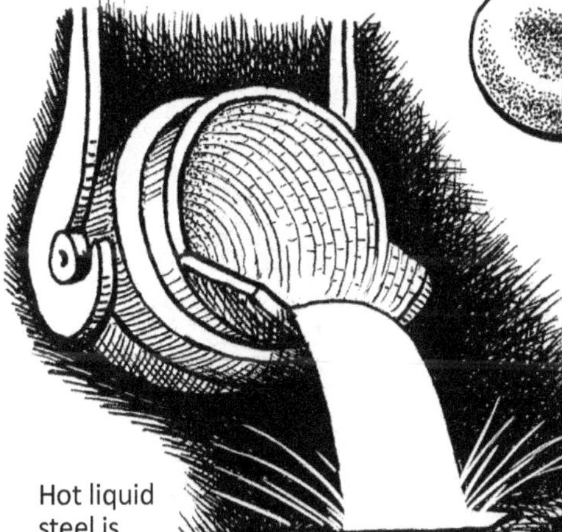

Hot liquid steel is poured into molds.

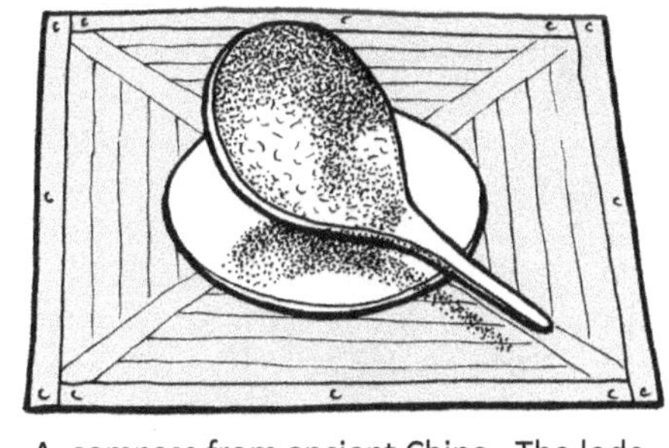

A compass from ancient China. The lodestone pointer resembles the constellation Ursa Major, which is always in the north.

Co 27

protons
32 neutrons
27 electrons
Atomic mass: 58.9

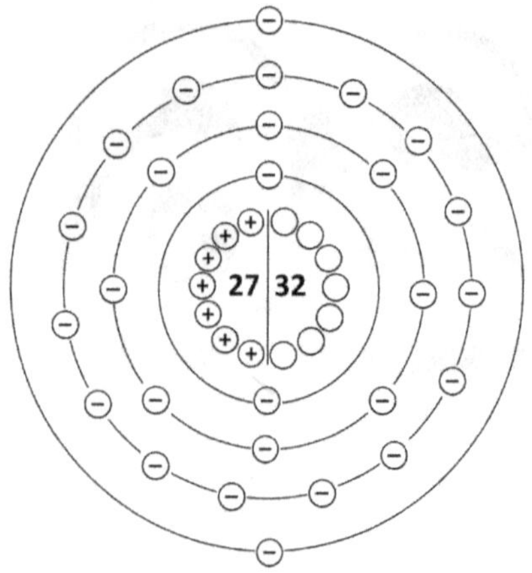

Cobalt

From the German word "kobold," meaning "evil goblin"

Cobalt has been used by craftsmen (as a blue pigment) since ancient times, but it did not receive its modern name until the 1700s, when a Swedish chemist named Georg Brandt officially declared it to be an element. He decided to use the name that German miners had associated with it: "kobold," meaning "goblin." The superstitious miners said that mischievous evil goblins lived in the dark mines, and could make miners sick. The mines had ores (rocks) that contained nickel and copper, but along with these came cobalt, bismuth and arsenic. When the rocks were heated to extract the nickel and copper, toxic fumes would make the miners sick. They guessed there was an unknown element in the ore, calling it the "evil goblin," but in the end it turned out to be the arsenic, not the cobalt, that was making them sick.

Cobalt is best known for its ability to make a bright blue pigment. Cobalt blue has been used for thousands of years in Asia and Europe to color glass, pottery and ceramics. The earliest known use was in Egypt 3,000 years ago.

Today, cobalt is mostly used to make batteries and magnets. Lithium cobalt oxide, $LiCoO_2$, is used in lithium-ion batteries in many electronic devices. Even nickel-cadmium (NiCad) batteries will use some cobalt to improve the oxidation of the nickel. The batteries used in electric cars need large amounts of cobalt. AlNiCo magnets combine aluminum, nickel, cobalt and iron. AlNiCo magnets were widely used in electronic devices such as motors, loud speakers, microphones, and guitar pickups, until the invention of rare earth magnets (containing neodymium, for example) which are stronger and smaller.

Steel alloys containing cobalt are very tough, and can be used to make parts for jet engines. The alloys are very resistant to corrosion and are non-toxic to the human body, so they can be used to make artificial joints.

If you bombard cobalt with neutrons, you can produce a heavier isotope called cobalt-60. This isotope is radioactive, but in a way that is very useful. It can be used by doctors for radiation therapy and for sterilization of medical equipment, and in agriculture to kill germs on food products.

Cobalt is found in the molecular structure of vitamin B12.

CoF_2
Cobalt (II) fluoride

Individual CoF_2 molecules get together to form this crystal structure. Samples of CoF_2 look like a bright pink powder.

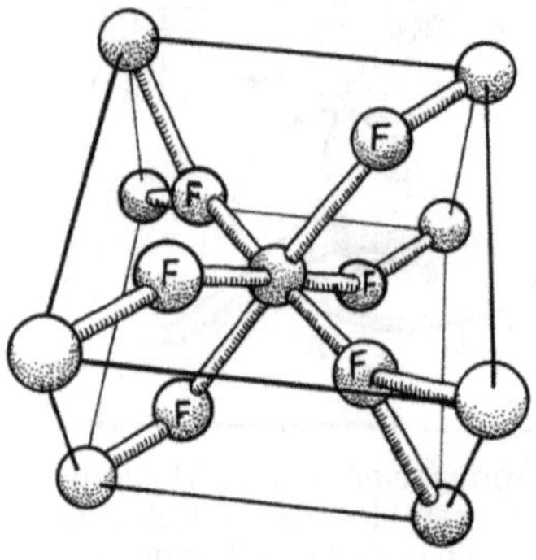

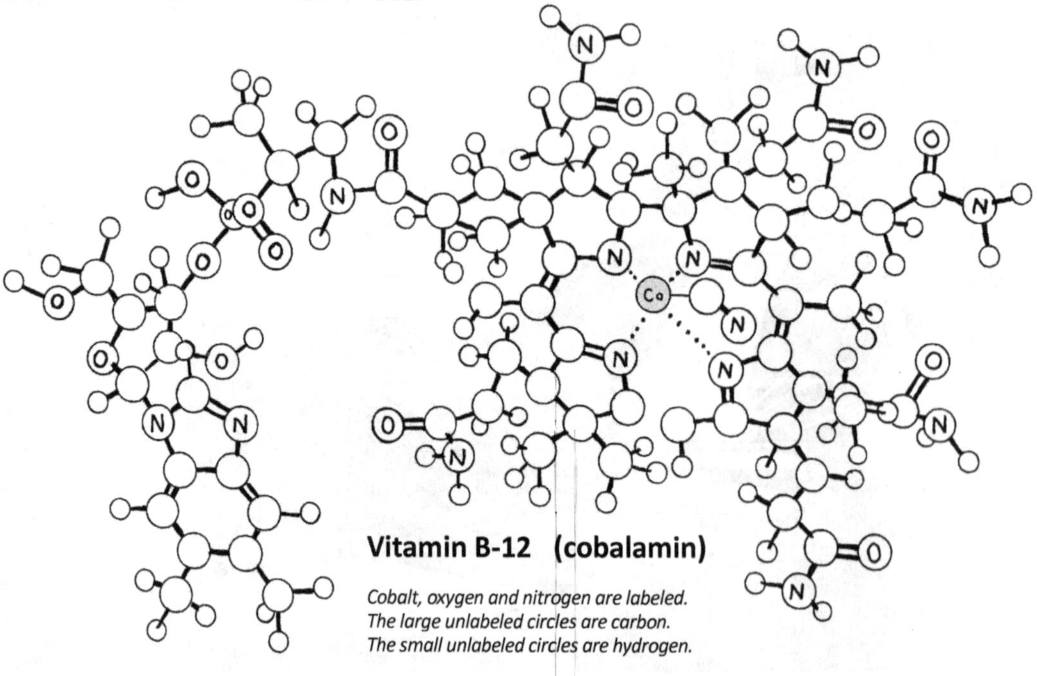

Vitamin B-12 (cobalamin)

Cobalt, oxygen and nitrogen are labeled.
The large unlabeled circles are carbon.
The small unlabeled circles are hydrogen.

27 Co

Cobalt

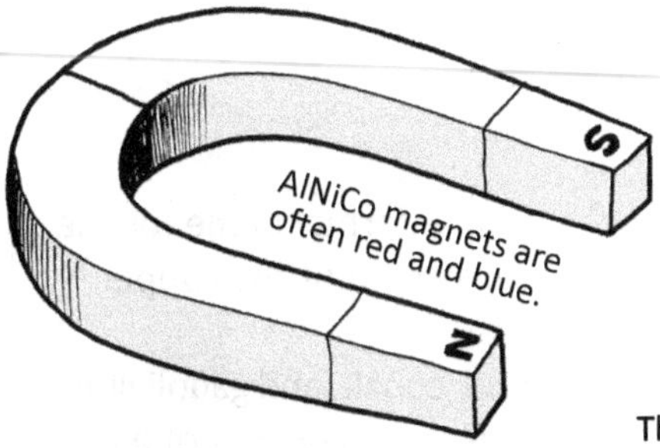

AlNiCo magnets are often red and blue.

This is how people in the 1700s imagined the kobolds who lived in the mines.

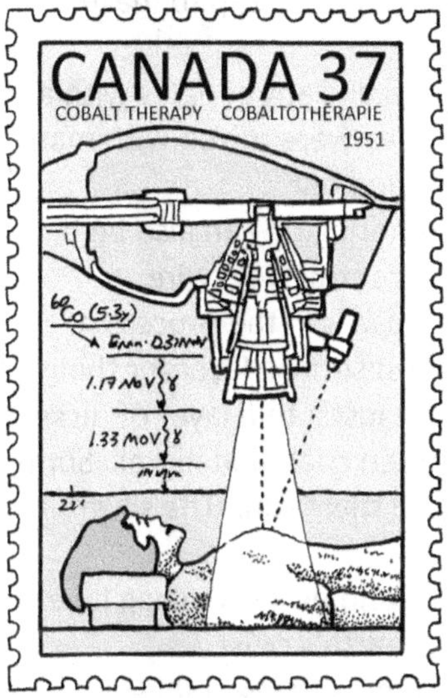

Canada designed a stamp to honor the inventors of cobalt-60 therapy.

This Roman vase is made of bright blue glass that was tinted with cobalt. (copper handles)

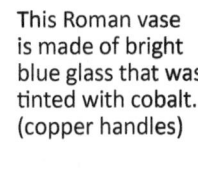

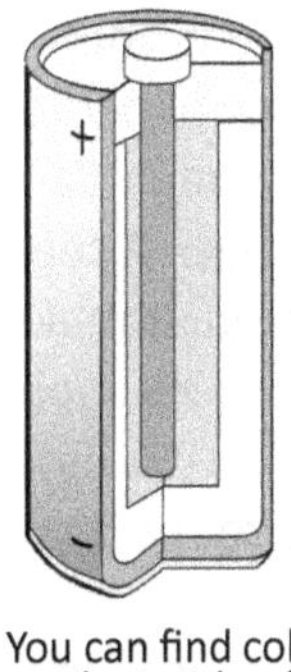

You can find cobalt on the inside of some batteries.

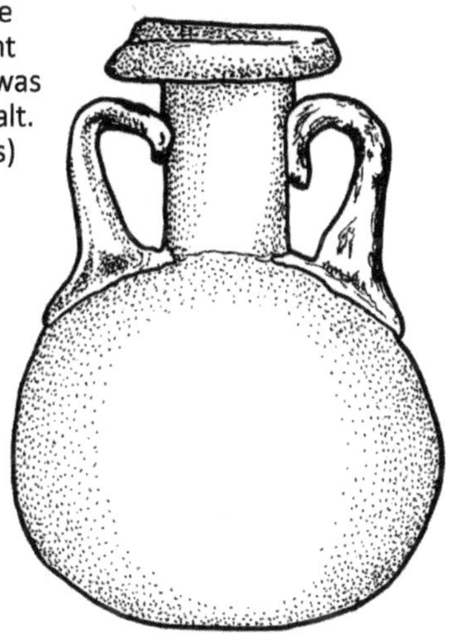

This porcelain vase was made in China in the 1300s. It is white with blue designs.

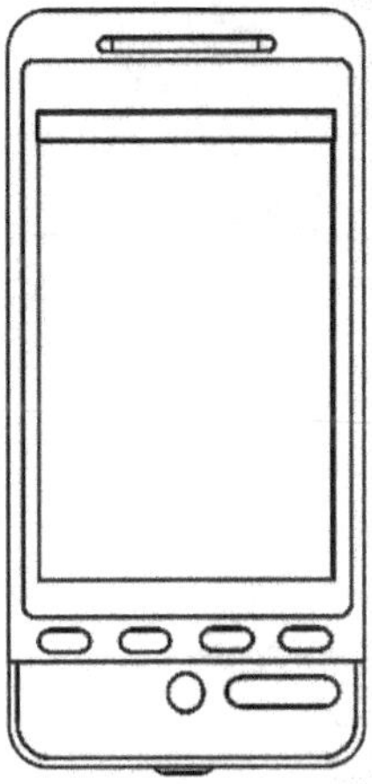

Rechargeable batteries in electronic devices often use cobalt.

Cobalt steel alloys are used to make artificial joints.

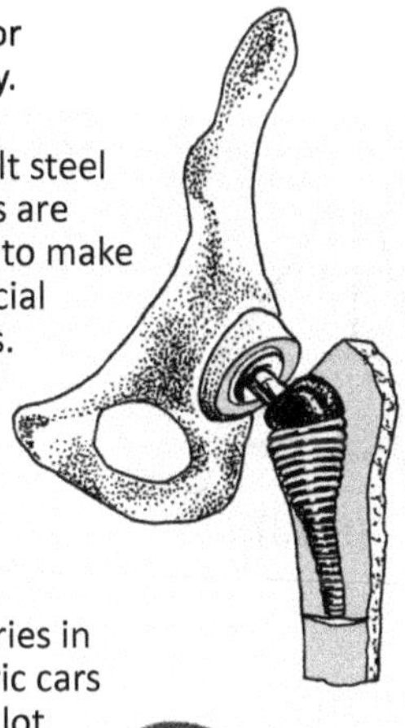

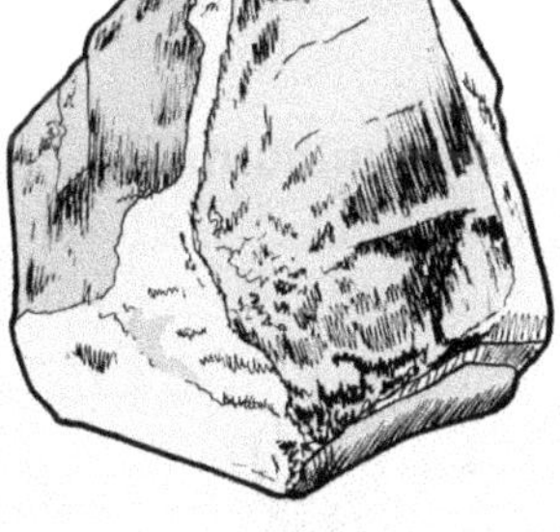

Shiny, gray cobaltite is a primary mineral ore of cobalt.

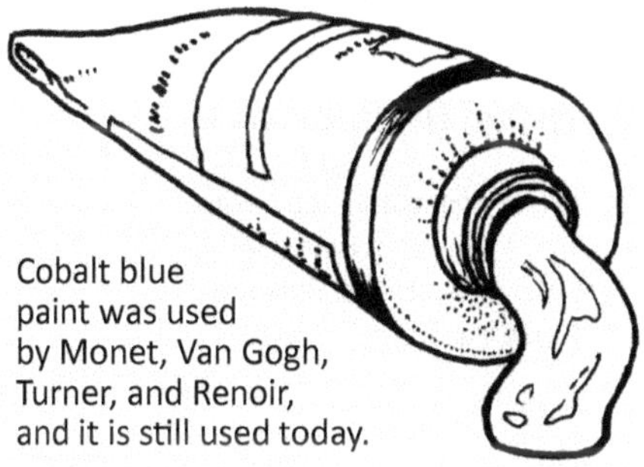

Cobalt blue paint was used by Monet, Van Gogh, Turner, and Renoir, and it is still used today.

Batteries in electric cars use a lot of cobalt.

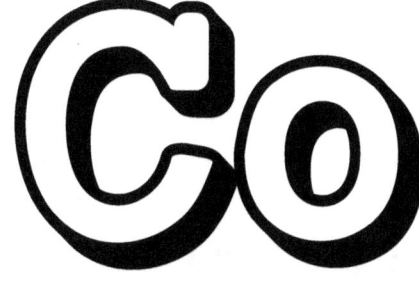

Ni

28

protons
30 neutrons
28 electrons

Atomic mass: 58.7

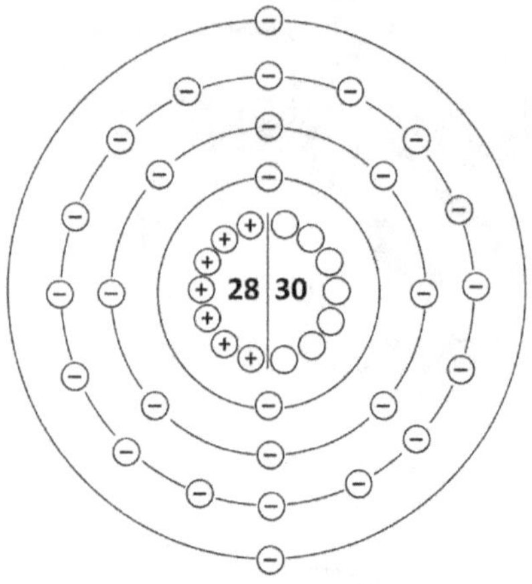

Nickel

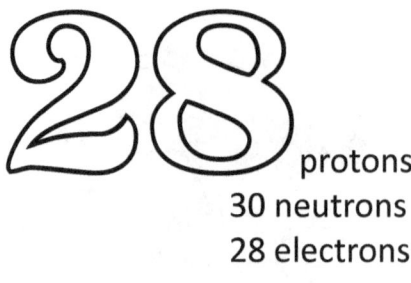

From a German word for the devil, "Old Nick"

Nickel, like cobalt, got its name from German mining mythology. "Old Nick" was an informal name for the devil, who the superstitious miners blamed when silvery nickel came out of the ore rocks, instead of the copper they were looking for. They would call the metal "Kupfernickel" (the devil's copper).

Nickel is one of only four elements that are magnetic at room temperature: iron, nickel, cobalt, and gadolinium. Magnets made of an alloy of aluminum, nickel, cobalt, and iron are known as AlNiCo magnets. They are not as strong as magnets made of rare earth metals such as neodymium, but they are much easier and cheap to make so they are still widely used. They are often painted red and blue.

Like iron, nickel was first discovered in meteorites. Iron-nickel meteorites provided ancient peoples with a source of almost pure metal. Later, these elements were discovered in ore rocks where they were mixed with many other elements. Through smelting (cooking the rocks) the metallic elements could be drained out.

The texture of nickel is hard but "ductile," meaning that it can be shaped by hammering. It can also be stretched to make very long wires. Guitar stings are made of nickel wire wound around a central steel wire.

Today, most nickel is used in the production of stainless steel alloys. A secondary use is in the process of electroplating other metals. Since nickel is shiny and resistant to corrosion, it makes a good surface layer for things like coins and jewelry. The American 5-cent coin, the "nickel," is actually 90% copper, with just a thin layer of nickel on the surface, though in the past it had a higher nickel content. Snaps on clothing are often coated in nickel. Some people have skin that is sensitive to nickel and develop a small rash where the snaps touch their skin. The silver keys on many musical instruments are electroplated with nickel.

Geophysicists speculate that the inner and outer core of the earth is made of iron and nickel, making it very abundant in the earth overall, despite it being relatively hard to find and extract from the earth's crust.

Nickel (and platinum) metal plates are used in the food industry to add hydrogen atoms to vegetable oils.

Nickel (II) fluoride

Nickel, like cobalt, can bond to fluorine to create a lattice structure. Nickel fluoride looks like a green powder.

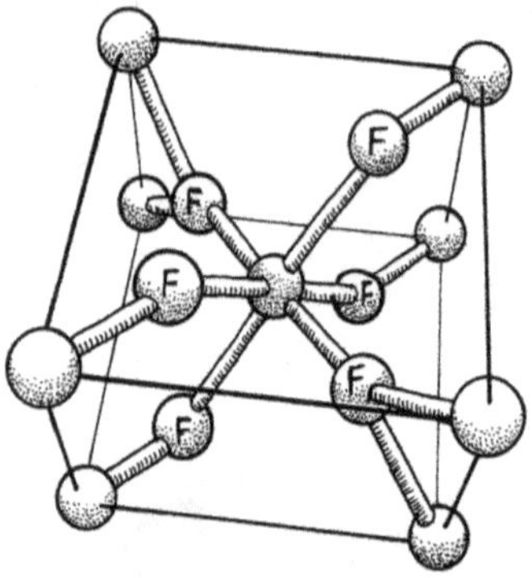

Label the Ni atoms and choose a color for each element.

Nickel-aluminum alloy "clusters"

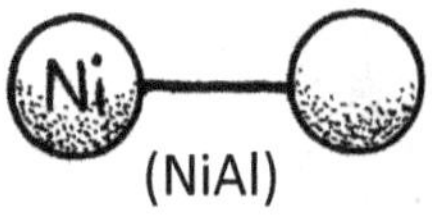

(NiAl)

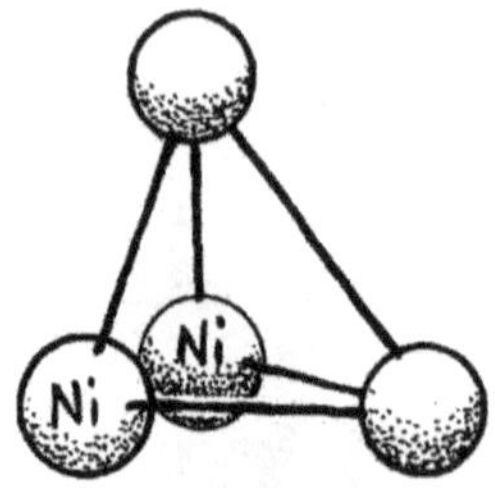

$(NiAl)_2$

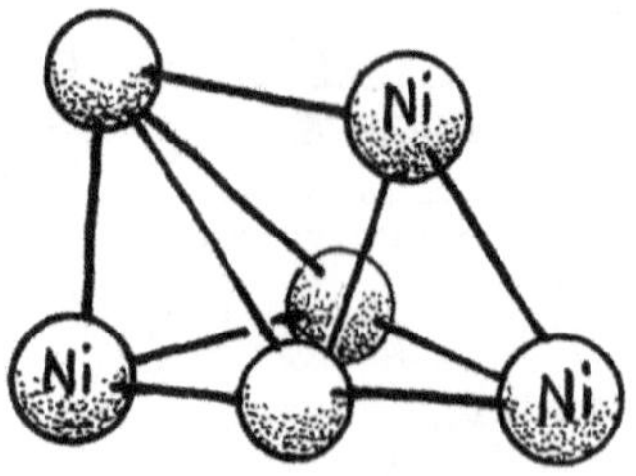

$(NiAl)_3$

The nickel atoms are labeled. The unlabeled atoms are aluminum. Choose a color for each element.

$(NiAl)_4$

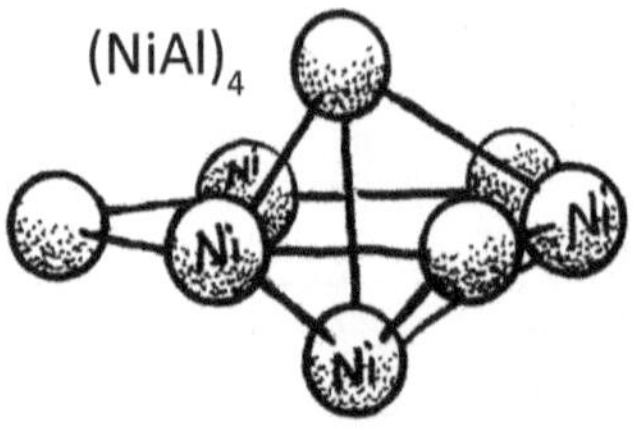

$(NiAl)_5$

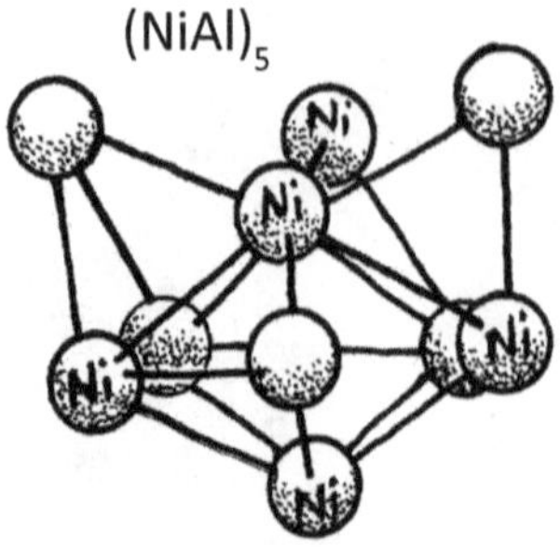

$(NiAl)_6$

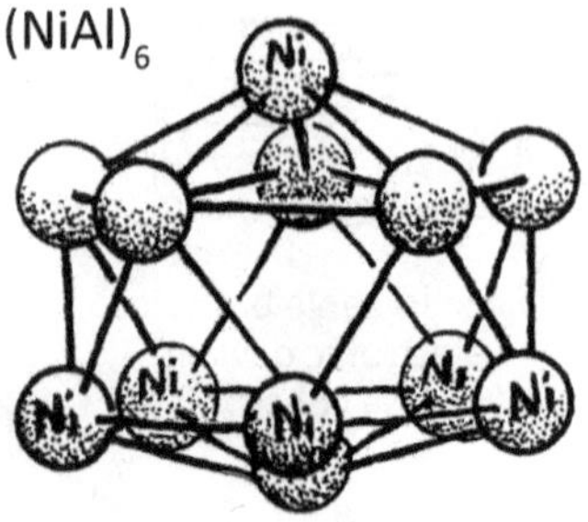

28 Ni

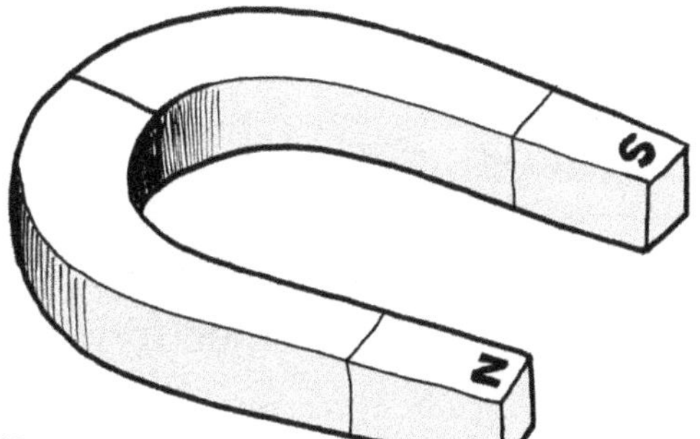

AlNiCo magnets are made of aluminum, nickel, cobalt, and iron.

Nickel

This was the first nickel issued in the US, known as the "shield nickel."

Scientists think that the solid inner core and the liquid outer core of the earth is made of iron and nickel.

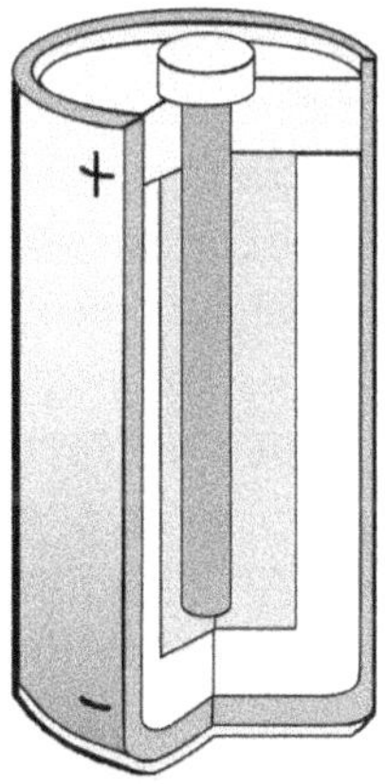

This is the standard US nickel.

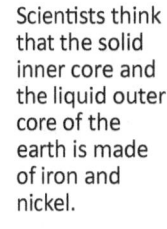

Snaps on clothing are often plated with nickel to make them shiny.

Rechargeable NiCad batteries use nickel and cadmium as electrodes.

Canada was the major source of nickel until a large supply was found on the island of New Caledonia.

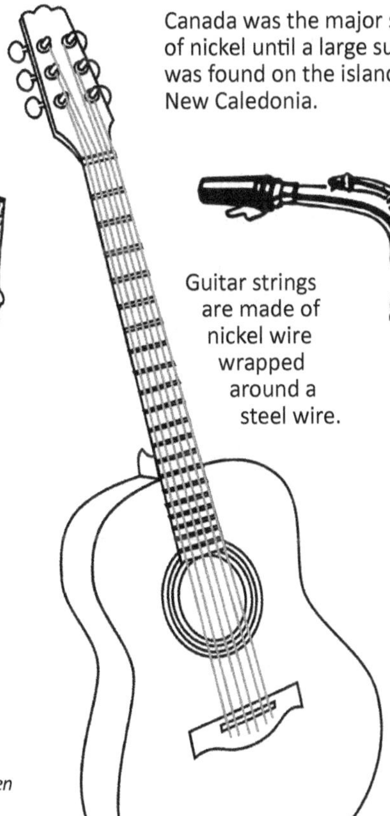

Guitar strings are made of nickel wire wrapped around a steel wire.

Nickel and platinum plates are used in the process of making hydrogenated vegetable oils.

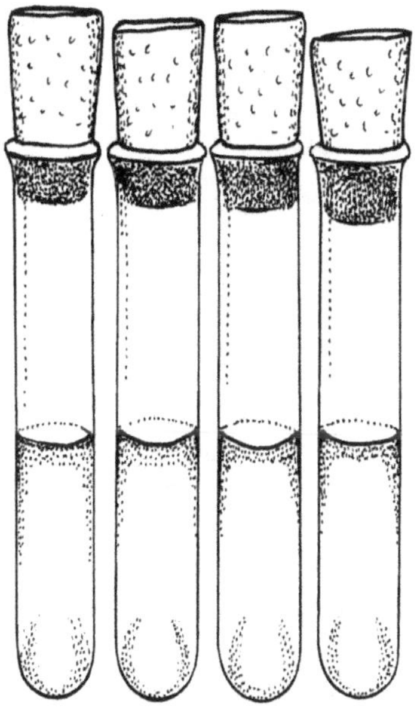

purple *blue* *light green* *dark green*

When nickel is dissolved in solutions of ammonia (NH_3), chlorine, or even water, you get bright colors.

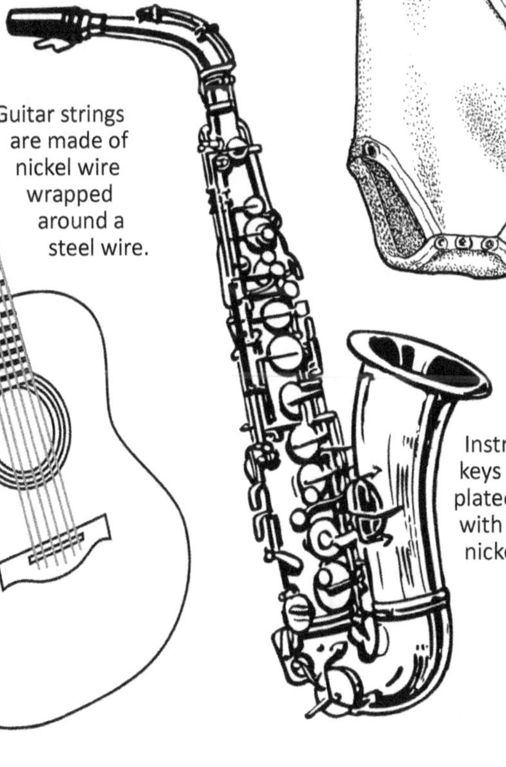

Instrument keys are plated with nickel.

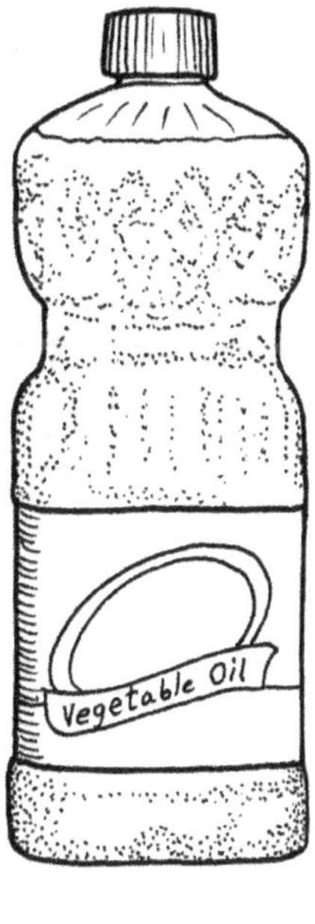

Cu 29

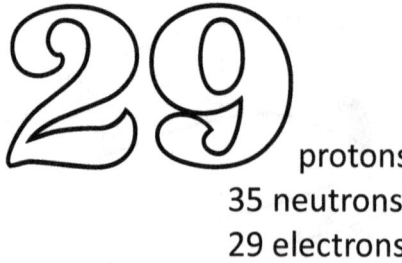

protons
35 neutrons
29 electrons

Atomic mass: 63.5

Copper

From the Latin word "cuprum" meaning "from Cyprus"

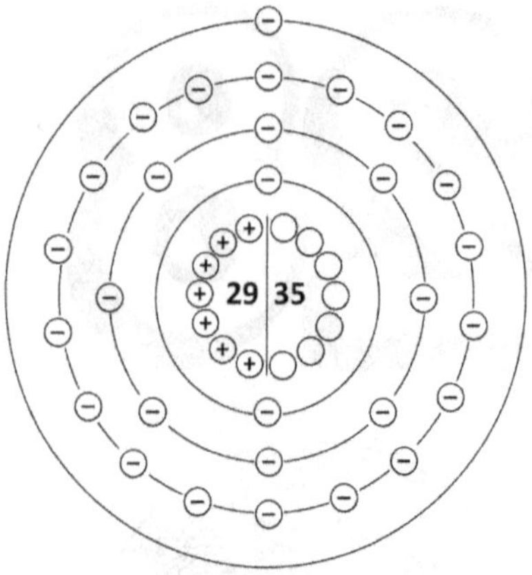

Copper has been used by many civilizations for thousands of years. It is one of the few metallic elements that can be found in its pure form in the crust of the earth, though most copper throughout history has been taken out of ore rocks that contain only a small amount of copper. Today, most copper comes from open pit mines that are digging through "porphyry copper" deposits, thought to have been formed by hot mineral fluids leaking up from far below the surface. Minerals that contain copper often have a green or blue color, such as malachite and turquoise. Pure copper tends to turn bluish green over time because it interacts with the oxygen in the air. The Statue of Liberty, which is made of copper, is now green.

More than 4,000 years ago, metal workers discovered that if you add some tin to the copper, it becomes harder, and therefore more suitable for tools and weapons. This combination (alloy) is called bronze. Much later, another alloy, brass, was invented by adding zinc to the copper. Brass is most well known today for its use in musical instruments that we call the brass instruments, such as the trumpet, trombone, and tuba.

The elements right below copper on the Periodic Table are silver and gold. All three elements, copper, silver, and gold have electron arrangements that give them similar properties. They are relatively soft and easy to hammer, or melt, into shapes. They also conduct electricity very well and can be made into electrical wires. You can find copper wire in just about any appliance or electric motor.

Copper kettles used to be very popular in homes, before the days of stainless steel and electric kettles. Until recently, copper pipes were the industry standard for plumbing in houses, and some builders still choose to install copper instead of PVC plastic. Thin sheets of copper are sometimes used to weatherproof roofs.

Copper is necessary for the formation of a few of our body's enzymes. Mollusks (snails, slugs, clams, squid) need more copper than we do because their blood uses a copper-based molecule to deliver oxygen (instead of iron, like most other animals). Copper is toxic to molds that grow on plants, so it is used to make agricultural fungicides. Copper is also toxic to many types of bacteria, so brass handles and doorknobs will carry fewer germs than steel.

$CuCl_2$
Copper chloride

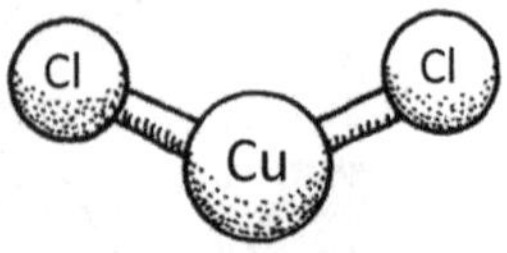

Cu_2O Copper (I) oxide

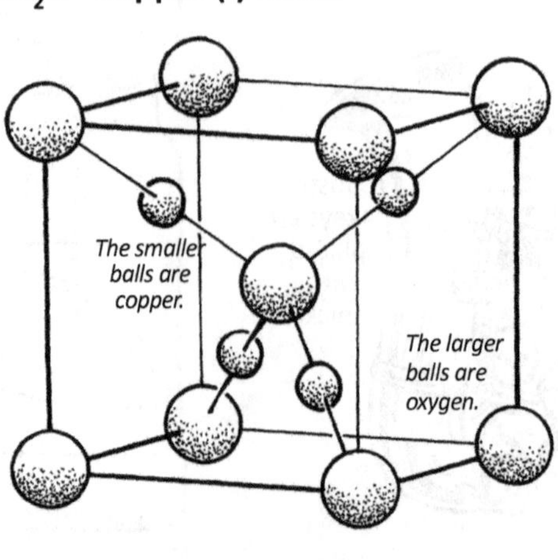

$CuCO_3$
Copper carbonate

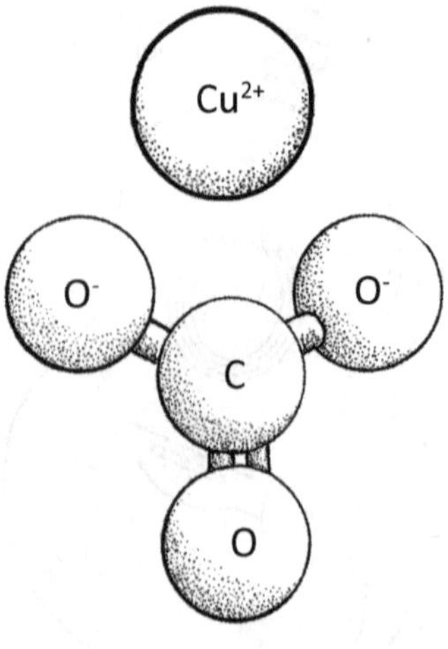

$CuSO_4$
Copper sulfate

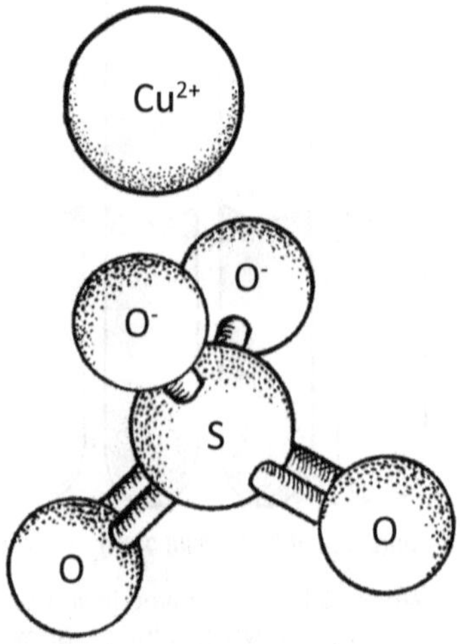

Most copper compounds can form crystal structures, though they are not shown here.

29 Cu

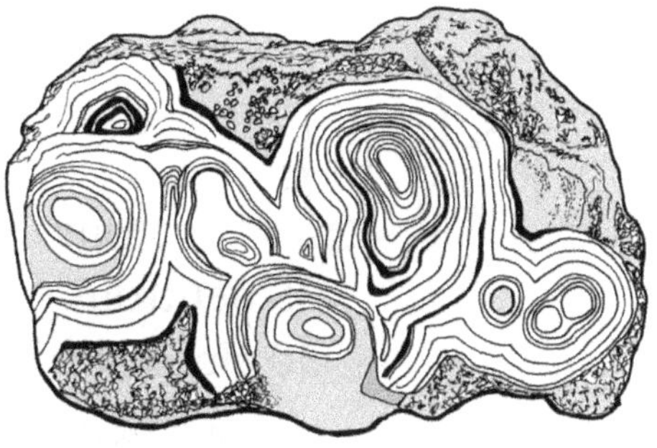

Malachite is bright green. $CuO_3 \cdot Cu(OH)_2$

Copper

This tuba is made of brass. Brass is an alloy of copper and can have 5% to 45% zinc.

The Statue of Liberty in New York City is made of copper. It is green because of copper's interaction with oxygen.

Copper kettle

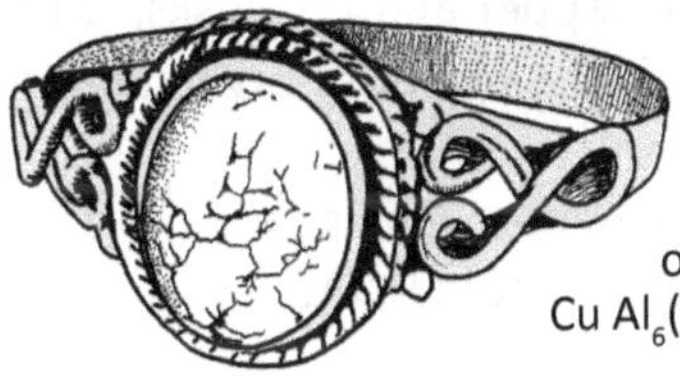

This ring displays a polished piece of turquoise. $Cu\,Al_6(PO_4)_4(OH)_8 \cdot 4H_2O$

A copper coin from ancient Rome.

Modern pennies are coated with copper.

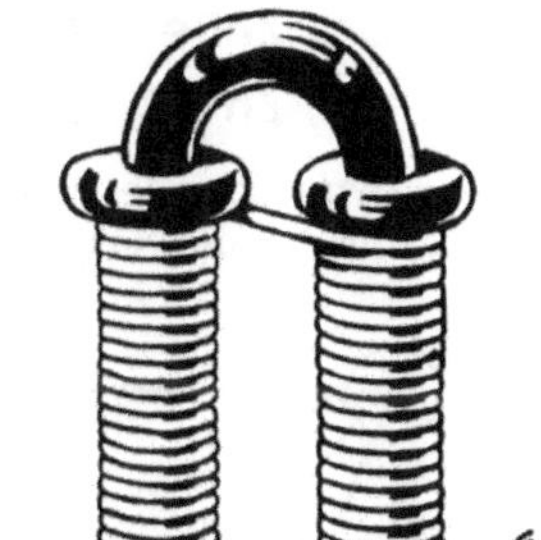

Copper wire is wound around metal rods to make electromagnets.

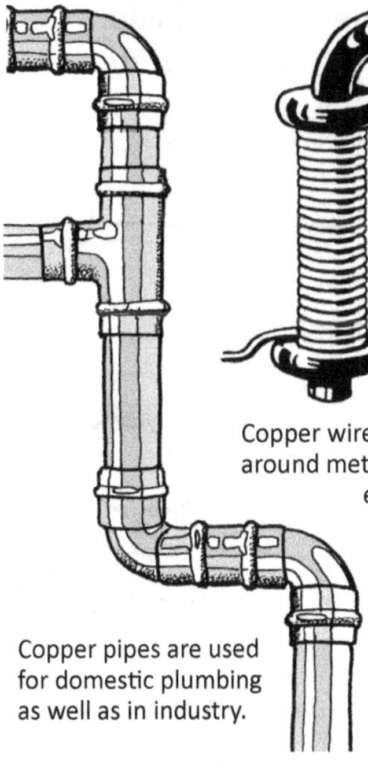

Copper pipes are used for domestic plumbing as well as in industry.

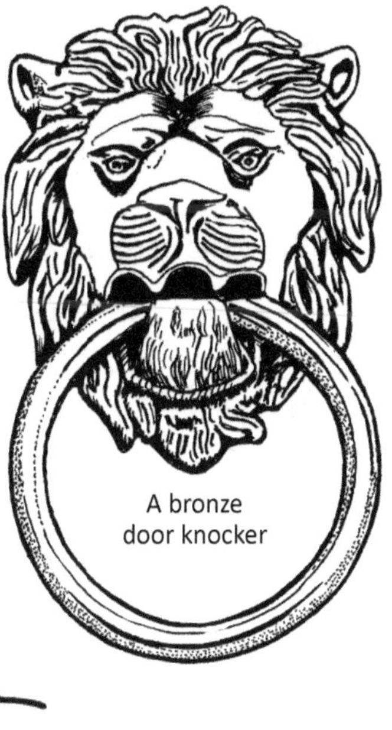

A bronze door knocker

Zn 30

protons
35 neutrons
30 electrons

Atomic mass: 65.4

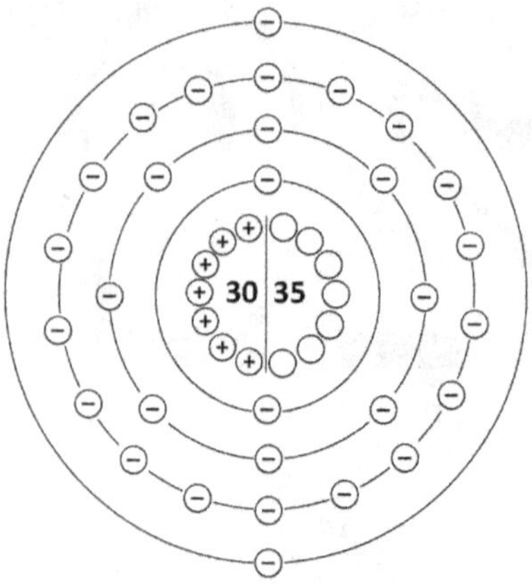

Zinc

From the German word "Zinke"

Zinc has been used by metal workers for over 3,000 years, but was not isolated as an element until 1746. Pure zinc is a shiny, silvery-gray metal. In nature, it is never found in pure form, but is bound to other elements such as sulfur or oxygen to form ore minerals, such as sphalerite. Zinc ores are often mixed with ores of copper and lead. These mixtures are likely what led to the discovery of brass and bronze.

Zinc is happy to form alloys with many other metals including aluminum, antimony, bismuth, gold, iron, lead, mercury, silver, tin, magnesium, cobalt, nickel, tellurium, and sodium. The most well-known alloy is brass, a combination of copper and zinc. Brass artifacts found in the Middle East date to about 1,400 BC. In modern times, people are most familiar with brass as the metal that trumpets, trombones and tubas are made of.

An important characteristic of zinc is its ability to combine with oxygen to form a protective surface coating. Metals can be weatherproofed by having a thin layer of zinc applied to them. This process is called "galvanization," named after Luigi Galvani, who did experiments with zinc in the late 1700s. Galvanized nails and screws are used to make structures that will be exposed to rain and snow.

Zinc oxide, ZnO, can absorb harmful ultraviolet light, so it can be used in sunscreen products and also in plastics to protect them from sun damage. The fact that ZnO is a white powder allows it to be used as a pigment for white paint. Zinc sulfide, ZnS, will glow after being exposed to light, and is used to make glow-in-the-dark products. Zinc chloride, $ZnCl_2$, is used in lumber to make the wood more resistant to weather and fire.

The first electric battery was made by Alessandro Volta in 1800, using a stack of copper and zinc disks, with wet pads between them. The pads were soaked with an electrolyte solution that would let electrons flow between the copper and zinc. This "Voltaic pile" was the forerunner of the modern dry cell battery.

Zinc is an important mineral in the body, as it is part of the structure of many enzymes. One of the most interesting zinc molecules is called the "zinc finger," a molecular structure that can grab and hold DNA.

$ZnCl_2$ Zinc chloride

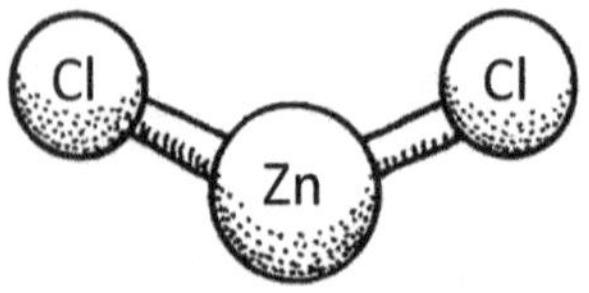

ZnO Zinc oxide

ZnO can also form a crystal structure (not shown here).

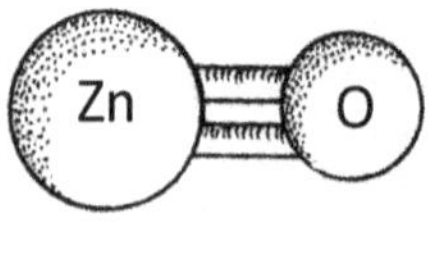

ZnS Zinc sulfide

ZnS can also form the structure shown for ZnSe.

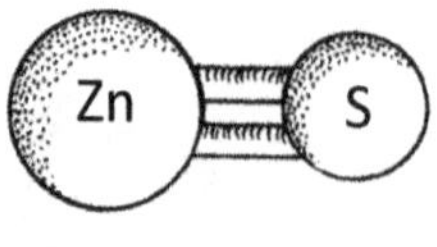

ZnSe Zinc selinide
ZnTe Zinc telluride

These molecules have the same structure. They are used in the semiconductor industry.

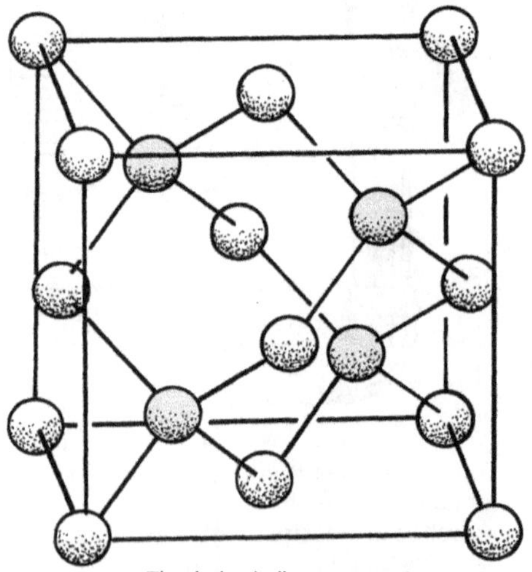

The darker balls represent zinc.

$ZnCO_3$ Zinc carbonate

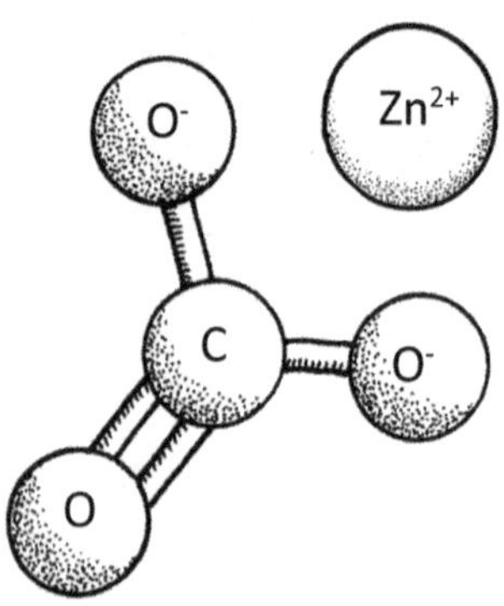

$ZnSO_4$ Zinc sulfate

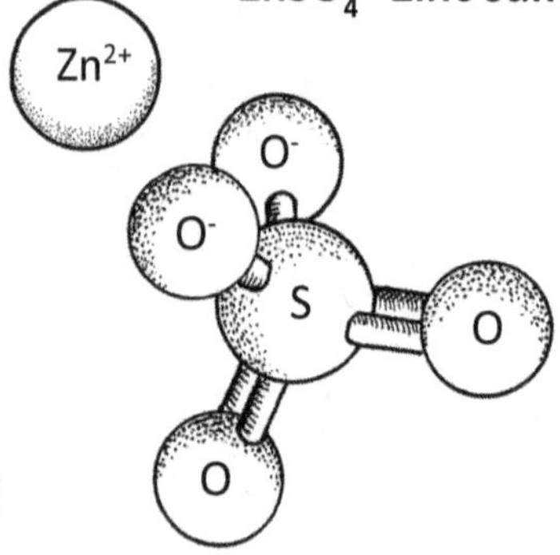

Red = oxygen
Black = carbon
Yellow = sulfur

30 Zn

Zinc

Brass instruments like this French horn are made of brass, which is an alloy of zinc and copper.

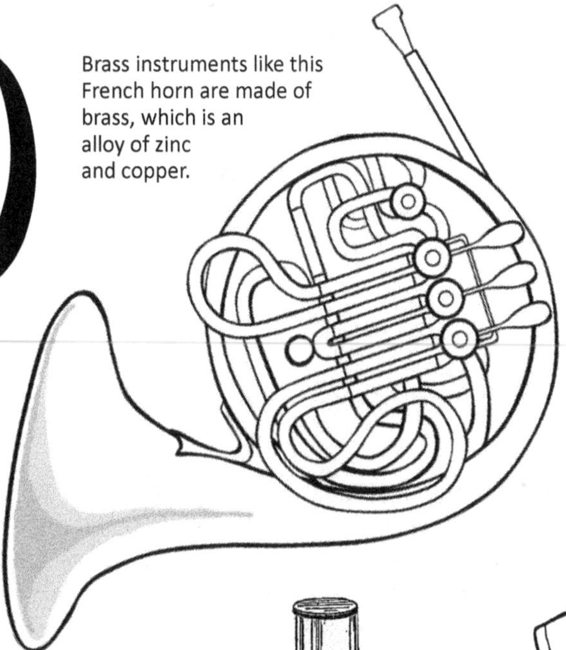

Zinc oxide is used as a white pigment in paint.

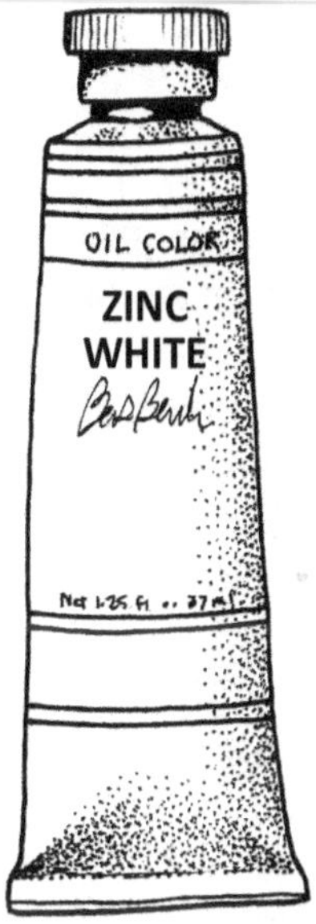

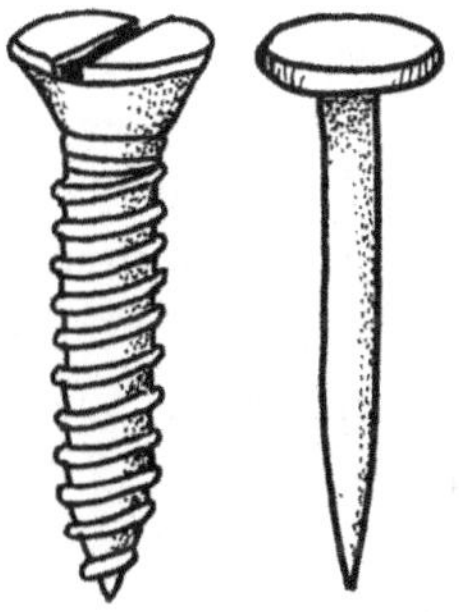

Galvanized nails and screws are coated with a thin layer of zinc to protect against rust.

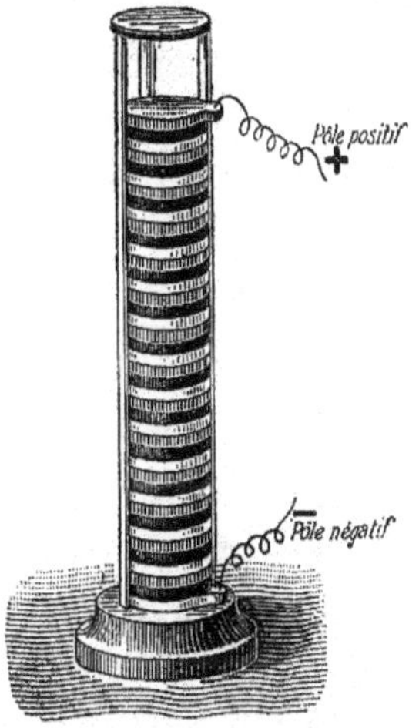

The "Voltaic pile" was a stack of copper and zinc discs. It was the first battery. Modern batteries still use zinc, though not as discs.

This T-shirt has a glow-in-the-dark design printed onto it. The GITD paint uses ZnS, zinc sulfide.

Zinc powder is combined with sulfur powder to make a safe explosive that will launch model rockets.

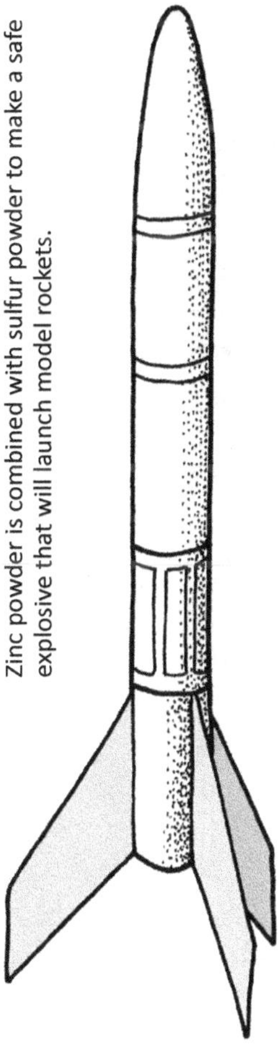

Pennies are made of zinc, with just a thin layer of copper on the outside.

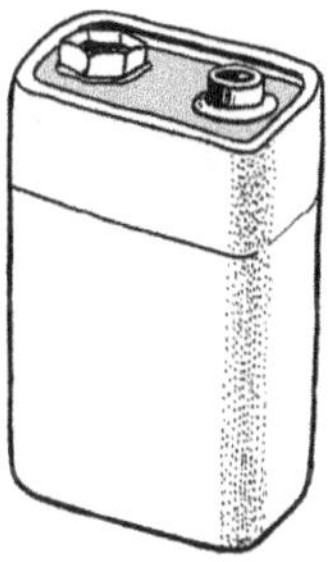

The "zinc finger" molecule has a zinc atom right in the middle. The molecule can grab and hold DNA.

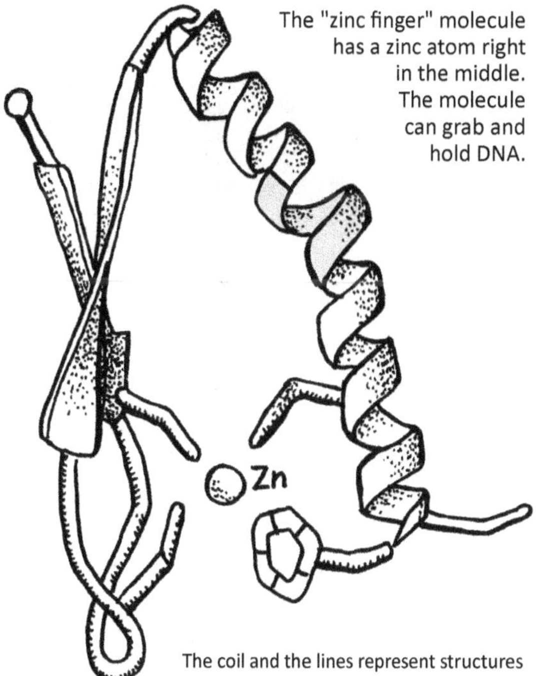

The coil and the lines represent structures made of amino acids (protein).

Oysters are the best dietary source of zinc.

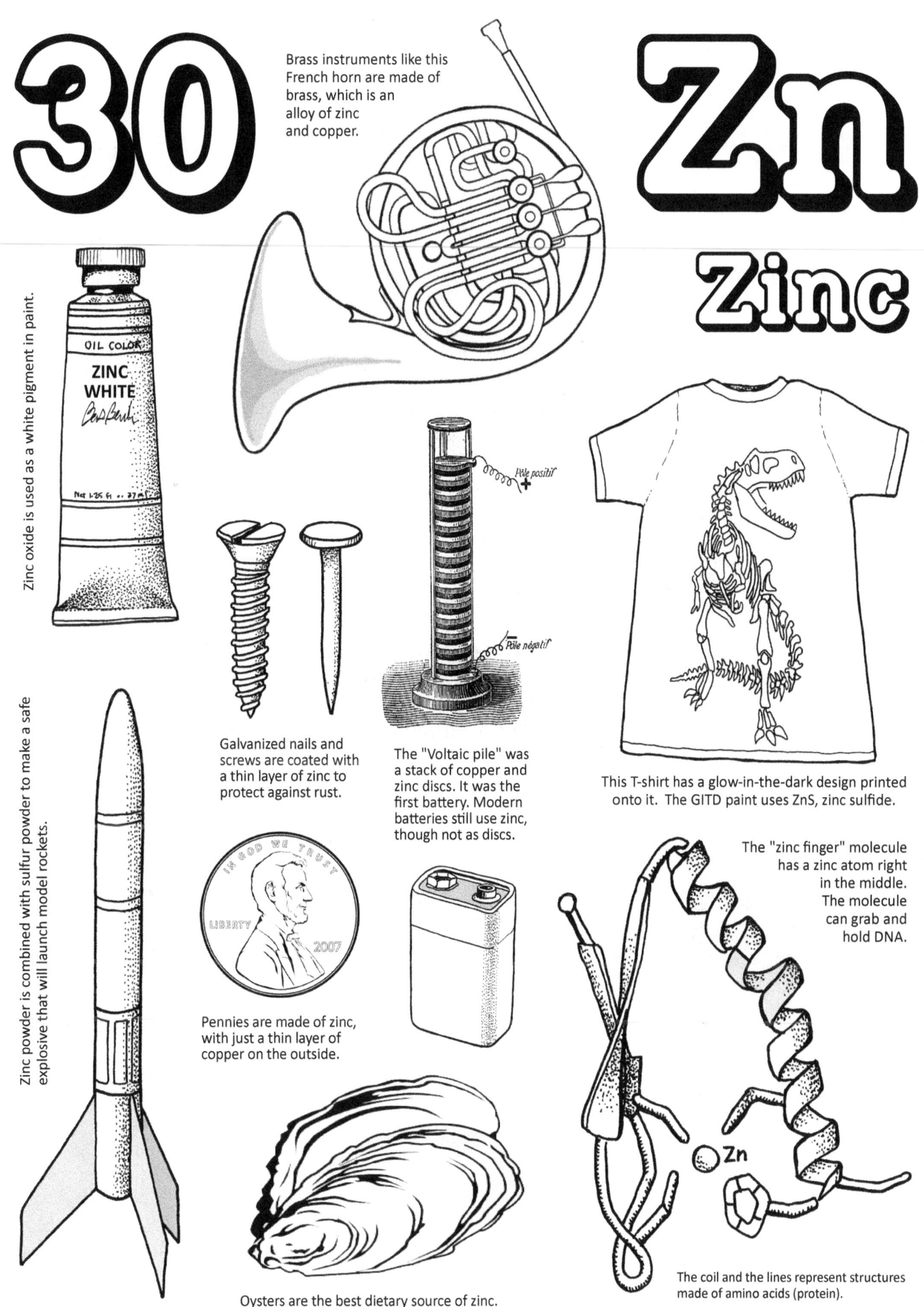

Ga 31

protons
39 neutrons
31 electrons

Atomic mass: 69.7

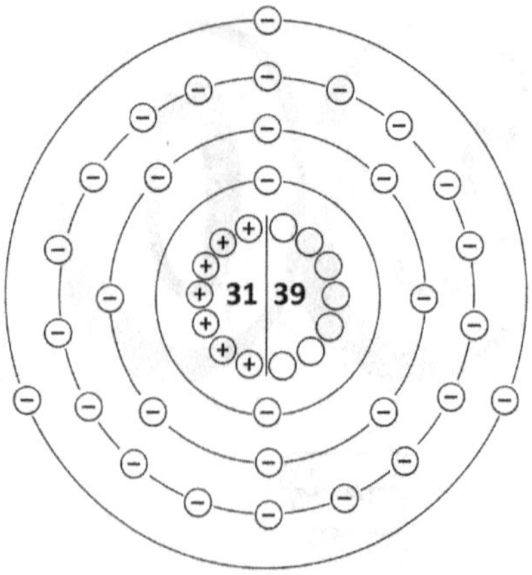

Gallium

From the Latin word for France, "Gallia"

Dmitri Mendeleyev, the inventor of the Periodic Table, made a prediction in 1871: a new element would soon be discovered. This element would be a soft metal with a low melting point, and would have an atomic mass of about 68. He even predicted that this new element would be discovered using a spectrometer, a machine that shows bands of colored light produced by elements when they are heated. He gave it a temporary name, "eka-aluminum," since it would be under aluminum in his table of elements. In 1875, a French chemist named Paul Emile Lecoq de Boisbaudran used a spectrometer to discover an element he decided to name "gallium" using the Latin name for France. The actual properties of gallium turned out to be very close to Mendeleev's predictions, which stunned the scientific world of that day, and made Mendeleyev famous.

Gallium has such a low melting point that it will become a liquid in your hand, just from your body heat. In the early 1900s, someone decided it would be funny to use gallium to make "disappearing spoons." At room temperature, the spoons looked like normal silverware, but when put into hot tea they quickly melted.

Before the 1960s, the main use of gallium was as a substitute for mercury in thermometers. Gallium was combined with indium and tin, forming an alloy called Galinstan.® ("Stan" is from the Latin word for tin, "stannum.")

Gallium is most useful in the world of technology when it is combined with other elements to make alloys. One of the most well-known alloys, gallium-arsenide, GaAs, is used in lasers. Gallium nitride, GaN, is used to make blue diode lasers that are used in Blu-ray disc players. GaAs and GaN are both used as semi-conductors, and can be found in computers, cell phones, car electronics, and fiber-optic devices. Aluminum-gallium-arsenide is used in high-power infrared lasers. Indium-gallium-nitride is used to make blue LEDs (light emitting diodes).

Gallium is not needed by the body, but it is also not toxic to the body. Various gallium compounds are being used as experimental medicines to treat infections and cancers. Radioactive isotopes of gallium, especially Ga-68, are used in PET scan diagnostic imaging.

Gallium's emission spectrum

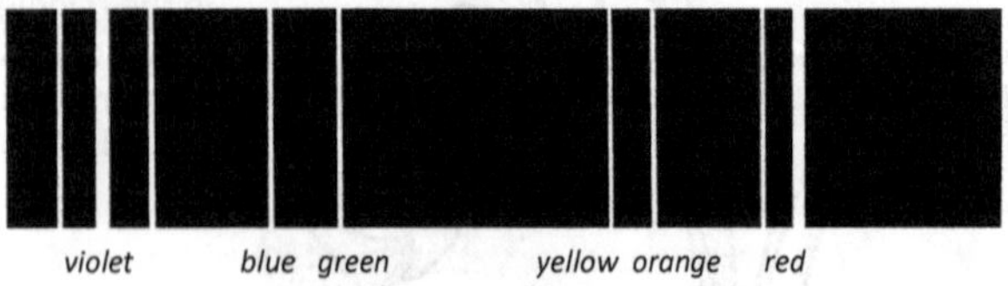

As Mendeleyev predicted, gallium was discovered with a spectrometer. Helium was the first element discovered this way.

$GaCl_3$
Gallium trichloride

Gallium trichloride is used by physicists to search for tiny particles called neutrinos.

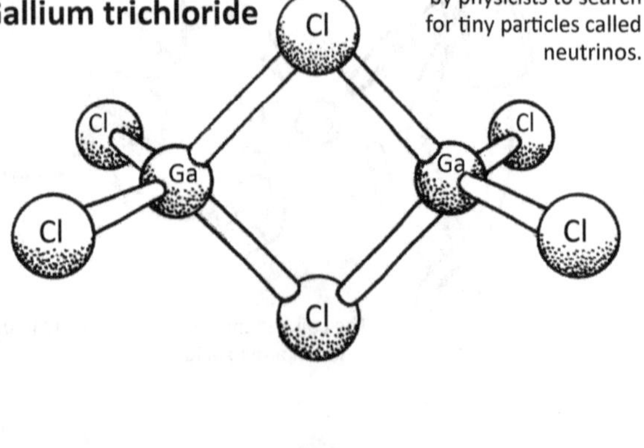

GaN Gallium nitride

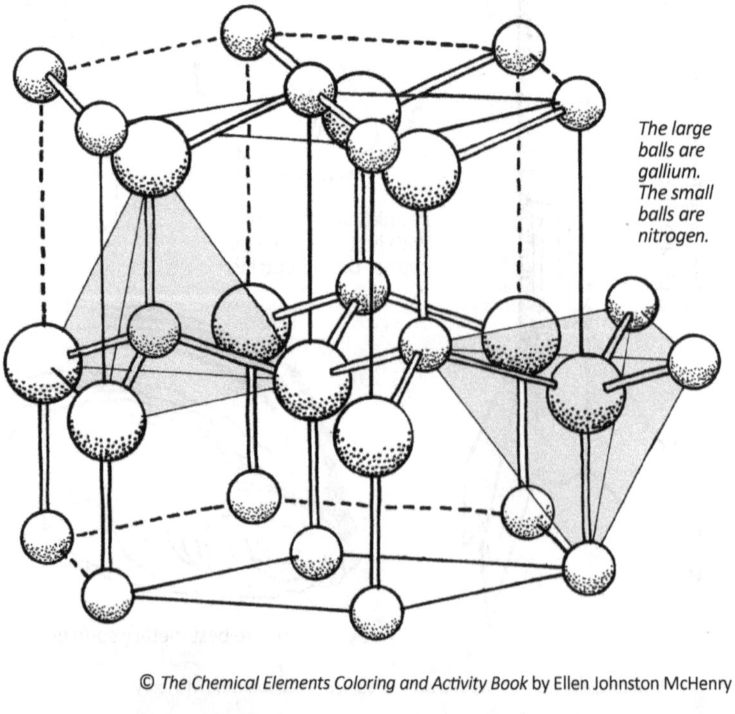

The large balls are gallium. The small balls are nitrogen.

31 Ga

Gallium is named after France.

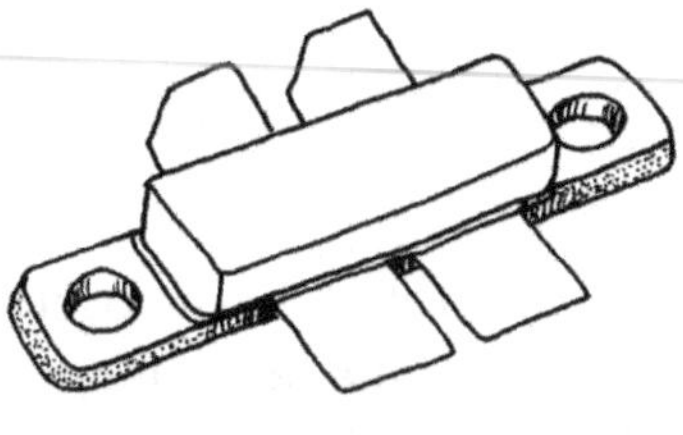

This GaN radio frequency device might not look like much, but it is at the heart of an industry that is worth millions of dollars.

Gallium

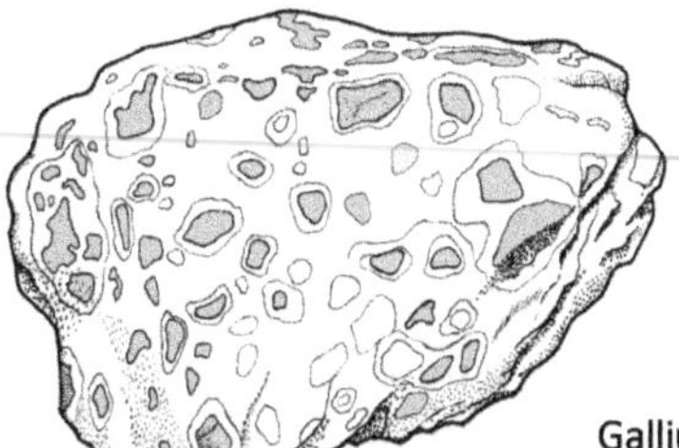

Gallium is found in ores of other minerals, like bauxite (aluminum).

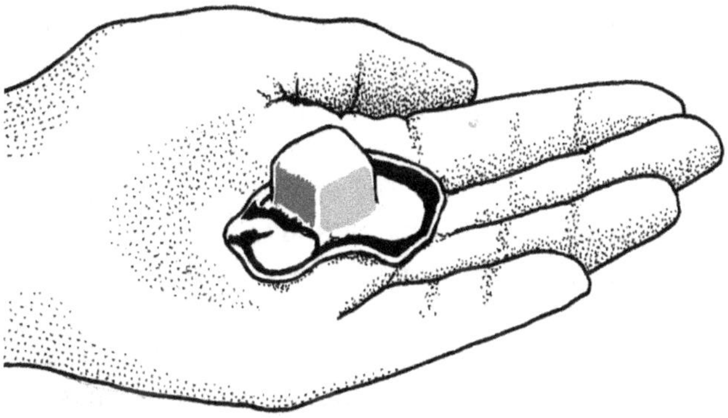

Gallium has such a low melting point that it will melt in your hand.

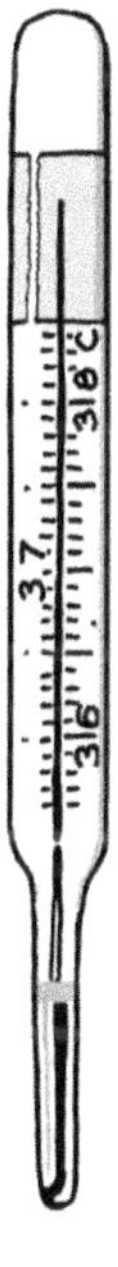

Gallium is used in thermometers, often as part of the alloy Galinstan (gallium, indium, tin).

Gallium is used to make blue LEDs.

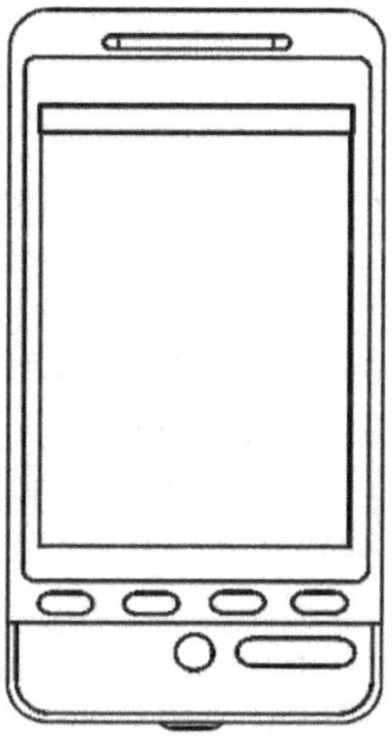

Gallium arsenide, GaAs, is in many of our devices.

Gallium alloys were used to make various parts for the Mars rovers.

The solar panels were made with thin layers of gallium arsenide on germanium.

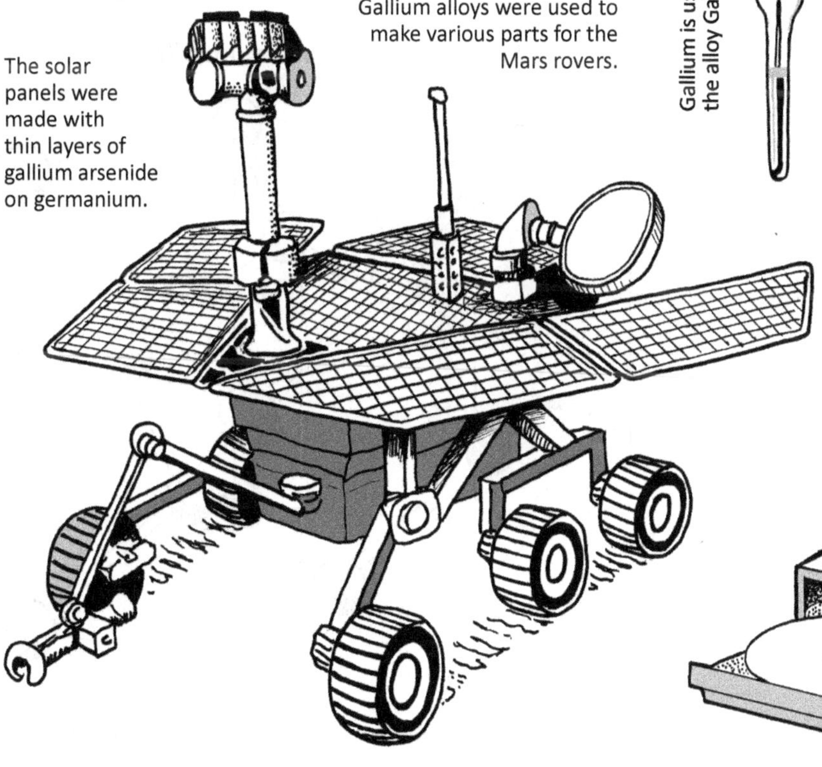

Years ago, gallium was used to make "disappearing spoons" (as a party trick).

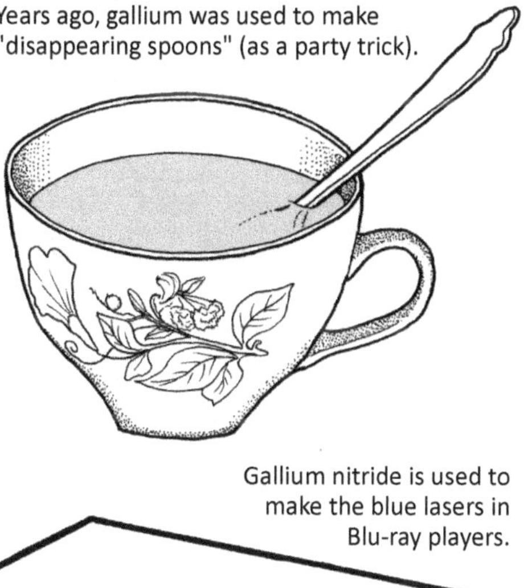

Gallium nitride is used to make the blue lasers in Blu-ray players.

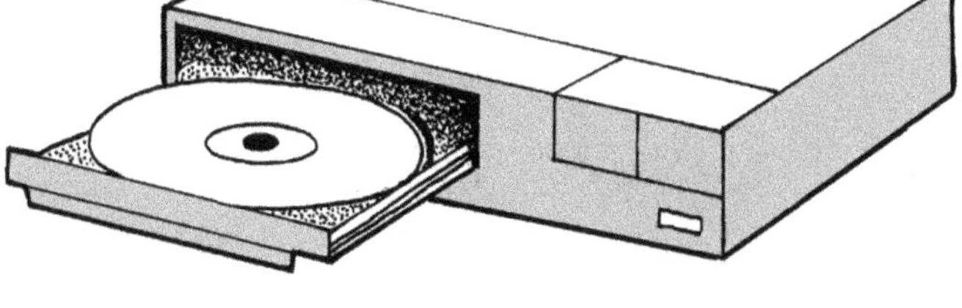

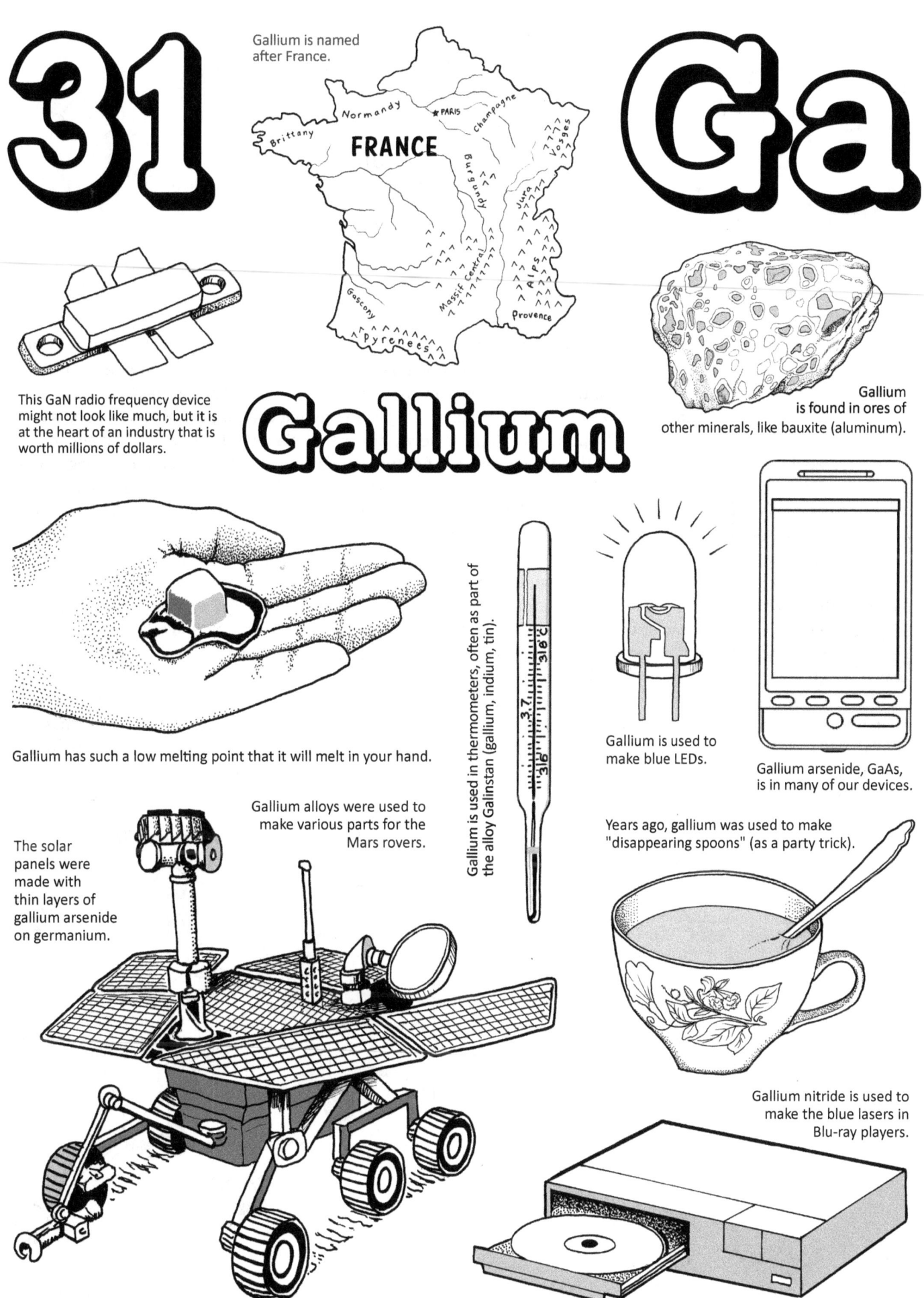

Ge 32

protons
41 neutrons
32 electrons

Atomic mass: 72.6

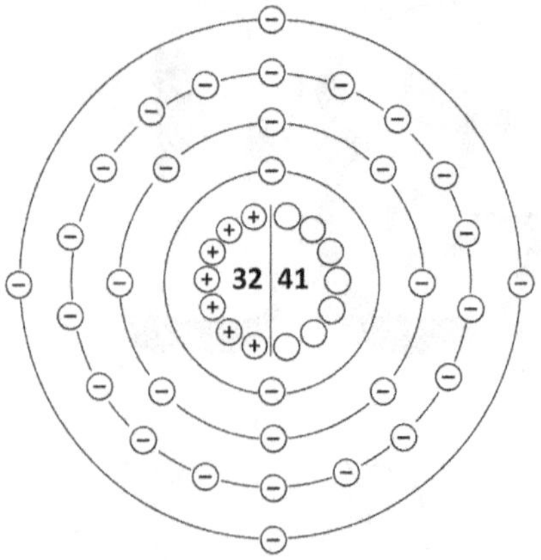

Germanium

Named after Germany

In 1869, two years before he predicted the discovery of gallium, Dmitri Mendeleyev predicted the discovery of a new element that would fit right below silicon on his Periodic Table of the Elements. He used the temporary name "eka-silicon" and gave estimates of this new element's mass, density, and chemical properties. In 1886, German chemist Clemens Winkler discovered a new element hiding in an ore that contained mostly silver and sulfur. He wanted to name this new element "neptunium" after the newly discovered planet, but this name had already been proposed for another element (not the one we now call neptunium, however), so he settled for naming it after his homeland, Germany. Today, germanium is most commonly extracted from sphalerite, the ore used for zinc.

Germanium is a lot like silicon and carbon, the elements right above it on the table. All of the elements in this column have four electrons in their outer shell. They would prefer to have eight instead of four, so they will arrange themselves into a lattice structure, allowing every atom to share its electrons with its neighbors.

In 1948, a great discovery was made about the germanium lattice structure: it was a "semi-conductor," meaning that it would only let electricity flow through it under certain circumstances. This was perfect for making a tiny electronic part called a "transistor." Transistors were then used to make the first radios and televisions.

Eventually, silicon was determined to be even more useful than germanium for this purpose, and since it was more abundant and less expensive, our modern devices use more silicon than germanium.

Germanium can be mixed with other elements to form alloys that are more useful to electronic industries than pure germanium is. When tiny amounts of other elements are added, this is called "doping." Germanium is doped with gallium, indium, arsenic, and antimony to make semi-conducting materials used in computers, phones, solar cells, and fiber optic systems. Germanium oxide is used in wide angle camera lenses.

Since germanium is transparent to infrared radiation it can be used in infrared optical devices such as remote controls and motion sensors.

GeO_2 Germanium dioxide

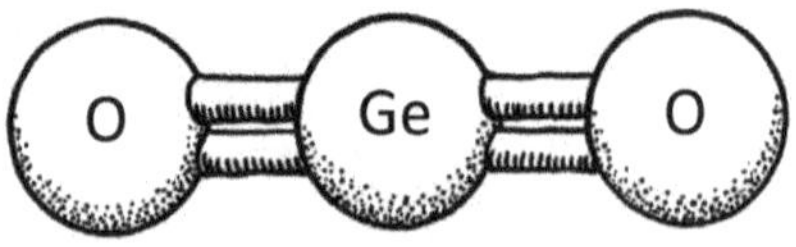

GeO_2 molecules will attach to each other to make a white powder.

GeH_4
Germane
(similar to methane)

GeH_4 burns in air to form GeO_2 and water.

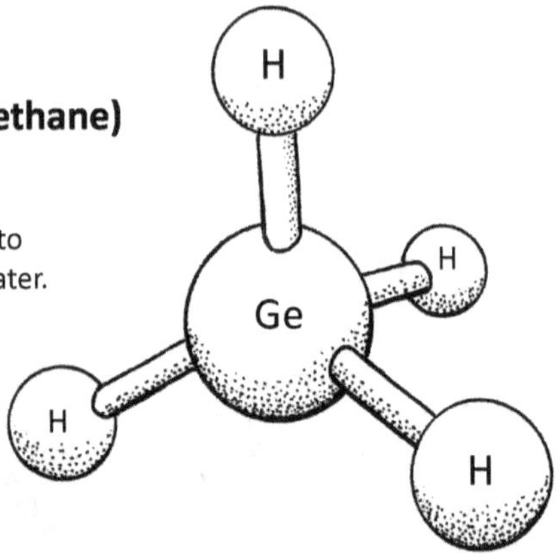

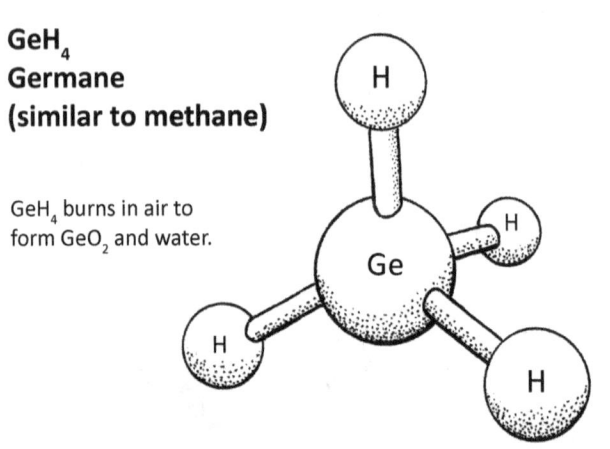

SiGe Silicon-germanium alloy

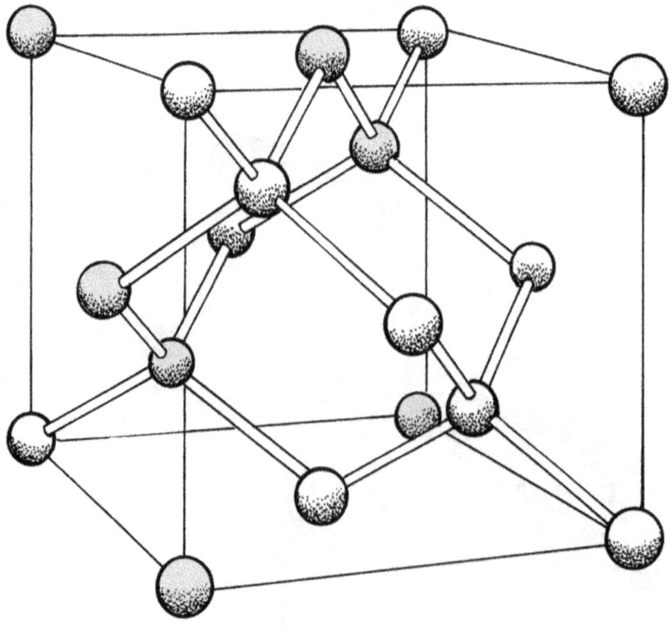

The darker balls represent germanium. The white balls represent silicon.

32 Ge

Germanium is named after Germany.

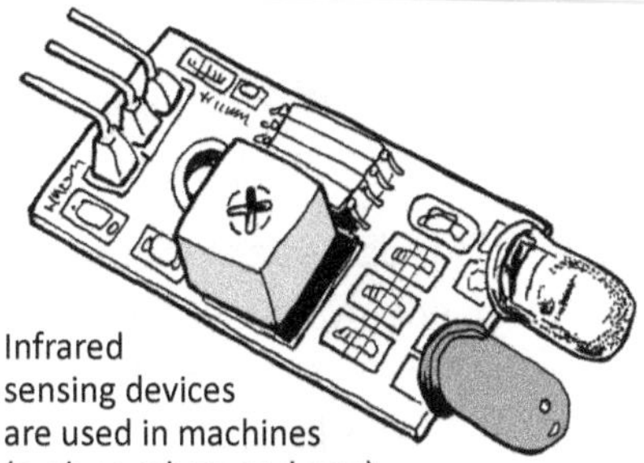

Infrared sensing devices are used in machines (such as robots and cars) to give them "vision."

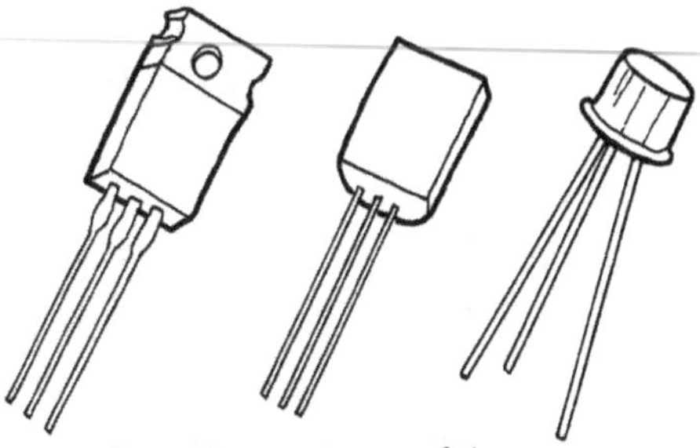

Transistors were one of the great inventions of the late 1900s.

Germanium

All of our tech devices probably have some germanium in them.

Wide angle lenses can be made of germanium compounds.

A "crystal radio" is easy to make.

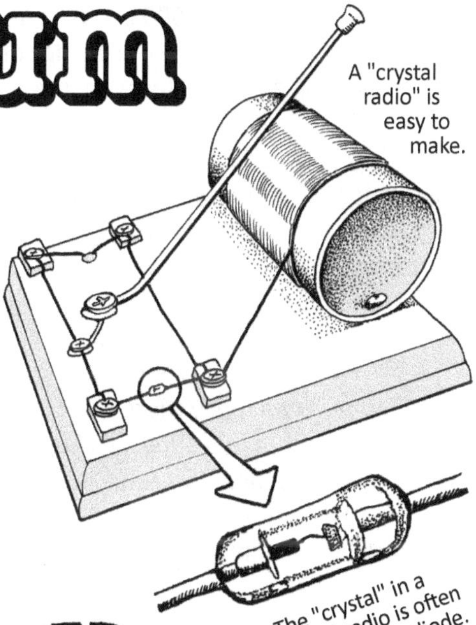

The "crystal" in a crystal radio is often a germanium diode.

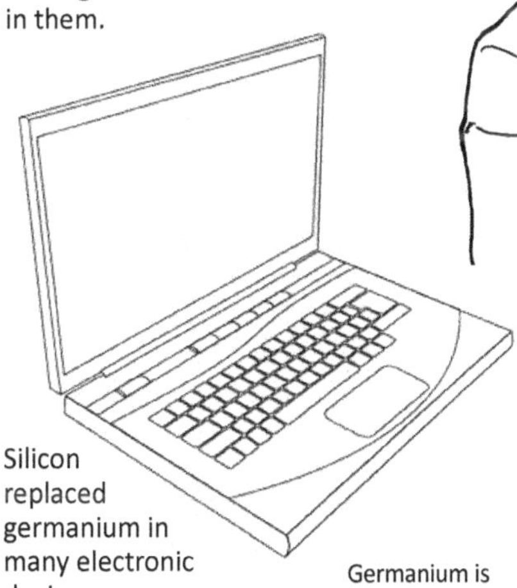

Silicon replaced germanium in many electronic devices.

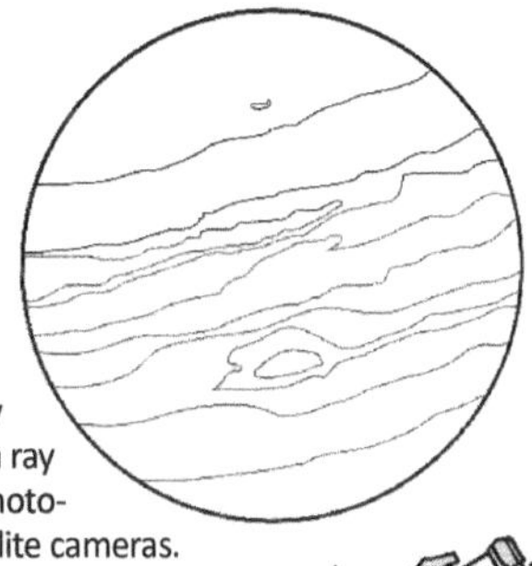

Germanium is one of the necessary elements for gamma ray spectrometry and photography used by satellite cameras.

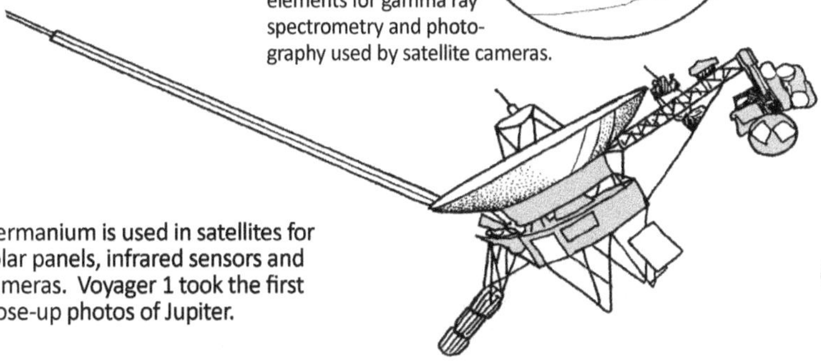

Germanium is used in satellites for solar panels, infrared sensors and cameras. Voyager 1 took the first close-up photos of Jupiter.

Germanium and gallium are used to make solar panels.

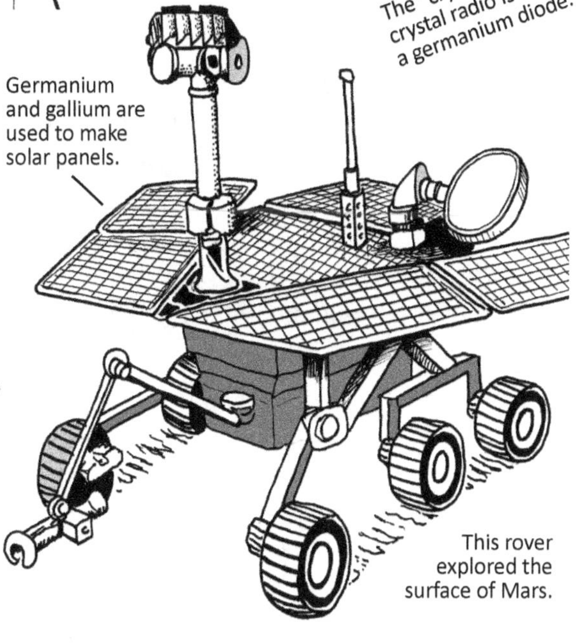

This rover explored the surface of Mars.

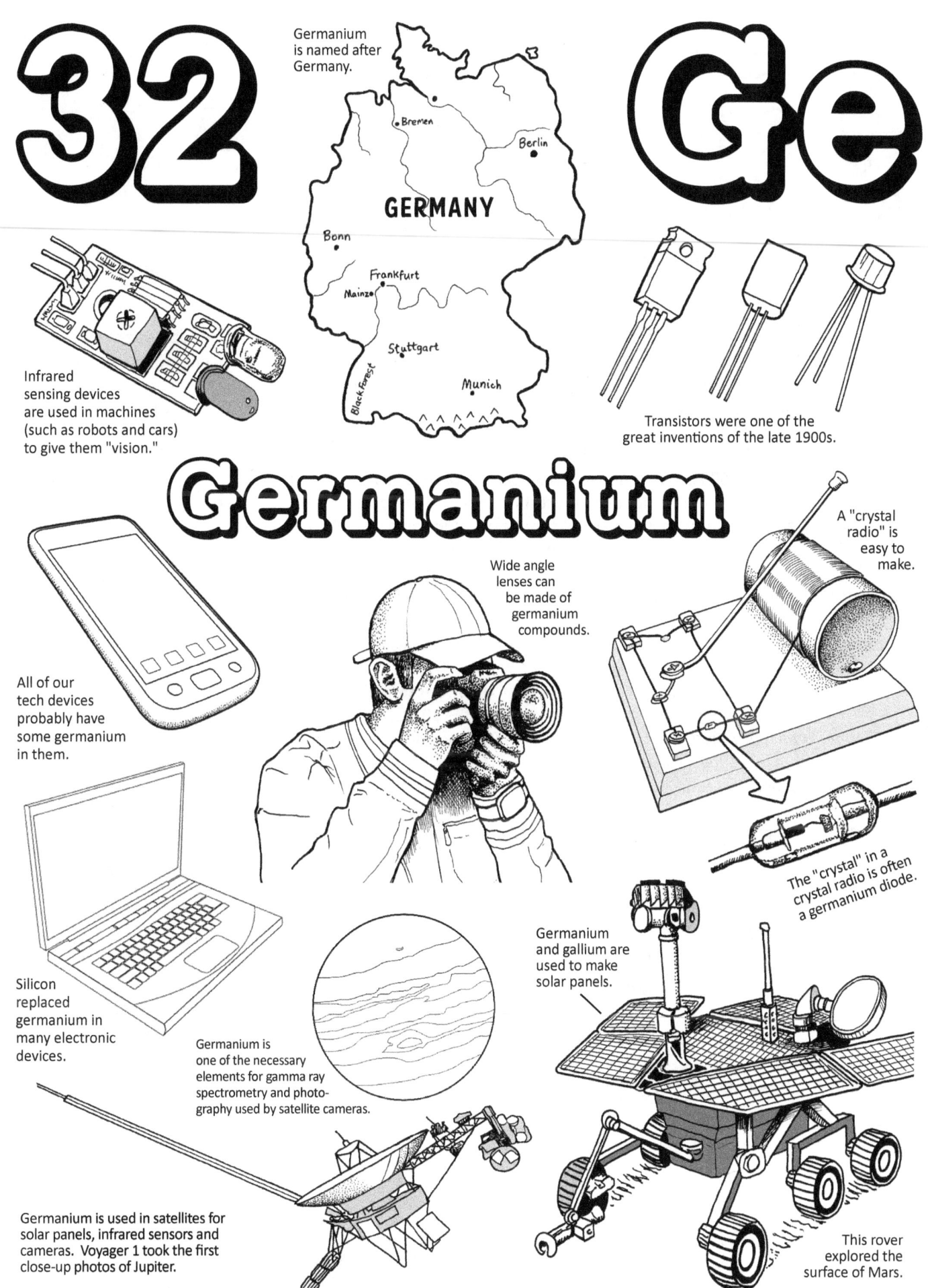

As 33

protons
42 neutrons
33 electrons
Atomic mass: 74.9

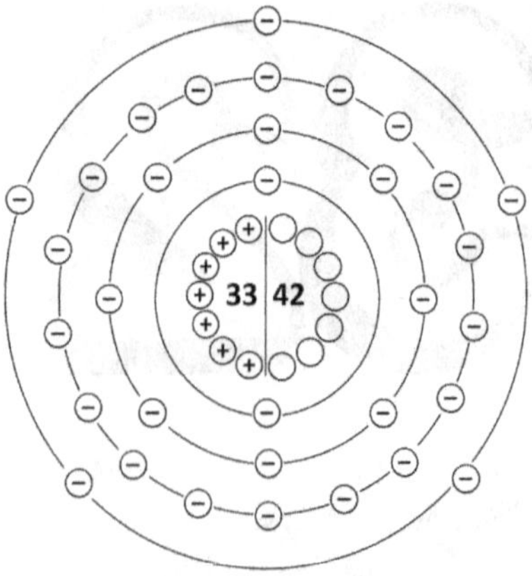

Arsenic

From an ancient Syrian word meaning "gold-colored"

Arsenic is poisonous, but it can also be very useful. Knowledge of arsenic compounds dates back thousands of years when people discovered it could be used as a bloodless way to get rid of a political enemy, but also as an additive to bronze to make it harder. An arsenic compound called "orpiment" could be used to make yellow or gold paint, and a white arsenic compound could be used by potters to make a white glaze for their clay pots.

Much later, in the 1800s, English women would mix white arsenic with vinegar and chalk and make a paste that they could either eat or rub on their skin to make them look pale, as being pale was in fashion at that time. Also during the 1800s, a copper-arsenic compound was used by chemist Carl Scheele to make a bright green pigment. This pigment was used not only by artists, but also to make wallpaper with green designs. Sadly, many people who lived in houses with green wallpaper got sick, and some (usually children) died.

Arsenic has long been used as rat poison. In the early 1900s chemists discovered how to use arsenic compounds to kill insects living in farm crops. Arsenic-based insecticides were completely phased out in the US in 2013, except for some (much-less-toxic) arsenic compounds used in cotton farming.

In the 1930s, chromated copper arsenic became popular as a wood preservative, protecting exterior wood surfaces from mold, algae, bacteria, and insect damage. These preservatives were banned in the US and Europe in 2004, but are still in use in other places around the world. Organic arsenic compounds (containing carbon atoms) are much less toxic than inorganic ones, and are fed to baby chickens and turkeys to make them grow faster.

From the 1700s until the discovery of penicillin in the 1930s, arsenic compounds were used by doctors as antibiotics. Even today there are still a few medicines (for uncommon diseases) that use small amounts of arsenic.

Since the late 1900s, gallium arsenide, GaAs, has been used in laser diodes, integrated circuit boards and solar panels. Gallium arsenide is particularly useful in devices that will go into space.

The largest use of arsenic today is as an alloy with lead. Car batteries contain lead that has small amounts of arsenic mixed into it. Lead-arsenic alloys can also be found in some bullets.

$CuHAsO_3$ Scheele's Green

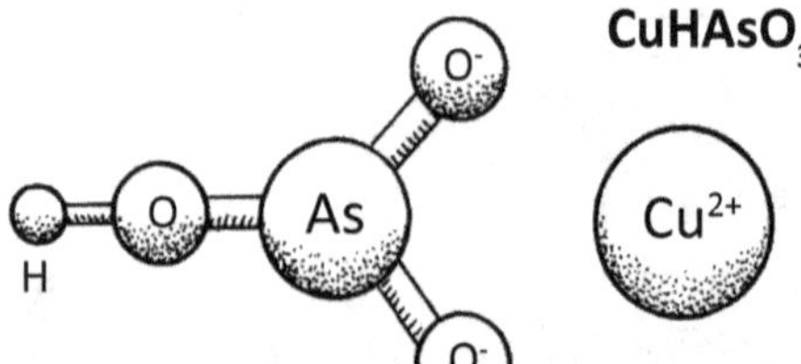

This is the famous toxic green pigment used on wallpaper in the 1800s.

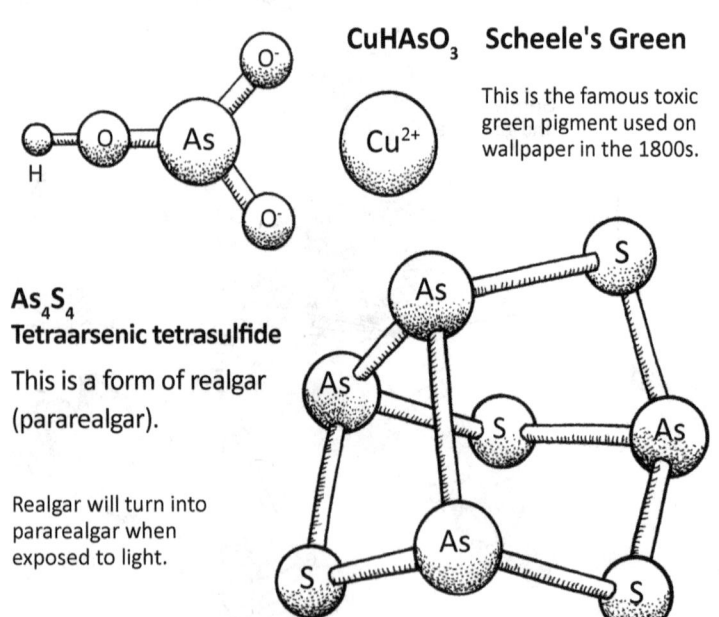

As_4S_4
Tetraarsenic tetrasulfide

This is a form of realgar (pararealgar).

Realgar will turn into pararealgar when exposed to light.

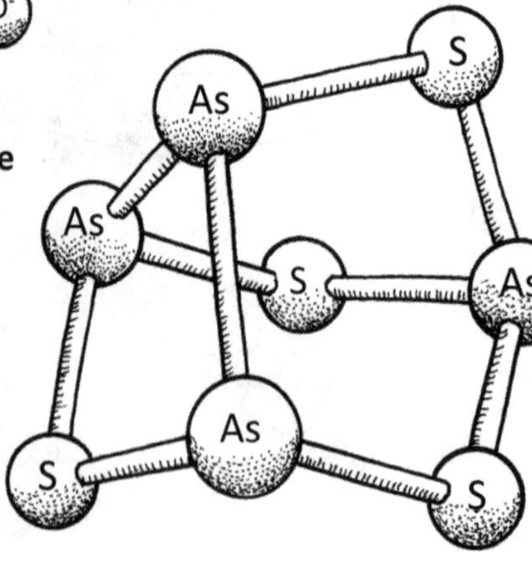

GaAs Gallium arsenide

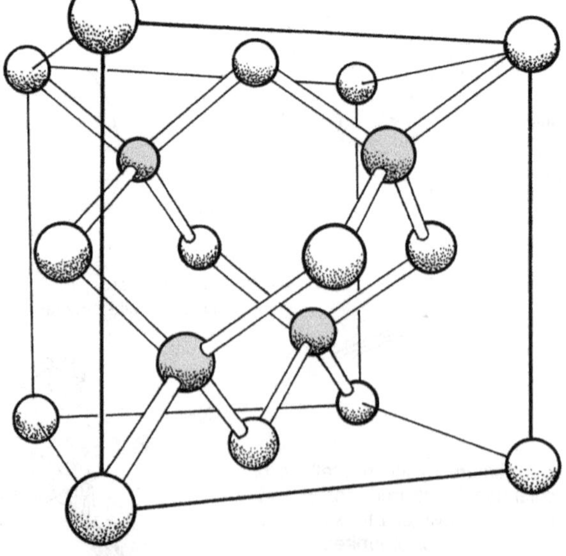

The darker balls represent arsenic. The white balls represent gallium.

33 As

Most people know arsenic as a rat poison.

Arsenic

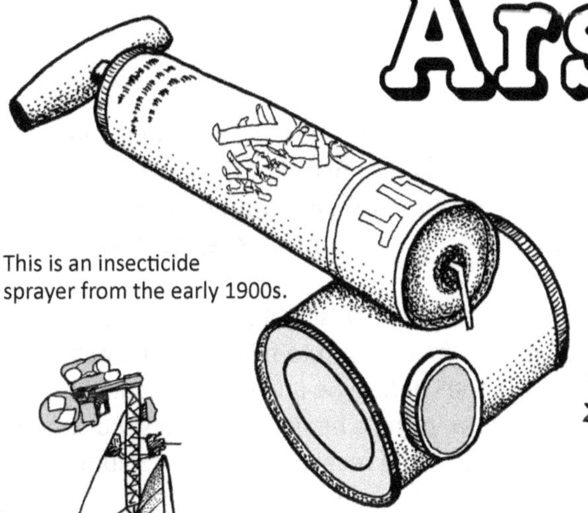

This is an insecticide sprayer from the early 1900s.

Arsenic has been used not only as a pesticide (to kill animals) but as an insecticide (to kill insects) in farm fields.

Realgar, As_4S_4, is often called "ruby arsenic" because the crystals are red.

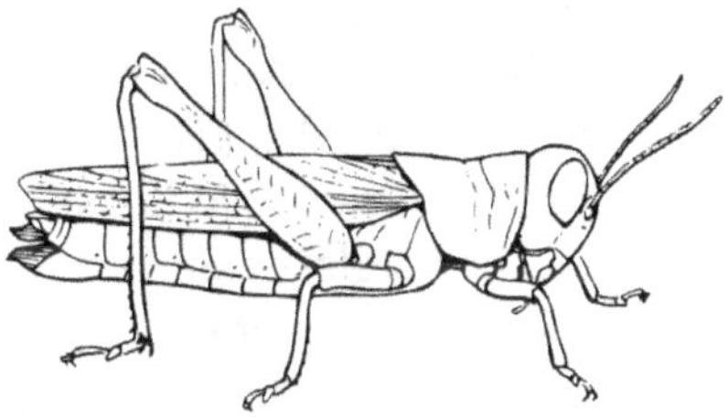

Orpiment, As_2S_3, can be yellow or gold.

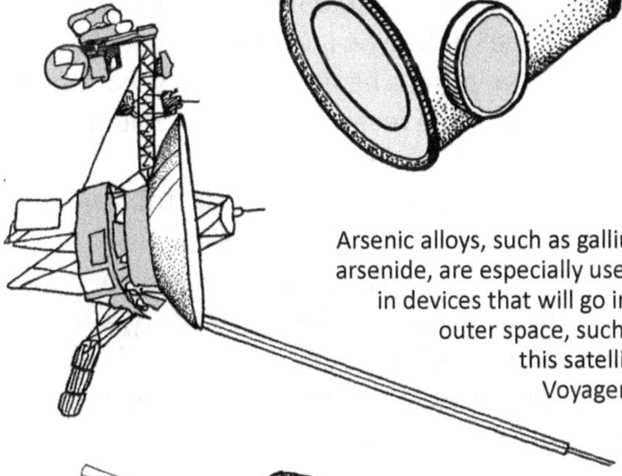

Arsenic alloys, such as gallium arsenide, are especially useful in devices that will go into outer space, such as this satellite, Voyager 1.

Arsenic wallpaper design. (Color everything green.)

Arsenopyrite, FeAsS, is shiny and gray.

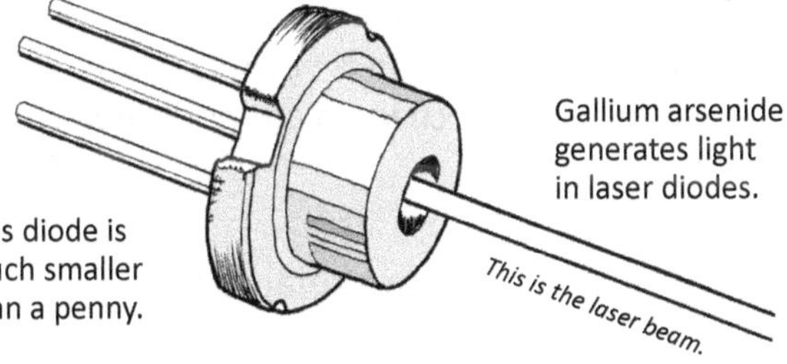

Gallium arsenide generates light in laser diodes.

This diode is much smaller than a penny.

Gallium arsenide is used in some solar panels. The panels on the roof of this house are blue.

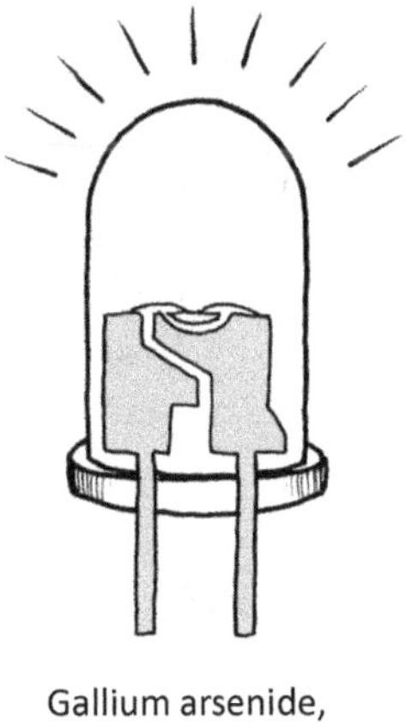

Gallium arsenide, GaAs, is used in LEDs.

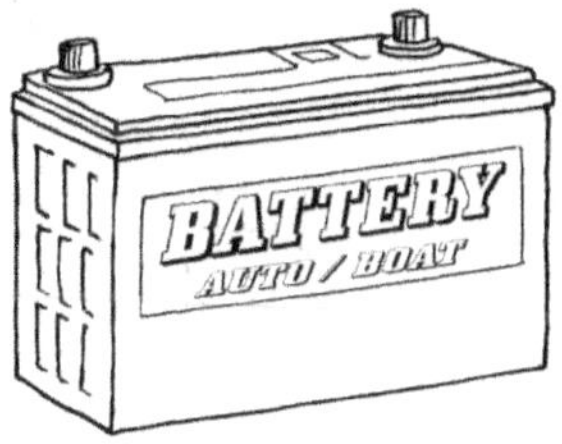

Lead-acid batteries contain arsenic.

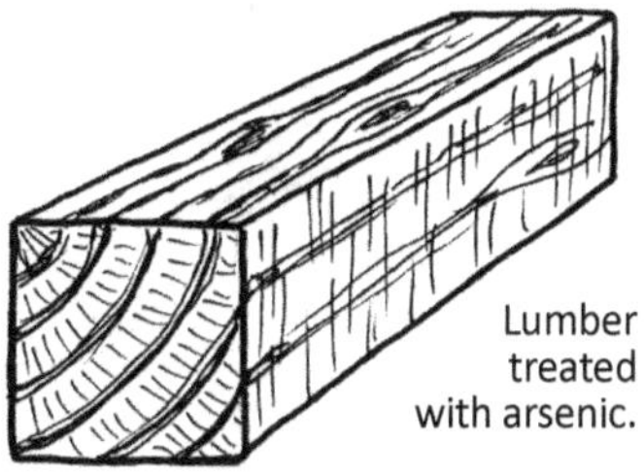

Lumber treated with arsenic.

Se 34

protons
45 neutrons
34 electrons

Atomic mass: 78.9

Selenium

Named using the ancient Greek word for moon, "Selene"

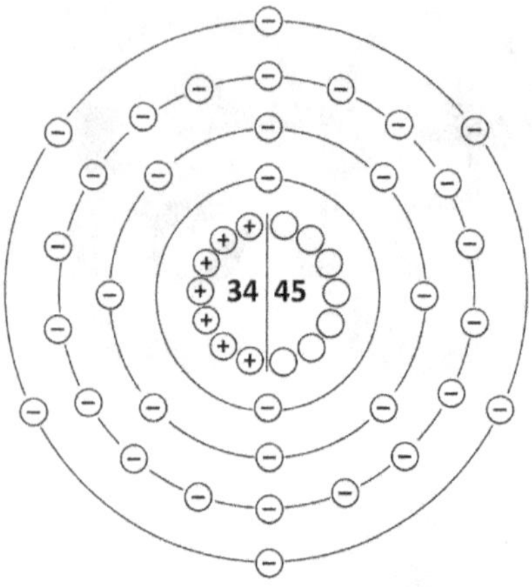

Selenium is named after the moon because it is located right above tellurium in the Periodic Table, and "tellus" is Latin for "earth." The Swedish chemists who discovered selenium thought at first that the element they were working with was tellurium because it had many of tellurium's characteristics. When it turned out to be a new element— a much smaller atom that would sit right on top of tellurium on the table—it just seemed appropriate to give the "earth" (tellurium) a "moon" shining down on it.

Selenium has a unique characteristic, setting it apart from any other element: it is light sensitive and will conduct electricity in direct proportion to how much light is shining on it. The first person to figure out a practical use for this property was Alexander Graham Bell, who invented a "photophone" before he invented the telephone. His photophone (in 1880) was the world's first wireless communication system, using a beam of light to send sound messages. Thin mirrors at either end could flex when hit by sound waves, and at the receiving end a selenium detector turned the sound waves into an electrical signal that went into an ear piece.

In the 1900s, selenium made possible the invention of other light-sensitive devices such as electric "eyes" (light sensors), light meters for cameras, solar cells, and (non-digital) photocopiers. Selenium can also be used as a semi-conductor and was a popular material in the electronics industry until silicon was discovered to be a better choice. Selenium was used to make tiny LEDs before the discovery of gallium arsenide. In the 21st century, a new use for selenium has been discovered, in flat-panel x-ray machines that produce digital images.

The largest use of selenium is in glass manufacturing, as selenium produces a bright red color when mixed with the silicon dioxide in glass. The steel industry also uses selenium in a few of its alloys. Selenium changes the texture of steel, making it easier to cut and shape into screws, bolts, and other hardware parts.

Selenium is needed in small amounts in animals and plants. In mammals, selenium is necessary to make an enzyme called glutathione, which can neutralize dangerous "free radicals" (electrically charged atoms or molecules that can damage or destroy good molecules).

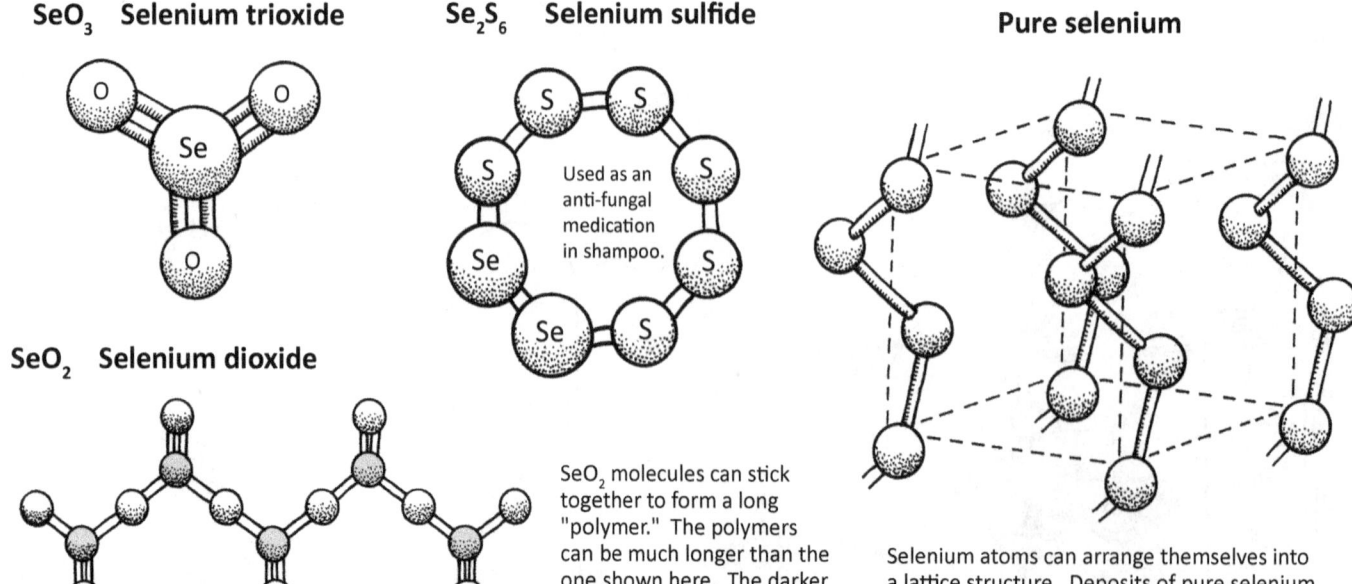

SeO_3 Selenium trioxide

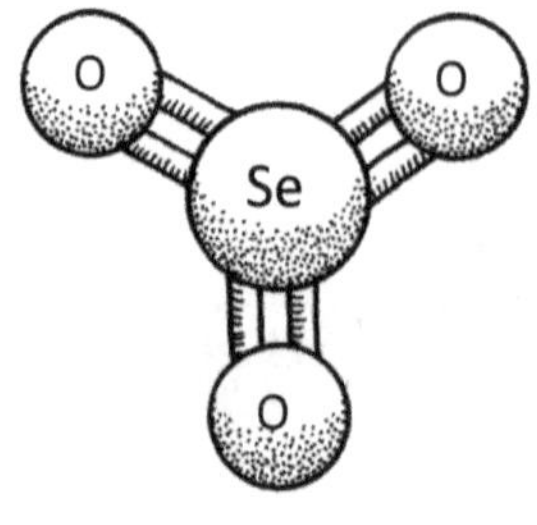

Se_2S_6 Selenium sulfide

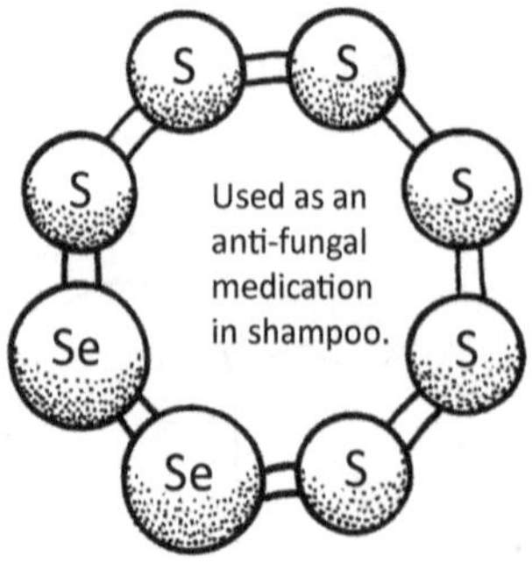

Used as an anti-fungal medication in shampoo.

Pure selenium

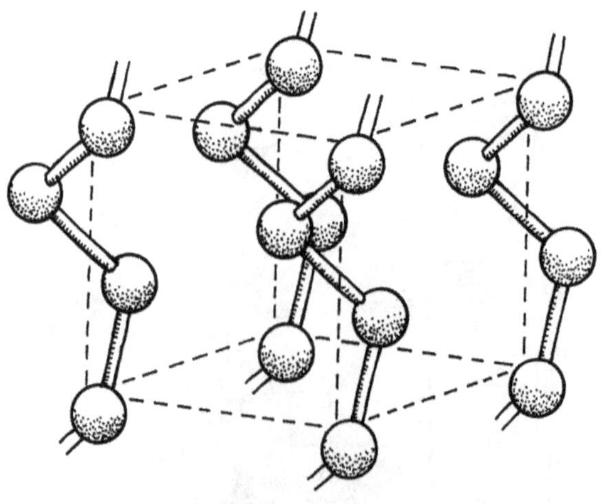

Selenium atoms can arrange themselves into a lattice structure. Deposits of pure selenium do exist, but are very rare.

SeO_2 Selenium dioxide

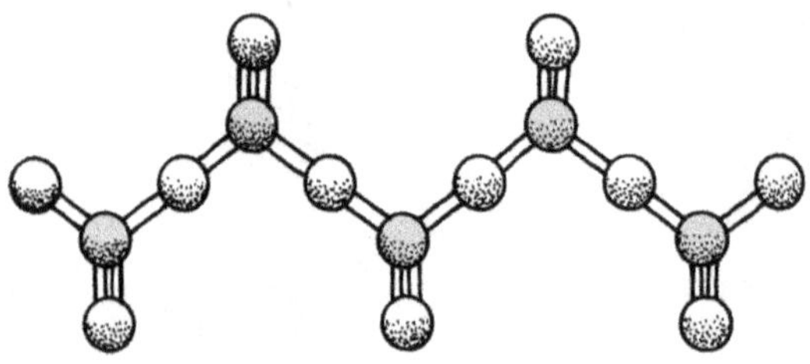

SeO_2 molecules can stick together to form a long "polymer." The polymers can be much longer than the one shown here. The darker circles represent Se atoms.

34 Se

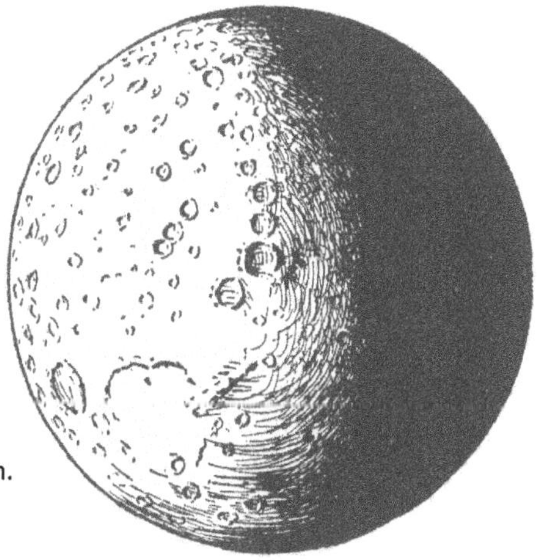

Selenium is named after the moon.

Selenium

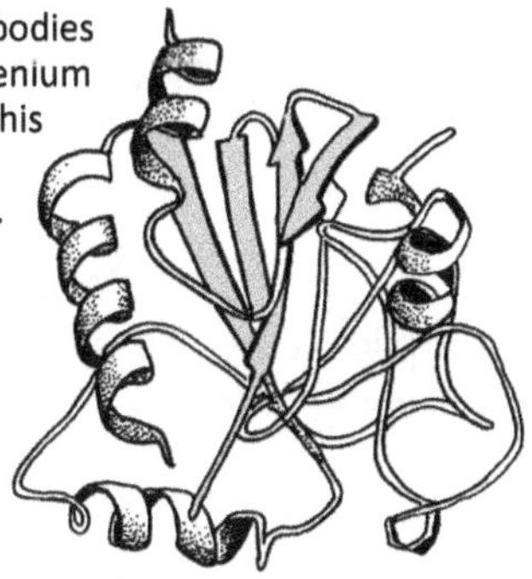

Our bodies need selenium to make this enzyme, glutathione.

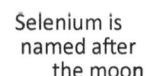

Selenium disulfide is the active ingredient in dandruff shampoo.

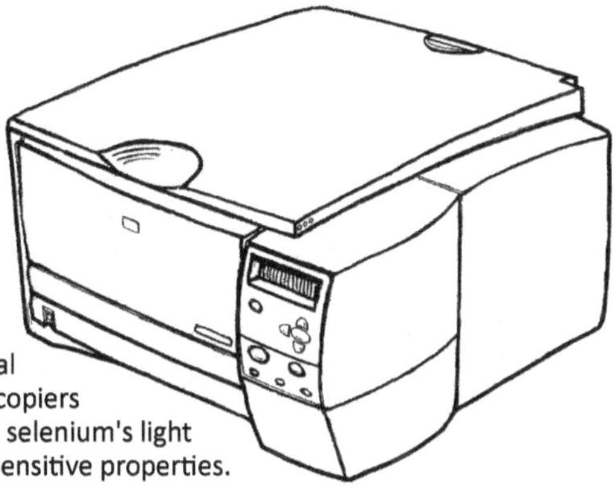

Traditional photocopiers used selenium's light sensitive properties.

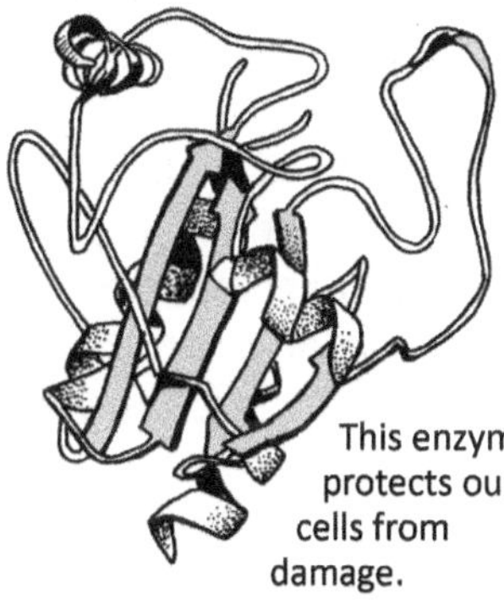

This enzyme protects our cells from damage.

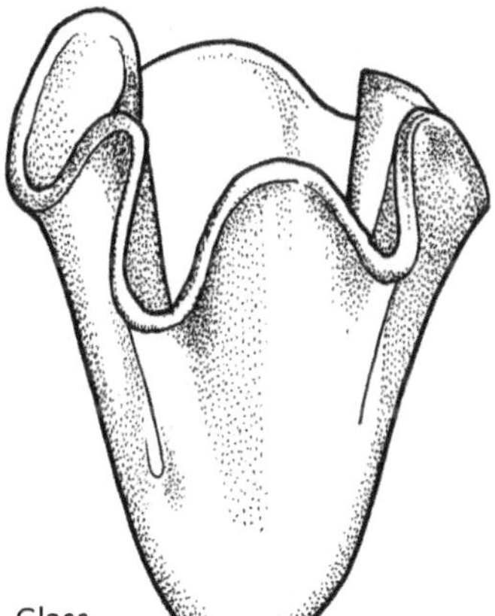

Glass can be colored bright red with selenium.

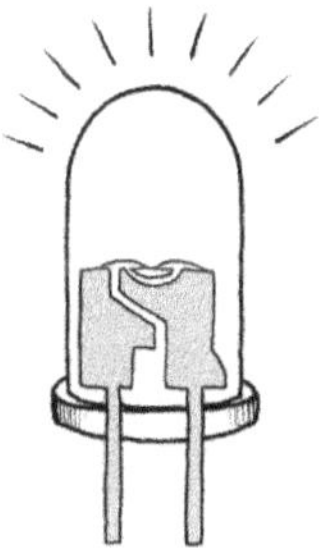

Zinc-selenide was used for blue LEDs before gallium nitride.

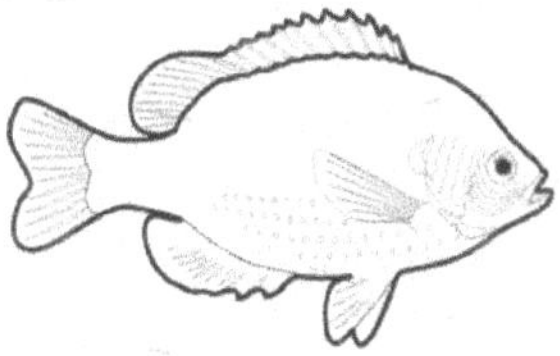

Fish is one of the best dietary sources of selenium. Other sources are eggs, nuts and beans.

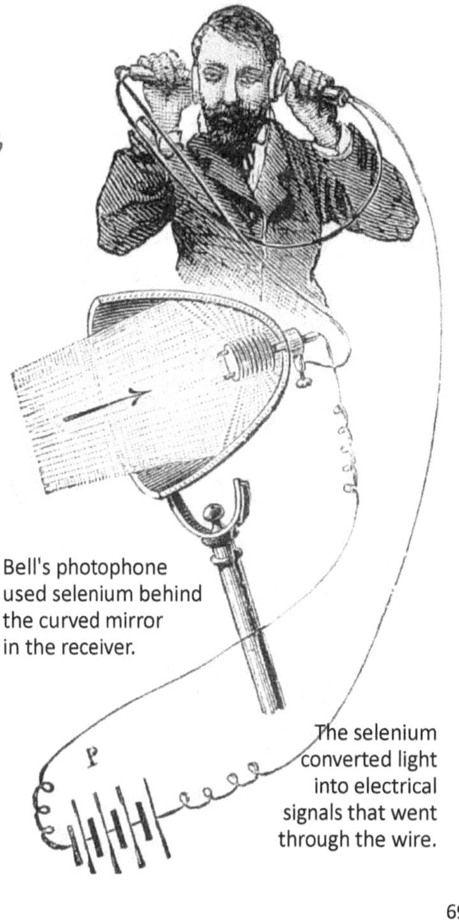

Bell's photophone used selenium behind the curved mirror in the receiver.

The selenium converted light into electrical signals that went through the wire.

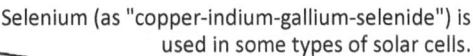

Selenium (as "copper-indium-gallium-selenide") is used in some types of solar cells.

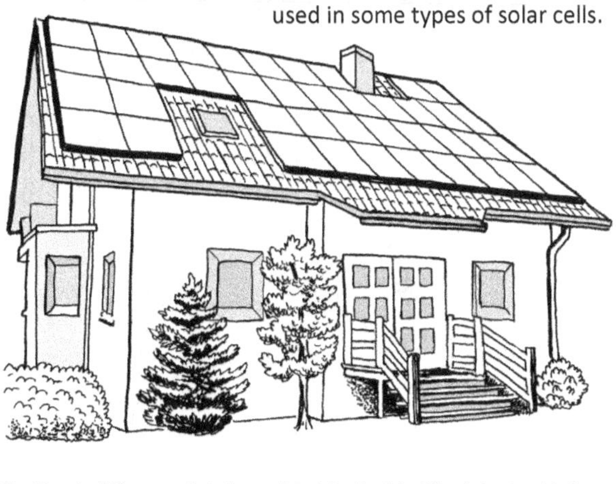

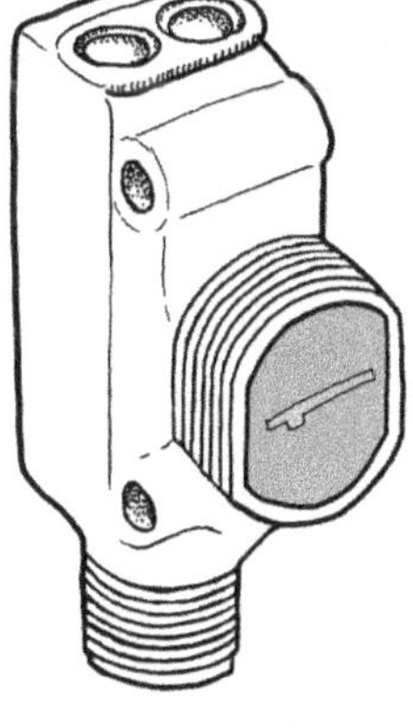

Selenium is used in photo sensors.

Br 35
protons
45 neutrons
35 electrons
Atomic mass: 79.9

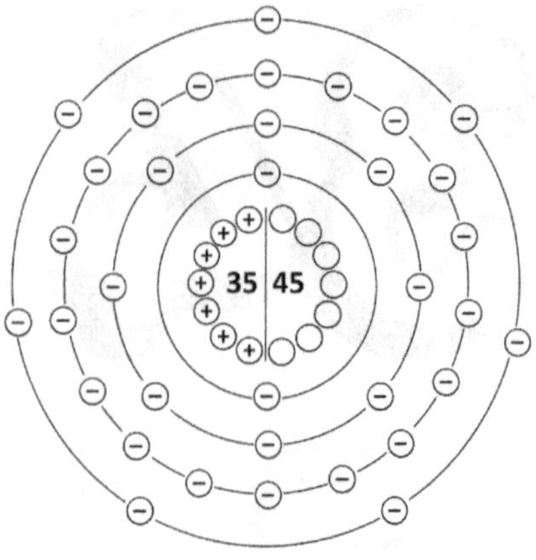

Bromine

The Greek word "bromos" means "stinky smell"

Bromine's claim to fame is that it is the only non-metallic element that is a liquid at room temperature. This red liquid element was first isolated by a German teenager, Carl Löwig, the summer before he started college, in 1825. He never got credit for his discovery, however, because the following year, a French scientist, Antoine Balard, produced the same red liquid from seaweed ash and was the first to publish his results. The name "bromine" was given to this new element, based on a Greek word meaning "stench." (The "fishy" smell along seashores is partly from bromine.)

Bromine is most often found in salty places that contain a lot of sodium and chlorine. The ocean contains a lot of bromine, as do salty lakes like the Dead Sea. Bromine is often a by-product of salt production.

Until the 1970s, most of the bromine produced was used by the gasoline (petrol) industry as an additive to prevent the build-up of lead inside the engine. The bromine atoms would attract lead atoms and form lead bromide, which would then be expelled with the exhaust fumes. Unleaded gasoline made this additive obsolete.

Other industrial uses for bromine have included water treatment, soil fumigation (to kill insect pests), zinc-bromide batteries, and as a flame retardant in fabrics and plastics. (Concerns about what bromine compounds might be doing to the ozone layer have put restrictions on industrial uses, especially fumigation of soil.)

In 1888, Bromo-Seltzer® was introduced as an over-the-counter medicine to treat indigestion and headache. It remained on the market until the 1970s but was eventually replaced by brands such as Alka-Seltzer®, which did not contain bromine. Bromine used to be found in many other medicines, such as sedatives and antiseptics, but these have now been replaced with formulas that do not contain bromine. However, some bromine compounds are still used to make medicines that treat brain diseases such as epilepsy and dementia.

A marine animal called the murex is able to extract bromine from the ocean water and use it to make a dark purple ink. Since ancient times, people have extracted this ink and used it to make purple dye for cloth. The city of Tyre (in modern-day Lebanon) was famous for its production of this dye. "Tyrian purple" was so expensive that only very wealthy people from royal families could afford it. Thus, the color purple became associated with royalty.

The earliest types of photographs used metal plates exposed to bromine vapor to make them light sensitive.

NaBr Sodium bromide

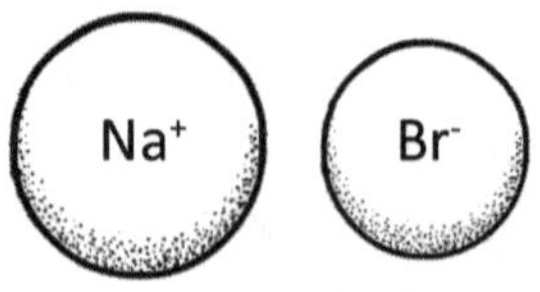

KBr Potassium bromide

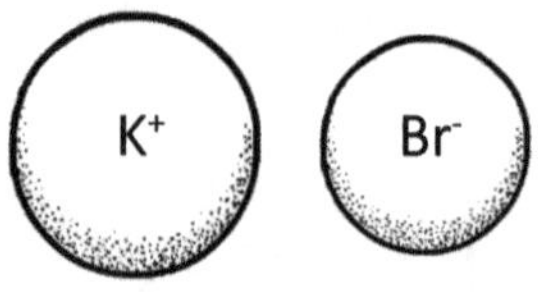

AgBr Silver bromide

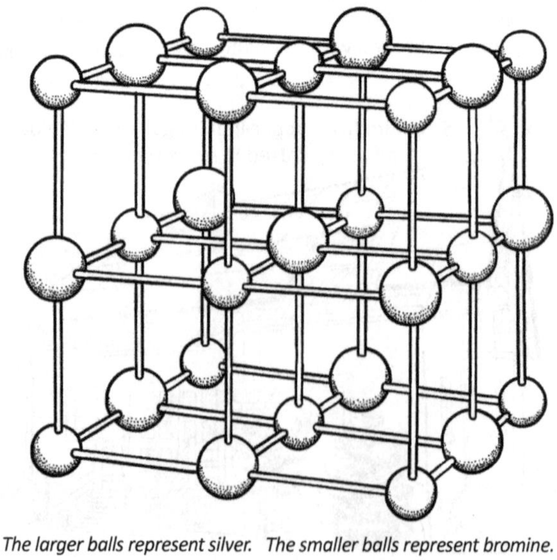

The larger balls represent silver. The smaller balls represent bromine.

$BrCH_3$ Bromomethane

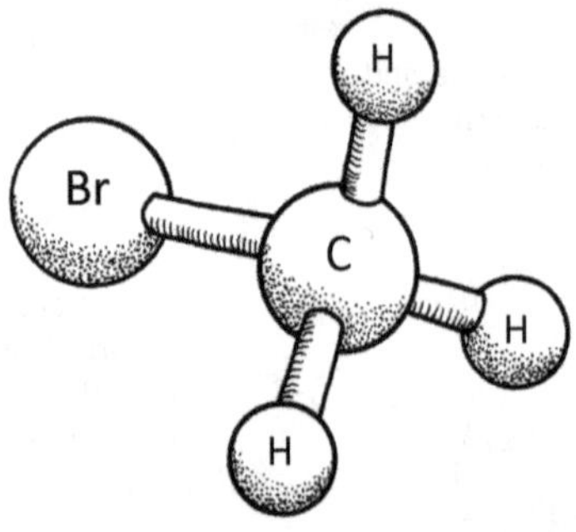

AgBr Silver bromide

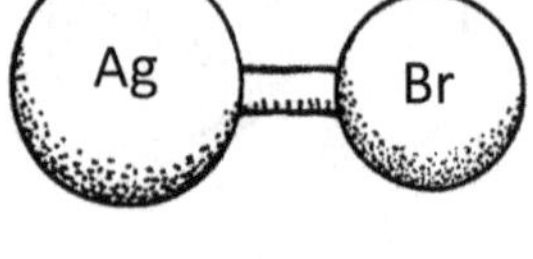

Silver bromide as a single molecule is shown above. These molecules can stick together to form the lattice shown on the right.

35 Br

This sea creature is called a murex.

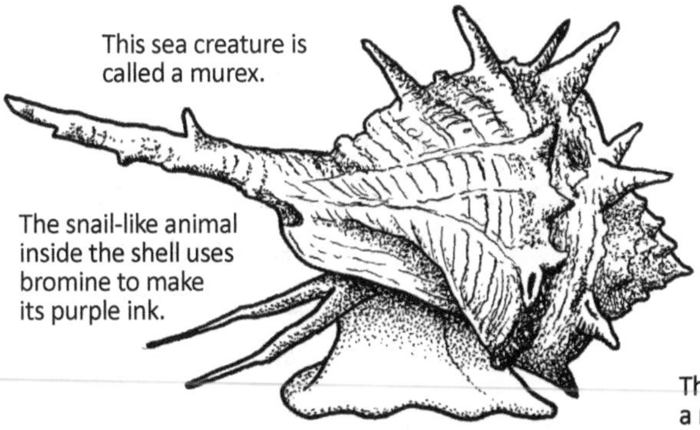

The snail-like animal inside the shell uses bromine to make its purple ink.

Bromine

In the 1970s, this was the largest industrial use of bromine.

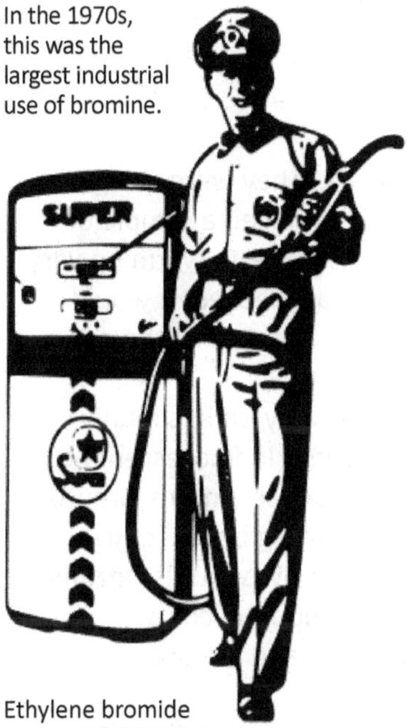

Ethylene bromide was an additive put into gasolines until the 1970s. The bromine atoms attracted lead atoms, forming lead bromide which was then expelled in the exhaust.

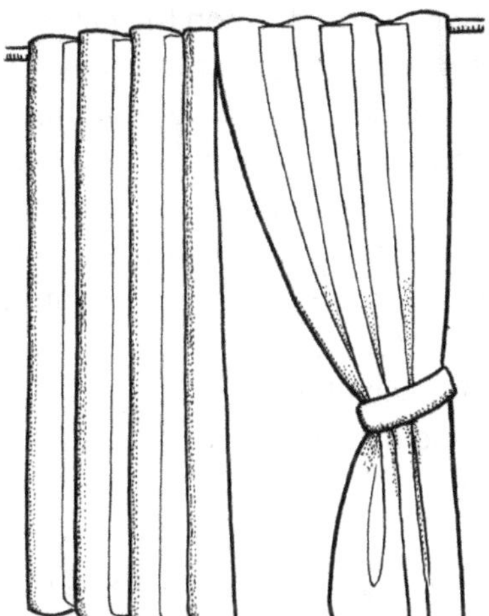

Fabrics and plastics can have bromine added to make them fire resistant.

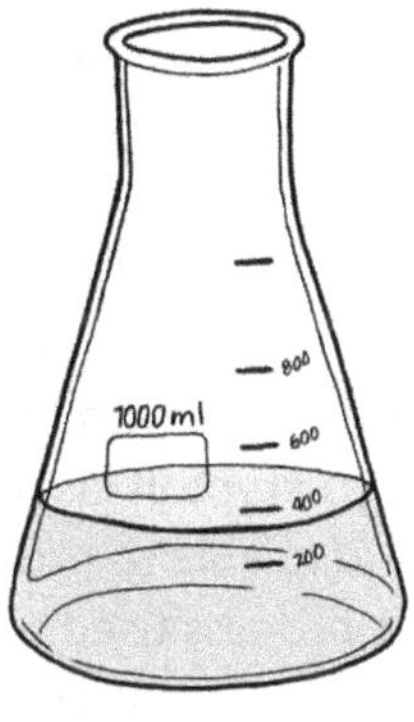

Pure bromine is a red liquid.

This king is wearing a robe dyed purple using murex ink.

Purple dye was very expensive.

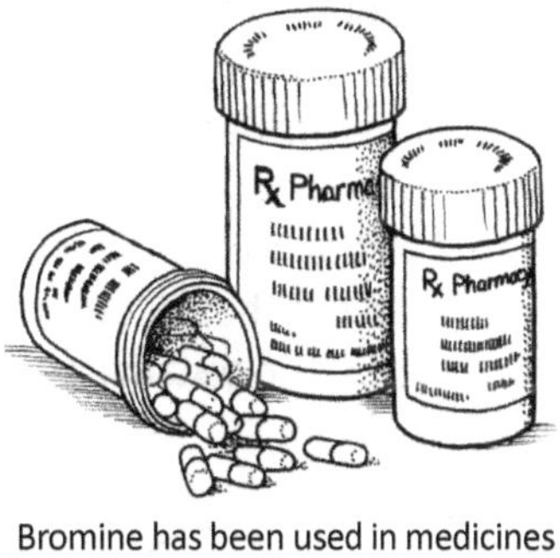

Bromine has been used in medicines that affect the nervous system, as it has a calming effect.

A blue Bromo-Seltzer bottle from the 1960s

A glass roof helps to light the room.

Bromine was an essential part of early photography. In this portrait studio, the cameras are the boxes sitting up high on the shelf. The man on the right is polishing a metal plate that will be exposed to bromine vapor to make it light sensitive. The plate will be put into a camera box and used as "film."

The fumigation tanks are red and green.

Until 1991, bromomethane was used to kill insects in soil, through a process called fumigation.

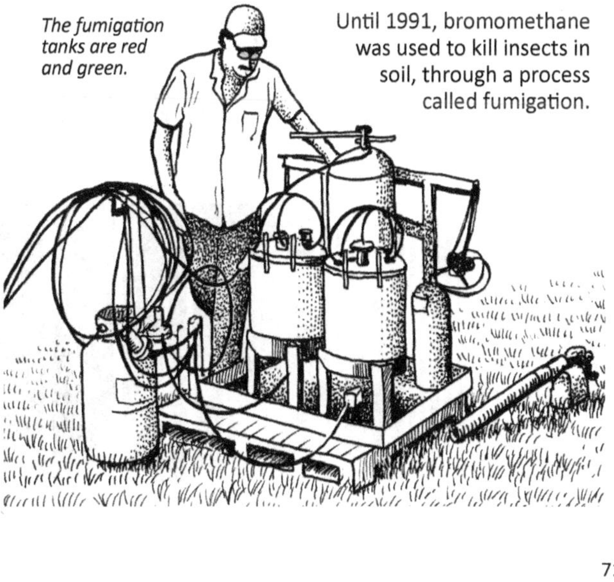

Kr 36

protons
48 neutrons
36 electrons

Atomic mass: 83.8

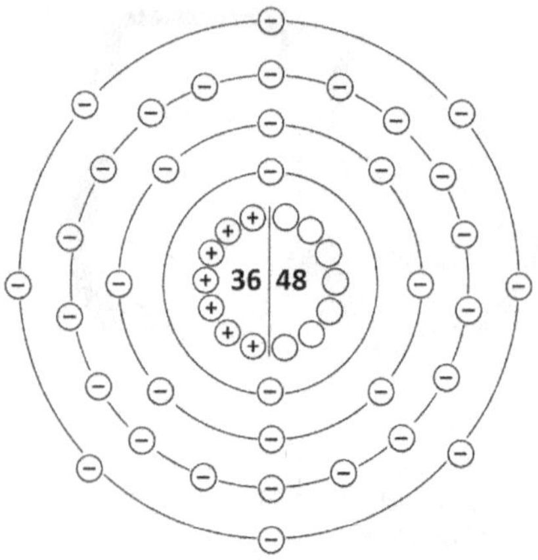

Krypton

The Greek word "kryptos" means "hidden"

Krypton was discovered on May 30, 1898, by Sir William Ramsay and Morris Travers. They were using a process called distillation to find out what type of gases are found in the air around us. They chilled a sample of air to a very low temperature so that all the gases turned to liquid. Then they allowed the sample to warm up very slowly, knowing that the gases all had different boiling points. When a gas's boiling point was reached, it would revert from liquid back into gas again, and the scientists could collect a pure sample of the gas. Most of the noble gases were discovered in this way. Neon was discovered just a few weeks after krypton.

The main use of krypton is in light bulbs and lasers. It is too expensive to be used widely, so it is found mainly in specialty bulbs, such as those used for outdoor lighting and high-speed photography. In fluorescent lights and gas discharge tubes it will be mixed with less expensive argon. In lasers, krypton can produce a more intense beam of light than other gases, so krypton lasers are used for laser light shows where very strong beams of light are needed. Krypton lasers are also able to produce ultra-violet light, which is invisible to our eyes, but is extremely useful for photolithography, the process used to print microscopic electrical circuits onto computer "chips."

Some isotopes of krypton (atoms with more or less than 48 neutrons) are radioactive and are used as tracers in medical imaging, especially in MRIs that look at the air passages in lungs.

Krypton is used as a propellant in the SpaceX Starlink satellite. Releasing a burst of gas in one direction causes the satellite to move in the opposite direction.

From 1960 to 1983, krypton was used to define the official length of a meter. The International Bureau of Weights and Measures defined a meter as "1,650,763.73 wavelengths of light emitted by the krypton-86 isotope." Before this, there had been an actual one-meter metal bar that all others were compared to. In 1983, the bureau changed the definition to "the distance that light travels in vacuum during 1/299,792,458 of a second."

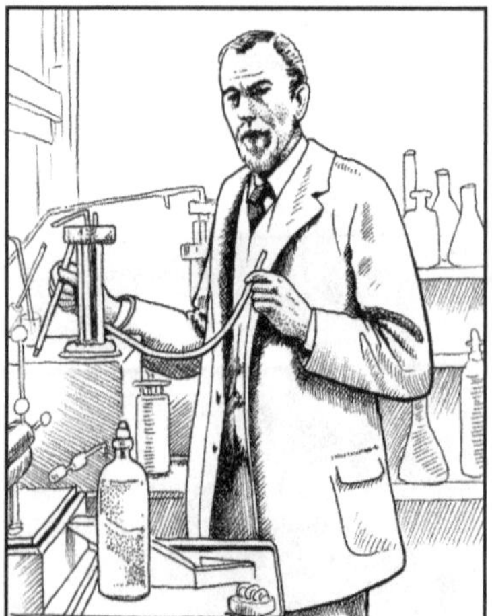

Sir William Ramsay in his lab.

KrF_2 Krypton difluoride

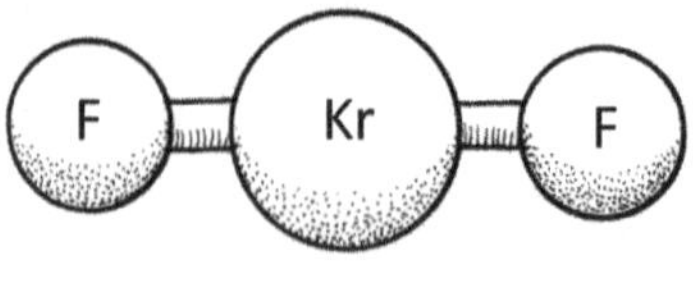

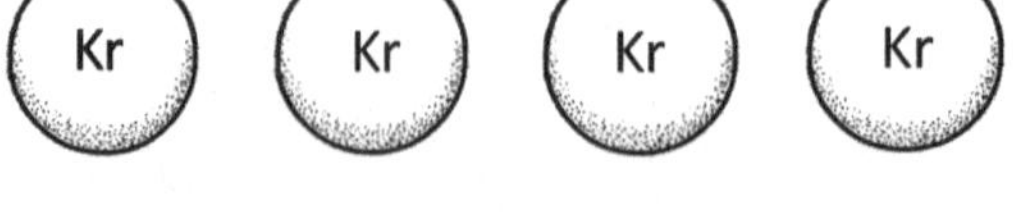

Normally, krypton does not form any molecules because it is a noble gas and does not react with other atoms. However, under extreme conditions, scientists can force krypton to make a molecule with two fluorine atoms.

If you look at burning krypton gas through a spectrometer, this is the light pattern you will see. It is called the emission spectrum.

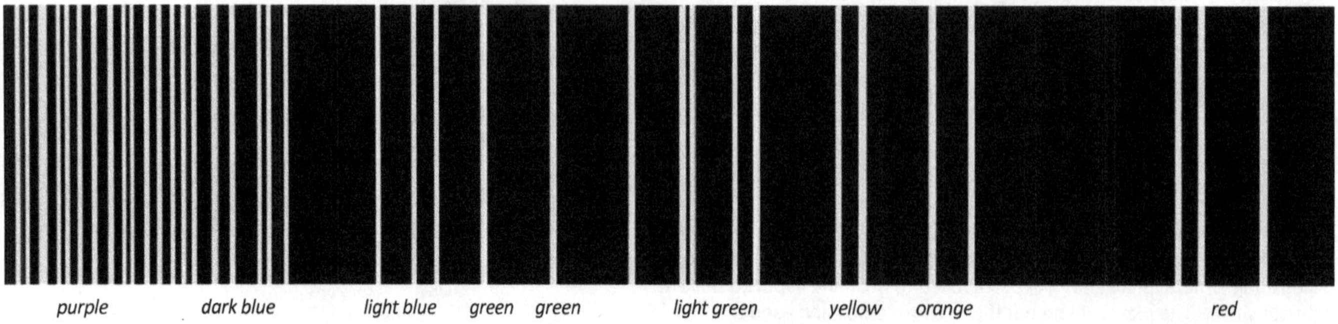

36 Kr

Krypton is used in flash bulbs for high speed photography.

Krypton

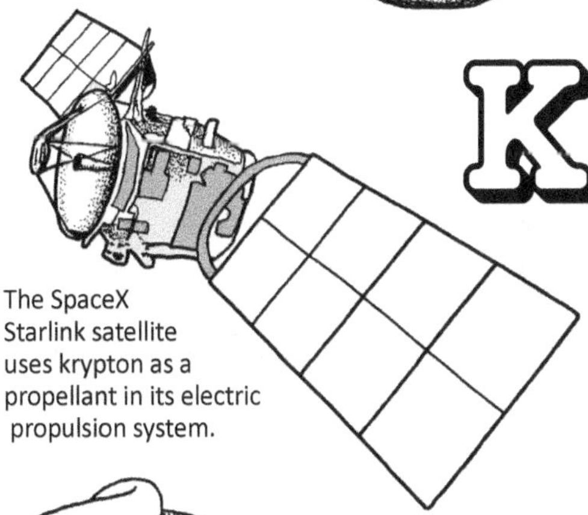

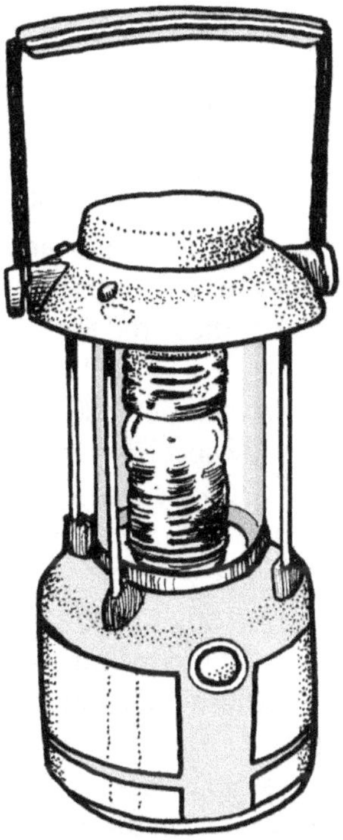

The SpaceX Starlink satellite uses krypton as a propellant in its electric propulsion system.

Krypton is mixed with argon to fill gas discharge lighting tubes.

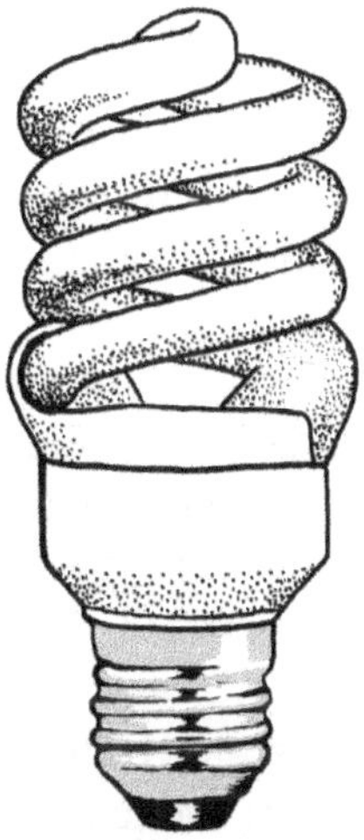

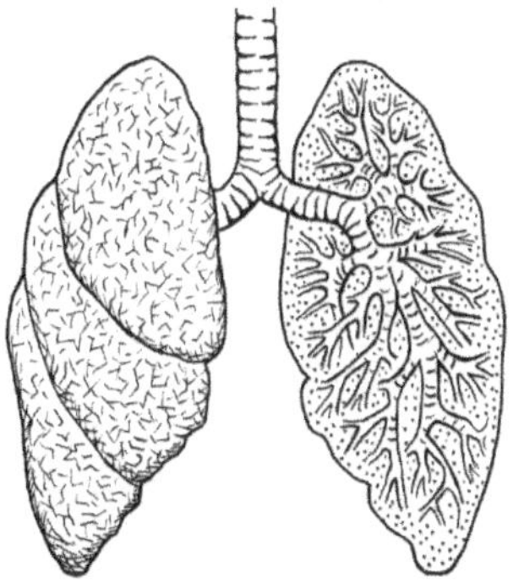

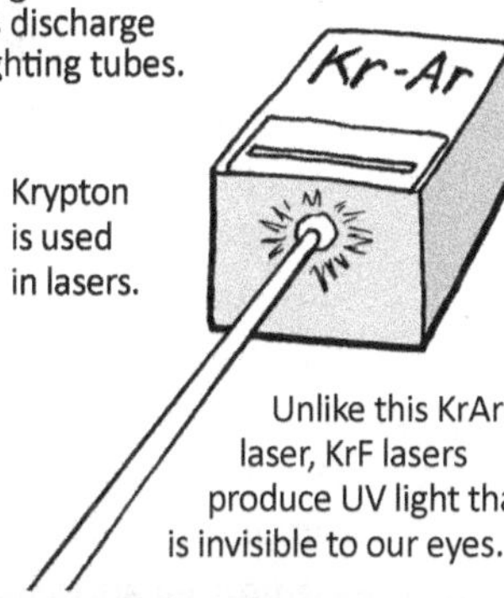

Krypton is used in lasers.

Krypton is used in bulbs that have to be very bright, like this emergency lantern. You might also find krypton (mixed with argon) in a fluorescent bulb.

A radioactive form of krypton is used with MRI to image airways.

Unlike this KrAr laser, KrF lasers produce UV light that is invisible to our eyes.

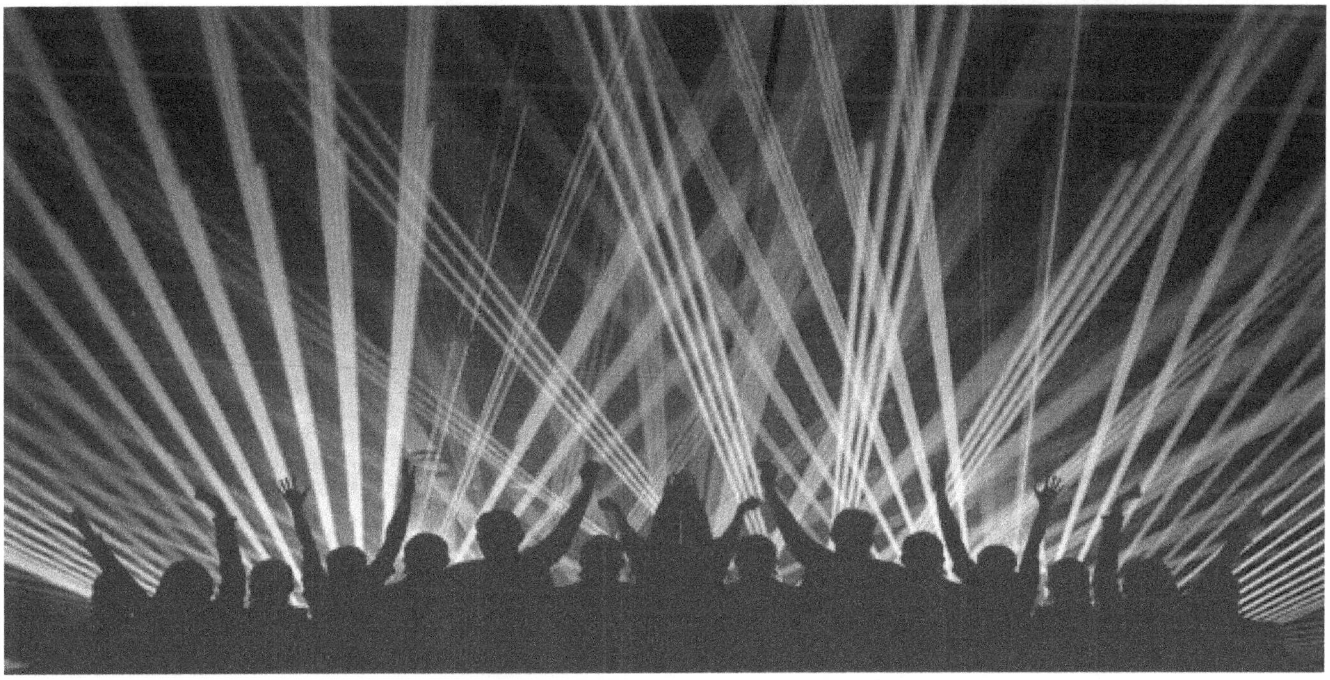

Krypton lasers provide a brighter light than lasers that use helium, neon, or argon, so they are used for laser shows. Add color to these laser lights.

Rb 37

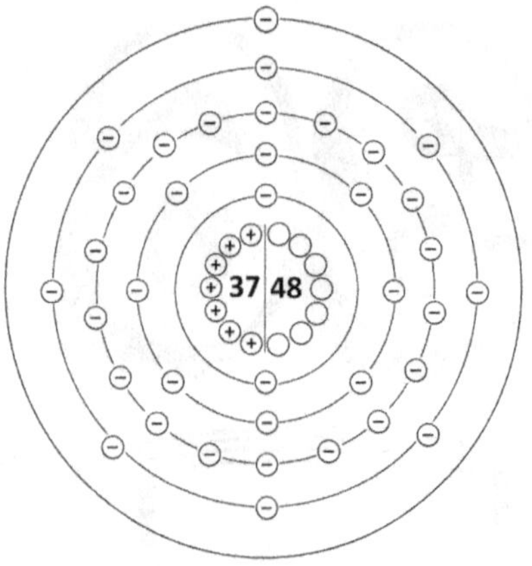

protons
48 neutrons
37 electrons

Atomic mass: 85.4

Rubidium

From the Latin word for deep red, "rubidus"

Rubidium was discovered 1861 by German chemists Robert Bunsen and Gustav Kirchhoff. They used their new invention, the spectrometer, to look at a sample of the mineral lepidolite. When they heated the mineral and looked at it through the spectrometer, they saw some bright red lines that were not associated with any element known at that time. They guessed correctly that they had discovered a new element. Because of those red lines they saw, they used the Latin word for "deep red" to name the new element.

Rubidium is chemically similar to potassium because they both have just one electron in their outer shell. Rubidium can therefore replace potassium atoms in many chemical compounds. The human body absorbs small amounts of rubidium and uses or stores the atoms as if they were potassium. Because of this, a radioactive isotope of rubidium, Rb-82, can be used in PET scans and are particularly useful for imaging the heart and brain. (PET stands for Positron Emission Tomography.) Normal cells retain the Rb-82 longer than damaged or sick cells.

Samples of pure rubidium are very reactive and will explode and burn in water, producing a bright reddish-purple flame. The ability of pure rubidium to grab other atoms is put to good use in vacuum tubes where the rubidium can take capture unwanted atoms of gas that might be inside the tube. Rubidium compounds (where rubidium is not by itself but is attached to other elements) can be used in fireworks to make purple sparks.

The structure of rubidium's electron shells makes it useful as a source of time-keeping in atomic clocks. Cesium is also used in atomic clocks, but rubidium is the better choice for smaller, less expensive clocks. In the early 1900s, it was discovered that rubidium could be used to make instruments called magnetometers that can detect magnetic forces. Recently, rubidium atoms (about 100 billion of them) have been put into very small magnetometers that can be used to measure the magnetic fields (and thus, activity) generated by the brain.

Rubidium is added to glass to make lenses for night vision goggles and to make specialty glass for the fiber-optic telecommunication industry.

RbCl Rubidium chloride

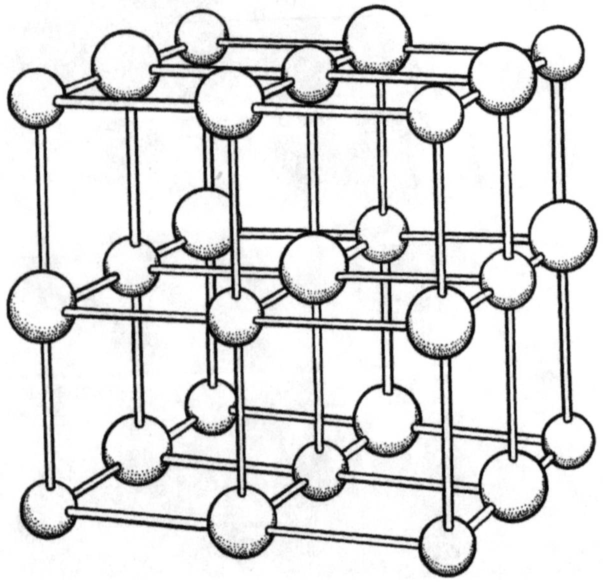

The larger balls are rubidium. The smaller balls are chlorine.

Rb_2S Rubidium sulfide

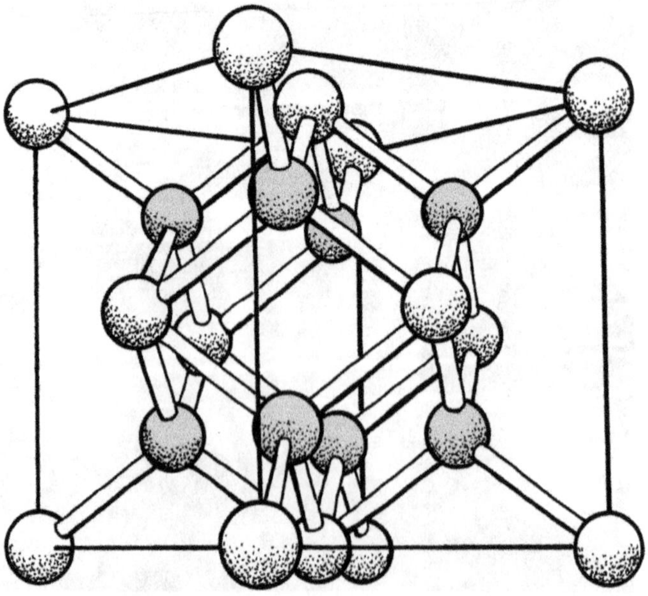

The white balls are sulfur. The gray balls are rubidium.

37 Rb

Pure rubidium burns in water!

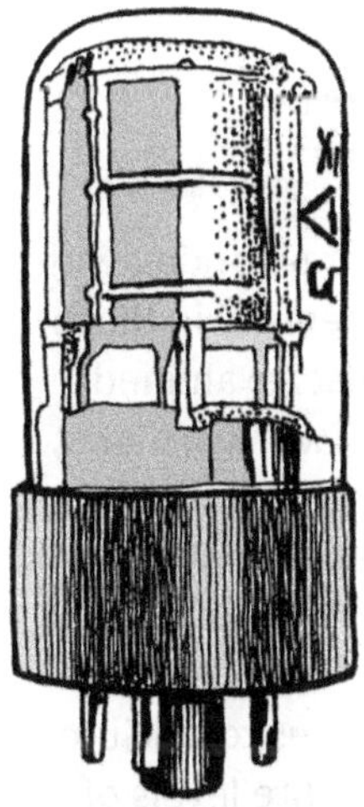

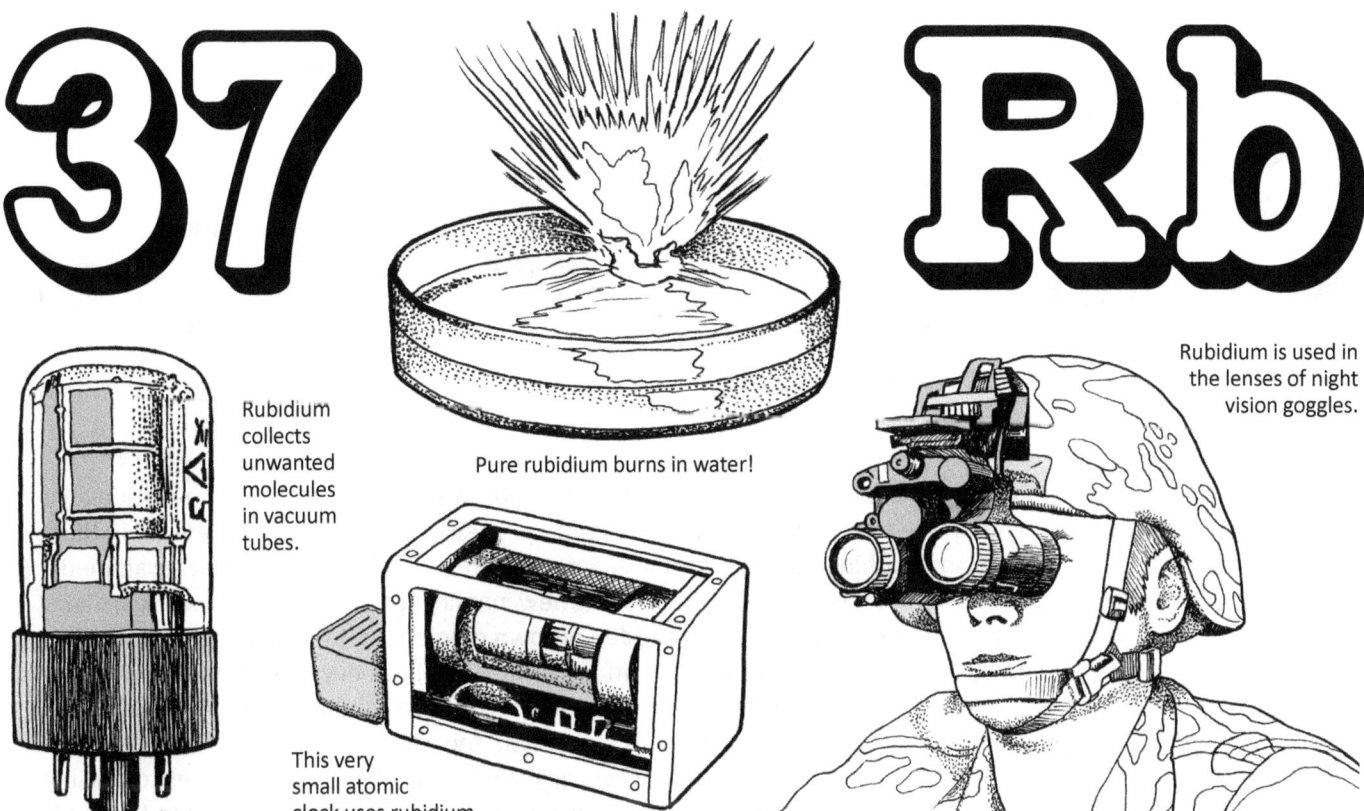

Rubidium collects unwanted molecules in vacuum tubes.

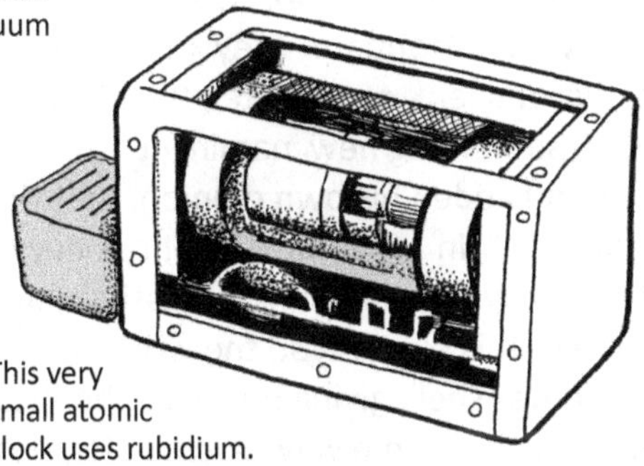

This very small atomic clock uses rubidium.

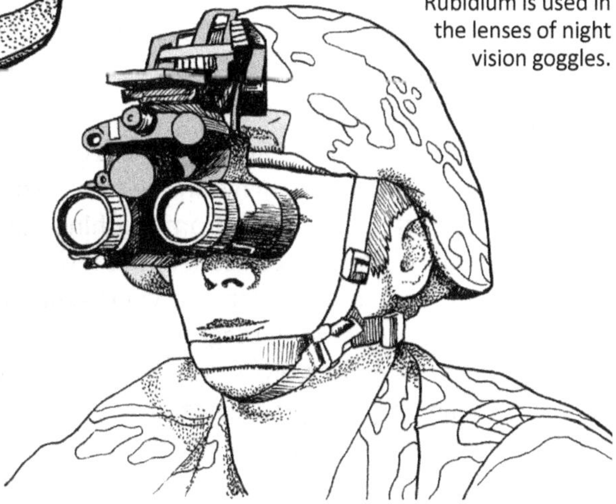

Rubidium is used in the lenses of night vision goggles.

Rubidium

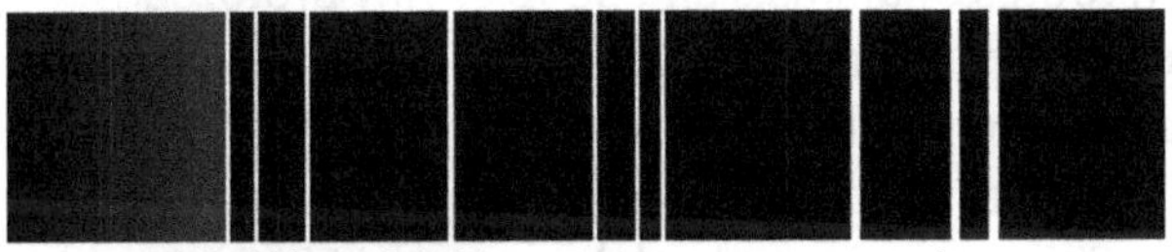

Rubidium's emission spectrum has several very bright red lines. Bunsen and Kirchhoff used a Latin word for "red" to name this new element.

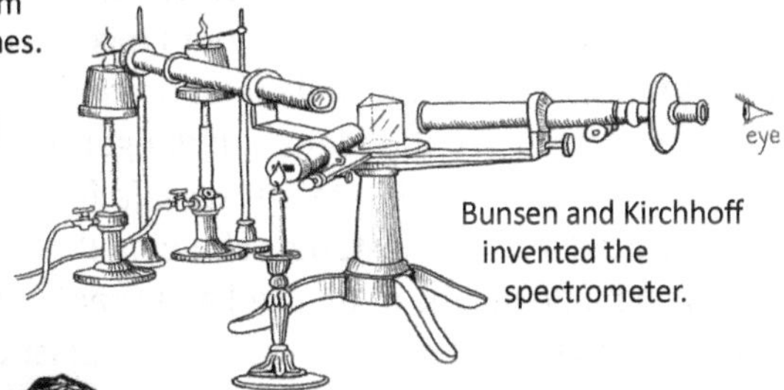

Bunsen and Kirchhoff invented the spectrometer.

Robert Bunsen

Gustav Kirchhoff

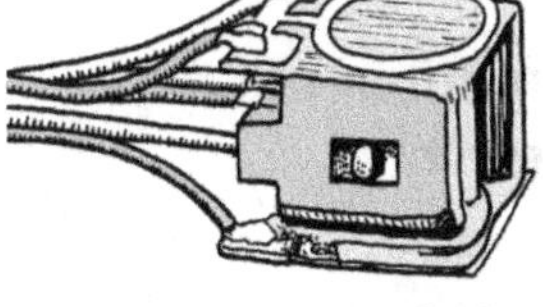

Rubidium atoms inside this tiny magnetometer (about the size of a penny) sense changes in the magnetic fields in the brain.

A source of Rb: purple lepidolite

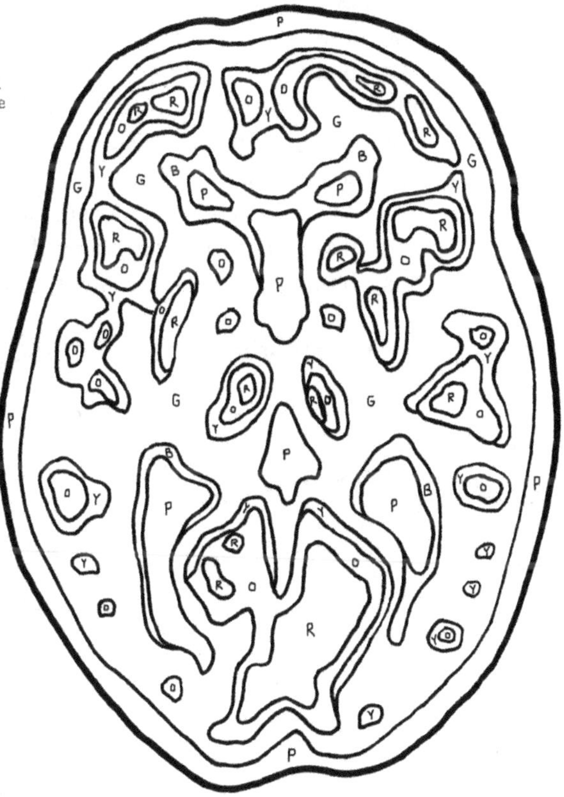

This is a PET scan image of the middle of a brain (top view). The colors show the activity level of each brain area. P=purple, B-blue, G=green, R=red, O=orange, Y=yellow.

Sr 38 protons

50 neutrons
38 electrons
Atomic mass: 87.6

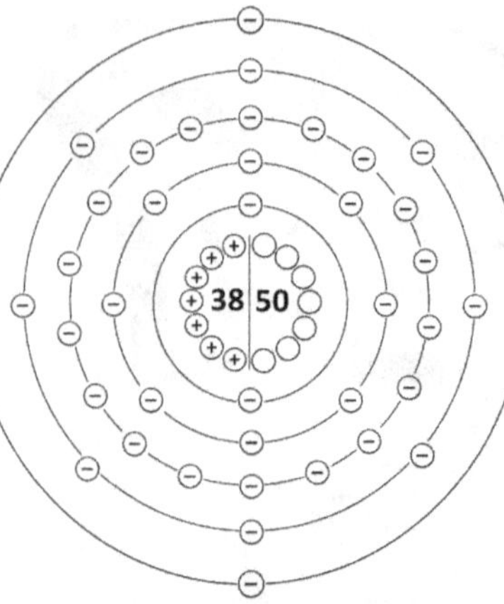

Strontium

Named after the Scottish town of Strontian

Many people were involved in the discovery of strontium. In 1790, two British chemists were working with mineral ores found in lead mines in the small Scottish town of Strontian. They knew there was a chemical in these ores that was acting differently from anything they had ever seen. A year later, two German scientists examined the Strontian mineral and concluded that it was, indeed, something new, naming it "strontianite." Two years later, a Scottish chemist confirmed the previous experiments and added his own opinion, that this mineral contained a hitherto unknown element, suggesting the name "strontites." In 1808, Sir Humphry Davy was able to isolate this new element using electrolysis (the method he used to discover sodium and potassium) and produced a sample of pure strontium metal. He suggested tweaking the name of this element so the word would end in "-ium."

Strontium, like calcium, has two electrons in its outer shell, giving it similar chemical properties to calcium. Plants and animals that use calcium will absorb strontium in the same way. Archaeologists can measure levels of strontium in old bones and use them as clues about past migration routes of animals and people. Tiny protozoans called acanthareans (a type of radiolarian) actually prefer to use strontium, rather than calcium, to build their shells.

In the late 20th century, the largest use of strontium was as an additive to the glass used in television screens. Cathode ray tubes, which produced the color pictures, had the unfortunate property of also producing x-rays. Strontium could absorb these x-rays and protect the viewer. (These CRT screens are obsolete; we use LCDs instead.)

In the 1800s, strontium was used in the extraction of sugar from sugar beets. Strontium carbonate would break apart the sugar molecules in the molasses without getting any of the strontium mixed into the sugar.

Strontium burns with a bright red flame and is non-toxic, making it ideal for use in fireworks and signal flares.

A radioactive isotope of strontium, Sr-90, is one of the most abundant and dangerous atoms produced by nuclear bombs. However, this same isotope can also be very useful. It is a relatively safe fuel for powering things that need to be "off the grid," such as lighthouses. Russia printed a series of stamps showing their lighthouses from the Soviet era, when Sr-90 was commonly used. Sr-90 is also made in labs to be used as a medicine against bone cancer.

Strontium aluminate is a compound that will glow in the dark. It can be put into plastic toys because it is non-toxic. It is a bit more expensive than other GITD compounds, though, so these toys are a bit higher in price.

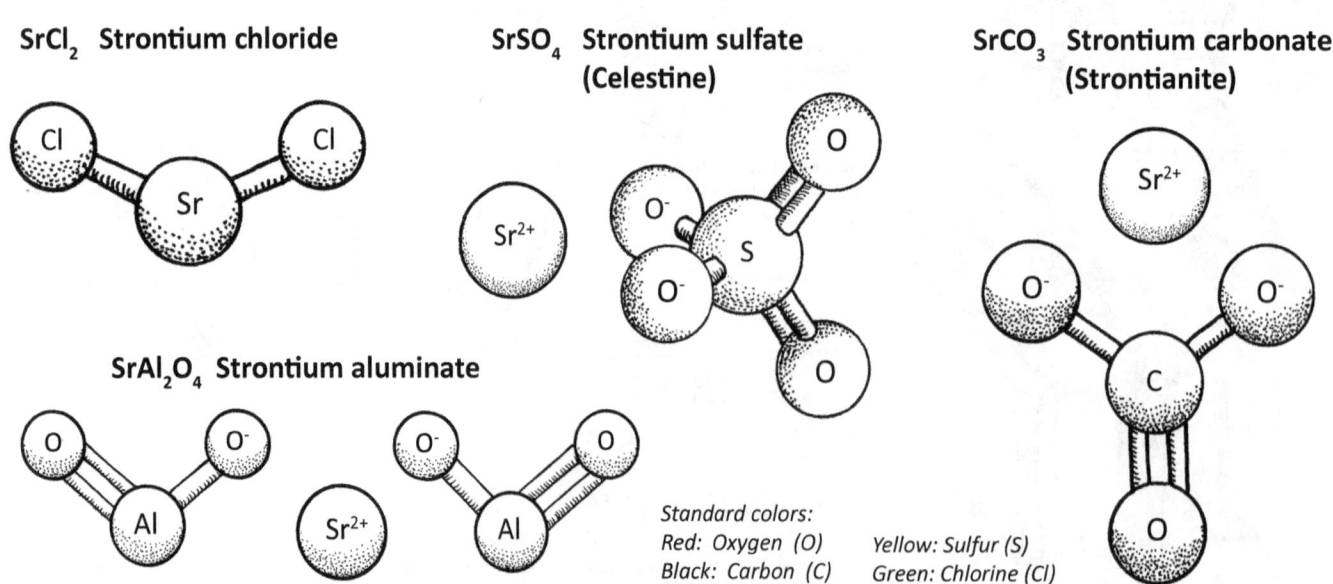

$SrCl_2$ Strontium chloride

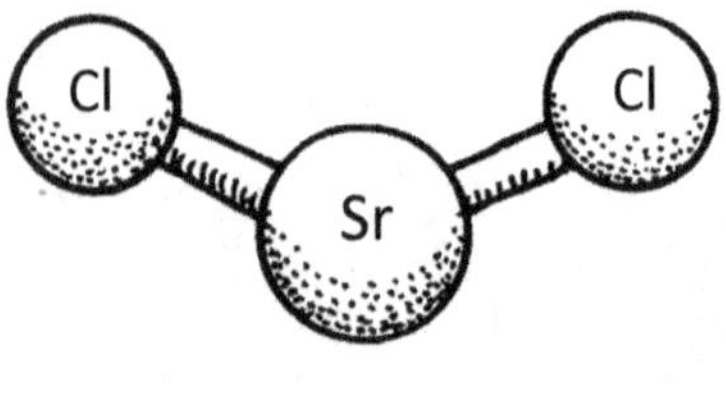

$SrSO_4$ Strontium sulfate (Celestine)

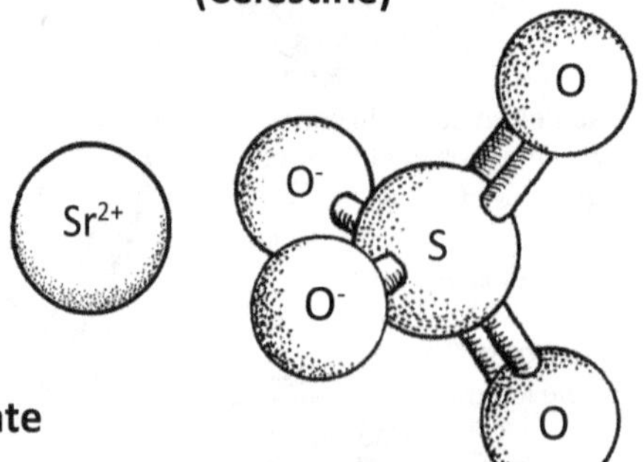

$SrCO_3$ Strontium carbonate (Strontianite)

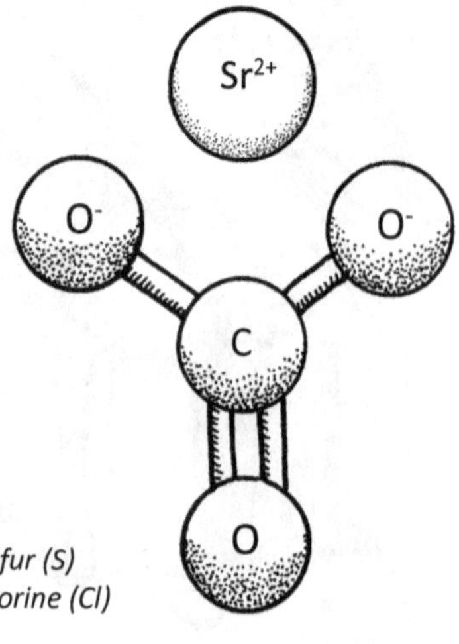

$SrAl_2O_4$ Strontium aluminate

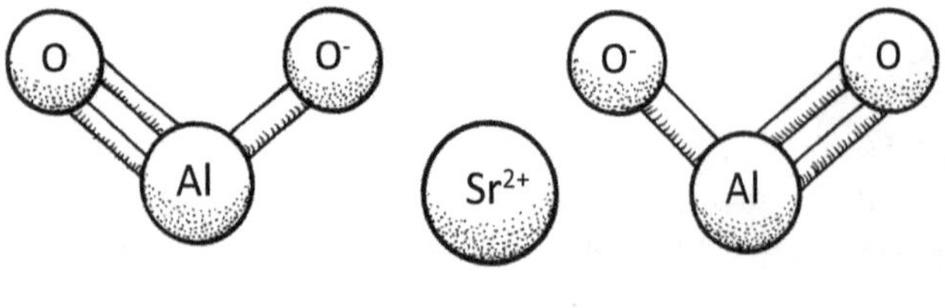

Standard colors:
Red: Oxygen (O)
Black: Carbon (C)
Yellow: Sulfur (S)
Green: Chlorine (Cl)

38 Sr

Strontium

Sr is used in red fireworks.

Russia used radioactive strontium to power light-houses in the mid-1900s.

Microscopic creatures called radiolarians can use Sr (in place of Ca) to build their shells.

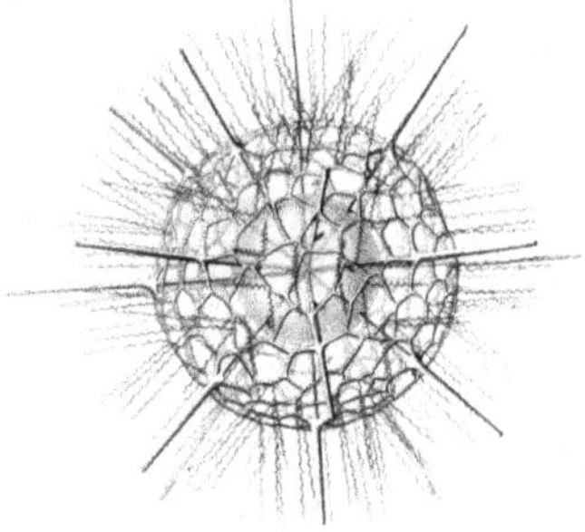

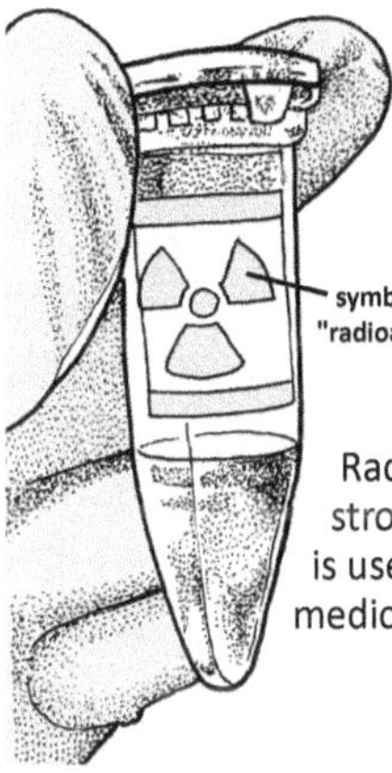

Radioactive strontium-90 is used as a medicine.

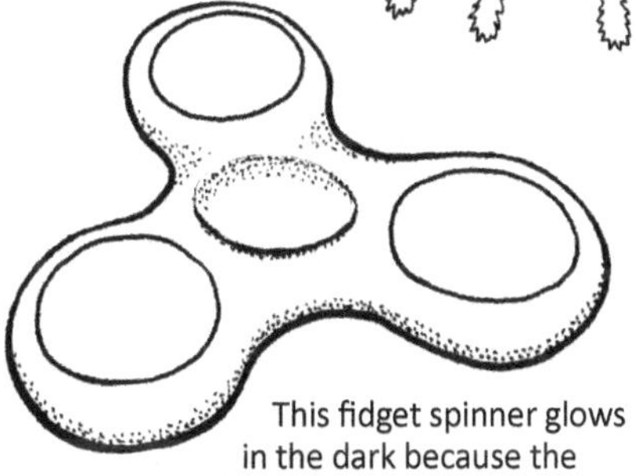

This fidget spinner glows in the dark because the plastic contains strontium aluminate.

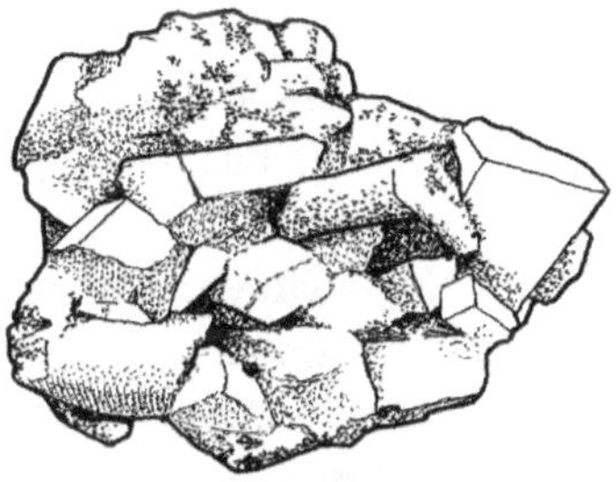

Strontianite is a primary source of strontium. (It can be white or yellowish white.)

Strontium provides the bright red color for signal flares.

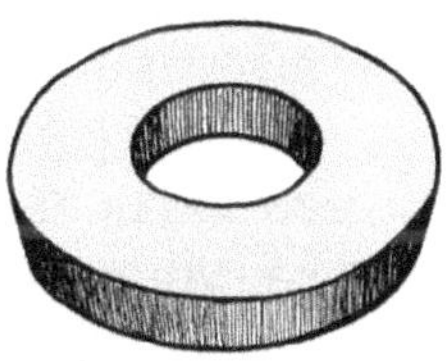

Strontium is used to make ferrite magnets found in loud speakers.

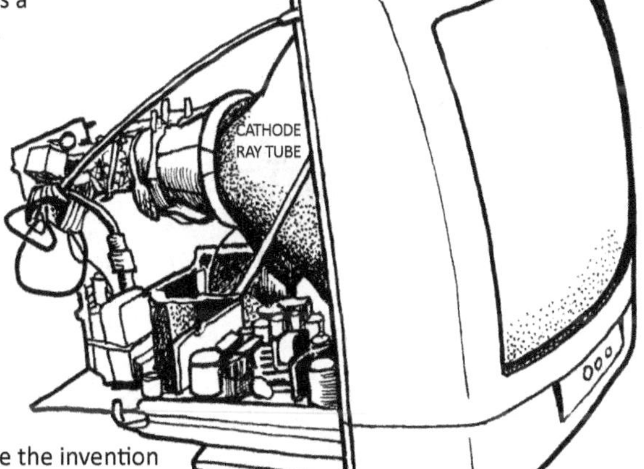

Before the invention of flat screens, the largest use of strontium was for glass television screens. The strontium absorbed harmful x-rays made by the cathode ray tube.

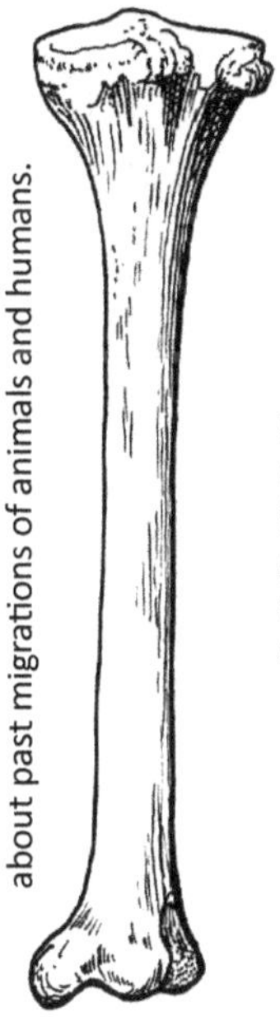

Archaeologists test bones for Sr, using it as a clue about past migrations of animals and humans.

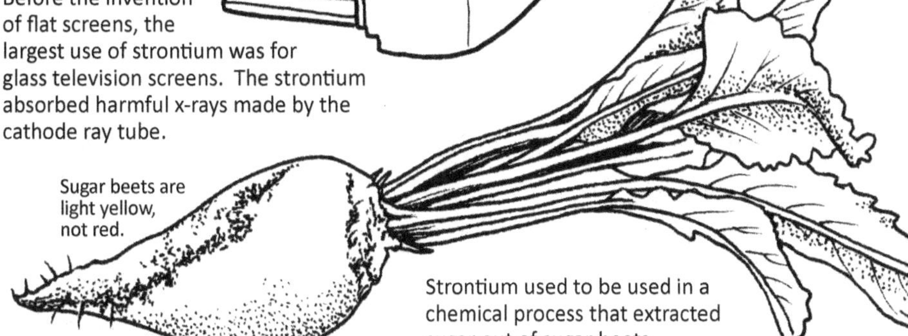

Sugar beets are light yellow, not red.

Sugar beets (not sugarcane) are the source of most sugar produced in the USA.

Strontium used to be used in a chemical process that extracted sugar out of sugar beets.

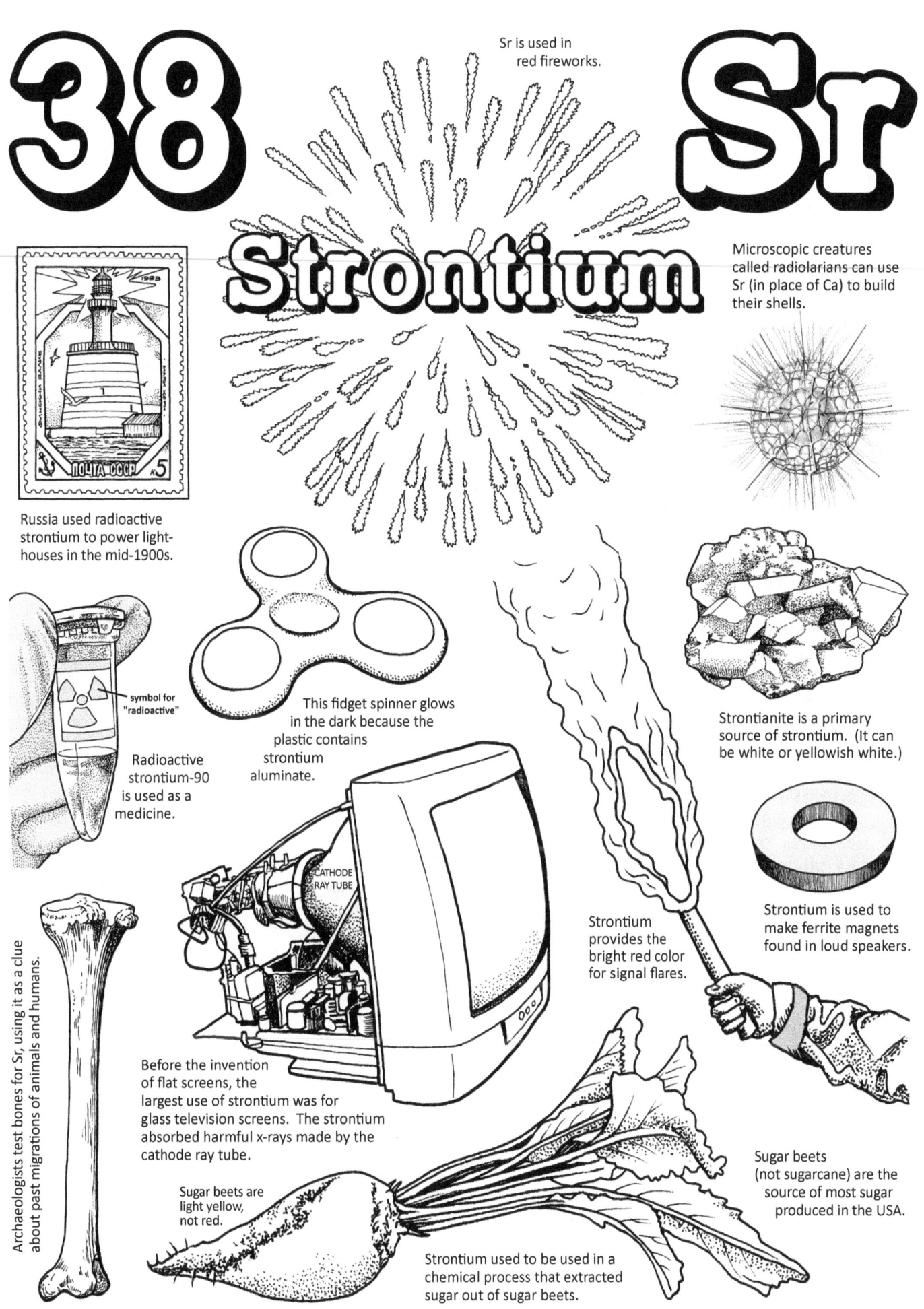

Y

39 protons

50 neutrons

39 electrons

Atomic mass: 88.9

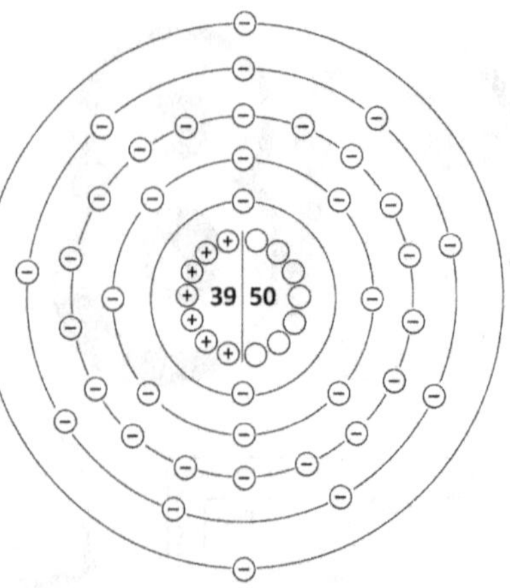

Yttrium

Named after the Swedish town of Ytterby

Yttrium *("it-ree-um")* is one of four elements on the Periodic Table named after the Swedish town of Ytterby. In 1787, an amateur Swedish geologist named Carl Arrhenius found a heavy black rock in an old mining quarry near Ytterby. The rock was very heavy, so he wondered whether it contained the newly discovered element, tungsten. He decided to name his mineral specimen "ytterbite" and he sent samples of it to several geologists and chemists, including Johan Gadolin, in Finland. Two years later, Gadolin published a complete analysis of this new mineral, so he usually gets credit for discovering yttrium. A pure sample of yttrium was not isolated until 1828. In 1843, another chemical analysis of the mineral turned up two more new elements: terbium and erbium.

Yttrium is never found by itself in nature, and is almost always mixed with elements 57-71, which are known as the "rare earth elements." Though yttrium is sometimes classified as a rare earth element it is hardly rare, being 400 times more abundant in the crust of the earth than silver.

Yttrium has had many industrial uses, and the uses keep changing to keep up with technology. One of the first uses was as an additive to strengthen Al and Mg alloys of steel. In the days when cathode ray tube televisions sets were being built, yttrium was used in combination with europium and terbium to produce the colors red and green. In 1987, researchers discovered that an yttrium alloy, yttrium barium copper oxide, was a "superconductor," meaning that at very low temperatures it lost all resistance to the flow of electricity. Superconductors are becoming more important in the 21st century. Yttrium is used in white LEDs and in specialized lasers used for eye surgery and for cutting metals. The lasers use man-made garnets (gemstones containing silicon, aluminum and yttrium) to produce the light.

Yttrium is added to lithium, iron and phosphorus to make safe, very high quality batteries that can be used in electric cars, submarines, and off-grid power systems. Yttrium has recently replaced thorium as a key ingredient in the mantles of gas lanterns (because of safety concerns about thorium's radioactivity).

Two important new uses for yttrium have been discovered in the 21st century. In 2009, researchers combined yttrium with indium and manganese to create a bright blue pigment, nicknamed "YInMn." Since about 2010, a radioactive form of yttrium, Y-90, has been used in medicines that fight cancers.

Y_2O_3 Yttrium oxide

Oxygen is usually red.

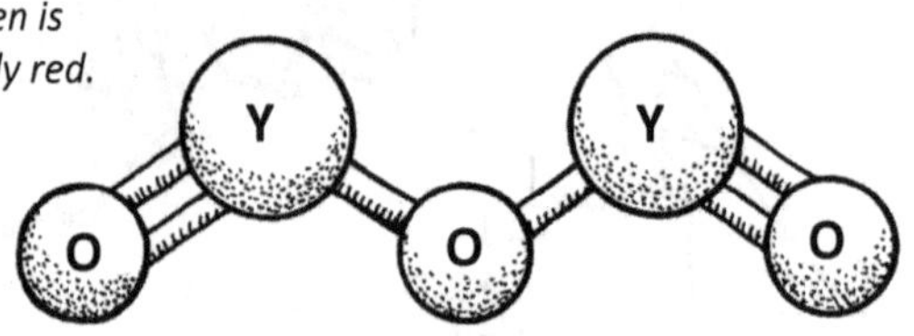

Y_2S_3 Yttrium sulfide

Sulfur is usually yellow.

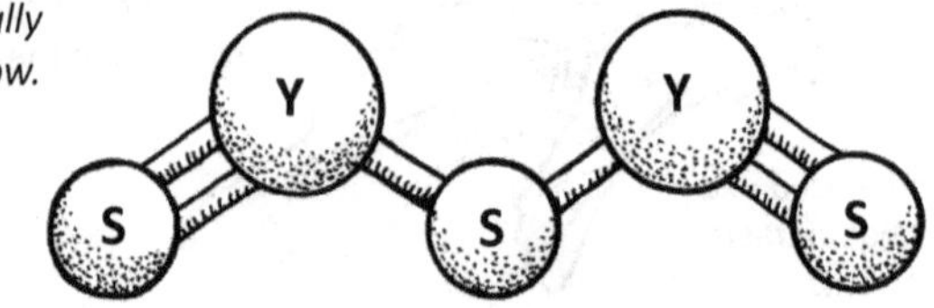

YF_3 Yttrium fluoride

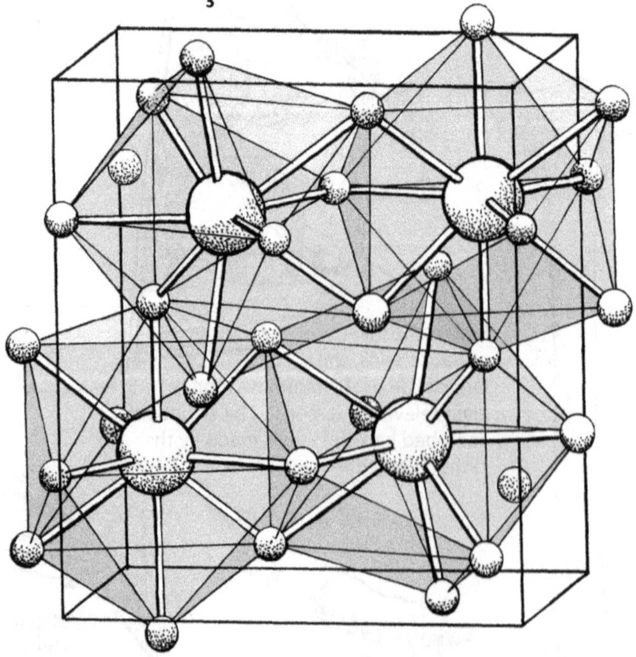

The larger balls represent yttrium, Y. The smaller balls represent fluorine, F.

39 Yttrium Y

Yttrium is one of four elements on the Periodic Table that is named after the small Swedish town of Ytterby. *("it-er-bee")*

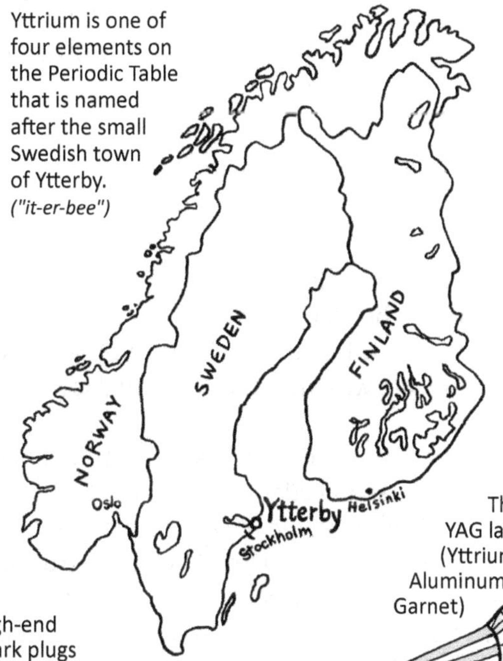

Yttrium can make lithium batteries more energy efficient.

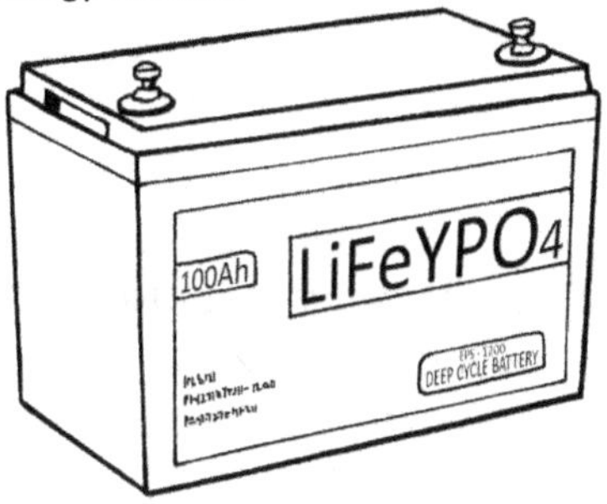

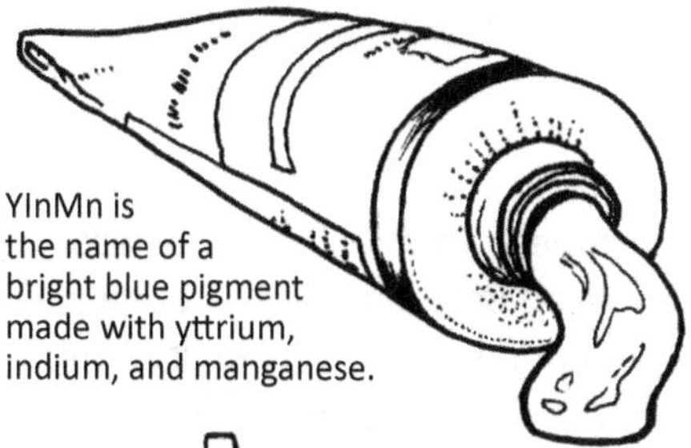

YInMn is the name of a bright blue pigment made with yttrium, indium, and manganese.

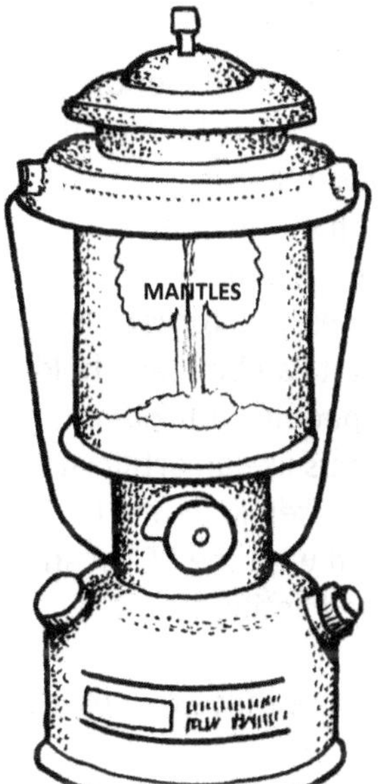

Yttrium has replaced thorium as the material used for mantles in gas lanterns.

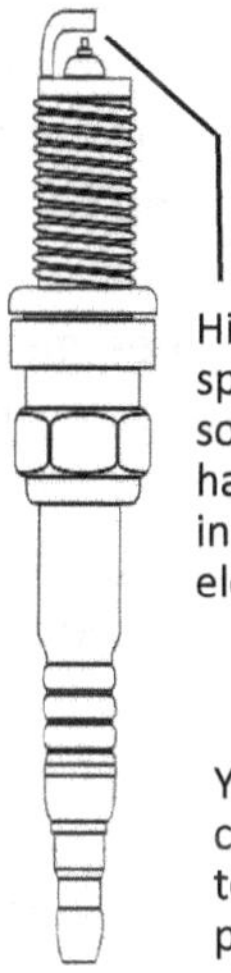

High-end spark plugs sometimes have yttrium in their electrodes.

This is a YAG laser. (Yttrium Aluminum Garnet)

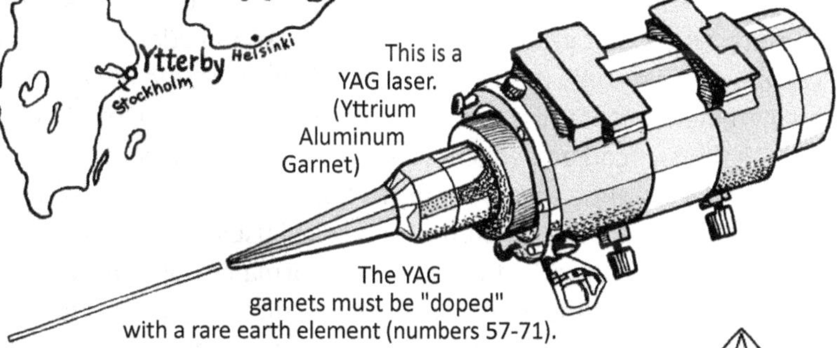

The YAG garnets must be "doped" with a rare earth element (numbers 57-71).

Before 1995, (when cubic zirconia became available), artificial garnets were popular as diamond substitutes.

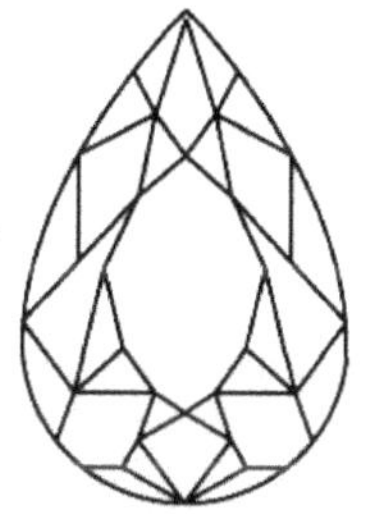

Yttrium was used in cathode ray tube televisions to produce colors.

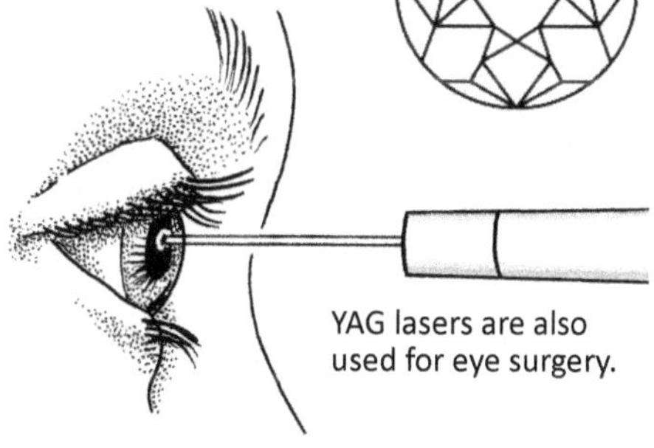

YAG lasers are also used for eye surgery.

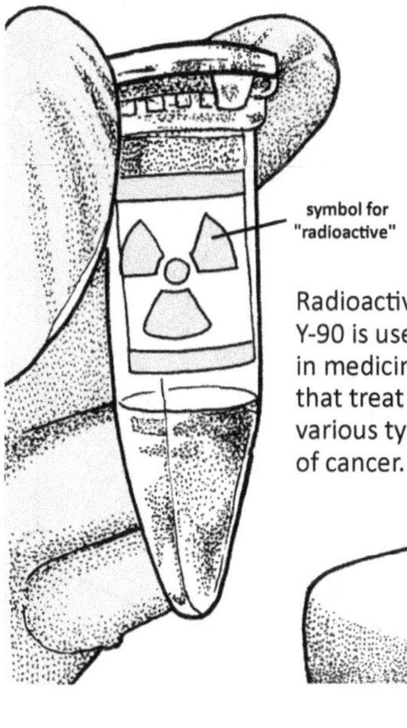

Radioactive Y-90 is used in medicines that treat various types of cancer.

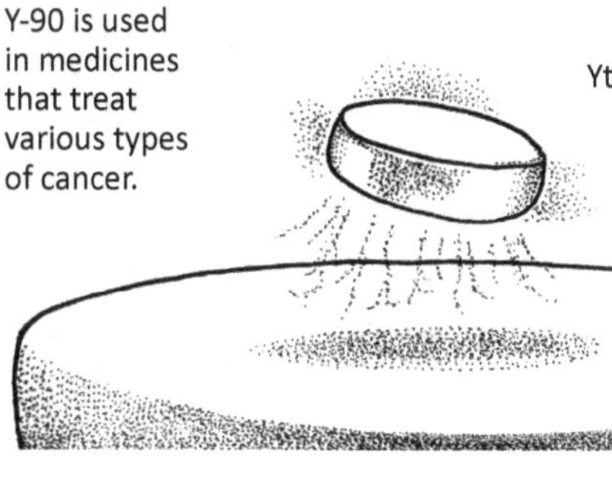

Yttrium barium copper oxide is a superconducting metal. At very low temperatures superconductors appear to float.

Yttrium is added to glass that is used in specialty camera lenses.

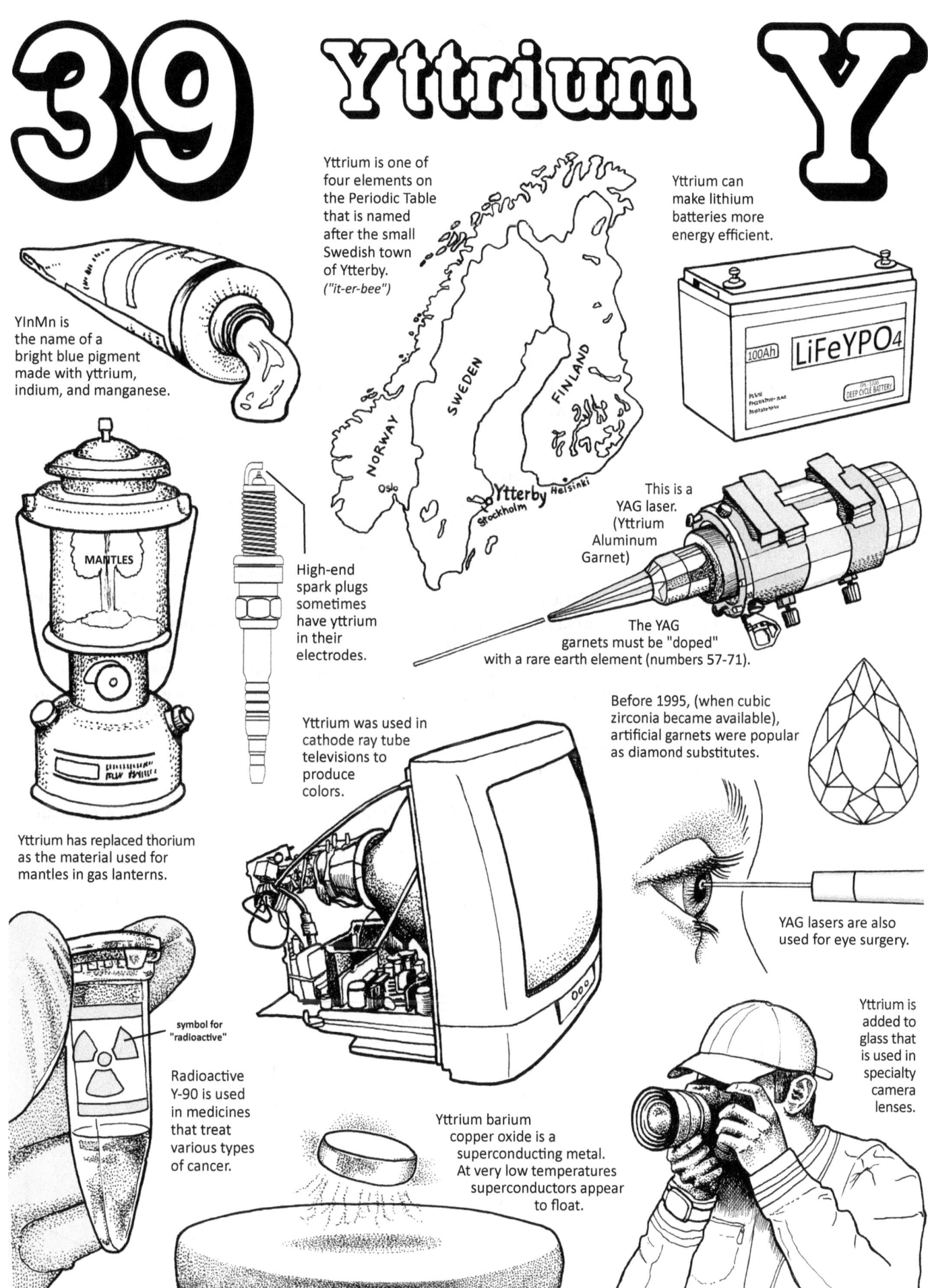

Zr 40 protons

51 neutrons
40 electrons
Atomic mass: 91.2

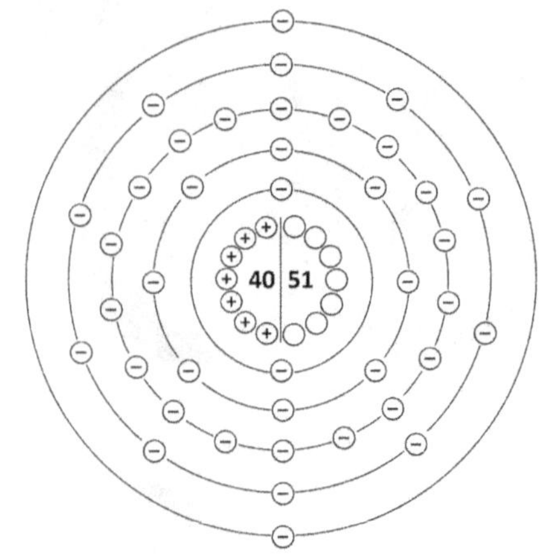

Zirconium

From the Persian word "zargun" meaning "golden"

Zirconium is found in zircon crystals, a type of mineral known since ancient times. Their primary use was always as gemstones. No one knew about the element zirconium until 1824 when Jacob Berzelius figured out a way to isolate a sample of pure zirconium metal by heating a sample of potassium zirconium fluoride inside an iron tube. It was not until 1925 that a chemical process was invented that could isolate zirconium on a large enough scale that it could be available for industrial uses.

Zircon crystals ($ZrSiO_4$) can be ground into tiny pieces and used to make abrasives (such as sandpaper or grinding wheels) or ground into a fine powder and added to ceramics to make them white instead of clear. A related crystal, cubic zirconia (also known as zirconium dioxide, ZrO_2), can also be used for these purposes, as well as being an ingredient in ceramics that need to be able to take a lot of heat, like crucibles and furnaces that melt metals.

Pure zirconium, which is a silvery metal, not a gemstone, can be added to steel alloys to make them harder and more resistant to heat. You might find zirconium steel alloys in pumps, jet engines, gas turbines, and rockets.

Zirconium has a number of medical uses. One of zirconium's unique properties is that it can bind and hold a biological waste product called urea. Normally, our kidneys take urea out of our blood, but people with kidney disease sometimes need a process called "dialysis" where blood is filtered by a machine that acts like an artificial kidney. Zirconium is a key ingredient in these machines. Zirconium-containing compounds are used in dental implants, and in artificial hips and knees. In nuclear medicine, a radioactive isotope, Zr-89, is used with the PET scan process (Positron Emission Tomography) to track microscopic infection-fighting molecules called antibodies.
In medical ultrasound, lead zirconate titanate (PZT) is used to generate sound waves.

Zirconium is sometimes used in antiperspirant deodorants, along with aluminum, because these metals are able to prevent sweat from leaving pores in the skin.

Zirconium alloys are used to make pipes that will be filled with radioactive nuclear fuel because the zirconium can prevent radioactivity from leaking out, under normal circumstances. Unfortunately, at very high temperatures, zirconium catches fire. Part of the Fukushima nuclear disaster in Japan involved zirconium coming into contact with burning hydrogen.

$PbZrO_3$ Lead zirconate

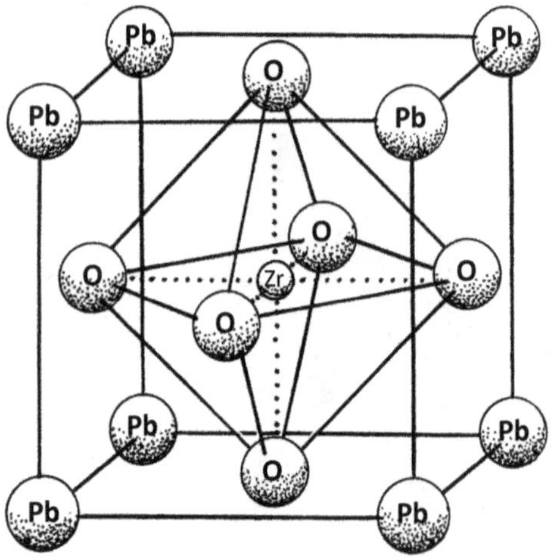

$ZrCl_4$
Zirconium chloride

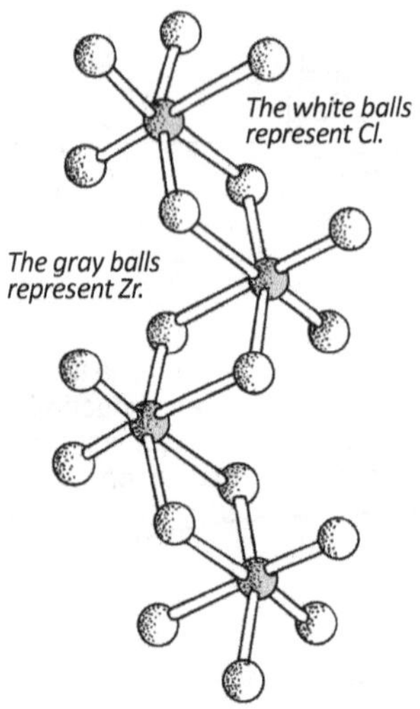

ZrO_2 Zirconium dioxide

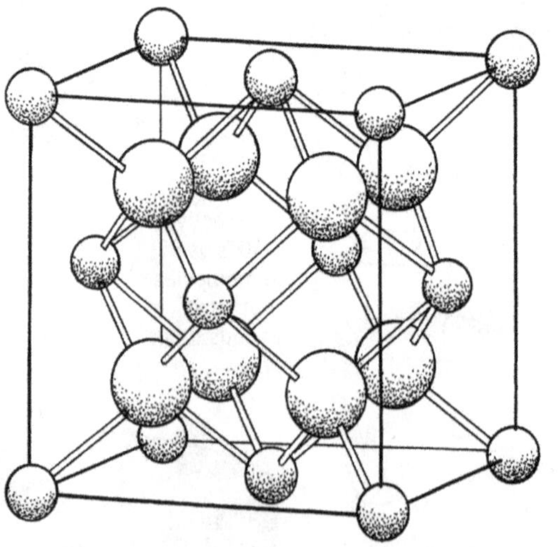

The larger balls represent O. The smaller balls represent Zr.

40 Zr

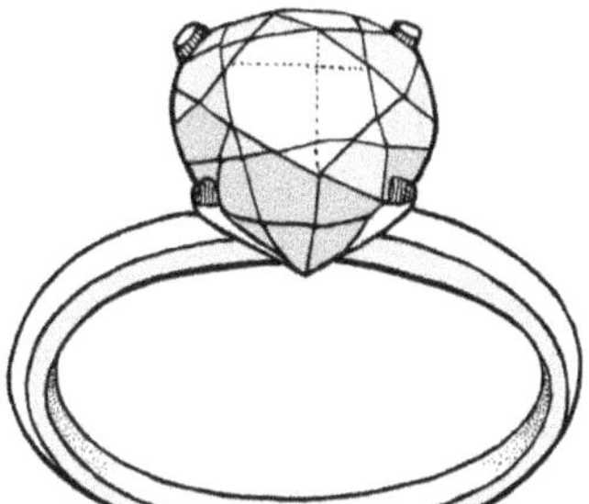

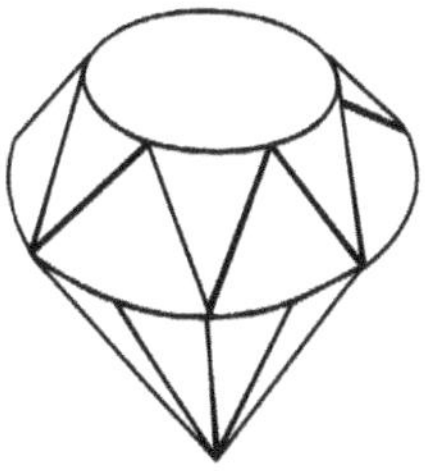

Crucibles are used in labs to "cook" things at very high temperatures.

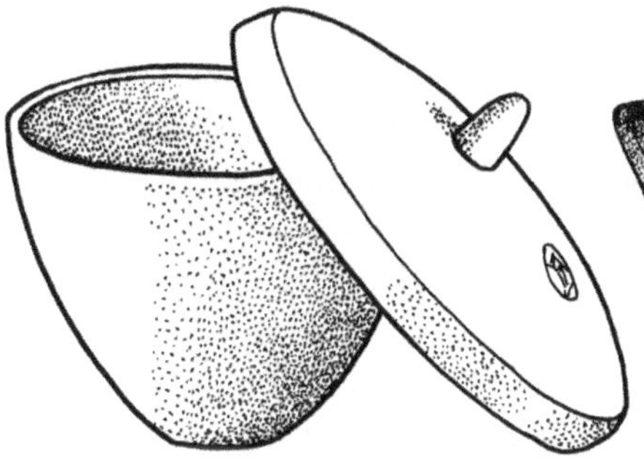

Cubic zirconia is used as a substitute for diamond because it is less expensive.

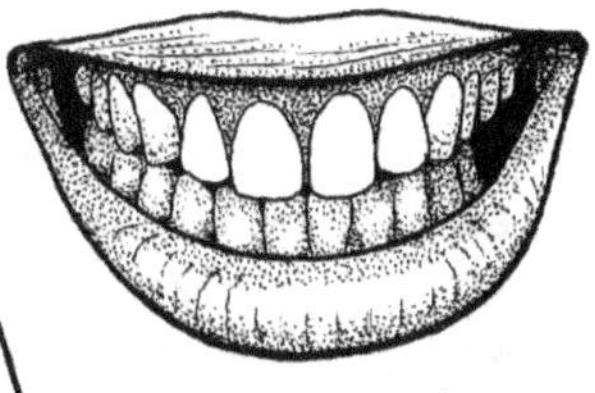

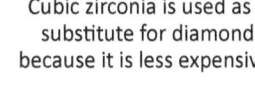

Zirconia is used in dental implants.

$ZrSiO_4$
Zircon crystal

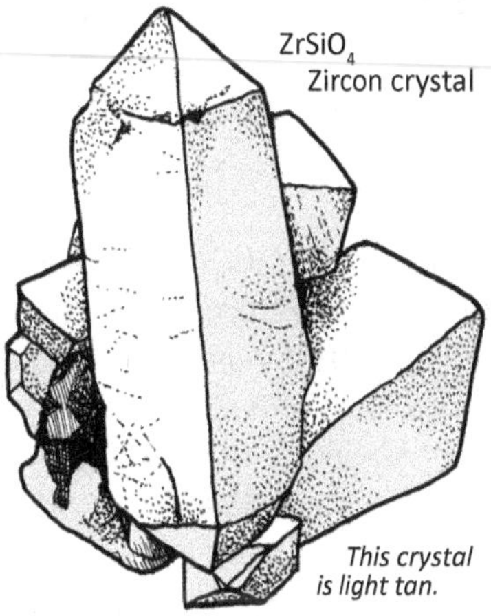

This crystal is light tan.

Zirconium is used (along with aluminum) in antiperspirants.

Zirconium

A tube of Zr-89 medicine

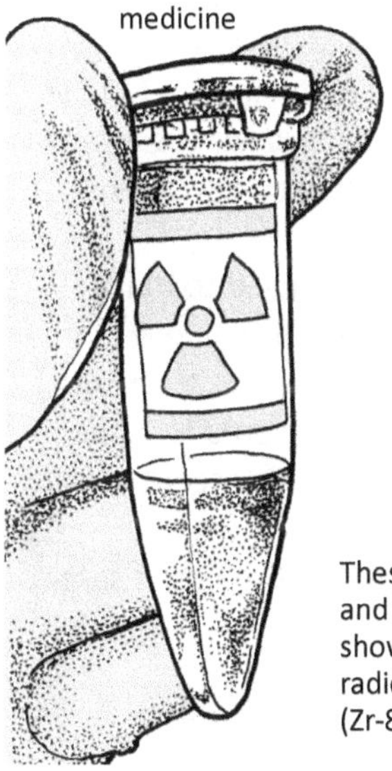

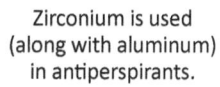

This lumpy Y-shaped thing is an infection-fighting molecule called an antibody.

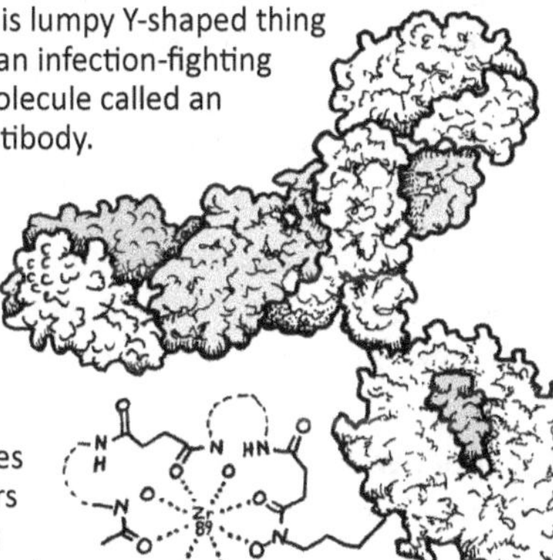

These lines and letters show the radioactive tag (Zr-89) added to the antibody.

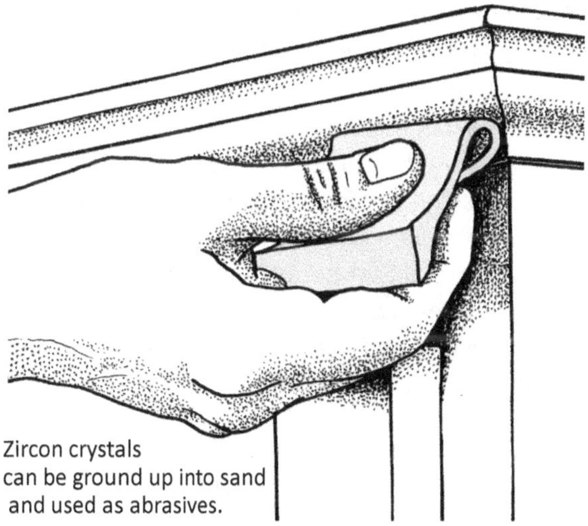

Zircon crystals can be ground up into sand and used as abrasives.

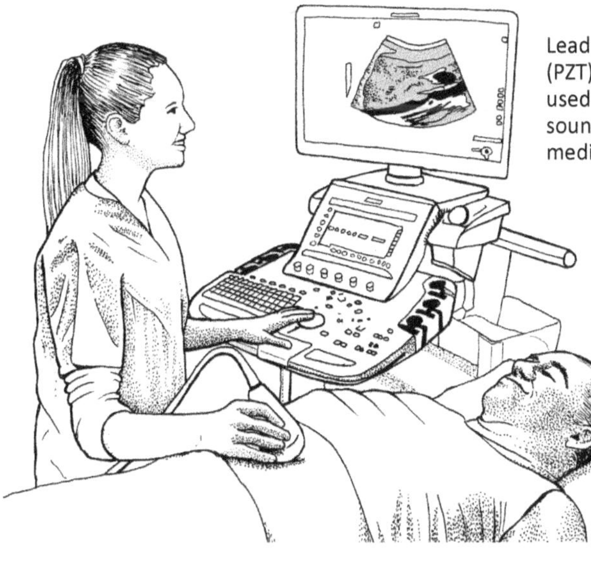

Lead zirconate titanate (PZT) ceramics are used to generate sound waves for medical ultrasound.

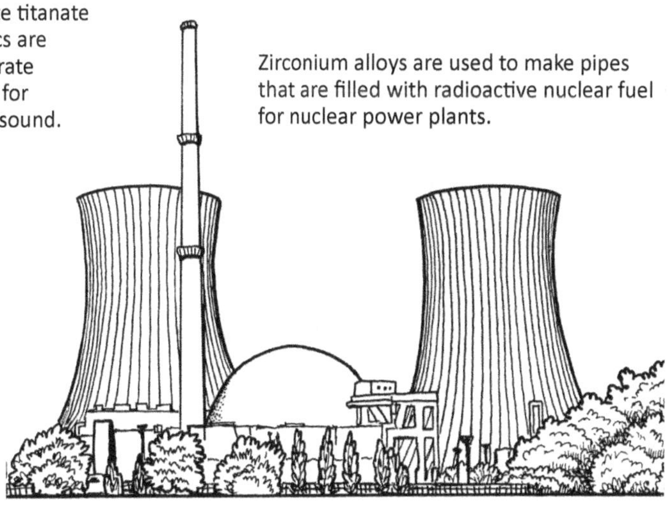

Zirconium alloys are used to make pipes that are filled with radioactive nuclear fuel for nuclear power plants.

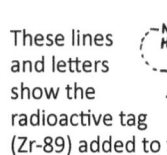

Nb 41

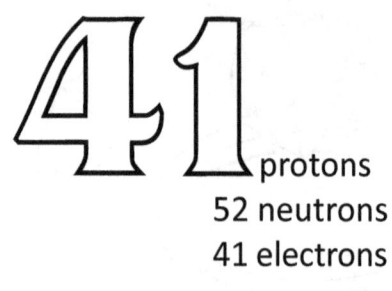

41 protons
52 neutrons
41 electrons
Atomic mass: 92.9

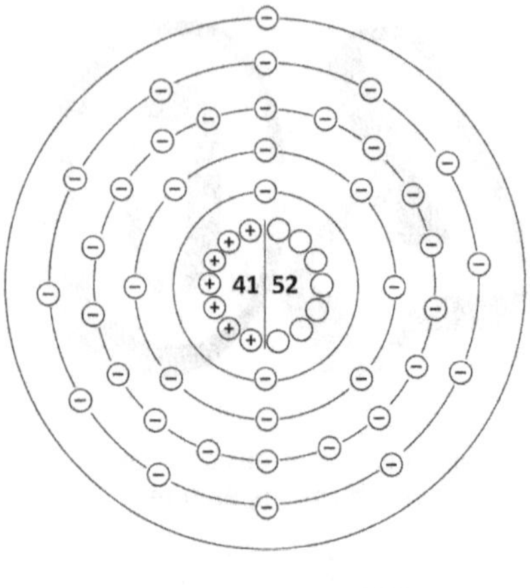

Niobium

Named after Niobe, from Greek mythology

For almost one hundred years, this element was known as columbium, not niobium. The mineral ores from which it was discovered had come from the state of Connecticut in America in the early 1700s. Samples were sent to chemists in England, and in 1801 it was announced that a new element had been discovered. The discoverer used a very old name for America, Columbia, to make a name for the new element. Only a few years later, another chemist did a second analysis of the mineral and declared that no new element had been discovered, only a new compound of the known element tantalum. In 1846, yet another chemist analyzed the minerals and came to the conclusion that the first analysis had not only been correct, but also that a second new element had been found. He chose the names niobium and pelopium, using the names Niobe and Pelops, children of the Greek god Tantalus, for whom tantalum had been named. In 1866, a chemist in Switzerland was finally able to isolate pure niobium, separating it from tantalum, and, in the process, proving that pelopium did not exist. American scientists continued to use the name columbium until 1949, when an international group of scientists decided the name should be niobium.

Niobium's primary use is as an additive to steel, but it can also be used as the primary metal in other alloys. C-103 is the name of an alloy that contains 89% niobium, 10% hafnium and 1% titanium. This alloy is very resistant to heat and has been used to make exhaust cones for rockets, even as early as the Apollo 15 mission. SpaceX uses niobium alloys in the upper stages of its Falcon 9 rocket.

When niobium is alloyed with tin or germanium, it forms a very useful material that can be used to make superconducting magnets. Superconducting niobium alloys are used to make giant electromagnets found in MRI machines and particle accelerators. (Particle accelerators are used for research on protons, neutrons and electrons.)

Niobium and its alloys have been found to be "hypoallergenic" and can be used as replacements for other types of metal that people have allergic reactions to, such as nickel. Niobium jewelry often has an iridescent quality and shimmers with rainbow colors.

Niobium is "inert" in the body, meaning it will not react with any body tissues or fluids. This makes it safe for use in parts that will be left inside the body, such as joint replacements and pacemakers.

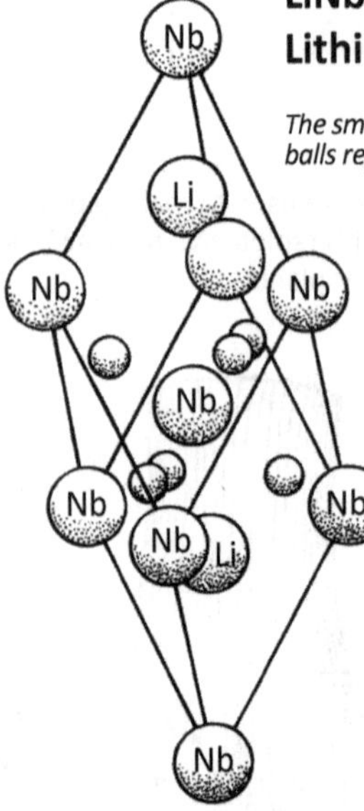

$LiNbO_3$ Lithium niobate

The small, unlabeled balls represent oxygen.

Lithium niobate is a man-made material, not found in nature. It is used in optical applications (things that use light).

Nb_2Cl_{10} Niobium pentachloride

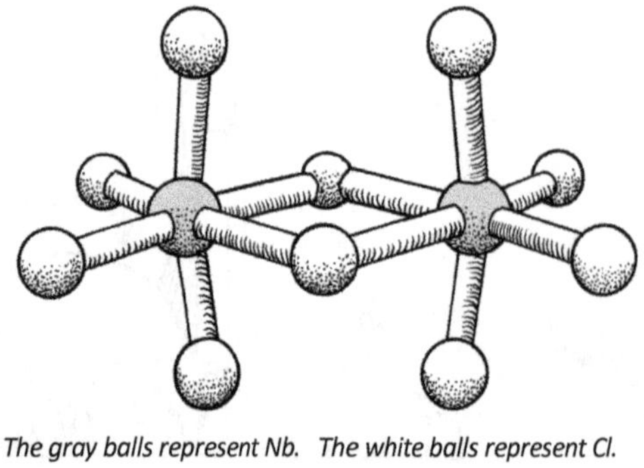

The gray balls represent Nb. The white balls represent Cl.

Niobium nitride is a superconducting material and can be used as an infrared light detector on satellites.

NbN Niobium nitride

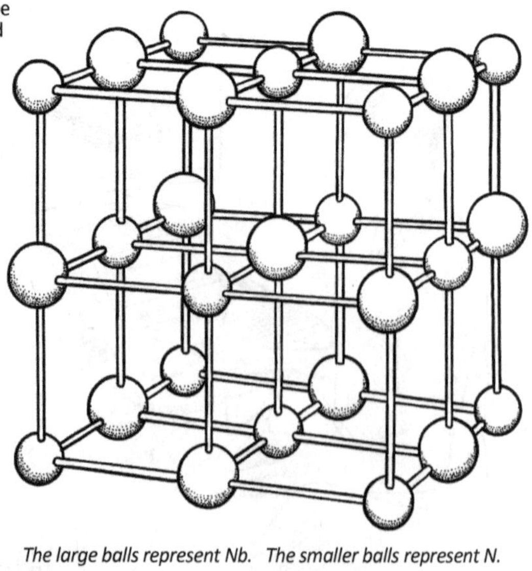

The large balls represent Nb. The smaller balls represent N.

41 Nb

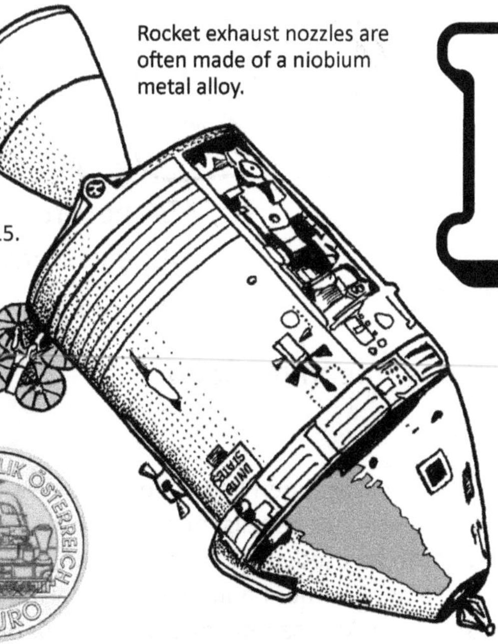

Rocket exhaust nozzles are often made of a niobium metal alloy.

This is Apollo 15.

Niobium jewelry shimmers with iridescent rainbow colors.

Collectible coins can be made of Nb. This one shimmers green.

Brazil is by far the largest producer of niobium metal. (Canada is second.)

Niobium

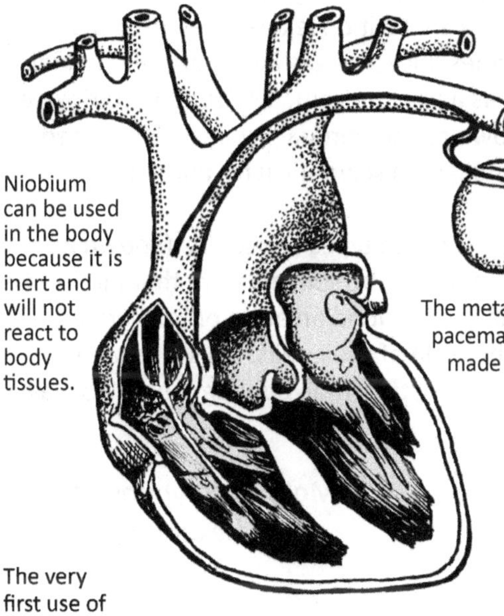

Niobium can be used in the body because it is inert and will not react to body tissues.

The metal cases of pacemakers can be made of niobium.

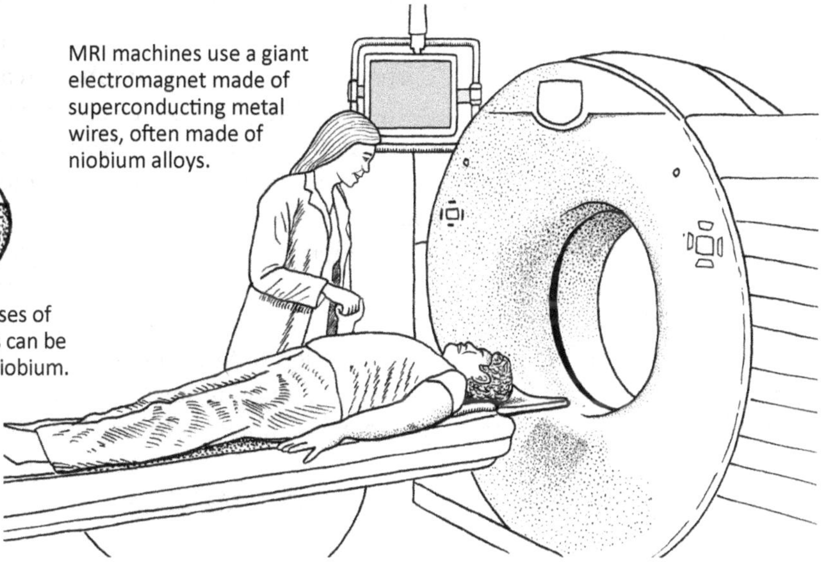

MRI machines use a giant electromagnet made of superconducting metal wires, often made of niobium alloys.

The very first use of niobium was in incandescent bulbs. Tungsten quickly replaced it.

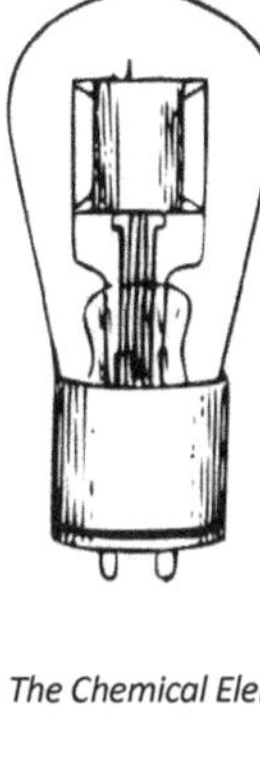

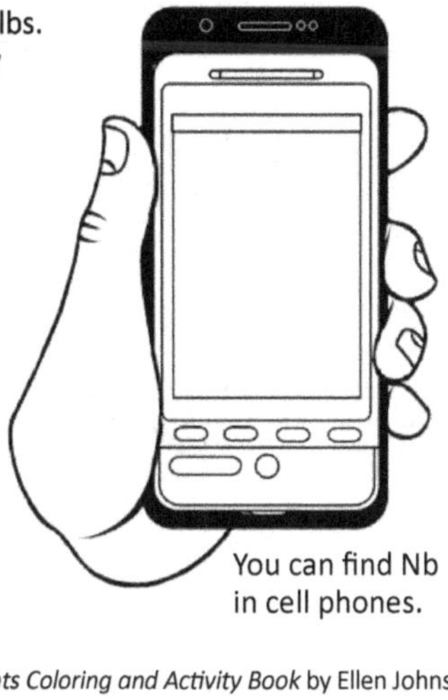

You can find Nb in cell phones.

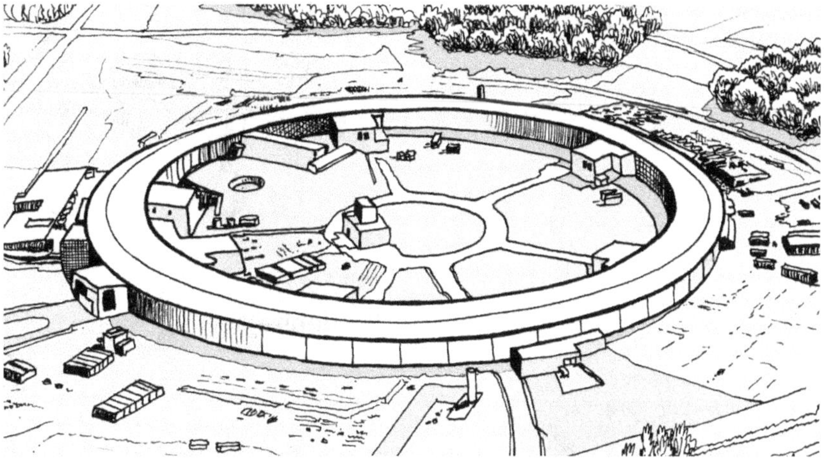

Niobium superconducting wire is used in the electromagnets in particle accelerators.

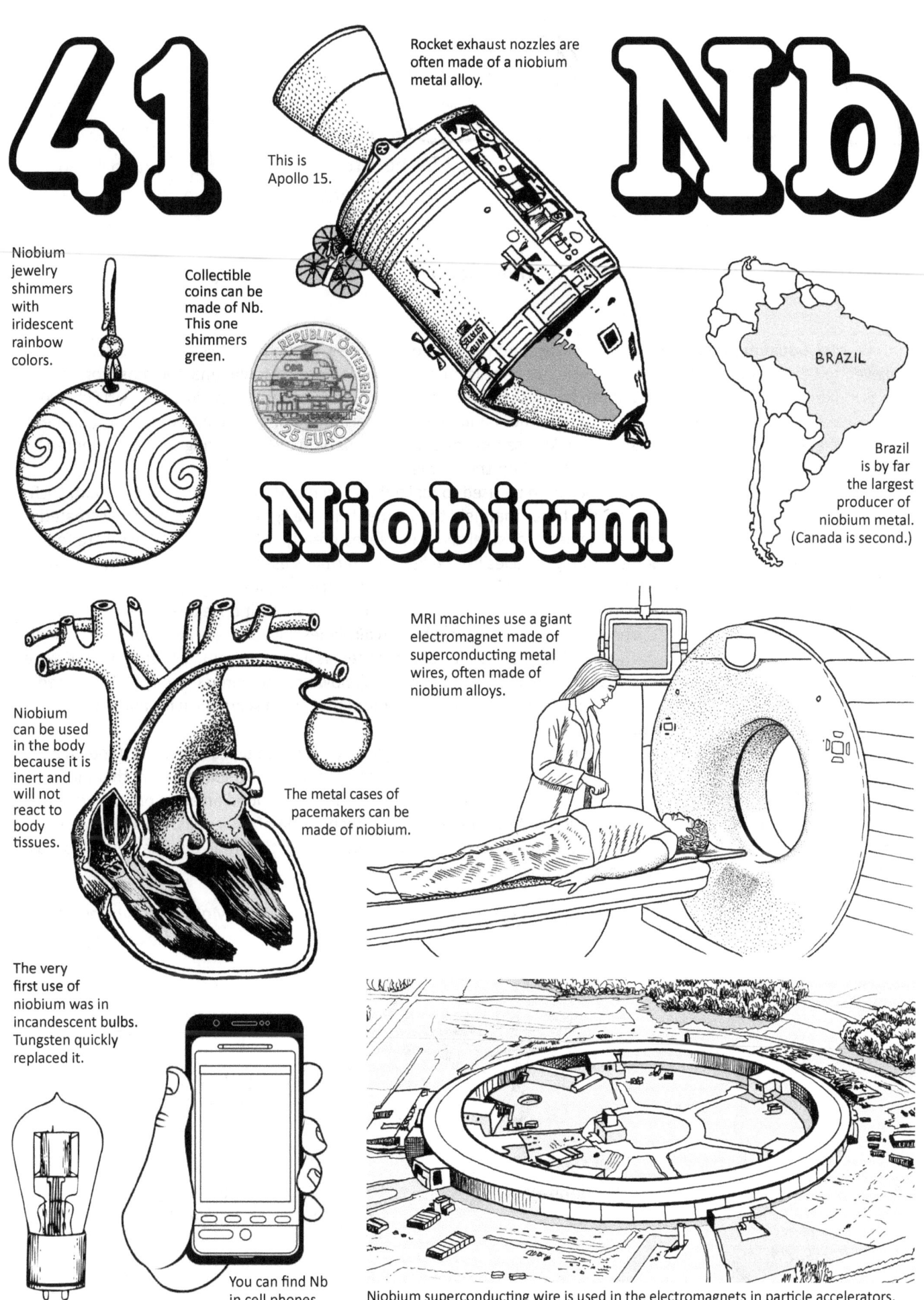

Mo 42

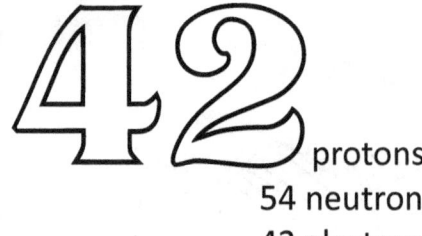

protons
54 neutrons
42 electrons
Atomic mass: 95.9

Molybdenum

From the Greek word "molybdos" meaning "lead"

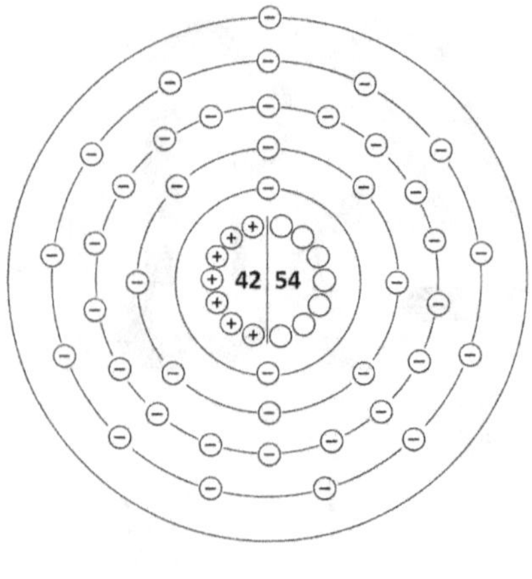

42
54

Molybdenum (*mo-LIB-den-um*) ores have been known for centuries, though they were often mistakenly believed to be either graphite (the stuff inside our pencils) or galena (a lead ore). Molybdena, the most common ore, is strikingly similar to graphite and can be used as a solid lubricant, just like graphite is. In 1778, Carl Scheele determined that molybdena contained a new element and named it molybdenum. Just a few years later, another Swedish chemist, Peter Hjelm, was able to make a sample of pure molybdenum metal. During the next hundred years, no practical or industrial use could be found for this element.

In the early 1900s, molybdenum began to be used to make things that had to withstand a lot of heat, such as the metal coils in electric heaters and the bases that held the filaments inside light bulbs. World War I created new opportunities for molybdenum, as the need for steel alloys increased greatly. Molybdenum alloys were found to be very strong and very light, perfect for big vehicles like tanks. The Germans used molybdenum steel to make the famous Big Bertha howitzer. Molybdenum could also be used as a replacement for tungsten in making steel tools. After the war, interest in molybdenum went down again, but in the late 1900s it once again began to play an important role in making steel. "Chromoly" steel is used for tubing in airplanes, bicycles, gas pipes, and gun barrels.

Today, molybdenum has many uses. Molybdenum disulfide is used as a dry lubricant. Molybdenum trioxide is used as an adhesive between metals and ceramics. Lead molybdate (known as the mineral "wulfenite") is used as a red pigment. Various molybdenum compounds are used in chemical process in scientific labs, and are used to make testing devices that sense pollution in air and water.

Molybdenum is at the heart of dozens of essential biological enzymes. Certain types of bacteria (known as "nitrogen-fixing bacteria") are able to take nitrogen gas out of the air and put it into the soil in a form that plants can use. They do this with an enzyme called nitrogenase, which uses a molecule built around an atom of molybdenum. Also, most forms of life make an enzyme called sulfite oxidase in the mitochondria of their cells. This enzyme is critical for the process of cellular respiration (burning sugars to harvest energy). Other molybdenum enzymes play a role in the transfer of oxygen atoms from one molecule to another.

Molybdenum trioxide
MoO_3

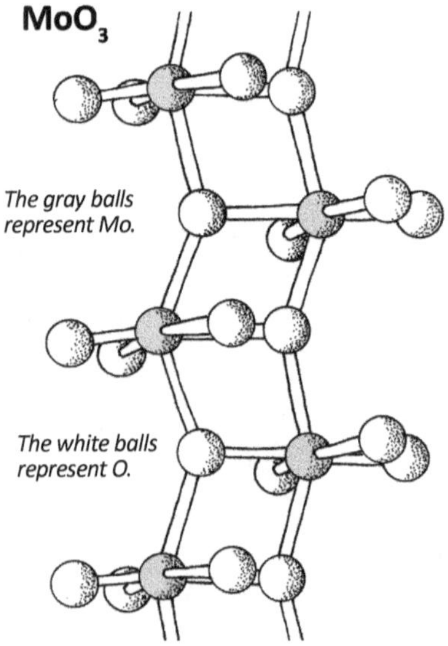

The gray balls represent Mo.
The white balls represent O.

$MoCl_5$ Molybdenum (V) chloride

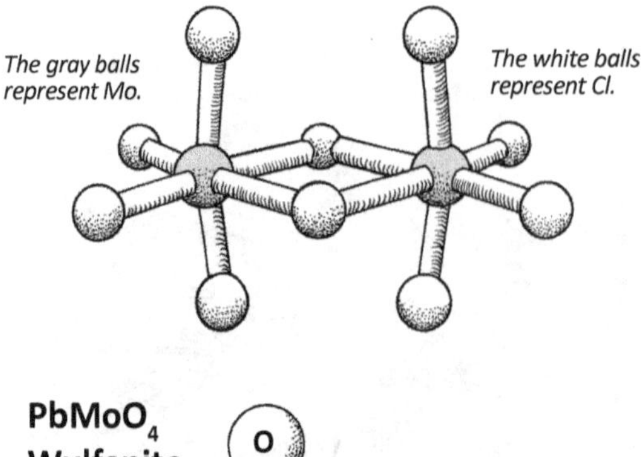

The gray balls represent Mo.
The white balls represent Cl.

MoF_6 Molybdenum hexafluoride

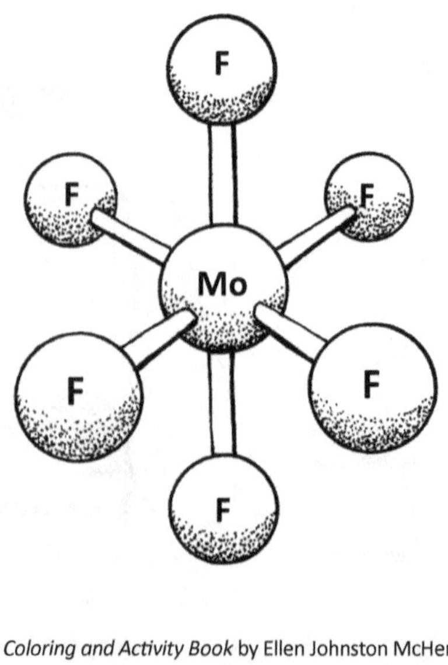

F
F
F
Mo
F
F
F

$PbMoO_4$
Wulfenite

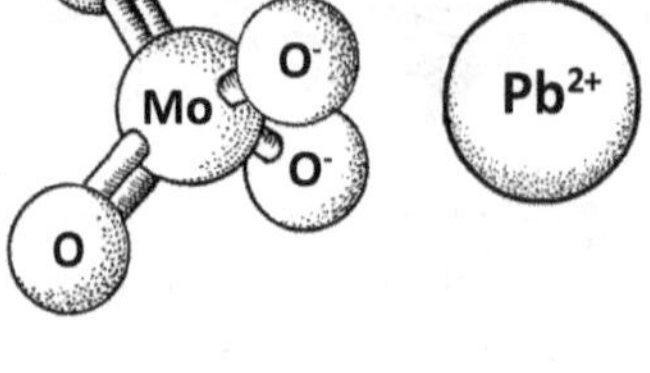

O
Mo
O-
O-
O
Pb2+

42 Mo

Wulfenite is often red.

Molybdenum

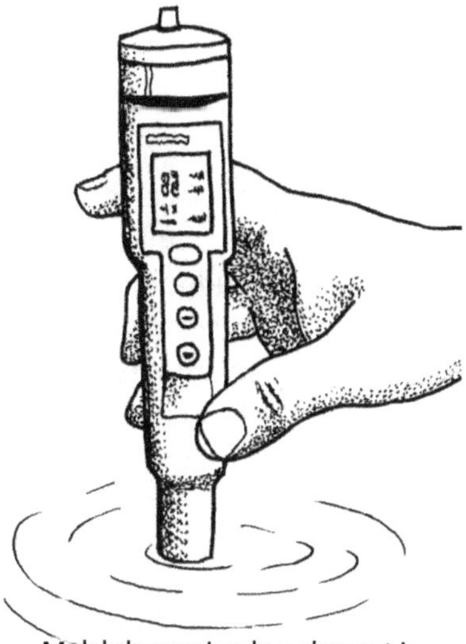

Molybdenum is a key element in chemical analysis instruments that test water and air quality.

Molybdenum disulfide (MoS_2) is used as a lubricant for machine parts.

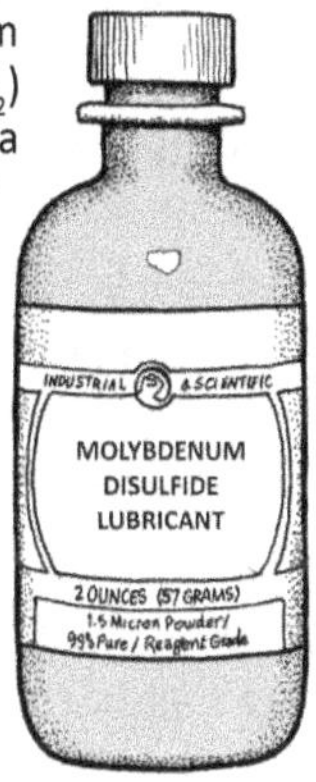

"Big Bertha" was the nickname for this howitzer built by Germany in WW 1.

This bicycle frame is made of "Chromoly" (Cr-Mo) steel.

Soybeans are one of the "legume" plants that grow little nodules on their roots to provide a place for nitrogen-fixing bacteria to live.

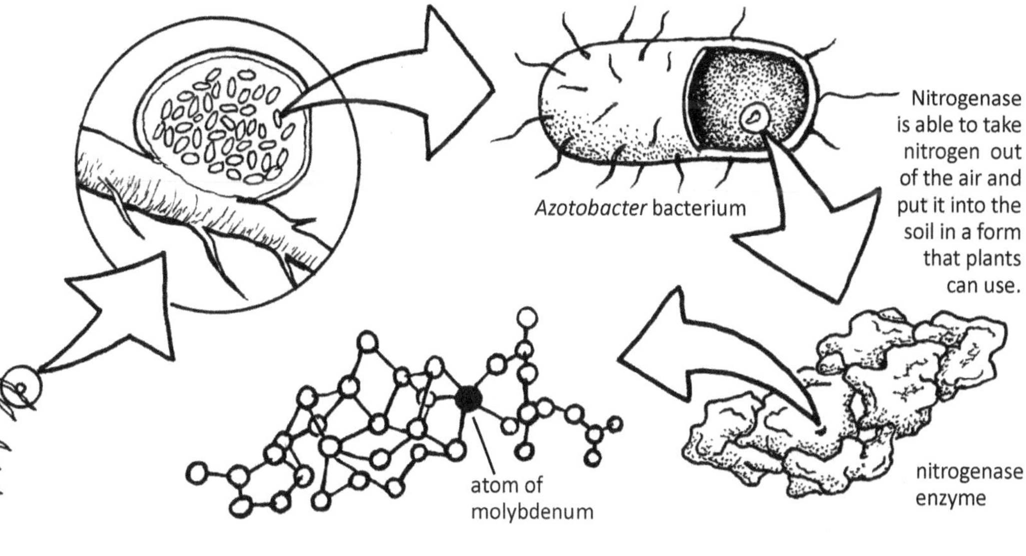

Nitrogenase is able to take nitrogen out of the air and put it into the soil in a form that plants can use.

Tc 43

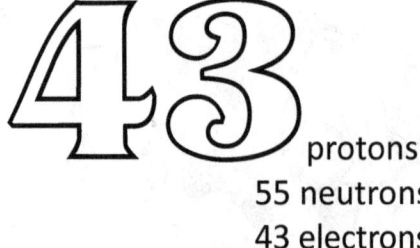

protons
55 neutrons
43 electrons
Atomic mass: 98

Technetium

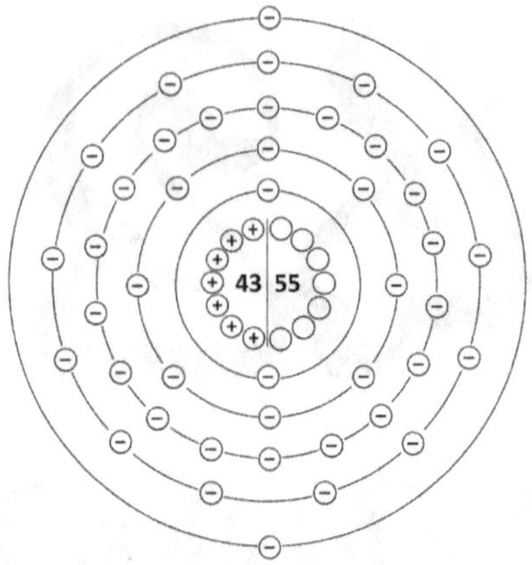

From the Greek word "teknetos" meaning "artificial"

The only place you find technetium (*tek-NEESH-ee-um*) in the natural world is in rocks that contain radioactive elements such as uranium and thorium. Large, unstable atoms, such as uranium, are gradually falling apart, and their pieces often include smaller radioactive atoms which will also eventually decay (fall apart) until a stable, non-radioactive atom is formed. All technetium atoms will eventually turn into molybdenum or ruthenium.

Technetium was officially discovered in 1936 by Carlo Perrier and Emilio Segrè in Palermo, Sicily (Italy). They had visited the Lawrence Berkeley National Lab in California (which would play a major role in element discovery in the future) and were allowed to take home some discarded parts that had become radioactive, including a sample of molybdenum. Since molybdenum is not radioactive, they guessed that there might be a new, radioactive element hiding in the sample. They were able to isolate Tc-95 and Tc-97. Later, they worked with Glenn Seaborg at the Berkeley lab and produced another isotope, Tc-99, which has proved to be incredibly useful to medical science.

One source of technetium is spent (worn out) fuel rods from nuclear reactors. The uranium in the rods decays to form many smaller atoms, including technetium. They are able to gather enough technetium atoms to make a solid lump of metal. Metallic Tc has many properties that are similar to the elements above, below, and next to it on the Periodic Table. Another way to make technetium is inside a machine called a particle accelerator where molybdenum atoms are bombarded with protons and neutrons. If a proton sticks to a molybdenum nucleus, that immediately turns it into technetium. The most useful form of technetium is "Tc-99m." ("M" stands for "metastable," a special type of isotope.) It is used widely in medial diagnostic testing because when it decays, it produces gamma radiation (similar to x-rays) that can be captured by a special camera. The gamma rays only last a short time so the exposure to harmful radiation is minimal. For example, a molecule called "Cardiolite" is used as a "tracer" to make images of the heart. Other Tc-based molecules are used to take pictures of many other body parts.

Tc-99m decays into Tc-99, an isotope with a very slow and steady decay rate that produces beta particles (electrons), making it useful to some industries for calibrating high-tech equipment.

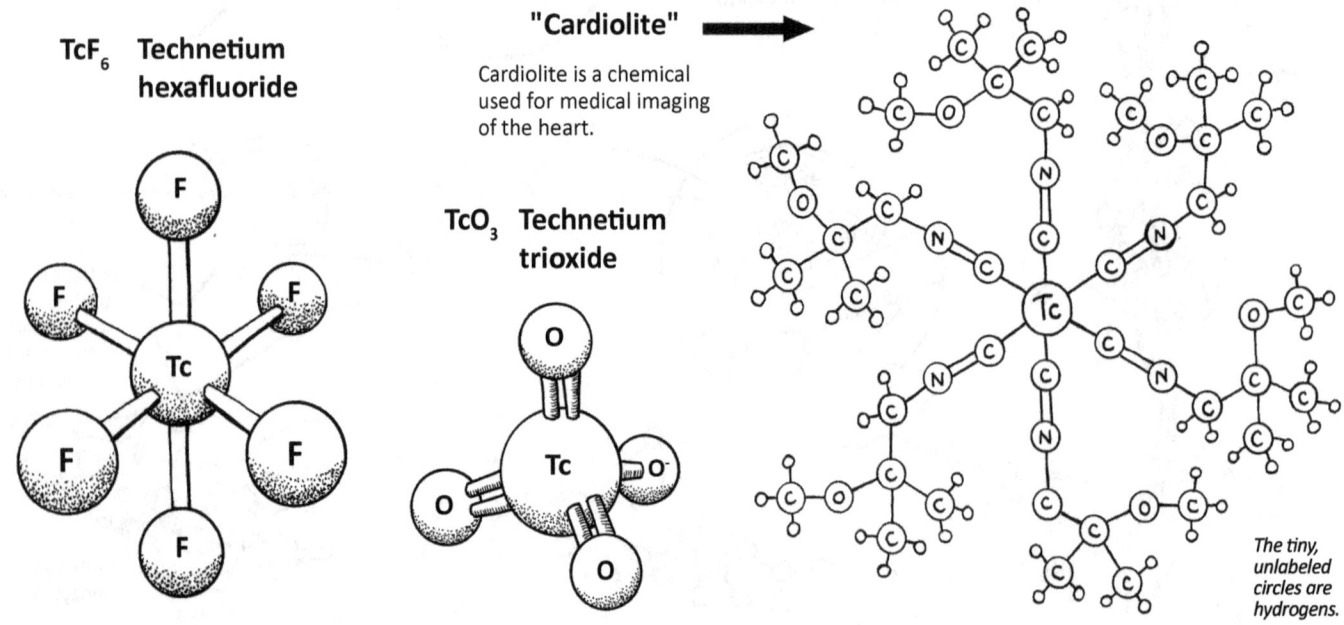

TcF_6 Technetium hexafluoride

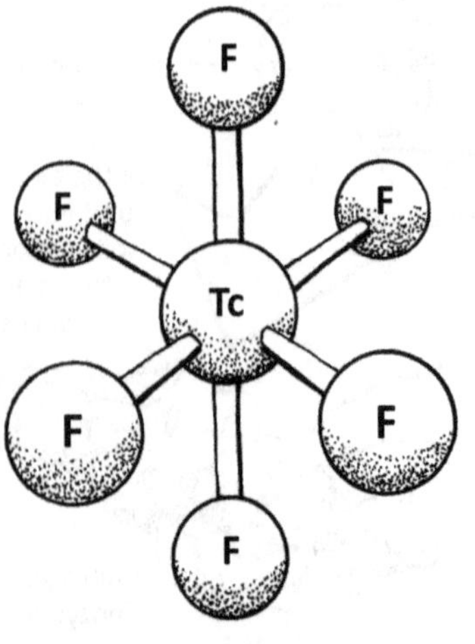

"Cardiolite"

Cardiolite is a chemical used for medical imaging of the heart.

TcO_3 Technetium trioxide

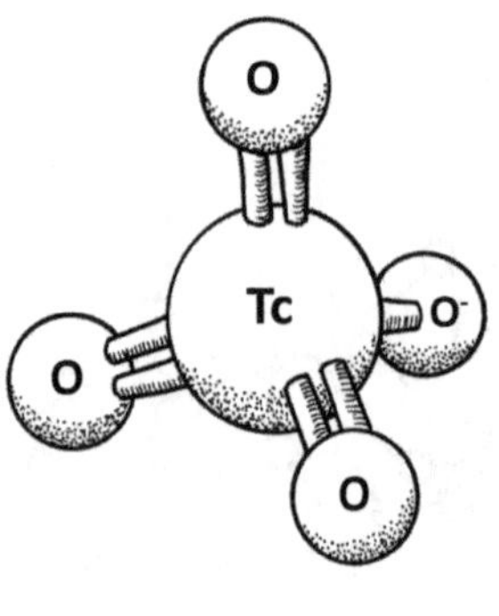

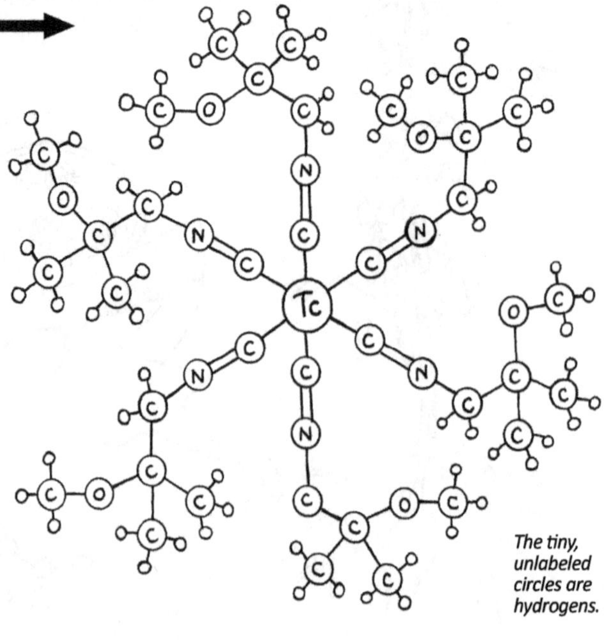

The tiny, unlabeled circles are hydrogens.

43

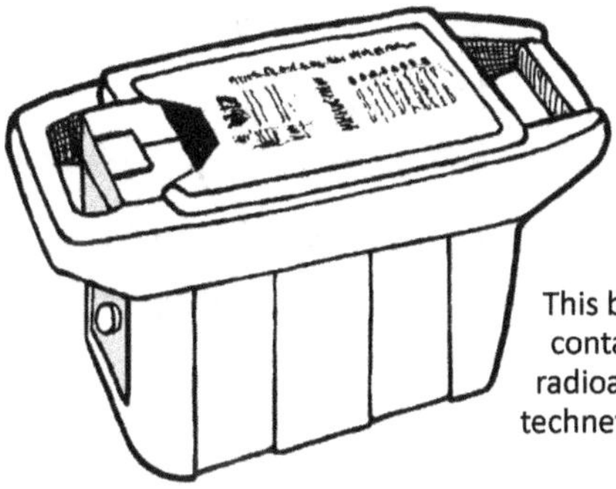

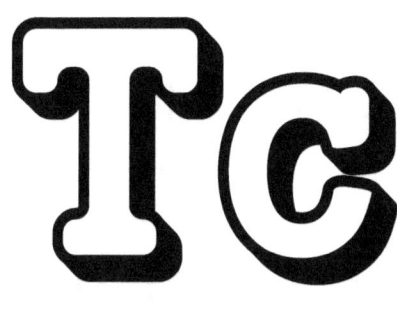

This blue box contains radioactive technetium.

Tc

Technetium

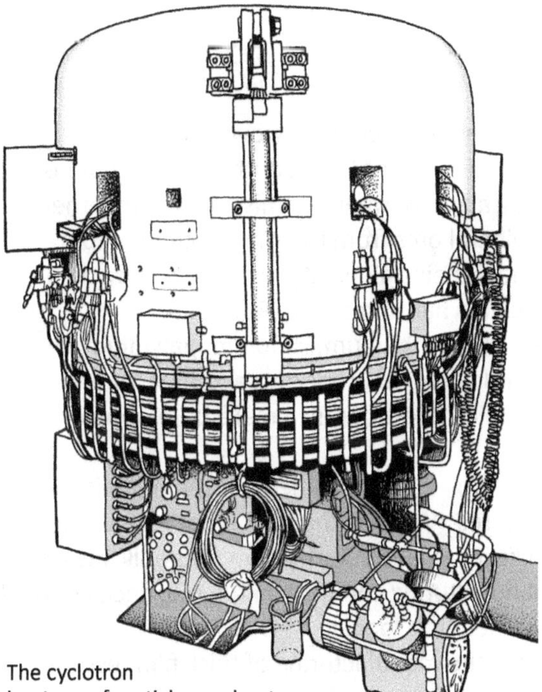

The cyclotron is a type of particle accelerator used for physics research and to make radioactive atoms for medical applications.

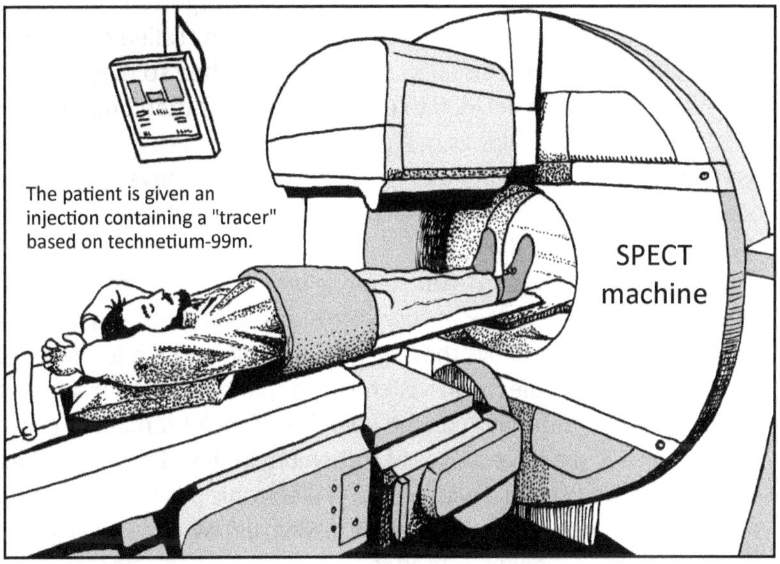

Tc-99m is used for tracers in over 50 different medical tests, including making diagnostic images of the brain and of internal organs.

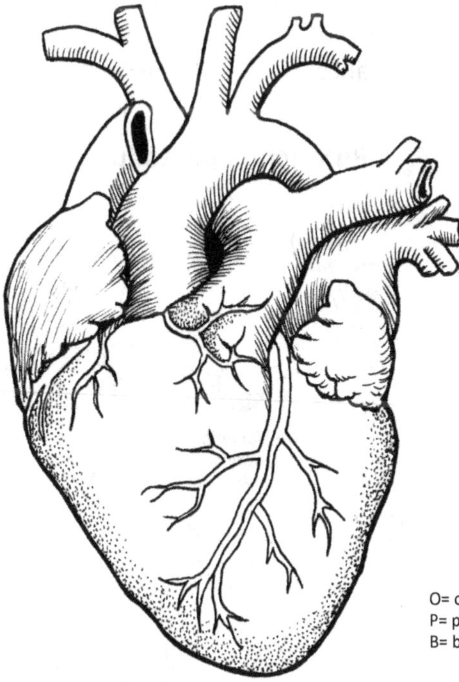

Cardiolite molecules will naturally stick to heart cells and make them visible to special cameras.

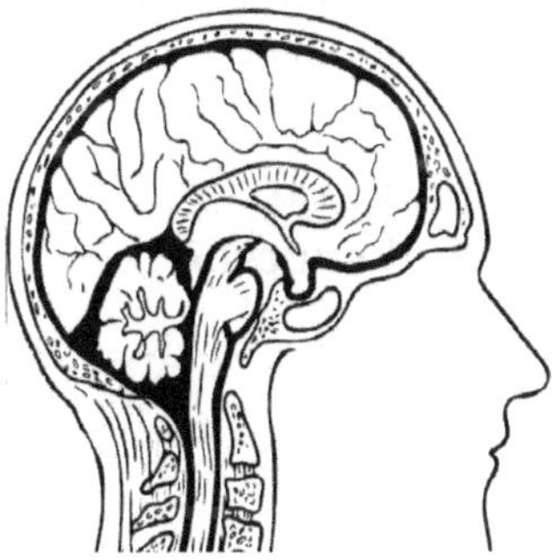

O= orange
P= purple
B= black

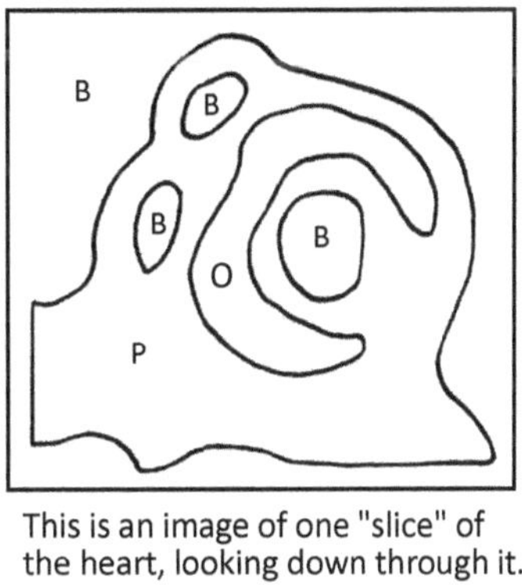

This is an image of one "slice" of the heart, looking down through it.

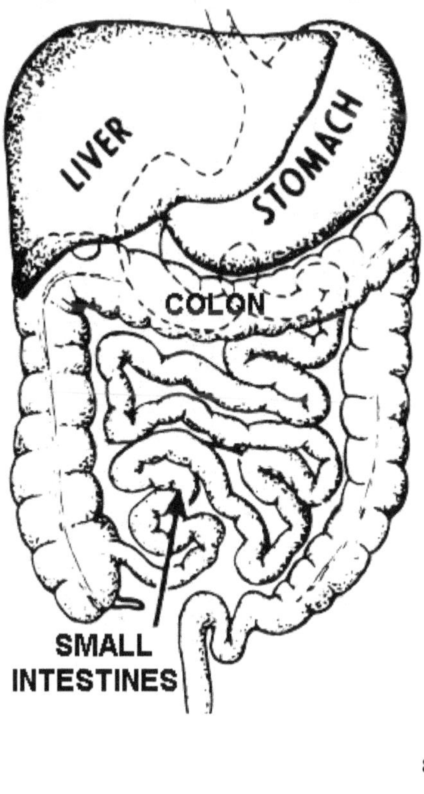

Ru 44

protons
57 neutrons
44 electrons

Ruthenium

Atomic mass: 101

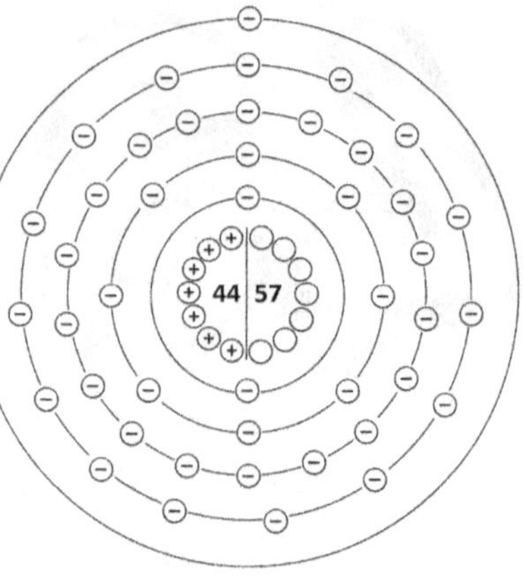

44
57

Ruthenia is an old name for a region in eastern Europe

Ruthenium was "almost discovered" twice before Karl Claus officially announced his discovery in 1844. In 1808, a Polish chemist published his work on a new element he found in mineral ores from South America, but his experiments were never verified and were thus forgotten. In 1827, a Swedish and German chemist worked together to discover three new elements, including one they called ruthenium, in mineral ores from the Ural Mountains, but the results of their experiments could not be verified so they had to withdraw their claims. When Karl Claus made his discovery in 1844, he decided to use their name, ruthenium, since his ancestors had come from Ruthenia.

Ruthenium belongs to a group of elements called the platinum group: ruthenium, rhodium, palladium, osmium, iridium, and platinum. These six elements have similar physical and chemical properties and they tend to occur together in minerals called "platinum ores." It takes complicated chemical procedures to make pure samples of these elements from the ores.

One of ruthenium's outstanding properties is its resistance to corrosion. It can be added to other metals to make alloys that are better at surviving harsh conditions and high temperatures. When ruthenium is added to titanium or to nickel-steel alloys they become durable enough to make things like jet engines. Ruthenium's durability is also useful in places like electrical switches and spark plugs. In 1944, a pen manufacturing company in the U.S. decided to use a coating of ruthenium on the gold tips of their fountain pens to make them last longer.

One of the largest uses of ruthenium in the 21st century has been in the manufacturing of thick film resistor chips which are used in a wide range of electronic products including many of the sensors found in cars. (Cars need sensors for oxygen, pressure, fuel/air mixing, airbag release, engine controls, and more.) During the manufacturing of these resistors, ruthenium is used to make a part called a photomask, in a process called extreme ultraviolet lithography.

Other uses of ruthenium include a biological stain called "ruthenium red," used to prepare biological samples for viewing under a microscope, and a powder used to lift fingerprints from crime scenes.

A radioactive isotope, Ru-106, is used by eye surgeons to treat a type of skin cancer that occurs inside the eye.

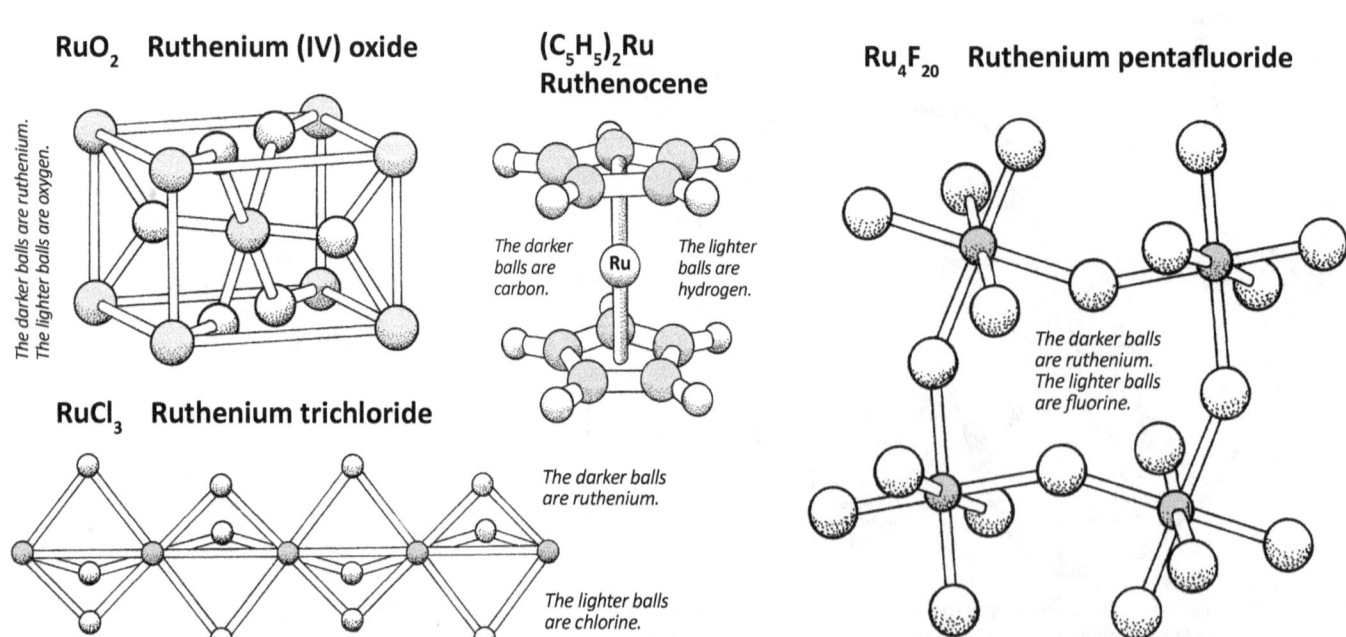

RuO_2 Ruthenium (IV) oxide

The darker balls are ruthenium. The lighter balls are oxygen.

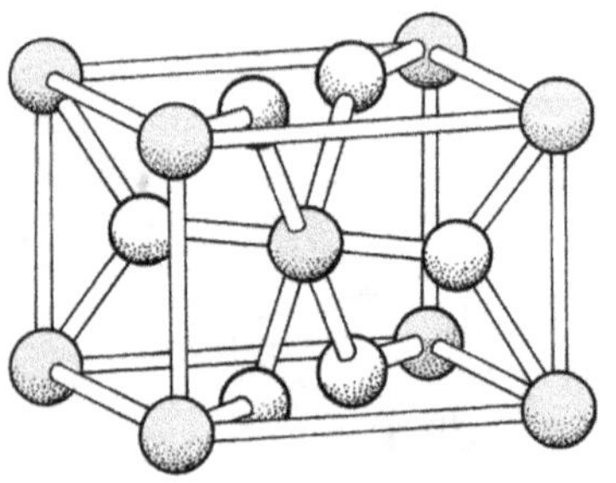

$(C_5H_5)_2Ru$ Ruthenocene

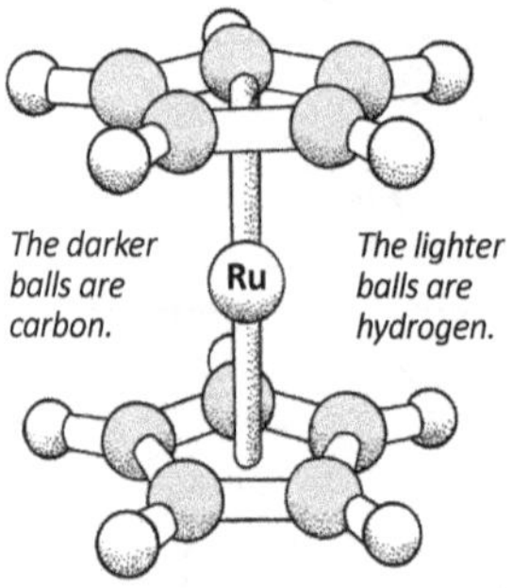

The darker balls are carbon.
Ru
The lighter balls are hydrogen.

Ru_4F_{20} Ruthenium pentafluoride

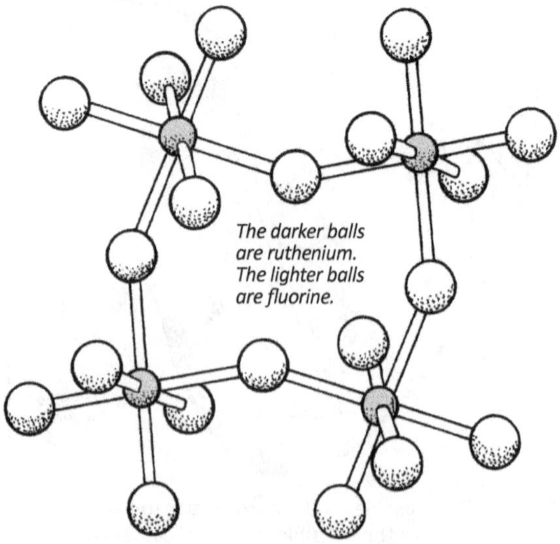

The darker balls are ruthenium. The lighter balls are fluorine.

$RuCl_3$ Ruthenium trichloride

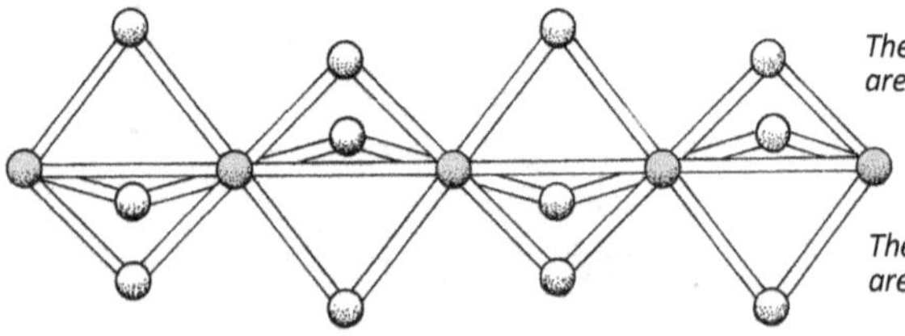

The darker balls are ruthenium.

The lighter balls are chlorine.

44 Ru

44 Ru Ruthenium	45 Rh Rhodium	46 Pd Palladium
76 Os Osmium	77 Ir Iridium	78 Pt Platinum

Platinum group

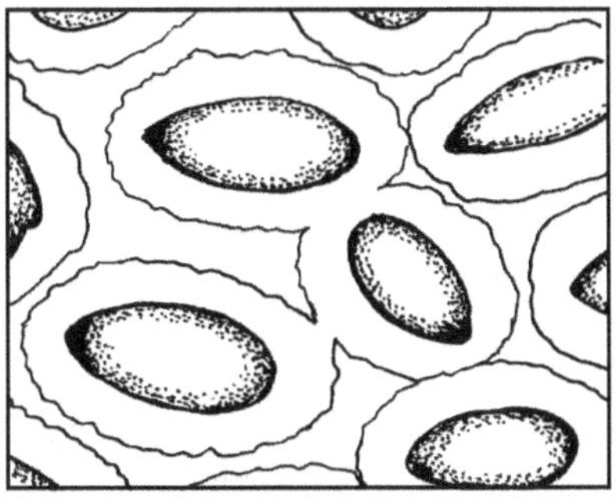

These seeds were stained with "ruthenium red" to reveal a sugar-based substance they produce. *(The rings around the seeds are bright pink.)*

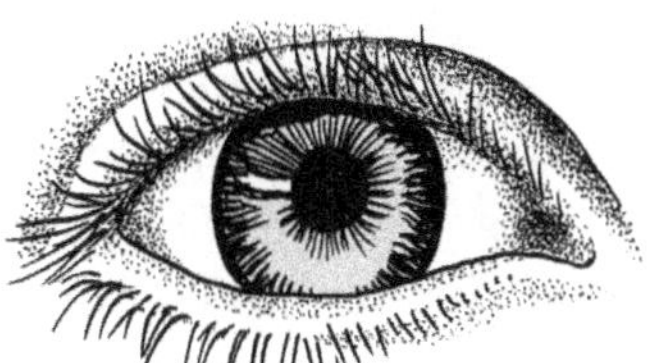

Radioactive ruthenium is used to treat cancer inside the eye.

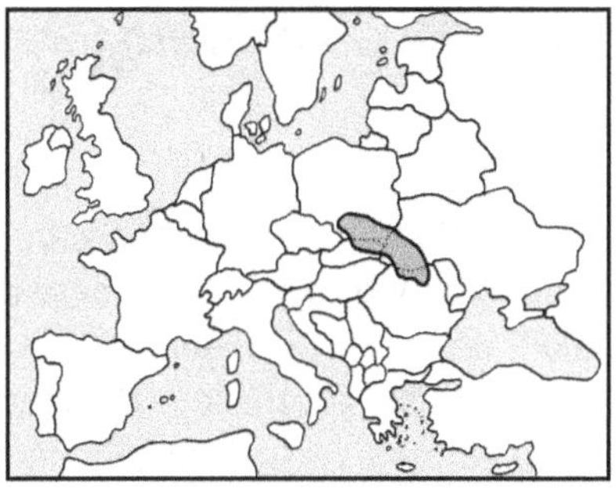

Ruthenium is named after an area of Europe that used to be called Ruthenia.

Ruthenium

Ruthenium can be used to make photomasks for a process called extreme ultraviolet lithography, which makes resistor chips.

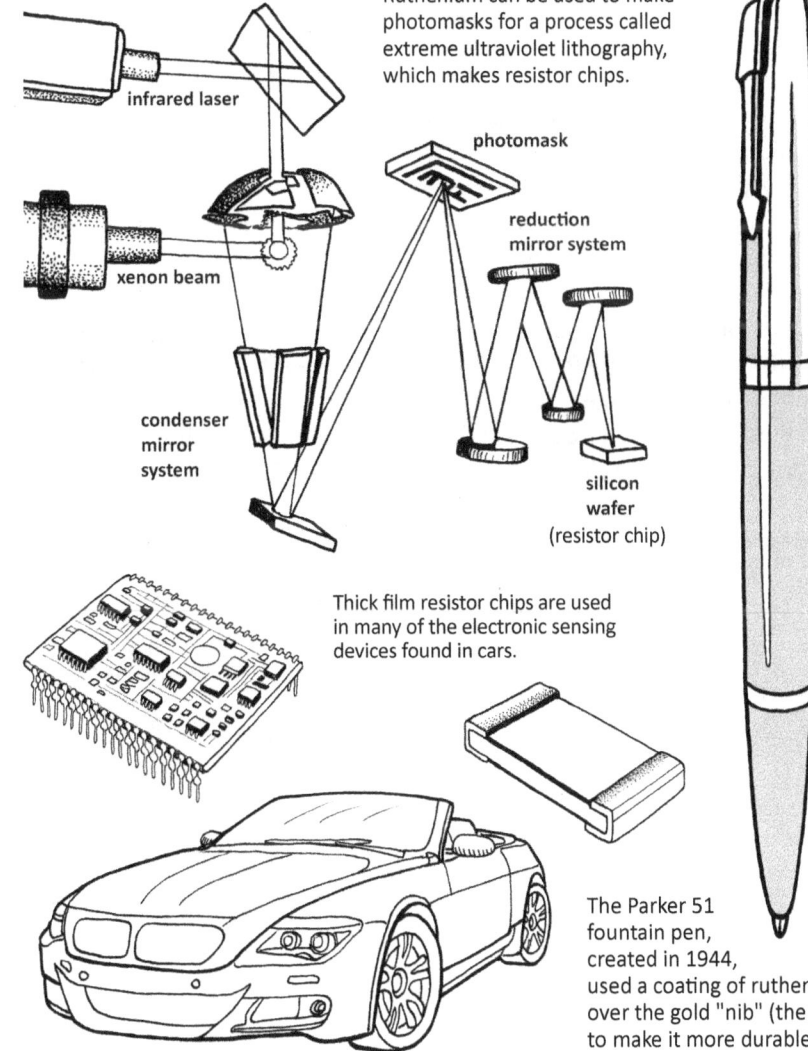

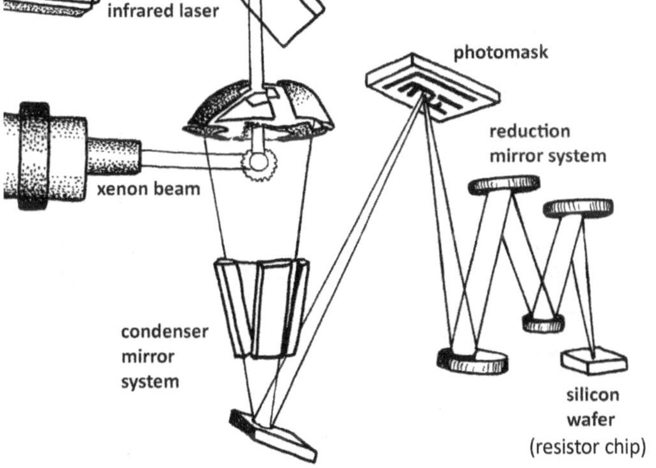

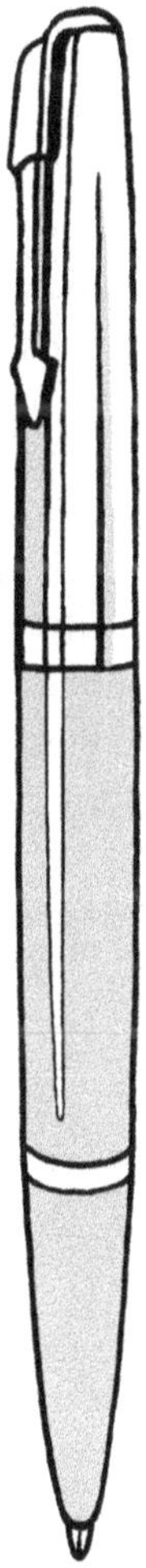

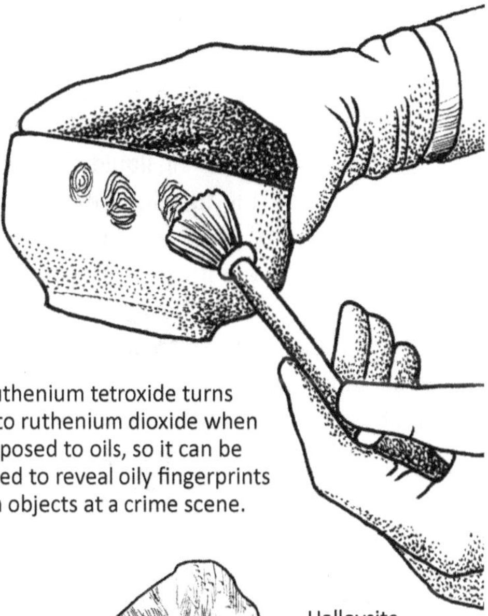

Ruthenium tetroxide turns into ruthenium dioxide when exposed to oils, so it can be used to reveal oily fingerprints on objects at a crime scene.

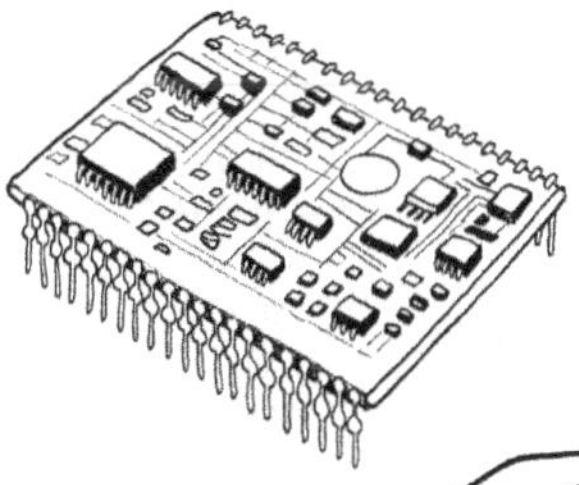

Thick film resistor chips are used in many of the electronic sensing devices found in cars.

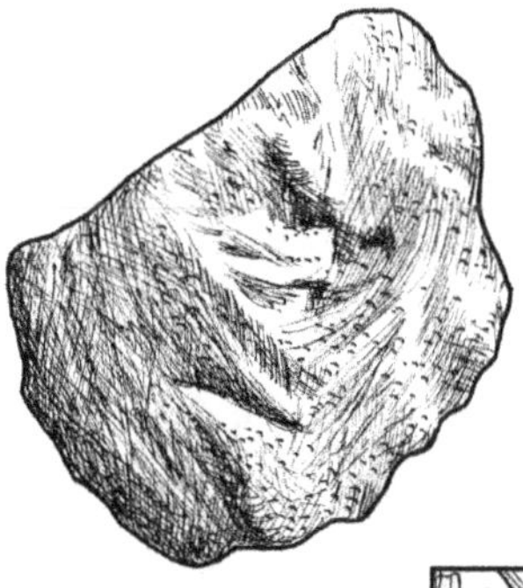

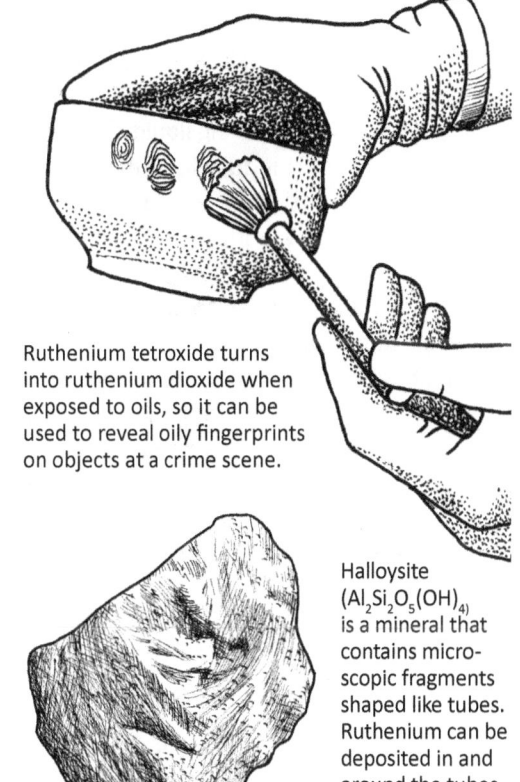

Halloysite ($Al_2Si_2O_5(OH)_4$) is a mineral that contains microscopic fragments shaped like tubes. Ruthenium can be deposited in and around the tubes.

The Parker 51 fountain pen, created in 1944, used a coating of ruthenium over the gold "nib" (the point) to make it more durable.

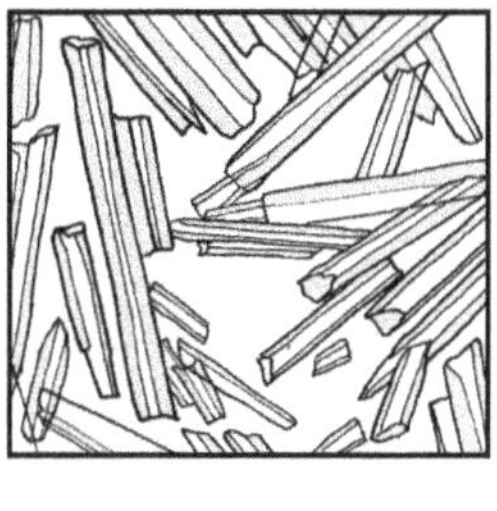

Ruthenium-filled Halloysite can be used as a catalyst in chemical reactions. Catalysts speed up reactions without being changed by them.

Rh 45

Rhodium

45 protons
58 neutrons
45 electrons
Atomic mass: 102.9

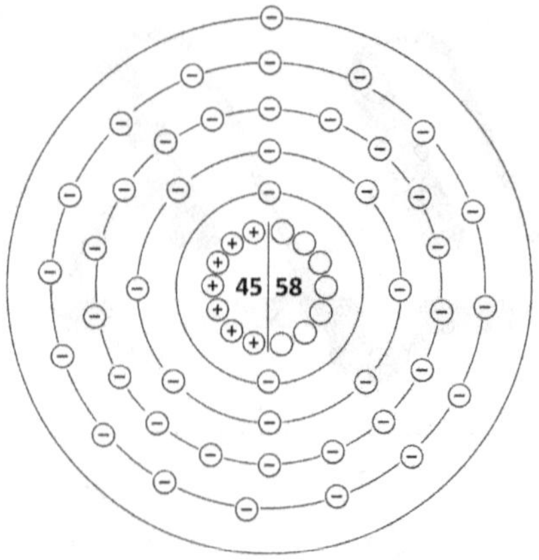

From the Greek word for rose, "rhodon"

Rhodium is one of the most rare elements on earth. It occurs only in very small amounts in minerals that contain a lot of platinum. These platinum ores also usually contain tiny amounts of ruthenium, palladium, osmium, and iridium. Because all of these elements are found together, a mining operation that produces platinum will also be able to supply these other elements. Rhodium is produced in such small amounts that it can only be used in thin layers, often as a surface coating [electroplated] on top of other metals. However, this is probably the best use for rhodium, since it is very resistant to corrosion and heat, but is difficult to cut and mold into shapes.

Rhodium was discovered by William Wollaston in 1803, just a few weeks after he had discovered palladium. Both elements were found in a piece of platinum ore that had come from South America. Wollaston used a series of acid solutions to pull the copper and lead out of the ore, then another series of chemical reactions to pull out platinum and palladium. After he was done, there was still something leftover: a rose-colored powder. He figured out that this powder contained a new element and named it rhodium using the Greek word for rose: "rhodon."

The first use that was found for this new element was making thermocouples that needed to withstand very high temperatures (1300-1800° C). It was then found that rhodium was also useful for making a very shiny protective coating on metals such as silver and gold. This trick is still used by jewelers today. They dip the jewelry into an electrically charged solution containing liquid rhodium. The jewelry glistens when new, but then dulls as the rhodium wears off.

The largest industrial use of rhodium began in 1975 when car manufacturers discovered that rhodium had the amazing ability to convert harmful exhaust fumes into nitrogen, carbon dioxide and water vapor, which are much safer to release into the environment. The part that does this is called the catalytic converter. (A "catalyst" is something that causes chemical changes.) Platinum and palladium are also used in catalytic converters.

Rhodium is used to coat metal parts inside furnaces and x-ray machines, and to coat and protect electrical switches and spark plugs. It is used in neutron detectors in nuclear power plants. The insides of searchlights and car headlights are sprayed with rhodium to make them super shiny so they reflect light more efficiently. Rhodium wires (coated with insulating plastic) can be found in heart pacemakers.

Rhodium is more expensive than gold or platinum, so it has been used to coat prestigious medals of honor. Paul McCartney (of the Beatles) received a rhodium-plated award from the Guinness Book of World Records in 1979.

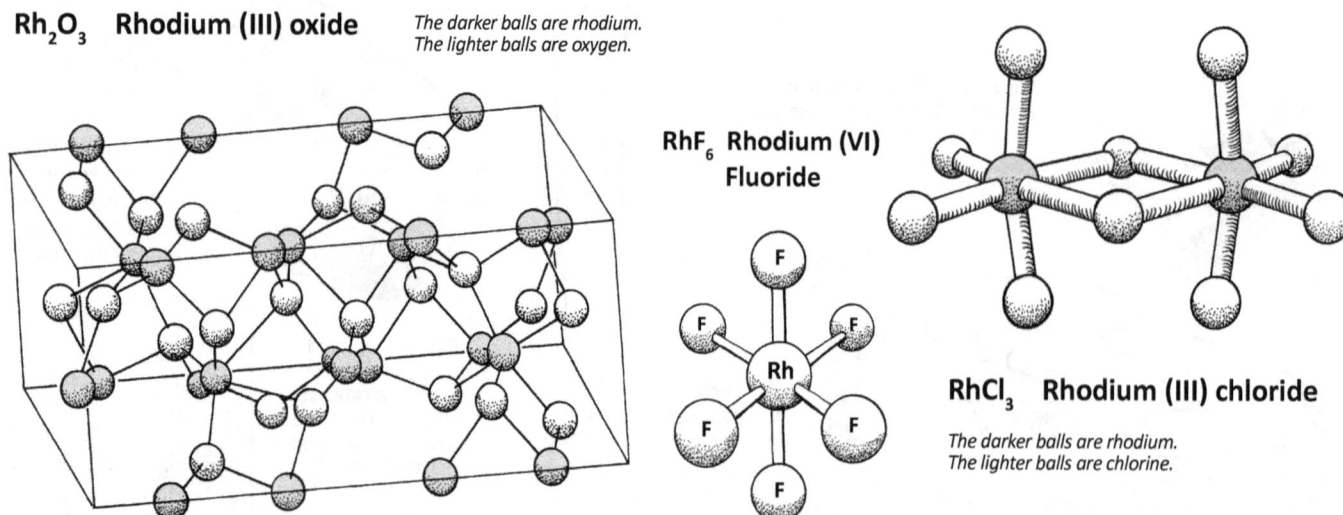

Rh_2O_3 Rhodium (III) oxide

The darker balls are rhodium.
The lighter balls are oxygen.

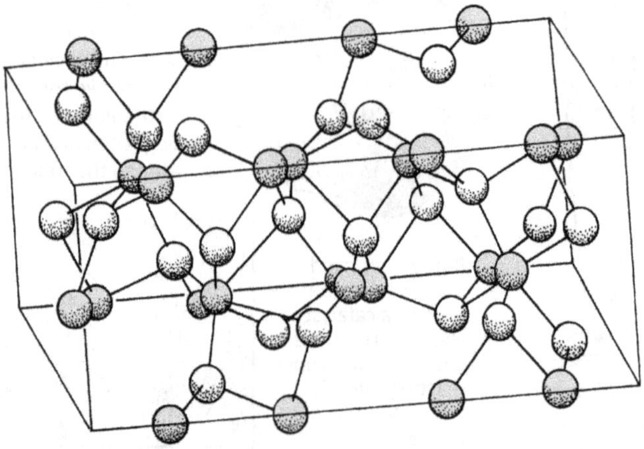

RhF_6 Rhodium (VI) Fluoride

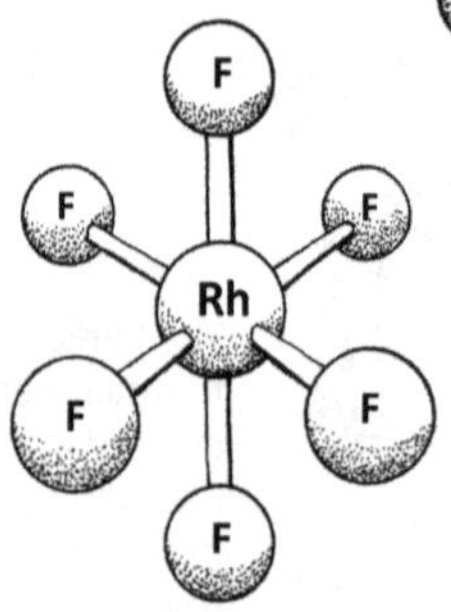

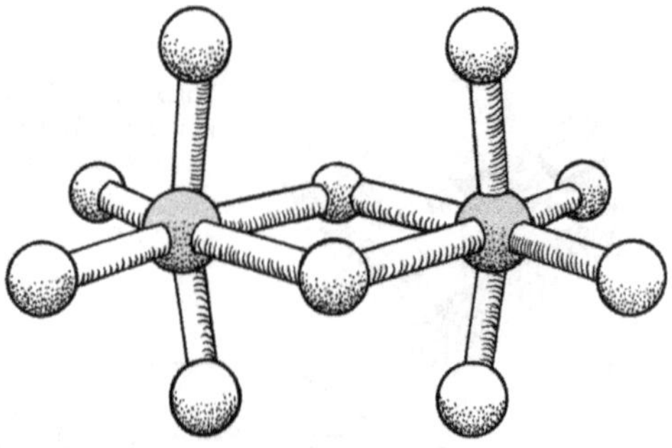

$RhCl_3$ Rhodium (III) chloride

The darker balls are rhodium.
The lighter balls are chlorine.

45 Rh

44 Ru Ruthenium	45 Rh Rhodium	46 Pd Palladium
76 Os Osmium	77 Ir Iridium	78 Pt Platinum

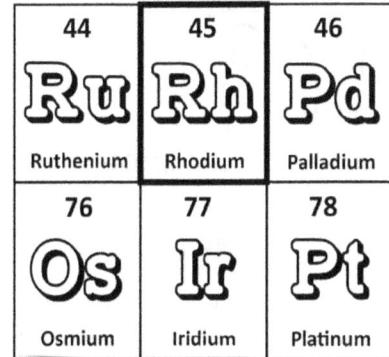

Platinum group

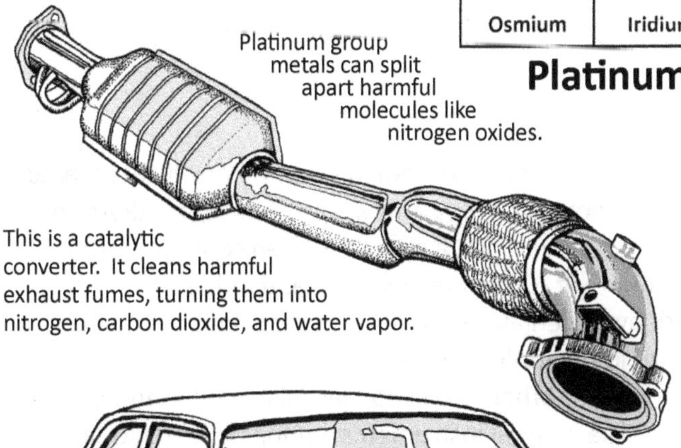

Platinum group metals can split apart harmful molecules like nitrogen oxides.

This is a catalytic converter. It cleans harmful exhaust fumes, turning them into nitrogen, carbon dioxide, and water vapor.

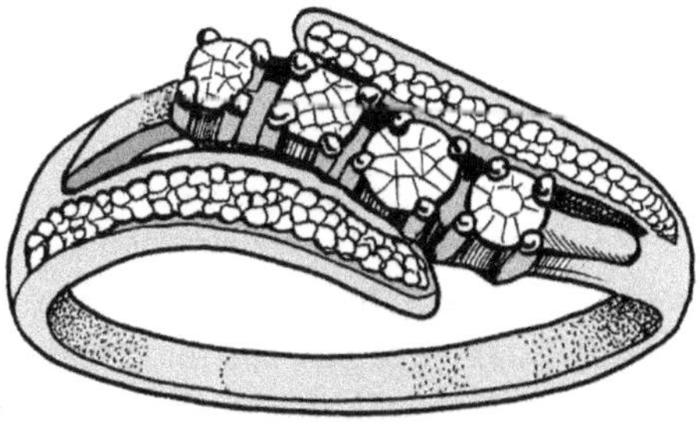

Jewelry is often electroplated with rhodium to make it look extra shiny.

Make each gemstone a different color.

Rhodium

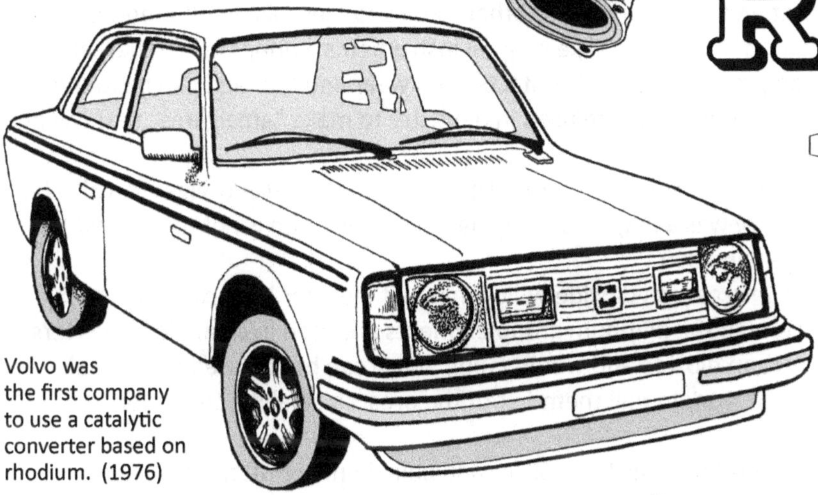

Volvo was the first company to use a catalytic converter based on rhodium. (1976)

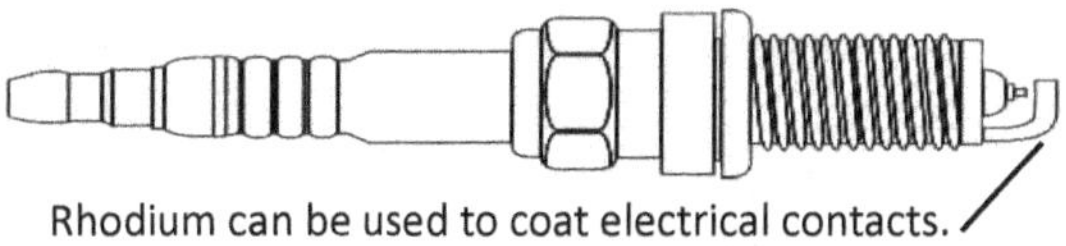

Rhodium can be used to coat electrical contacts.

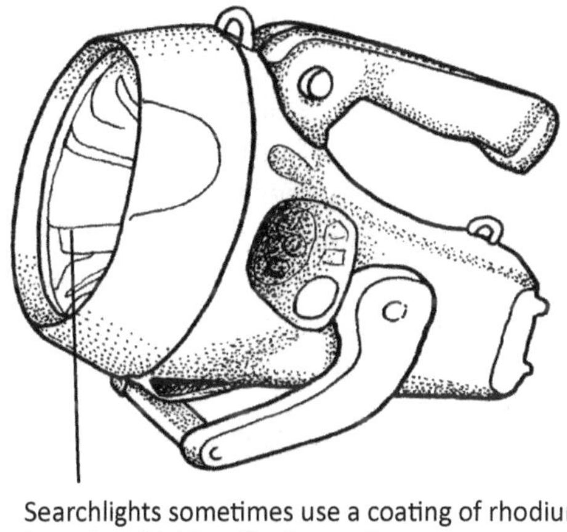

Searchlights sometimes use a coating of rhodium to make a highly reflective surface.

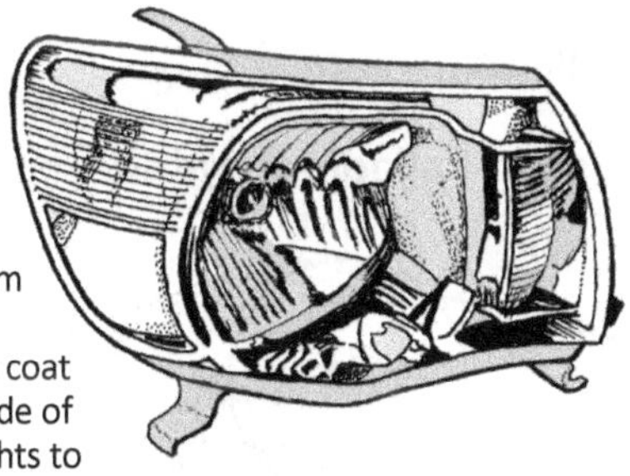

Rhodium can be used to coat the inside of headlights to make them more reflective.

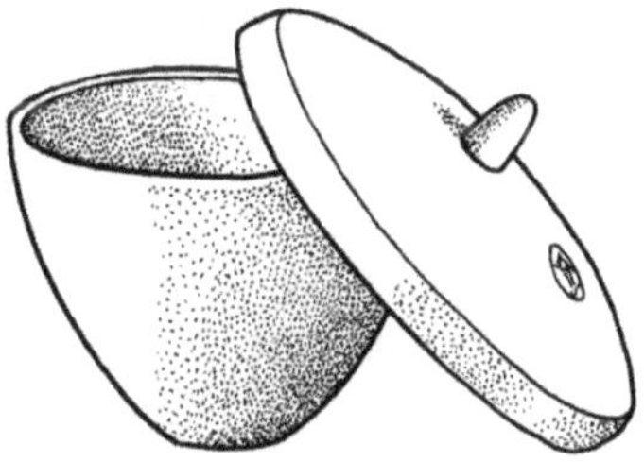

This crucible is made of an alloy containing rhodium.

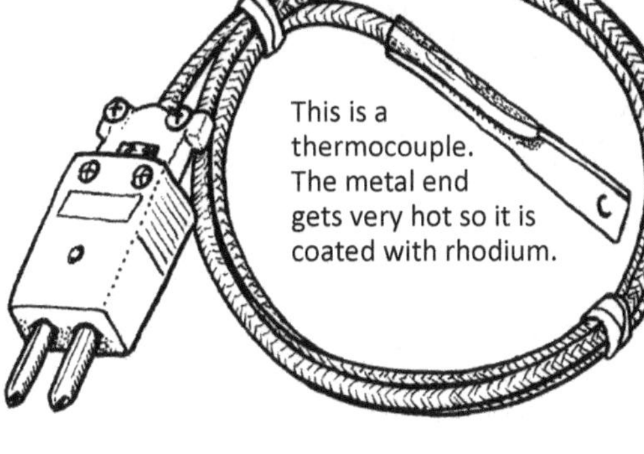

This is a thermocouple. The metal end gets very hot so it is coated with rhodium.

This Canadian coin looks almost black, though it is made of silver and is coated with rhodium.

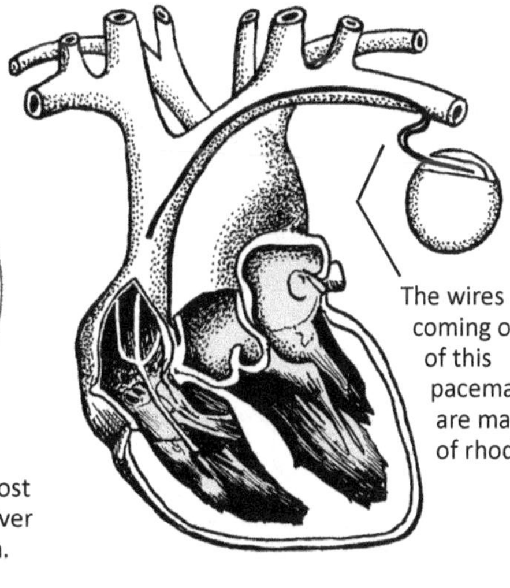

The wires coming out of this pacemaker are made of rhodium.

Pd 46

protons
60 neutrons
46 electrons

Atomic mass: 106.4

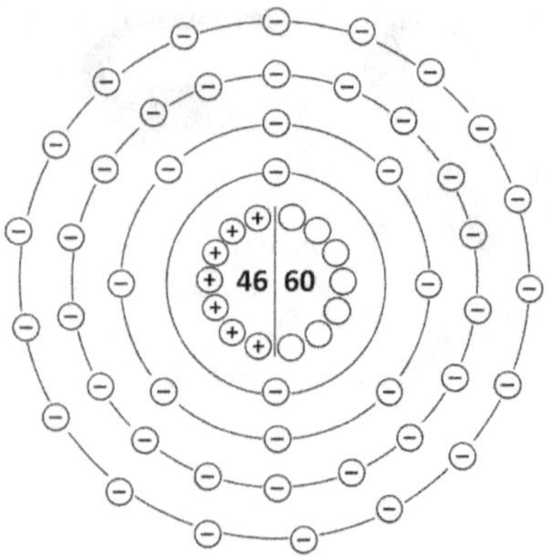

Palladium

Named after the asteroid "Pallas"

Discovered by William Wollaston in 1803, palladium *(pah-LAY-dee-um)* was named after the newly discovered asteroid, Pallas. (The asteroid was named after a figure in Greek mythology.) Wollaston would discover rhodium soon after palladium because the mineral ore he was working with was a platinum ore, a rock that often contains both rhodium and palladium. Some platinum ores also have small amounts of ruthenium, osmium and iridium.

One of palladium's unique features is its ability to absorb hydrogen atoms, almost like a sponge holds water. This feature is being researched as a way to store hydrogen for use in things like fuel cells.

Palladium is resistant to corrosion and heat and can be added to other metals to make alloys. It is added in small amounts to metals such as steel and titanium to make things like surgical instruments, springs in watches, and nibs (points) for pens. Like rhodium, palladium can protect electrical contacts from wear and tear and can be found in places like aircraft spark plugs. Other specially formulated alloys are used in dentistry to make "amalgams" that will be part of dental implants that will replace teeth.

"White gold" is an alloy of gold and palladium and is popular for making jewelry. (Nickel and silver can also be used to make white gold.) In the 20th century, jewelry was a major use of palladium. Palladium is not often used for coins, but Russia did use it for a commemorative coin in 1989.

Palladium chloride turns carbon monoxide into carbon dioxide which makes it ideal for both carbon monoxide detectors and catalytic converters in cars. Several of the platinum group elements are used in catalytic converters since they can turn harmful gases into nitrogen, carbon dioxide and water vapor, which can be released into the air. Platinum group elements are rare and it takes a lot of work to pull them out of the ores, so catalytic converters are recycled to reclaim the platinum group elements.

In the 21st century, a major use of palladium is for capacitors. Silver and palladium are often combined to make these tiny parts that can hold electrical charges like miniature batteries.

Palladium is sometimes used in the manufacturing of professional-quality flutes.

$PdCl_2$ Palladium chloride

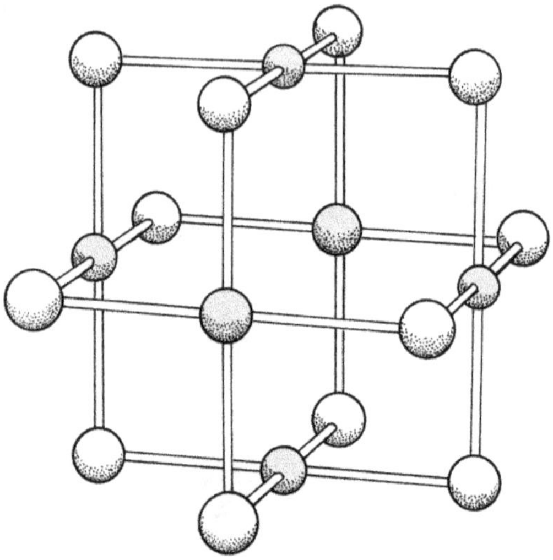

The darker balls represent palladium.
The lighter balls represent chlorine.

PdH Palladium hydride

Palladium has the amazing ability to attract and hold hydrogen atoms, almost like a sponge. It can absorb 900 times its own volume of hydrogen!

The combination of these two elements is technically an alloy. Researchers are investigating whether palladium can be used to make hydrogen storage tanks.

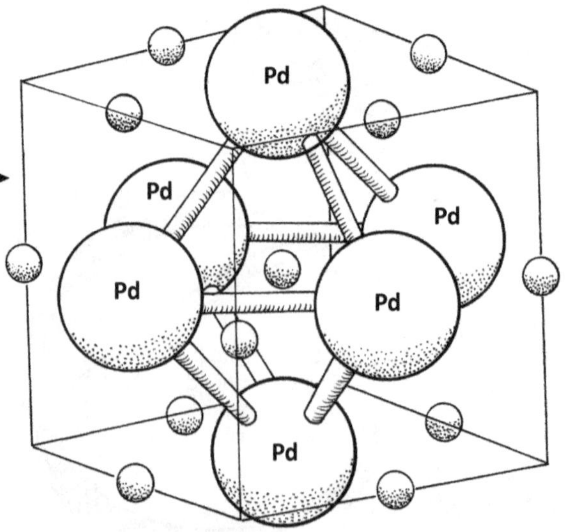

The tiny balls represent hydrogen atoms.

46 Pd

44	45	46
Ru Ruthenium	**Rh** Rhodium	**Pd** Palladium
76	77	78
Os Osmium	**Ir** Iridium	**Pt** Platinum

Platinum group

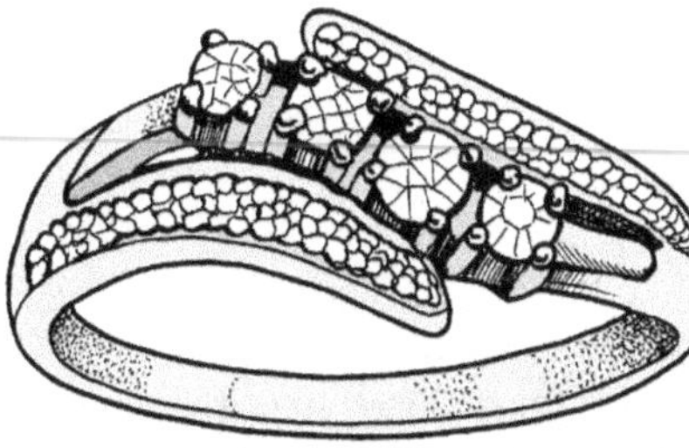

Palladium is mixed with gold to make "white gold."

Make the gemstones on this ring four different colors.

Palladium is used to make professional quality flutes.

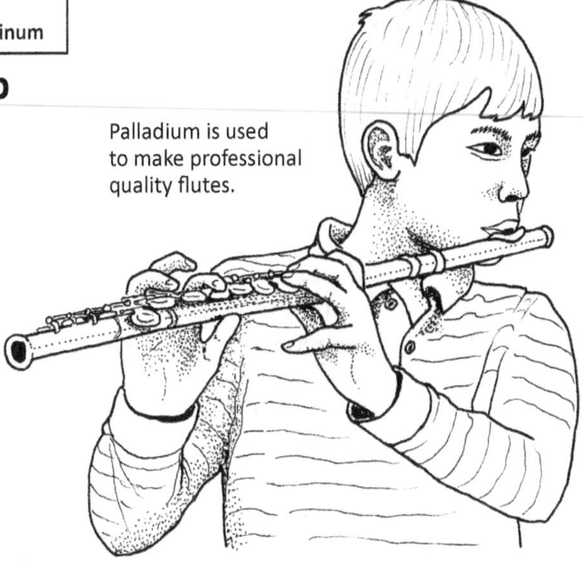

Palladium

Palladium is used inside catalytic converters in cars. Like rhodium, palladium can turn harmful exhaust fumes into less harmful gases.

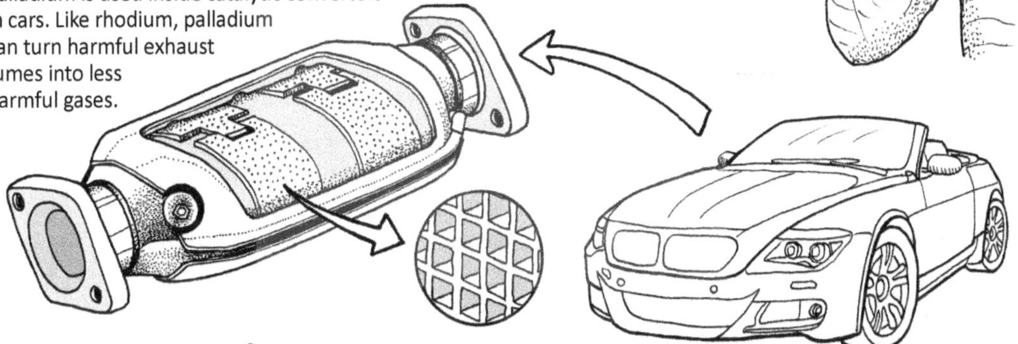

Springs inside non-digital watches often contain palladium.

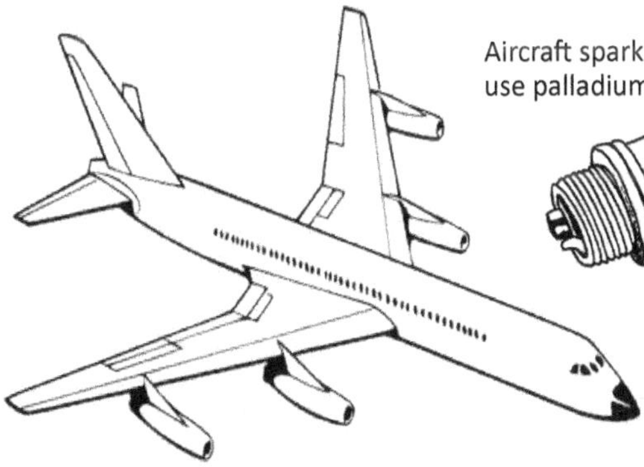

Aircraft spark plugs use palladium.

Palladium can be used to make durable pen points (nibs).

Pd is used to make carbon monoxide detectors to keep us safe in our homes.

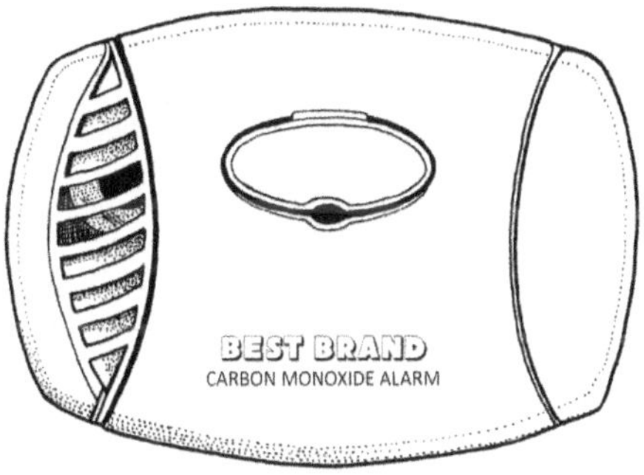

Palladium and silver are used in ceramic capacitors.

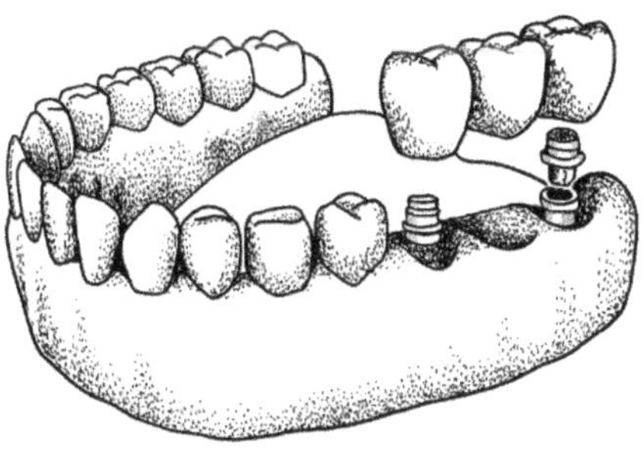

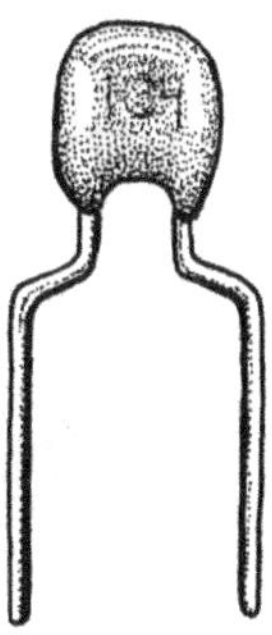

Surgical tools are sometimes made using Pd alloys.

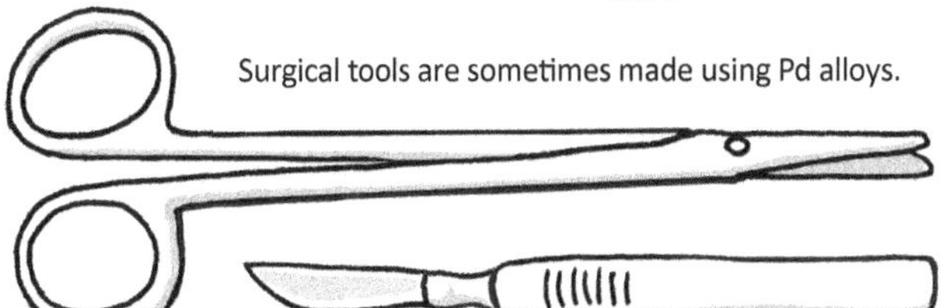

Palladium is used in small amounts in dental amalgams to make the dental implants resistant to corrosion.

Ag 47

protons
61 neutrons
47 electrons
Atomic mass: 107.8

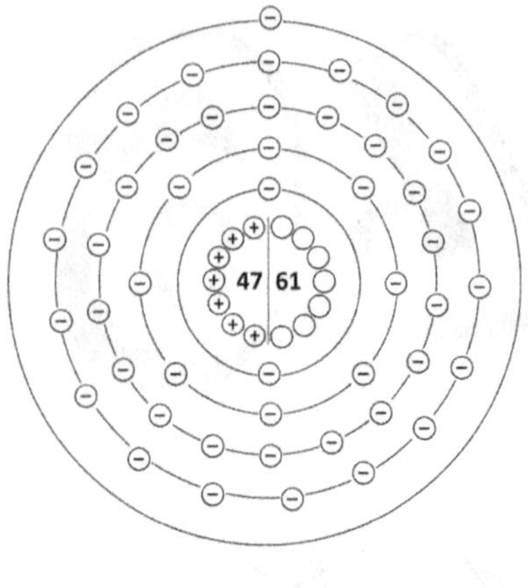

Silver

"Silver" is from the old Anglo-Saxon word "seolfor"

The symbol, Ag, comes from the Latin word for silver, "argentum"

Silver is one of the metals that has been known since ancient times and has been found all over the world. The ancients discovered that silver was too soft to make good weapons, so they used it for decorative purposes such as jewelry, tableware and coins. Silver is still used for all of these purposes today. Unfortunately, silver has a tendency to tarnish (turn gray) and must be polished occasionally. Expensive jewelry is often made tarnish-proof by the addition of some platinum, palladium, or gold.

Silver is sometimes found in its pure form in nature, but most often it is found mixed into ore rocks that primarily contain lead or copper. The silver must be melted out of the ore, collected, and purified. The Greeks had learned how to do this is by the 7th century B.C., and their silver mines in Laurion, near Athens, provided the funds to build their navy of trireme ships. This navy fought the famous "battle of Salamis" against King Xerxes of Persia.

In modern times many more uses have been found for silver, many of them related to health or to the electronics industry. Though some people in the past discovered silver's antibiotic (germ-killing) properties and put silver coins into their rain barrels to keep the water from going bad, today we have further exploited silver's ability to slow the growth of bacteria, putting it into bandages, medicines, and fabric for athletic clothing. Silver diamine fluoride is used by dentists in cavity prevention. Silver nitrate has been used to prevent eye infections and used to be part of standard care for newborn babies. Colloidal silver can be used to disinfect swimming pools.

Silver conducts electricity better than any other metal, and it can also be drawn out into wires that are only a few atoms wide. It is useful to the electronics industry not only for tiny wires, but also for capacitors (which hold an electrical charge like a battery) when layered with palladium.

Silver bromide is light-sensitive and was used to make photographs from the mid 1800s till the late 1900s.

Silver is basically non-toxic and can be used in small amounts on food items. In some cultures, it is rolled into an extremely thin foil that is used to decorate cakes and cookies.

Silver fulminate is an incredibly unstable molecule and is used to make (harmless) "bang snaps" for holidays.

$AgNO_3$ Silver nitrate

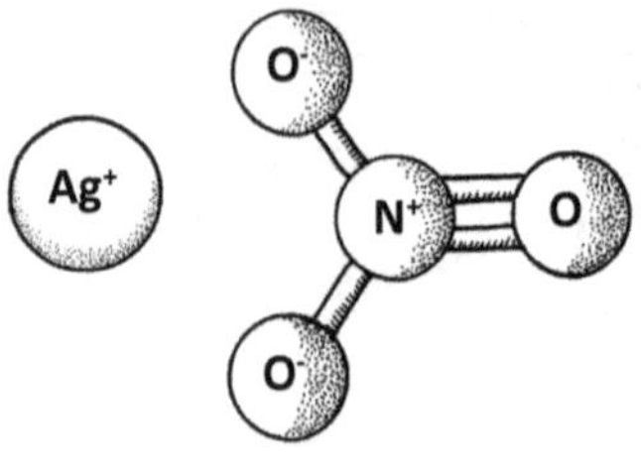

AgCNO Silver fulminate

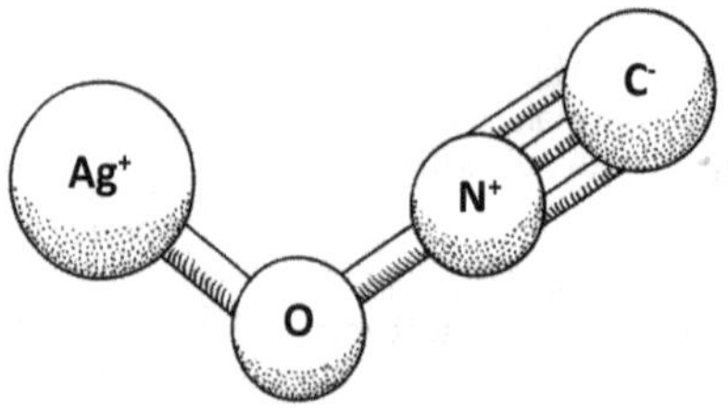

AgS_2 Silver sulfate

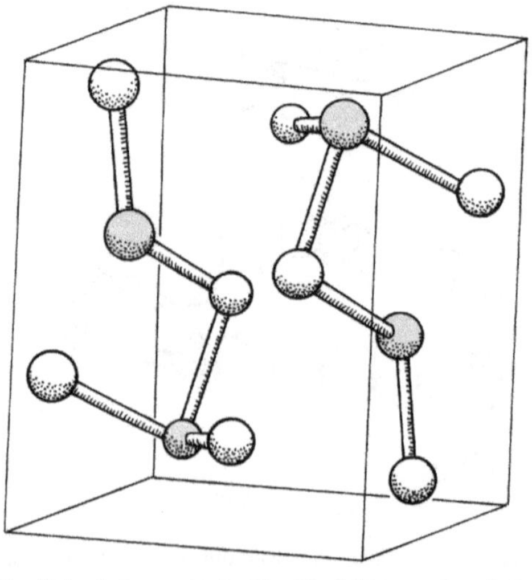

The darker balls represent sulfur. The lighter ones are silver.

$Ag(NH_3)_2$ Silver diamine

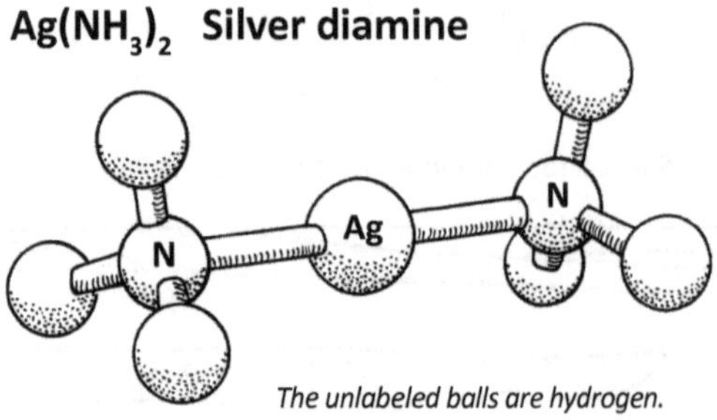

The unlabeled balls are hydrogen.

AgF_2 Silver fluoride

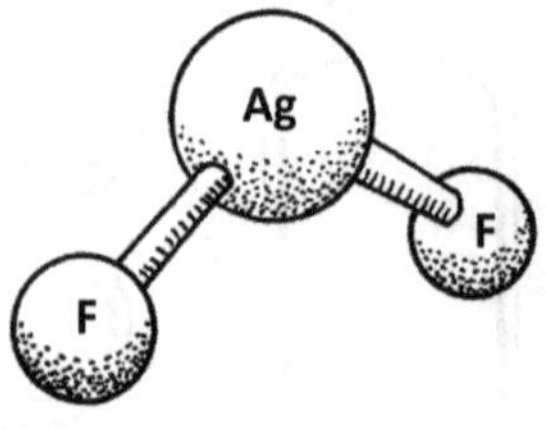

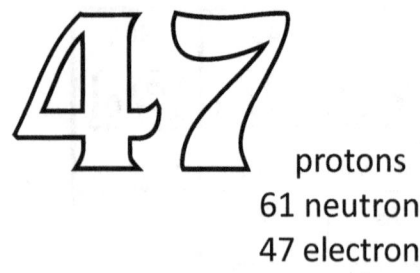

47 Ag Silver

"Bang snaps" are made of silver fulminate, an unstable molecule that falls apart with a "BANG"! when you throw it.

American "silver eagle"

Ancient Greek tetradrachm coin from about 300 B.C.

Ancient Egyptian silver statue of Horus the falcon god from 500 B.C.

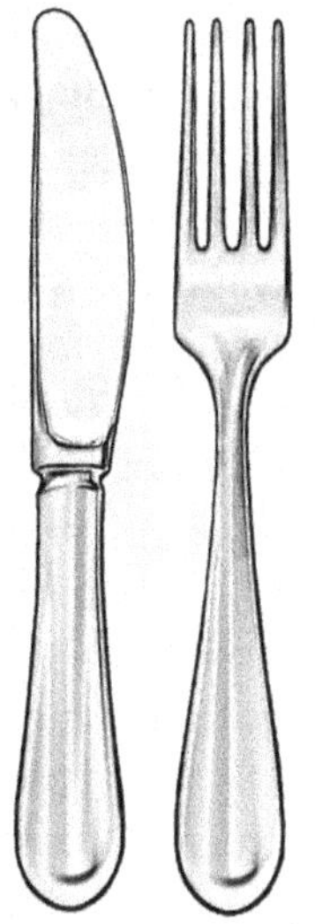

Real "silverware" is often made of a silver alloy.

In past centuries, silver coins were placed in rain barrels to keep the water fresh. Silver discourages the growth of bacteria.

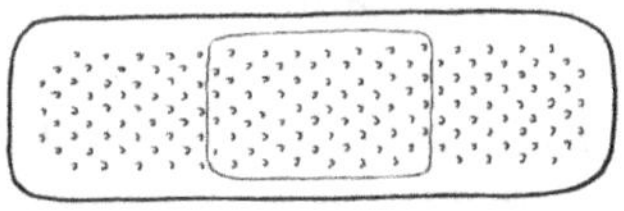

Some bandages include silver to fight infection.

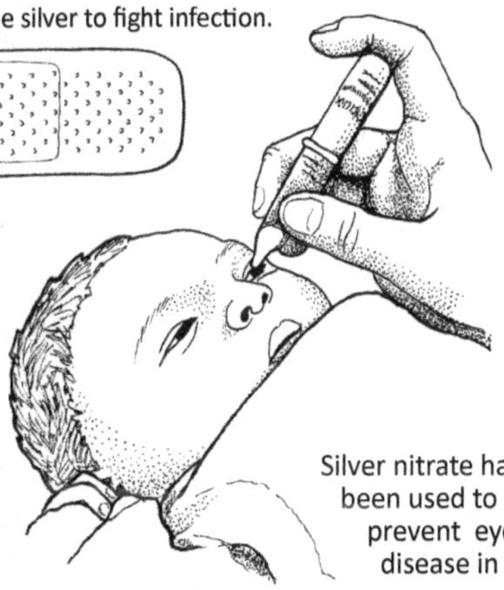

Silver nitrate has been used to prevent eye disease in babies.

These sails have red and white stripes.

The Laurion silver mines provided the money needed to build the Greek navy of trireme ships that won the battle of Salamis against King Xerxes of Persia.

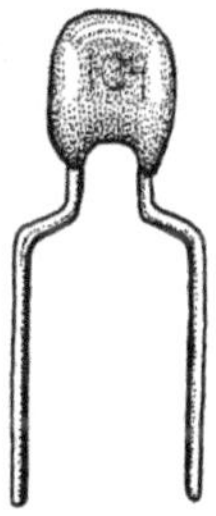

Palladium and silver are used in ceramic capacitors.

The Laurion silver mines near Athens, Greece.

Silver bromate is light-sensitive and was used to prepare metal plates for capturing a photograph.

Cd 48

protons
64 neutrons
48 electrons

Atomic mass: 112.4

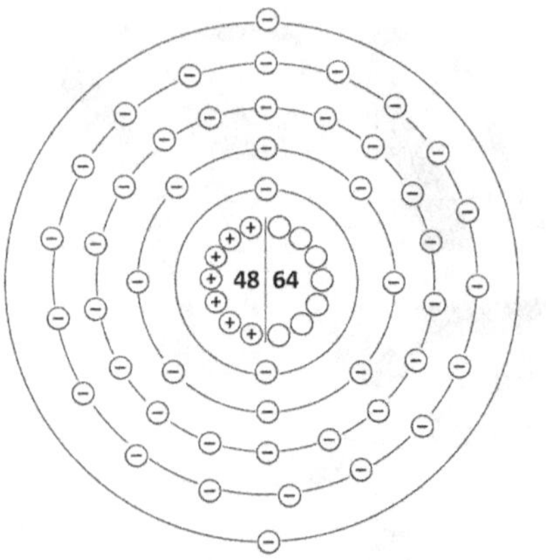

Cadmium

Named after the Greek hero, Cadmus, founder of Thebes

Cadmium was discovered by a German chemist in 1817 while he was studying samples of a mineral called calamine, $ZnCO_3$. Some of his samples glowed yellow when heated and others did not. He guessed that some of his samples contained a hidden element. He found a way to isolate the element and named it cadmium, from the Greek word "kadmea," an old name for calamine ore. Zinc ores continue to be the major source of cadmium.

One of cadmium's outstanding properties is that it is toxic. Because it is poisonous, its uses are limited. One of the oldest uses for cadmium is as pigment for paint. Cadmium yellow (cadmium sulfide) is bright and fade-proof and was highly valued by artists. Cadmium red (cadmium selenide) is another common cadmium paint. Cadmium pigments can also be used to color glass. (Since cadmium is toxic, these paints should only be used by adult artists.)

The largest use of cadmium today is in batteries. Nickel-cadmium (Ni-Cad) battery packs can be found in toys and tools that are rechargeable. The batteries are often bundled together and wrapped in bright yellow plastic.

Cadmium is resistant to corrosion and is sometimes used to coat steel for the aircraft industry.

Cadmium is very good at absorbing neutrons so it is used in fuel rods in nuclear reactors. When they want to slow down the fission process, they can expose the cadmium rods and catch many free neutrons that would otherwise go on to hit more atoms and cause more fission. Cadmium acts like "brakes" on the fission process.

In the late 20th century, cadmium was used to make the phosphors in black and white television sets, and the blue and green phosphors in color sets. It was also used as a photoconductive coating on the drums in photocopiers.

Cadmium is an ingredient in some alloys that have a very low melting point. One of these alloys is called Wood's metal. This metal is very useful for making the fire-sensitive valve in sprinkler systems. If fire hits the sprinkler, the metal valve melts and immediately water starts pouring out of the sprinkler system.

Helium-cadmium lasers are a common source of ultra-violet laser light. These lasers can be used on microscopes to do experiments that require UV light.

Cadmium telluride is a light sensitive substance and is used in solar cells in solar panels.

CdO Cadmium oxide

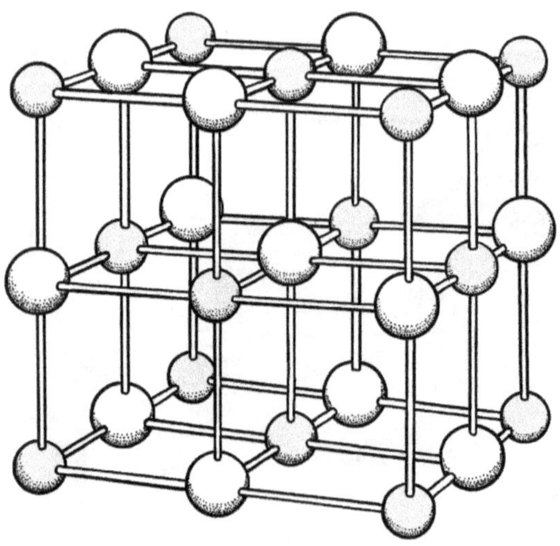

The darker balls are oxygen. The lighter balls are cadmium.
Does this pattern look familiar? (It is the same as NaCl, RbCl and NbN.)

$CdSO_4$
Cadmium sulfate

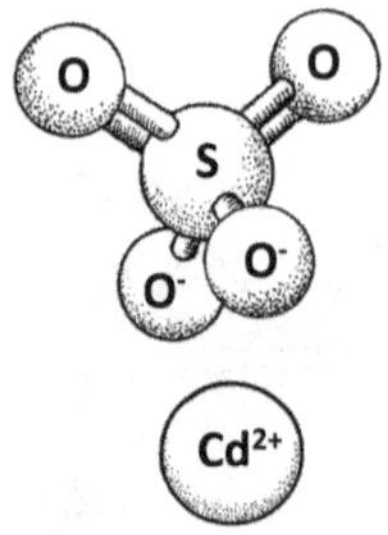

$CdCl_2$
Cadmium chloride

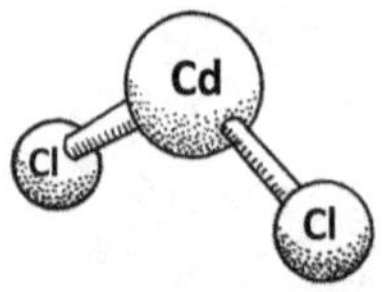

CdSe Cadmium selenide

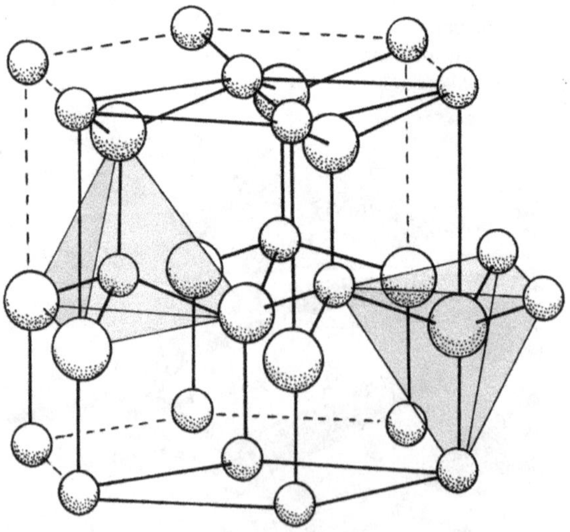

The larger balls are Se, the smaller ones are Cd. CdSe is used in orange paint, and to make windows transparent to infrared light.

48 Cd

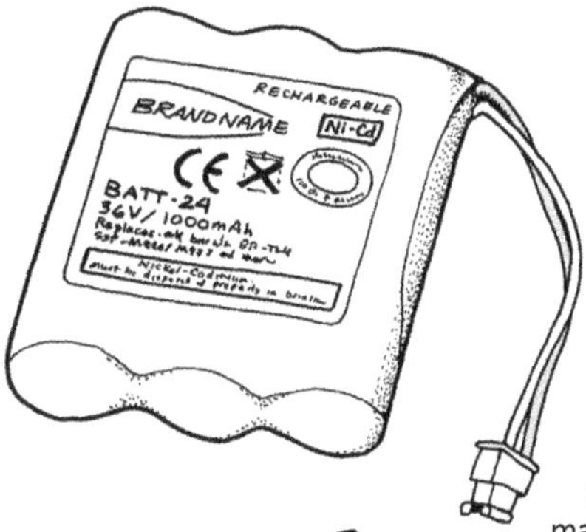

Cadmium and nickel are used to make Ni-Cd batteries, often bundled together to make rechargeable units.

Rechargeable Ni-Cd battery units are often found in remote control toy vehicles.

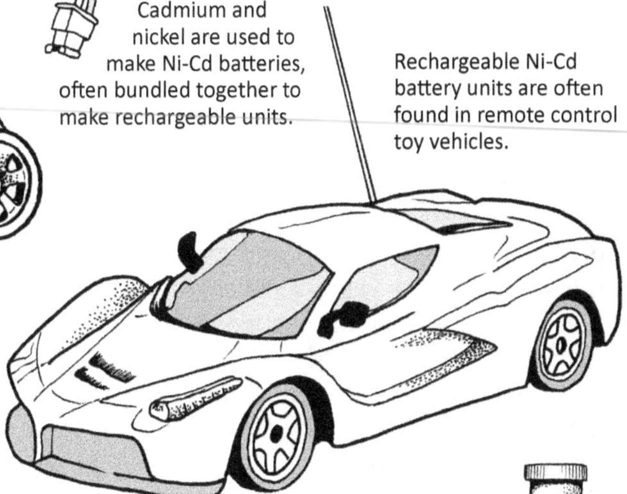

Calamine ores were named after the Greek mythological hero, Cadmus, shown here fighting a serpent.

Cadmium

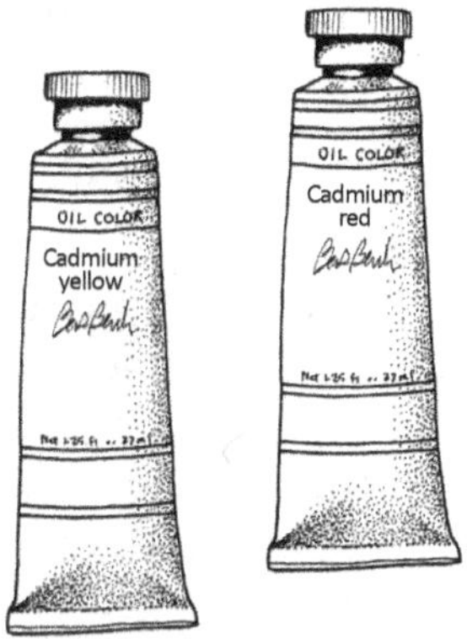

Cadmium compounds are used to make yellow, orange, and red paints for professional artists.

He-Cd (helium-cadmium) lasers are used for optical data storage, holography, lithography, spectroscopy, and more.

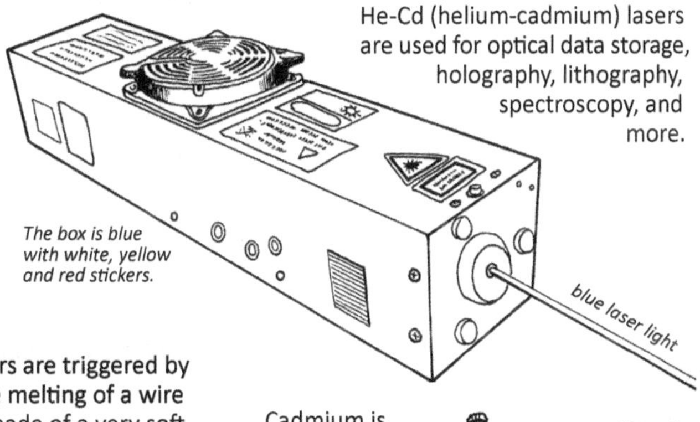

The box is blue with white, yellow and red stickers.

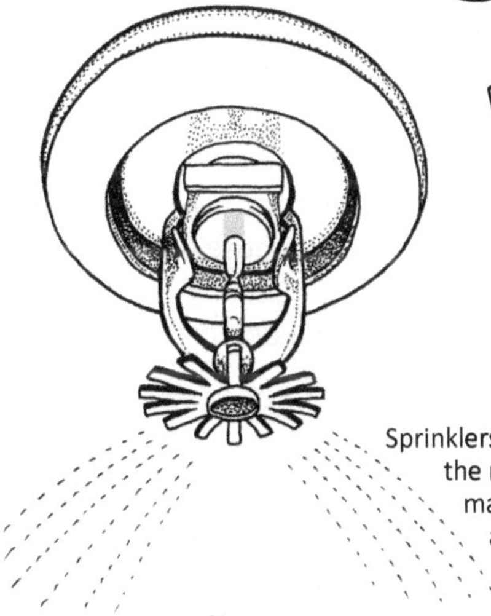

Sprinklers are triggered by the melting of a wire made of a very soft alloy of cadmium and bismuth.

Cadmium is used in control rods in nuclear reactors. It can slow down the fission process.

Aircraft steel can be coated with Cd.

Cadmium telluride is used in solar cells.

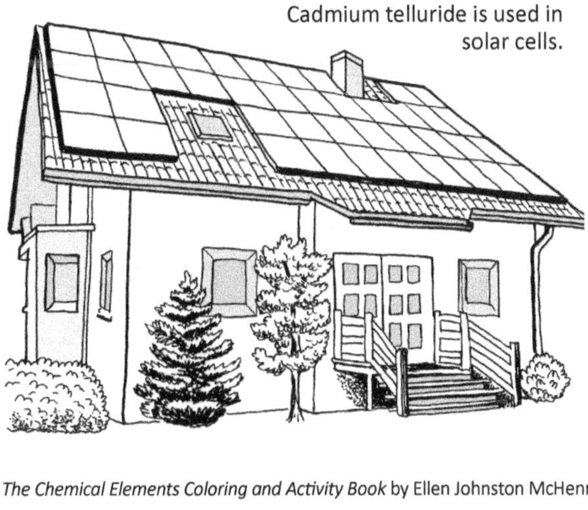

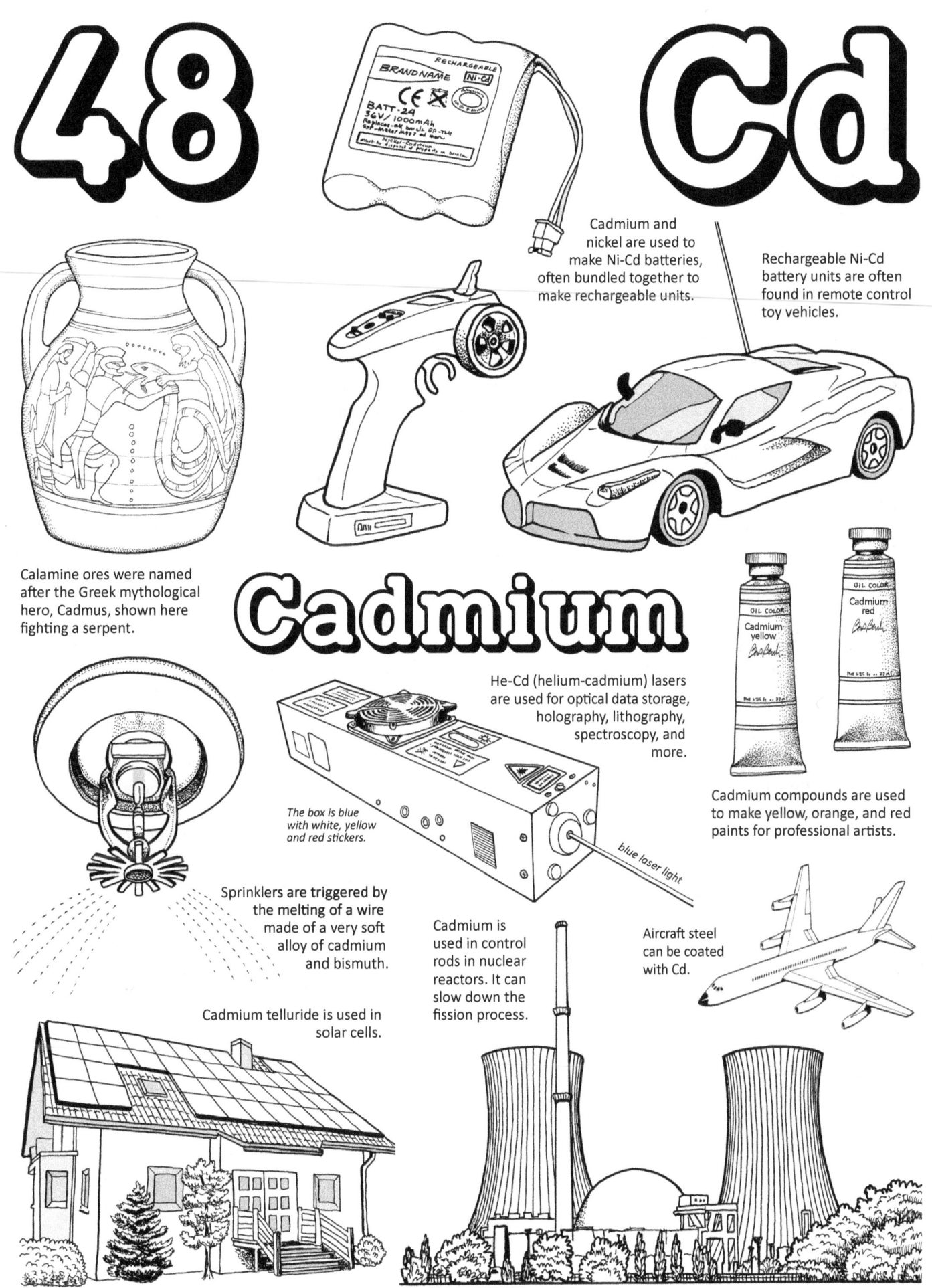

In 49

Indium

protons
66 neutrons
49 electrons

Atomic mass: 114.8

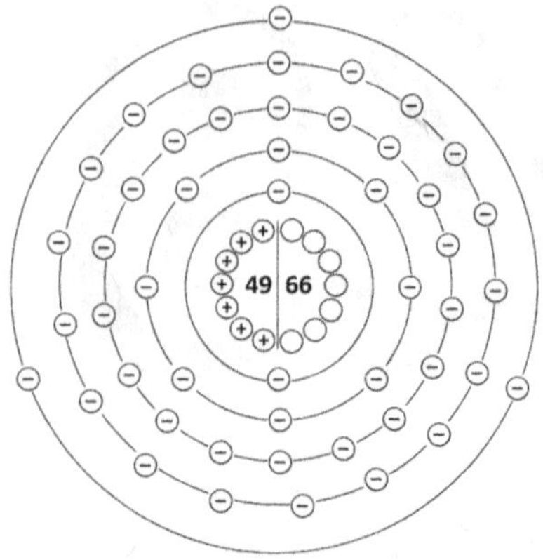

Named after the indigo line in its spectrum

Indium is found primarily in zinc ores and is a product of zinc mining. It was discovered in 1863 by two German chemists who were attempting to discover all the elements hiding in zinc ore rocks. They were using a spectrometer to look at the light coming from heated samples, and were expecting to find the bright green lines of thallium. Instead, they saw some bright bluish-purple (indigo) lines they had not seen before. Because of these indigo blue lines, they named the new element indium. The following year they managed to isolate samples of pure indium and by 1867 they had made enough of it to exhibit a one-pound bar of pure indium at the 1867 World's Fair in Paris.

Indium is a very soft metal and can be easily cut by hand with a knife. Because it is so soft, indium wire can be used to seal small gaps around lids of vacuum chambers. The soft indium will fill very tiny cracks where air might leak in.

Indium tin oxide (abbreviated as ITO) is used to make electrically conductive transparent coatings on touch-sensitive screens. It is also used on the windshields of airplane cockpits to keep frost from forming. Other uses include gas detectors and anti-reflective coatings on eyeglasses and other lenses.

Indium arsenide, InAs, and indium antimonide, InSb, are used in low-temperature transistors. Indium gallium arsenide, InGaAs, and indium gallium nitride, InGaN, are used in LEDs and lasers. Copper indium gallium selenide (CIGS) is a main ingredient in the "2nd generation" thin-film flexible solar cells, but can also be found in large solar panels.

Copper indium sulfide, $CuInS_2$, has the right properties for a number of uses in various electronics industries and is being studied for possible use in bio-imaging, solid-state lighting and solar cells. It is less toxic than other compounds currently being used for these purposes.

Indium is good at absorbing free neutrons, and is often combined with cadmium and silver to make control rods for nuclear reactors. When the control rods are placed between the fuel rods, the fission reaction slows down.

In 2009, researchers at Oregon State University combined indium with yttrium and manganese to make a bright blue pigment called "YInMn blue." It is non-toxic and fade-resistant.

Radioactive isotopes of indium are used in medical imaging to track proteins and white blood cells.

$CuInS_2$ Copper Indium Sulfide

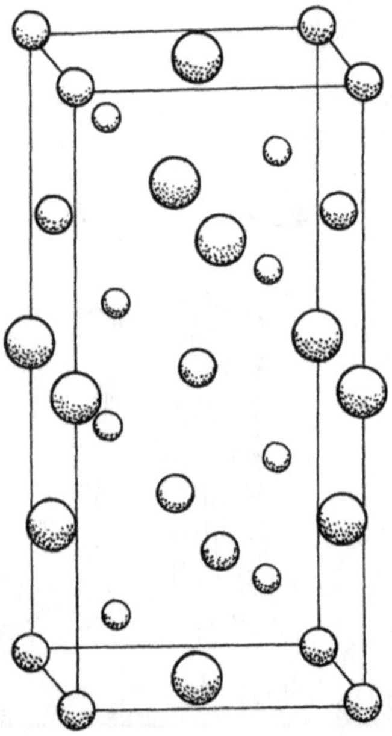

The large balls are indium. The medium balls are copper. The small balls are sulfur.

CIGS Copper indium gallium selenide

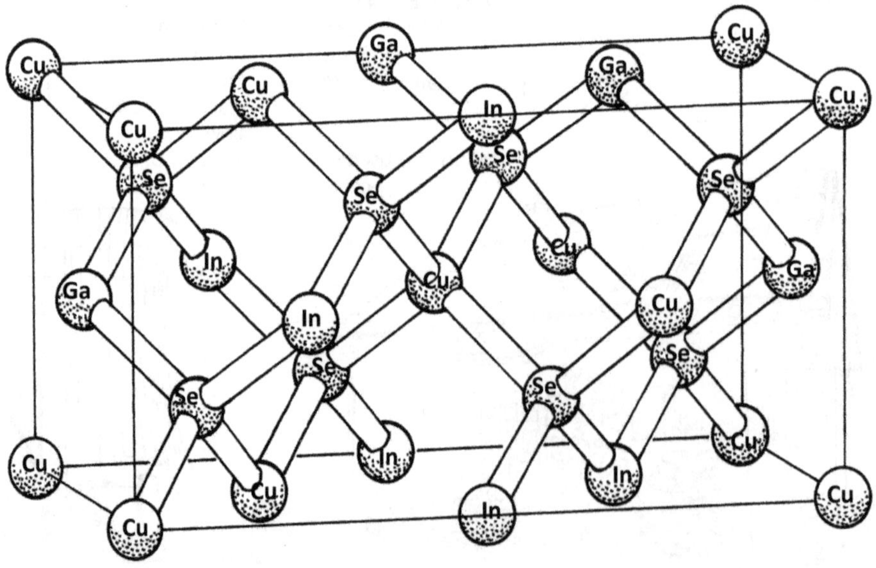

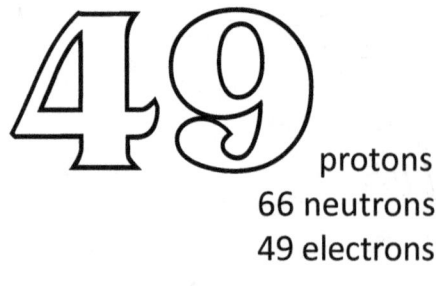

49 In

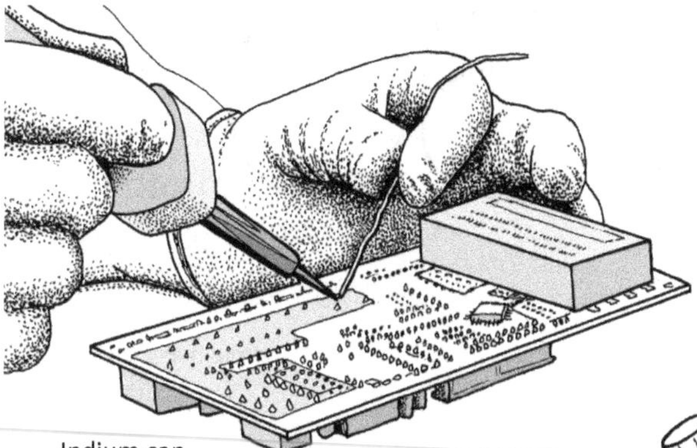

Indium can
replace lead in solder. (Indium is less toxic.)

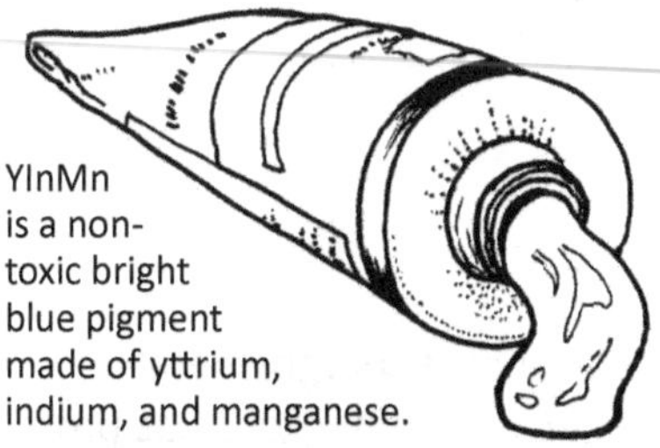

YInMn
is a non-
toxic bright
blue pigment
made of yttrium,
indium, and manganese.

Indium

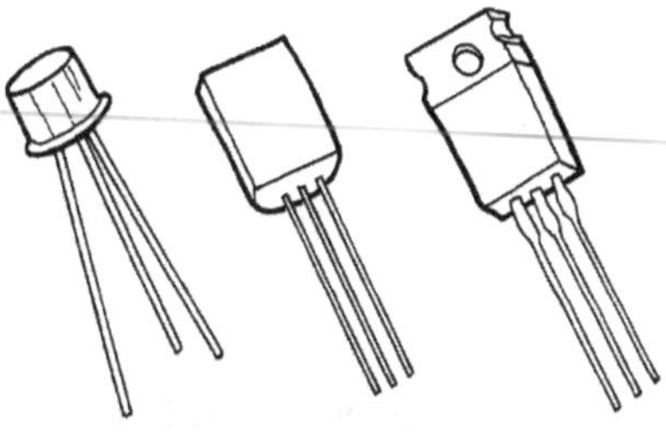

InAs and InSb are used in transistors.

Indium's emission spectrum has bright indigo lines.

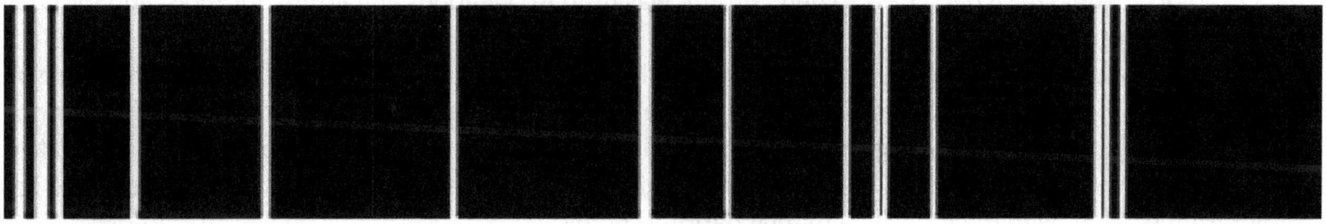

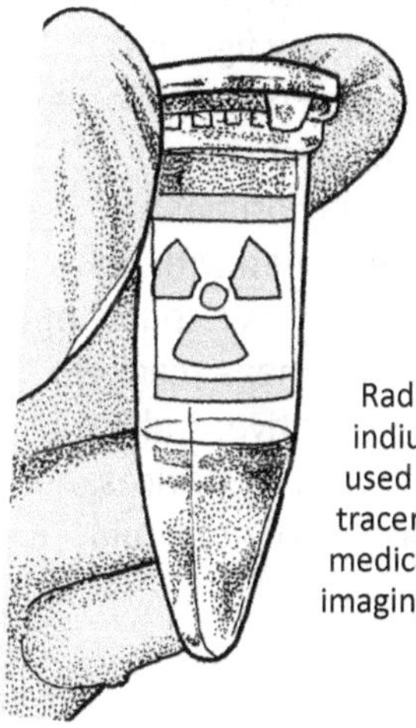

Radioactive
indium-111 is
used as a
tracer in
medical
imaging.

Indium is
used in LEDs.

ITO can be an
anti-reflective
coating on glasses.

Sprinklers are
triggered by
the melting of
a wire made of
a soft alloy that
often includes
indium.

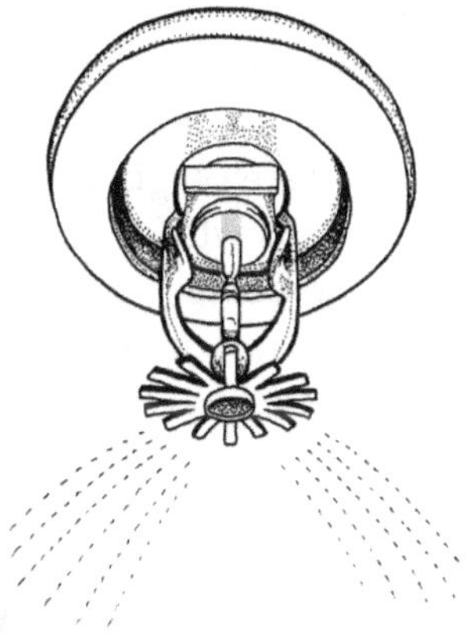

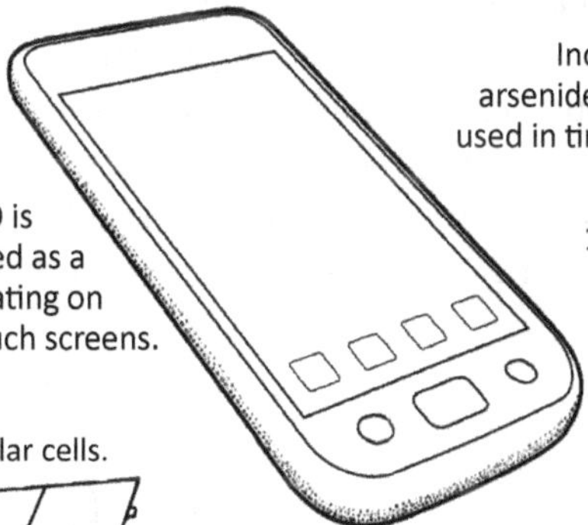

ITO is
used as a
coating on
touch screens.

Indium gallium
arsenide (InGaAs)is
used in tiny lasers.

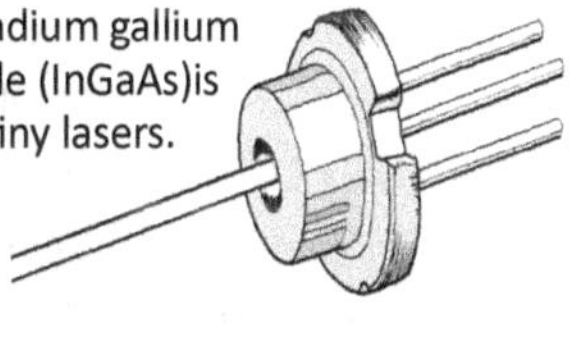

Indium tin oxide
(ITO) is used for
defrosting airplane
windshields.

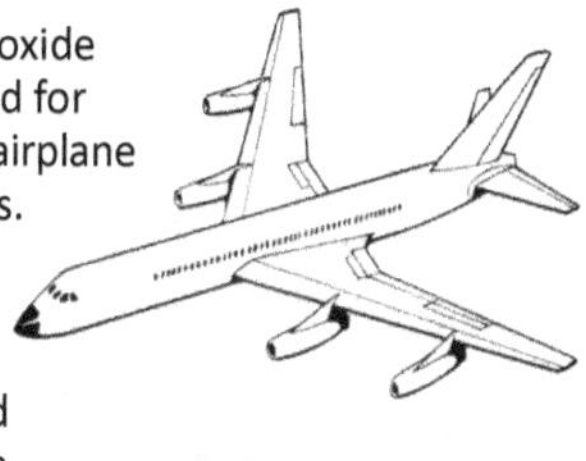

Copper indium gallium selenide is used in solar cells.

Indium (mixed
with cadmium
and silver) is
used to slow
down nuclear
reactions.

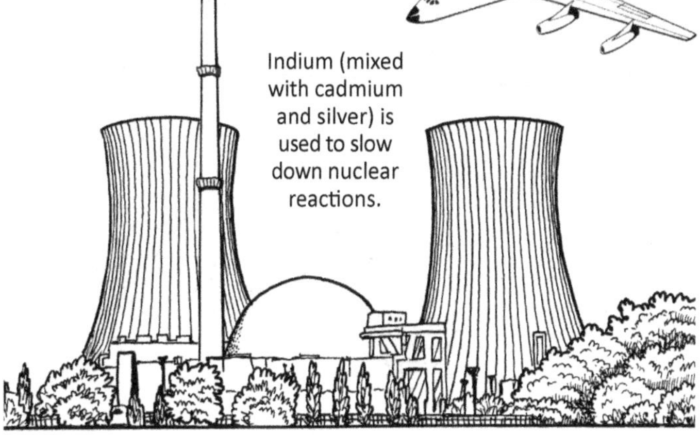

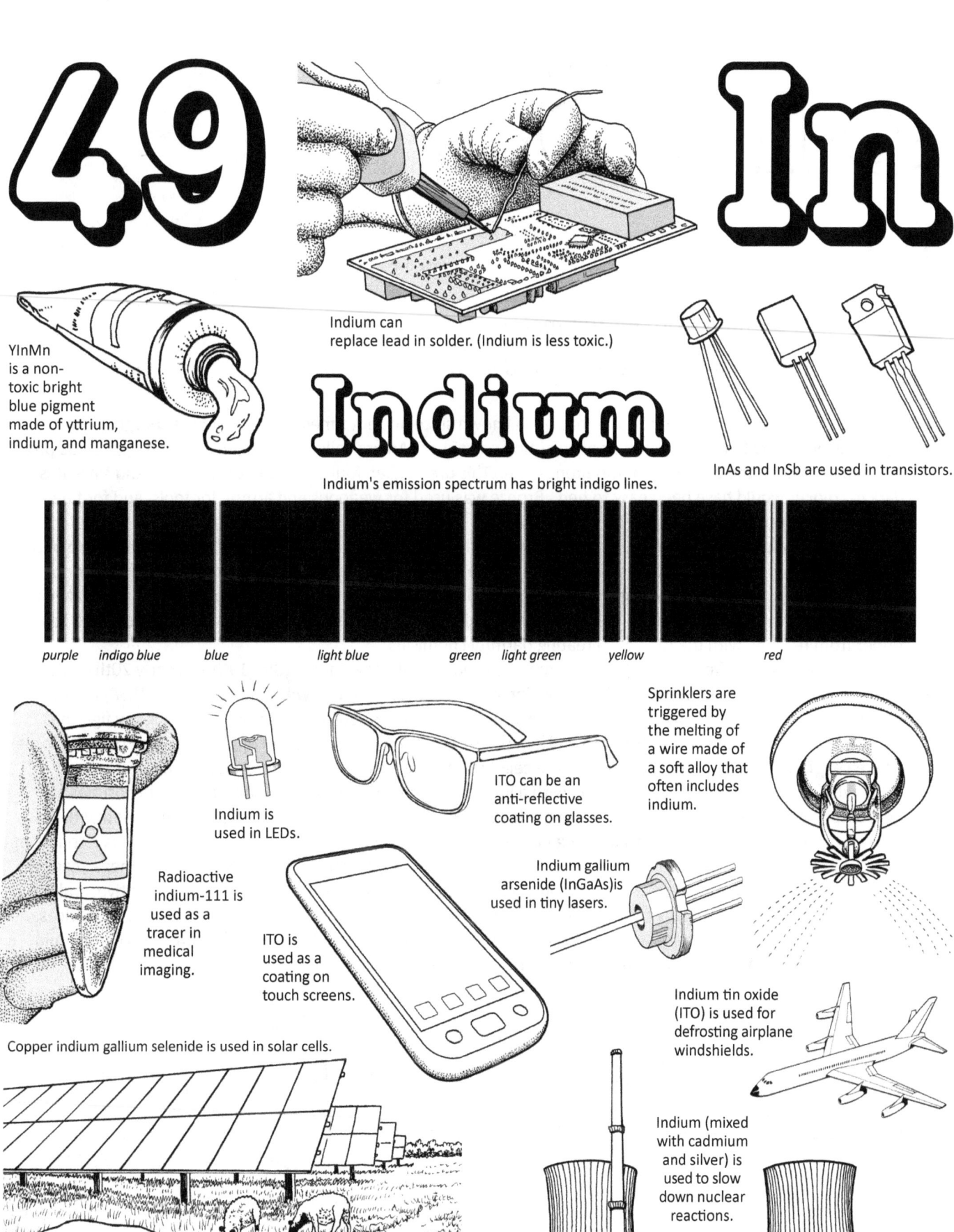

Sn 50

protons
69 neutrons
50 electrons

Atomic mass: 118.7

Tin

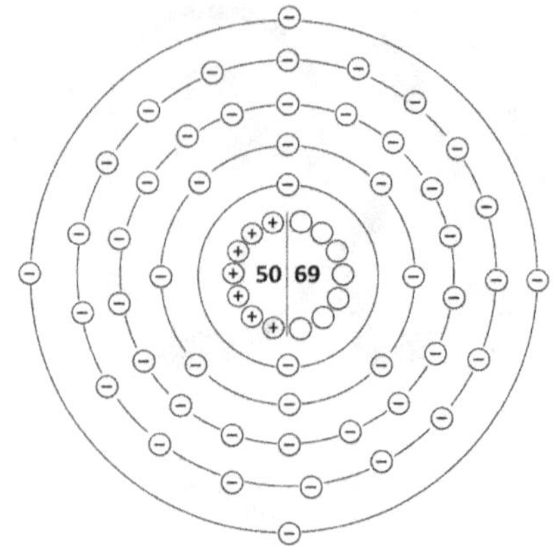

"Tin" is the old German word for this metal

The symbol, Sn, comes from the Latin word for tin, "stannum"

Tin is one of the seven elements that has been known since ancient times. About 5,000 years ago, metal workers discovered how to make bronze, an alloy of tin and copper. An ore called cassiterite, SnO_2, tin oxide, was probably the source of tin at the beginning of the Bronze Age. This ore is often found along river channels, and since it is very dark in color, it would have been easy to find. Bronze was used for weapons and armor, for tools, and for fine art such as jewelry and sculpture.

Today, half the tin supply in the world is used for solder (in a tin-lead alloy). The next most common use of tin is in electroplating other metals, such as steel. "Tin cans" are actually steel cans that are coated with a thin layer of tin. The first tin-plated food cannister was produced in London in 1812. A century later, tin cans of food were taken on expeditions to both the North and South Poles. Copper cooking pans are often coated with a thin layer of tin to keep the copper from reacting with the food and creating harmful chemicals.

Tin and tin alloys (often tin-lead) were commonly used to make toys in the 18th, 19th and early 20th centuries. Tin soldiers were made famous by writer Hans Christian Anderson when he wrote a fairy tale (in 1838) about a tin soldier who falls in love with a paper ballerina. In the late 20th century, plastic replaced tin for making toy figurines. Another use for tin-lead alloys during the 18th and 19th centuries was the construction of pipes for pipe organs. Some pipes were very large and reached all the way to the ceiling. A mix of 50% tin, 50% lead produces the best sound.

Pewter, an alloy consisting of mostly tin with a little antimony and copper, was popular in past centuries for making kitchenware and utensils. Craftsmen also made lanterns and "pie safes" from sheets of tin punched with holes.

Tin compounds can show up in surprising places. Dental products may contain tin fluoride, SnF_2, as a source of fluoride to prevent tooth decay. Niobium-tin, Nb_3Sn, is used to make super-conducting coils of wire for mega-magnets found in things like particle accelerators. A compound called tributyltin oxide is used as a wood preservative. Indium tin oxide is a transparent, electrically conductive compound that is used to make touch-sensitive screens for devices.

Pure molten tin is used to make "float glass." A small "lake" of hot liquid tin will support a layer of hot liquid glass. Gravity causes the glass to flow out and to form a uniformly flat sheet on top of the tin.

SnO_2 Tin oxide

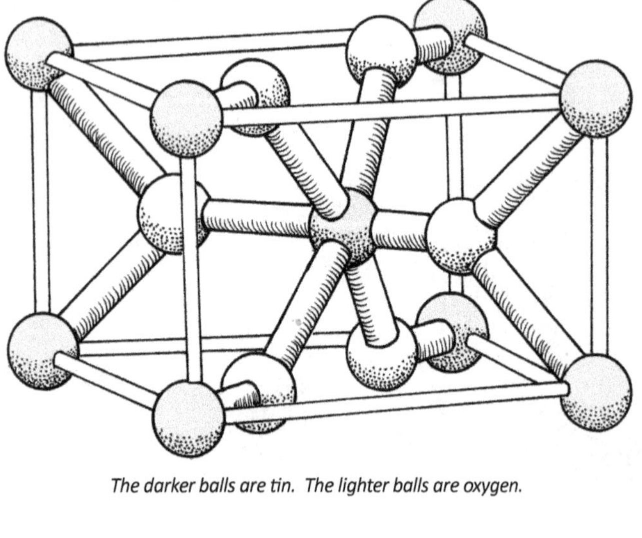

The darker balls are tin. The lighter balls are oxygen.

SnS_2 Tin sulfide

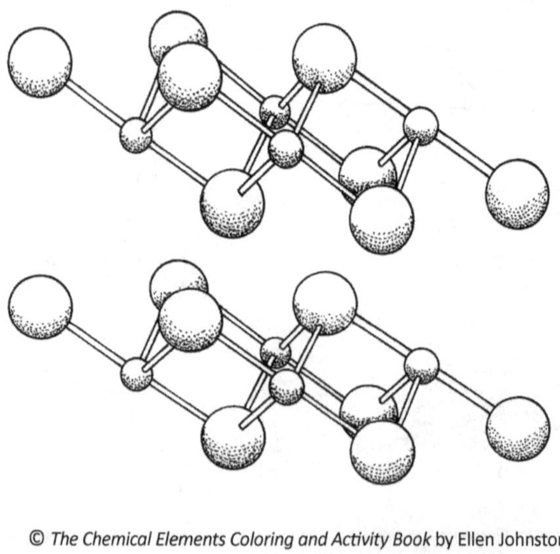

The small balls are tin. The larger balls are sulfur.

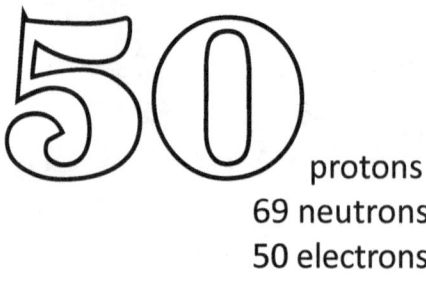

50 Sn

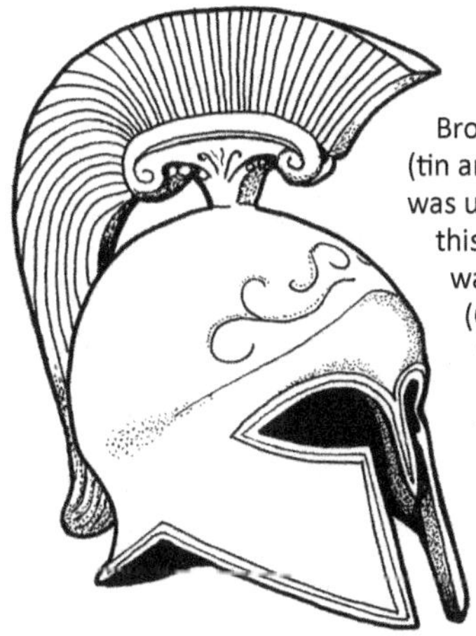

Bronze (tin and copper) was used to make this Greek war helmet. (Over the years the copper has tarnished and looks green in places.)

Terra Nova Expedition, 1910-1913

A member of an Antarctic expedition poses while eating beans from a tin can. (The picture was taken as an ad for Heinz products.)

Tin cans were invented in 1812.

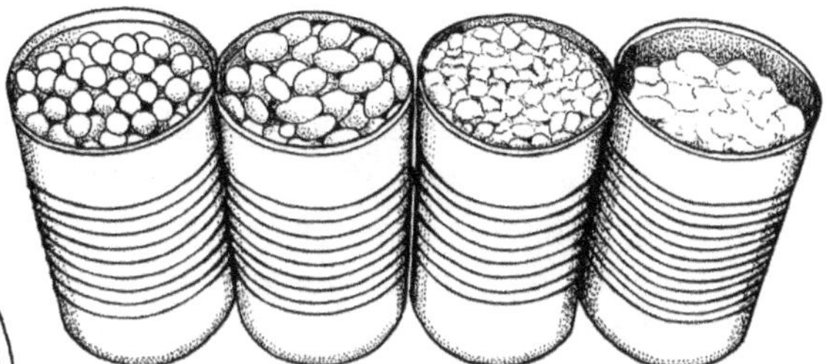

Modern "tin cans" are made of steel, and plated with a thin layer of tin.

Punched tin is a form of art that produces items that are both useful and decorative.

Tin

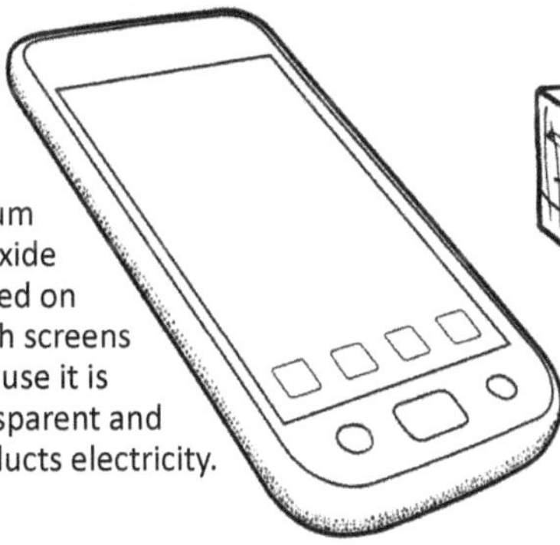

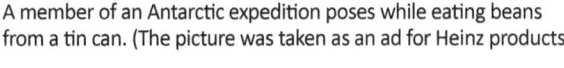

Indium tin oxide is used on touch screens because it is transparent and conducts electricity.

A tin compound is used as a wood preservative.

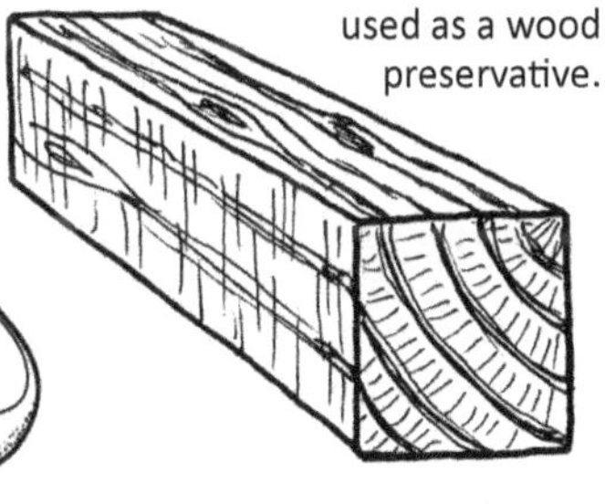

This door knocker is made of bronze (tin and copper).

Tin soldiers were made of tin alloys, often containing lead. Since lead is toxic, toys like this are now made of plastic.

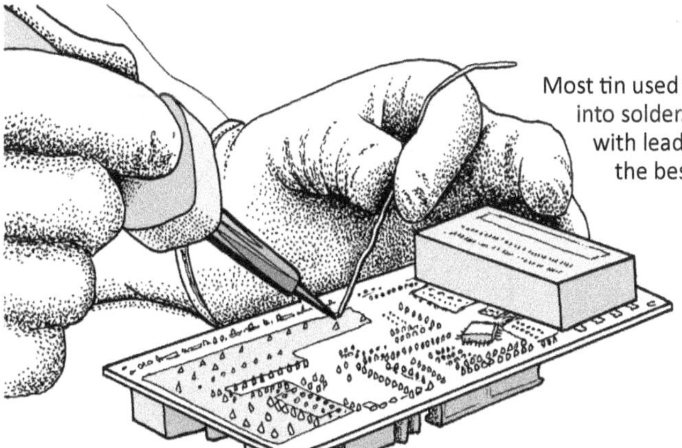

Most tin used today goes into solder. Tin mixed with lead makes the best solder.

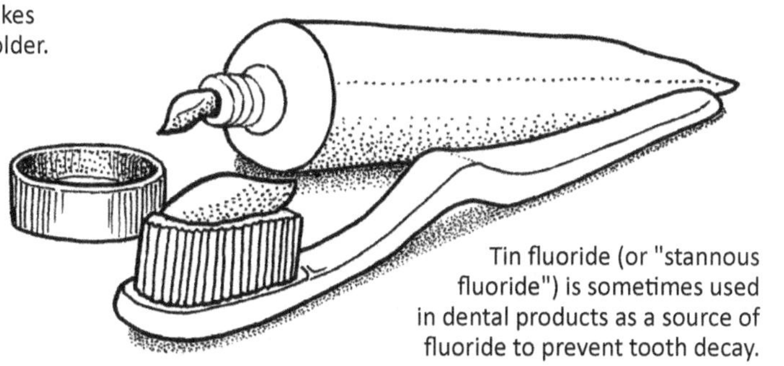

Tin fluoride (or "stannous fluoride") is sometimes used in dental products as a source of fluoride to prevent tooth decay.

Sb 51

Antimony

51 protons
71 neutrons
51 electrons

Atomic mass: 121.7

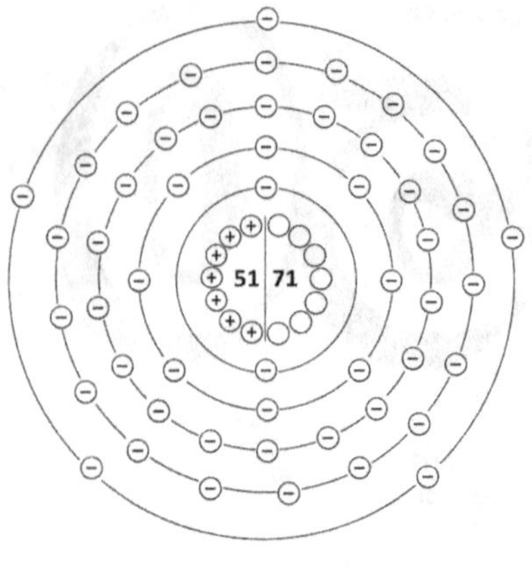

The name is from a Greek word

The symbol is from the Latin "stibium"

No one knows for sure where the name "antimony" came from. This element has been used for so many centuries, that the origins of its name have been lost in time. Some people say it is from the Greek word "anti-monos," meaning "not alone," because it is always found in ores with other elements, but others say it is from the Greek word "anti-monachos" meaning "monk-killer," suggesting that during the Middle Ages monks who dabbled in chemistry may have had bad experiences with this element. Greek writers also used the words "stibi" and "stimi" when writing about this element, and these words were almost certainly borrowed from old Egyptian or Arabic words. The Egyptians also called it "msdmt." They did not write the vowels so we don't know how to pronounce it. All the ancient words for antimony were connected to its primary use at that time: as a eye-shadowing cosmetic. Both men and women would use it around or under their eyes as decoration and as protection from the glare of the sun. The Egyptians also used an antimony compound ($Pb_2Sb_2O_7$) to make a yellow pigment. European painters in the 1700s discovered this pigment and named it "Naples yellow." (Naples yellow sold today might not contain antimony.)

hieroglyphs for "msdmt"

The symbol "Sb" was first used by the Swedish chemist Berzelius in the early 1800s, as an abbreviation for the word "stibium," a Latinized version of "stibi." The mineral stibnite, Sb_2S_3, was named using the word "stibi."

In the modern world, antimony has many uses. Antimony trioxide, Sb_2O_3, is used to make fire-resistant compounds that can be added to plastics or fabrics (especially children's clothing) and sprayed onto car seats. Antimony is part of a chemical reaction during the manufacturing of PETE plastics. Antimony can remove air bubbles from glass during the manufacturing of glass screens. In the semiconductor industry it is added to silicon wafers.

When alloyed with lead, antimony can be found in solder, lead-acid batteries, bullets, and safety matches. A mixture of lead, antimony, and tin is used to make pipes for large pipe organs. An alloy of indium and antimony is used to make infrared radiation detectors.

Antimony compounds are used by veterinarians to cure diseases (especially in cattle) caused by protozoans.

Sb_2O_3 Antimony trioxide

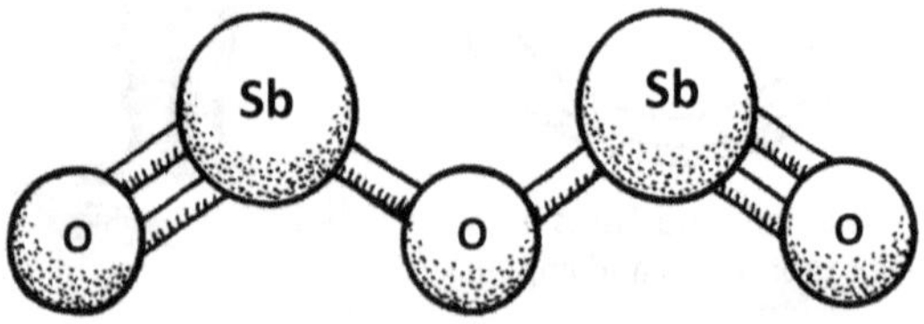

Sb_2O_3 molecules can also form a lattice shape.

AsSb (stibarsen, a natural mineral)

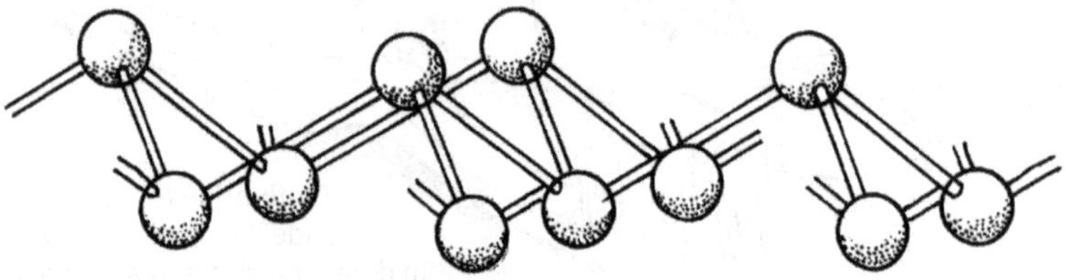

The balls can represent either antimony or arsenic, in any order.

InSb Indium antimonide

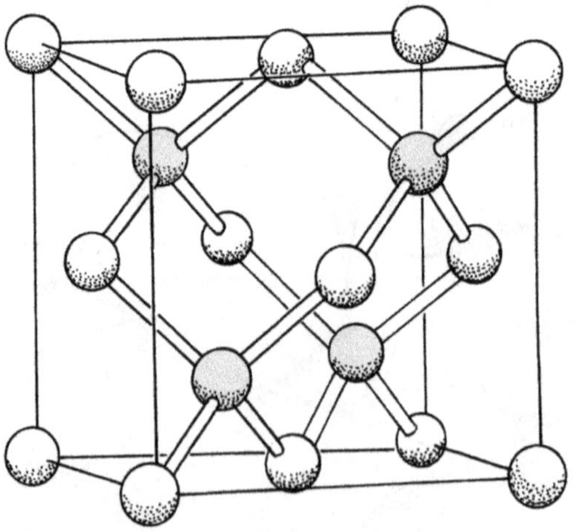

The darker balls are antimony. The lighter balls are indium.

51 Sb

Stibnite, Sb_2S_3

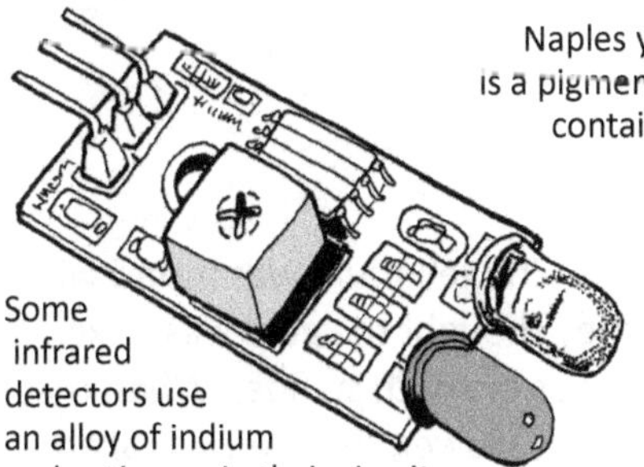

Some infrared detectors use an alloy of indium and antimony in their circuits.

Naples yellow is a pigment that contains Sb.

Antimony

Egyptians, like many people groups around the world, used antimony compounds to draw dark lines around their eyes. They called it "kohl."

Sb_2O_3 is a key ingredient in fire-retardants that are used in children's clothing and on car seats.

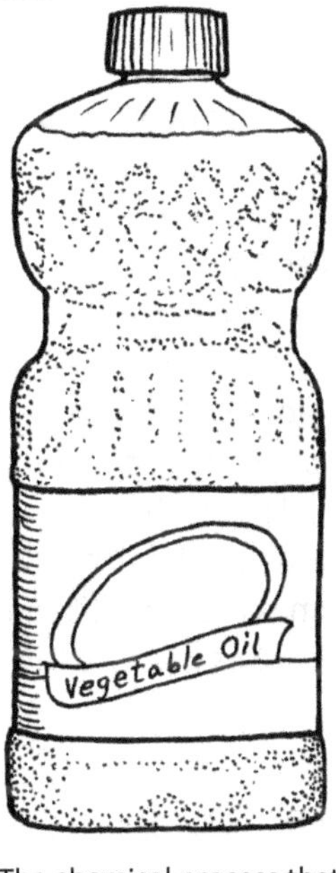

The chemical process that produces PETE plastic bottles uses Sb as a catalyst

Antimony can be found in the tips of some safety matches.

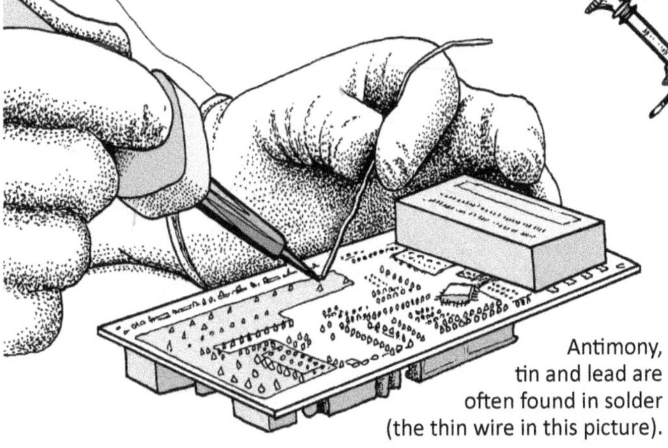

Antimony, tin and lead are often found in solder (the thin wire in this picture).

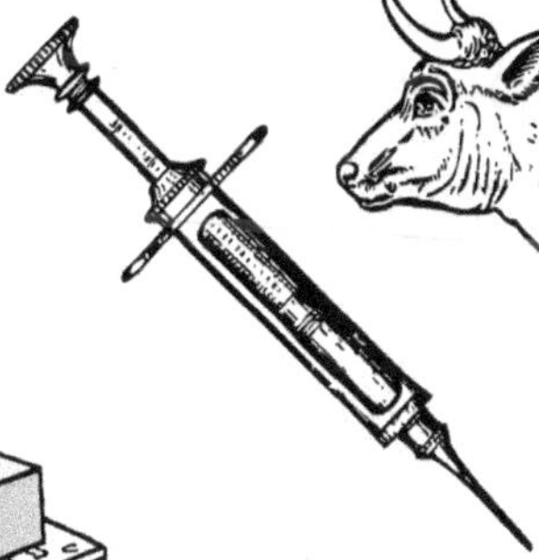

Antimony compounds (such as lithium antimony thiomalate) are used to treat some skin diseases found in cattle.

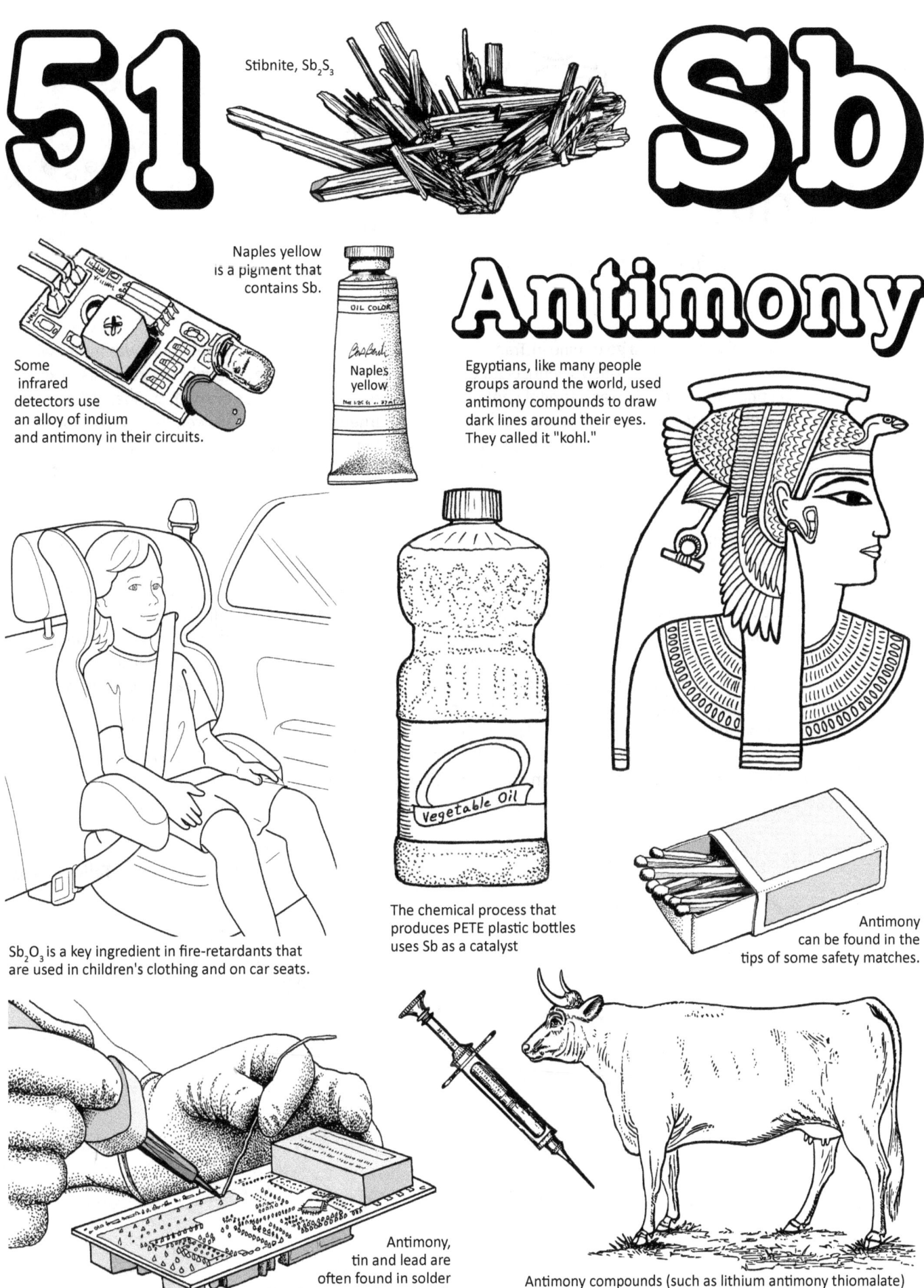

Te 52

protons
76 neutrons
52 electrons

Atomic mass: 127.6

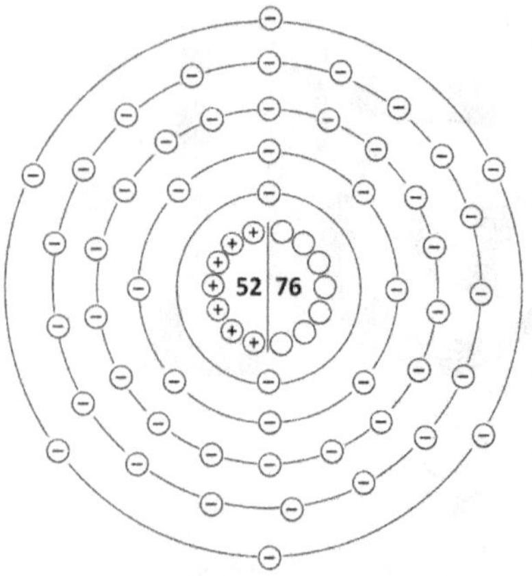

Tellurium

Named using the Latin word for earth, "tellus"

Tellurium was discovered in a gold mine in Transylvania (now part of Romania). The tellurium was mixed with the gold to create an ore rock that mystified scientists for quite a few years. At first they thought the gold was mixed with antimony. Testing showed that it was definitely not antimony, so the next guess was bismuth sulfide. When that was shown to be incorrect and several more years of testing had not turned up a definite answer, the ore was called "metallum problematicum" (problem metal). Finally, this mysterious element was isolated, and, in 1798, a German chemist named Martin Klaproth chose the name "tellurium" for this new element, although he did not take credit for its discovery. (If you read about selenium, you will remember that it was named after the moon because it sits on top of the earth, "tellus," in the Periodic Table.)

Since tellurium is in the same column as sulfur and selenium, this means that it shares some chemical characteristics with these elements and therefore can sometimes be used as a substitute for them. For example, tellurium can replace sulfur to vulcanize (toughen) rubber. Tellurium can replace selenium in many semi-conducting alloys.

Although tellurium is used to make steel and copper alloys, the most well-known tellurium alloys are CdTe and HgCdTe. Cadmium telluride, CdTe, is used to make efficient solar panels. If some of the cadmium in CdTe is replaced with zinc, it can be used to detect x-rays. Mercury cadmium telluride, HgCdTe, is used for detecting various wavelengths of infrared radiation. By adjusting the amount of cadmium in the mix, this compound can be used in various applications including military airplanes, night vision goggles, telescopes, and satellites.

Tellurium is added to glass to make optical glass fibers used in communication cables. Tellurium is added to BaO_2 to make "delay powder" in blasting caps for dynamite detonators. TeO_2 is used to make DVD and Blu-ray discs.

The only biological application for tellurium is the identification of the bacteria that causes the disease diphtheria. Tellurium is added to the growing medium in petri dishes because this is the only type of bacteria that will grow in it.

TeO_2 Tellurium dioxide

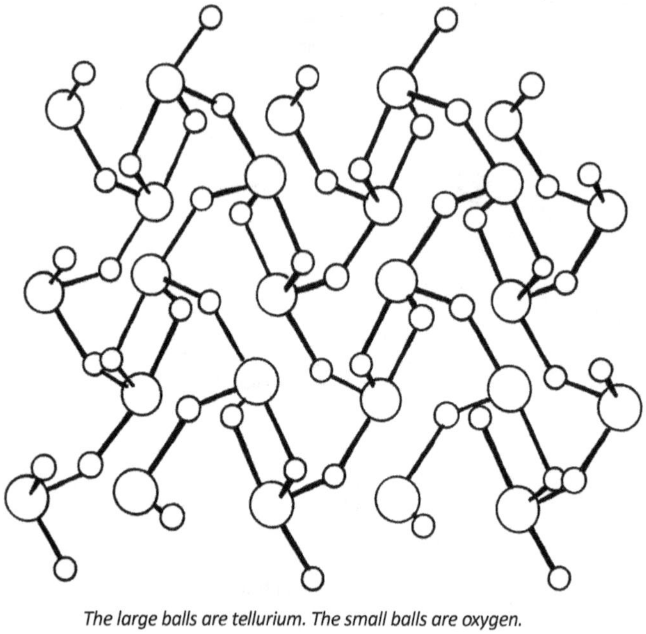

The large balls are tellurium. The small balls are oxygen.

H_2TeO_3 Tellous acid

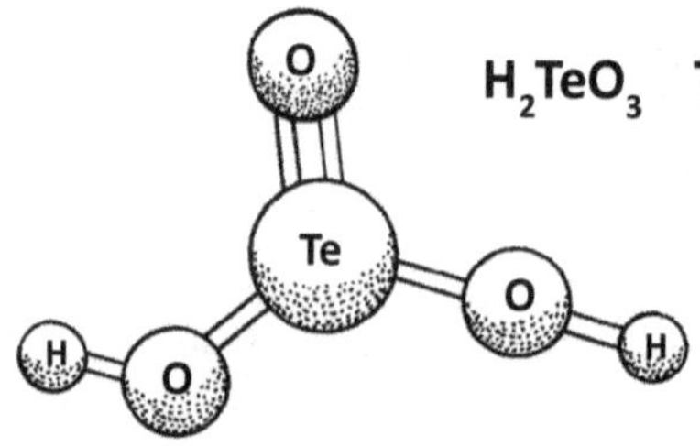

$AuAgTe_4$ Sylvanite

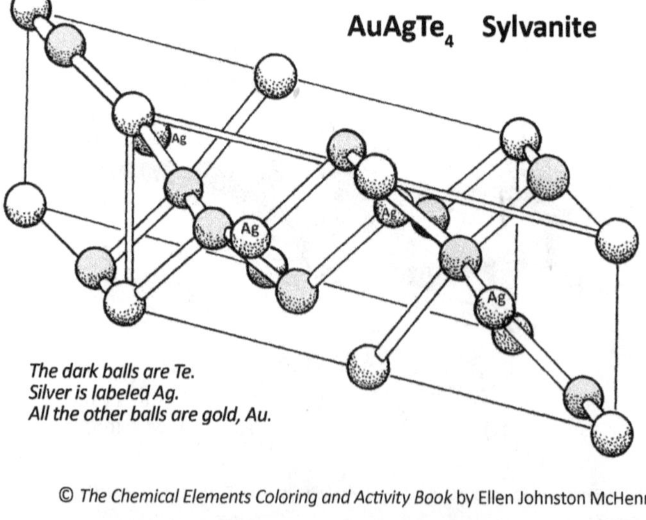

The dark balls are Te.
Silver is labeled Ag.
All the other balls are gold, Au.

52 Te

Tellurium is named after the earth, using the Latin, "tellus."

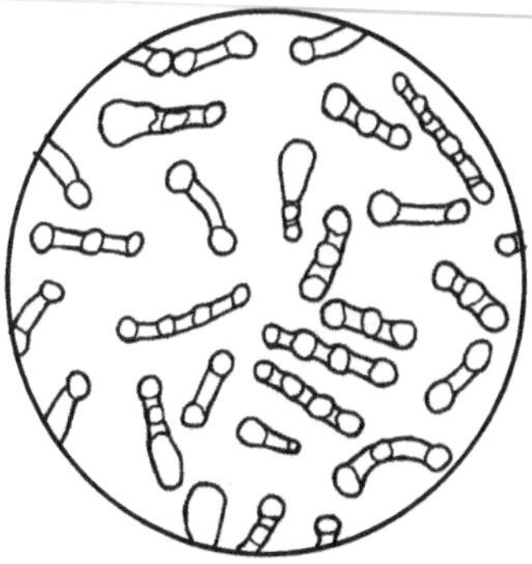

Tellurium

Only diphtheria bacteria will grow on a gel that contains Te.
(These bacteria were stained; the circles look dark purple.)

Tellurium can be used in place of sulfur to vulcanize rubber.

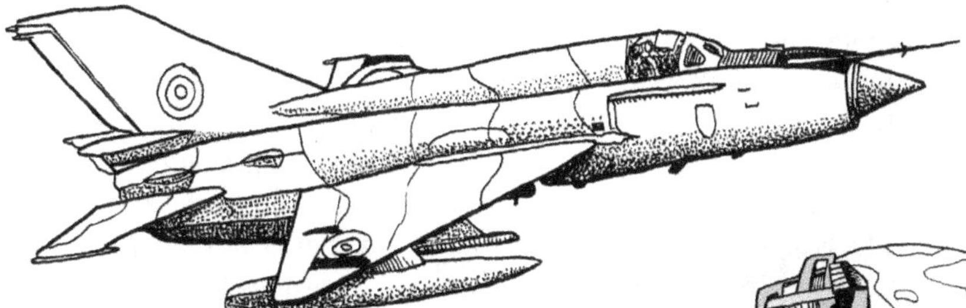

HgCdTe, is used in night vision equipment on military planes.

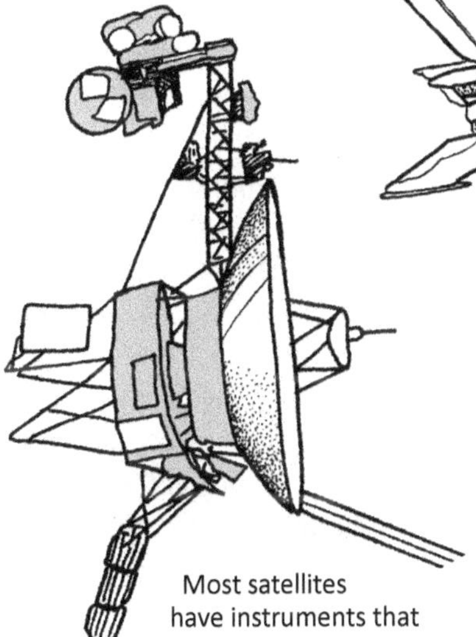

Most satellites have instruments that detect infrared radiation. HgCdTe is often used in the detecting mechanism.

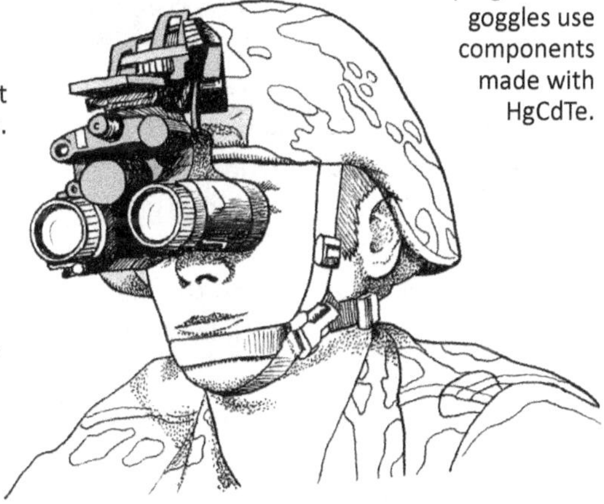

Military night vision goggles use components made with HgCdTe.

Tellurium oxide is used to make DVD and Blu-ray discs.

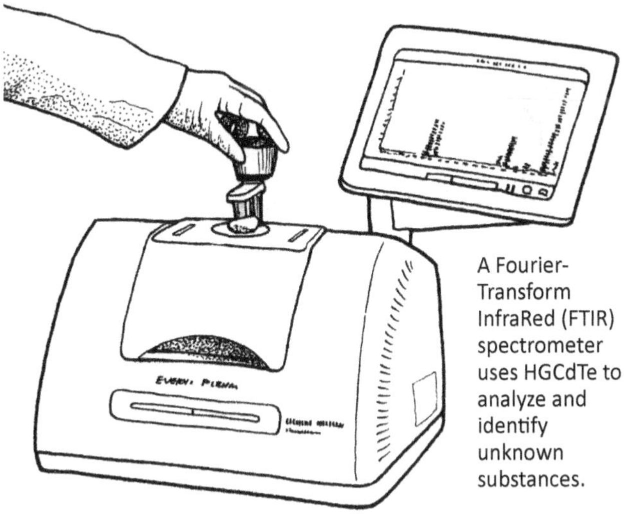

A Fourier-Transform InfraRed (FTIR) spectrometer uses HGCdTe to analyze and identify unknown substances.

CdTe (cadmium telluride) solar panels are highly efficient.

I

53

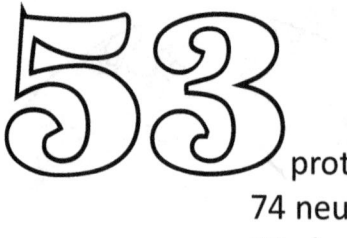

protons
74 neutrons
53 electrons
Atomic mass: 126.9

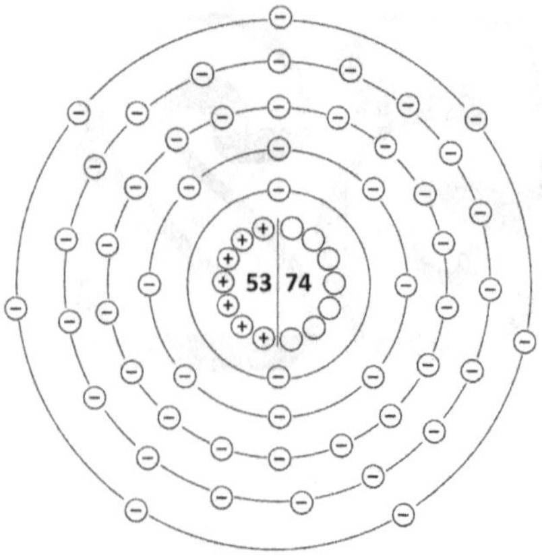

Iodine

Named using the Greek word for purple, "iodes"

Iodine was discovered in 1811 by French chemist Barnard Courtois *(kur-TWAH)*, who was trying to extract sodium and potassium from seaweed ash. He accidentally added too much sulfuric acid and a cloud of purple smoke began to fill the room. The purple gas condensed onto all the objects in the room, covering them with a thin layer or purple stuff. What was the purple stuff? Courtois sent samples of it to several famous chemists. After studying it for several years they all came to the same conclusion, that it was a new element, and someone suggested naming it using the Greek word for purple. Then several of these scientists began to fight about who was going to receive credit for the discovery. However, in the end, they did the right thing and named Courtois as the official discoverer of iodine.

Humphry Davy was one of the chemists who studied this new element, and he pointed out that it was very similar to chlorine and bromine and might therefore be useful for killing germs. It was not until 1908 that iodine was officially introduced into operating rooms as a pre-surgical sterilization technique. Iodine penetrates bacteria and begins to destroy its amino acids, causing the bacteria to die very quickly. Iodine solutions are still used today to sterilize skin wounds. Lugol's solution is an iodine solution that is also handy for detecting the presence of starch, as it will cause starch molecules to turn black. Iodine solutions have been used to detect counterfeit paper money.

Potassium iodide, KI, is light-sensitive and was one of the chemicals used to make photographic film. It can also be used to sterilize water. AgI, silver iodide, is also photosensitive but has a fascinating use not related to light. Many countries have "cloud seeding" programs where silver iodide is released into the atmosphere (often by airplanes) over areas where they would like it to rain or snow. Airports sometimes use this technology to try to prevent fog around runways.

In the body, the thyroid gland needs iodine to produce its hormone, thyroxine. People who do not get enough iodine in their diet can develop an iodine deficiency disease called goiter, which produces a visible lump on the thyroid.

Iodine-131 is one of the most common radioactive isotopes produced by nuclear bombs (as nuclear "fall out") but when produced artificially in a lab, it can be useful in medical imaging and for treating some thyroid diseases.

IBX acid (2-iodoxybenzoic acid)

This molecule is used as a source of oxygen atoms in chemical reactions that need a supply of oxygen.

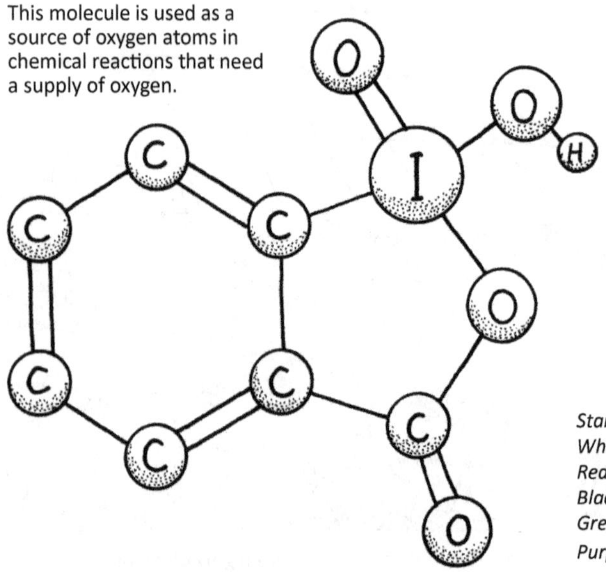

Erythrosine (Red #3)

A very common artificial food coloring.

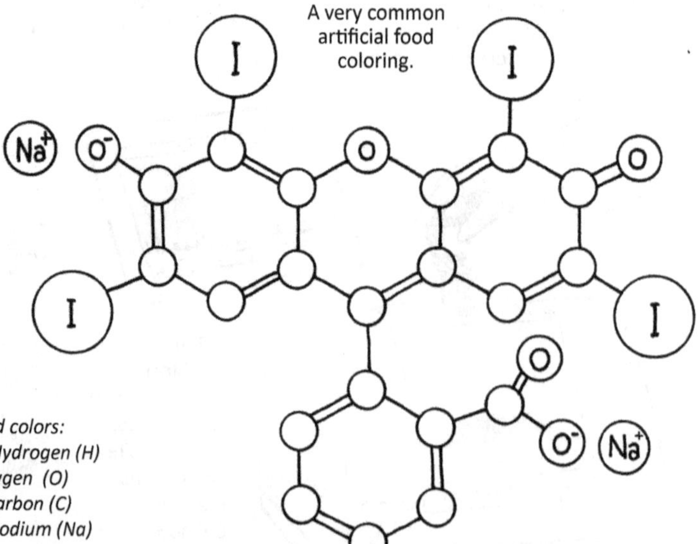

Standard colors:
White: Hydrogen (H)
Red: Oxygen (O)
Black: Carbon (C)
Green: Sodium (Na)
Purple: Iodine (I)

The unlabeled balls are carbon.

53 I

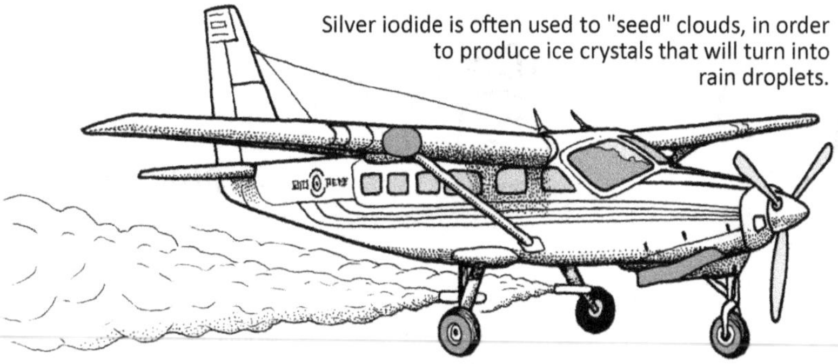

Silver iodide is often used to "seed" clouds, in order to produce ice crystals that will turn into rain droplets.

Iodine

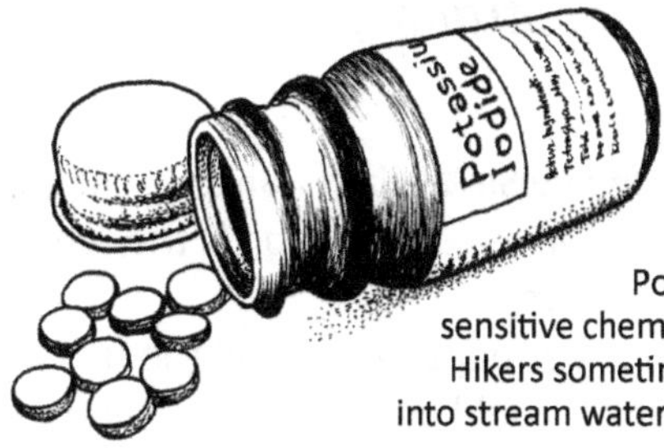

Iodine is added to table salt to prevent iodine deficiency diseases.

Potassium iodide is a light-sensitive chemical that also kills germs. Hikers sometimes buy KI tablets to put into stream water to make it safe to drink.

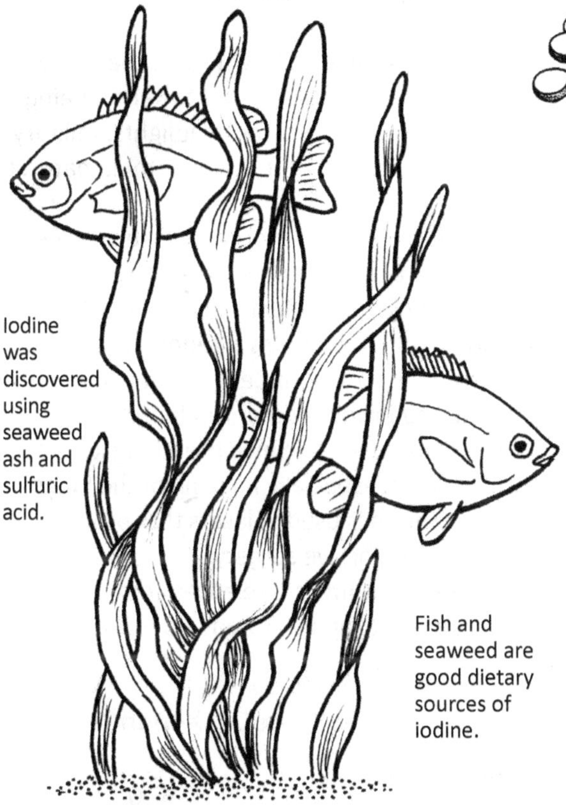

Iodine was discovered using seaweed ash and sulfuric acid.

Fish and seaweed are good dietary sources of iodine.

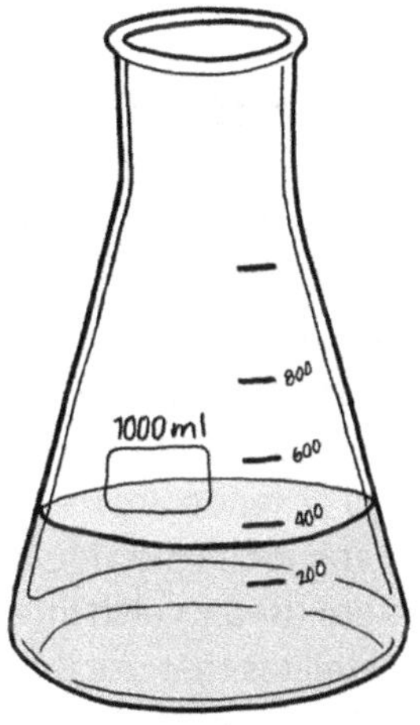

Iodine solutions look orange, but when dropped onto starch they turn black.

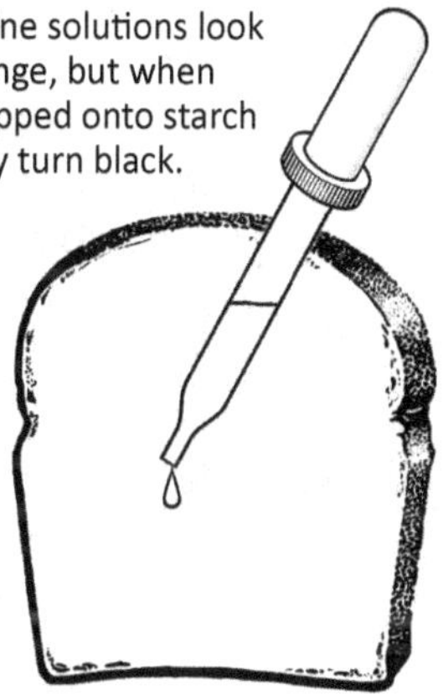

Iodine is a key ingredient in several classic chemistry demonstrations including the "chemical clock" and "golden rain."

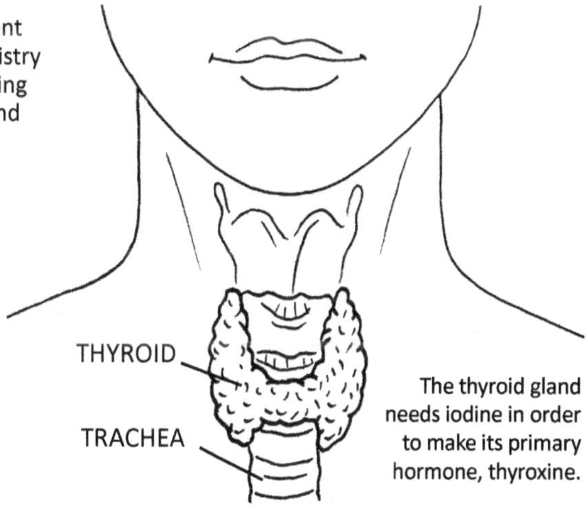

The thyroid gland needs iodine in order to make its primary hormone, thyroxine.

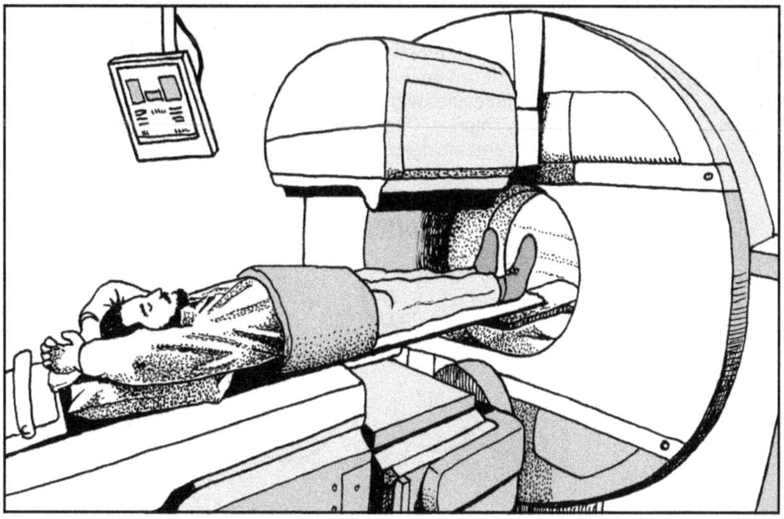

Radioactive isotopes of iodine (I-131, I-123, I-125) are used in medical imaging.

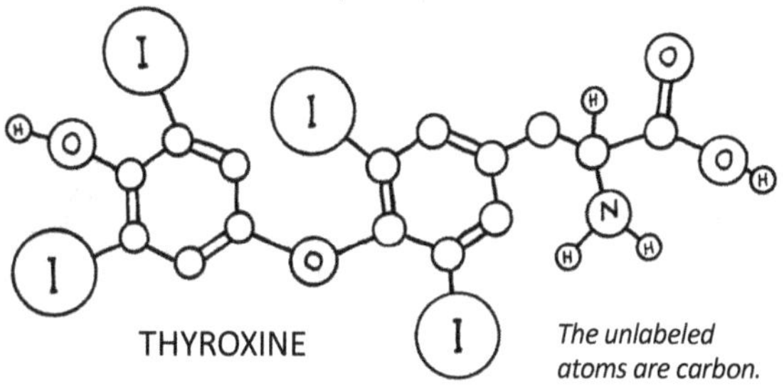

THYROXINE

The unlabeled atoms are carbon.

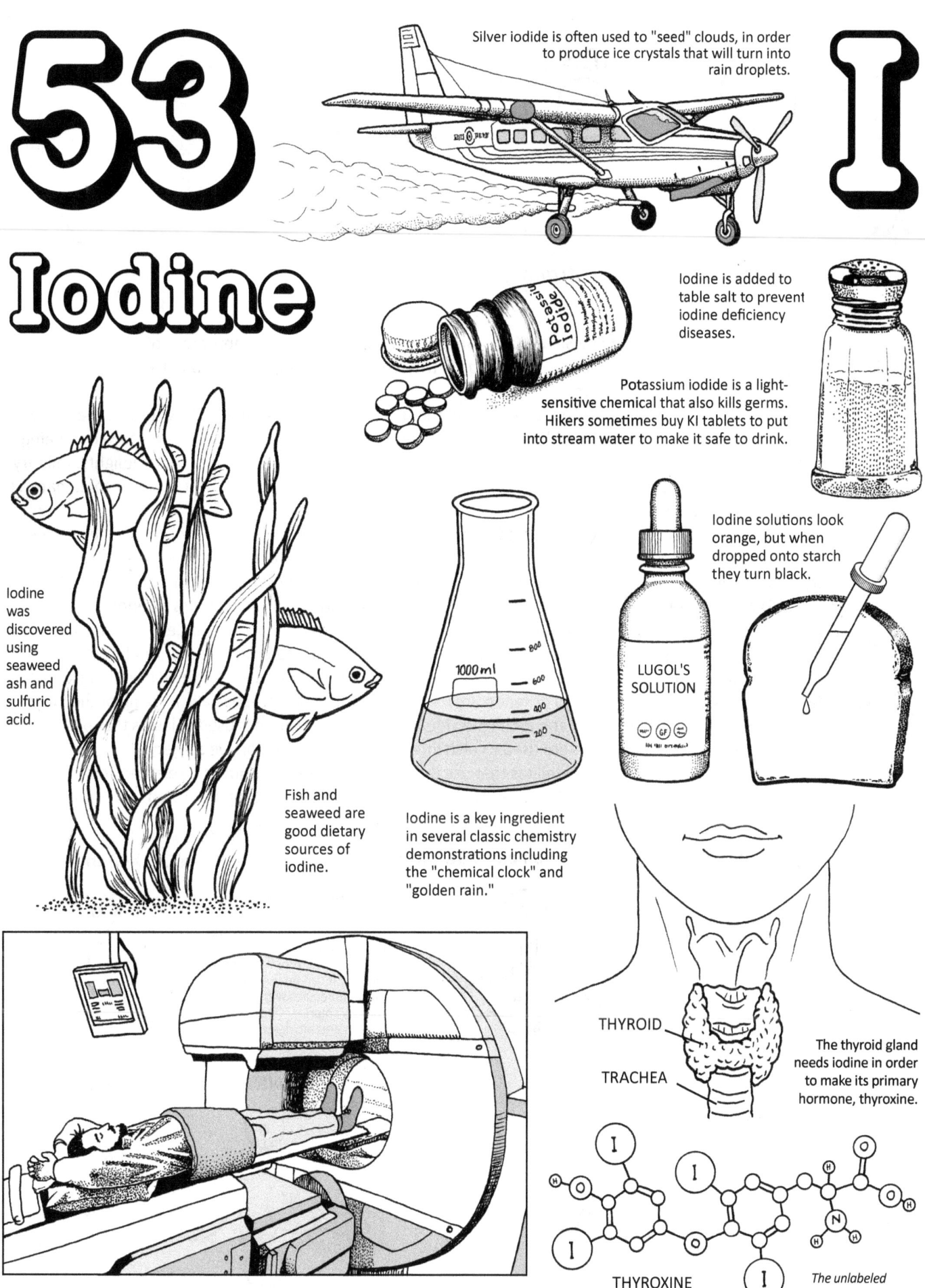

Xe 54

54 protons
77 neutrons
54 electrons

Atomic mass: 131.3

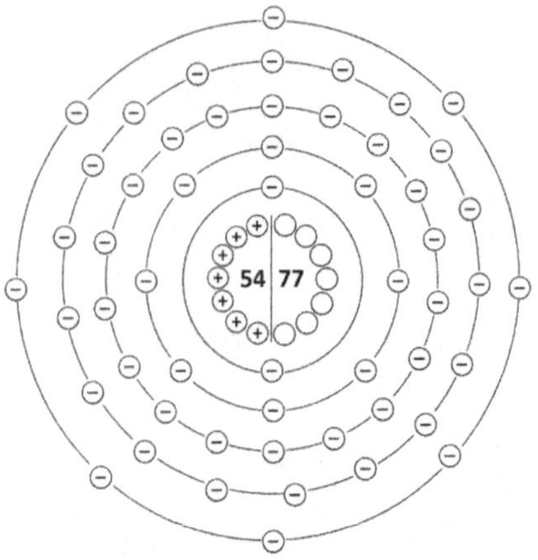

Xenon

Named using the Greek word for strange, "xeno"

Xenon *(ZEE-non)* was discovered by Sir William Ramsay and Morris Travers in September of 1898, just a few weeks after they had discovered krypton and neon. Xenon is the rarest of all the gases found in Earth's atmosphere. Only one atom in 20 million is a xenon atom.

In the 1930s, an American engineer began exploring the use of xenon as a source of bright light. He found that passing an electric current through it could produce a very intense flash of white light. This led to xenon being used in all kinds of specialty light bulbs, including strobe lights, sodium lamps, arc lamps, HID car headlights, military flashlights, and the projectors used for IMAX films. Xenon bulbs generate wavelengths of light that resemble natural sunlight and are used in solar simulators.

In 1939, an American doctor began to investigate the cause of a strange medical phenomenon seen in deep sea divers that produced symptoms similar to drunkenness. Guessing that it might be the air they were breathing from their tanks, he began experimenting with various mixtures of gases, seeing what effect they had on the nervous system. To his surprise, his research led him to conclude that xenon might be useful as a general anesthetic during surgery. By 1951, enough experiments had been done on mice that surgeons felt confident about using the procedure on humans. Xenon has another effect on the body; it can stimulate the production of red blood cells. Athletes, especially distance runners, sometimes use xenon to enhance their performance. Since breathing xenon does no harm to the lungs, radioactive isotopes of xenon are used in medical imaging, especially imaging of the lungs.

Plasma screens use both xenon and neon in the individual plasma cells. Xenon is used in lasers that produce ultraviolet light (excimer lasers). These lasers are used to etch integrated circuits, and for eye surgery.

Xenon is the preferred propellant in satellites that use ion propulsion. Xenon atoms are released from the satellite, acting as a force that pushes the satellite forward. Unlike many other gases, it can be stored under pressure at room temperature, and therefore does not need extreme refrigeration.

XeF_4 Xenon tetrafluoride

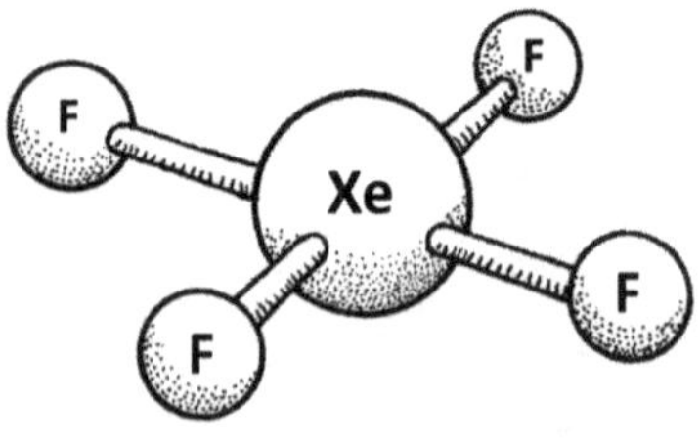

$XeOF_4$ Xenon oxytetrafluoride

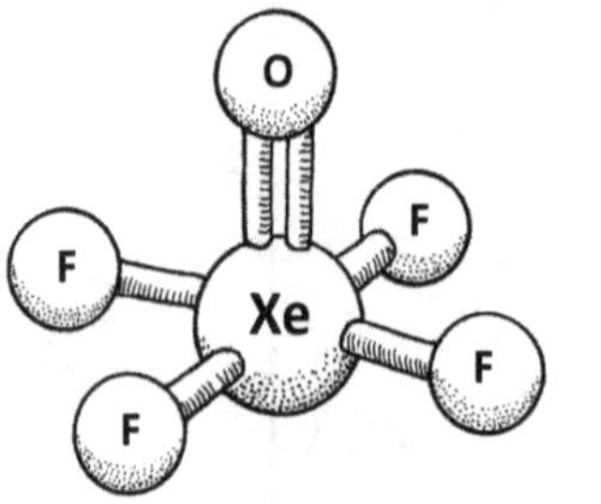

Under normal circumstances, xenon is "inert" and will not combine with other atoms to form molecules. However, if it is heated, pressurized or ionized, it can be forced to make molecules with atoms such as chlorine, fluorine, oxygen and nitrogen.

XeO_4 Xenon tetroxide

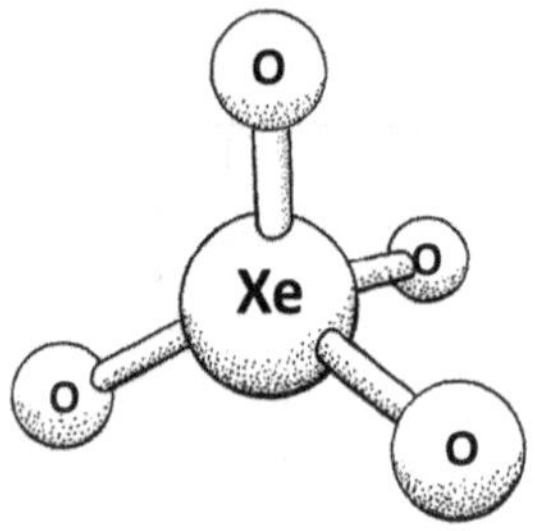

H_2XeO_4 Xenous acid

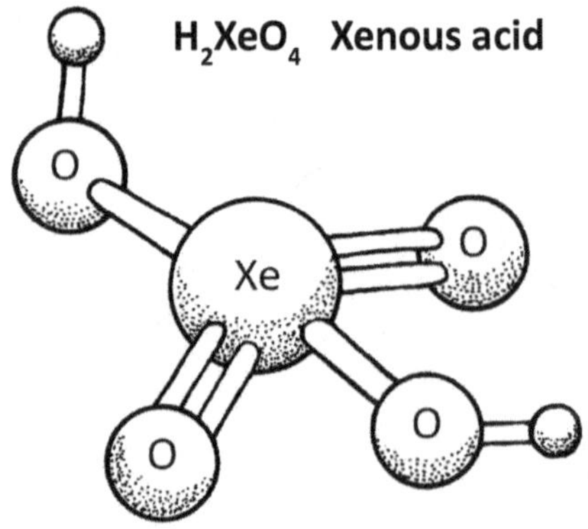

XeCl Xenon chloride

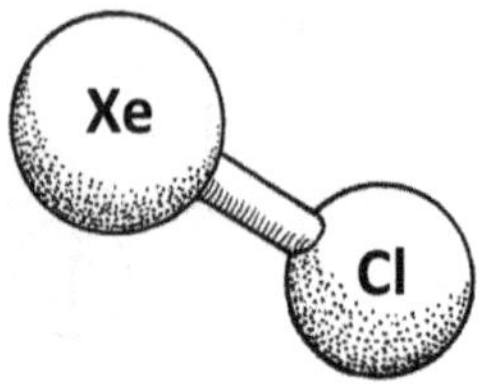

$Xe(NH_3)_2$ Xenon nitrate

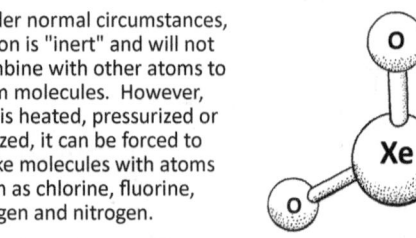

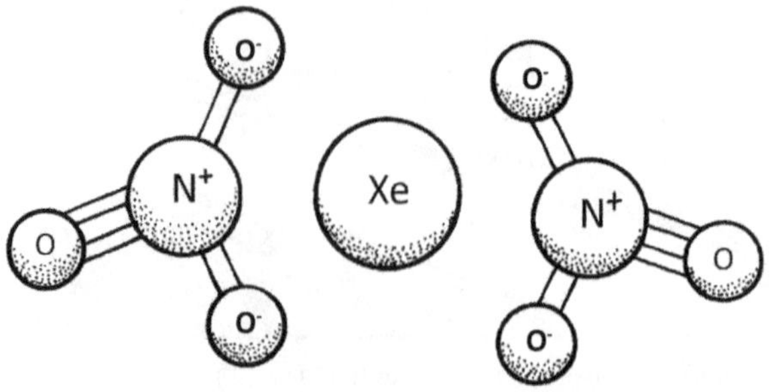

54 Xenon Xe

When electricity passes through xenon gas, it produces white light that has a spectrum similar to natural sunlight.

Satellites that need to travel long distances are often equipped with ion propulsion systems that use xenon gas.

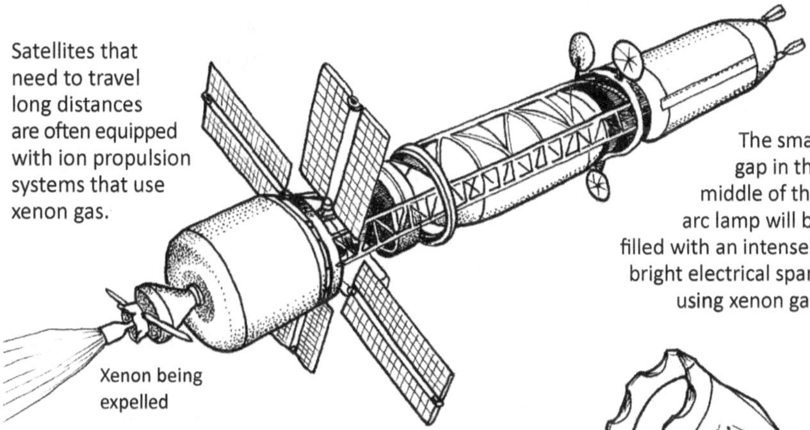

Xenon being expelled

The small gap in the middle of this arc lamp will be filled with an intensely bright electrical spark using xenon gas.

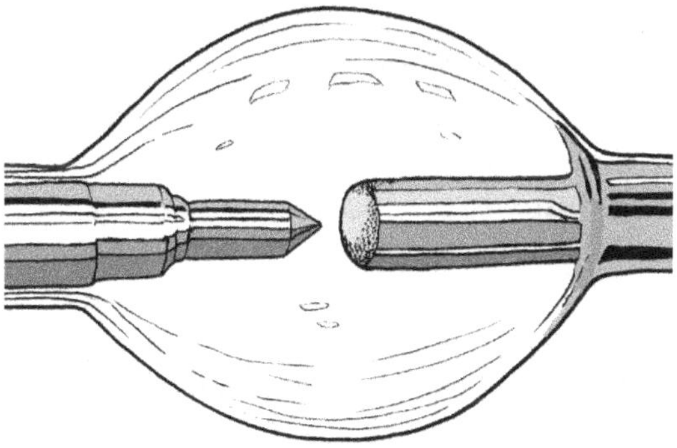

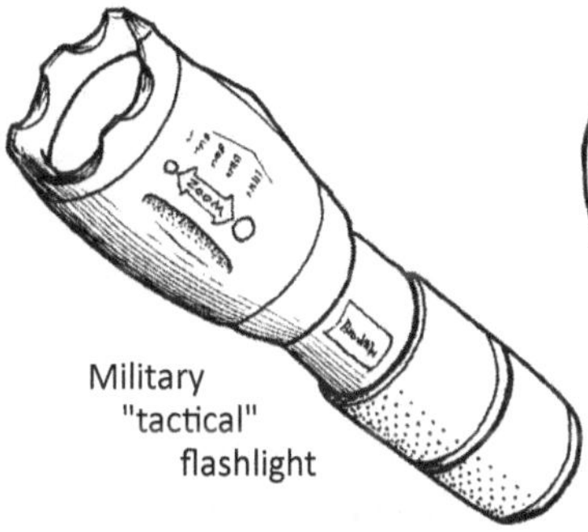

Military "tactical" flashlight

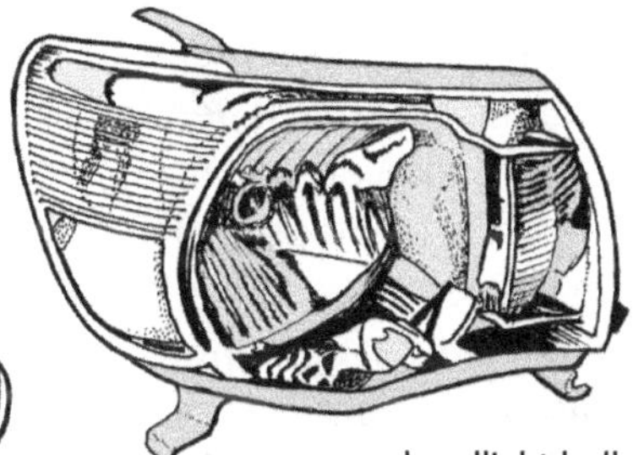

car headlight bulbs

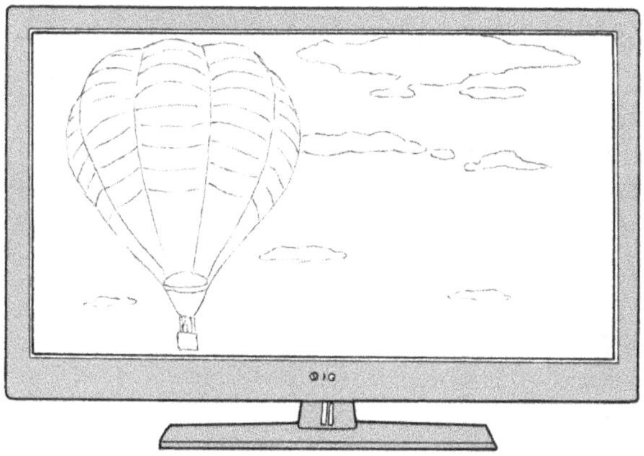

Plasma screens use xenon and neon.

One use for xenon excimer lasers is eye surgery.

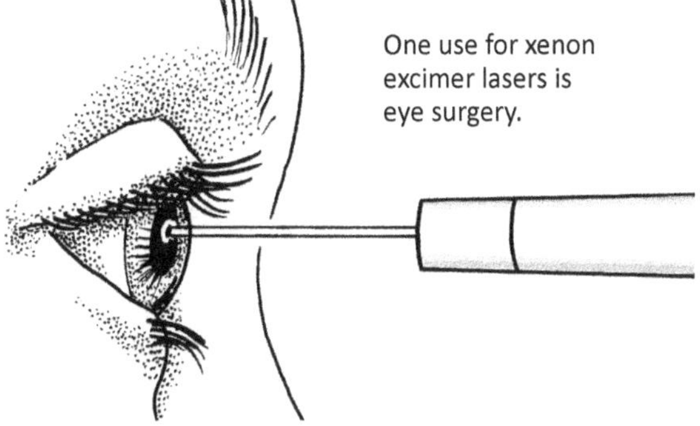

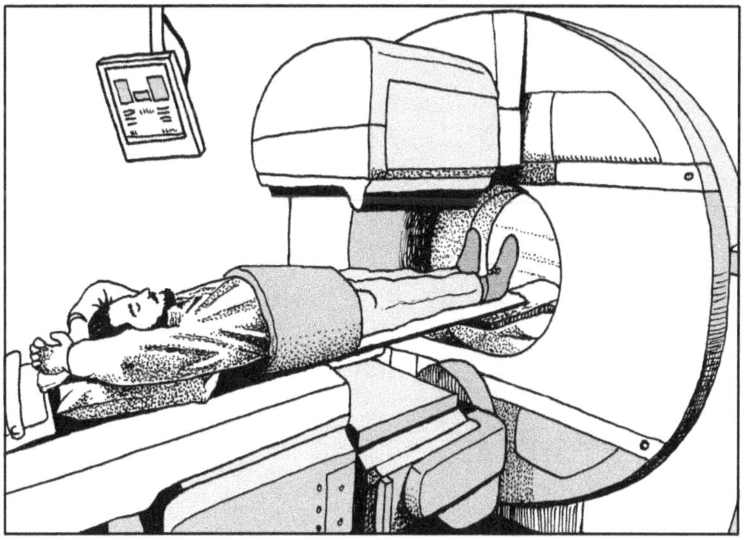

Radioactive isotopes of xenon are used for medical imaging.

Cs 55

protons
78 neutrons
55 electrons

Atomic mass: 132.9

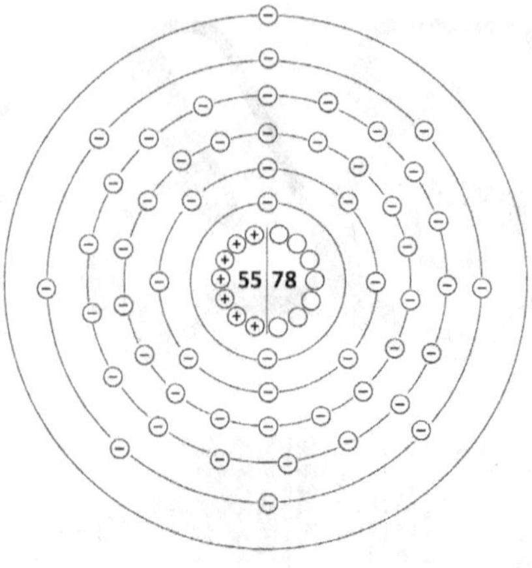

55
78

Cesium

Named using the Latin word for sky-blue, "caesius"

Cesium (or caesium) was discovered in 1860 by Robert Bunsen and Gustav Kirchhoff while examining a sample of mineral water from a well in Germany. They heated the sample and looked at it through their new invention, the spectrometer. They had used their spectrometer to look at many elements and they knew the unique patterns that each one produced. So when they saw a new pattern that featured many light blue lines, they knew they had discovered a new element. They named it using the Latin word for light blue. Bunsen and Kirchhoff would use their spectrometer to discover several other elements, including rubidium, the element right above cesium on the table. These two elements share many chemical properties, including being so reactive that they explode and burn if they come into contact with water. Pure cesium is a soft, silvery metal that will melt into a liquid at only 29° C (83° F).

Cesium is always found in compounds, never alone. Pollucite ($CsAlSi_2O_6$) is the mineral ore most often used in industry to produce pure cesium. Most of the world's pollucite occurs in Manitoba, Canada. The only other significant ore is pezzottaite, which can be used to make a pinkish-red gemstone called "raspberry beryl."

The first industrial use for cesium came in the 1920s when vacuum tubes were invented. Cesium was used as a "getter" that would get rid of unwanted gas molecules in the tube by catching and holding them. Now that vacuum tubes have been replaced by microchips, the largest use of cesium is in the petroleum industry where it is used in drilling fluids. Though pure cesium is dangerous, cesium compounds are considered to be fairly safe and non-toxic.

Extremely accurate atomic clocks (that lose less than one second every million years!) use cesium atoms as their source of time-keeping. All atoms vibrate (or "oscillate"). Cesium oscillates 9,192,631,770 times per second. Or, rather, a second is defined as the time it takes for a cesium atom to oscillate 9,192,631,770 times. GPS satellites and the entire Internet depend on atomic clocks to provide the highly accurate time keeping they need.

Cesium is used in photoelectric cells since it can turn light into electrical current. Cesium compounds are used in machines that detect x-rays and gamma rays, and in dosimeters that keep track of exposure to radiation. Radioactive cesium can be manufactured in nuclear labs, and Cs-137 is used as a source of gamma rays for machines used to measure the density of rock formations and the thickness of materials, and for treatment of certain cancers.

Solutions of cesium chloride and cesium sulfide are used in microbiology as a fluid in centrifuges (machines that spin test tubes) to isolate samples of viruses, parts of cells, and pieces of DNA.

CsF Cesium fluoride

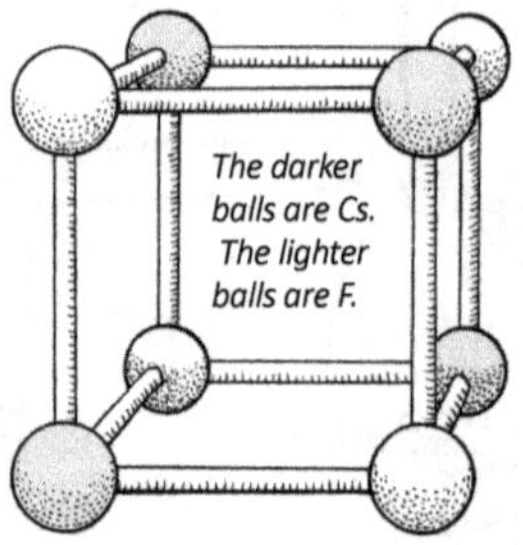

The darker balls are Cs. The lighter balls are F.

CsCl Cesium chloride

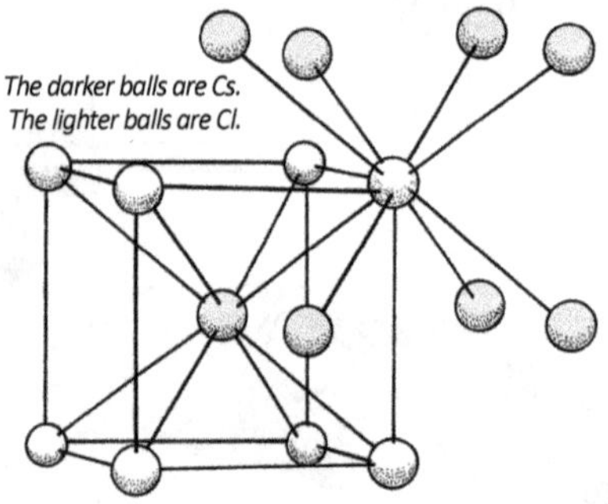

The darker balls are Cs.
The lighter balls are Cl.

Cs_2O Cesium oxide

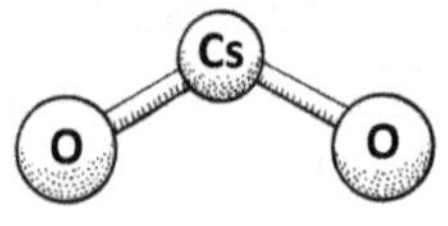

Cs
O
O

$Cs_{11}O_3$ Cesium "suboxide"

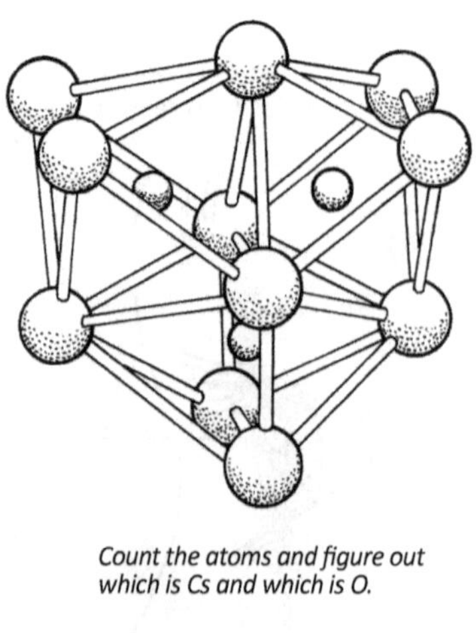

Count the atoms and figure out which is Cs and which is O.

CsN_3 Cesium azide

The darker balls are Cs.
The lighter balls are N.

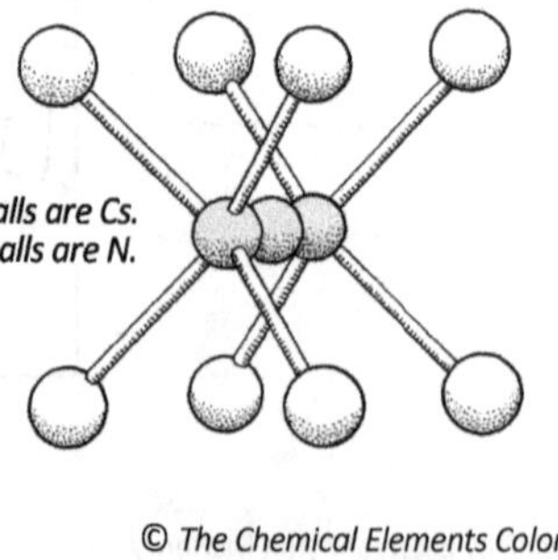

Cs_2CO_3
Cesium carbonate

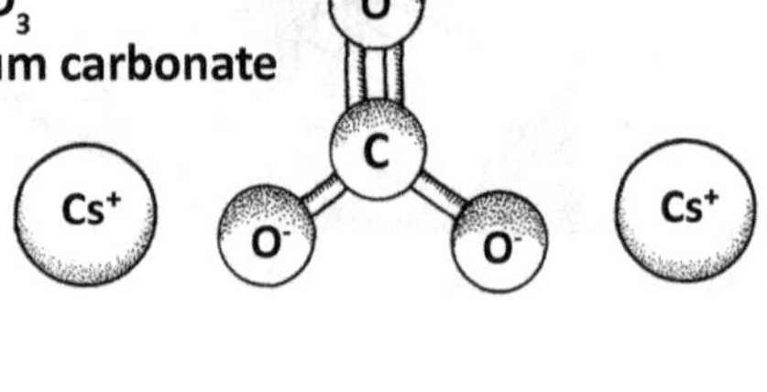

O
C
Cs+
O-
O-
Cs+

55 Cs

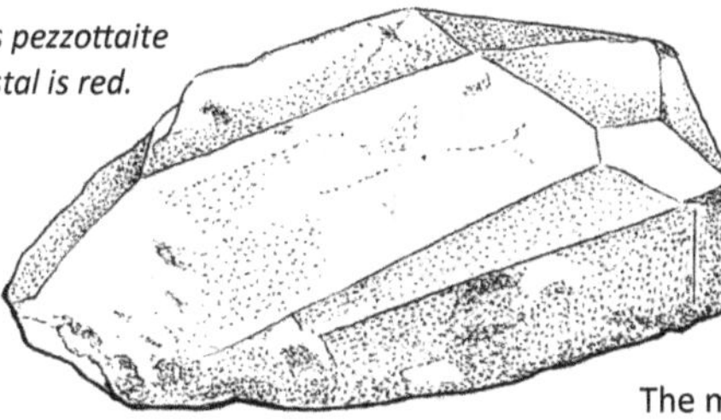

This pezzottaite crystal is red.

The mineral pezzottaite $(Cs(Be_2Li)Al_2Si_6O_{18})$ is one of the ores from which cesium can be extracted.

Cesium

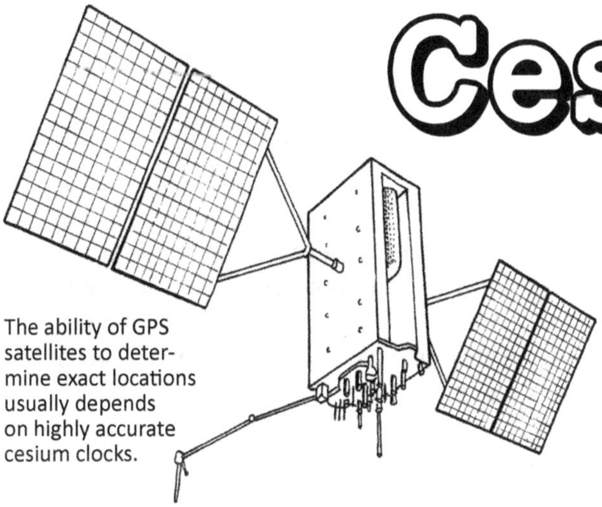

The ability of GPS satellites to determine exact locations usually depends on highly accurate cesium clocks.

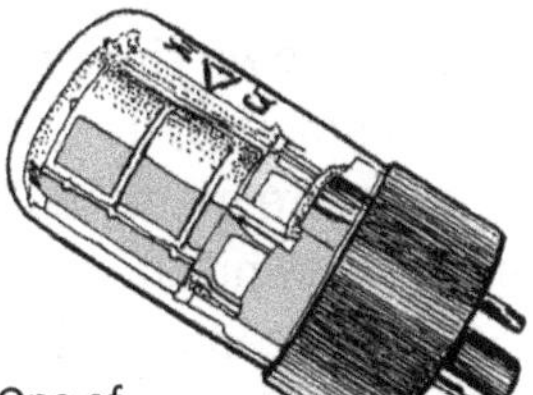

One of the first uses for cesium was in vacuum tubes as a "getter" to remove unwanted gases.

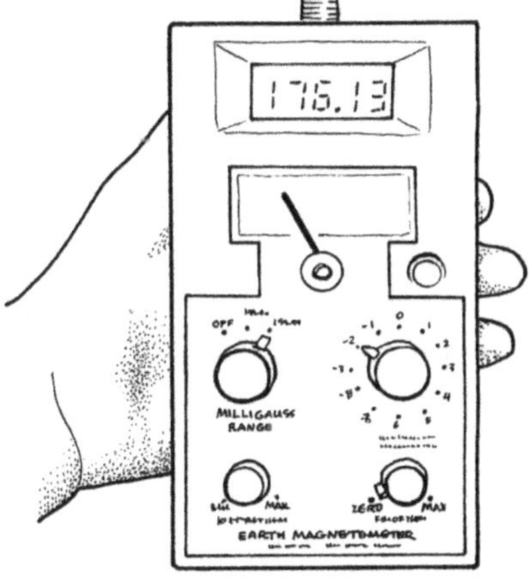

Magnetometers, which measure magnetism, may use cesium.

The first cesium clocks were very large. The clock shown here is in Canada and belongs to the National Research Council (NRC).

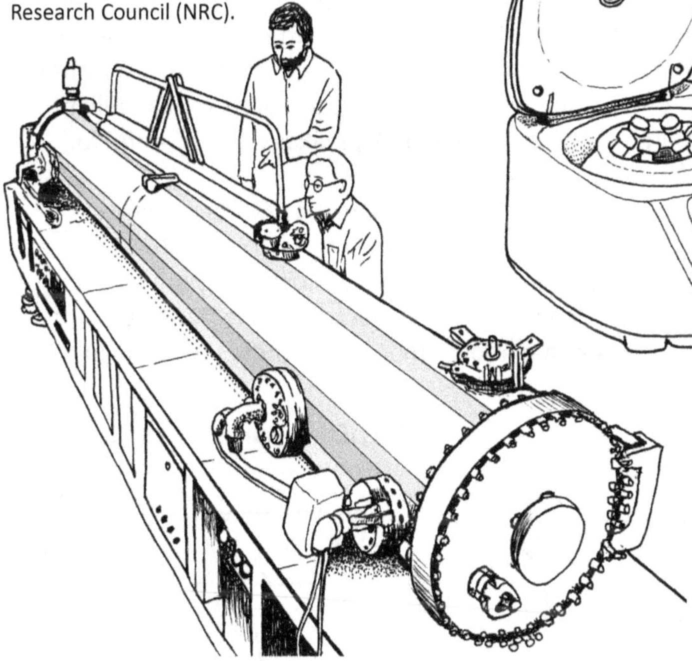

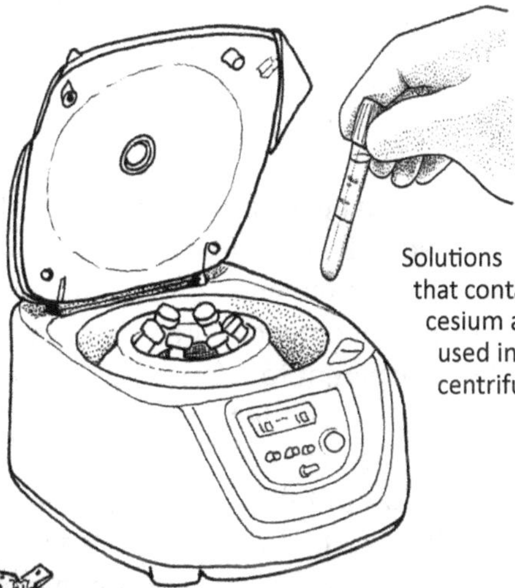

Solutions that contain cesium are used in centrifuges.

Radioactive cesium is used to track contamination from nuclear accidents (such as Fukushima).

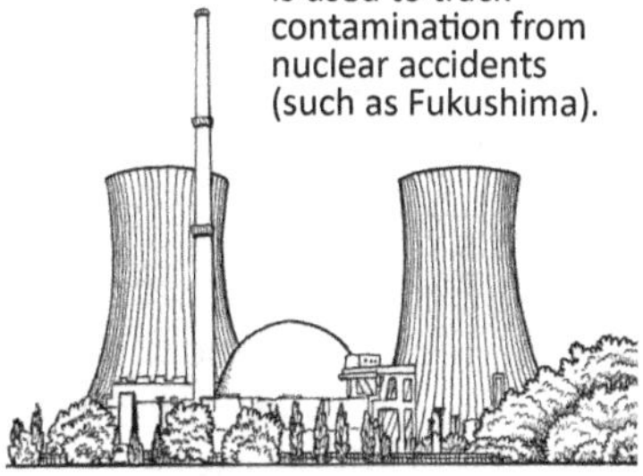

Gustav Kirchhoff

Robert Bunsen

Cesium was the first element discovered by Robert Bunsen and Gustav Kirchhoff using their spectrometer, a machine that separates light into a pattern of colored lines.

The spectral emission pattern of cesium

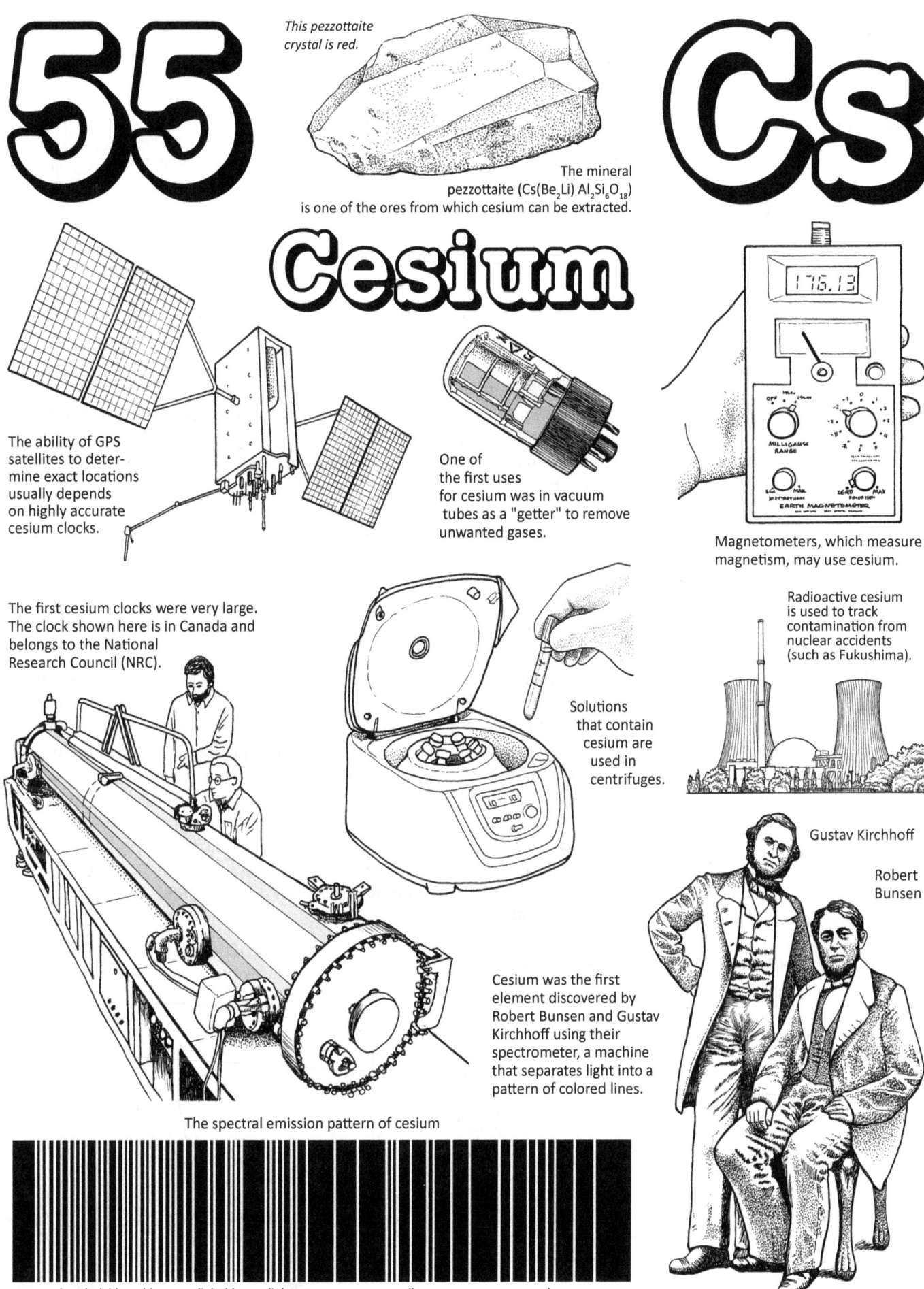

Ba 56

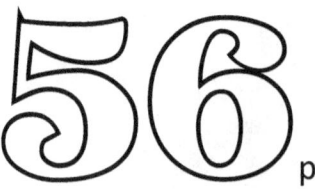

protons
81 neutrons
56 electrons
Atomic mass: 137.3

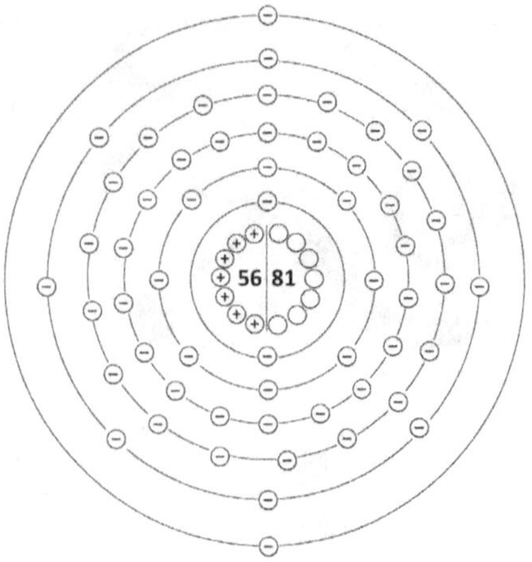

Barium

Named using the Greek word for heavy, "barys"

Minerals containing barium have been known since the Middle Ages. In 1602, a chemist wrote down his observations of "baryte," a volcanic rock from Bologna, Italy, that would glow for years after being exposed to light. This fascinated chemists and several attempts were made to discover the substance in the rocks that caused this phenomenon. Two German chemists were able to produce barium oxide, but still were not able to identify the new element it contained. Pure barium was not produced until 1808 when Sir Humphry Davy pulled the barium out of a solution using electricity. Barium is right below magnesium, calcium and strontium, and therefore shares similar chemical properties. It is a very reactive silvery metal that is never found alone in nature, only in compounds.

Barium forms compounds with many other atoms. Some of these compounds are toxic, but others are very safe. Barium carbonate, $BaCO_3$, is dangerous and is sometimes used as a rat poison. (Humans are at less risk of barium carbonate poisoning since we are so much larger than rats and would have to eat quite a lot of it.) Barium carbonate is useful to the glass and ceramic industries where it is used in various ways, including in clay and glazes, and as a coloring agent. It is also used by the oil drilling industry as an ingredient in the liquid "slurry" that goes down the shaft to keep the drill shaft lubricated. Barium sulfate, $BaSO_4$, is non-toxic and can be safely swallowed because it won't dissolve in water. Barium sulfate drinks are given to patients who need x-ray images of their intestines.

Barium oxide, BaO, is used on electrodes in fluorescent lamps (to aid the release of electrons) and as a desiccant, absorbing excess moisture in places that need to be very dry. Barium chloride, $BaCl_2$, can be used to "soften" water, helping to remove atoms such as calcium, magnesium and iron. Barium nitrate, $Ba(NO_3)_2$, will burn with a brilliant bright green color and is used in fireworks and emergency flares. Barium fluoride, BaF_2, makes crystals useful for making specialized optics that need to be transparent to infrared light waves.

Some barium compounds are used in very high-tech applications. Barium titanate, $BaTiO_3$, has electrical properties and can be used in capacitors and microphones. Barium is an ingredient in a material called YBCO, yttrium barium copper oxide. Only small amounts of this substance have been made and mostly they are used for research into superconductivity, when a substance loses all resistance to the flow of electrons. The significance of YBCO is that it will become superconducting at a higher temperature than most other superconductors, though this "high" temperature is still very cold: -196° C (-321° F).

$BaCO_3$ Barium carbonate

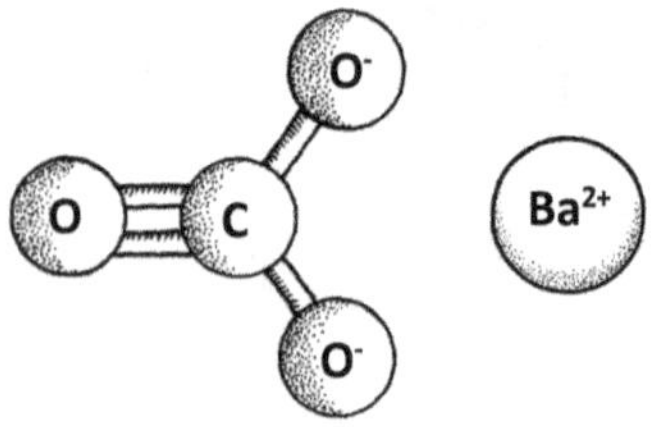

$BaSO_4$ Barium sulfate

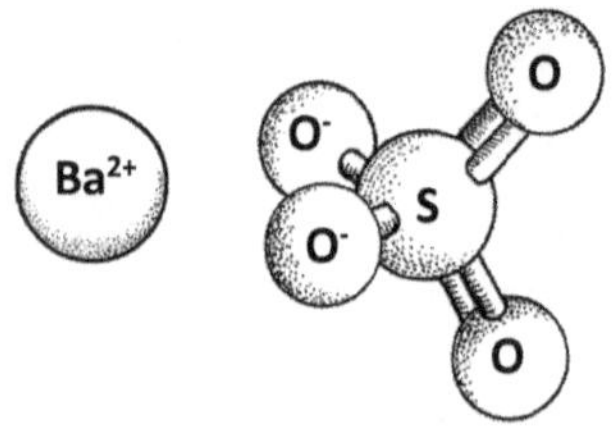

BaF_2 Barium fluoride

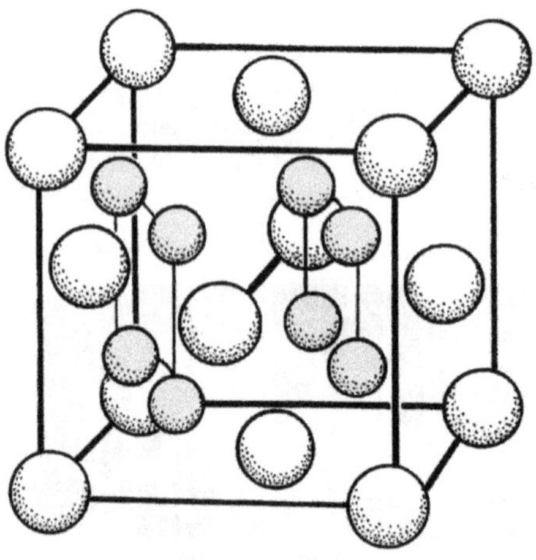

The darker balls are F. The lighter balls are Ba.

$BaTiO_3$ Barium titanate

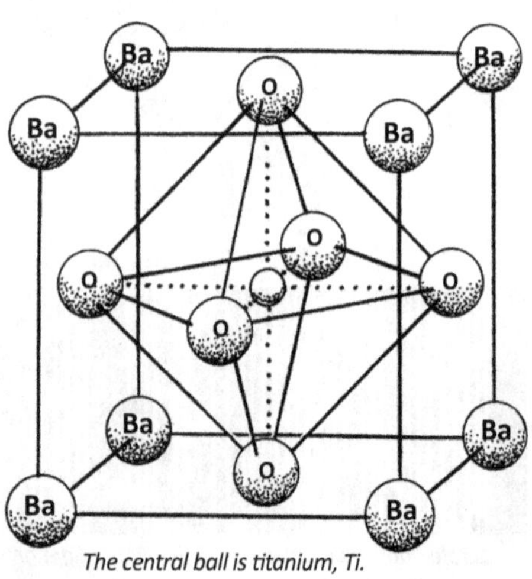

The central ball is titanium, Ti.

56 Ba

Reddish brown or tan.

This mineral formation, called the desert rose, is made of $BaSO_4$.

Barium

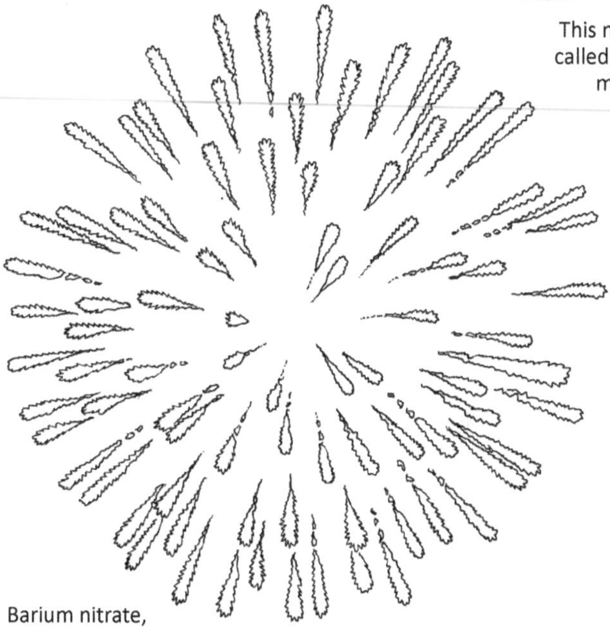

Barium nitrate, $Ba(NO_3)_2$, burns bright green and is used in fireworks.

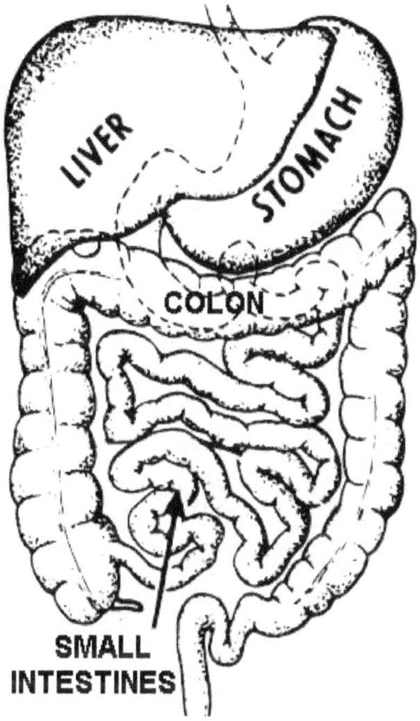

Barium sulfate is the primary ingredient in a solution given to patients who need to have x-rays of their intestines.

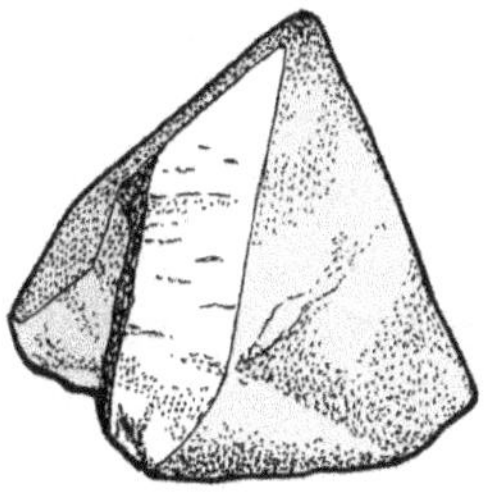

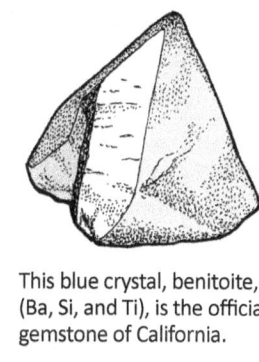

This blue crystal, benitoite, (Ba, Si, and Ti), is the official gemstone of California.

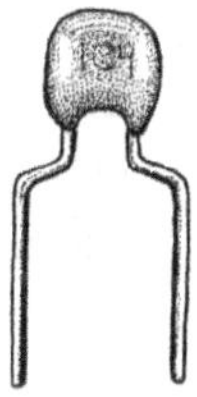

Capacitors can contain $BaTiO_3$.

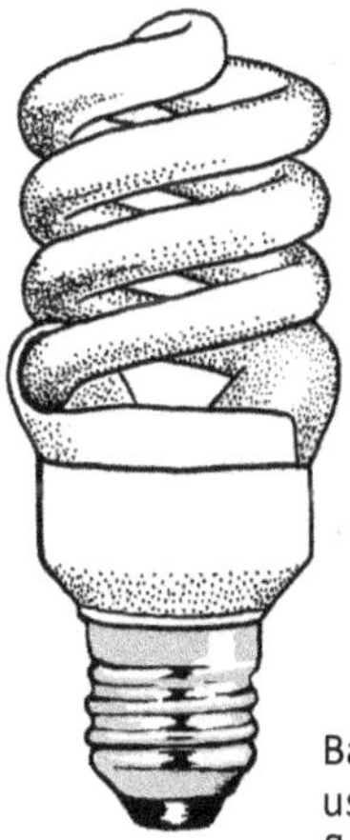

Clay that is high in barium sulfate is used to make pottery called Jasperware. This teapot is dark blue with white designs.

Barium oxide, BaO, is used on electrodes inside fluorescent lights.

Baryte rocks were discovered around volcanoes in Italy. After exposure to light or heat they will glow for years.

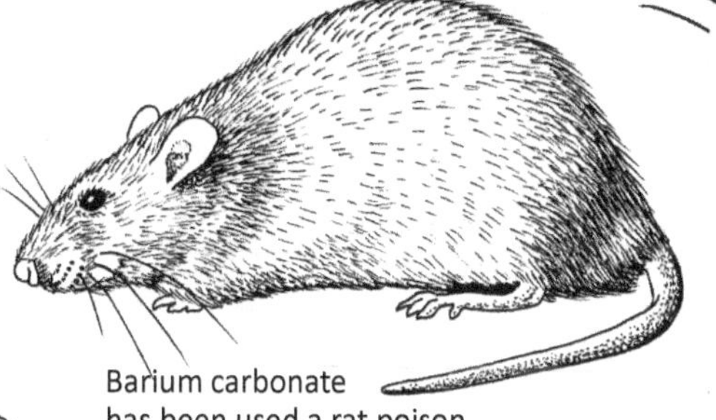

Barium carbonate has been used a rat poison.

YBCO (yttrium barium copper oxide) is being studied for its superconducting properties. Superconductors lose all resistance to electrical flow at very low temperatures.

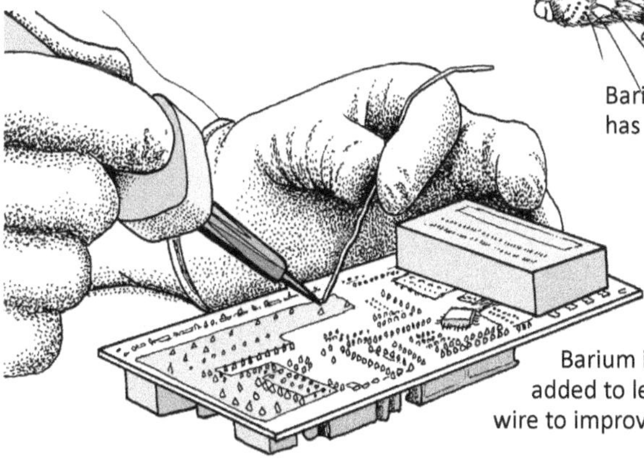

Barium is sometimes added to lead soldering wire to improve its texture.

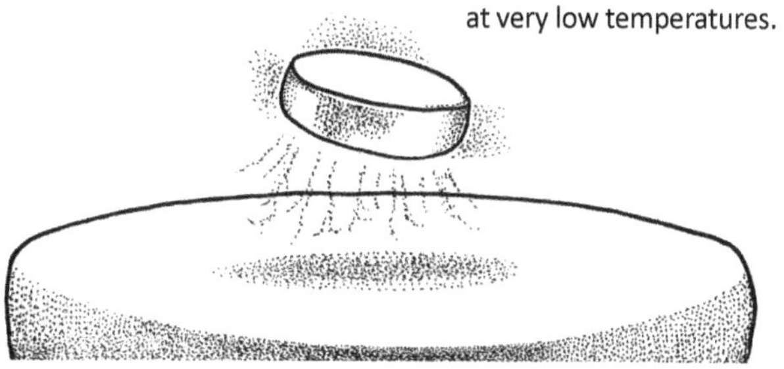

La 57

57 protons
82 neutrons
57 electrons
Atomic mass: 138.9

Lanthanum

Named using the Greek word for hidden, "lanthanein"

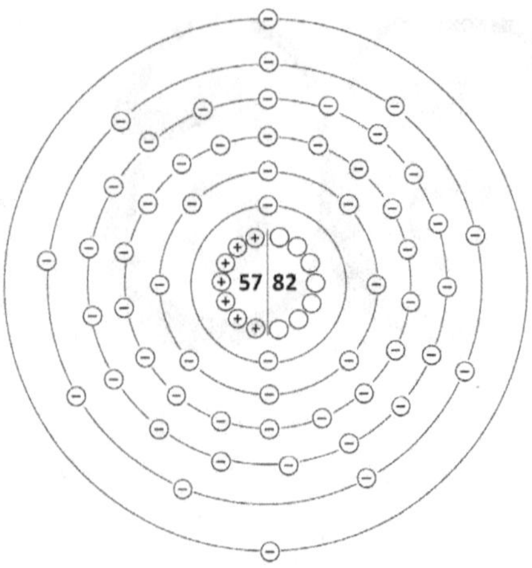

The discovery of lanthanum is a complicated story involving many scientists, and its story is intertwined with the stories of other rare earth elements (numbers 58 to 71). These elements are very difficult to isolate, and often what the scientists thought was one new element turned out to be a mixture of two. Cerium had just been discovered in a mineral sample from Sweden, and as the scientists continued to study the sample, they were surprised to find out that another new element had been there all that time. Since this second element seemed to have been "lying hidden" all that time, they named it using the Greek word lanthanein, which means to lie hidden. Though lanthanum is classified as a "rare earth" element, it is not all that rare. It is three times more abundant than lead.

The first commercial use of lanthanum was in 1886, when lanthanum oxide was combined with zirconium oxide to made mantles for gas lanterns. The light it produced was too green, however, limiting its success.

One of lanthanum's unusual characteristics is that friction will make it produce sparks. Because of this, lanthanum is a key ingredient in the "flint" that makes sparks in lighters and other fire-starting devices. In another application related to sparking, lanthanum is alloyed with tungsten to make electrodes for gas-metal welding machines.

Another characteristic of lanthanum is known for is the ability to capture phosphate (PO_4) molecules. Although phosphorus is essential for all life, too much of it in one place is dangerous. For example, algae cells will multiply too fast if put in a high-phosphorus environment. Pool cleaning products often contain lanthanum compounds that remove phosphates from the water so algae won't grow. To treat lakes and ponds, clay high in lanthanum can be safely added. Another place you don't want too much phosphorus is in your blood. Doctors can use lanthanum carbonate as a medicine to bind and hold excess phosphorus in the blood.

Lanthanum isn't one of the elements that our bodies need, but neither is it highly toxic. The only form of life that actually needs lanthanum is a rare species of bacteria that lives around hot vents called fumaroles, which produce poisonous gases. These bacteria can live in conditions that would kill other forms of life.

Lanthanum is sometimes added to steel to make it easier to work with. Lanthanum oxide is added to glass that will be used for specialized lenses such as those found in telescopes and other high-tech optical devices.

One of the most recent uses for lanthanum is in "nickel-metal-hydride" batteries for hybrid (partially electric) cars. Each battery can contain as much as 30 pounds (15 kg) of lanthanum.

LaF_2 Lanthanum difluoride

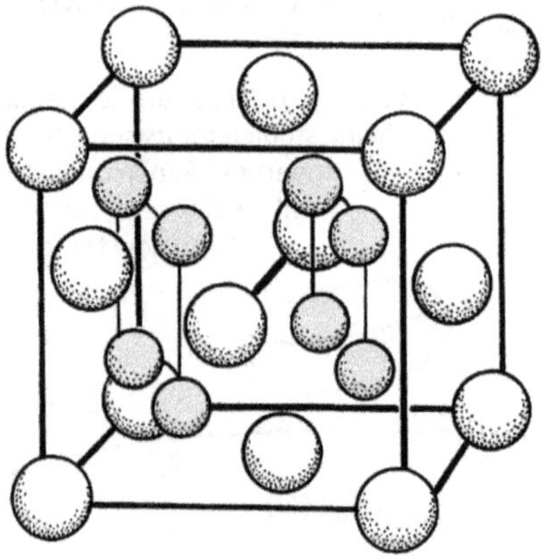

The darker balls are La. The lighter balls are O.

La_2O_3 Lanthanum oxide

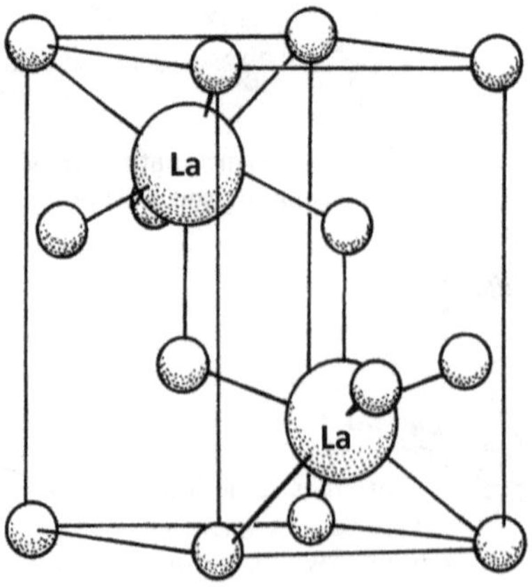

LaB_6 Lanthanum hexaboride

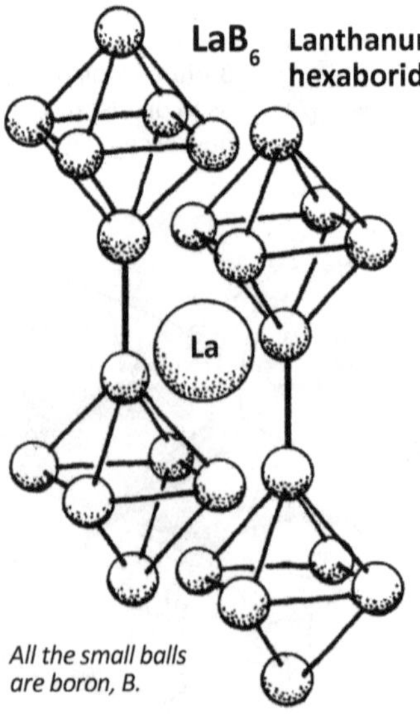

All the small balls are boron, B.

57 La

Lanthanum is one of the metals used in the "flints" in fire starters.

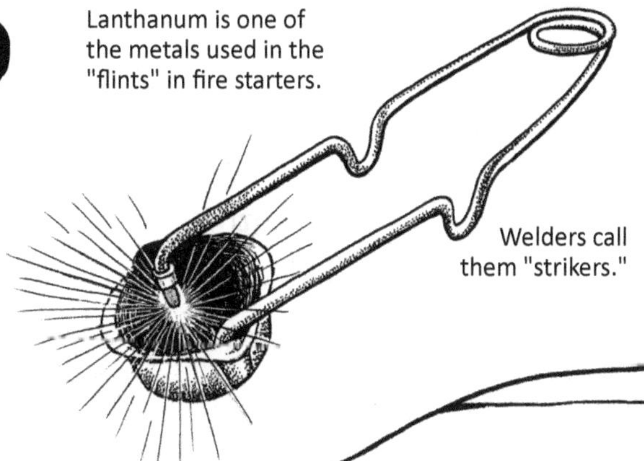

Welders call them "strikers."

This laser engraver uses lanthanum hexaboride.

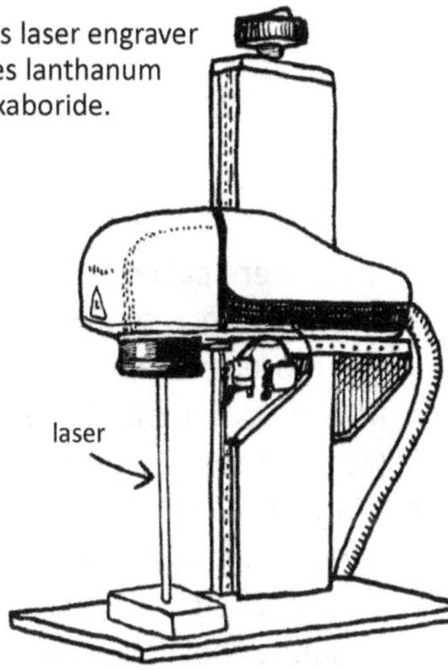

Rechargeable batteries in hybrid cars contain as much as 30 pounds (15 kg) of lanthanum.

Lanthanum

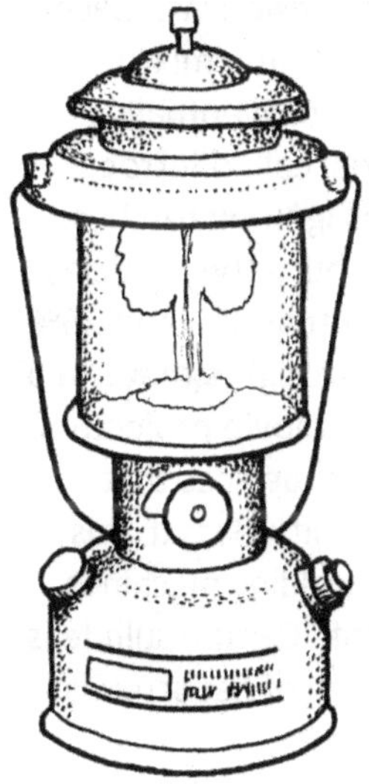

green algae cells

Lanthanum is an ingredient in cleaning solutions that prevent the growth of algae in pools.

Lanthanum is added to glass that will be used in telescopic lenses.

Early gas lanterns used La_2O_3 in their mantles (the glowing part).

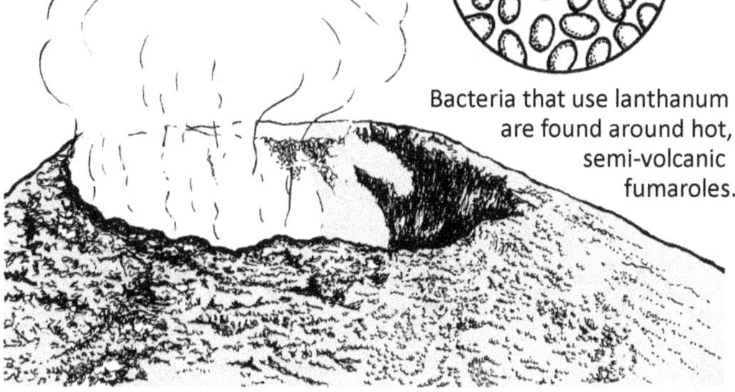

Bacteria that use lanthanum are found around hot, semi-volcanic fumaroles.

This gas-metal welding equipment uses an electrode that is made of an alloy of tungsten and lanthanum.

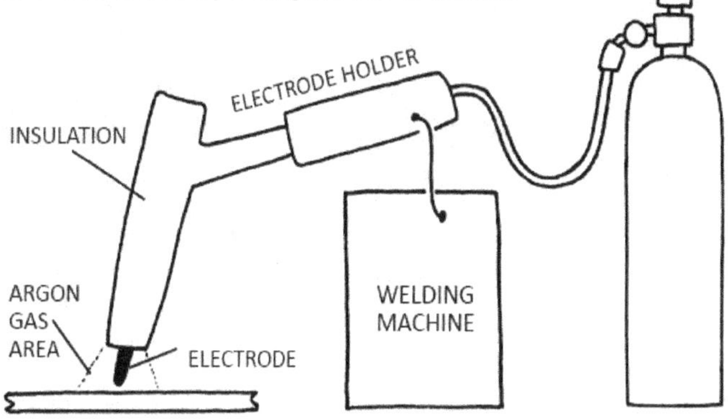

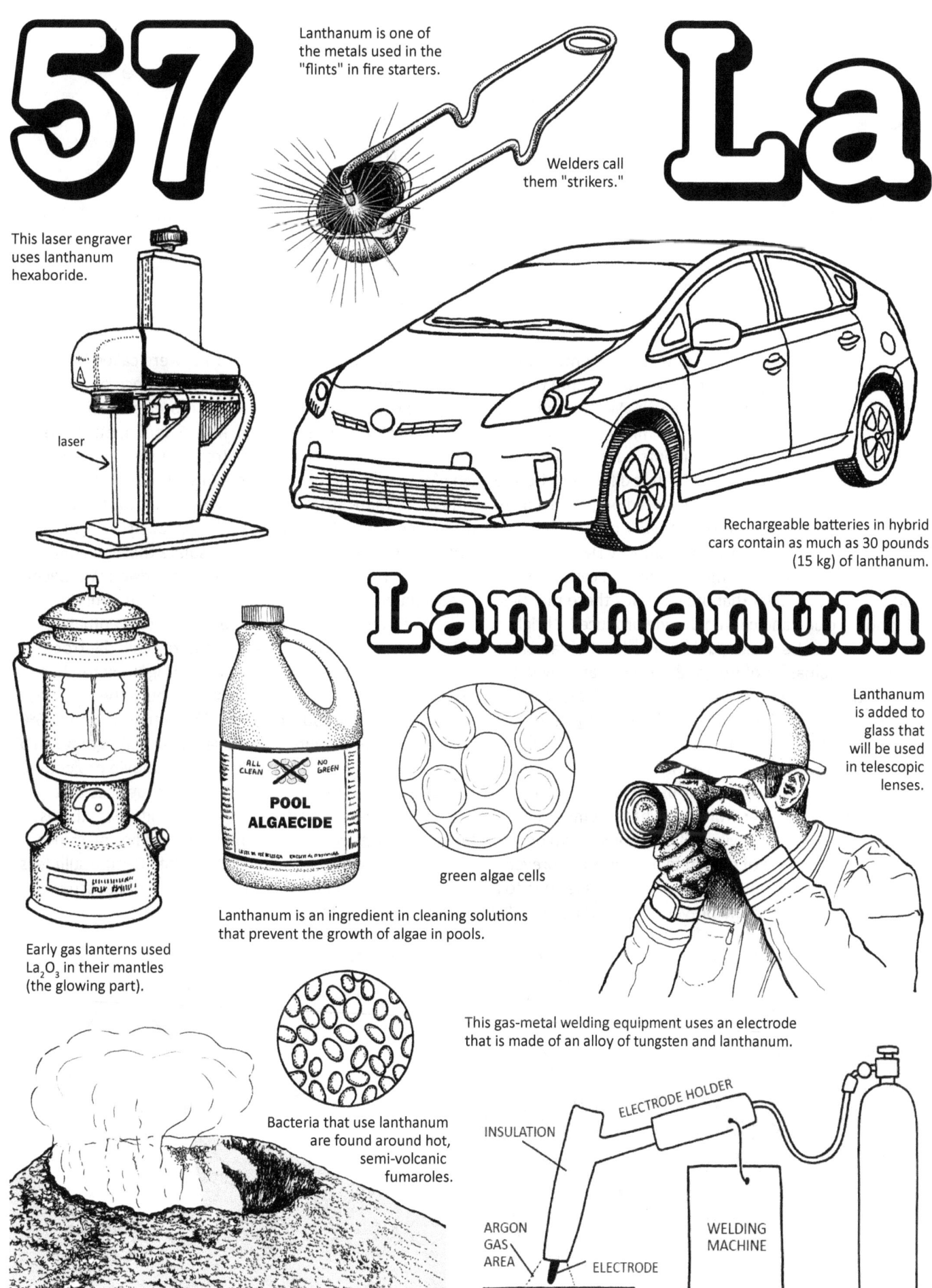

Ce 58

protons
82 neutrons
58 electrons

Atomic mass: 140.1

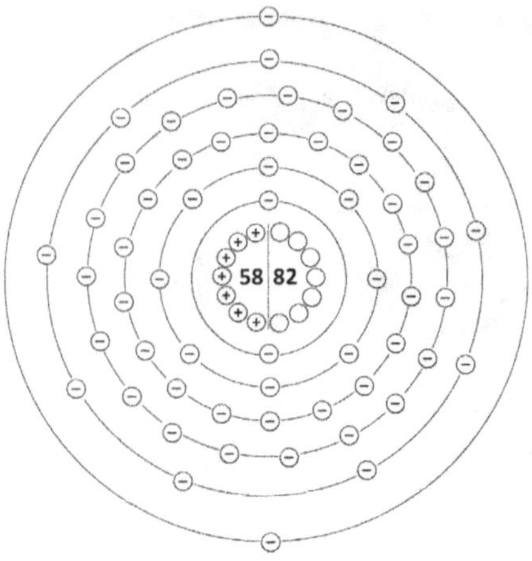

Cerium

Named after the asteroid Ceres

Cerium was the first "rare earth" element to be discovered. These elements (57 to 71) were called "rare" when they were discovered because at the time they did indeed seem rare compared to the abundance of other minerals being mined from the earth. We now know that they are not rare, and that there is more cerium in the earth than tin or lead. The original mineral sample in which cerium was discovered came from Sweden, but since then, sources of rare earth elements have been found all over the world. Rare earth ores (rocks or sand) always contain a mix of these elements. It is difficult to isolate each element, and often when chemists thought they had discovered one new element it turned out to be a mixture of two or more. Three scientists get credit for the discovery of cerium in 1803, although none of them were able to produce a sample of pure cerium metal. It was not until the invention of electrolysis by Humphry Davy in the late 1880s that scientists were able to isolate these elements.

One of the first commercial uses of cerium was in mantles for gas lanterns. Austrian chemist Carl von Welsbach had originally used lanthanum oxide for these mantles but the light was too green. Then he found that a mixture of thorium oxide and cerium dioxide produced a bright white light, exactly what his customers wanted. Eventually, thorium had to be removed due to its level of radioactivity. Welsbach went on to create a material called ferrocerium, a combination of iron and cerium that proved to be very useful for making sparking devices igniting torches. Like lanthanum, cerium will make sparks if it is scratched. The sparks from the cerium ignite and burn the particles of iron. Another alloy often used in sparking devices, "mischmetal," is about 50% cerium, 25% lanthanum, and 25% other rare earth elements such as samarium, europium and neodymium. Cerium is also used in arc lights that use a bright spark jumping across a small gap. These bulbs are very bright and were especially useful in movie projectors.

Cerium oxide has many uses in various industries. It is commonly used in catalytic converters, devices that remove toxic molecules from the exhaust fumes of cars. It is used to polish glass, especially high-quality glass that is being used for optical devices. It is combined with other metals to make electrodes for gas metal arc welding machines.

The addition of cerium to pigments can prevent them from darkening when exposed to light. Cerium sulfide is used as a non-toxic red pigment. Cerium was added to glass for television screens (until the invention of flat screens) because the cerium could prevent the glass from darkening due to bombardment with electrons.

Ce_2O_3 Cerium oxide

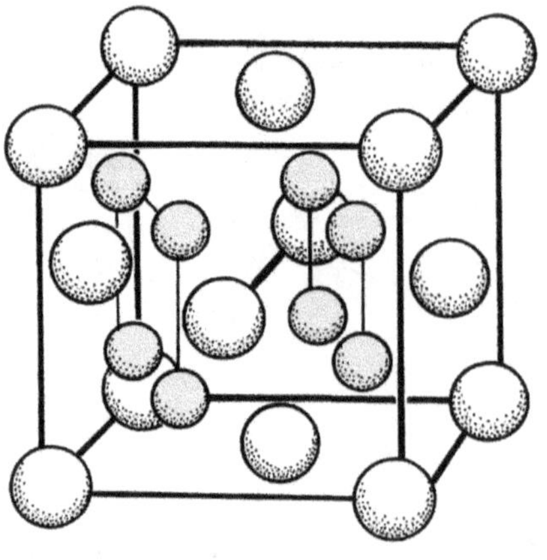

The darker balls are Ce. The lighter balls are O.

Monazite crystal structure

Monazite sand is one of the ores from which rare earth elements are extracted. The sand always has many rare earth elements from the lanthanide series.

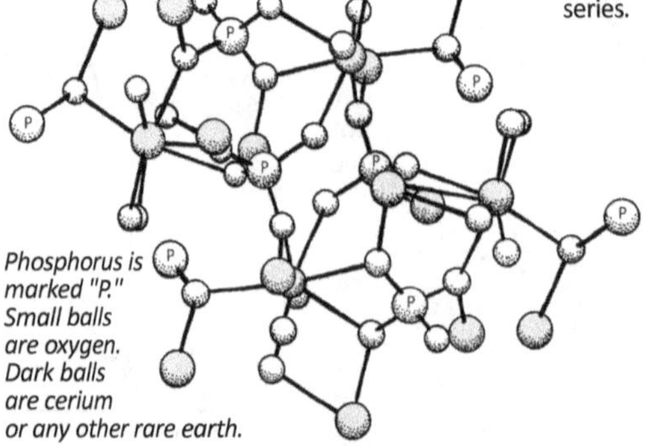

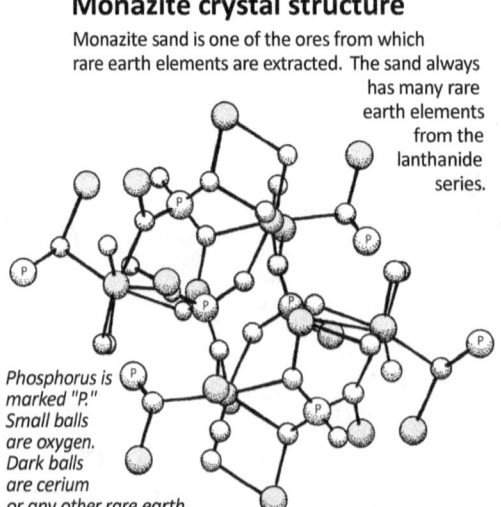

Phosphorus is marked "P." Small balls are oxygen. Dark balls are cerium or any other rare earth.

CeB_6 Cerium hexaboride

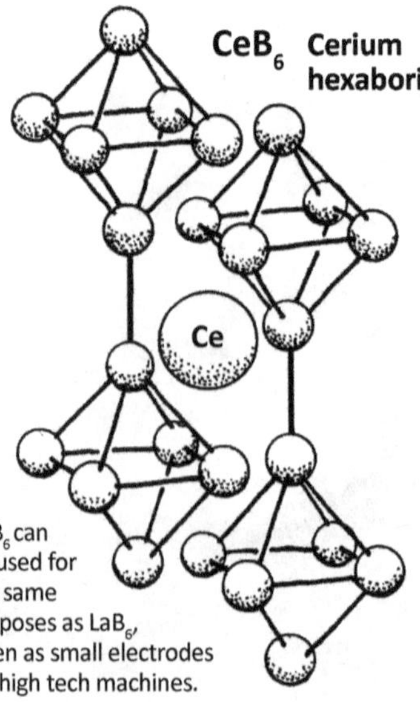

CeB_6 can be used for the same purposes as LaB_6, often as small electrodes for high tech machines.

58 Ce

Cerium was named after the newly discovered asteroid, Ceres.

Cerium

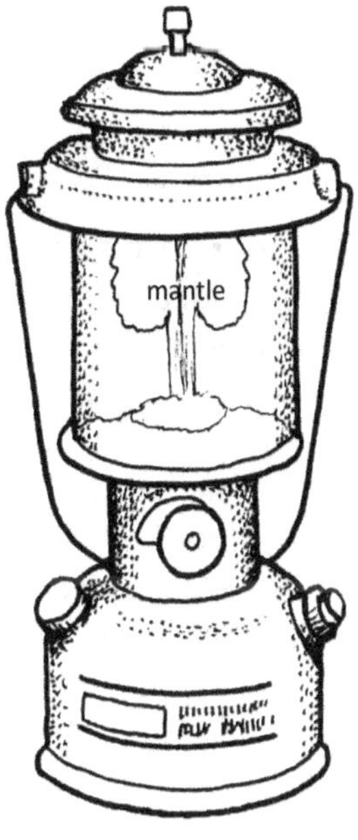

Cerium replaced lanthanum in the mantles of gas lanterns.

Cerium is one of the ingredients used to make the inside lining of self-cleaning ovens.

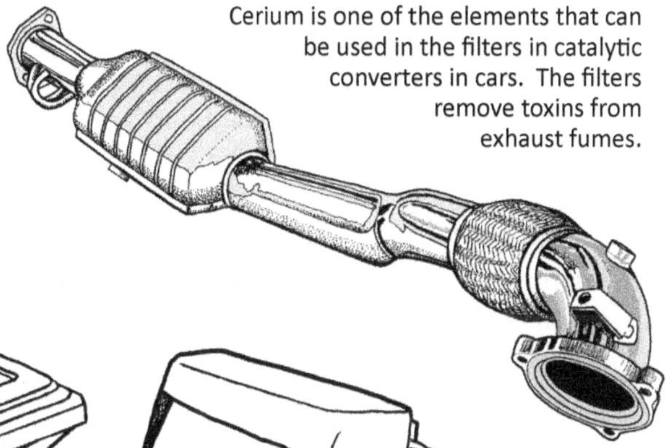

Cerium is one of the elements that can be used in the filters in catalytic converters in cars. The filters remove toxins from exhaust fumes.

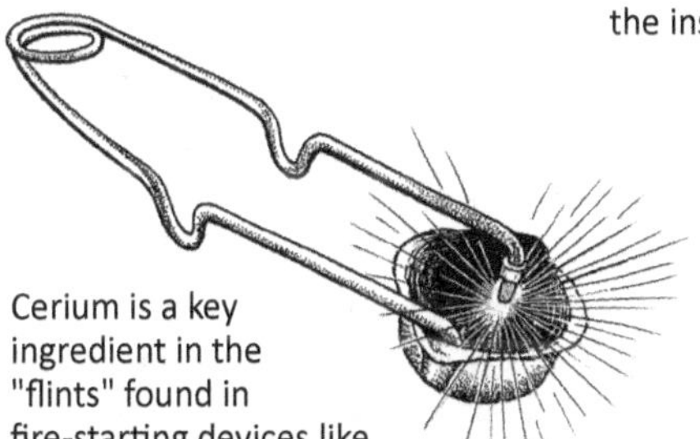

Cerium is a key ingredient in the "flints" found in fire-starting devices like this striker used by welders.

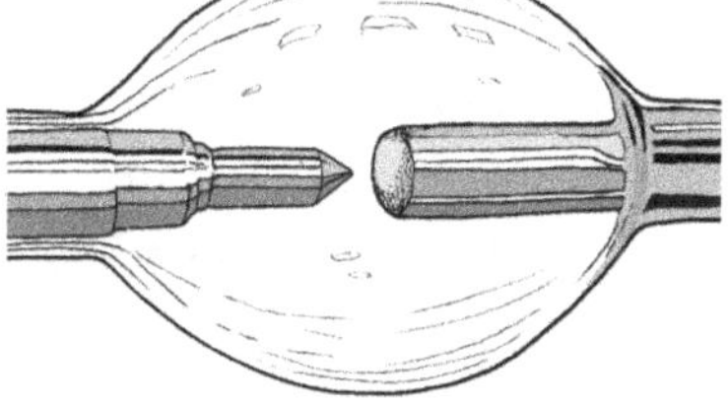

"Carbon arc" bulbs can use cerium in the metal rods that create a bright spark.

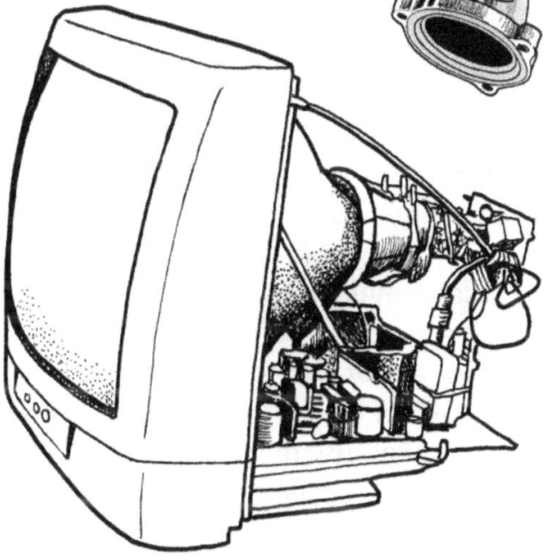

Cerium was added to the glass used in TV screens to keep the glass from darkening over time.

Cerium oxide is used to polish glass.

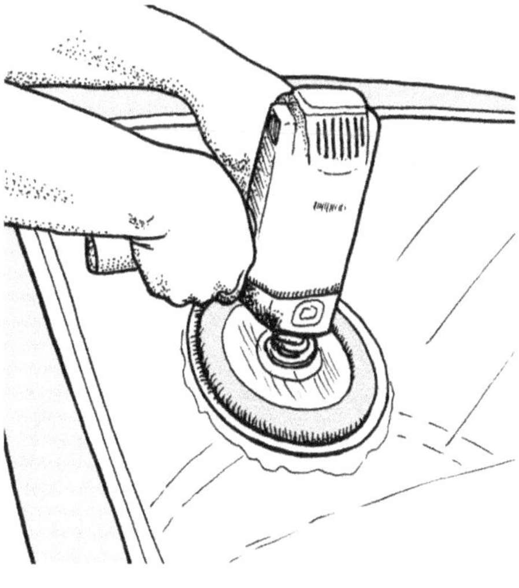

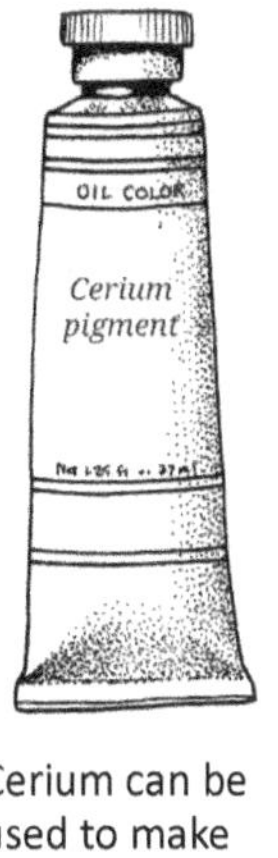

Cerium can be used to make pigments.

This gas-metal welding equipment uses an electrode that is made of an alloy of tungsten and cerium.

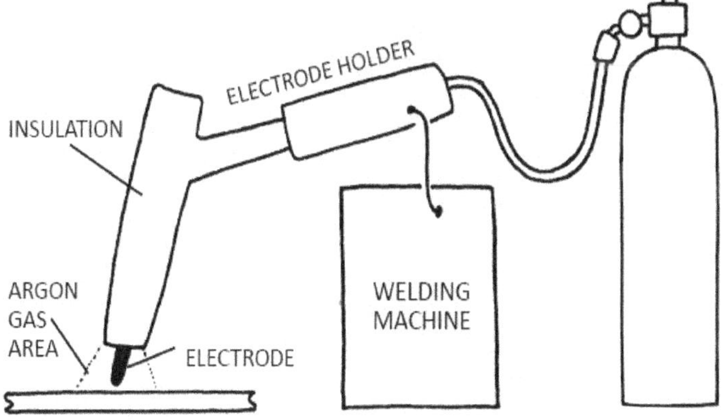

Pr 59

protons
82 neutrons
59 electrons

Praseodymium

Atomic mass: 140.9

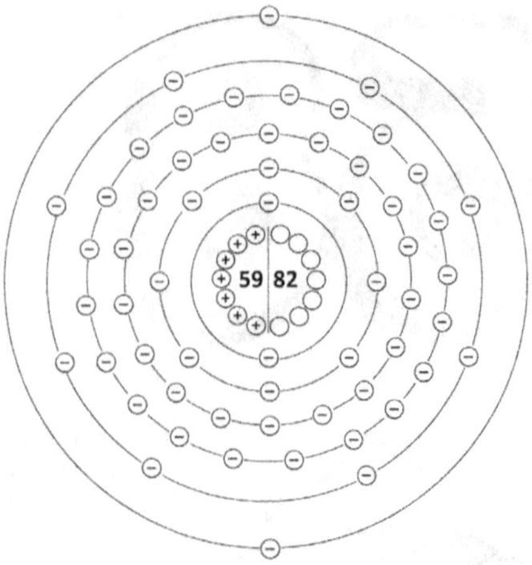

From Greek "praseos" (green) and "didymos" (twin)

The elements in the "lanthanide series" (numbers 57-71) are also called "rare earth" elements. If you watch those electron rings as we go through the lanthanides, you will notice that every time an electron is added it will go into the second ring from the outside, not the outermost ring. That second ring will get pretty crowded by the time we get to element 71! These rare earth elements turned out not to be as rare as people thought when they were first discovered. Despite the fact that you have probably never heard of praseodymium, there is more of it in the crust of the earth than gold, silver, tin or mercury. The main source of praseodymium (or any of the rare earths) is monazite sand. Carl von Welsbach, the discoverer of praseodymium, was one of the first scientists to realize that this sand held so many of the rare earth elements. Welsbach discovered neodymium at the same time as praseodymium because what he thought was one element turned out to be two. If praseodymium metal is dropped into hydrochloric acid, it will turn the solution light green, explaining the use of "praseos" in its name.

Praseodymium, like neodymium, is strongly magnetic. It is combined with other metals to make very stable and strong magnets for things like motors, loud speakers, and printers. When praseodymium is alloyed with nickel, it can create a super magnet that is able to slow down the vibration of atoms until they almost reach absolute zero temperature (-273° C). Another useful alloy can be made when praseodymium is combined with magnesium, and a few other metals, to make a material strong enough to be used in aircraft engines.

Despite its being named after the color green, when praseodymium is added to glass, the glass turns yellow. Praseodymium is combined with some other elements to make "didymium glass" which is used to make glasses for welders and glass blowers. (Didymium glass is typically purple, not green or yellow.) The didymium glass absorbs yellow and orange light produced by burning sodium. In another light-related application, praseodymium is put into fluorescent lamps to activate phosphors that will glow red, green or blue. It is also used in the fiber optics industry.

Because the rare earth elements (lanthanides) are so similar, any of them can be used to make mischmetal, the "flint" that makes sparks in fire-starting devices. Any of the rare earths, including praseodymium, can be used in the electrodes found in "carbon arc" bulbs, which are used in movie projectors and other devices that need a very strong, bright white light.

Pr_2O_3 Praseodymium (III) oxide

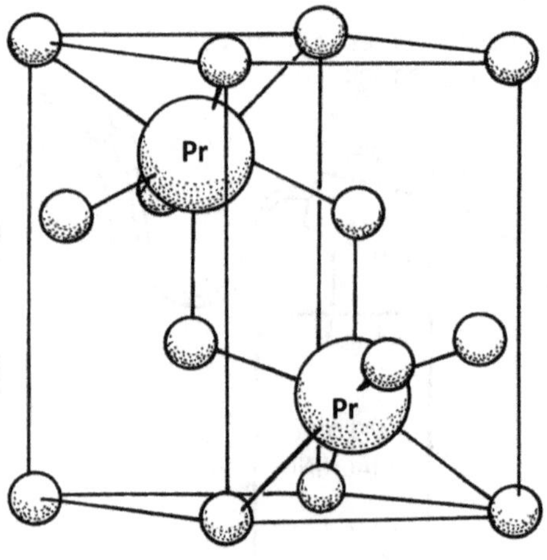

The unmarked balls are oxygen, O.

PrF_3 Praseodymium fluoride

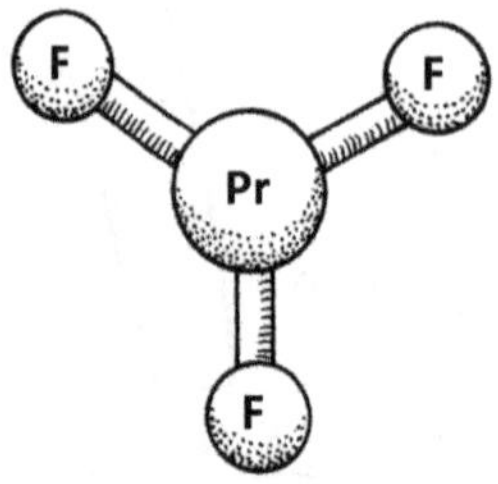

Monazite crystal structure

Monazite sand is one of the ores from which rare earth elements are extracted.

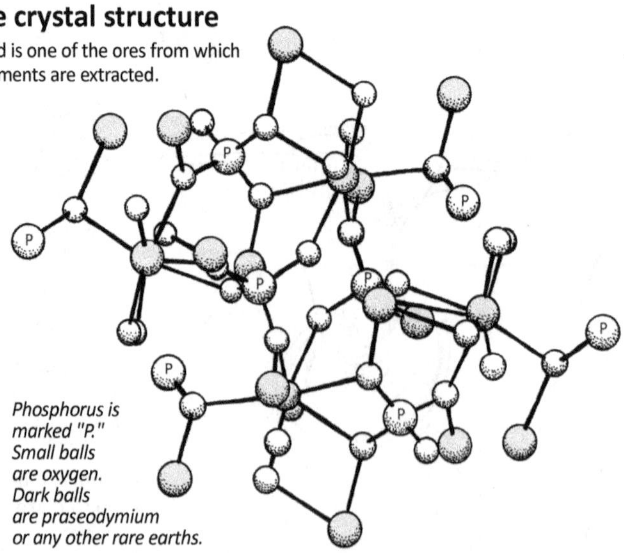

Phosphorus is marked "P." Small balls are oxygen. Dark balls are praseodymium or any other rare earths.

59 Pr

Didymium glasses have purple lenses and are used by glass blowers and welders because they absorb bright yellow light.

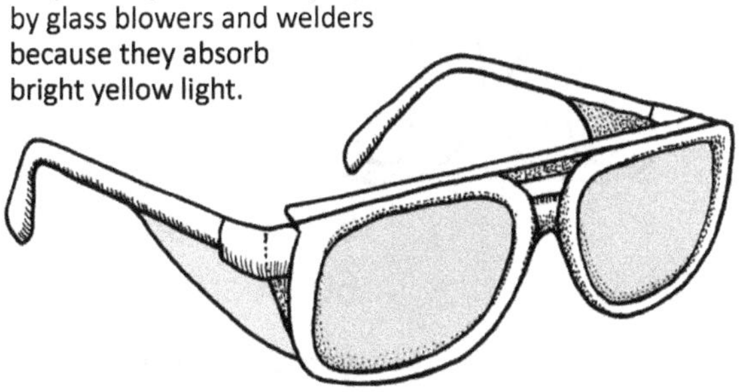

Praseodymium

Praseodymium and magnesium are used to make an alloy that is very strong and heat resistant and can be used in aircraft engines.

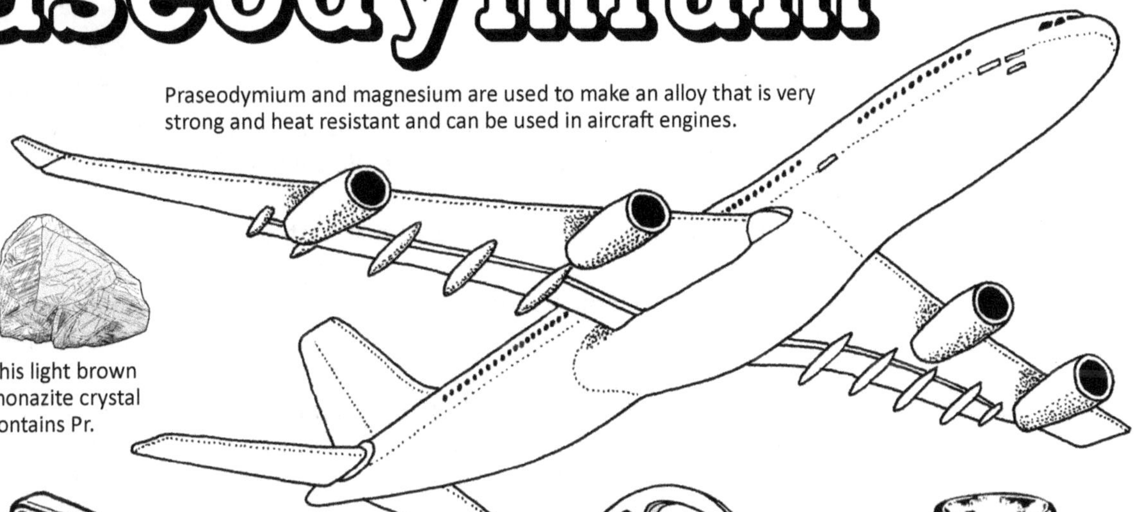

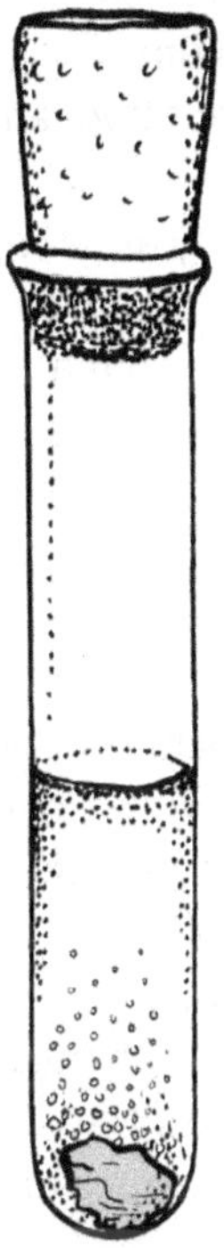

This light brown monazite crystal contains Pr.

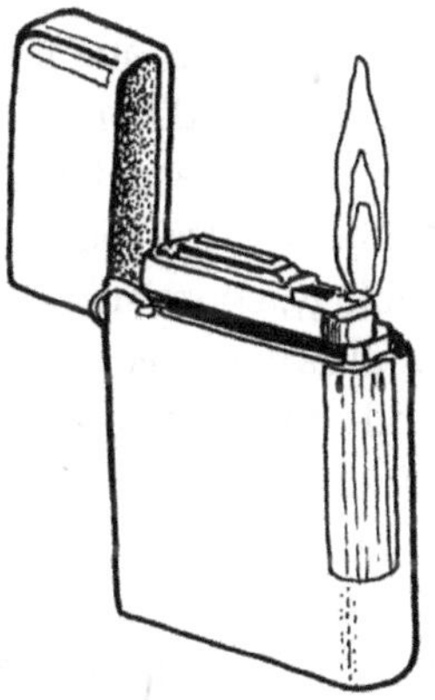

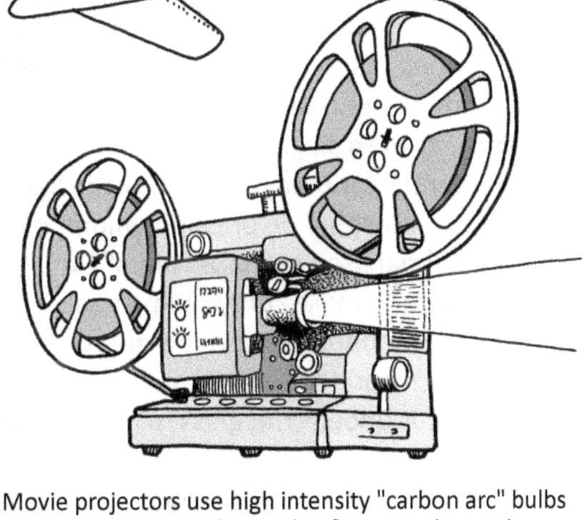

If you put a piece of praseodymium in hydrochloric acid, the solution turns light green.

Praseodymium is often one of the rare earth elements used in the "flints" in lighters.

Movie projectors use high intensity "carbon arc" bulbs that contain electrodes made of rare earth metals.

Praseodymium is used to make yellow glass.

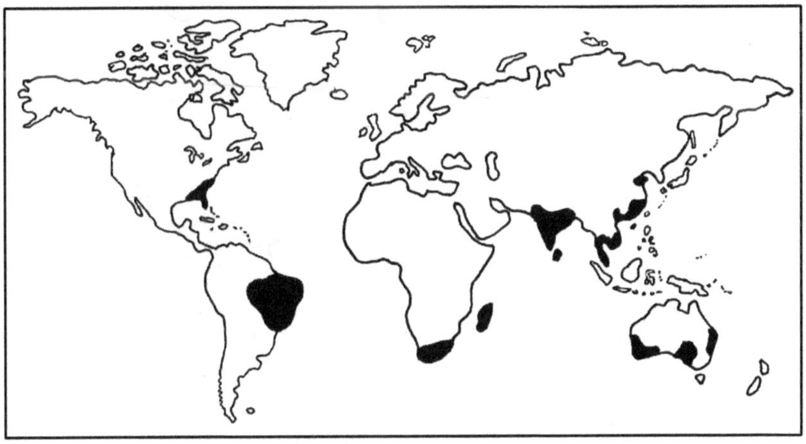

Monazite sand is only found in a few places around the world.

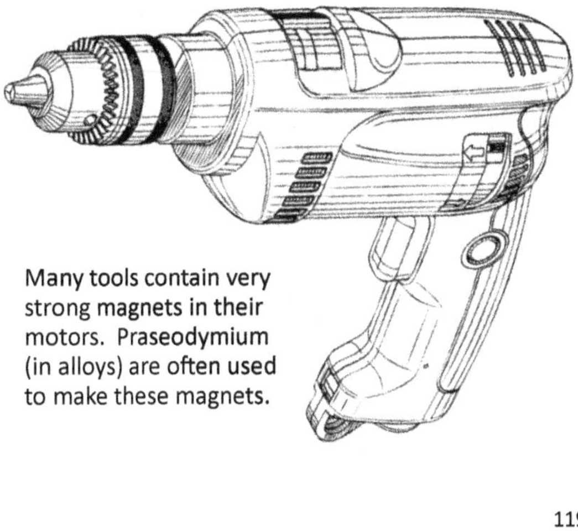

Many tools contain very strong magnets in their motors. Praseodymium (in alloys) are often used to make these magnets.

Nd 60

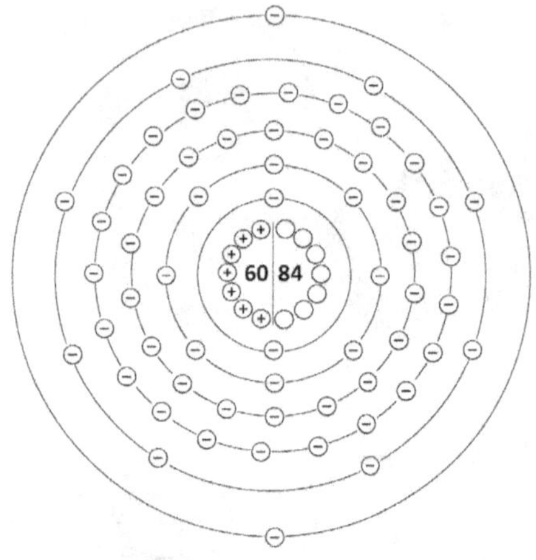

60 protons
84 neutrons
60 electrons

Atomic mass: 144.2

Neodymium

From Greek words "neo" (new) and "didymos" (twin)

Neodymium was discovered at the same time as praseodymium. The year was 1885. Carl von Welsbach, one of the world's first experts on rare earth minerals, thought he had discovered a new element: didymium. However, upon further investigation he found that his mineral sample contained not one new element, but two, and he named them praseodymium and neodymium.

Neodymium can be found in two mineral ores: monazite and bastnasite. Monazite is found mainly as sand, though larger crystals show up once in a while. Bastnasite is a type of rock named after the town in Sweden where it was first identified. Extracting rare earth elements from mineral ores is a long and complicated process, no matter which ore you use. These ores always contain many of the rare earth elements so extraction of neodymium always produces many other elements as well.

Like all of the rare earth elements (numbers 57- 71), neodymium limits the number of electrons in its outer shell. This arrangement enhances its magnetic properties and makes it ideal for alloys (such as $Nd_2Fe_{14}B$) that are used to make small, but extremely strong, magnets. Tiny-but-strong magnets are necessary for making electronic devices such as computers, phones, microphones, headphones, pickups on electric guitars, and many machines used in technical industries. Neodymium magnets are strong enough to lift an object 1,000 times their own weight.

Neodymium-yttrium-aluminum-garnet lasers (Nd-YAGs) are used to create laser beams for "optical tweezers." Microscopic things like bacteria, DNA, or even single atoms, can be trapped and held at the focal point of the laser light beams. Optical tweezers are essential for many branches of research in biology, chemistry and physics.

Neodymium can be used with, or instead of, praseodymium in many applications. For example, neodymium is an ingredient in mischmetal, the material used to make "flints" for lighters and other fire-starting devices. Neodymium is combined with praseodymium to make "didymium glass" which is used to make filters for telescopes and protective glasses for glass blowers and welders because it is able to absorb yellow and orange light.

When neodymium oxide (Nd_2O_3) is added to glass, it makes a most unusual effect: the glass will change color depending on what type of light it is under. The glass might look reddish-purple in daylight, but blue under fluorescent lights.

Nd_2O_3 Neodymium oxide

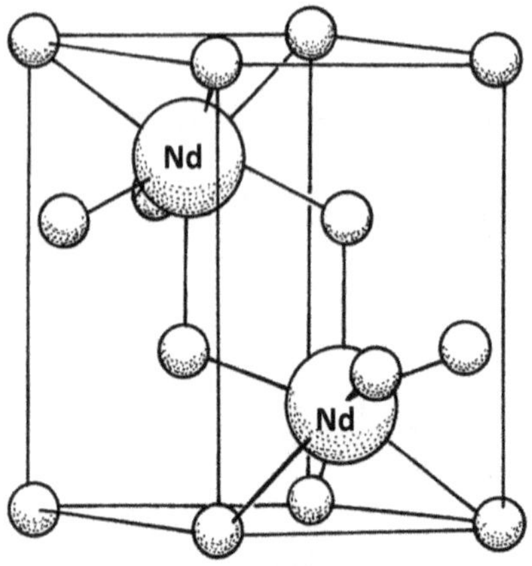

The darker balls are Nd. The lighter balls are O.

$NdCl_3$ Neodymium chloride

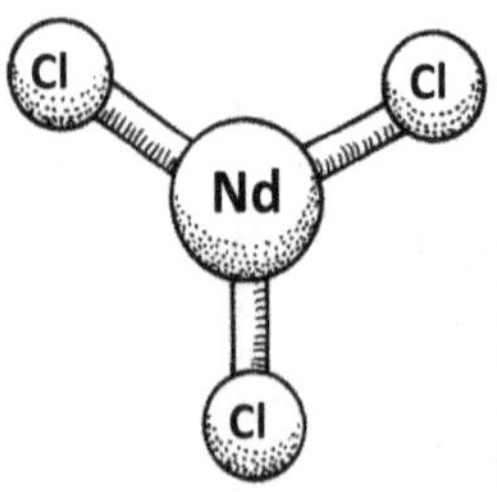

Monazite crystal structure

Monazite sand is one of the ores from which rare earth elements are extracted.

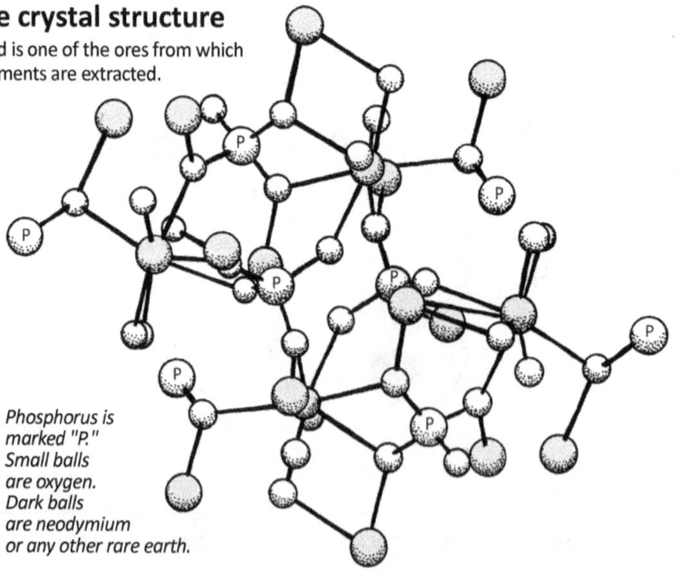

Phosphorus is marked "P." Small balls are oxygen. Dark balls are neodymium or any other rare earth.

60 Nd

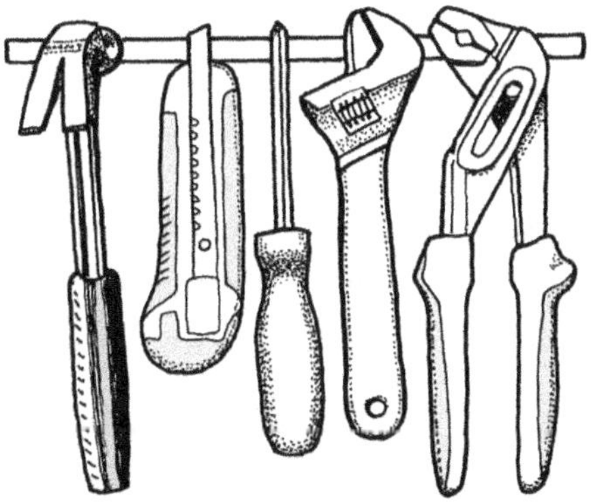

Neodymium magnets can hold 1,000 times their weight! Here, a long bar magnet holds heavy tools.

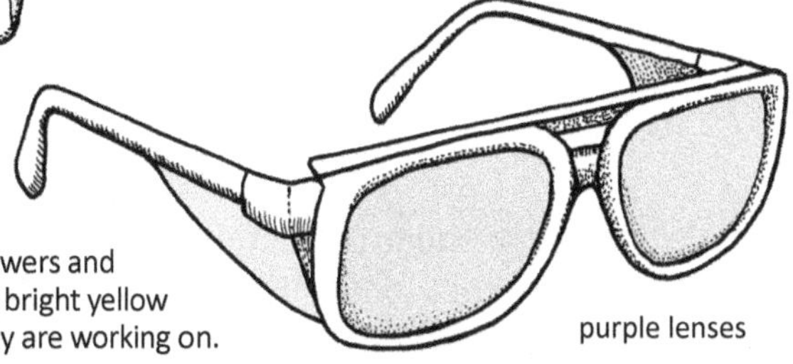

Didymium glasses are used by glass blowers and welders because they absorb annoying bright yellow light that makes it hard to see what they are working on.

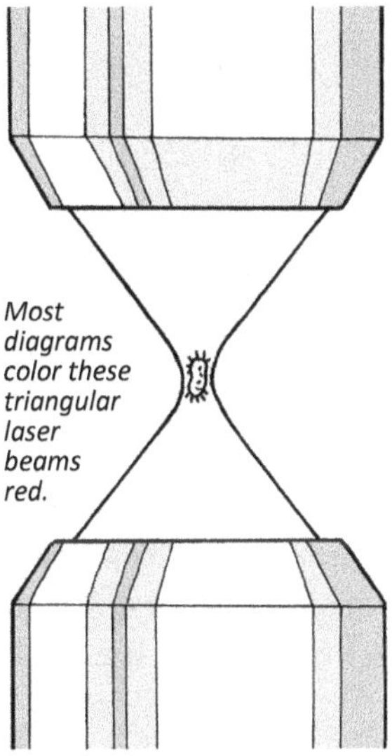

Most diagrams color these triangular laser beams red.

Nd-YAG lasers are used in "optical tweezers" that can trap and hold microscopic things from bacteria and DNA down to single atoms.

Neodymium

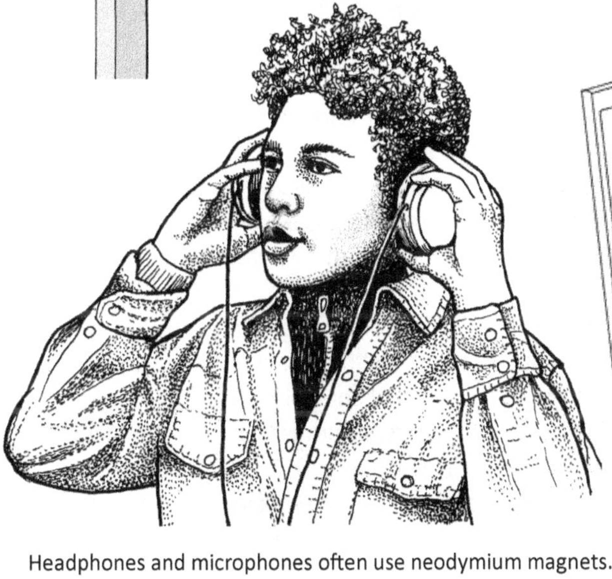

Headphones and microphones often use neodymium magnets.

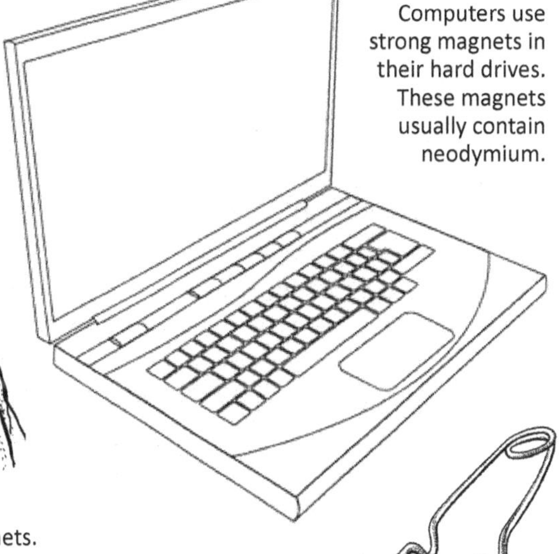

Computers use strong magnets in their hard drives. These magnets usually contain neodymium.

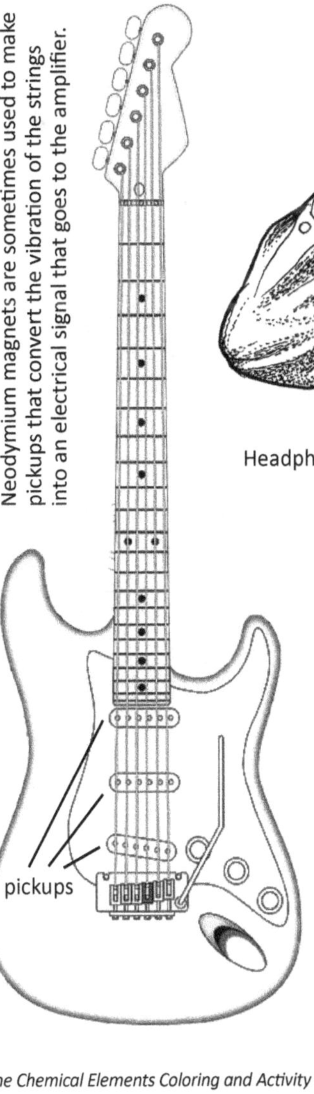

Neodymium magnets are sometimes used to make pickups that convert the vibration of the strings into an electrical signal that goes to the amplifier.

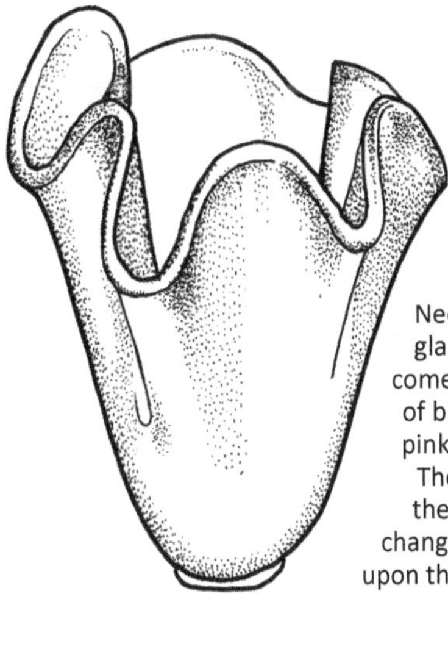

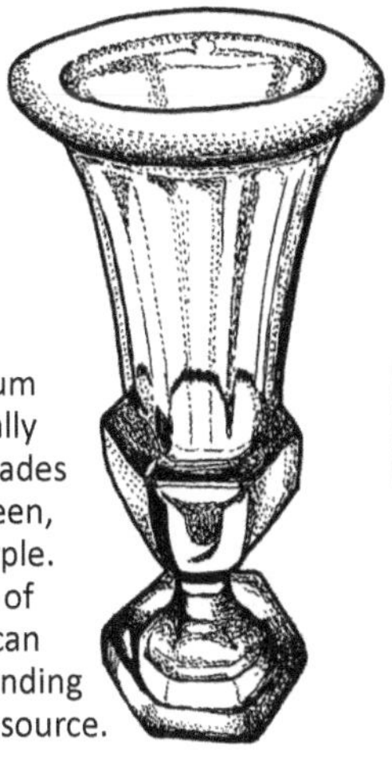

Neodymium glass usually comes in shades of blue, green, pink or purple. The color of the glass can change depending upon the light source.

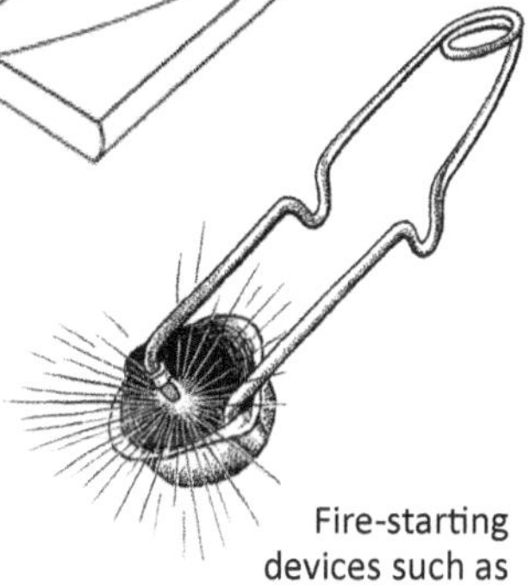

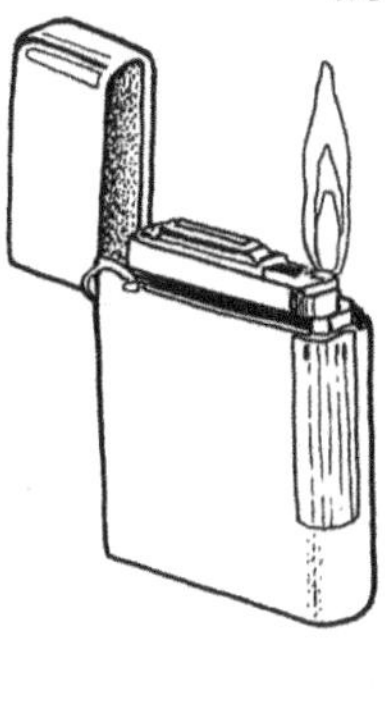

Fire-starting devices such as welding sparkers and lighters have "flints" made of rare earth alloys that often include Nd.

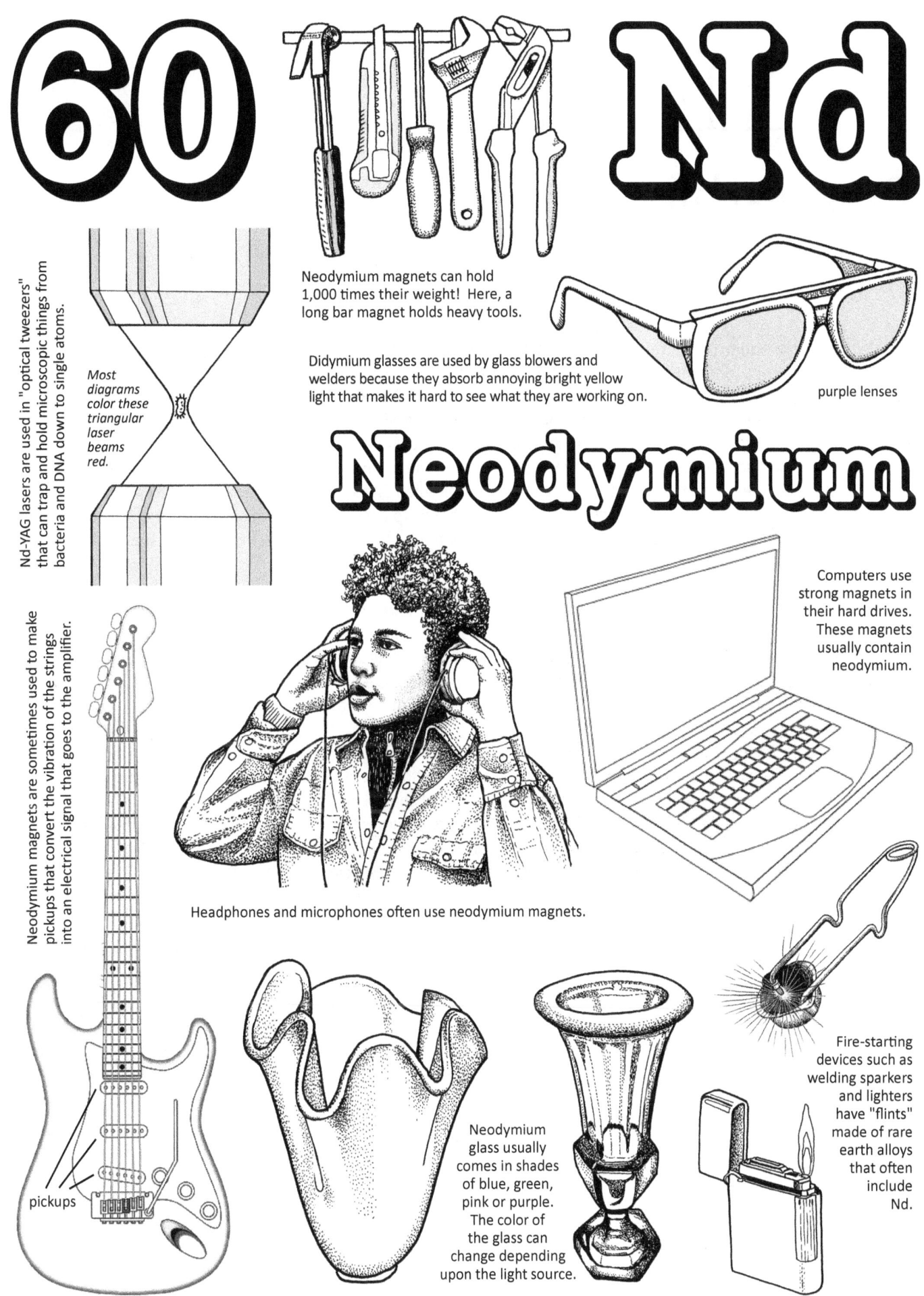

Pm 61

61 protons
84 neutrons
61 electrons

Atomic mass: 145

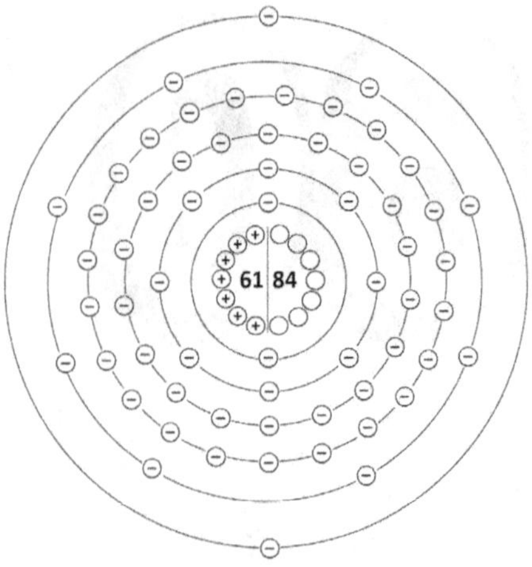
61 84

Promethium

Named after Prometheus, the Greek god who gave fire

Like technetium, promethium is a man-made element, produced in labs that work with radioactive elements. The existence of element 61 was recognized as early as 1902, but no one was able to produce an actual sample until 1944. Three scientists working at what is now called Oak Ridge National Lab (in Tennessee) were examining the smaller radioactive atoms produced by the splitting of uranium atoms. This was right at the end of World War 2, so the reason they were researching uranium was to help create atomic weapons. They were not able to announce their discovery until 1947. They were going to name the new element "clintonium" after the name of the lab at that time (Clinton Lab) but the wife of one of the scientists suggested naming it after the Greek god Prometheus who was said to have brought the technology of fire to early humans. This was right after two atomic bombs had been dropped on Japan to end the war, so the naming of this element was to suggest "the benefits and the possible misuse of mankind's intellect."

In 1963, promethium fluoride was created, and was allowed to react with lithium ions so that the fluoride atoms left the promethium and joined with the lithium to form molecules of lithium fluoride. This left the scientists with enough pure promethium metal that they were able to do experiments to determine its properties. There are several isotopes of promethium (atoms with more or less than 84 neutrons) but most decay quickly and turn into either neodymium or samarium. The most stable isotope, Pm-145, has a half-life of about 17 years, which means that after 17 years half of your original sample will be gone and in another 17 years half of that remaining half will be gone. After 68 years you will only have 1/16 of your original sample left.

The only useful isotope of promethium is Pm-147, which does not emit deadly gamma rays like many radioactive atoms do. The "beta" radiation it emits (essentially electrons) has an extremely short range and is easily absorbed by other elements. It can be used to make glow-in-the-dark paint that glows by using the energy coming out of the promethium atoms. The Pm-147 has a half-life of only 2.6 years, so it has to be used in places that only need a good strong glow for a few years. The Apollo lunar rover, used in 1971 and 1972 on missions 15, 16 and 17, had control panels painted with luminous promethium paint (promethium chloride mixed with zinc sulfide). During the mid 1900s, Pm-147 was used as a safer replacement for radium paint, which they had used for decades to make the faces of watches glow in the dark, not realizing how dangerous the radium was.

$PmCl_3$ Promethium (III) chloride

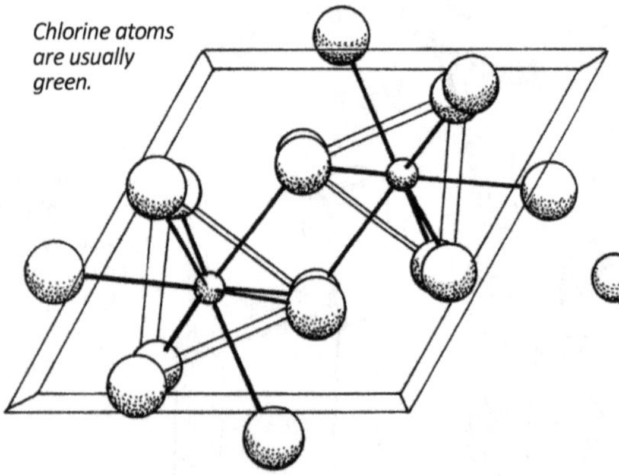
Chlorine atoms
are usually
green.

The larger balls are Cl. The smaller balls are Pm.

Pm_2O_3 Promethium (III) oxide

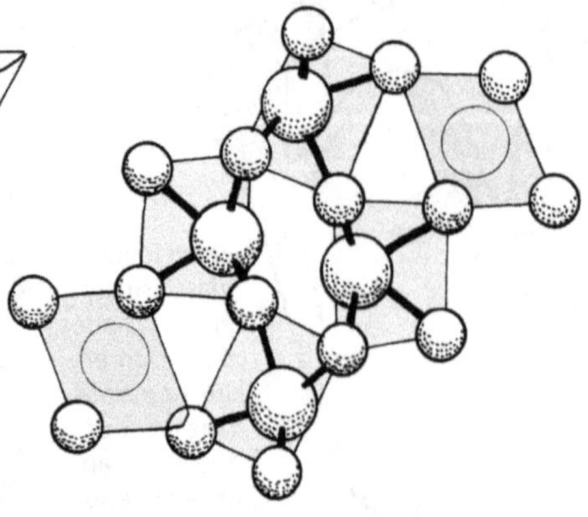

The larger balls are Pm. The smaller balls are O.

PmF_3 Promethium fluoride

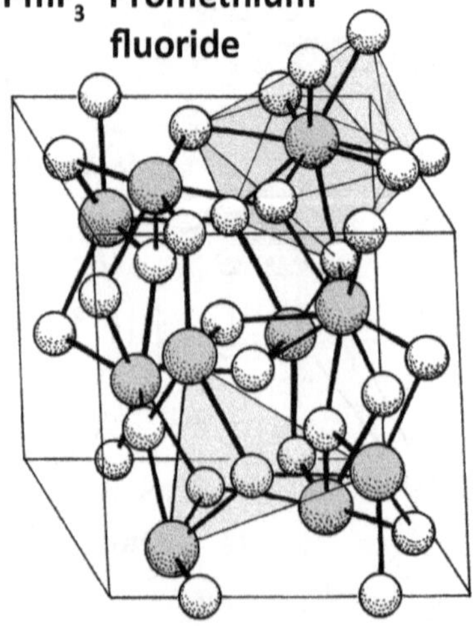

The dark balls are Pm. The light balls are F.

61 Pm

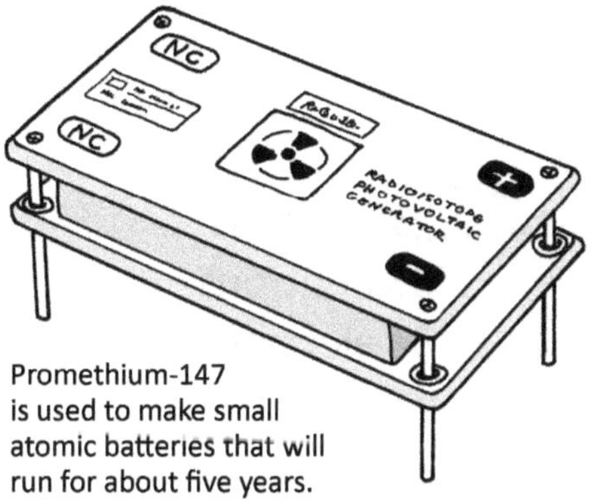

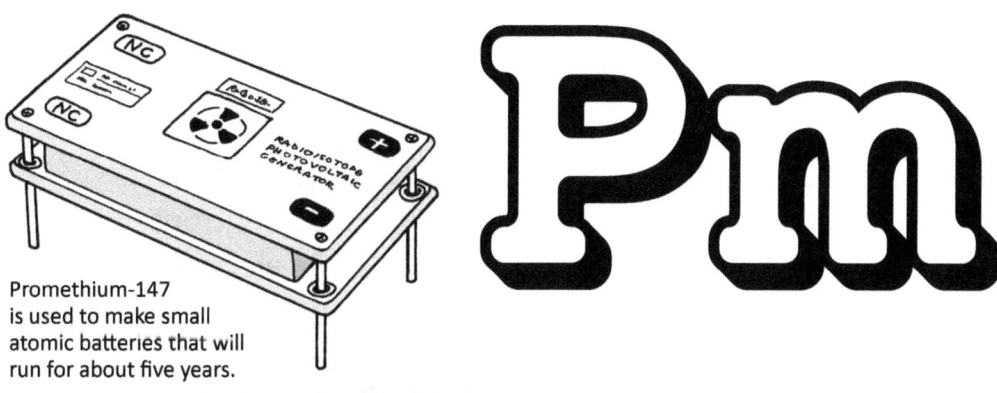

Promethium-147 is used to make small atomic batteries that will run for about five years.

Promethium was discovered in 1944 as a waste product produced during experiments with uranium atoms.

Promethium

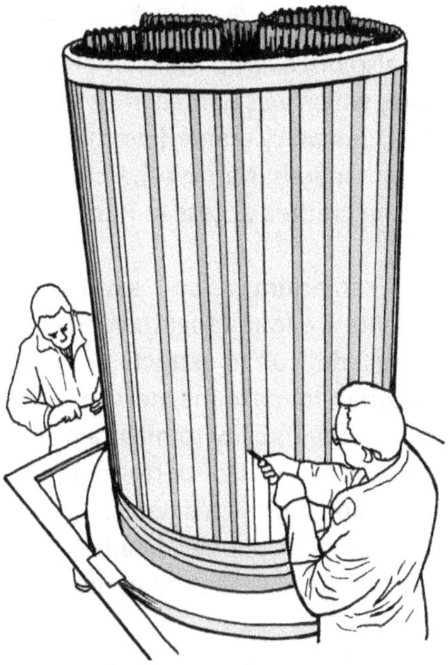

The nuclear reactor at the lab that discovered promethium (now Oak Ridge National Lab) used a core of graphite to slow down the free neutrons that were hitting uranium atoms. (This is a more modern graphite core, smaller than the one used at Oak Ridge.)

In a way, many people helped to discover promethium, though they did not know it at the time. Many non-scientists were employed at the top-secret facility in Oak Ridge, Tennessee, where scientists were enriching uranium in order to develop the atomic bomb. This woman was taught to push buttons in a certain order. She did not know that the lab was making a nuclear weapon.

This drawing was inspired by a photograph taken in 1944 at Oak Ridge.

Promethium's radioactivity provided the energy for the glow-in-the-dark paint that was used to mark buttons on the lunar rover used by Apollo missions 15, 16 and 17.

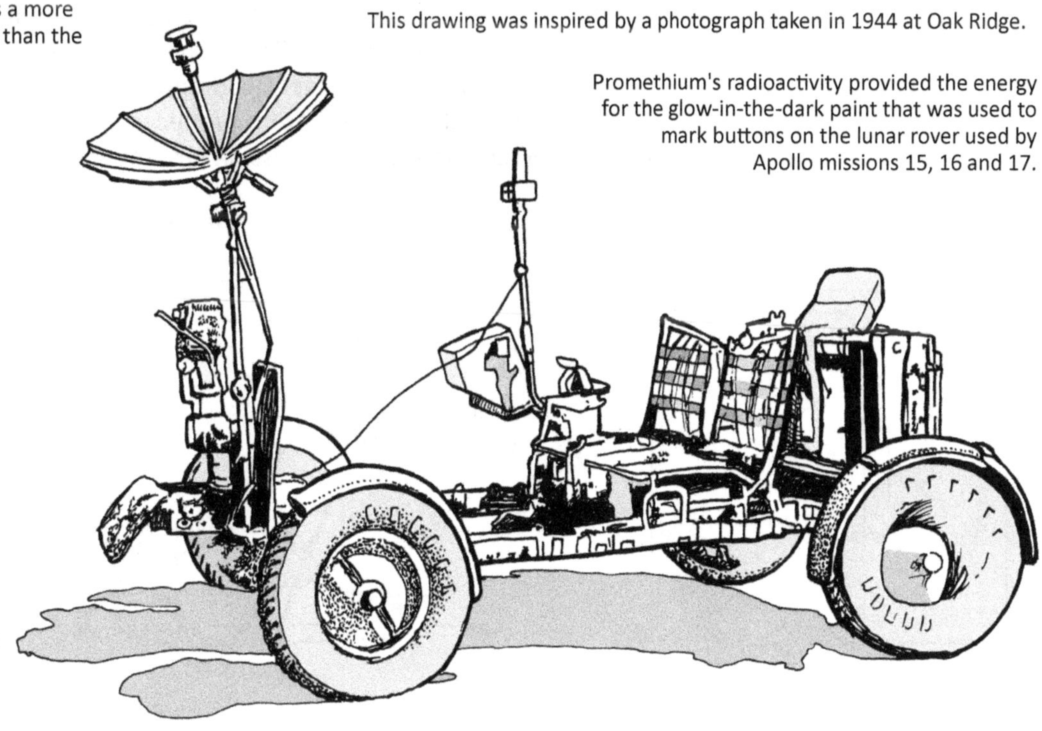

Promethium replaced radium in luminescent paint after they discovered how dangerously radioactive radium is.

Sm 62

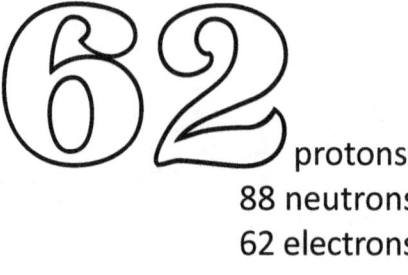

protons
88 neutrons
62 electrons
Atomic mass: 150.3

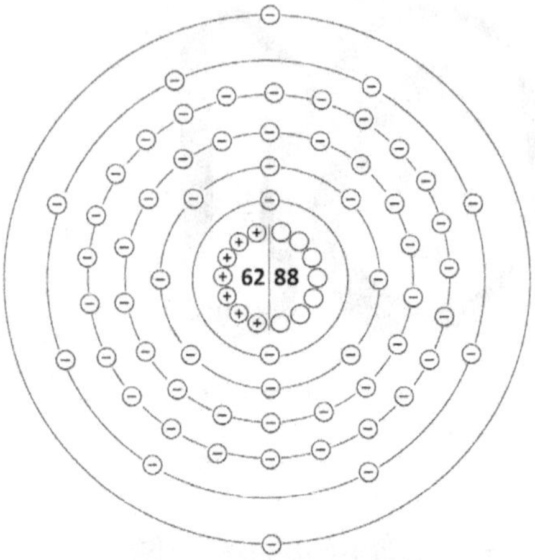

Samarium

Named for Russian mining engineer Vasili Samarsky-Bykhovets

Samarium was discovered in 1879 by Paul-Émile Lecoq de Boisbaudran, who also discovered the elements gallium and dysprosium. Samarium is actually named after its mineral ore, samarskite, which was named after Vasili Samarsky-Bykhovets, the chairman of the Russian Corps of Mining Engineers. In 1839, Samarsky granted permission to two German geologists to come into Russia, into the Ural Mountains, to look for new minerals. Naming the new mineral they found after Samarsky was the geologists' way of saying thank you for being granted access to Russian mining areas. Samarium was the first element to be named after a person.

Like all the elements in the lanthanide series (the "rare earth" elements), pure samarium is a soft, silvery metal that tarnishes (loses its shininess) when exposed to air. All the rare earth elements are found together in ore minerals such as monazite sand and a rock called bastnasite. They are not rare at all, just difficult to extract.

Like neodymium, samarium is highly magnetic. Samarium is combined with cobalt to make magnets that are almost as strong as neodymium magnets (NdFeB), but are superior to neodymium in their ability to maintain their magnetism even at high temperatures. Samarium-cobalt magnets ($SmCo_5$) can be found in high-performance motors and engines, such as railroad locomotive traction motors, industrial generators, and motors for boats.

SmCo magnets are also used in places where temperature is not an issue, such as microphones, headphones, loudspeakers, cell phones, and "noiseless" pickups for electric guitars. SmCo magnets were used in the motor of the "Solar Challenger," one of the world's first solar-powered electric airplanes.

Compounds of samarium are important in chemical industries, often as catalysts. A catalyst is a "helper" that encourages chemical reactions to occur faster than they would otherwise. A catalyst is not itself changed by the reaction, so it can be used over and over again. Some samarium compounds assist in the breakdown of plastics, including PCBs, which are considered to be environmental pollutants.

Samarium is one of the metals that is used to "dope" crystals that are used in specialty lasers. Calcium fluoride crystals doped with samarium were used to make one of the world's first solid-state lasers at IBM in 1961.

Radioactive samarium-153 is used in nuclear medicine (as "Quadramet") to kill cancer cells.

Samarium is very good at capturing and holding free neutrons produced by nuclear fission, so it is used in control rods in nuclear reactions. The control rods slow down the fission process when fission is going too fast.

Sm_2O_3 Samarium oxide

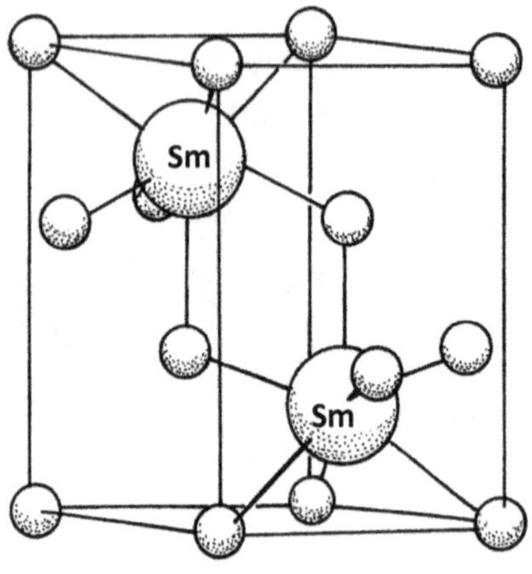

The unmarked balls are oxygen, O.

SmI_2 Samarium iodide

Iodine atoms are often colored purple or pink, although real crystals of SmI_2 are green.

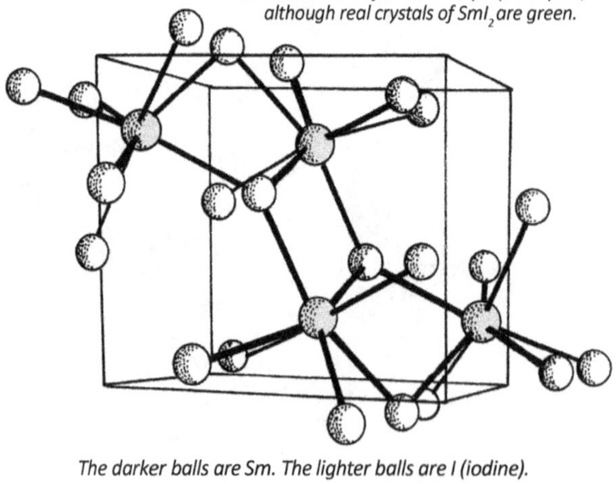

The darker balls are Sm. The lighter balls are I (iodine).

SmB_6 Samarium hexaboride

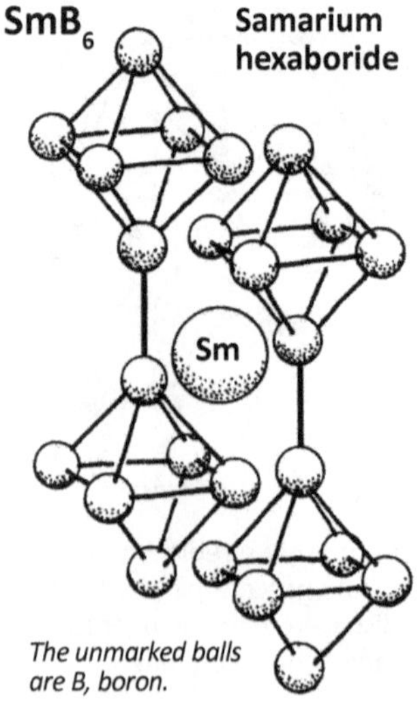

The unmarked balls are B, boron.

62 Sm

Samarskite crystals are black or dark gray.

Solar airplanes need very strong magnets. Samarium-cobalt is often a good choice because you can keep the magnets small and not add extra weight.

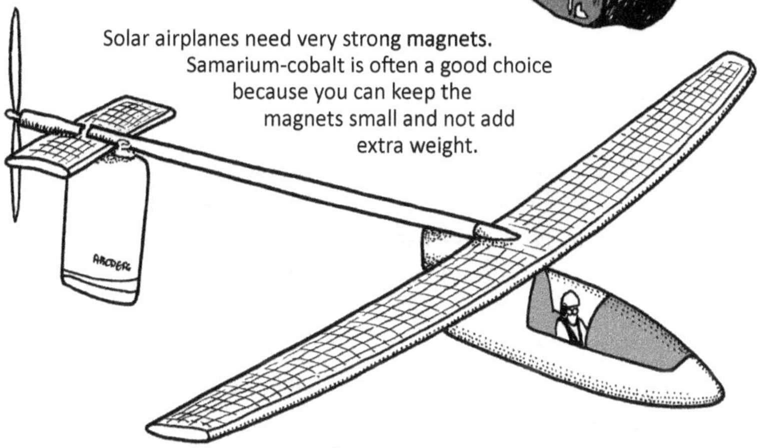

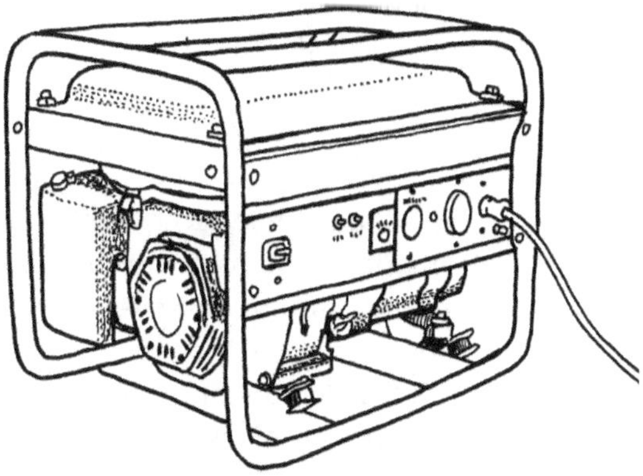

Heavy duty generators sometimes use samarium-cobalt magnets because they work at high temps.

Samarium

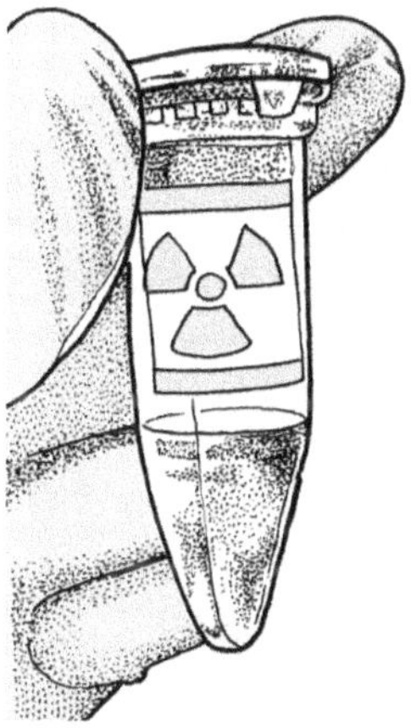

Samarium-153 is used in nuclear medicine to kill cancer cells.

You might find tiny SmCo magnets in a cell phone.

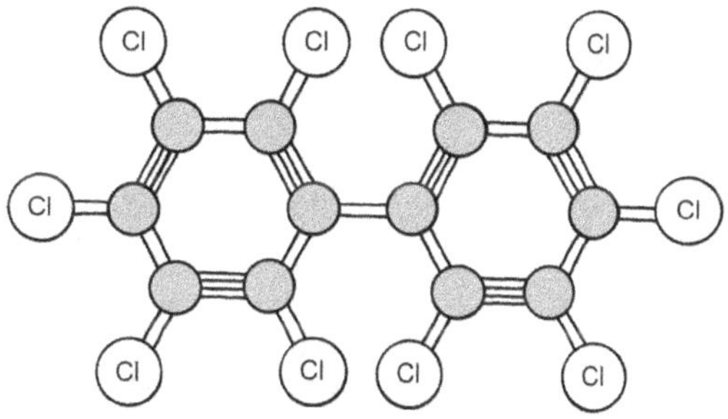

PCB plastics are a major environmental concern. Samarium compounds help to get rid of them safely. *(The darker circles represent carbon atoms.)*

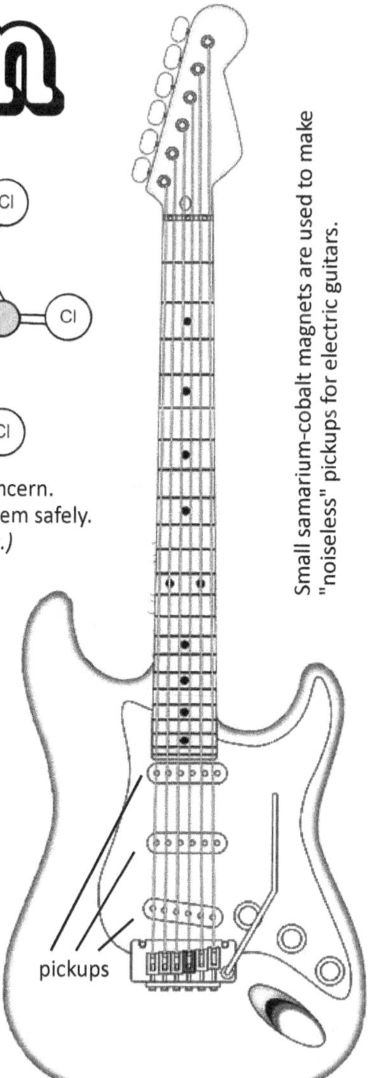

Small samarium-cobalt magnets are used to make "noiseless" pickups for electric guitars.

Samarium can slow down free neutrons.

Samarium is used to make control rods for nuclear reactors.

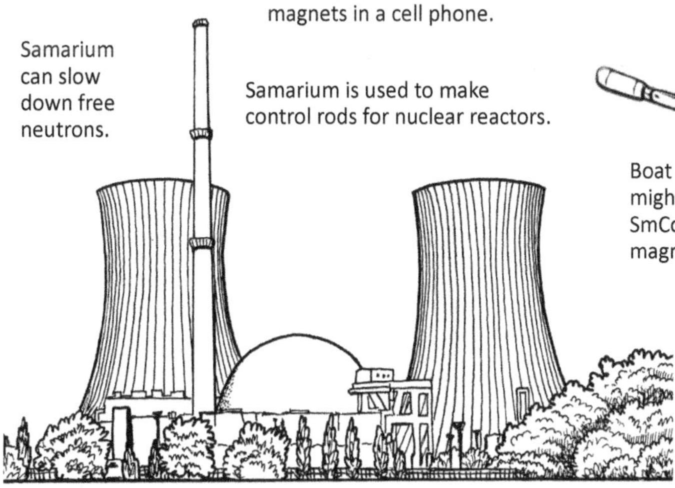

Boat motors might use SmCo magnets.

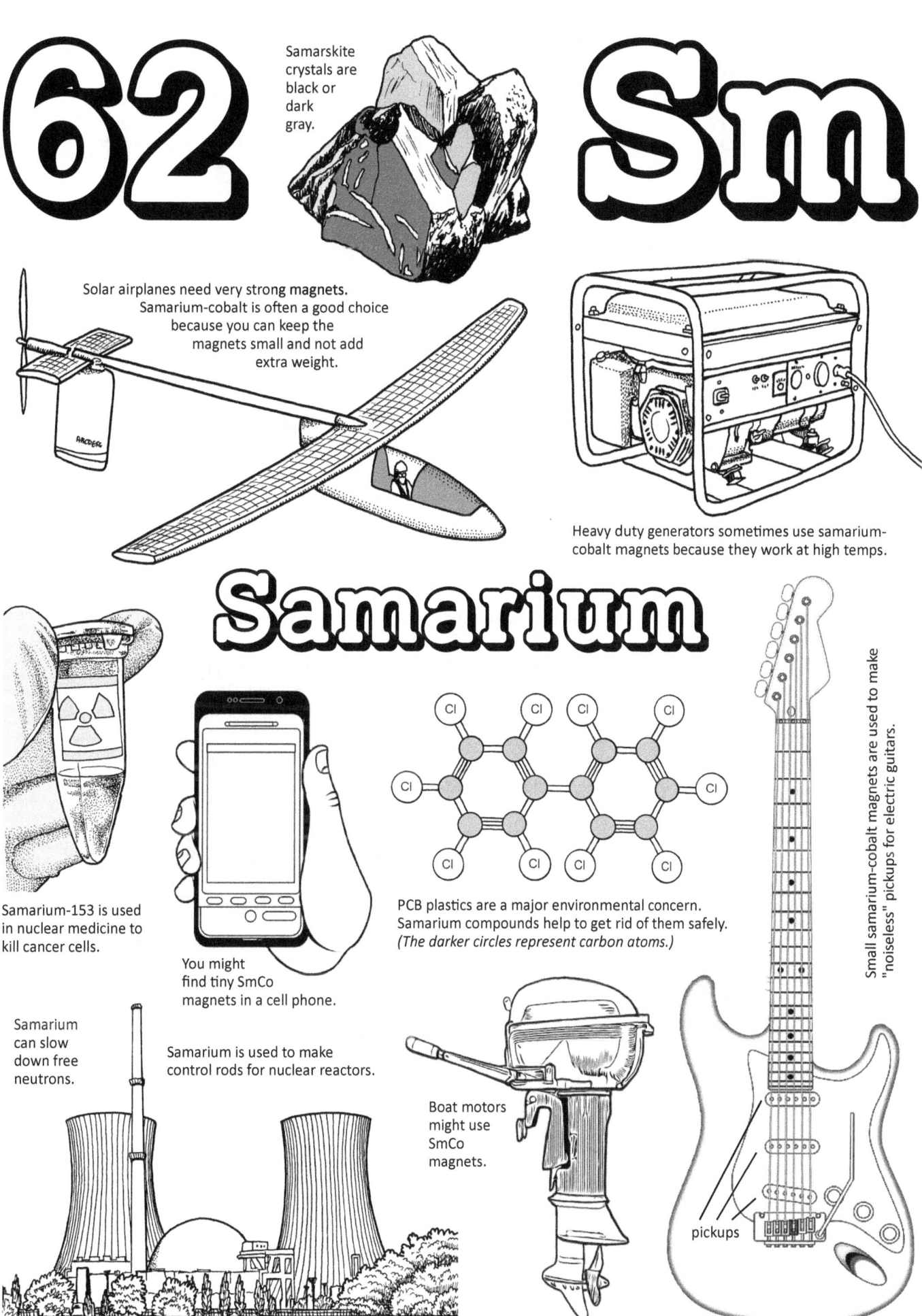

Eu 63

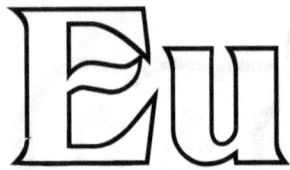

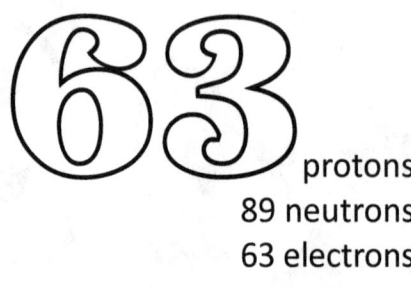

protons
89 neutrons
63 electrons
Atomic mass: 151.9

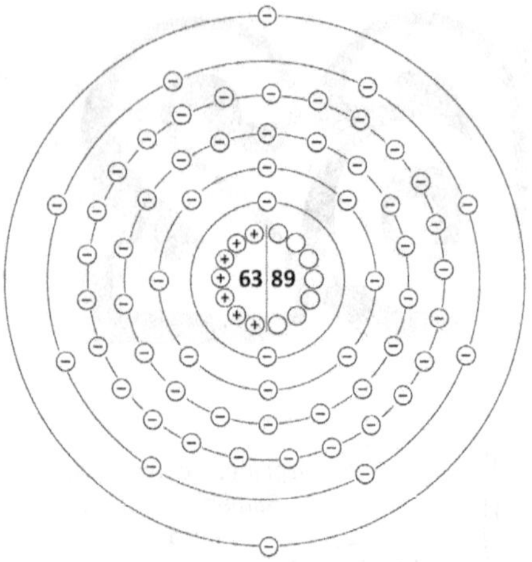

Europium

Named after Europe

In 1892, Paul-Émile Lecoq de Boisbaudran was using a spectrometer to analyze samples of minerals that contained samarium, and he saw some spectral lines that suggested there might perhaps be something else in his sample besides samarium. However, he was unable to isolate it to prove its existence. Another French chemist, Eugène-Anatole Demarçay, was able to find and extract this hidden element in 1901, and was thus declared to be its discoverer. He decided to name this new element after Europe.

Pure europium is the most reactive of all the rare earth elements. Its reactivity is often compared to lithium. Samples of europium metal must be kept sealed in tubes that protect them from coming into contact with oxygen or water. If you drop a piece of europium into a test tube of water, it will turn the water bright yellow and produce bubbles of hydrogen gas. ($2\ Eu + 6\ H_2O \rightarrow 2\ Eu(OH)_3 + 3\ H_2$)

Europium's claim to fame is being able to fluoresce. This means that when molecules containing europium are energized with certain forms of electromagnetic radiation (often including ultraviolet light), they release light. In the 1960s, when color televisions were being developed, it was discovered that when crystals of yttrium orthovanadate (YVO_4) were "doped" with europium (had europium atoms added to the crystals), they would give off a very bright red light. This allowed an amazing improvement in the brightness of color TVs. When europium is put into other compounds it can make shades of blue, green, yellow, or orange. Most famously, europium compounds are used in euro bank notes, to detect counterfeiting. If you hold a genuine euro note under a UV light, you will see bright red stars or other patterns that are not visible under ordinary light. The process of printing with europium pigments is just difficult enough that most counterfeiters won't be able to do it.

Other rare earth elements can also glow. Terbium makes bright yellowish-green light. When red and blue europium compounds are combined with greenish-yellow terbium compounds, the result is a very clean white light. The insides of fluorescent bulbs are coated with these "phosphors" made of powdered Eu and Tb compounds.

Europium is used as a dopant (additive) for glass used in certain types of lasers.

In 2015, it was discovered that europium may be useful in quantum memory chips.

Eu_2O_3 Europium oxide

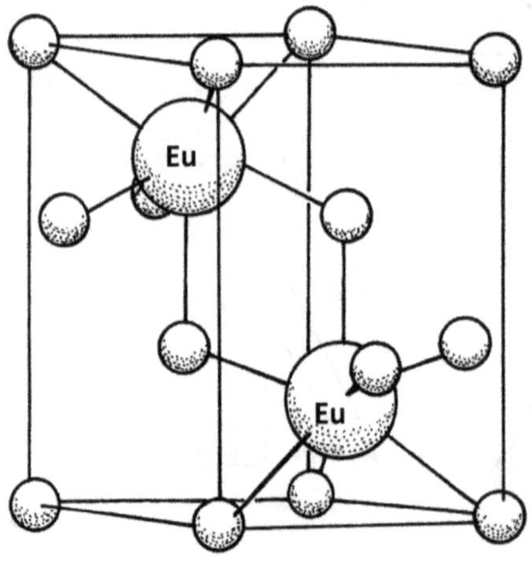

The darker balls are Eu. The lighter balls are O.

Bastnäsite The main mineral ore of europium.

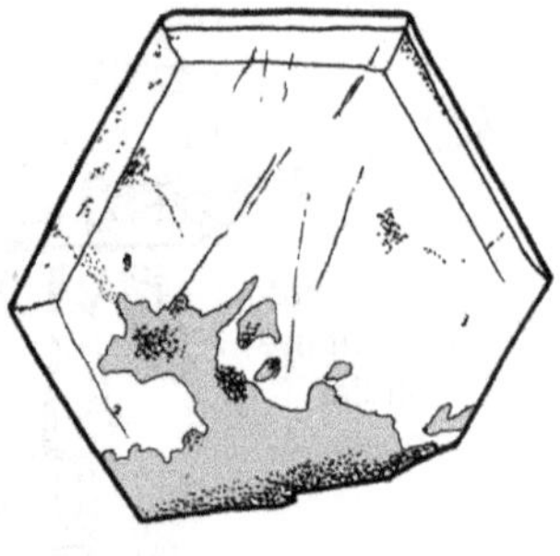

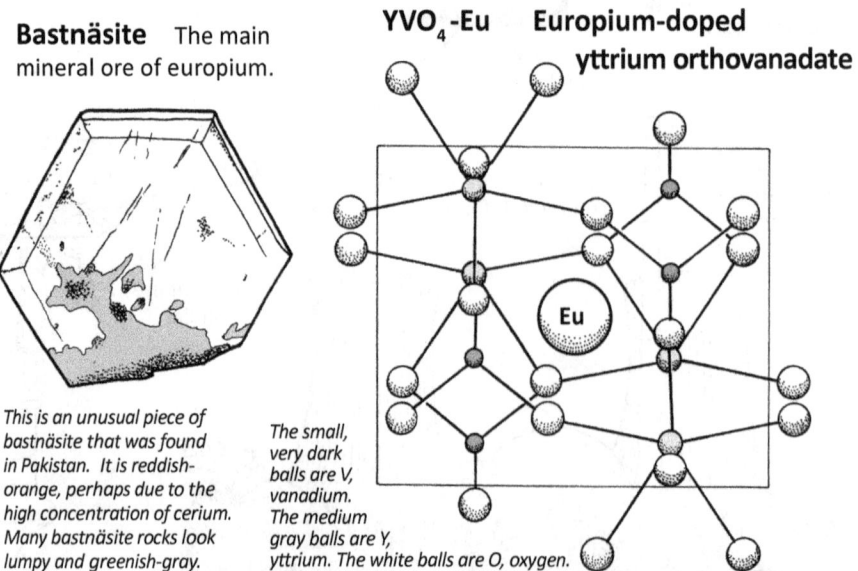

This is an unusual piece of bastnäsite that was found in Pakistan. It is reddish-orange, perhaps due to the high concentration of cerium. Many bastnäsite rocks look lumpy and greenish-gray.

YVO_4-Eu Europium-doped yttrium orthovanadate

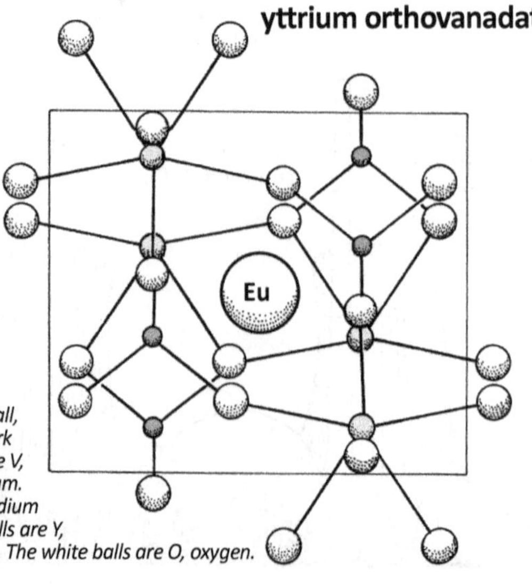

The small, very dark balls are V, vanadium. The medium gray balls are Y, yttrium. The white balls are O, oxygen.

63 Eu

Europium was named after Europe, where it was discovered.

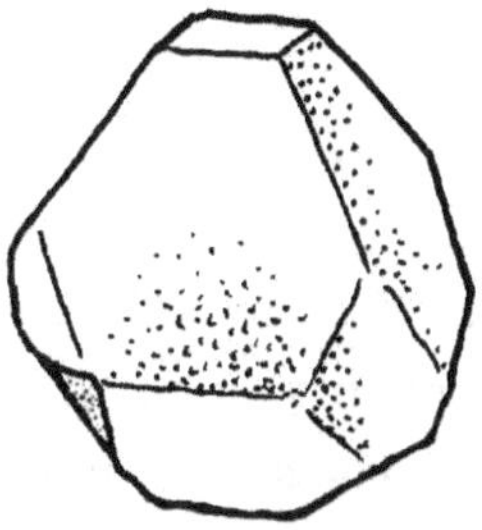

Crystals of fluorite, CaF_2, that contain small amounts of europium will look blue under ultraviolet light. (This is an example of fluorescence.)

The white light produced by helical fluorescent bulbs is made by powdered phosphors of both europium and terbium.

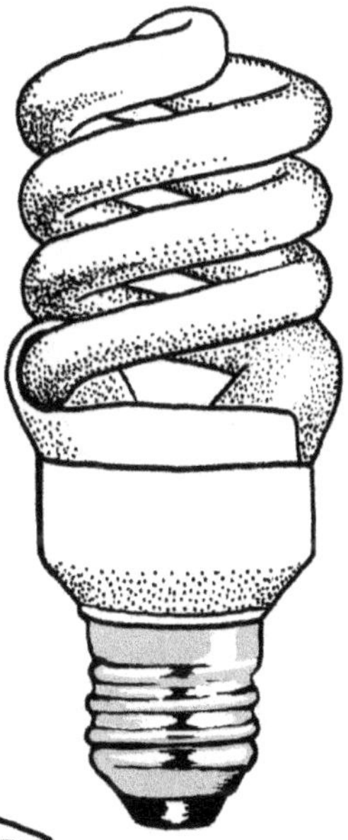

Europium

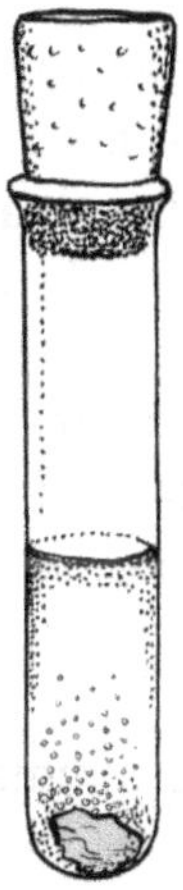

If pure europium is put into water, the solution will turn yellow.

Euro bank notes are printed using europium to make designs that can only be seen under UV light.

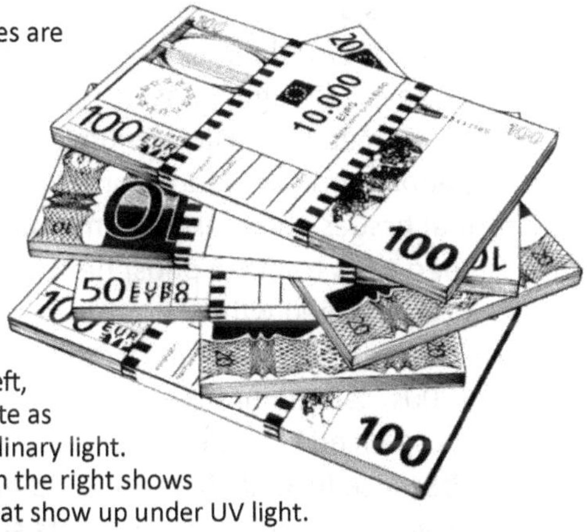

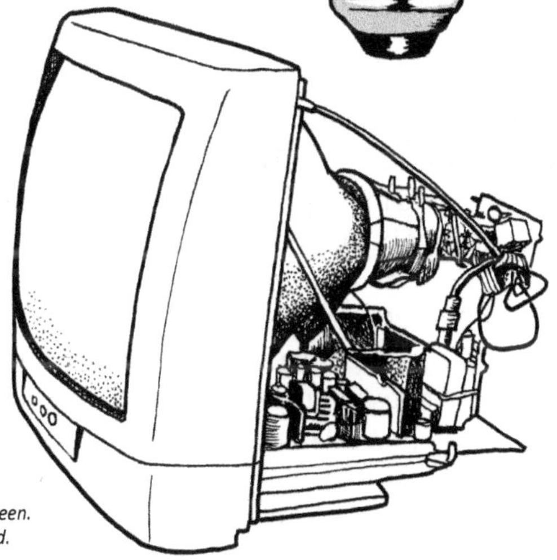

Below, on the left, is a 50 euro note as seen under ordinary light. The diagram on the right shows the patterns that show up under UV light.

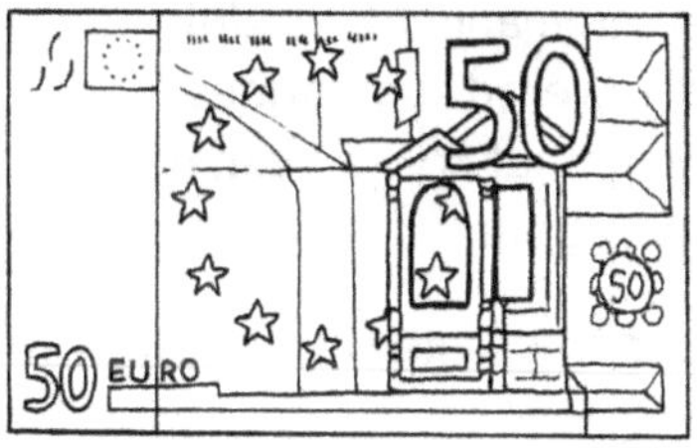

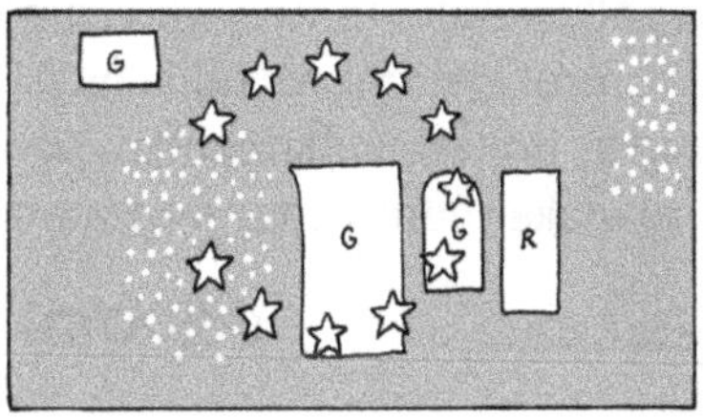

G stands for green. R stands for red. The 6 lower stars are green, the 5 upper stars are red. The dots are red.

CRT (cathode ray tube) television sets made from the 1960s until about the year 2000 used yttrium-orthovanadanite doped with europium to create their bright red colors.

Small amounts of europium are added to powders of zinc sulfide or strontium aluminate to create a wide range of pigments that glow in the dark.

(Use bright (or neon) pastel colors to color the pigment powders in these jars.)

Gd 64

protons
93 neutrons
64 electrons
Atomic mass: 157.25

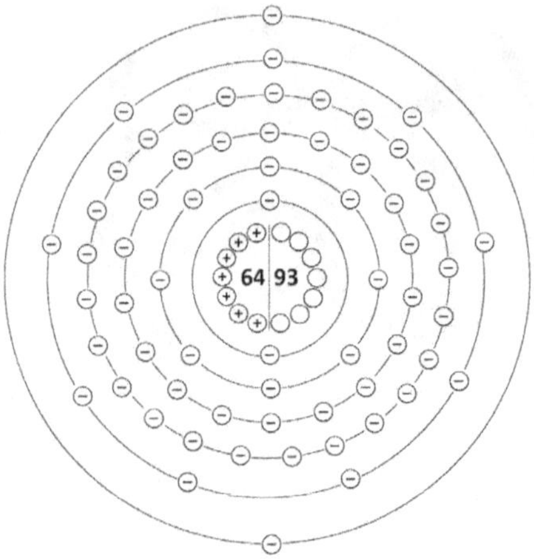

Gadolinium

Named after Finnish chemist Johan Gadolin

Gadolinium was named after the mineral ore in which it was found, gadolinite, which contains a number of rare earth elements, plus iron, beryllium, silicon and oxygen: $(Ce,La,Nd,Y)_2FeBe_2Si_2O_{10}$. As you can see, gadolinium isn't in the official chemical formula. This is because it is found in such small quantities.

The scientist for whom gadolinite was named, Johan Gadolin, didn't name or discover either the mineral or the element. He did find what he called "yttria" in 1792, but gadolinite was given that name by someone else in 1800. In 1880, the Swiss chemist Jean Charles Galissard de Marignac observed strange new spectroscopic lines coming from a sample of gadolinite and guessed correctly that it must be a new element. In 1886 he was able to produce pure gadolinium metal from gadolinite mineral ore, and he named the element after the ore.

Like most of the rare earth elements in the lanthanide series, gadolinium is magnetic. Its magnetism, however, is unique among the rare earths because it changes at 20° C (68° F). Below this temperature, gadolinium acts like a regular magnet (ferromagnet). Above this temperature the magnetism decreases significantly, becoming "paramagnetic," allowing it to respond to magnetic fields even though it has lost its own strong magnetism. Many substances that are not "magnetic" (even glass) can be paramagnetic. Gadolinium's magnetic properties make it particularly useful in MRI scans. Solutions containing gadolinium compounds are injected into patients before they go into the scanner. The gadolinium collects in abnormal tissues and makes them visible on the (magnetic) scans.

Like many of the rare earths, gadolinium has been used to make phosphor powders that produce light in fluorescent bulbs. Gadolinium makes green light, so it is combined with phosphors that make red and blue light. The combination of red, green, and blue light produces white light. The ratio of the phosphors can be adjusted to make the light look "warmer" or "colder." In the late 1900s, gadolinium green was used in televisions.

A radioactive isotope of gadolinium, Gd-153, is produced in nuclear reactors. It emits dangerous gamma rays, but does not create long term disposal problems because its half life is only 240 days. Gd-153 can be used in machines that calibrate high-tech equipment, as well as various medical instruments such as bone density scanners and portable x-ray units. These portable units are especially helpful to vets who care for large animals. Gadolinium can also absorb free neutrons. It has been used in the control systems found in nuclear powered submarines.

Gadolinium gallium garnets ($Gd_3Ga_5O_{12}$) are used as imitation diamonds. Yttrium garnet crystals can have gadolinium added to them (a process called "doping") to make them better for laser applications.

Gd-doped cerium oxide crystal, CeO_2

Gadolinium atoms have been "doped" into this crystal of cerium oxide, replacing two atoms of cerium. The result is an awkward "blank spot" right were the X is, making the crystal unstable in this area. Far from being a bad thing, this makes the crystal useful for opto-electronics (ex: lasers).

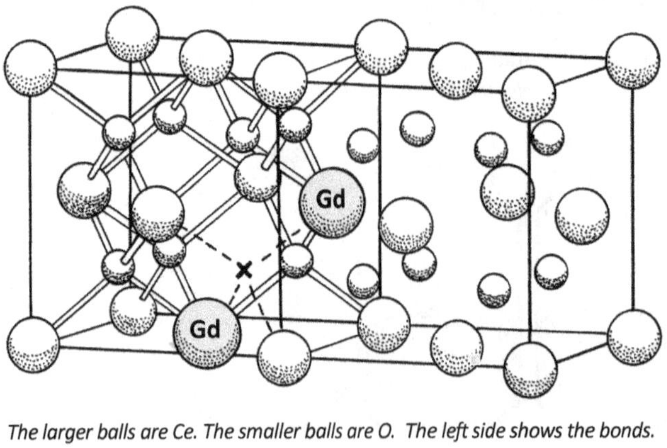

The larger balls are Ce. The smaller balls are O. The left side shows the bonds.

Gd(DOTA) contrast agent used in MRI

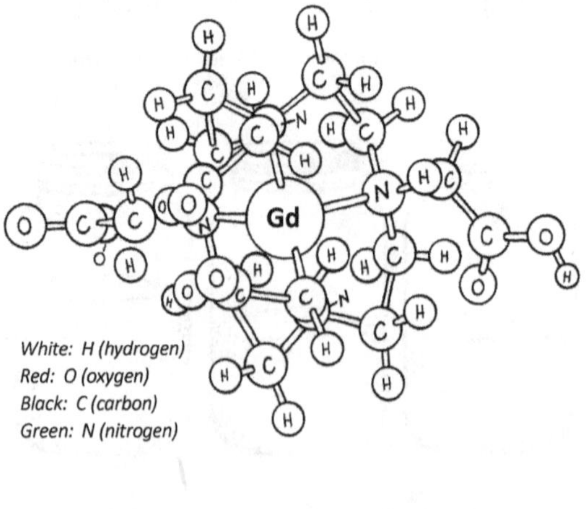

64 Gd

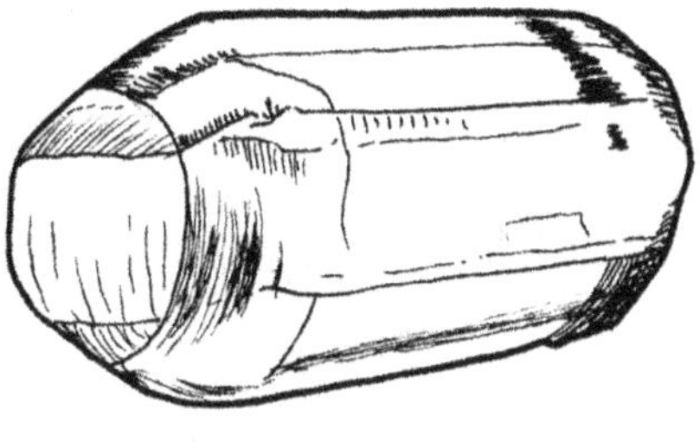

This purple yttrium garnet crystal has been "doped" with Gd to make it more useful in lasers.

Gadolinium

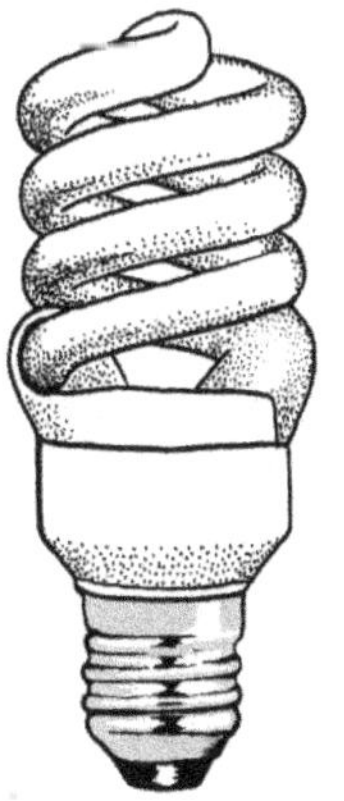

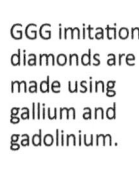

Gadolinium is used in phosphor powders that produce green light. Fluorescent bulbs use a mix of red, green and blue phosphors to make white light.

GGG imitation diamonds are made using gallium and gadolinium.

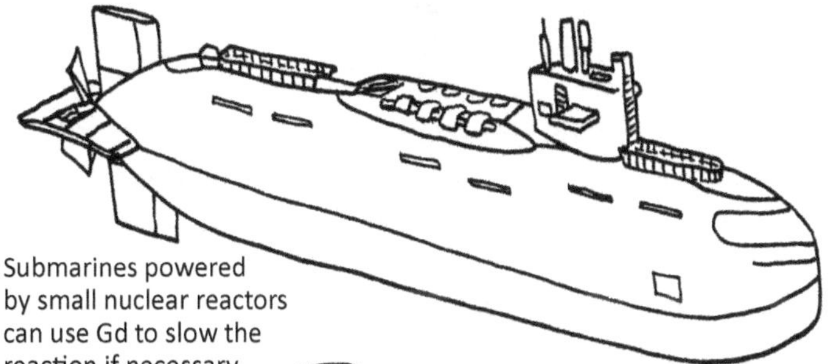

Submarines powered by small nuclear reactors can use Gd to slow the reaction if necessary.

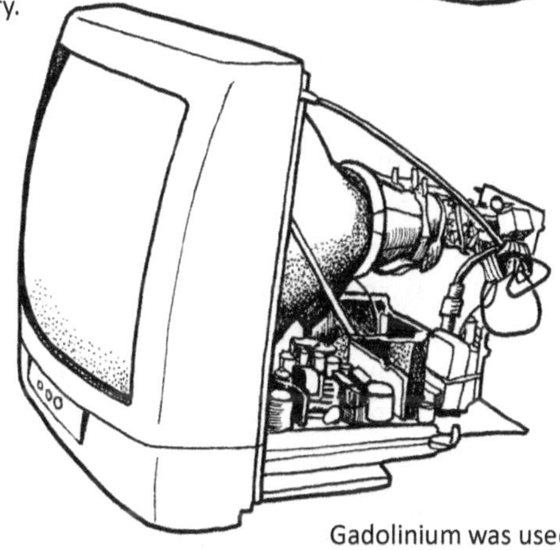

Gadolinium was used in CRT televisions to make green light.

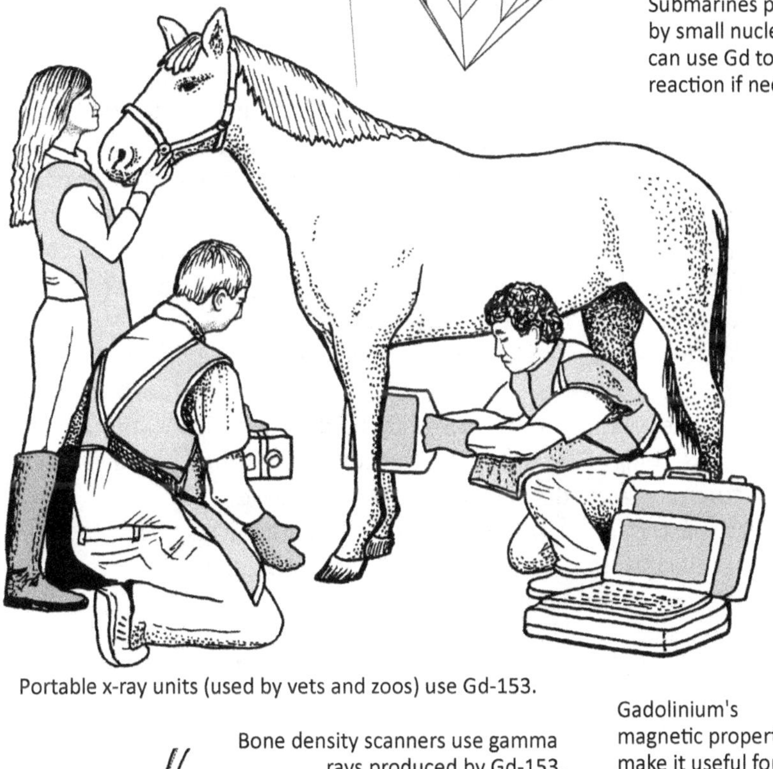

Portable x-ray units (used by vets and zoos) use Gd-153.

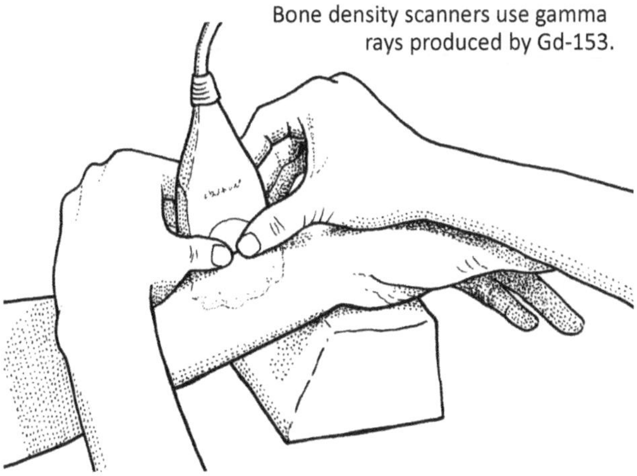

Bone density scanners use gamma rays produced by Gd-153.

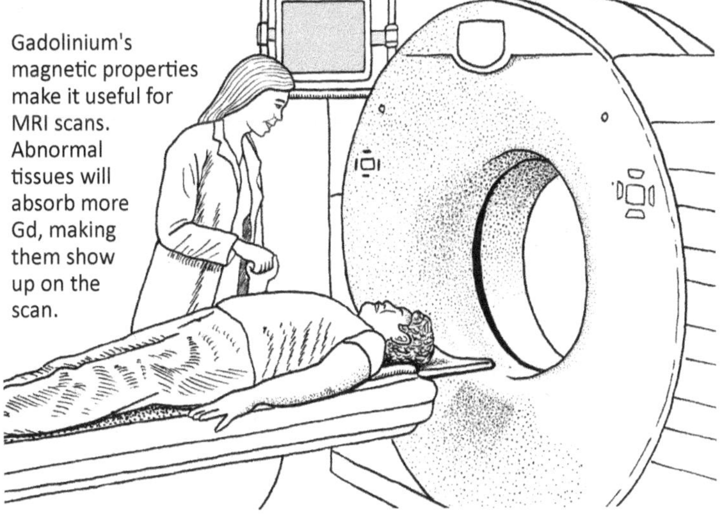

Gadolinium's magnetic properties make it useful for MRI scans. Abnormal tissues will absorb more Gd, making them show up on the scan.

Tb 65

protons
94 neutrons
65 electrons
Atomic mass: 158.9

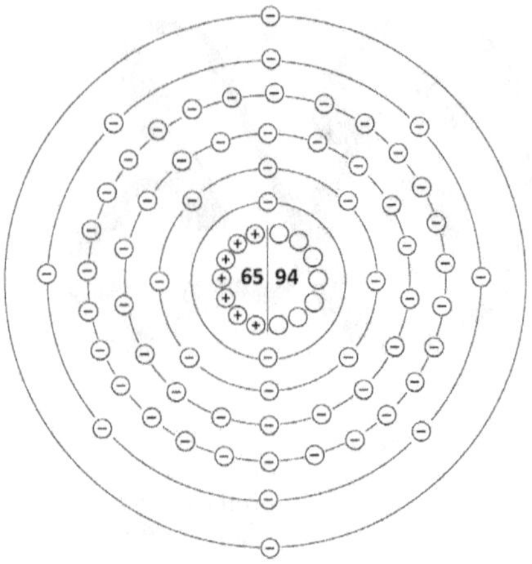

Terbium

Named after the Swedish town of Ytterby

Terbium was discovered in 1843 by Swedish chemist Carl Gustaf Mosander and named after the small village of Ytterby, Sweden. Just outside this village was a small above-ground mining operation where interesting mineral ores were being discovered. Many of the rare earth elements in the lanthanide series were discovered using rocks from the Ytterby quarry. Mosander had isolated yttrium oxide, Y_2O_3, from an Ytterby rock, and discovered terbium as an "impurity" in his sample. He was not able to extract the terbium, however. This was only made possible years later when chemists began using electricity to isolate elements. Today, most terbium is extracted from monazite sand, but it can also be found in euxenite and bastnäsite.

Mosander determined that there were actually three oxides present in his sample, which he called yttria, erbia, and terbia. The only difference between erbia and terbia was that one produced a pink color and the other was colorless. As other scientists begin to investigate these new oxides, there was some confusion between erbia and terbia, and the result was that the names got switched. Mosander's erbium is now our terbium.

Terbium has properties very similar to gadolinium. It is used to make phosphor powders that glow bright green. Therefore, it is usually combined with powders of europium, which make red and blue. The red, blue, and green light blends together to make white light. This might seem strange since you probably have experience combining paints, and you know that red, green and blue paint won't make white paint. However, light doesn't act like paint. (It took several famous physicists to figure this out, so don't worry if it seems confusing!)

Terbium can be "doped" (added) into crystals such as calcium fluoride and calcium tungstenate, which are used in solid state devices (electronics based on microchip processors). It can also be added to other metals to create an alloy called Terfenol-D®, which will change its shape depending on the magnetic fields around it. Materials like this are used to make actuators, which are used to make things like sonar for boats and submarines, and even consumer products such as the SoundBug® which can turn any large, flat surface into a loudspeaker.

In biology, terbium is used as a fluorescent marker to detect round endospores made by bacteria. In environmental science, terbium can be added to molecules that detect nitrogen-based pollutants produced by factories that make nitrogen-based products such as dyes, pesticides and explosives.

Tb-Dy-Fe alloy Terfenol-D®

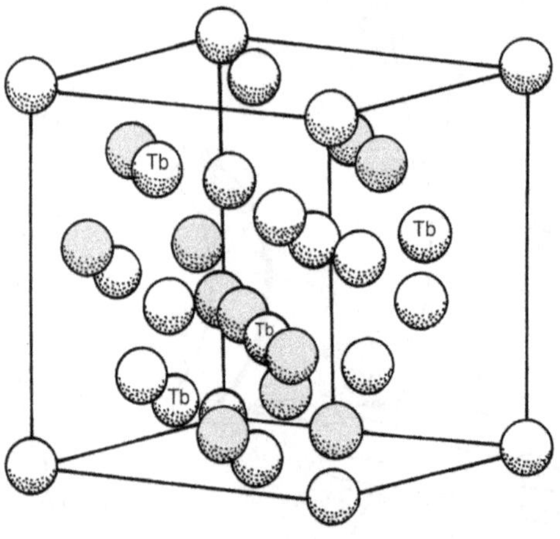

The darker balls are Dy. The lighter balls are Fe.

Nitrotoluene (a pollutant)

This molecule will stick to a terbium indicator molecule and cause it to glow.

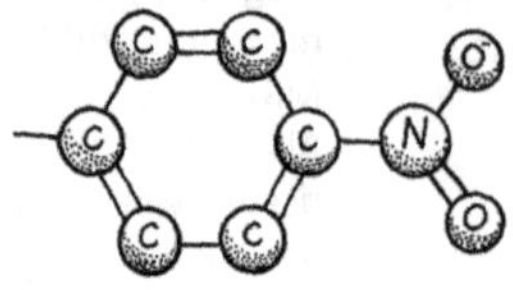

Dipicolinic acid

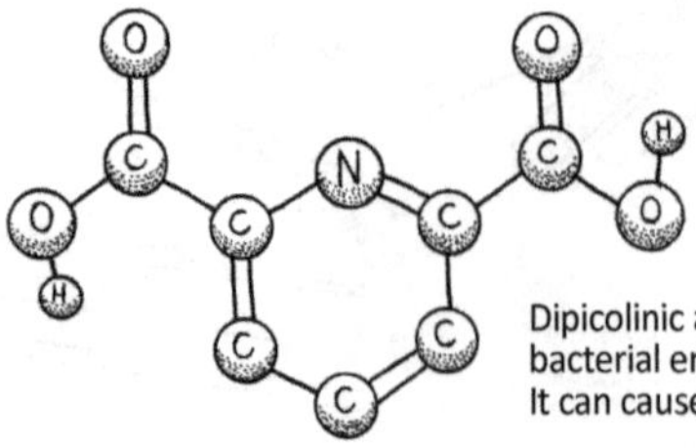

Dipicolinic acid is made by bacterial endospores. It can cause terbium to glow.

Tb MOF (Metal Organic Framework)

When dipicolinic acid sticks to this molecule, the Tb will glow green.

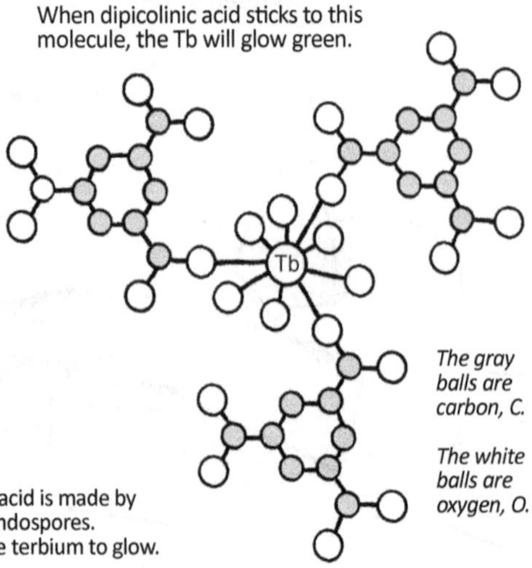

The gray balls are carbon, C.

The white balls are oxygen, O.

65 Tb

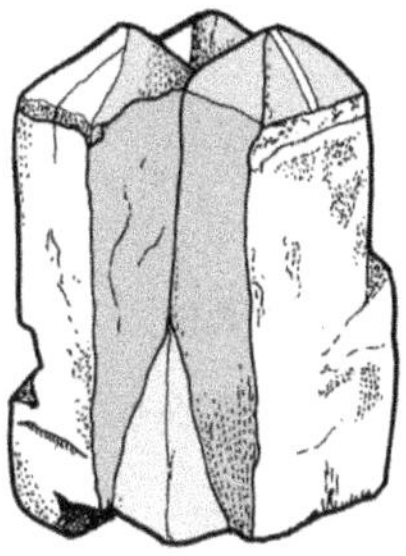

Xenotime

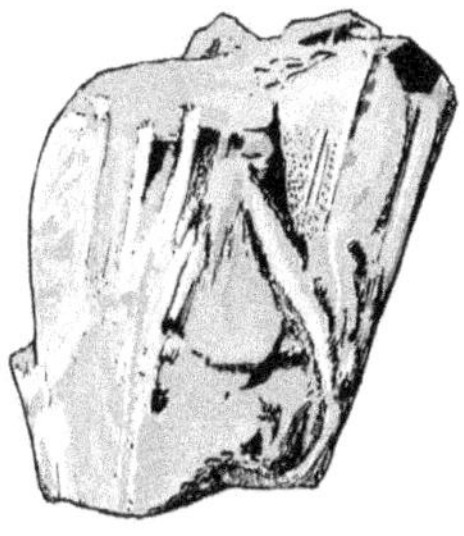

Euxenite *(yuke-sen-ite)*

Mineral ores that contain small amounts of terbium are dark gray, brown, or tan.

Terbium

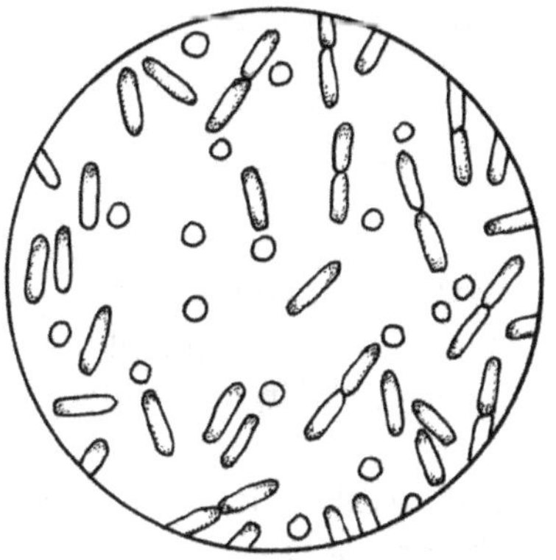

Terbium is used to make bio-markers that will cause bacterial endospores to glow green. *(Round endospores are green and rod-shapes are red.)*

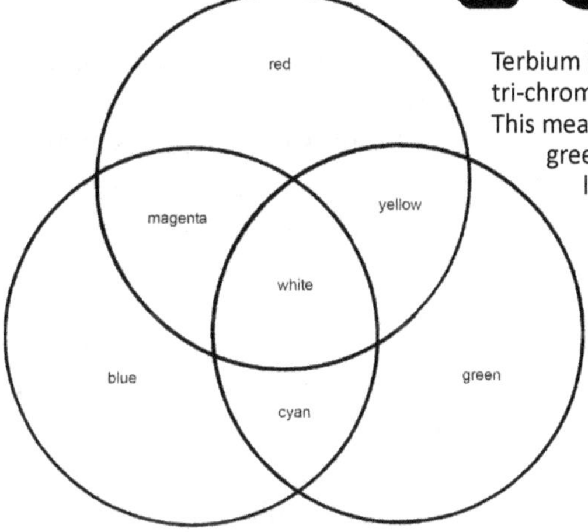

Terbium is used in tri-chromatic lighting. This mean that red, green, and blue light are used to make white light.

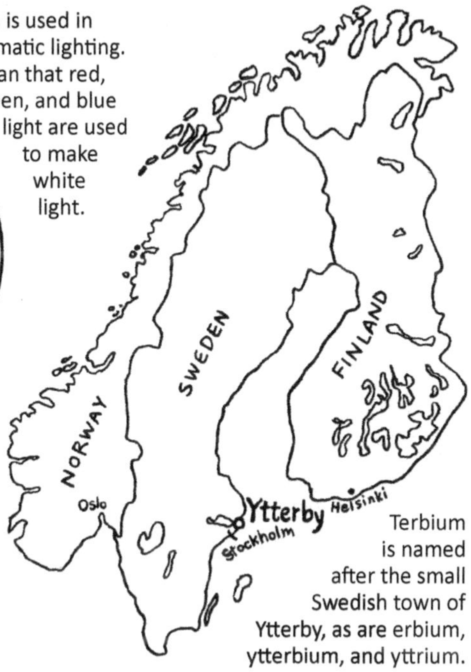

Terbium is named after the small Swedish town of Ytterby, as are erbium, ytterbium, and yttrium.

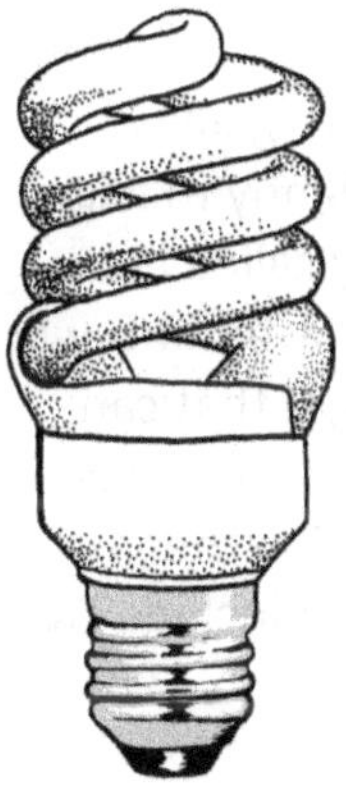

The inside of fluorescent bulbs are coated with phosphor powders that glow red, green or blue. These colors combine to make white light.

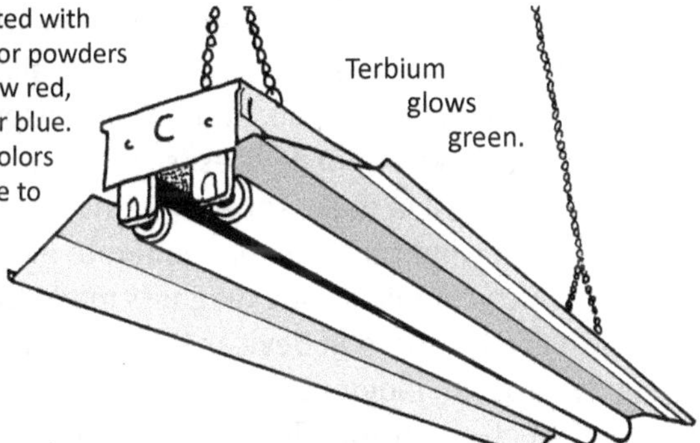

Terbium glows green.

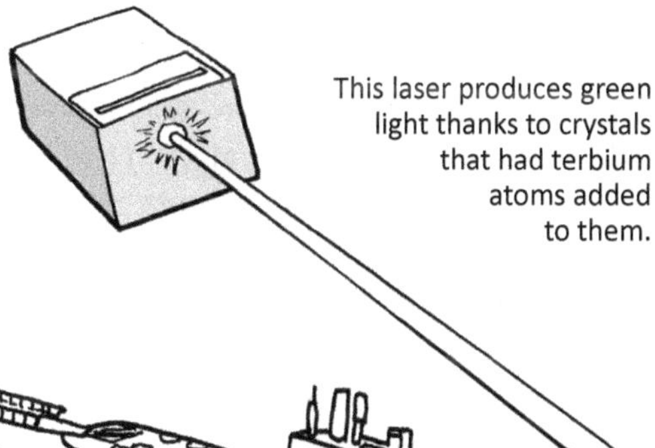

This laser produces green light thanks to crystals that had terbium atoms added to them.

The SoundBug® device uses Terfenol-D (Tb, Fe, and Dy) to turn any large, flat surface into a loudspeaker.

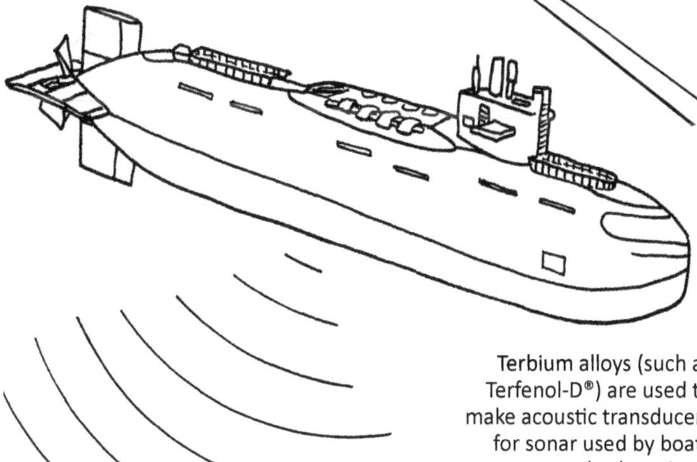

Terbium alloys (such as Terfenol-D®) are used to make acoustic transducers for sonar used by boats and submarines.

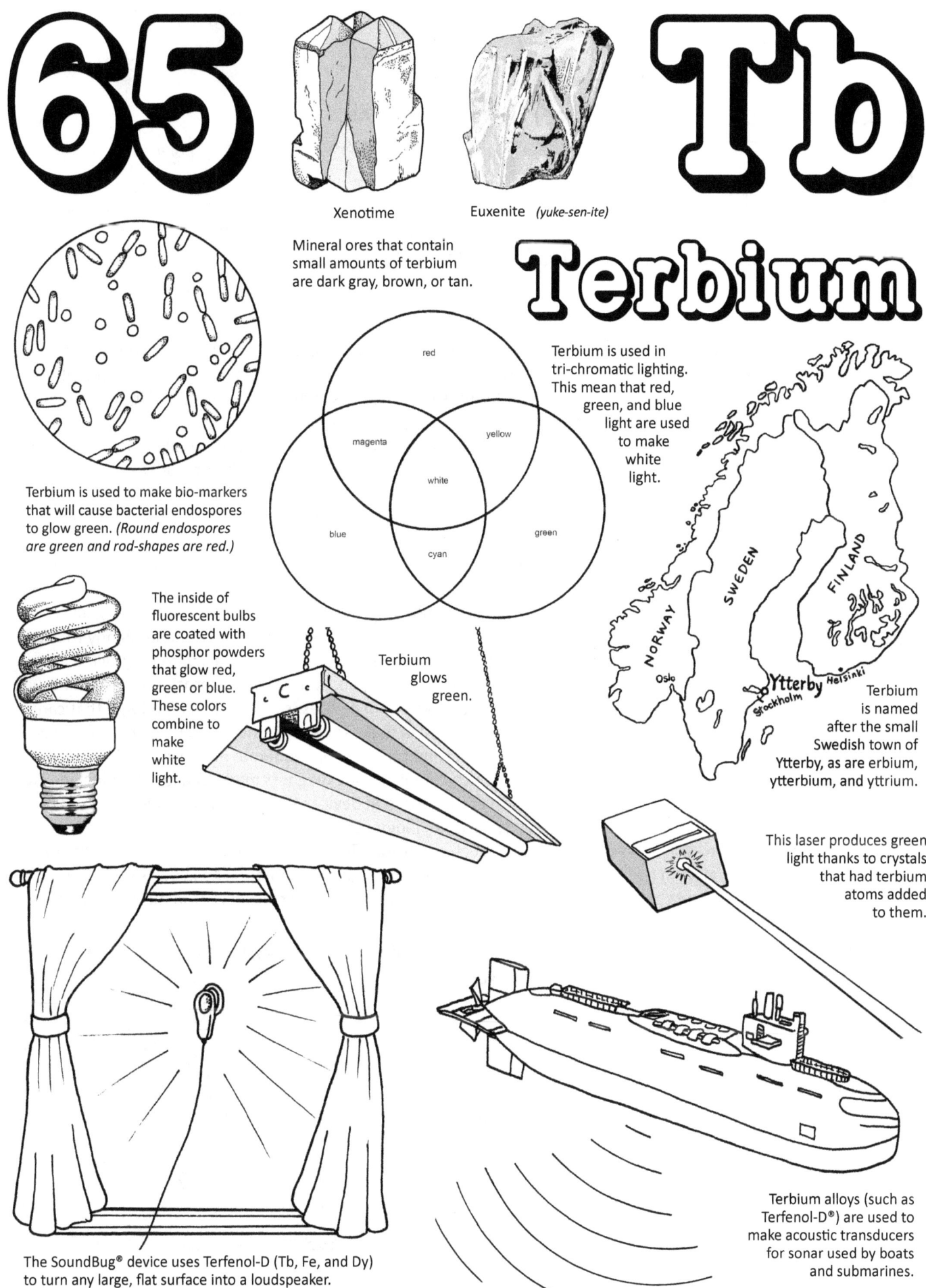

Dy 66

protons
97 neutrons
66 electrons

Dysprosium

Atomic mass: 162.5

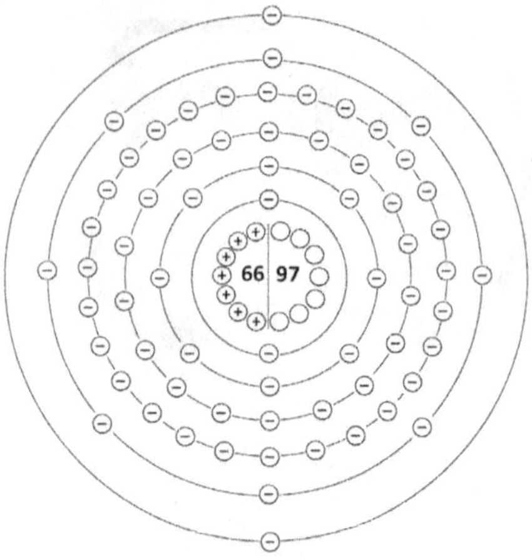

Named using Greek word "dysprositos" ("hard to get")

Dysprosium was discovered by Paul Émile Lecoq de Boisbaudran in 1886 while working with mineral ores containing erbium, holmium and thulium. He knew there was yet another new element lurking in this mineral ore and wanted to be its discoverer. It took him over 30 attempts to isolate it, so when he finally found it, he decided its name should reflect his experience, and used a Greek word meaning "hard to get."

A very pure sample of dysprosium was not produced until the 1950s, when a procedure called "ion exchange" was invented. The process allows the rare earth elements to bind to elements such as chlorine or lithium, then in a second step, the chlorine or lithium "ions" are pulled off, and the rare earths find themselves all alone. This method of extraction is used to pull dysprosium out of monazite sand and clay. Southern China has the richest sources of dysprosium and produces 99% of the world's supply.

Like the other rare earth elements in the lanthanide series, pure dysprosium metal is soft and shiny. It is always combined with other elements to make compounds or alloys. It is similar enough to neodymium that it can replace some of the neodymium in NdFeB magnets. When combined with terbium and iron, the resulting alloy has an interesting property called "magnetostriction." This means that the metal will change shape slightly to align with magnetic fields around it. The Naval Ordnance Lab (NOL) was the first to create this alloy, so they called it Terfenol-D® ("Ter" for terbium, "Fe" for iron, "NOL" for the name of the lab, and "D" for dysprosium). Terfenol-D will cause vibration to result from magnetism, so it was used to make audio devices (such as the SoundBug®) that can make any flat surface into a loudspeaker. Terfenol-D is also used in make sonar systems for boats and submarines.

Dysprosium is found in very strong magnets used in wind turbines that generate electricity, and in the motors of electric cars. Hard drives in computers might also contain strong magnets made of alloys that include Dy.

Like the other rare earth elements, dysprosium can be found in devices that make light. Dysprosium is most often found in metal halide bulbs, in the compound dysprosium iodide, DyI_3.

Dysprosium can be "doped" (added) to crystals such as calcium fluoride to make them useful as radiation detectors, such as those found in dosimeters that measure how much radiation someone has been exposed to.

Dysprosium can capture or slow down free neutrons, so it can be used in control rods in nuclear reactors.

Dy_2O_2 Dysprosium oxide

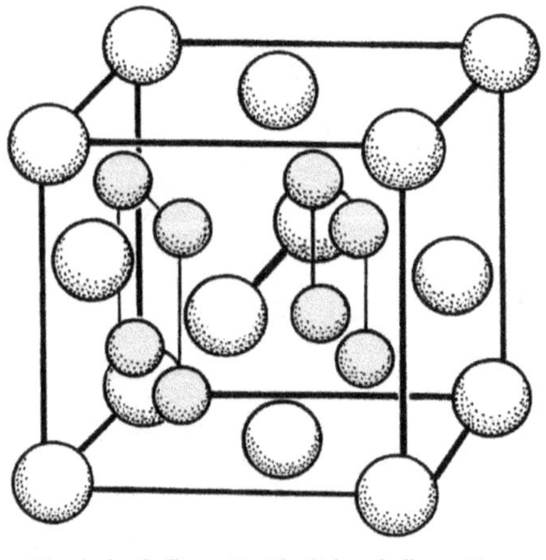

The darker balls are Dy. The lighter balls are O.

DyI_3 Dysprosium iodide

Used in high intensity light bulbs.

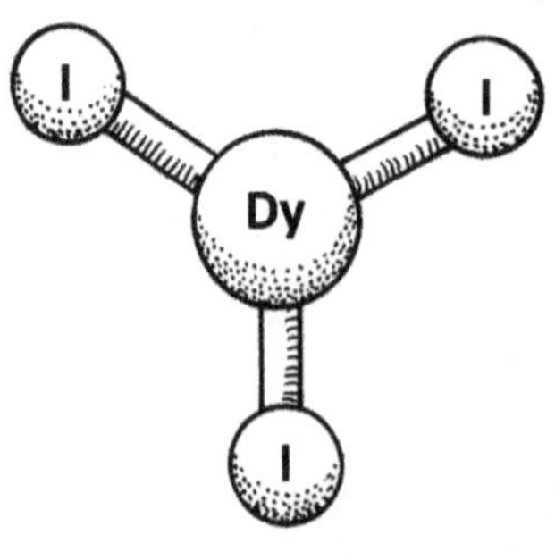

Dysprosium will make similar molecules when it combines with fluorine, bromine, or chlorine.

Tb-Dy-Fe alloy Terfenol-D®

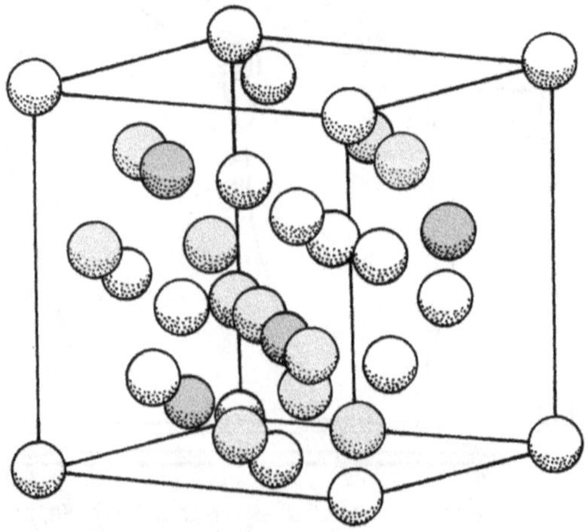

Darkest balls are Tb, lighter gray balls are Dy, white is Fe.

66 Dy

The SoundBug® device uses the magnetic properties of Terfenol-D® (terbium, iron, and dysprosium) to turn any large, flat surface into a loudspeaker.

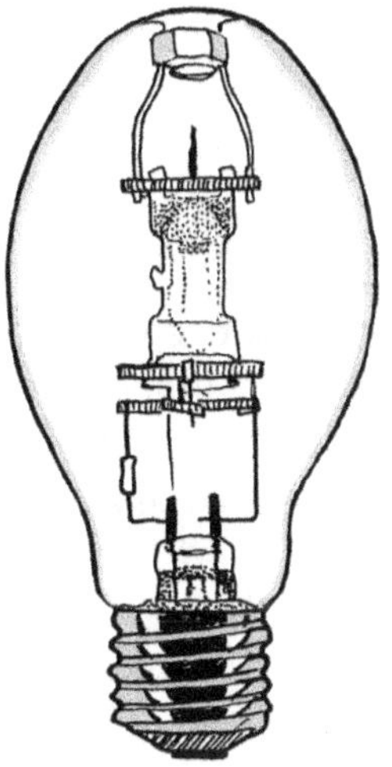

Dysprosium iodide can be found in metal halide bulbs, and in some fluorescent bulbs.

Dysprosium

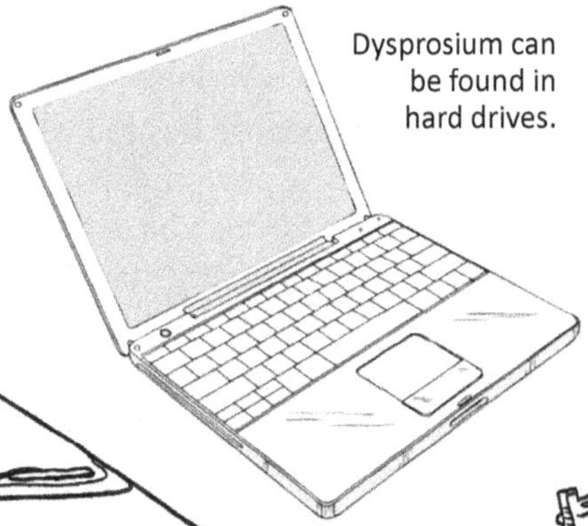

Dysprosium can be found in hard drives.

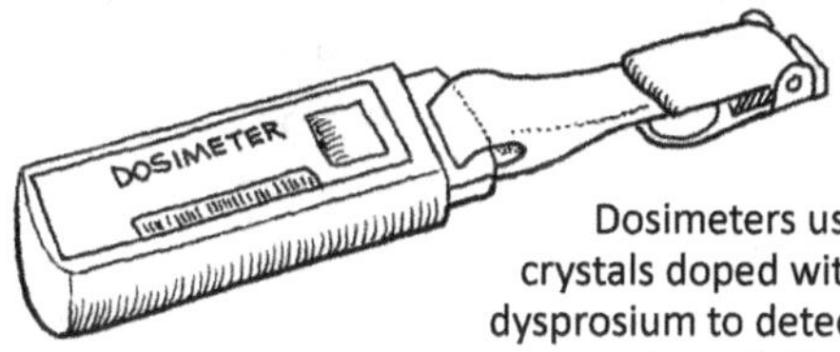

Dosimeters use crystals doped with dysprosium to detect exposure to radiation.

Dy can be found in metal alloys used to make magnets for motors in electric cars.

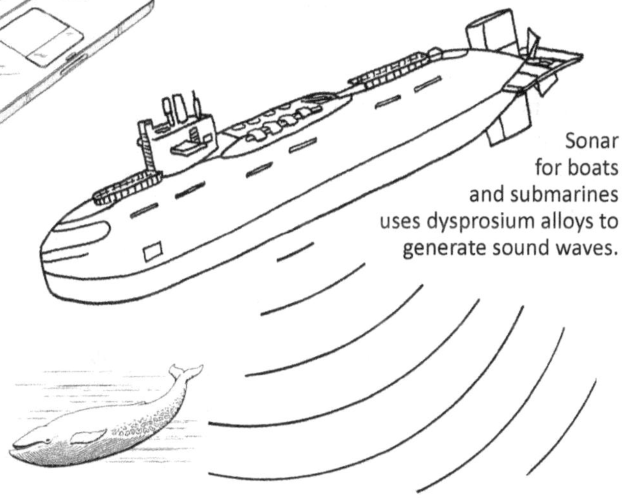

Sonar for boats and submarines uses dysprosium alloys to generate sound waves.

Dysprosium magnets are used in wind turbine generators.

Dysprosium is used in control rods in nuclear reactors because it can absorb the energy of free neutrons and slow down reactions.

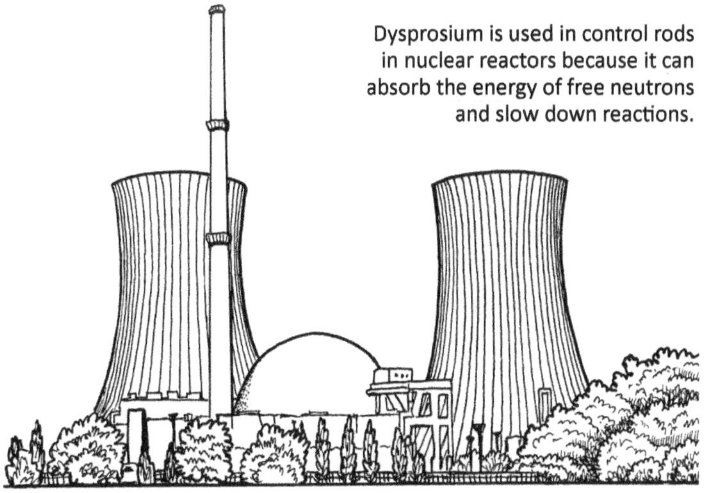

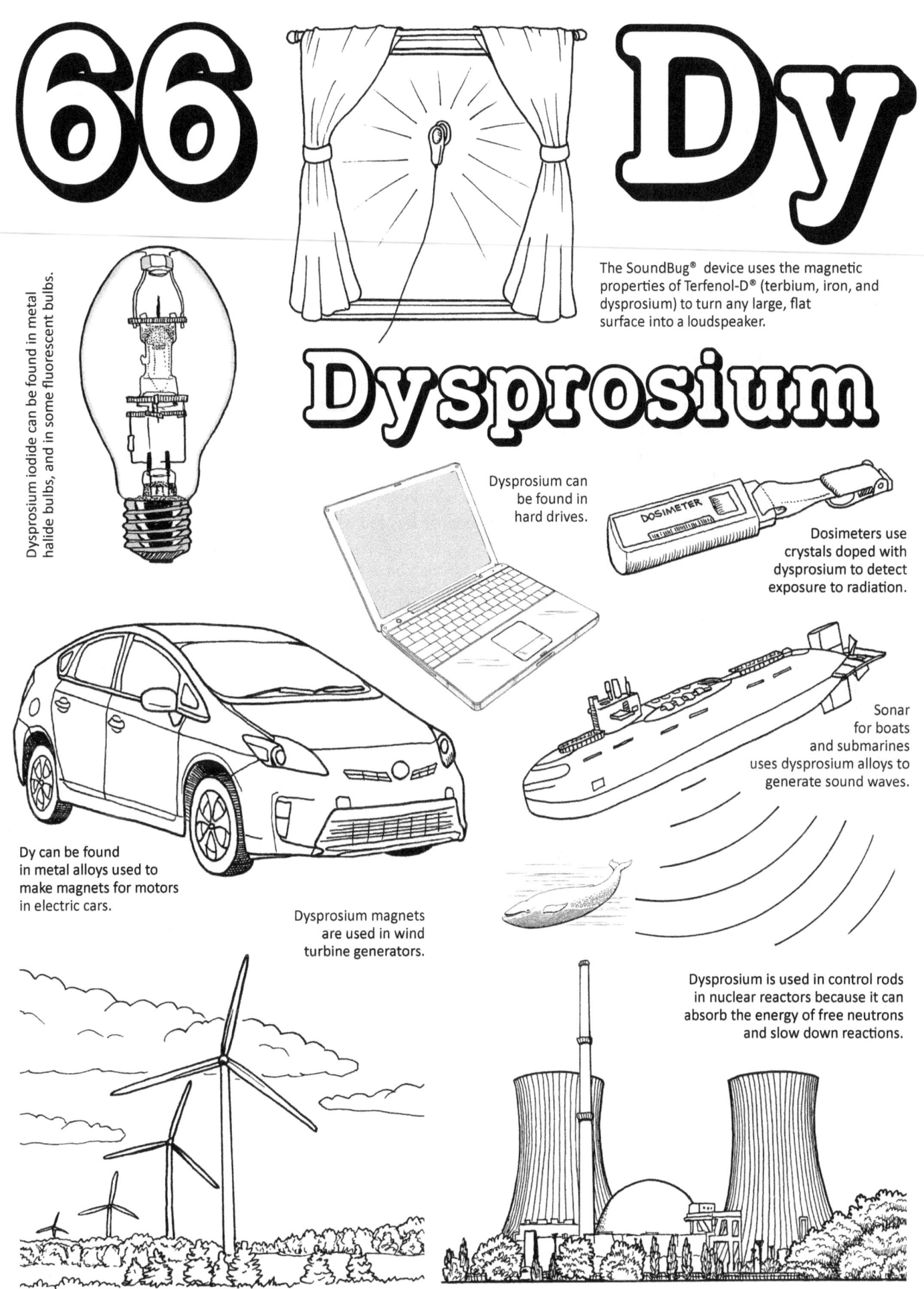

Ho 67 protons

98 neutrons
67 electrons

Atomic mass: 164.9

Holmium

Named after Stockholm, Sweden

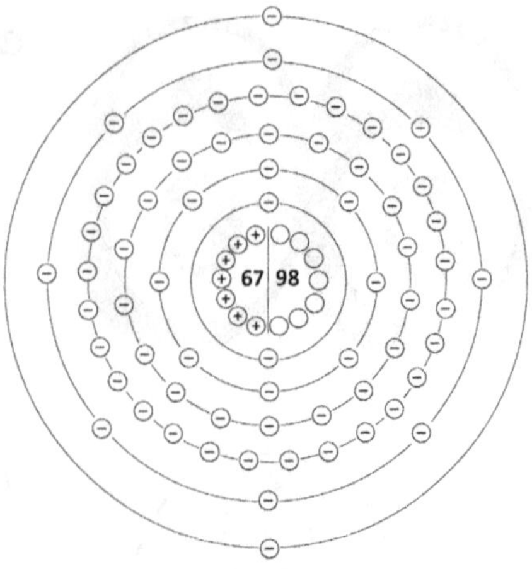

Holmium was discovered in 1879 by Per Teodor Cleve, the Swedish chemist who also discovered thulium. Like the discoveries of the other rare earth elements of the lanthanide series, he was working with an oxide mineral that was known to contain a rare earth (in this case, erbium), and he suspected that there was another element still hiding in the ore. When he found the new element he named it after his birthplace, Stockholm, using the Latin word for the city, Holmia. Soon after that, he discovered a second element in the ore and named it thulium, using the ancient Latin word for Scandinavia: Thule.

Holmium is the most magnetic element on the Periodic Table, with its strongest magnetism occurring at very low temperatures. It can't be used in its pure form because it is too soft. Small amounts of holmium are mixed into magnetic metal alloys that are used to make extremely powerful magnets for machines such as MRI scanners.

Magnetism isn't holmium's only useful quality; it can also be used to color glass red or yellow. Glass colored with holmium produces certain wavelengths of light so consistently that these pieces of glass can be used to calibrate spectrometers for accuracy. Holmium can also provide red color for cubic zirconia, a gemstone often used as a substitute for diamond. Holmium oxide powder will look either pink or yellow, depending on the light source.

Yttrium iron garnet (YIG) and yttrium aluminum garnet (YAG) crystals can be "doped" with holmium to make them more useful for solid state lasers. Holmium-YAG lasers are used in dentistry and in some surgical operations, especially in removal of mineral deposits ("stones") in the kidneys. The lasers produce wavelengths of light beyond the visible spectrum The energy in the laser beam is high enough that the laser can cut through tissue better than even a sharp scalpel, and it can also cauterize (seal) blood vessels so they don't bleed.

Like some of the other rare earth elements, holmium can absorb fast-moving free neutrons, so it can be put into control rods in nuclear reactors. The control rods are lowered when they want the reaction to slow down.

Holmium is one of the less abundant rare earth elements (with the even numbered elements being more abundant than the odd ones) but it is still more abundant than silver, gold or mercury. Like the other rare earths, most holmium is extracted from monazite sand or clay, using a process called "ion exchange."

Ho_2O_3 Holmium oxide

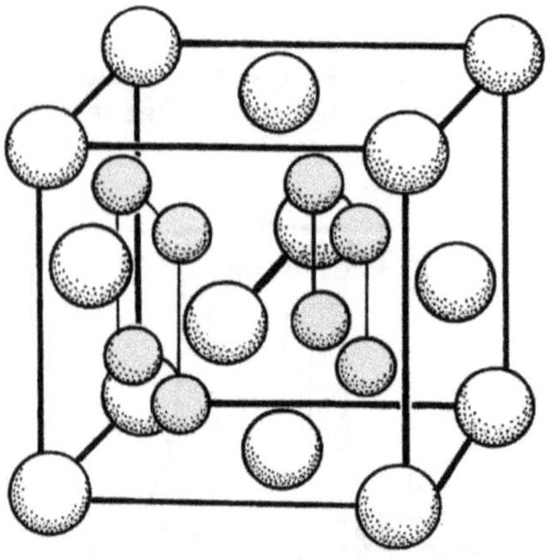

The darker balls are Ho. The lighter balls are O.

$HoCl_3$ Holmium chloride

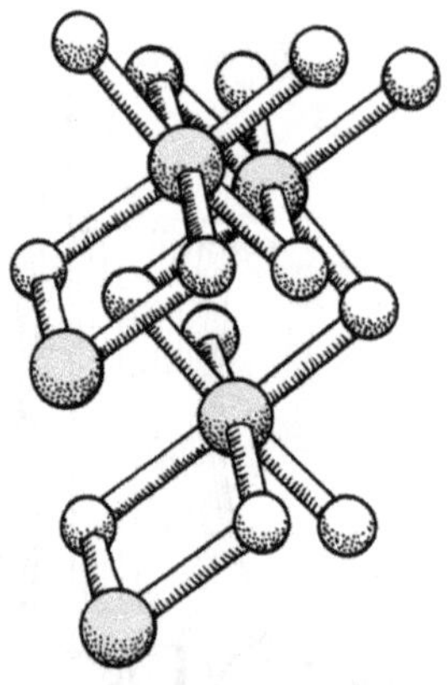

The darker balls are Ho. The lighter balls are Cl.

Monazite crystal structure

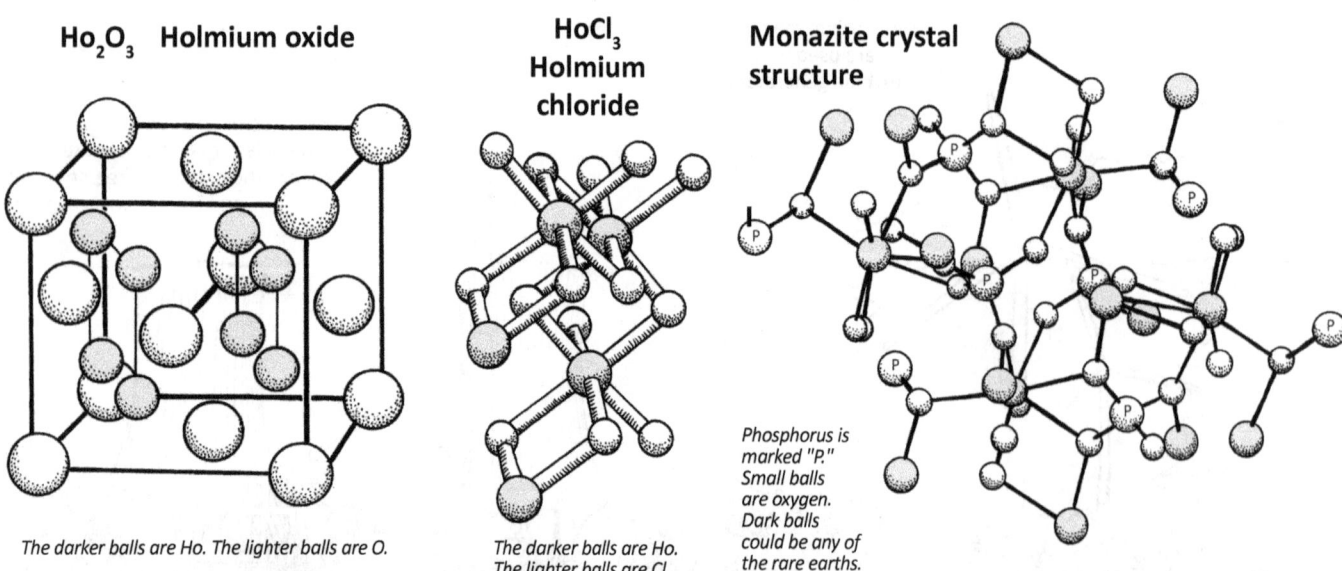

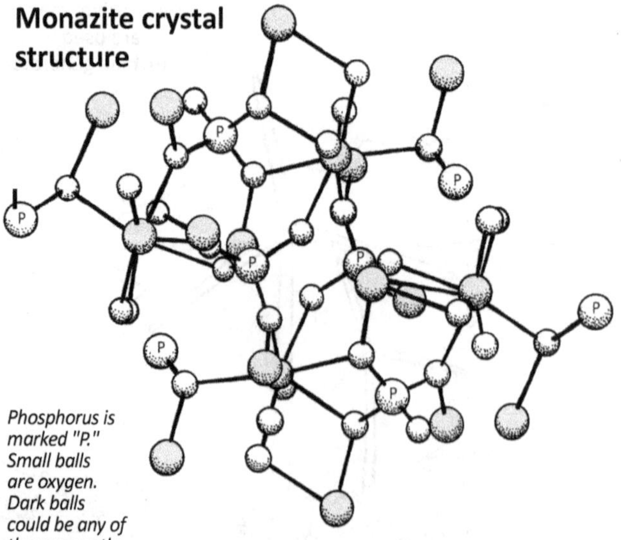

Phosphorus is marked "P." Small balls are oxygen. Dark balls could be any of the rare earths.

67 Ho

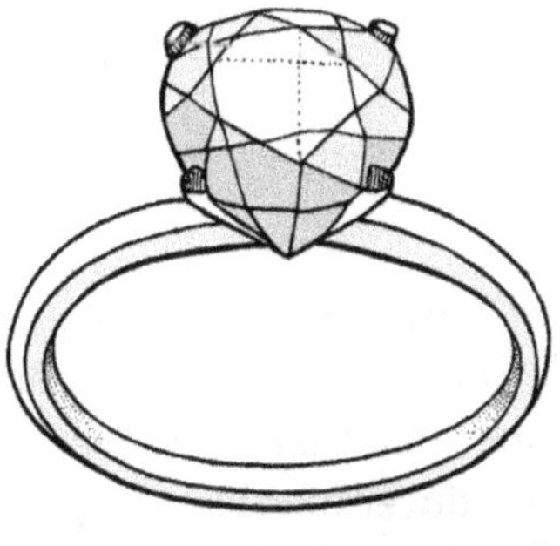

Holmium in cubic zirconia (ZrO_2) can add a red tint to the crystal.

Stockholm, Sweden, has many historic buildings.

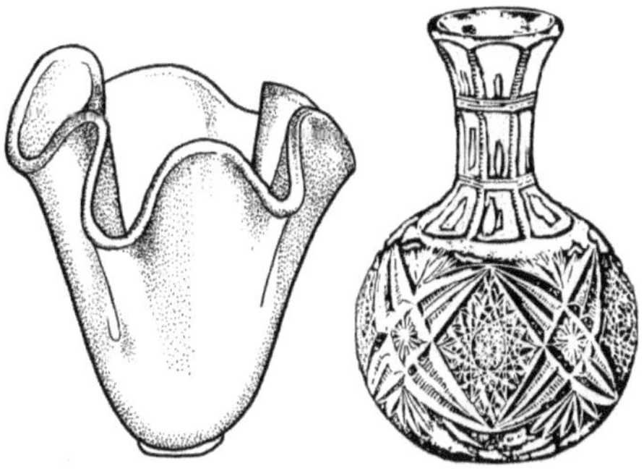

Holmium can color glass red or yellow.

Holmium

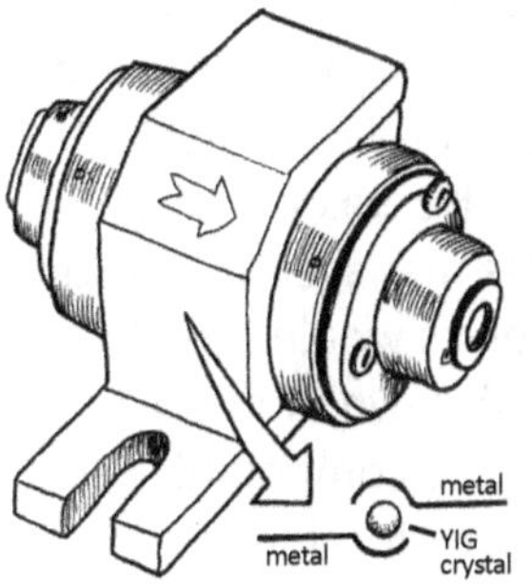

This optical isolator uses a tiny (the size of a pinhead) Ho-YIG crystal as a "one way valve" for laser light.

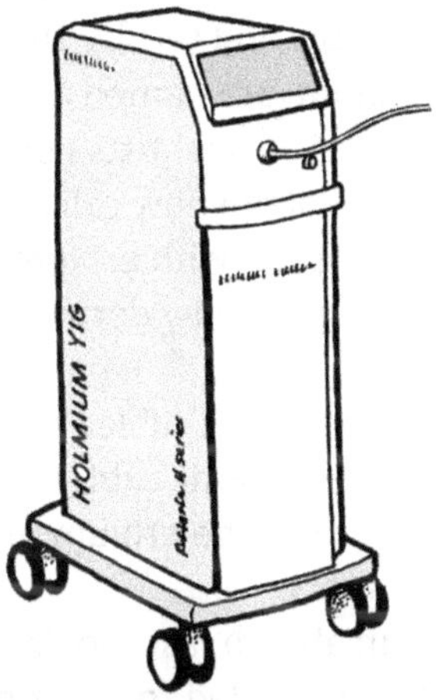

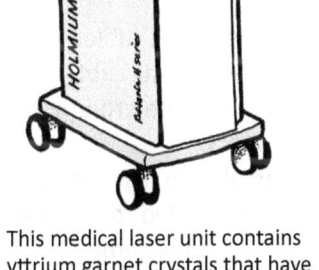

This medical laser unit contains yttrium garnet crystals that have been "doped" with holmium.

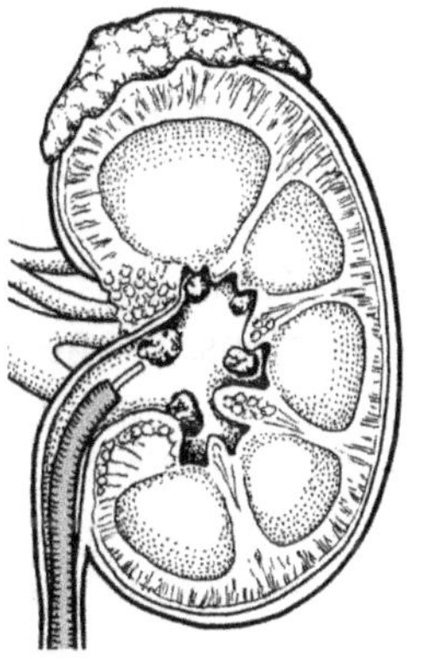

Ho:YIG lasers are often used to remove mineral stones from kidneys.

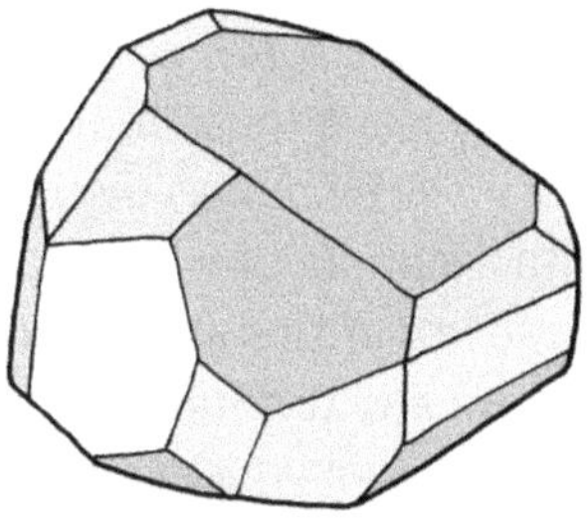

This yttrium-iron garnet (YIG) crystal is deep reddish-purple.

Ho_2O_3 looks yellow in daylight, pink under fluorescent light.

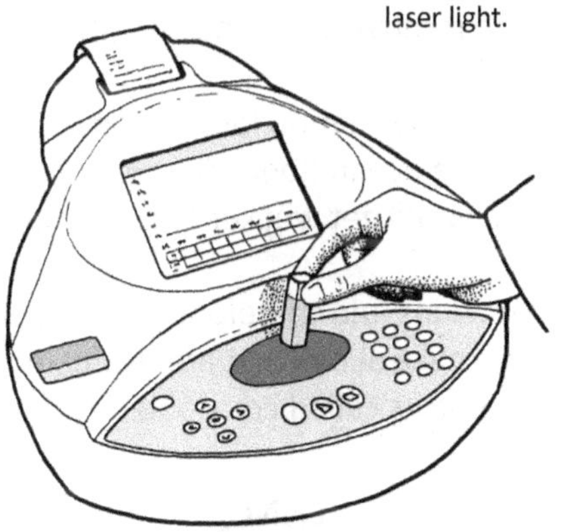

Holmium glass is used to calibrate (check for accuracy) spectrophotometers that use light to analyze samples.

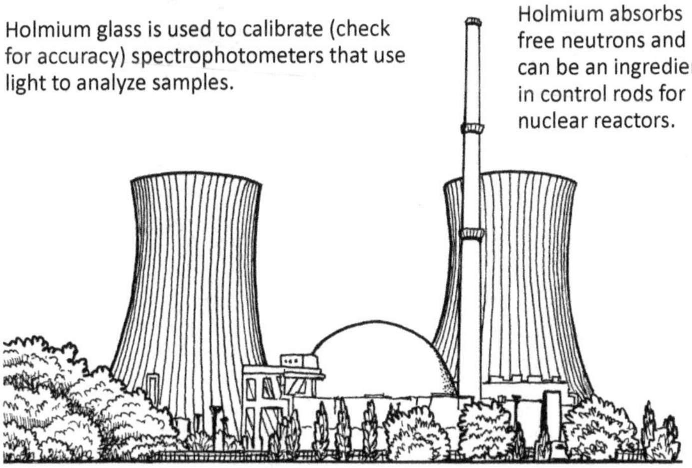

Holmium absorbs free neutrons and can be an ingredient in control rods for nuclear reactors.

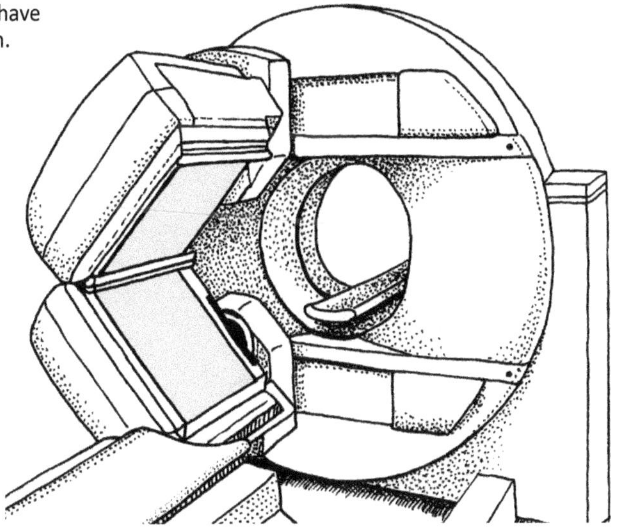

MRI scanners use extremely powerful magnets containing holmium.

Er 68

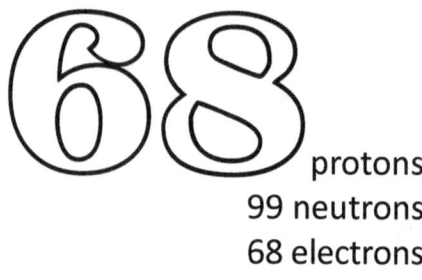

protons
99 neutrons
68 electrons

Atomic mass: 167.25

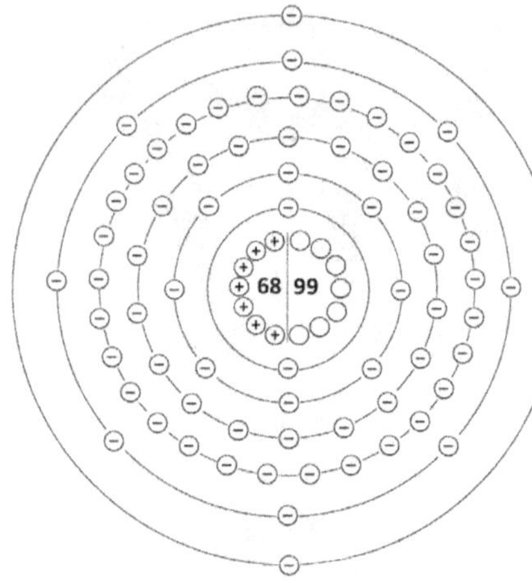

Erbium

Named after the Swedish town of Ytterby

Erbium is strongly associated with the color pink. Compounds and solutions containing erbium are pink, and they can be used to add pink color to glass, ceramics and cubic zirconia, ZrO_2. Like all of the rare earth elements in the lanthanide series, erbium isn't really rare. There are more erbium atoms in our planet that there are silver, gold or mercury, or even some of the gases like neon and helium. Currently, the biggest producer of erbium is southern China, where they have large amounts of a special type of clay that is very high in these elements. However, erbium can also be extracted from monazite sand found in Florida, Brazil, and India, and from the mineral ores xenotime and euxenite. The process of extracting and isolating each element from the clay is called "ion exchange." It requires a series of chemical reactions where the elements are offered a better situation than the one they are in, so they leave those molecules and join others. Eventually, in the last step, the rare earth element is left by itself.

Erbium was discovered at the same time as terbium, by Carl Gustaf Mosander in 1843. He was working with a sample of what he called "yttria" (Y_2O_3), but found that there were several impurities. He managed to isolate these impurities as oxide compounds and named them erbia and terbia. The elements themselves would then be erbium and terbium. All three, yttrium, erbium and terbium, are named after the small Swedish town of Ytterby, the site from which the mineral samples were taken. As other scientists begin to investigate these compounds using spectrometry, they accidentally switched the names, so what we now call erbium was Mosander's terbium.

Yttrium aluminum garnet (YAG) crystals can be "doped" with erbium to make them more useful for solid state lasers used for medical applications. Er:YAG lasers are used by dermatologists to remove scars and wrinkles, and dentists use them for some types of dental surgeries.

Erbium is added to glass fibers that are used in fiber-optic cables that can carry information long distances. All the continents are connected to the Internet through fiber-optic cables that lie on the ocean floor. The role of erbium in these optical fibers is to amplify the signal. Without the erbium, the signal would get weaker and weaker as it traveled the thousands of miles between the continents.

Like many other rare earth elements, erbium can absorb fast-moving free neutrons and therefore can be used in control rods in nuclear reactors. Erbium can also be added to other metals to make alloys that are useful in certain industries. Erbium-nickel, Er_3Ni, is used to make ultra-low temperature refrigerators called cryocoolers.

Er_2O_3 Erbium oxide

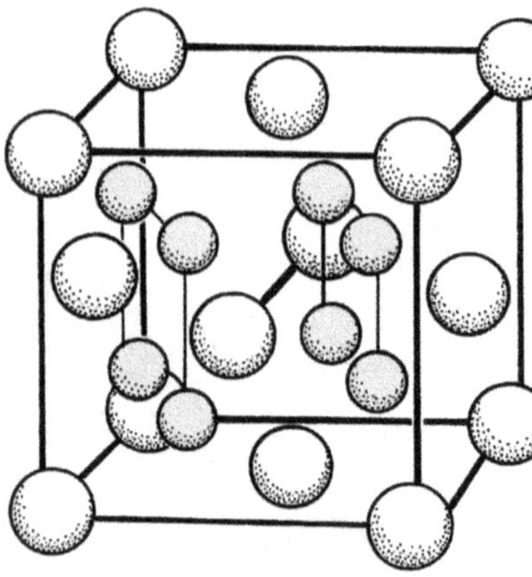

The darker balls are Er. The lighter balls are O.

$ErCl_3$ Erbium chloride

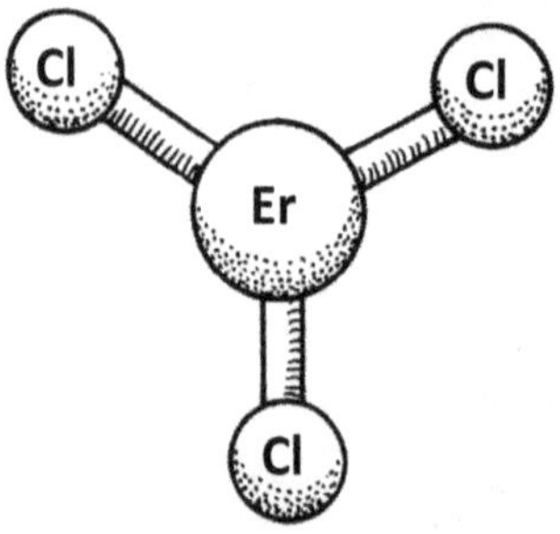

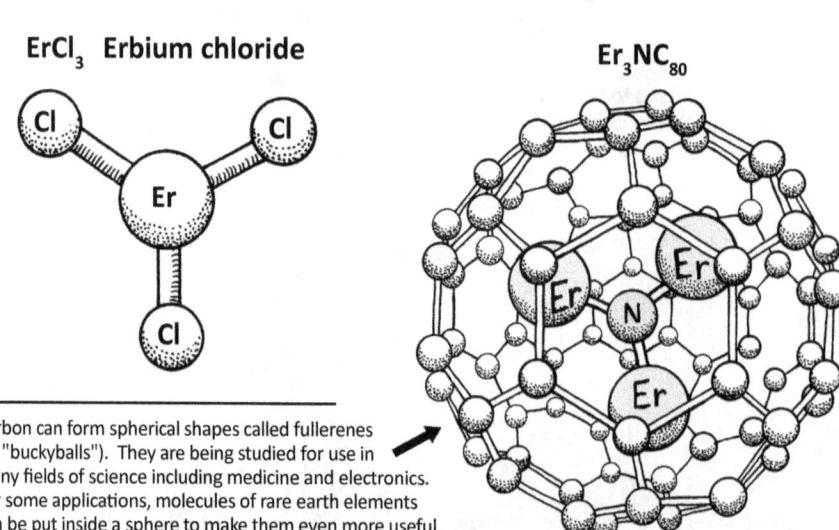

Carbon can form spherical shapes called fullerenes (or "buckyballs"). They are being studied for use in many fields of science including medicine and electronics. For some applications, molecules of rare earth elements can be put inside a sphere to make them even more useful.

Er_3NC_{80}

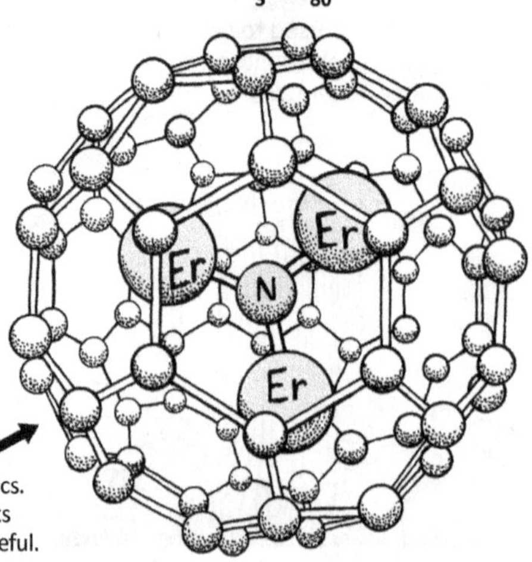

68 Er

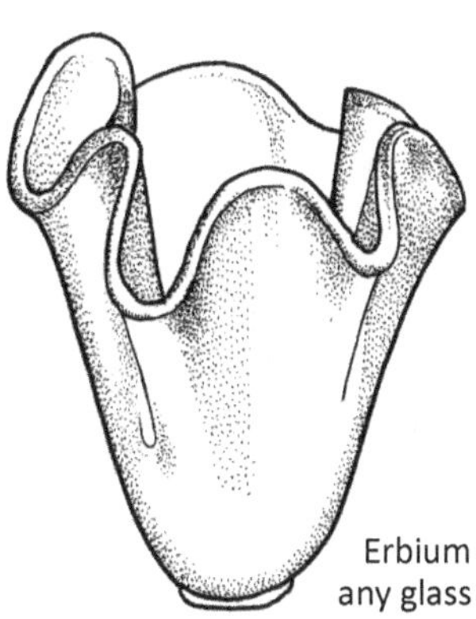

Erbium can color any glass pink.

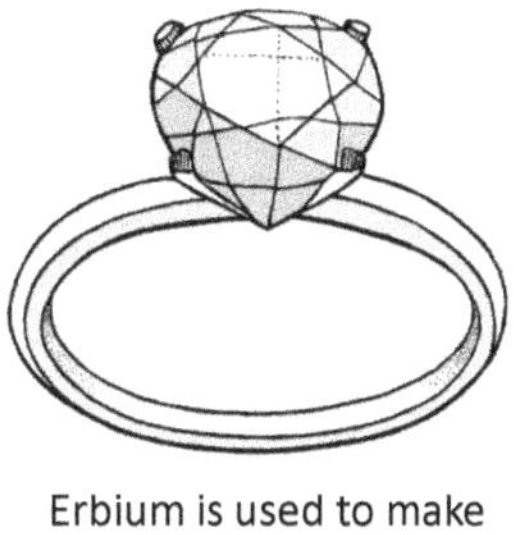

Erbium is used to make pink cubic zirconia.

Erbium

Erbium is named after the small Swedish town of Ytterby, as are terbium, ytterbium, and yttrium.

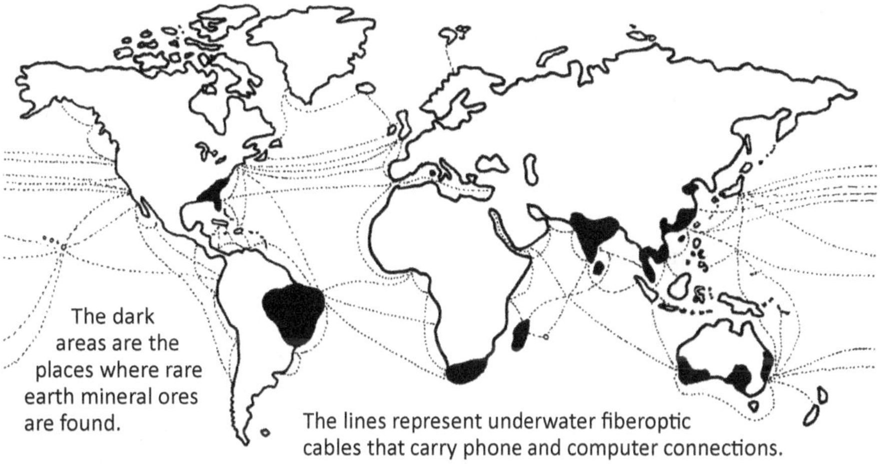

The dark areas are the places where rare earth mineral ores are found.

The lines represent underwater fiberoptic cables that carry phone and computer connections.

This ship will lay fiberoptic cables onto the seafloor.

Some of the fibers in these cables contain erbium.

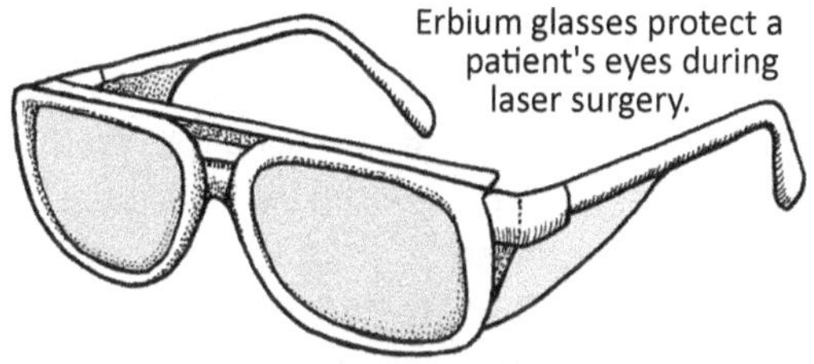

Erbium glasses protect a patient's eyes during laser surgery.

Sharks like to bite underwater cables. Shark-proofing the cables is a major challenge. Sometimes the cables are wrapped in Kevlar®, a super-tough plastic made by DuPont, which is many times stronger than steel.

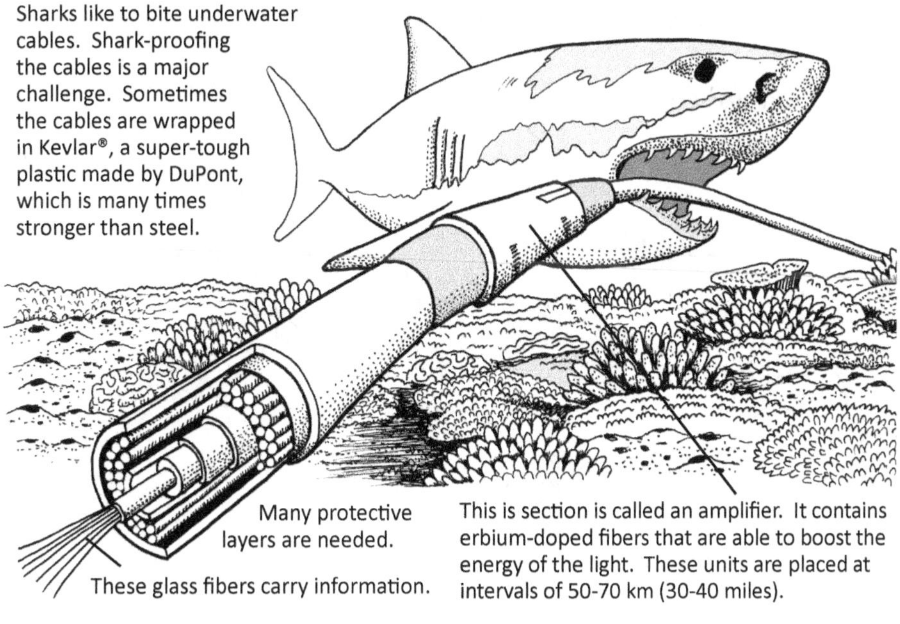

Many protective layers are needed.

These glass fibers carry information.

This is section is called an amplifier. It contains erbium-doped fibers that are able to boost the energy of the light. These units are placed at intervals of 50-70 km (30-40 miles).

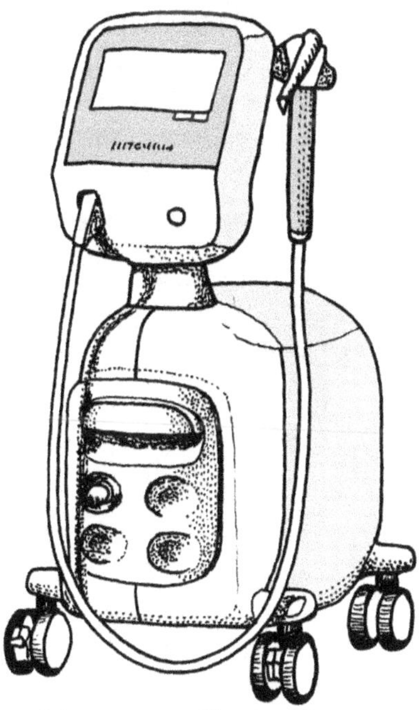

Er:YAG lasers are used for dental surgery and for removing scars and wrinkles on skin.

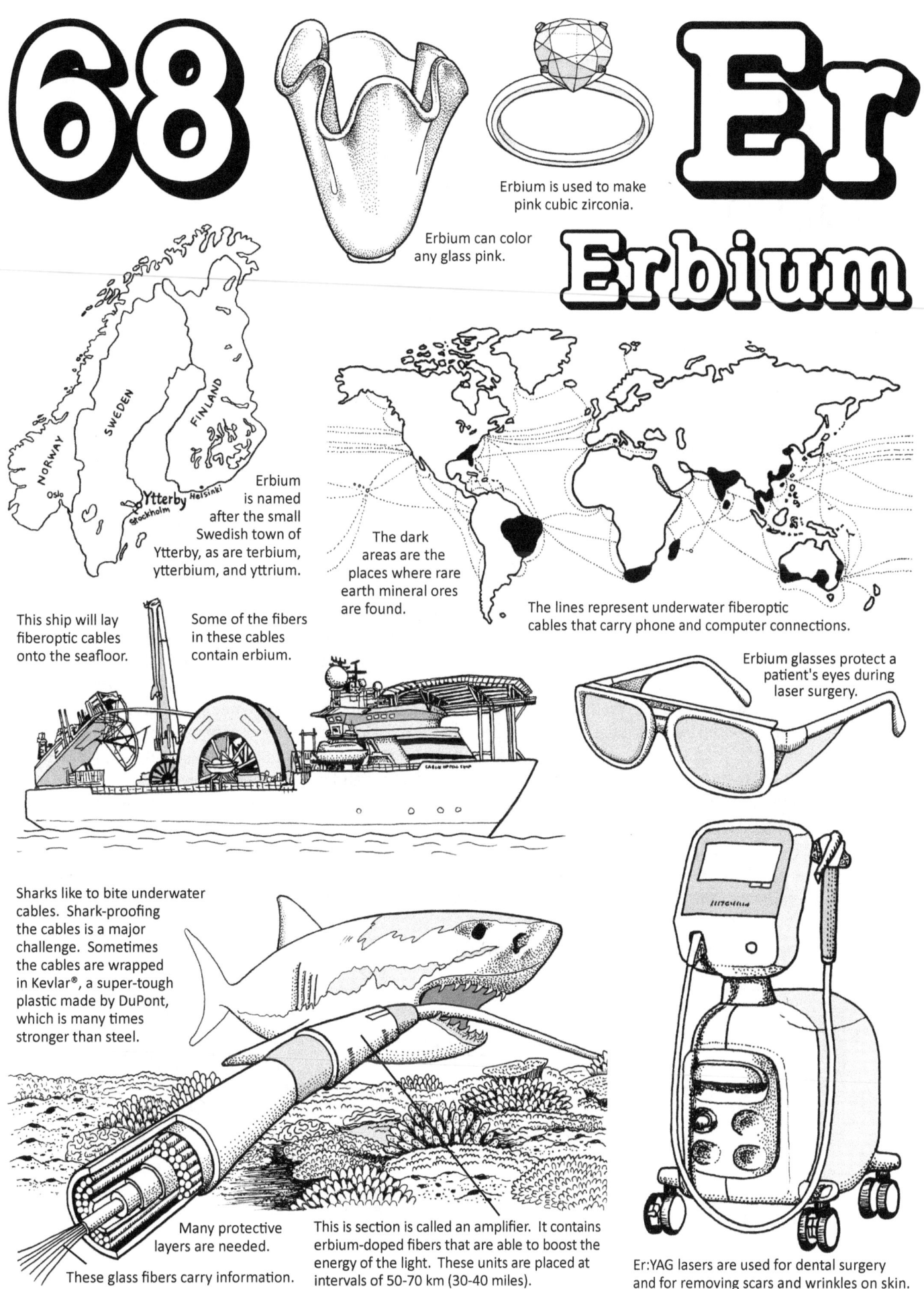

Tm 69

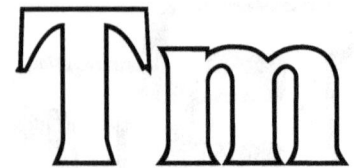

protons
100 neutrons
69 electrons

Atomic mass: 168.9

Thulium

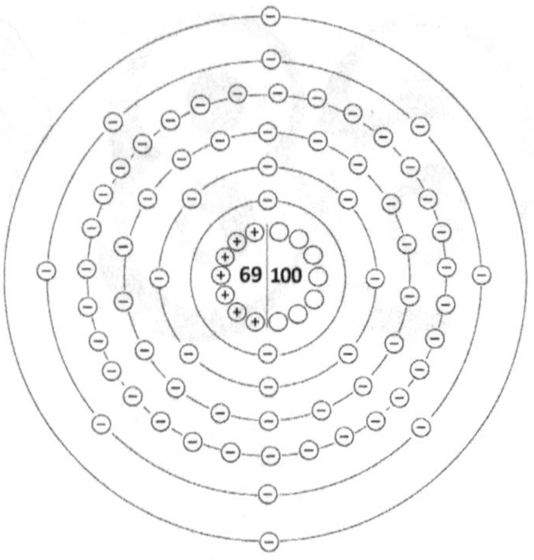

Named using an ancient word for Scandinavia, "Thule"

Thulium is the least abundant of all the rare earth elements in the lanthanide series. Even so, there are still more thulium atoms in the crust of the earth than atoms of gold, silver, or mercury. Thulium was discovered at the same time as holmium, by Swedish chemist Per Teodor Cleve in 1879. Cleve was working with a sample of erbium oxide and trying to isolate impurities that he knew were not erbium. He managed to extract two mineral oxides, a brown one and a green one. The brown one he named "holmia," after Stockholm, and the green one was called "thulia," using a very old name for Scandinavia, Thule.

Like all the other rare earth elements, thulium can be found in monazite sand. It also occurs in gadolinite, xenotime and euxenite. Thulium was added to the list of known elements long before anyone had actually seen a sample of it in its pure form. It was not until the 1950s, with the invention of the ion exchange process, that someone was able to produce a piece of very pure thulium metal.

Thulium will glow blue under ultraviolet light. For this reason, thulium is put into Euro bank notes to create a design that will show up only under UV light. Counterfeiters trying to make euro notes will find it difficult to imitate the effects of thulium. Thulium's ability to glow is also useful in making dosimeters that measure the levels of radioactivity that someone has been exposed to. Dosimeters use crystals that have been "doped" with thulium.

Thulium-doped crystals are also used in lasers, usually in combination with holmium and chromium. Lasers known as Ho,Cr,Tm:YAG are used for weather radar, military infrared equipment, and in medical devices. The wavelength produced by thulium lasers is ideal for medical procedures on skin, such as removal of scars or tattoos.

Thulium isn't naturally radioactive, but nuclear labs can bombard it with neutrons and create a radioactive isotope called Thulium-170. Tm-170 is a relatively safe source of x-rays and can be used in portable x-ray units. These units are not only useful in medicine and dentistry, but also in mechanical industries where they need to track down defects in metal structures and electrical components.

Thulium is often added to the metal alloys used for arc lighting. The light is produced by a spark that jumps from one piece of metal to the other. The color of the light comes from the hot metals. Thulium's contribution is light in the green part of the spectrum, an area seldom seen coming from other metals.

Tm_2O_3 Thulium oxide

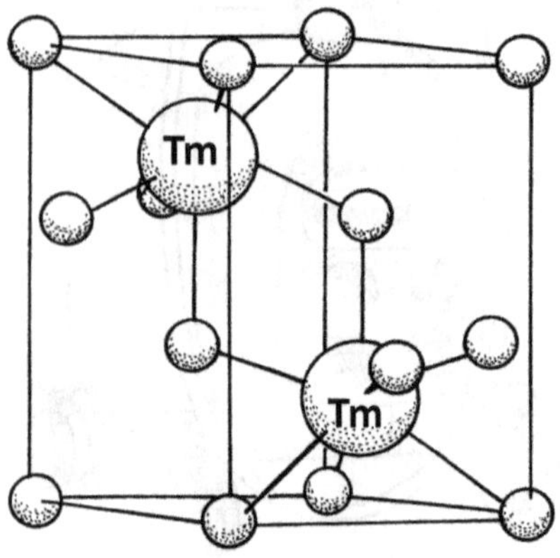

The small balls are O.

Thulium-doped CaF_2 crystal

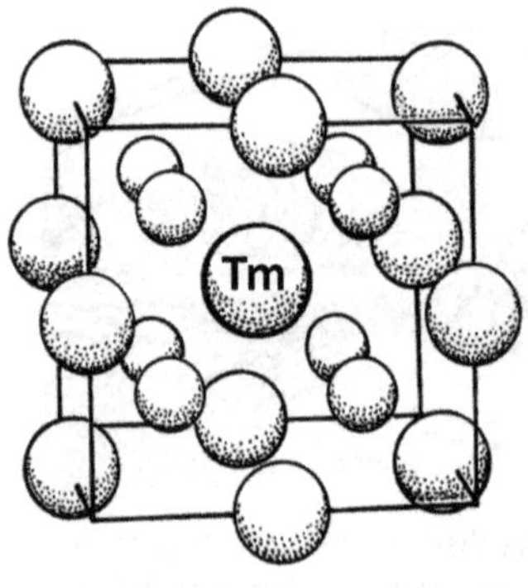

The smaller balls are F, fluorine. The larger balls are Ca, calcium.

Monazite crystal structure

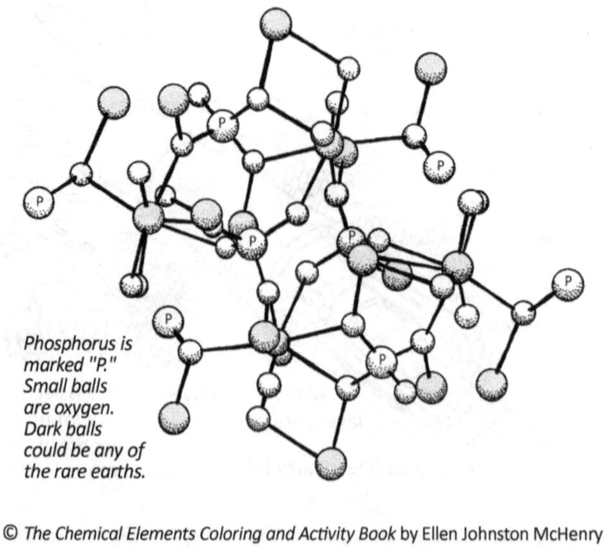

Phosphorus is marked "P." Small balls are oxygen. Dark balls could be any of the rare earths.

69 Tm Thulium

THULE is an old name for SCANDINAVIA.

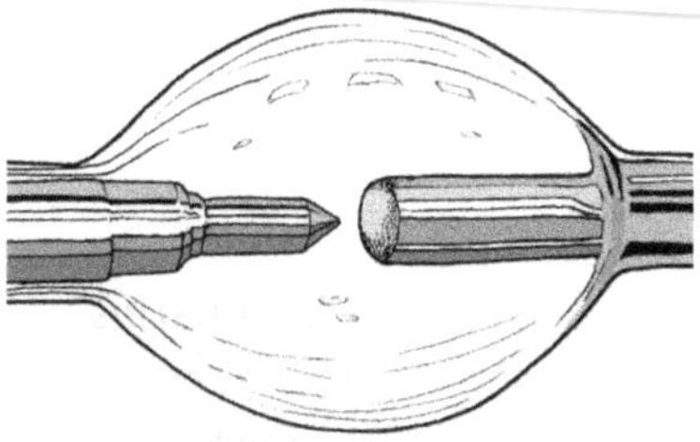

Thulium is used in arc light bulbs because it can produce light in the green part of the spectrum, unlike most other metals suitable for arc lights.

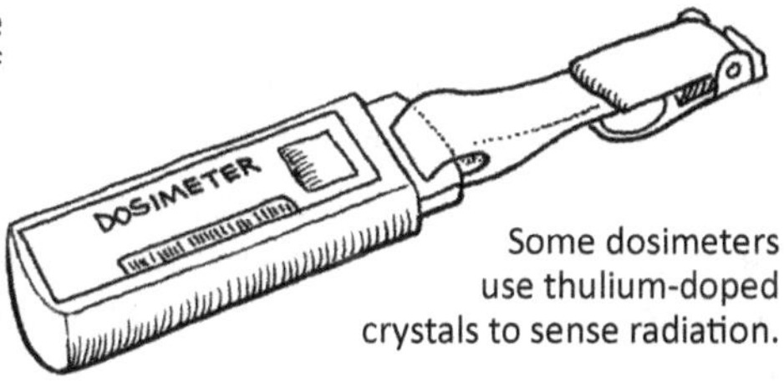

Some dosimeters use thulium-doped crystals to sense radiation.

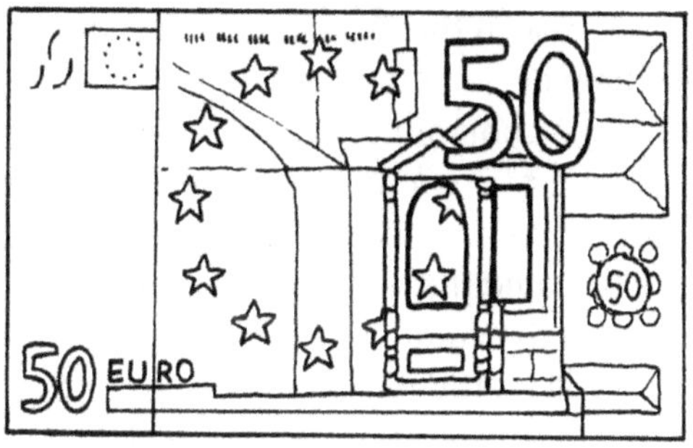

Thulium is found in euro bank notes. It causes them to glow blue under ultraviolet light.

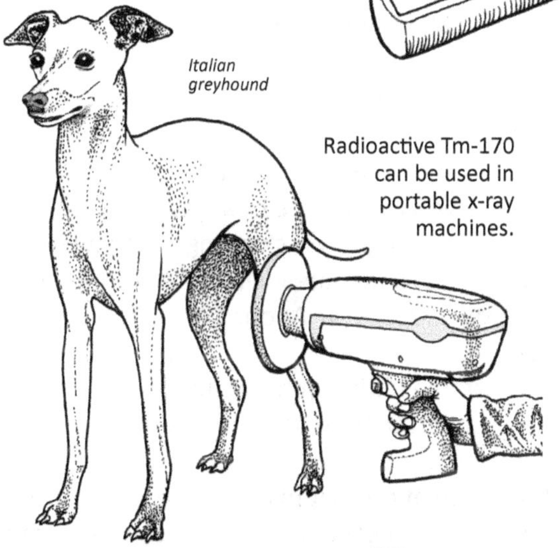

Radioactive Tm-170 can be used in portable x-ray machines.

Like the other rare earth elements, one of the minerals in which Tm can be found is monazite. This large crystal is light brown. However, most monazite occurs as sand.

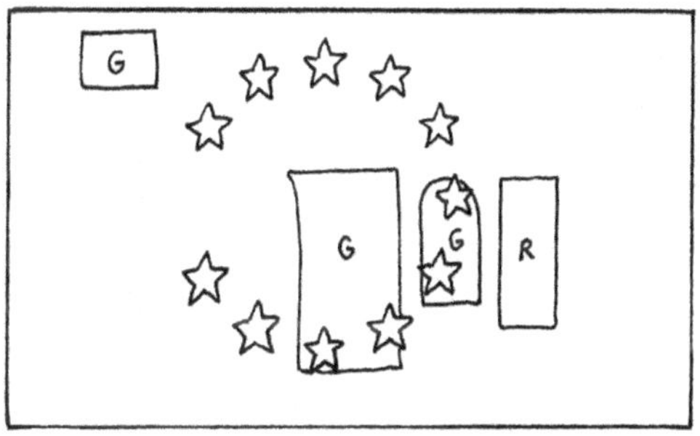

G stands for green, R stands for red. The 6 lower stars are green, the 5 upper stars are red. The background is blue.

Lasers that use thulium-doped crystals are perfect for medical procedures that involve skin.

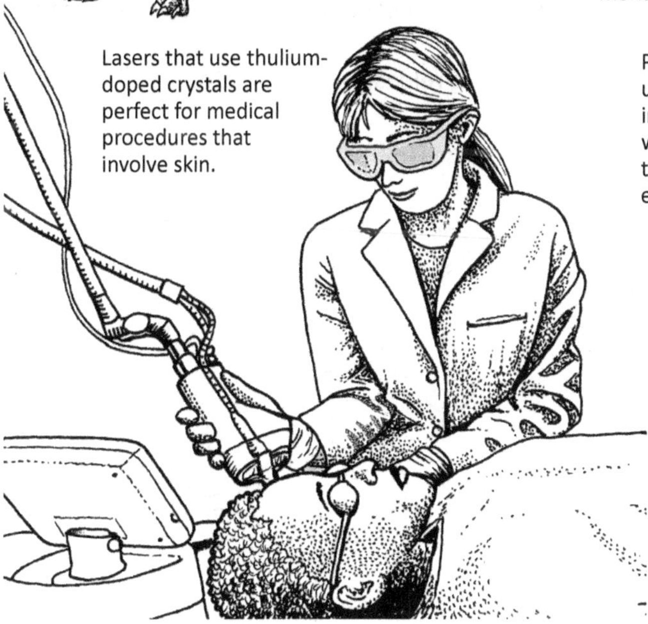

Portable x-ray units are also used by non-medical industries. Here, workers use it to inspect equipment.

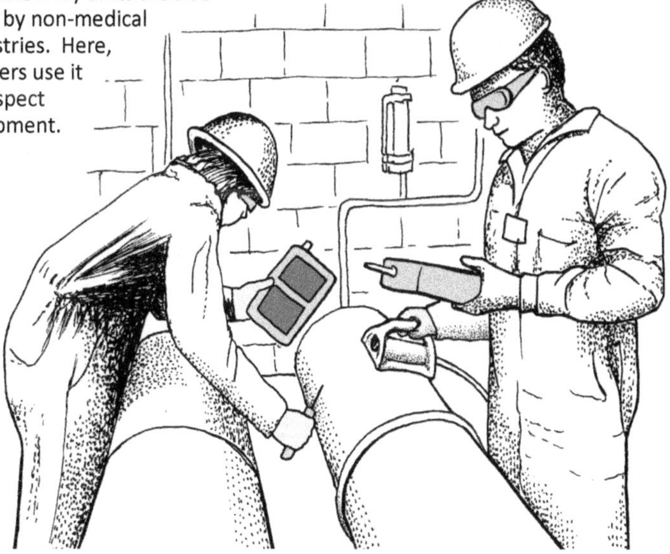

Yb 70

protons
103 neutrons
70 electrons

Atomic mass: 173

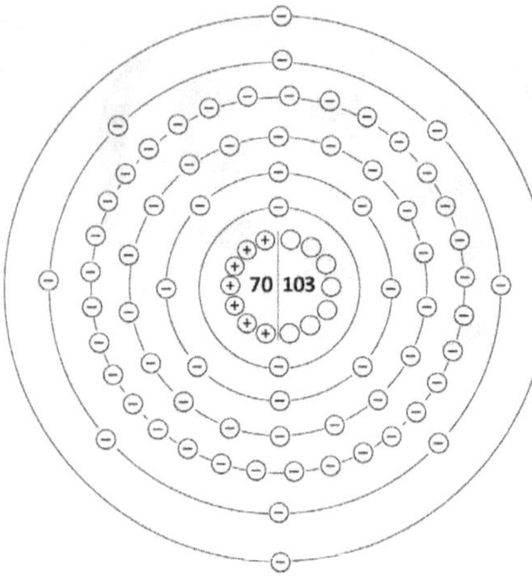

Ytterbium

Named after Ytterby, Sweden

Ytterbium *(i-TER-bee-um)* was discovered several times. The first time was by Swiss chemist Jean Charles Galissard de Marignac in 1878. He was working with a sample of erbia, Er_2O_3. He suspected that there was something else in the sample, and managed to isolate its oxide form, naming it ytterbia. In 1907, French chemist Georges Urbain decided to take a second look at ytterbia and to his surprise, his chemical tests revealed that ytterbia wasn't a single compound, but a combination of two, which he called neoytterbia and lutecia. (Lutetia is the old Roman name for Paris.) What Urbain didn't know was that two other chemists were working on the same experiments at that same time. One of these chemists, Carl von Welsbach (who discovered neodymium and praseodymium) wanted to name the new elements aldebaranium and cassiopeium. An international committee finally decided to give Urbain the credit for discovery, but also to honor Marignac's work, and they officially named the elements ytterbium and lutetium. (Urbain was one of the three people on the committee. Was that cheating?)

The largest use of ytterbium is in solid state lasers. Yb is "doped" into a crystal (often YAG, yttrium aluminum garnet) to make the crystal better for producing a certain wavelength of light. These lasers are able to do extremely precise cutting, and can make "cuts" that are so shallow that only microscopic amounts of material are removed. This is ideal for cleaning very old paintings, stone monuments, or even the surface of airplanes. These lasers can also be used in dentistry, for medical surgery, and for engraving letters or patterns on rings and jewelry.

Ytterbium atoms are being used to make new and improved atomic clocks that are even more accurate than cesium clocks. The Yb atoms must be cooled down to almost absolute zero (-273.15° C). About 10,000 Yb atoms are trapped by a laser beam while another beam is passed through them.

Radioactive Yb atoms can be used as a source of gamma rays in portable radiography testing units.

Military researchers are experimenting with using ytterbium as a substitute for magnesium compounds in decoy flares. Military airplanes can shoot decoys to confuse heat-seeking missiles that are trying to track the plane by using infrared sensors. Ytterbium is able to produce even more infrared radiation so it might be a better decoy.

Ytterbium's electrical properties change when subjected to high levels of mechanical stress, so can be used in machines that monitor earthquakes and large explosions.

YbF_3 can be used for dental fillings. It has the added benefit of slowing releasing fluorine atoms which can be absorbed by teeth, making them stronger and healthier.

Yb_2O_3 Ytterbium (III) oxide

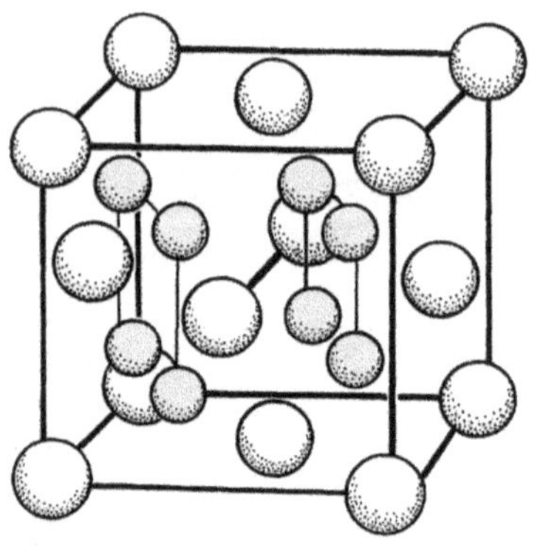

The darker balls are Dy. The lighter balls are O.

YbF_3 Ytterbium fluoride

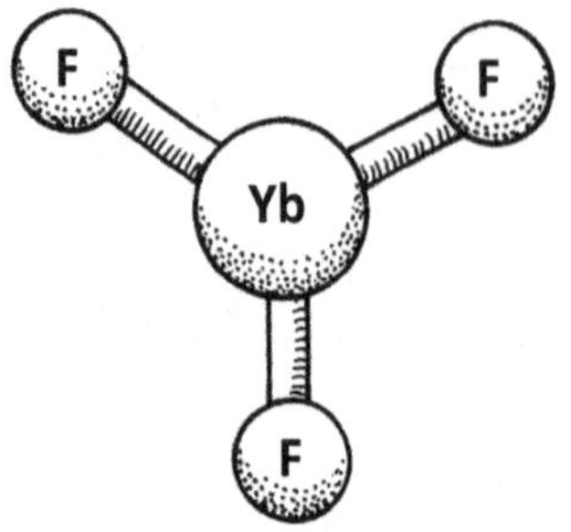

YbO Ytterbium (II) oxide

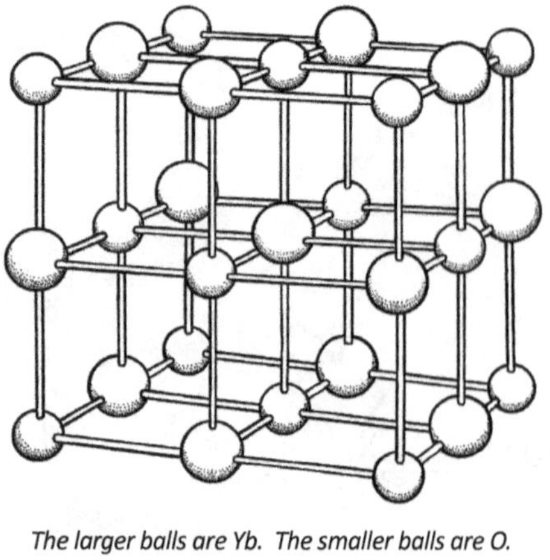

The larger balls are Yb. The smaller balls are O.

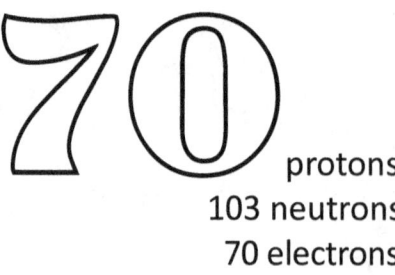

70 Yb

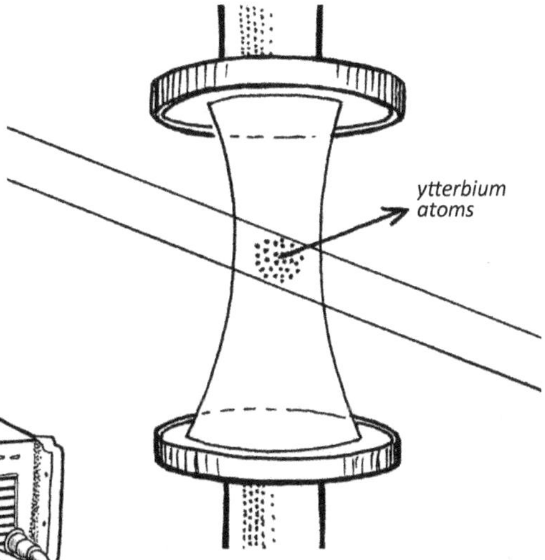

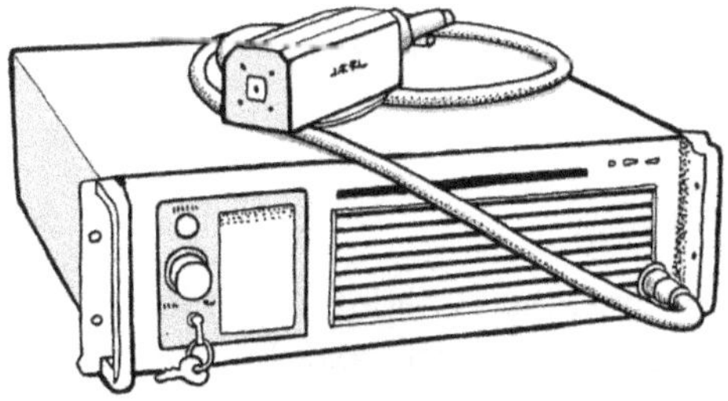

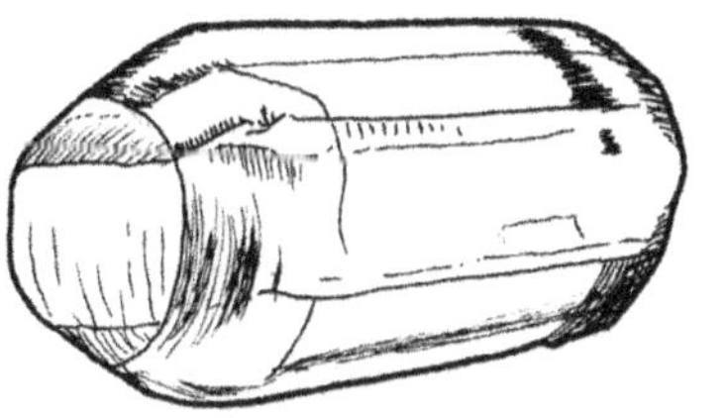

The newest atomic clock uses atoms of ytterbium suspended (trapped) in a laser beam.

Yb-doped YAG lasers are used in industries for engraving, cutting, welding, or surface cleaning (ablation).

Ytterbium

Yb: CALGO Ytterbium-doped Calcium Aluminum Gadolinium Oxide crystal

Ytterbium might replace magnesium to make burning decoys that produce intense heat.

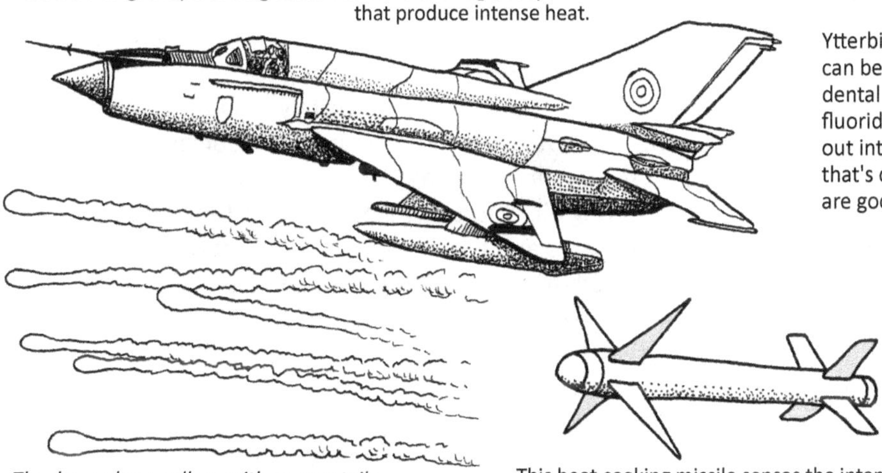

The decoys burn yellow, with orange tails.

Ytterbium fluoride can be used for dental fillings. If fluoride ions leak out into the tooth that's okay, they are good for teeth!

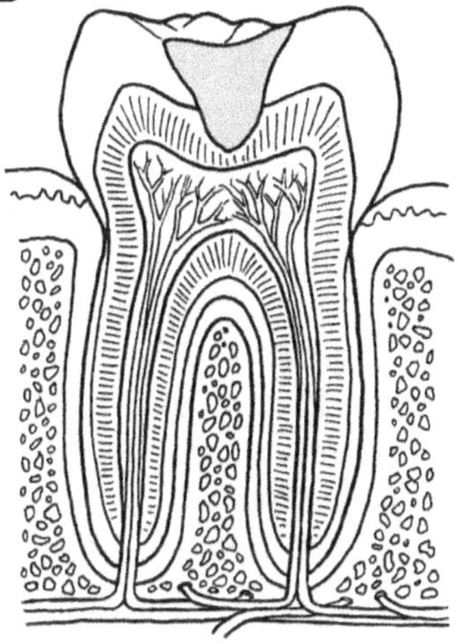

This heat-seeking missile senses the intense infrared radiation of the decoys and is drawn off course, away from the airplane.

Yb-doped YAG (yttrium aluminum garnet) lasers can be used to clean old paintings because their precision allows for the removal of just dirt, not paint.

Radioactive ytterbium is useful as a source of gamma rays for portable radiography units used to inspect equipment.

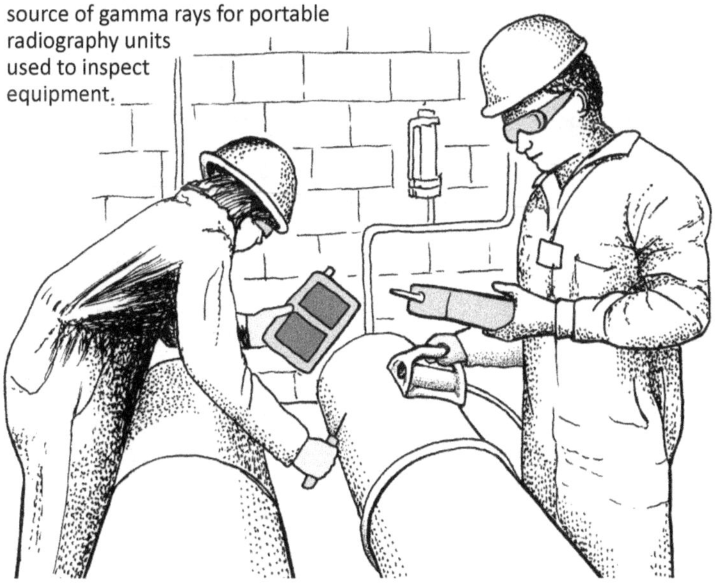

Lu 71

Lutetium

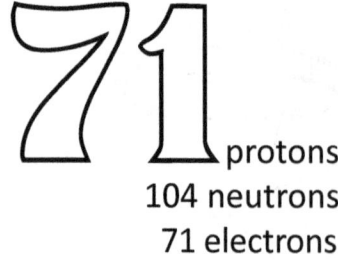

71 protons
104 neutrons
71 electrons

Atomic mass: 174.9

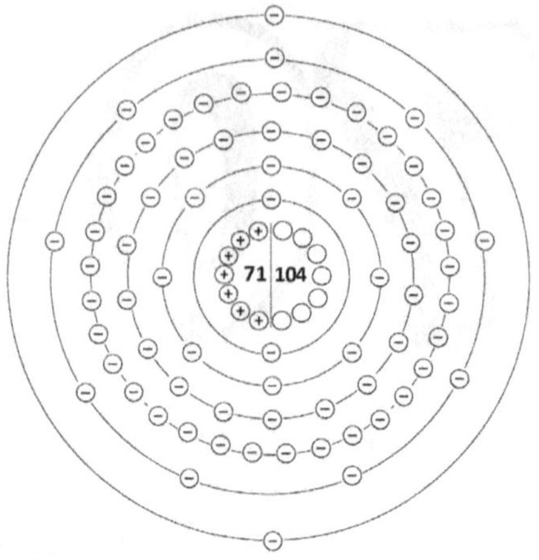

Named after Paris, using its old Latin name, Lutetia

The story of lutetium started with the discovery of ytterbium by Swiss chemist Jean Charles Galissard de Marignac in 1878. Several other chemists were also working with the ytterbium ore at this time, and they found that "ytterbium" was actually a mixture of two new elements. An argument arose as to which of these scientists had been the first to discover these new elements. Georges Urbain wanted to name the elements neoytterbium and "lutecium" (using the old Latin name for Paris, Lutetia). Carl von Welsbach wanted to name them after stars and chose the names aldebaranium and cassiopeium. An international committee (which happened to include Urbain) had to settle the dispute, and in 1909 the names ytterbium and lutetium were finally chosen.

Like the other rare earth elements in the lanthanide series, lutetium is found primarily in monazite sand. The process of extraction is difficult and involves the use of many chemicals, so the rare earth elements tend to be expensive. However, for most applications, only small amounts of these elements are needed, so a little goes a long way.

Lutetium is an ingredient in alloys that are used to make metal plates that act as catalysts in the refining of petroleum. Crude oil (petroleum as it comes out of the ground) is made of very long chains of carbon atoms with hydrogens attached. The chains are broken apart into smaller chains of various lengths. Longer chains become waxes. Medium-sized chains are turned into diesel fuel, jet fuel, and heating oil. Shorter chains with 8 to 10 carbons become gasoline (petrol) for cars. Very short chains of only a few carbon atoms are used by factories that make plastics. The smallest molecule produced is natural gas, CH_4. The role of the catalyst is to cause the chains to break into smaller pieces. Lutetium alloys are also used in catalysts that do the reverse—put small molecules together to make larger ones. This process is used to make plastics and other polymers.

Lutetium tantalate ($LuTaO_4$) is both white and dense and is ideal for making the "film" for x-ray machines; it has phosphors (usually containing other rare earth elements) added to it that will glow when exposed to x-rays. Another use for lutetium in medical imagining is in positron emission tomography (PET) scans where cerium-doped lutetium oxyorthosilicate (LSO) is used to detect positrons.

Radioactive Lu-177 is used in a medicine that treats tumors found in glands and nerves.

Lutetium aluminum garnet (LuAG) is used in solid state lasers, in optical fibers and switches, and in infrared sending devices such as heat seeking missiles and night vision equipment.

$LuSiO_4$ Lutetium oxyorthosilicate A crystal that can be doped with cerium and used for detection of positrons.

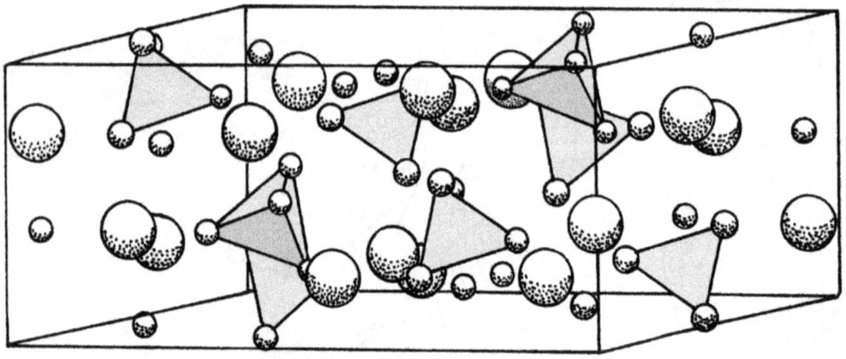

The large balls are Lu, lutetium. The small balls are O, oxygen. You can't see the silicon atoms because they are inside the triangle shapes.

LuO_8 Lutetium oxide

Created in a lab using Lu and H_2O molecules.

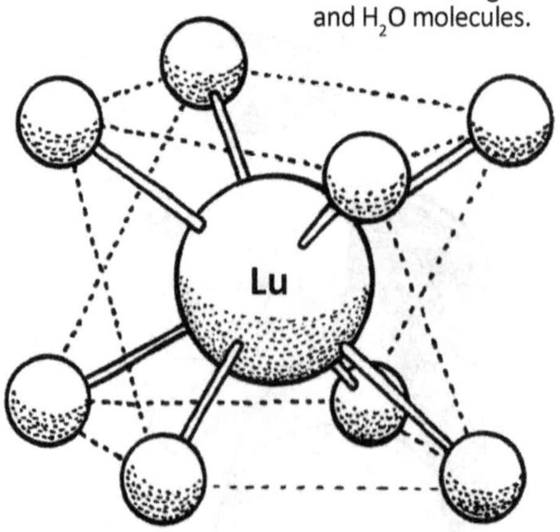

All the unmarked balls are O, oxygen.

71 Lu

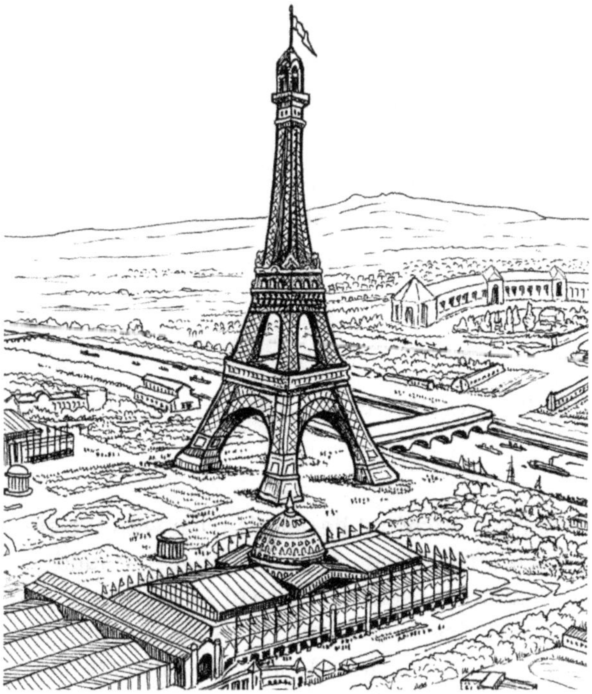

The Eiffel Tower was built in 1889 for the World's Fair in Paris. This drawing shows the tower and its surroundings in 1890.

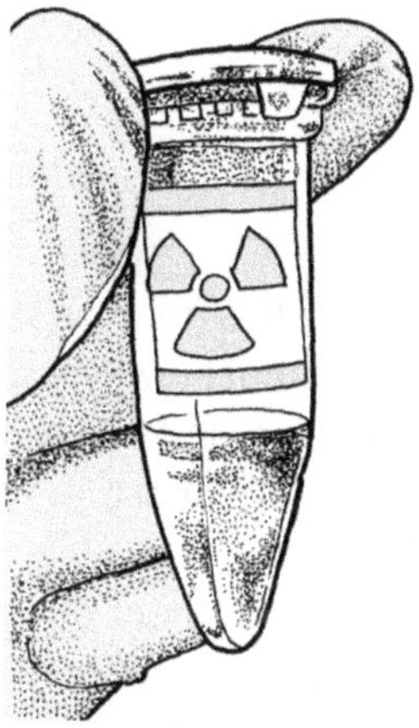

Lu-177 is used in radioactive medicines.

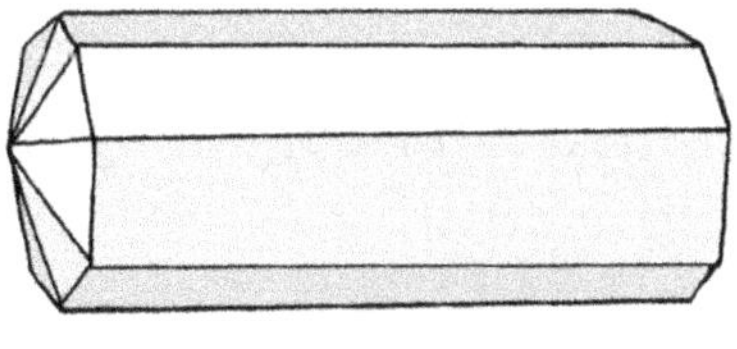

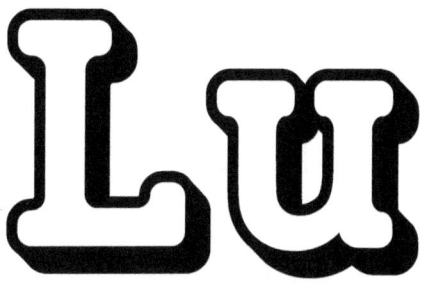

Lutetium aluminum garnet ($Al_5Lu_3O_{12}$), or LuAG, is used in lasers, optical switches , optical fibers, and infrared sensing devices

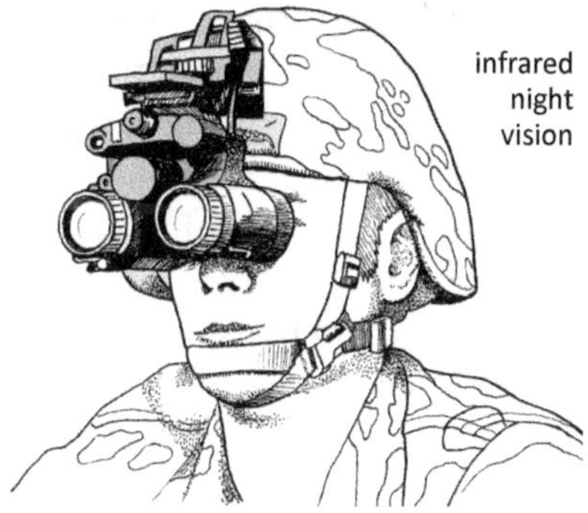

infrared night vision

Lutetium

Monazite crystals are a major source of lutetium. It is difficult to separate lutetium from ytterbium.

Lutetium is used in alloys that function as catalysts for petroleum "cracking." Long chains of carbon atoms are broken into shorter chains, most of which are used as fuel.

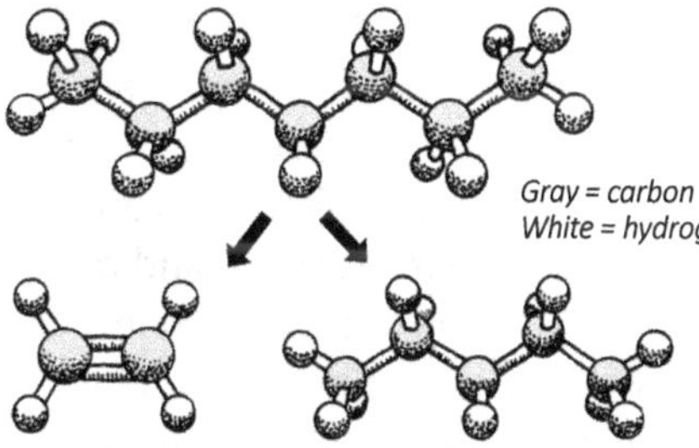

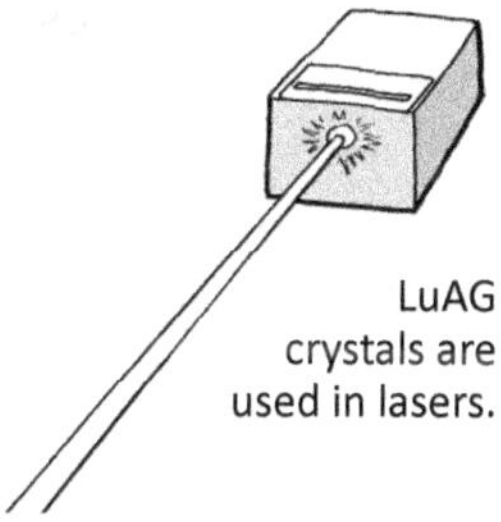

LuAG crystals are used in lasers.

$LuTaO_4$ is useful as a "host" material for phosphors that glow when exposed to x-rays.

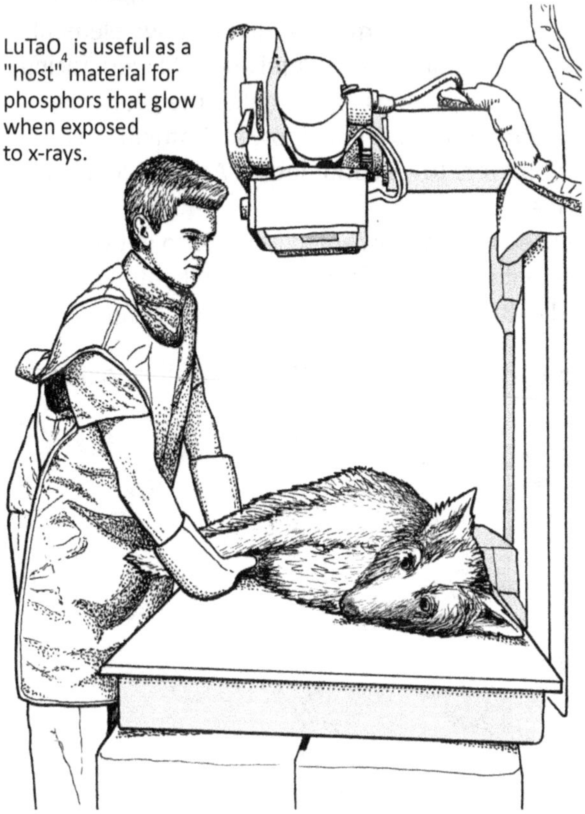

Cerium-doped lutetium orthosilicate is used to detect positrons (positively charged electrons) in PET scanners.

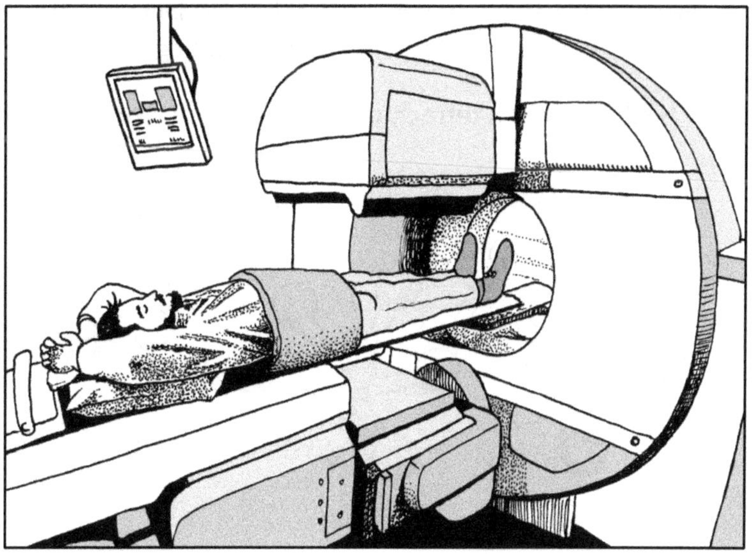

Hf 72

protons
106 neutrons
72 electrons

Atomic mass: 178.4

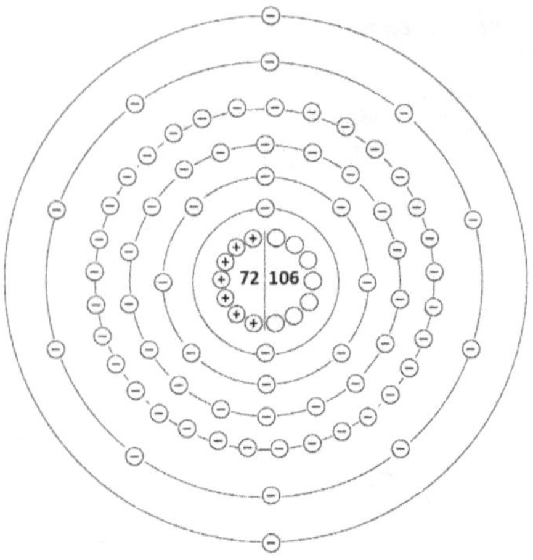

Hafnium

Named after Copenhagen, using its old name, Hafnia

Even people who have studied chemistry may not be very familiar with the element hafnium. It is not considered to be a rare earth element, but is often found in the same places around the world. It is almost always found inside zircon crystals, $ZrSiO_4$. The zircons are usually the size of sand grains and are part of a mix called heavy mineral sand. Other elements found in heavy mineral sands are tungsten, titanium, thorium and iron. The best deposits of this ore are found in Brazil, Malawi, and western Australia, particularly in an area called Mt. Weld.

Hafnium sits right under zirconium in the Periodic Table, which means they have the same arrangement of electrons in their outer shell and therefore are chemically very similar. Hafnium can take zirconium's place in any molecule. Up to 5% of the $ZrSiO_4$ molecules in a zircon can be $HfSiO_4$. One difference between the two elements, however, is what happens when they are bombarded with fast-moving free neutrons. Hafnium "catches" the neutrons and adds them to its nucleus. Zirconium doesn't interact at all with the neutrons and they pass right through it. In a nuclear reactor, you will find hafnium in the control rods, and zirconium in the metal case around the control rods.

Dmitri Mendeleyev, inventor of the Periodic Table, predicted the discovery of an element below zirconium on the table, but this element turned out to be so difficult to find that it was not officially discovered until 1923, using x-ray spectroscopy. The place it was discovered was a lab in Copenhagen, Denmark, so it was named after this location, but using its old Latin name, Hafnia. Hafnium was the last non-radioactive element to be discovered.

Hafnium is used in alloys with iron, titanium, niobium, tantalum, and other metals. Since these alloys are extremely resistant to heat, they are used to make nozzles for rocket thrusters. The main engine of the Apollo lunar module was made of an alloy that was 10% hafnium. Hafnium carbide is the toughest material made of two elements.

In 2007, hafnium made the news as being part of new, improved microprocessors. Intel and IBM were racing to reduce the size of the "gates" through which electrons traveled in their microchips. The smaller the gates, the more transistors could be put into the microchips, and more chips meant more processing power. Hafnium oxide was part of the solution to reducing the size of the gates from 90 nanometers to 45 nanometers. As of 2020, new computers are being made with gates that are only 5 nanometers wide.

Hafnium is very good at grabbing gas molecules so it can be used in incandescent light bulbs to get rid of the few nitrogen and oxygen molecules that accidentally get into the bulbs. Hafnium's resistance to heat makes it suitable for using as an electrode in "plasma cutting," a type of metal cutting that creates extreme temperatures.

$HfCl_4$ Hafnium tetrachloride

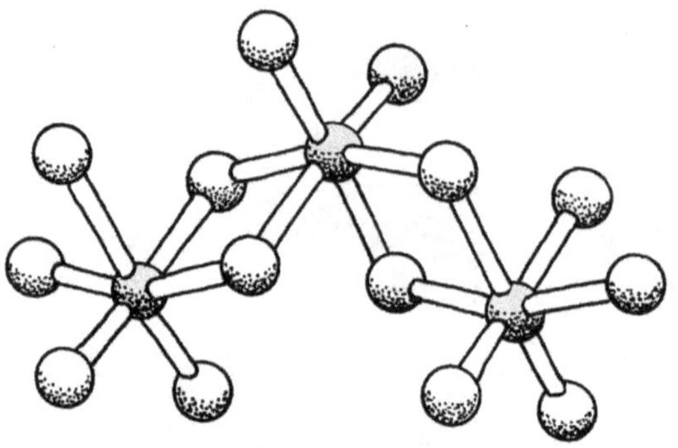

The gray balls are hafnium. The white balls are chlorine.

HfC Hafnium carbide

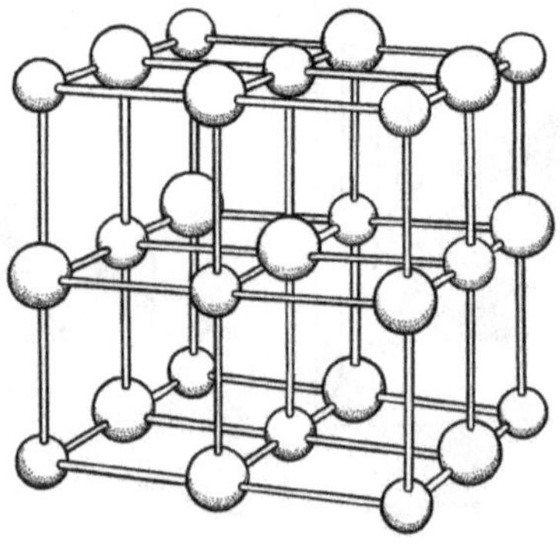

The larger balls are Hf. The smaller balls are C.

HfO_2 Hafnium oxide

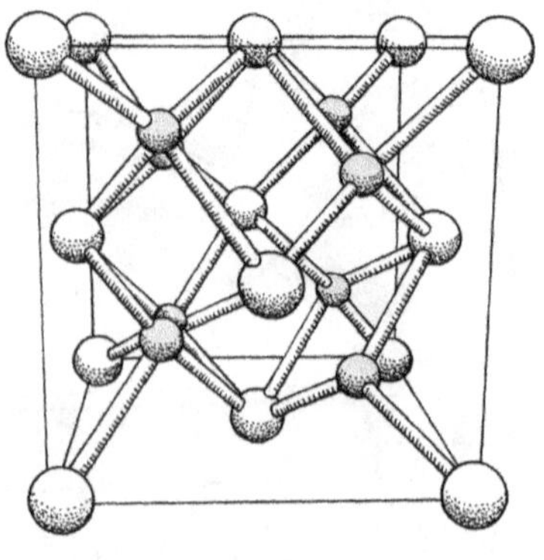

White balls are Hf. Gray balls are O.

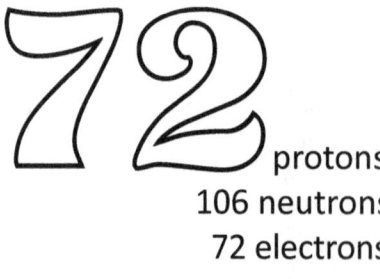

72 Hf

Hafnium was named after Copenhagen, Denmark.

Hafnium

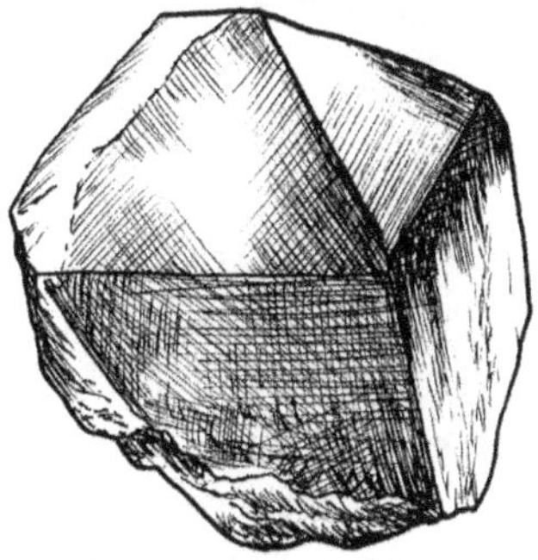

This reddish-brown crystal is a zircon ($ZrSiO_4$). Up to 5% of the zirconium atoms have been replaced with hafnium atoms.

The Rundetaarn in Copenhagen. The roof of this building was used as an observatory; the building behind it housed a large science library.

pink *green* *red*

These garnet crystals contain a small amount of Hf.

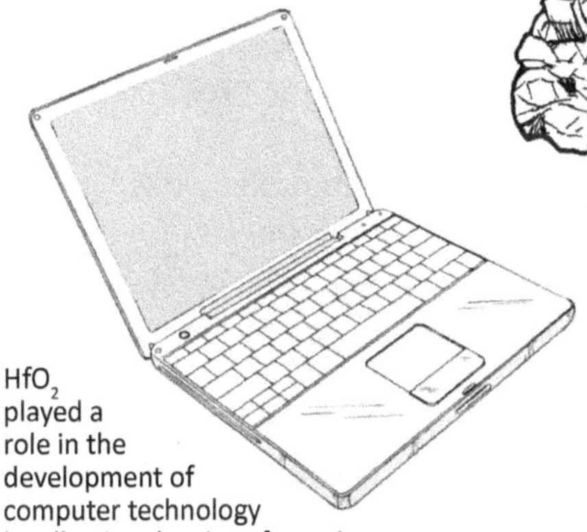

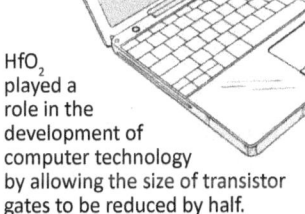

HfO_2 played a role in the development of computer technology by allowing the size of transistor gates to be reduced by half.

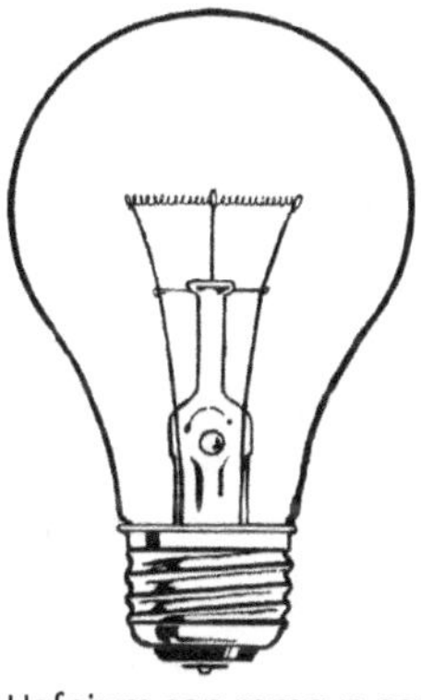

Hafnium can remove gas molecules that are not supposed to be in bulbs.

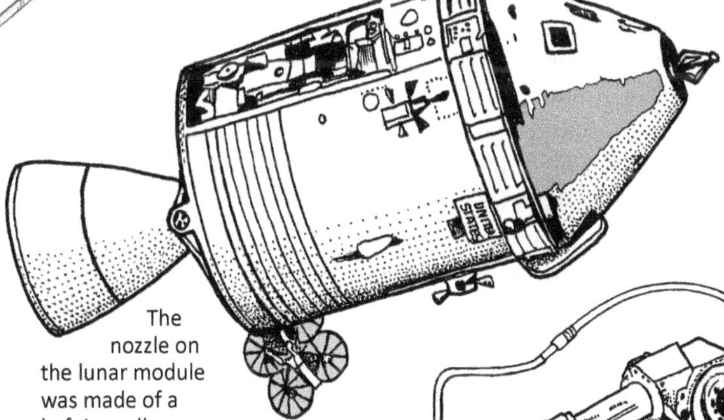

The nozzle on the lunar module was made of a hafnium alloy.

Diagrams of transistor gates look like this. The goal is to make the gap as small as possible.

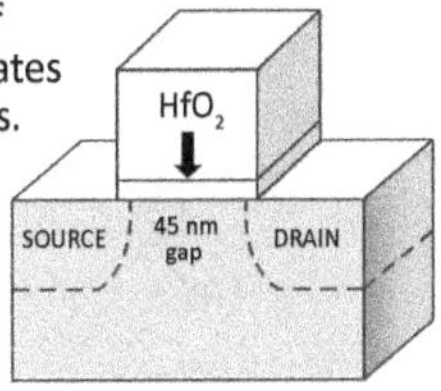

Hafnium absorbs neutrons in control rods used in nuclear reactors.

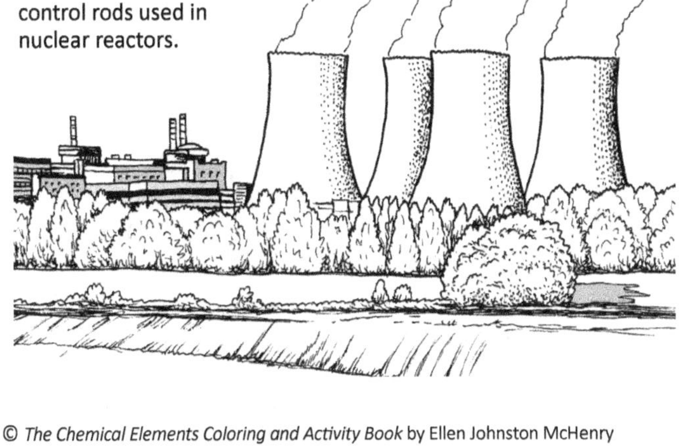

Hafnium can take a lot of heat, so it is used to make metal tips (electrodes) for plasma cutters. Showers of sparks shoot out as this robot arm uses hot plasma to cut metal.

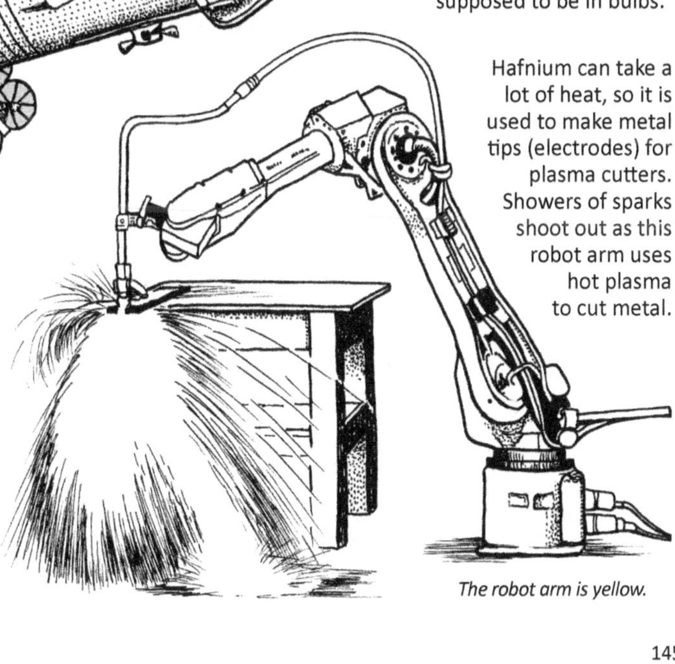

The robot arm is yellow.

Ta 73

protons
108 neutrons
73 electrons

Tantalum

Atomic mass: 180.9

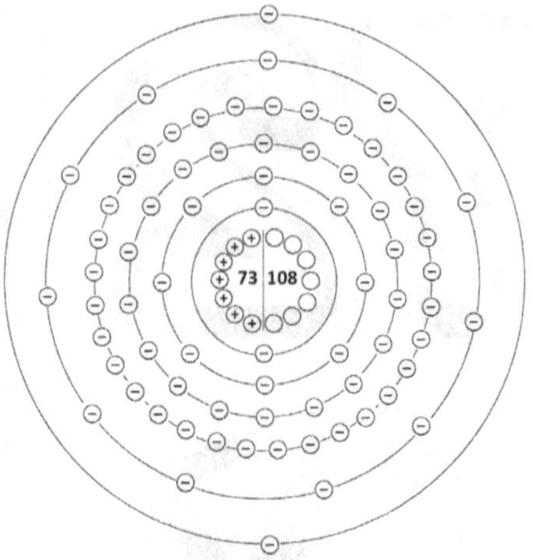
73
108

Named after Tantalus, from Greek mythology

Tantalum was discovered in 1802 by Anders Ekeberg, in mineral samples from Sweden and Finland. Since tantalum sits right below niobium on the Periodic Table, they have many similar chemical properties, and this led to quite a bit of confusion. Chemists were not absolutely sure about the identity of these elements until the 1860s. The name "tantalum" had been chosen by Ekeberg because some of this metal's chemical properties reminded him of the story of Tantalus in Greek mythology. Tantalus is remembered for the punishment he received from Zeus for his crime of stealing food from the divine banquet table. Zeus ordered that Tantalus be eternally condemned to standing in a pool of water with ripe fruit hanging over his head. If he bent down to drink the water, it would drain away, and if he tried to pick the fruit, the branch would rise and take the fruit out of his reach. (The element niobium was named after the daughter of Tantalus, Niobe.)

The mineral ores from which tantalum is extracted occur mainly in Australia and Africa. In Central Africa, especially in the Democratic Republic of the Congo, there has been much fighting over control of the ores and many people work in smuggling operations.

Pure tantalum is a "ductile" metal, which means it can be stretched and shaped, often into thin wires. It was used in very early light bulbs, before they discovered that tungsten was even better for this purpose. Tantalum is also very resistant to corrosion so it used to make tubes and pipes for industries that process harsh chemicals. Tantalum alloys are durable even when exposed to intense heat, so they can be used to make parts for jet engines, nuclear reactors, missiles, and tanks.

The largest industrial use of tantalum is in the electronics industry, mostly in making capacitors and resistors. Tantalum is useful in capacitors primarily because it can combine with oxygen to form an oxide compound. Tantalum capacitors, which store electrical charge like a battery, are used in many electronic products including computers, tablets, and cell phones, and in the electronic parts found in cars.

Tantalum alloys can also be used for decorative purposes, such as jewelry, watches, and coins. Kazakhstan minted a coin made of silver and tantalum, with images of the Apollo-Soyuz and the International Space Station.

Some alloys of tantalum can be used to make artificial joints. The advantage of tantalum alloys is that they are not magnetic and are therefore don't pose a safety risk to patients needing an MRI scan.

$Ta(CH_3)_5$ Pentamethyltantalatum

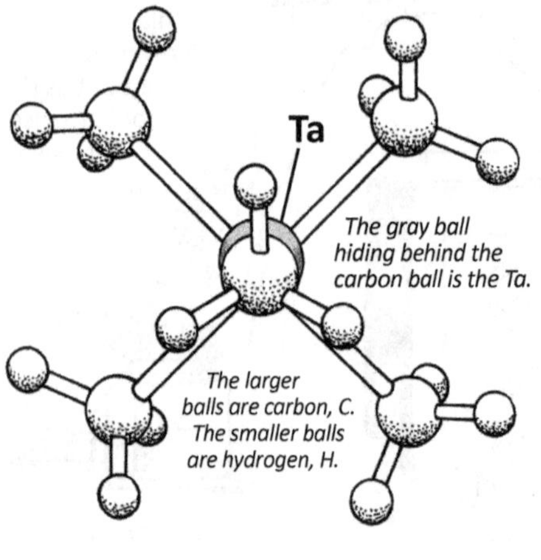
Ta
The gray ball hiding behind the carbon ball is the Ta.
The larger balls are carbon, C. The smaller balls are hydrogen, H.

TaS_2 Tantalum (IV) sulfide

a semiconductor material

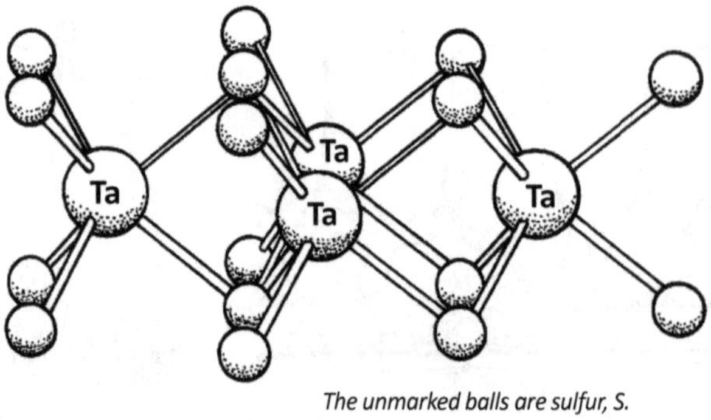
Ta
Ta
Ta
Ta

The unmarked balls are sulfur, S.

$LiTaO_3$ Lithium tantalate

a material with useful optical properties

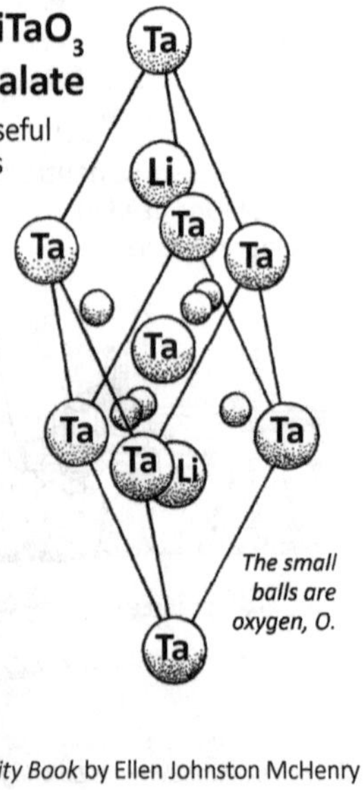
Ta
Li
Ta
Ta
Ta
Ta
Ta
Ta
Ta
Li
Ta
The small balls are oxygen, O.

73 Ta

In Greek mythology, Tantalus was punished by Zeus by being tempted with fruit that he could not reach.

Tantalum

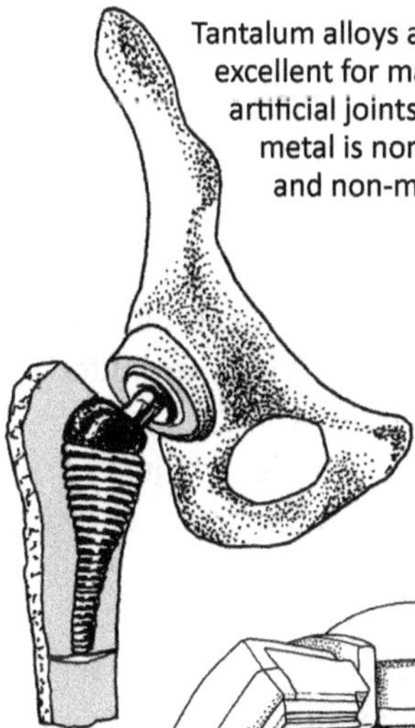

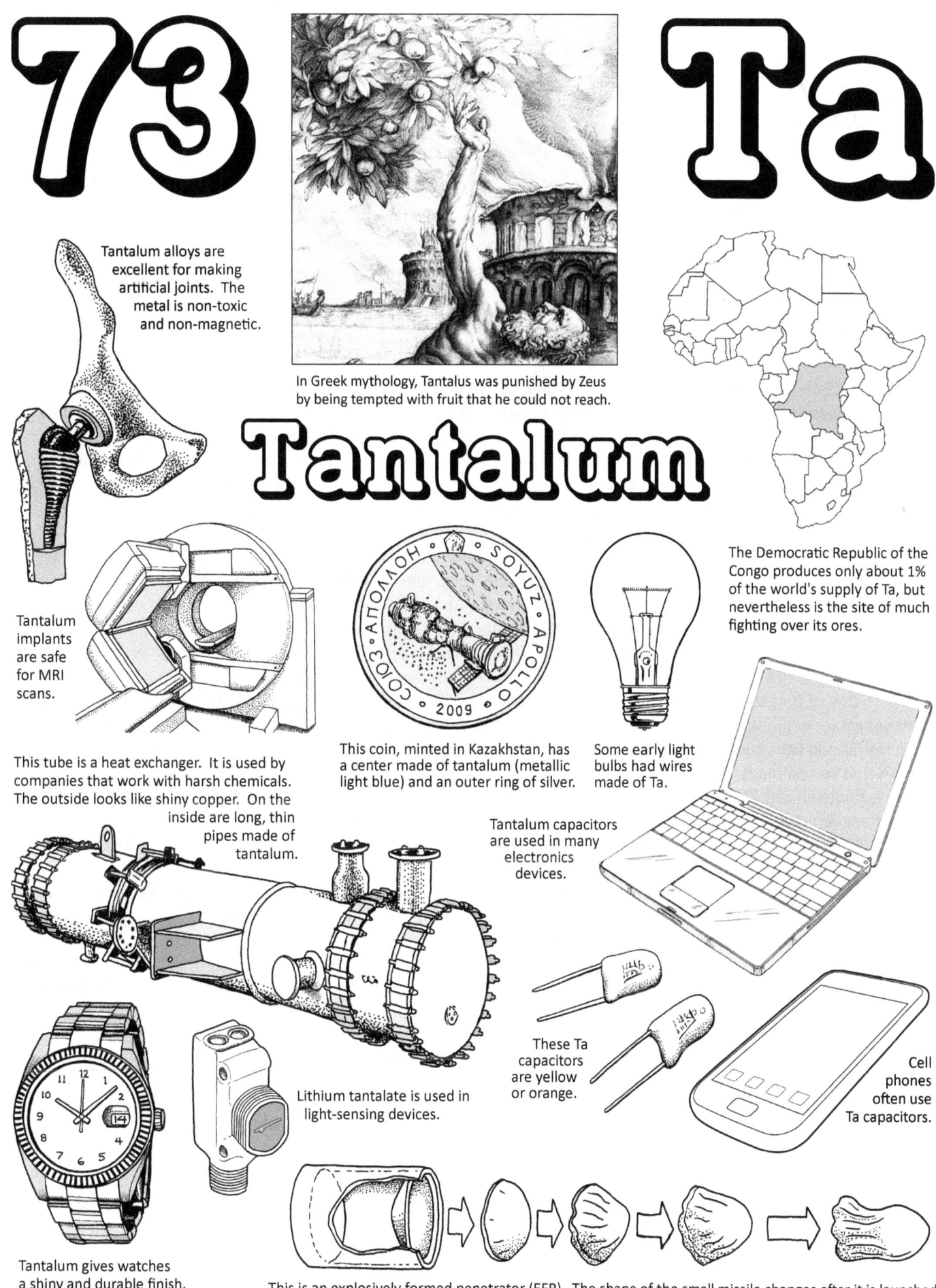

Tantalum alloys are excellent for making artificial joints. The metal is non-toxic and non-magnetic.

The Democratic Republic of the Congo produces only about 1% of the world's supply of Ta, but nevertheless is the site of much fighting over its ores.

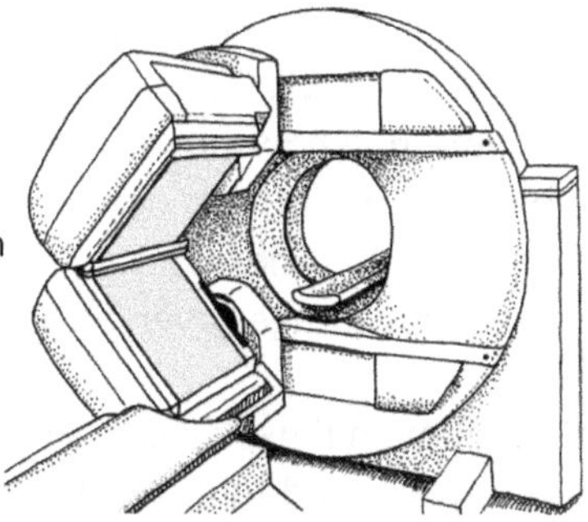

Tantalum implants are safe for MRI scans.

This coin, minted in Kazakhstan, has a center made of tantalum (metallic light blue) and an outer ring of silver.

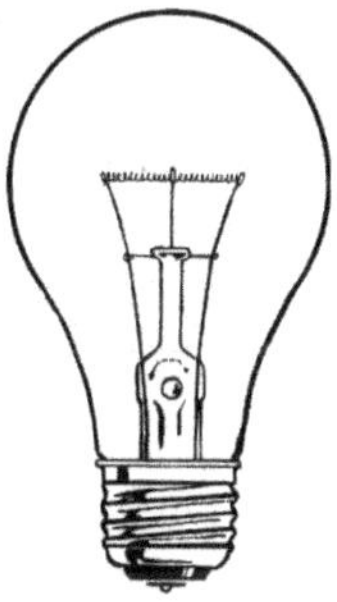

Some early light bulbs had wires made of Ta.

This tube is a heat exchanger. It is used by companies that work with harsh chemicals. The outside looks like shiny copper. On the inside are long, thin pipes made of tantalum.

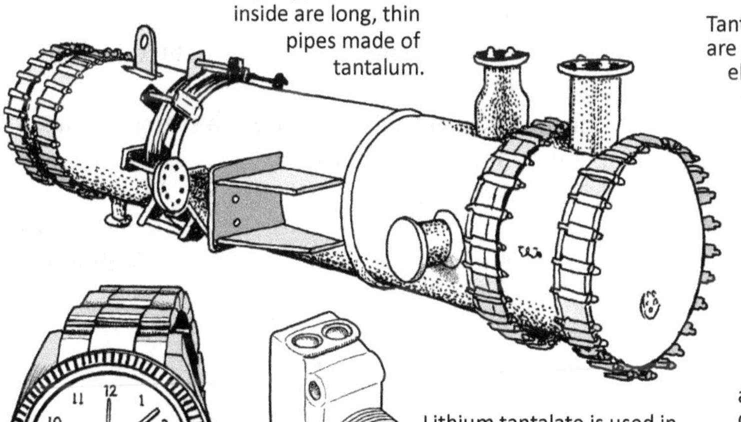

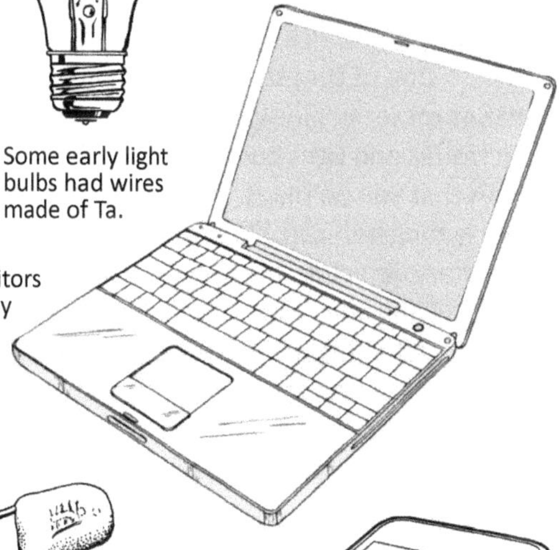

Tantalum capacitors are used in many electronics devices.

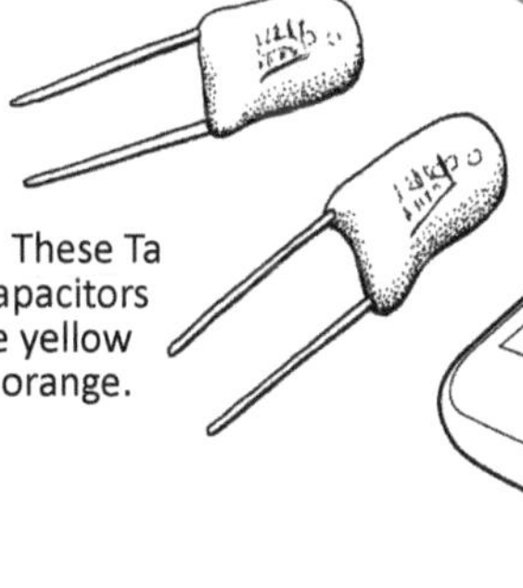

These Ta capacitors are yellow or orange.

Cell phones often use Ta capacitors.

Tantalum gives watches a shiny and durable finish.

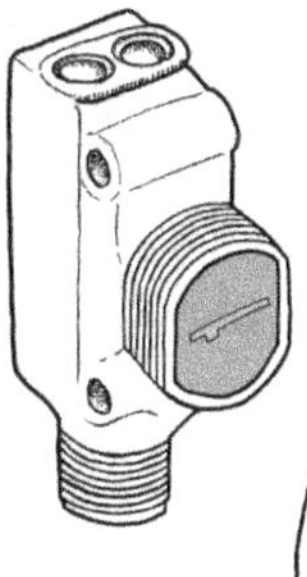

Lithium tantalate is used in light-sensing devices.

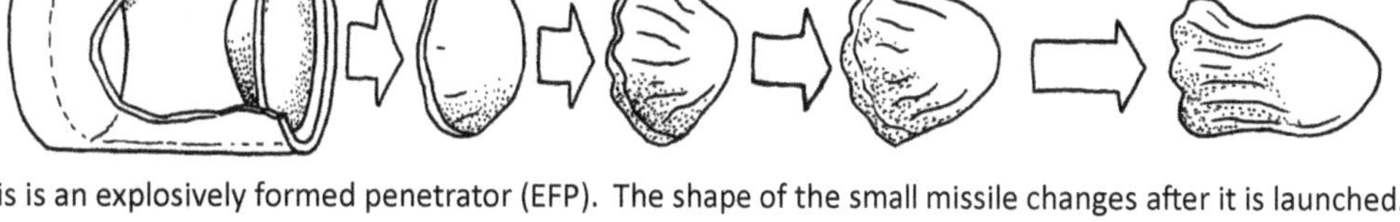

This is an explosively formed penetrator (EFP). The shape of the small missile changes after it is launched.

W 74

Tungsten

protons
110 neutrons
74 electrons

Atomic mass: 183.8

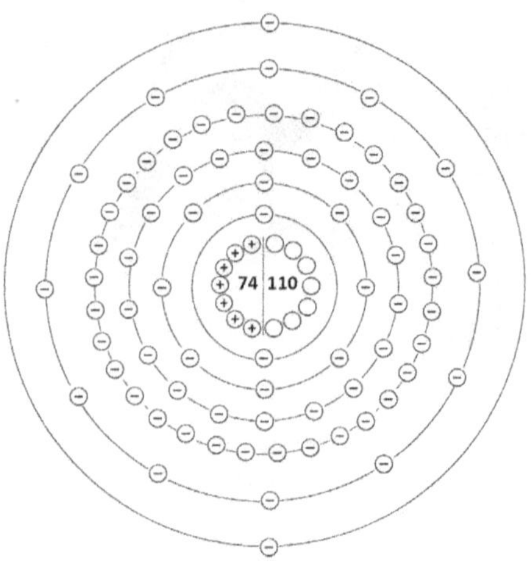

Tungsten means "heavy stone" in Swedish.
The symbol, W, is for "wolfram," another name for this element.

In 1546, Georg Agricola was experimenting with a mineral ore he called "lupi spuma" which is Latin for "wolf froth." He was amazed at how much tin was consumed in the process of trying to extract this ore, and it reminded him of a wolf's large appetite. This name was translated into German in 1747, and became "wolf rahm."

Wolframite is a dark gray or black mineral and is very heavy. It is so heavy that it can replace lead in just about any application. Since lead has been discovered to be toxic, using non-toxic tungsten is often the better choice. Tungsten has replaced depleted (used) uranium as counterweights in ships, airplanes, helicopters and race cars. Tungsten can also be used as a substitute for molybdenum, because it sits right below it on the Periodic Table. Elements that occur in the same column will share many chemical properties because they will have the same arrangement of electrons in their outer shell. For example, tungsten disulfide, WS_2, and molybdenum disulfide, MoS_2, can be used almost interchangeably. Both are used as dry lubricants and for removal of sulfur from petroleum. Recently, it has been discovered that both of these compounds can form sheets of molecular hexagons in the same way that carbon forms sheets of graphene. Sheets of WS_2 and MoS_2 may have safety advantages over graphene, while being just as useful in the world of electronics and nanoscience.

One of tungsten's outstanding qualities is its high boiling point. It won't melt even if exposed to the intense heat of rocket engines, so it can be used for engine nozzles. Other hot places where you'll find tungsten are welding electrodes and light bulb filaments. Tungsten is also very tough and resistant to wear and corrosion, so it is used in alloys that will be made into almost any metal part you can think of that has to be tough. The most famous tungsten alloy is tungsten carbide, used for drill bits, saw blades, knives, and other cutting tools. Tungsten is also used widely in military vehicles and weapons, including grenades and anti-tank penetrators.

Tungsten oxides are used for coating ceramics, inside fluorescent lighting, and as radiation detectors. The Voyager satellites had cosmic (gamma) ray detectors that used tungsten in their sensors.

A lesser-known use of tungsten is in C strings for cellos. These strings are said to produce a stronger tone.

Tungsten has a density that is similar enough to gold that counterfeiters have tried to make fake gold bars by putting an outer layer of gold over an inner core of tungsten.

WS_2 Tungsten disulfide

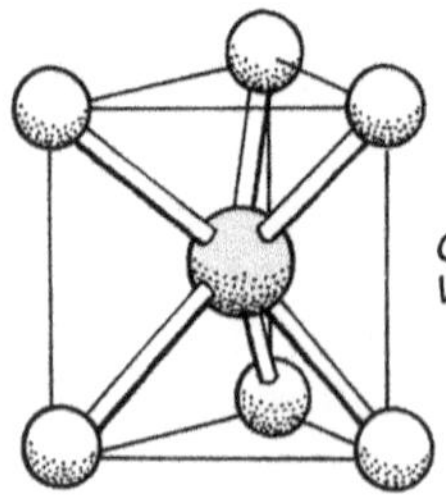

Gray balls are W. White balls are S.

Hexagonal sheet of WS_2

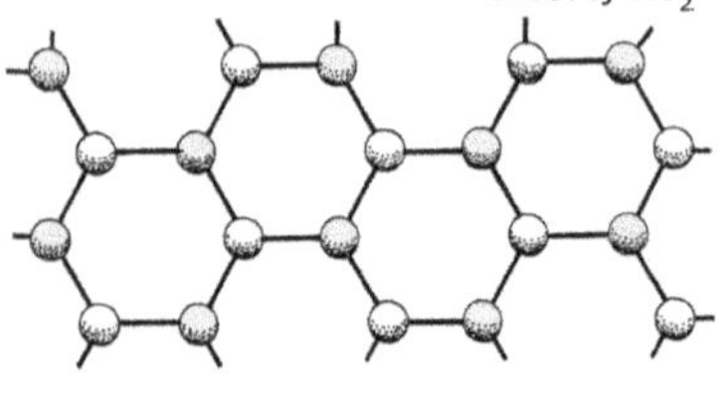

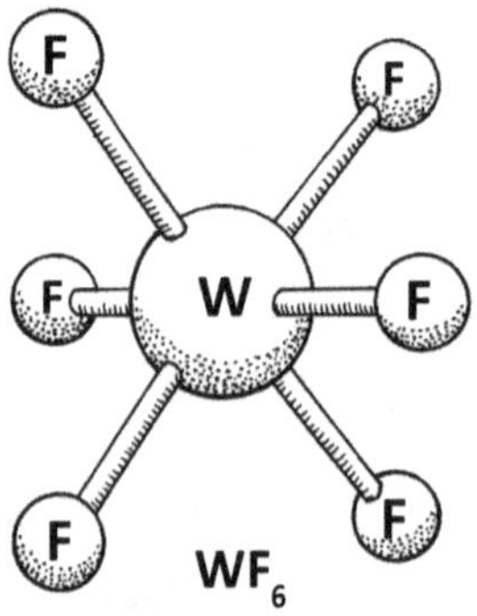

WF_6
Tungsten hexafluoride

This is a colorless, toxic glass.

WO_3
Tungsten (VI) oxide

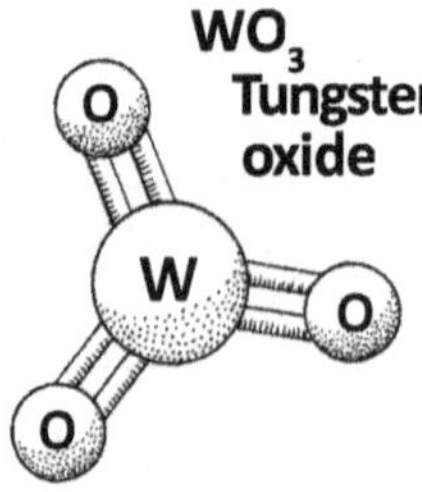

Gray is W. White is Cl.

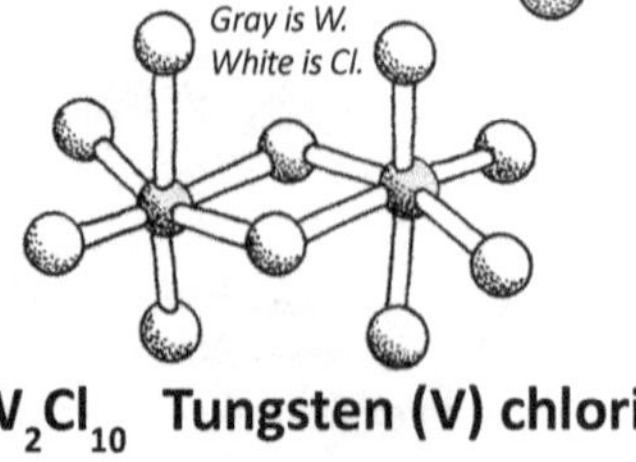

W_2Cl_{10} Tungsten (V) chloride

W_6Cl_{18}
Tungsten (VI) trichloride

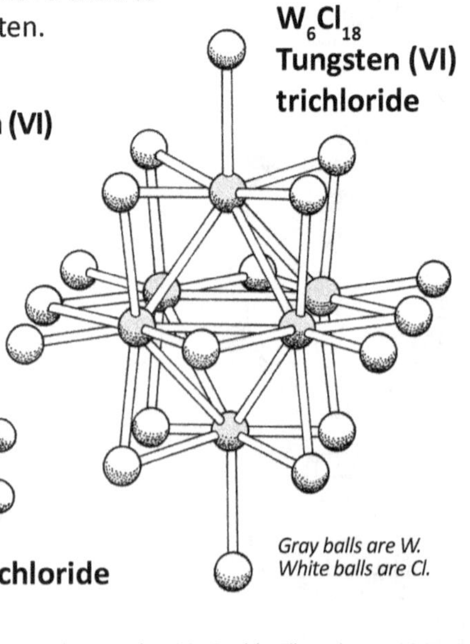

Gray balls are W. White balls are Cl.

74 W

Tungsten

In many parts of the world tungsten is still known by its old name, wolfram, derived from the name of its mineral ore, wolframite. Georg Agricola gave this name to the ore in 1546. He though the ore acted like a wolf, as it "ate" much tin during the extraction process.

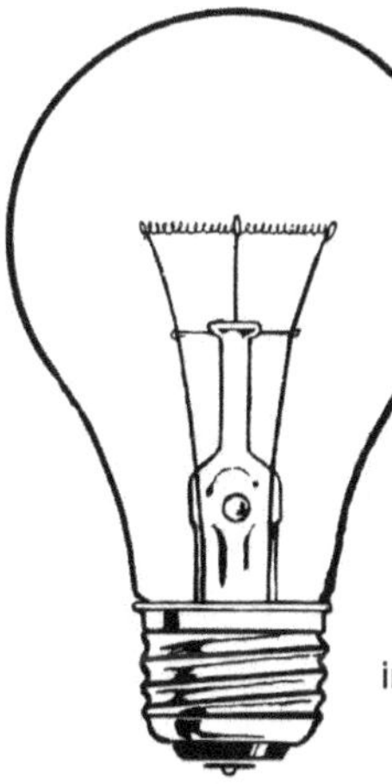

Tungsten wire is the glowing filament in incandescent light bulbs.

Tungsten carbide drill bits can cut into other metals.

Wolframite is dark gray.

Tungsten carbide is used to make saw blades.

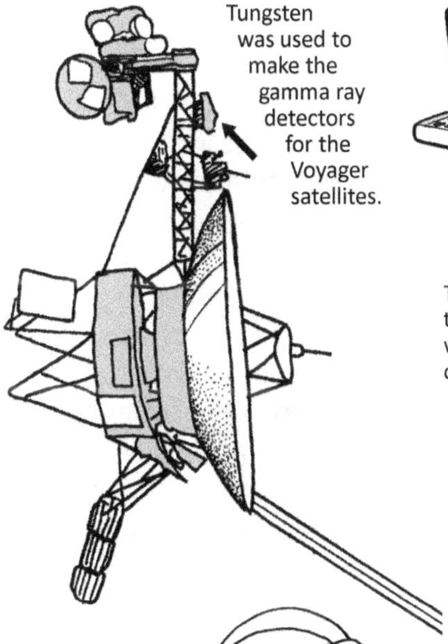

Tungsten was used to make the gamma ray detectors for the Voyager satellites.

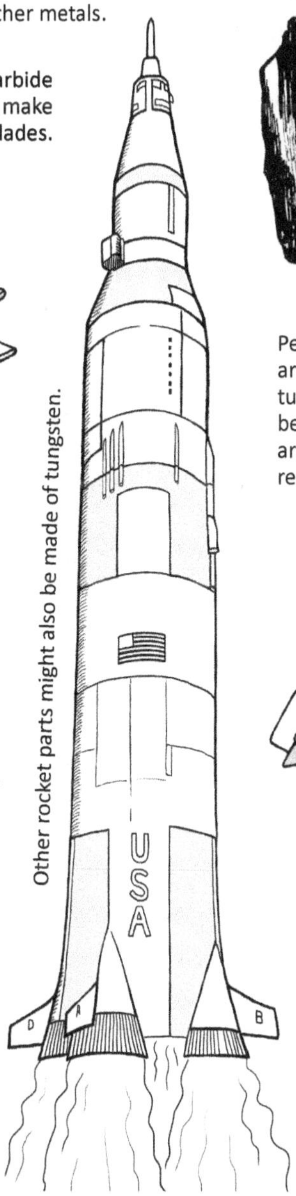

Other rocket parts might also be made of tungsten.

Tungsten has a very high melting point. W alloys are used to make rocket nozzles.

These bars are tungsten painted with a layer of gold.

Tungsten has been used to counterfeit gold because their densities are so similar.

Penetrating weapons are often made of tungsten alloys because they are so heat resistant.

This anti-tank penetrator is red and black.

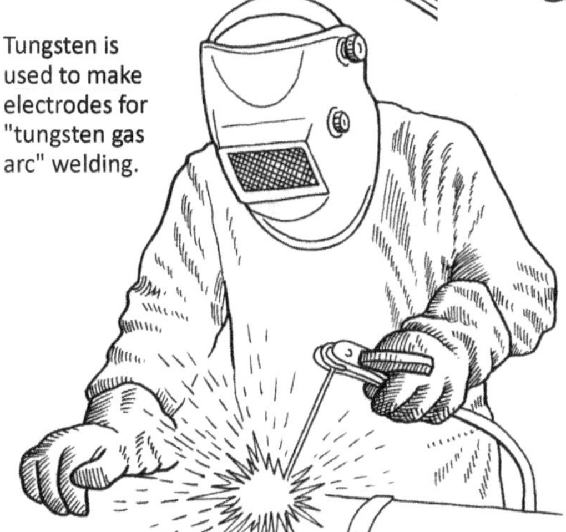

Tungsten is used to make electrodes for "tungsten gas arc" welding.

Tungsten disulfide is a dry lubricant.

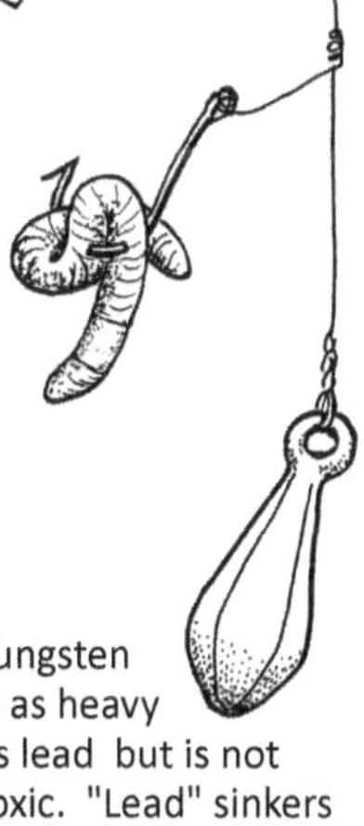

Tungsten is as heavy as lead but is not toxic. "Lead" sinkers are often made of a tungsten alloy.

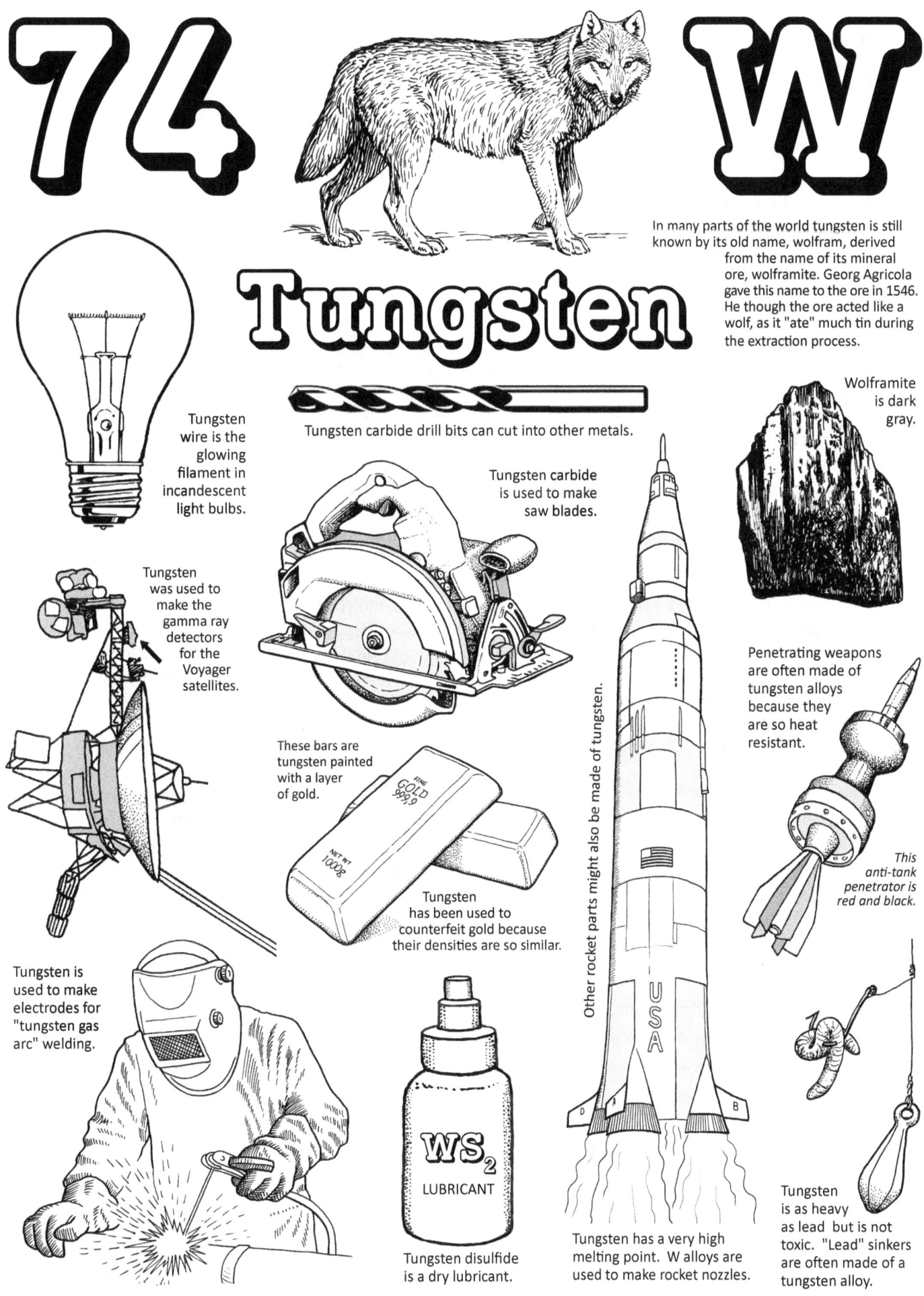

Re 75

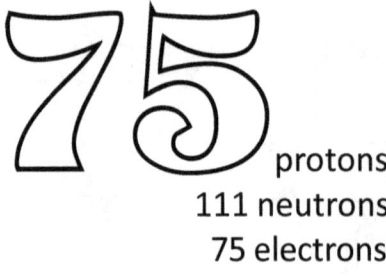

protons
111 neutrons
75 electrons

Atomic mass: 186.2

Rhenium

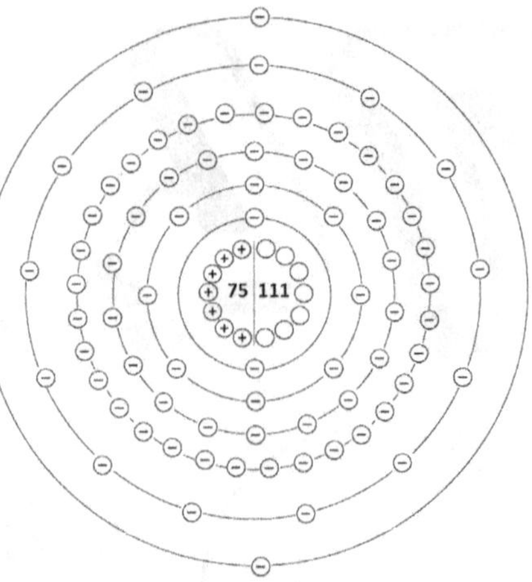

Named after the Rhine River

Rhenium is one of the rarest elements. There is less rhenium in the earth than any of the "rare earths." If fact, it is the least abundant solid element that is not radioactive. It occurs as an impurity in molybdenum ores, although rarely a piece of rhenium sulfide is found around the edges of a volcano (Kudriavy, Kuril Islands, 1994). The discovery of rhenium was not until the 1920s, when three German chemists managed to isolate 1 gram from 660 kilograms of molybdenite ore. They decided to name it after a famous landmark in Germany, the Rhine River. Today, the main producers of rhenium metal are Chile, the US, Peru, and Poland. Chile has the largest resources, with rhenium being an impurity found in their copper ores.

Rhenium is not very useful as a pure element, although some metal-producing companies do sell pure rhenium wires and rods for special applications, such as heating coils in mass spectrometers, ion gauges and flash units for photography. Mostly, rhenium is added to other metals to make "superalloys" which can withstand an immense amount of heat and are resistant to corrosion. Superalloys usually contain large amounts of nickel, iron, and cobalt, and have small amounts of other elements such as rhenium, molybdenum, titanium, chromium, aluminum, hafnium or tantalum. Superalloys are used to make things like jet engines and turbines used in power plants.

When rhenium is added to tungsten, the alloy is called rheni-tung. This alloy has many uses in the areas of electronics, welding, heating, and in the aerospace industry. Temperature gauges that use rheni-tung wires can measure temperatures up to 2200° C. Rheni-tung can also generate x-rays when exposed to the right kind of high energy beam, so these wires can be found in x-ray machines.

When rhenium is alloyed with platinum, it becomes useful as a catalyst for the refining of petroleum. The long chains of carbon atoms in the crude oil need to be broken into smaller chains to make products such as wax, diesel fuel, heating oils, gasoline (petrol) and chemicals for making plastics, paints, and dyes. The catalyst must be a metal that will not bond to any of the petroleum molecules, just encourage them to break apart. Metal catalysts, if put into the right chemical situation, can do the reverse and help small molecules join together. Rhenium alloys can be catalysts for chemical process that produce molecules for detergents, medicines, and plastics.

Radioactive isotopes of rhenium can be produced in a lab and are useful for treating liver and skin cancers. Radioactive rhenium is short-lived and will cease to be radioactive in a matter of days or weeks.

ReO_2 Rhenium (IV) oxide

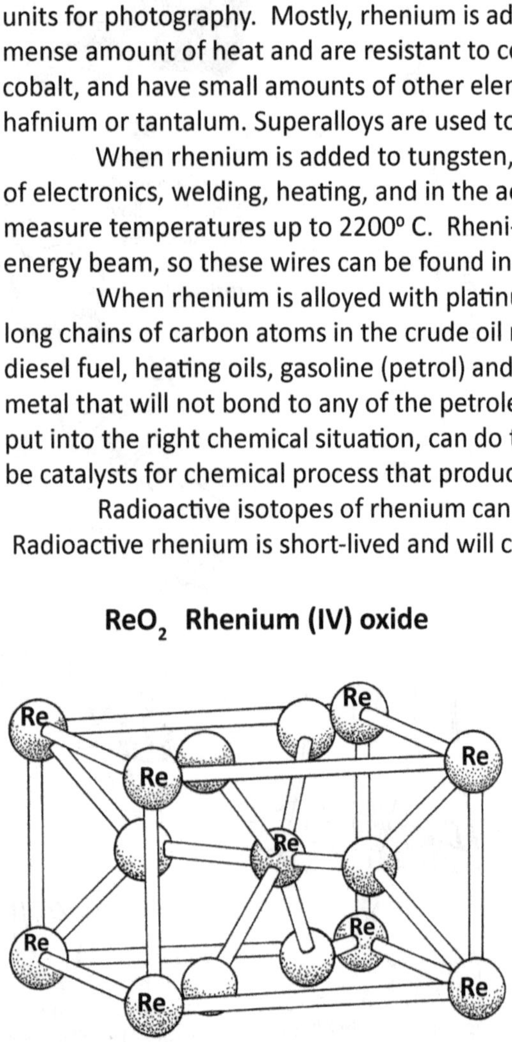

The unmarked balls are oxygen, O, and are always red.

ReS_2 Rhenium disulfide

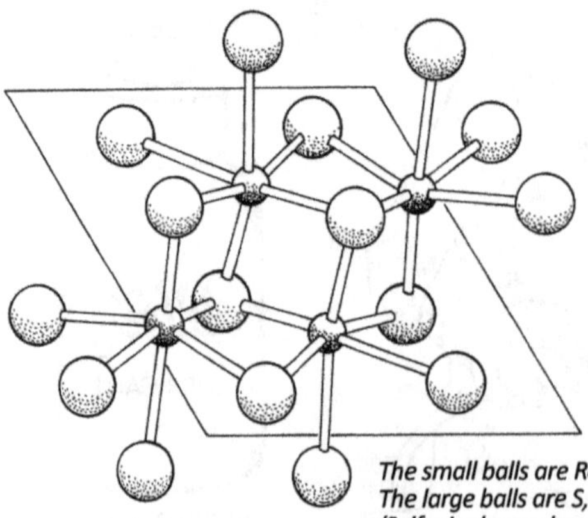

The small balls are Re.
The large balls are S, sulfur.
(Sulfur is always shown as yellow.)

$H_4Re_2O_9$ Perrhenic acid

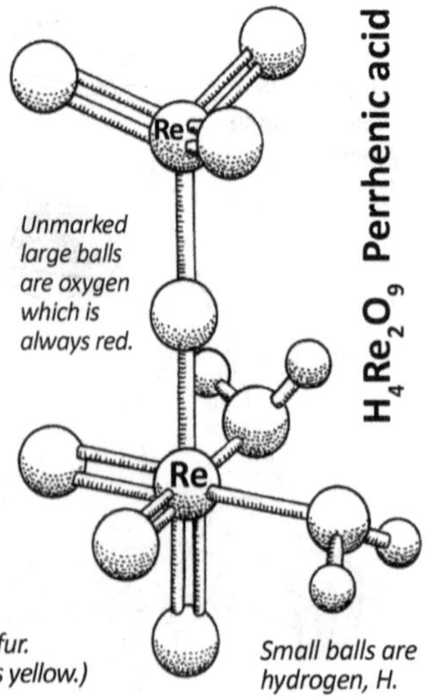

Small balls are hydrogen, H.

75 Re

Rhenium

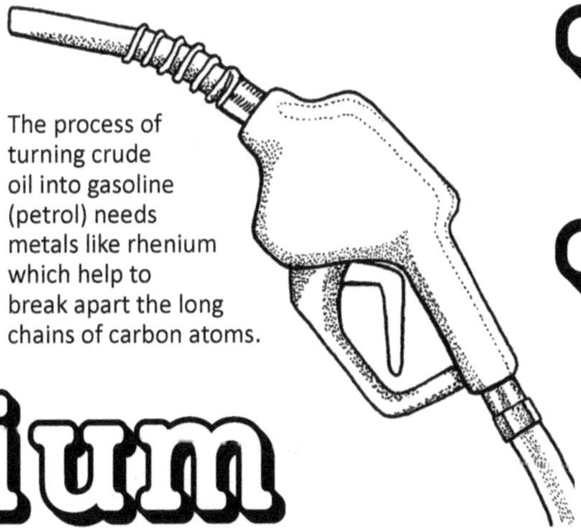

The process of turning crude oil into gasoline (petrol) needs metals like rhenium which help to break apart the long chains of carbon atoms.

Rhenium is named after the Rhine River.

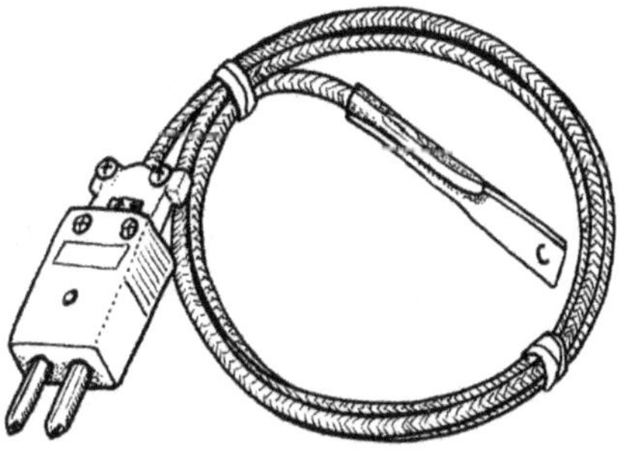

Rheni-tung is used for wires and parts used in heaters and thermocouples.

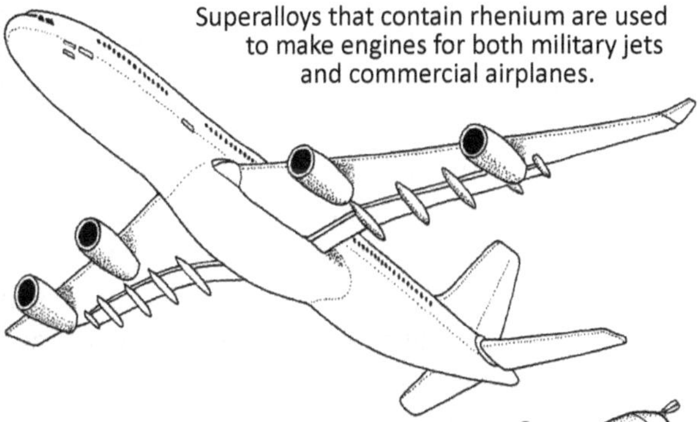

Superalloys that contain rhenium are used to make engines for both military jets and commercial airplanes.

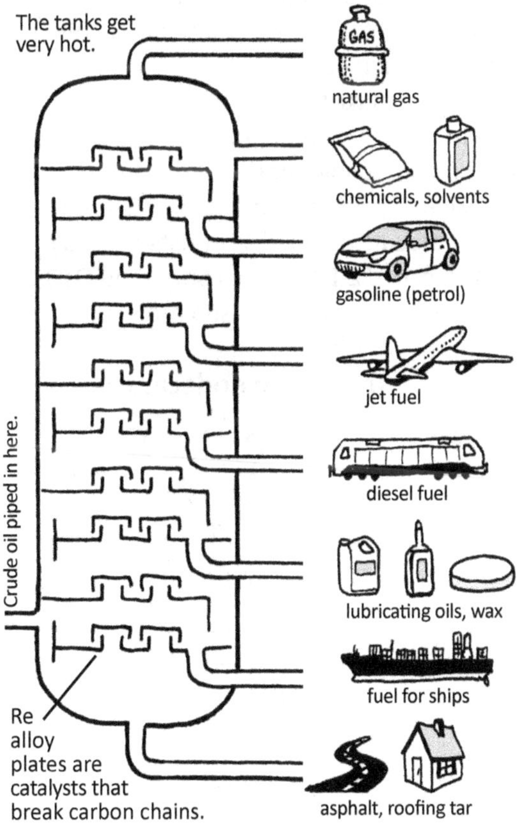

Chile is the leading producer of rhenium., but Peru also has rhenium ores.

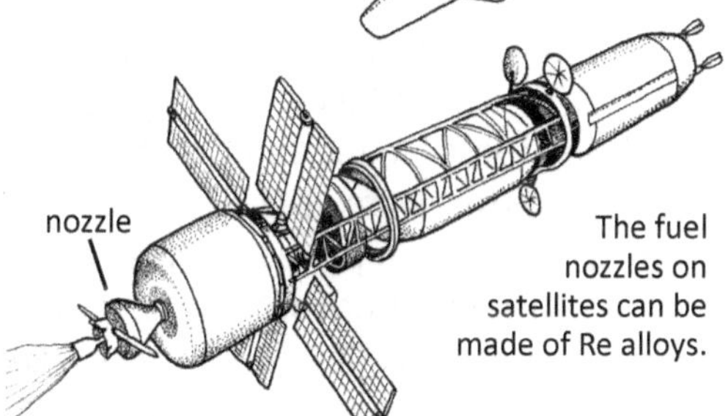

The fuel nozzles on satellites can be made of Re alloys.

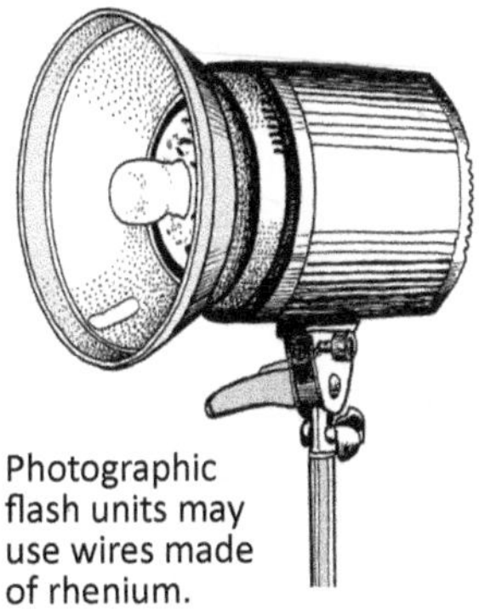

Photographic flash units may use wires made of rhenium.

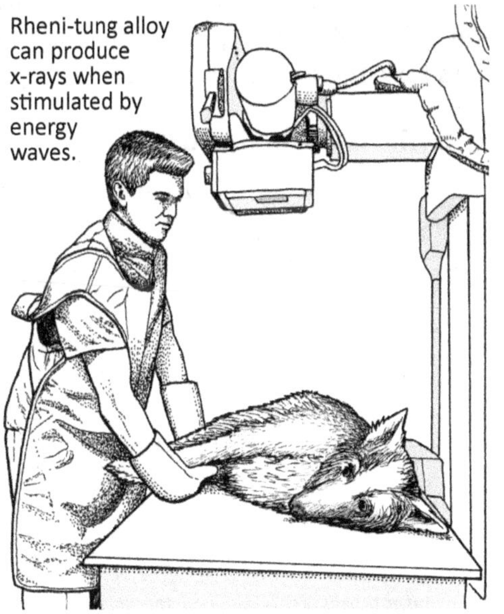

Rheni-tung alloy can produce x-rays when stimulated by energy waves.

Os 76

protons
114 neutrons
76 electrons

Atomic mass: 190.2

Osmium

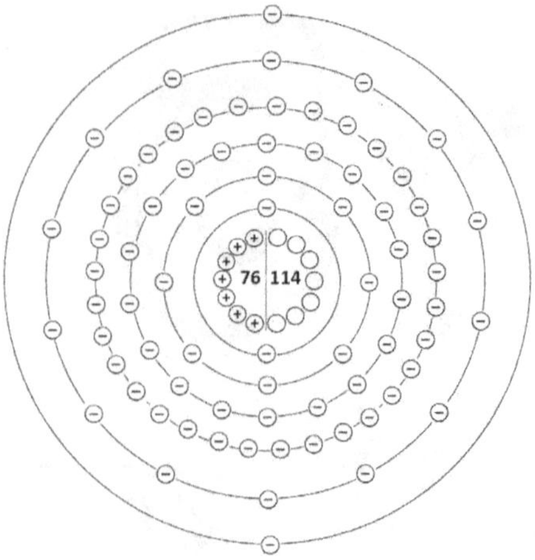

Named using the Greek word for smell, "osme"

Osmium is a member of the "platinum group" which also includes ruthenium, rhodium, palladium, iridium, and platinum. All of these metals have similar properties and many of them can be used for the same purposes. They are usually found as impurities in copper or nickel deposits, and because it takes a lot of work to extract them from the ores, the pure metals are expensive to purchase.

Osmium and iridium are very often found together and can be difficult to separate. They were discovered together in 1803 by Smithson Tennant, in a black residue left behind after a piece of platinum was dissolved in a mixture of hydrochloric acid and nitric acid. Tennant did more experiments with the residue and one of the results was a yellow solution that smelled terrible. It was this smell that caused him to choose the name osmium for the new element, because "osme" is Greek for "smell."

Osmium holds the record for being the most dense element. We all know that lead is dense and very heavy, but osmium is twice as dense as lead. Pure osmium is also brittle and has a very high melting point, which makes it hard to work with. Osmium is combined with other metals, often iridium or platinum, to make useful alloys. Since these alloys are expensive, they show up only in small things like pen points and electrical contacts.

One early use for an osmium alloy was as a filament in light bulbs. In 1902, Carl Auer von Welsbach started a manufacturing company to make "Oslamps." Only a few years later, the company began using tungsten instead. Another early use for osmium was as a catalyst in a chemical process that could pull nitrogen out of the air and turn it into NH_3 which can be used to fertilize plants. Again, replacements were soon found for osmium so it is no longer used for this purpose. In the 1940s and 1950s, osmium alloys were used to make "needles" for phonographs. In the 1990s, osmium mirrors were used by the space shuttles as part of their ultraviolet ray detection equipment, but since then osmium has been replaced by other elements due to osmium's tendency to attract destructive oxygen atoms.

Although pure osmium is not toxic, it quickly binds with oxygen atoms to form osmium tetroxide (OsO_4) which is very toxic. Despite its toxicity, OsO_4 is useful to scientists. It sticks to phospholipid molecules that make cell membranes, so it is used as a stain in electron microscopy. (In 2020, a researcher used it to find unfossilized nerve cells in dinosaur bones!) Its ability to stick to certain parts of the DNA molecule enabled scientists to study the shape of DNA in great depth and they discovered "right-handed" DNA, also called Z-DNA. The compound OsO_2 is less toxic but is also less useful. $KOsO_4$ is used as a catalyst by chemists who work with carbon-based molecules.

OsO_4 Osmium tetroxide

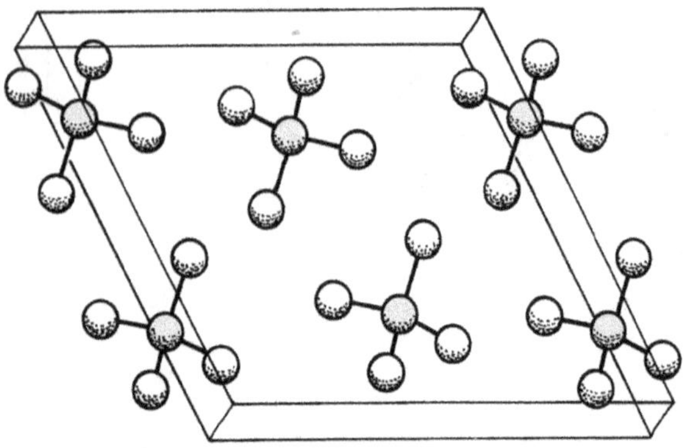

The gray balls are Os, osmium. The white balls are O, oxygen.

OsF_6
Osmium hexafluoride

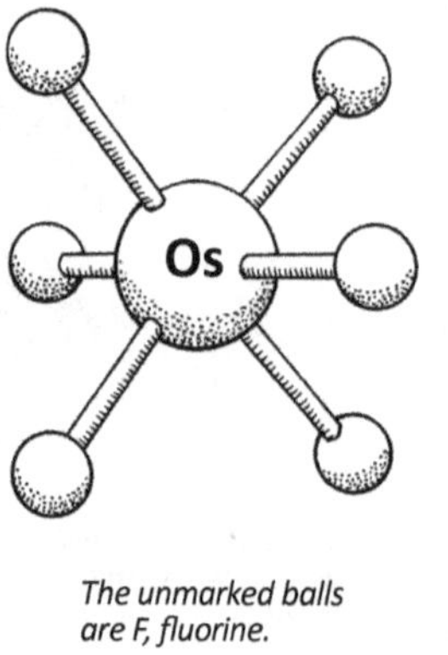

The unmarked balls are F, fluorine.

$Os_3(CO)_{12}$
Triosmium dodecacarbonyl

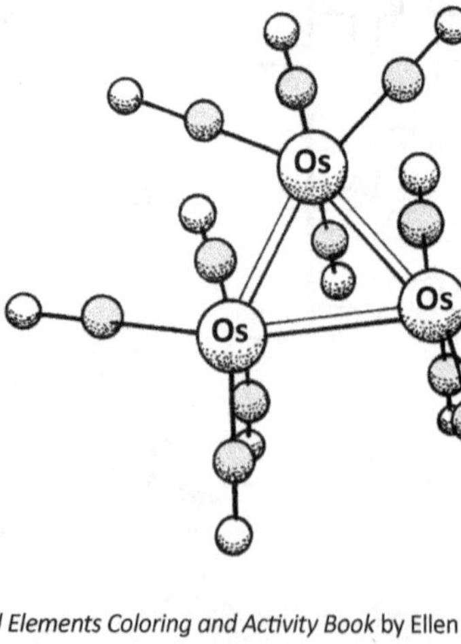

Gray balls are C, carbon. White balls are H, hydrogen.

76 Os

44 Ru Ruthenium	45 Rh Rhodium	46 Pd Palladium
76 Os Osmium	77 Ir Iridium	78 Pt Platinum

Platinum group

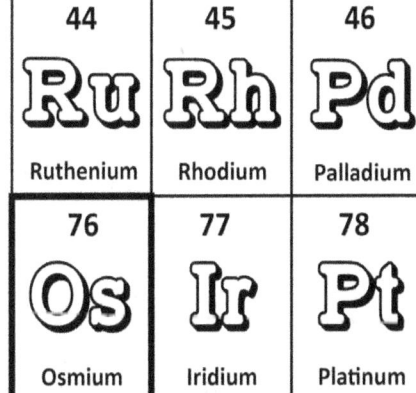

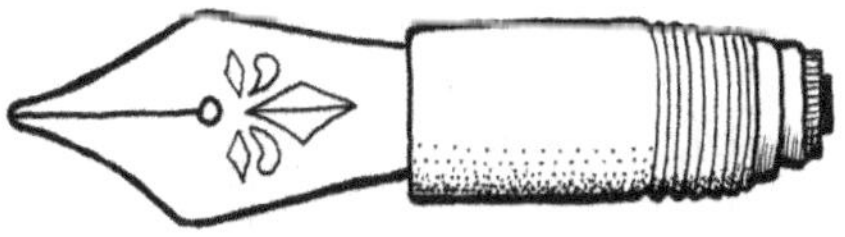

Osmium alloys are used to make pen points.

Osmium

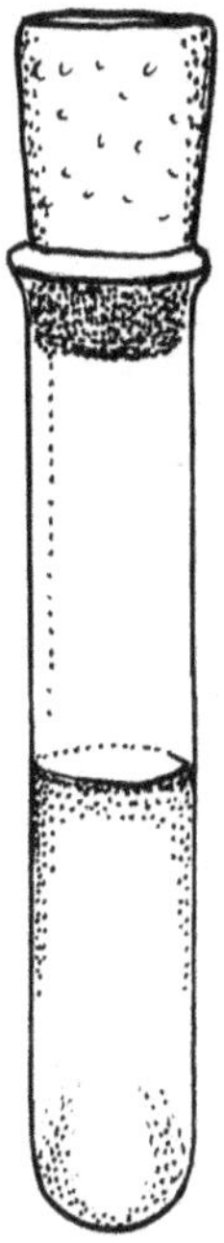

Osmium was discovered as a smelly yellow solution.

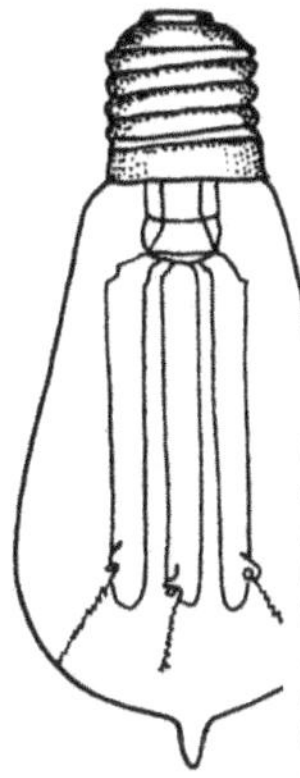

One early use for Os was as a filament in light bulbs.

This stamp honors the work of Carl von Welsbach.

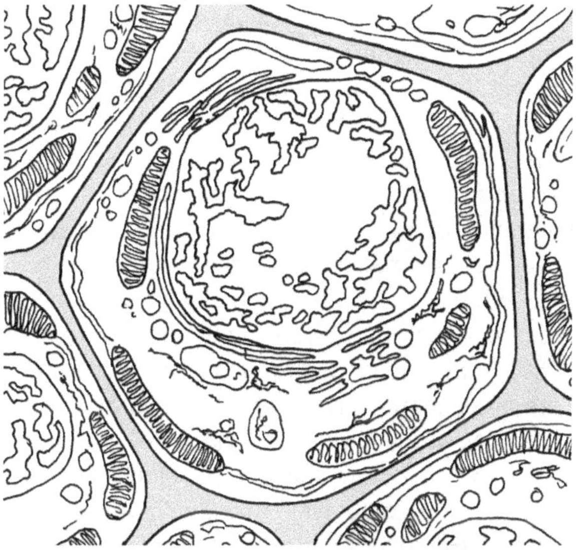

Osmium tetroxide, OsO_4, is used to stain cells so that their membrane shows up when viewed under an electron microscope.

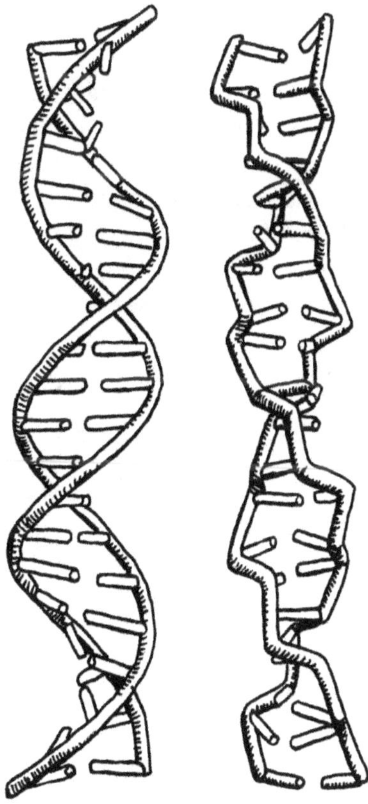

The "rungs" on the DNA "ladder" are drawn using 4 colors.

left-handed DNA *right-handed DNA*

Osmium was a key in the discovery of "right-handed" DNA, or Z-DNA.

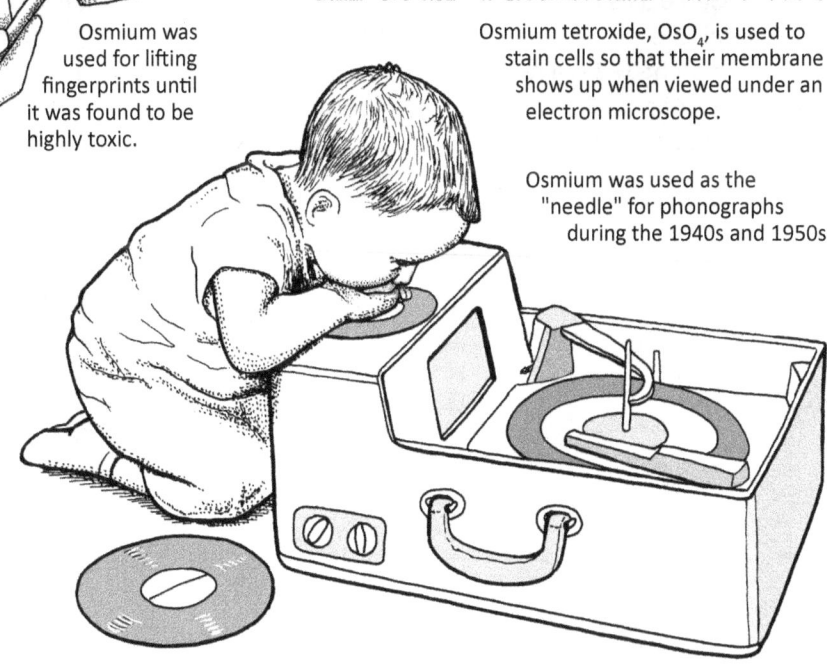

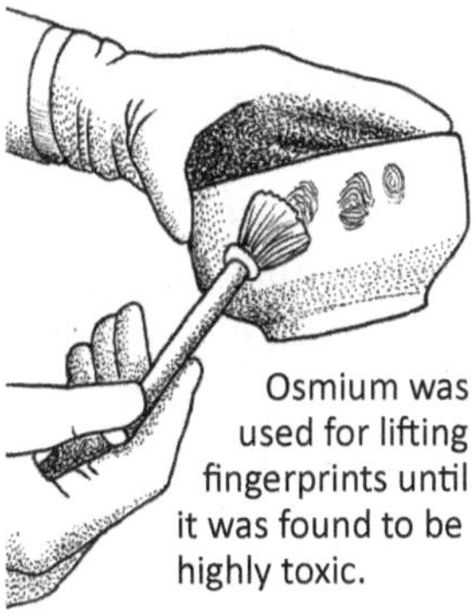

Osmium was used for lifting fingerprints until it was found to be highly toxic.

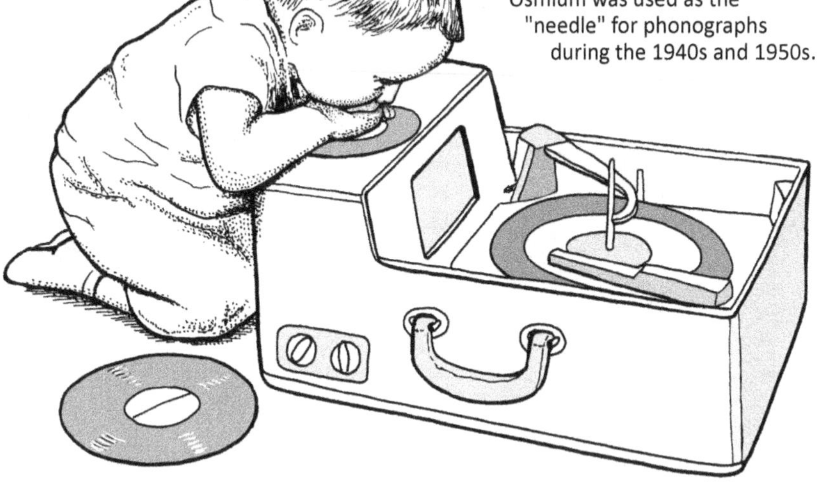

Osmium was used as the "needle" for phonographs during the 1940s and 1950s.

Ir

77 protons
116 neutrons
77 electrons

Atomic mass: 192.2

Iridium

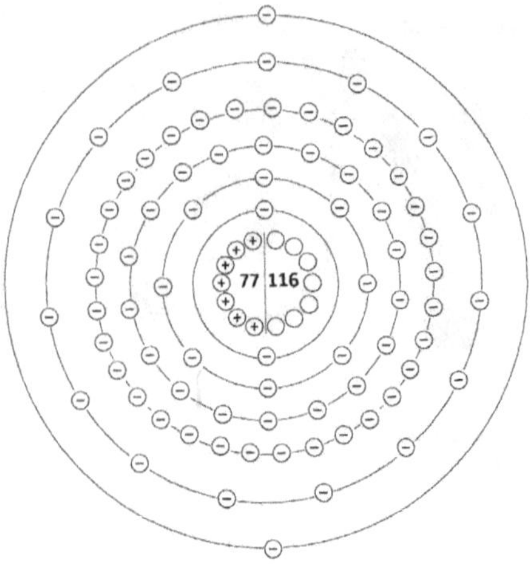

Named after Iris, Greek goddess of the rainbow

Iridium and osmium were discovered by Smithson Tennant in 1803. He had been dissolving platinum in "aqua regia," a mixture of hydrochloric acid and nitric acid. The reaction left some black residue and he wondered what it was. By doing more experiments on the residue, he found that it contained two new elements, osmium and iridium. Iridium was named after Iris, the Greek goddess of the rainbow, because its compounds were of various colors. Tennant's knowledge of iridium was through its compounds; iridium metal was not seen until 1813, when a new invention, the electric battery, made it possible to obtain samples of elements that never occur alone in nature.

By the 1840s, enough iridium had been produced that could be used to make "nibs" (points) for ink pens, and certain cannon parts that received a lot of wear and tear. In 1889, iridium and platinum were used to make an object that received no wear and tear at all: the official measuring rod for the length of 1 meter.

In 1933, ruthenium and iridium were combined to make thermocouples that were used in thermometers that could measure temperatures up to 2,000° C. Iridium can be combined with other members of the platinum group to make very durable alloys used for small parts for machines. Since iridium is the most corrosion-resistant element, iridium alloys are used to make pipes that will sit in salt water for a long time, and for machines that are used in industrial processes that involve harsh chemicals or high heat. Some iridium alloys can be used to make crucibles (melting pots) for melting and forming garnet crystals (such as YAG) used in lasers. Iridium is also perfect for electrical contacts, so you will find it in places like heavy-duty spark plugs.

Iridium is used for a number of high-tech applications. Specimens for scanning electron microscopy (SEM) are often coated with a very thin layer of iridium so that electrons will bounce off and create the SEM image. And speaking of thin layers, the Chandra X-ray Observatory satellite has mirrors that are coated with a microscopically thin layer of iridium, and perfectly flat to within a few atoms. These mirrors reflect x-rays onto a focal point inside the telescope tube, where they will form an image of the astronomical object that is sending out the rays.

Geologists find more iridium in some geological layers than in others and wonder if this is connected to an astronomical event. Meteorites have a higher concentration of iridium than any natural rock.

Iridium-192 is a radioactive isotope produced by nuclear labs for scientific and medical purposes. It is used as a source of gamma radiation for radiography units that inspect structures, and for cancer therapy.

Iridium is used by particle physicists to make "antiprotons," a type of antimatter.

IrO_2 Iridium tetroxide

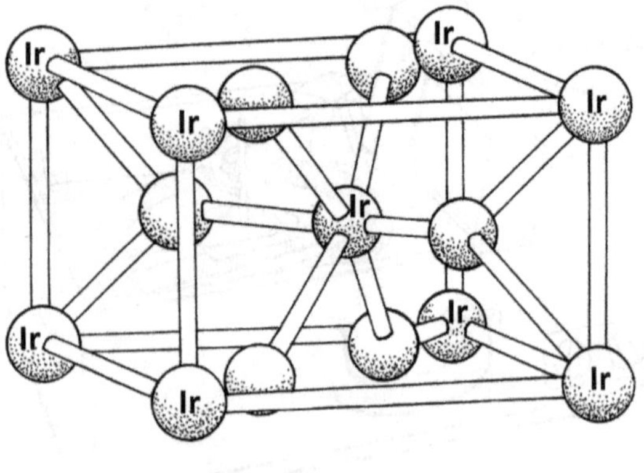

All the unmarked balls are O, oxygen.

IrF_6 Iridium hexafluoride

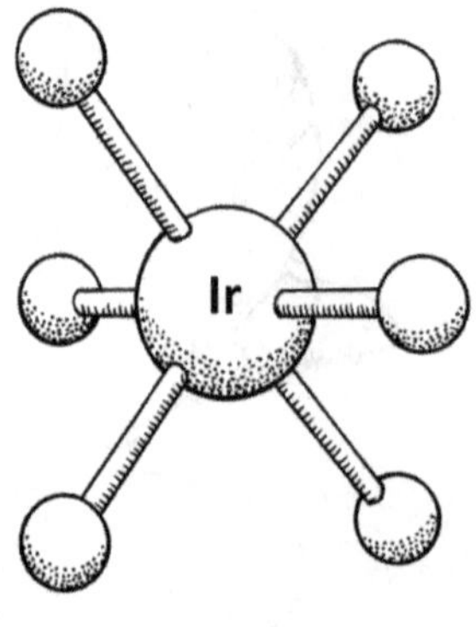

All the unmarked balls are F, fluorine.

IrS_2 Iridium disulfide

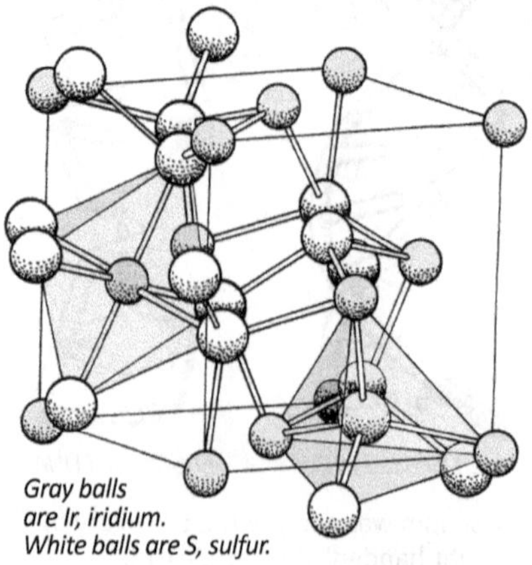

Gray balls are Ir, iridium. White balls are S, sulfur.

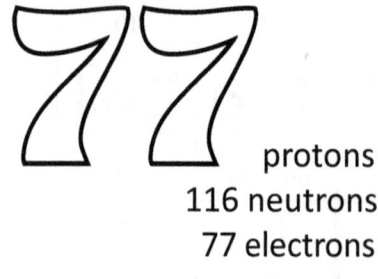

77 Iridium Ir

44 Ru Ruthenium	45 Rh Rhodium	46 Pd Palladium
76 Os Osmium	77 Ir Iridium	78 Pt Platinum

Platinum group

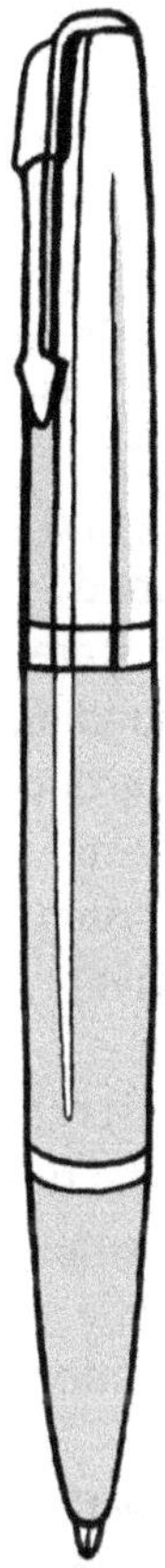

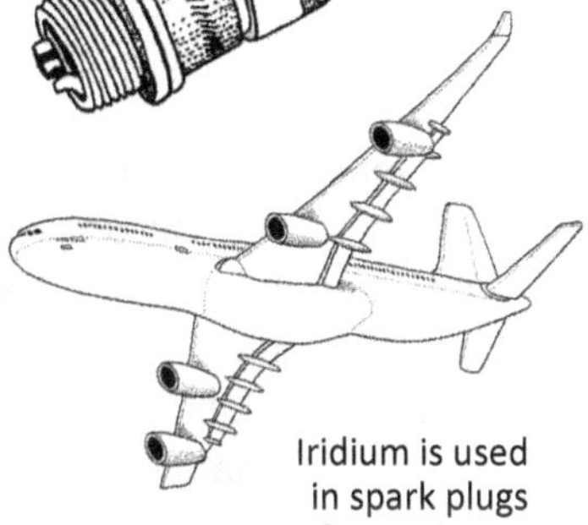

Iridium is used in spark plugs for airplanes.

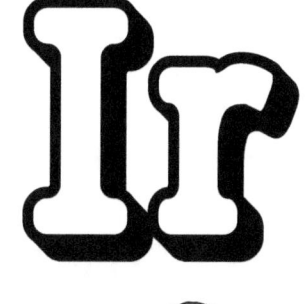

Iris, the Greek goddess of the rainbow. She was a messenger, like Mercury.

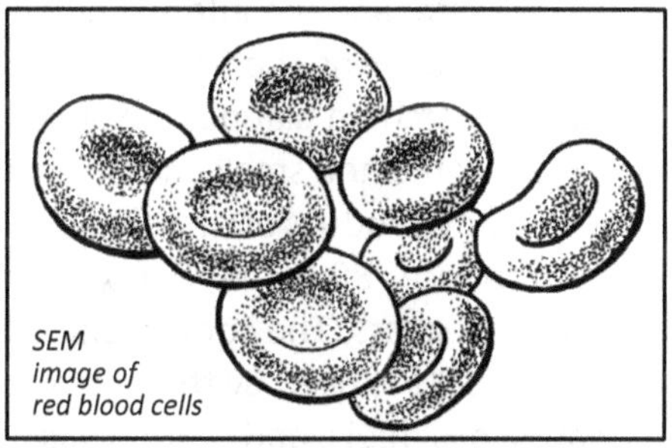

SEM image of red blood cells

SEM (scanning electron micrograph) images require the specimens to be coated with a thin layer of metal, often iridium.

Iridium crucibles can withstand the heat needed to melt other metals. Garnet crystals (such as YAG) are made in crucibles.

Starting in 1943, fountain pen nibs (tips) were made of gold-coated iridium or ruthenium.

The solar panels tend to look blue.

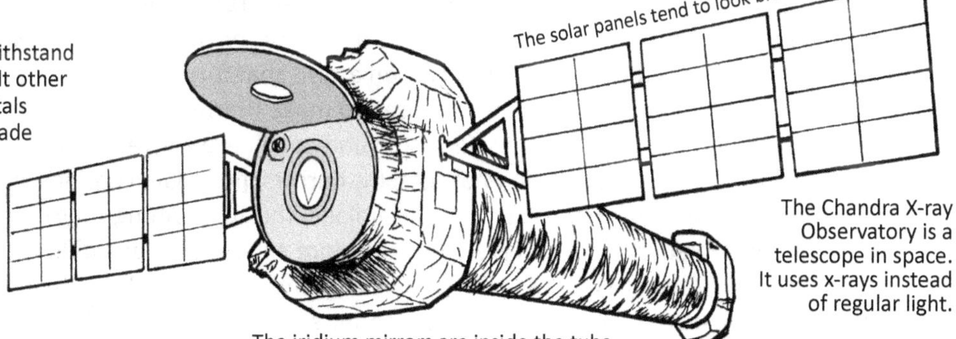

The Chandra X-ray Observatory is a telescope in space. It uses x-rays instead of regular light.

The iridium mirrors are inside the tube.

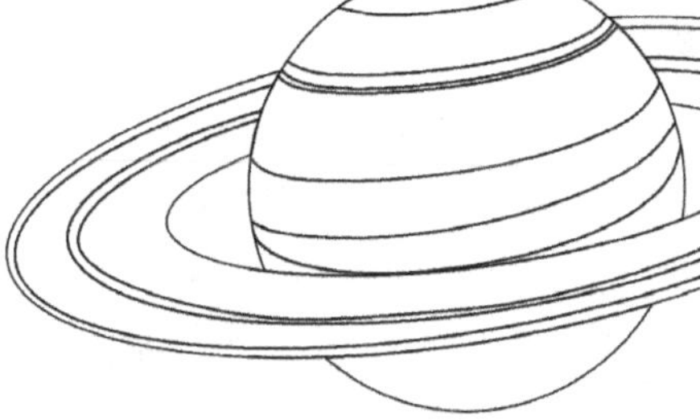

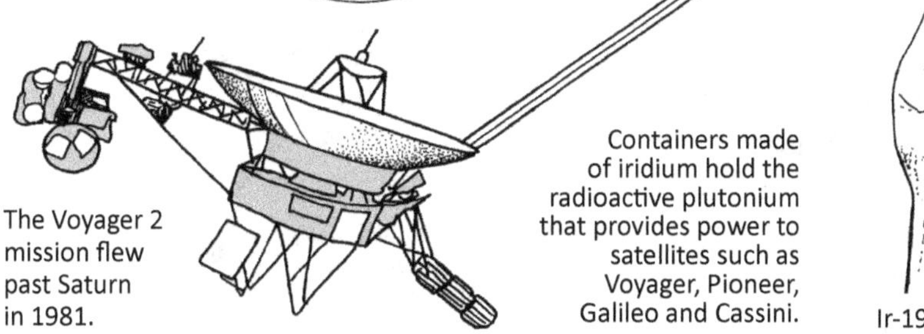

The Voyager 2 mission flew past Saturn in 1981.

Containers made of iridium hold the radioactive plutonium that provides power to satellites such as Voyager, Pioneer, Galileo and Cassini.

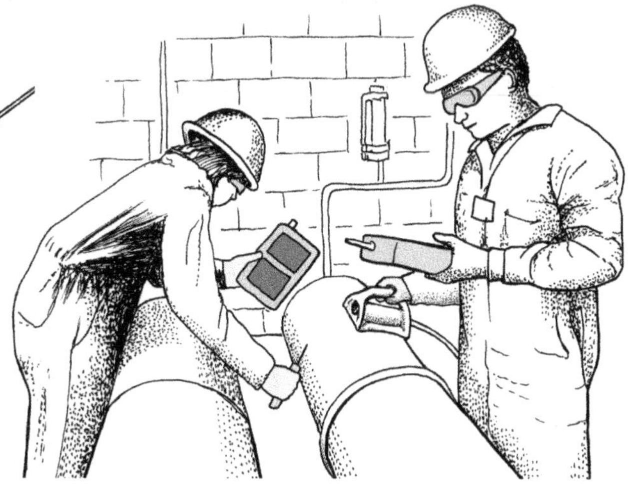

Ir-192 is used in gamma radiography units that inspect equipment.

Pt 78

protons
117 neutrons
78 electrons

Atomic mass: 195.0

Platinum

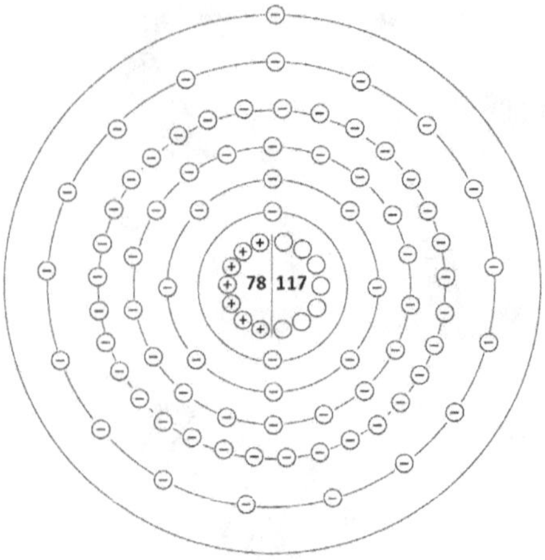

Name means "little silver" (platina) in Spanish

In 1743, a young Spanish scientist named Antonio de Ulloa joined an eight year tour of exploration to the area we now call Ecuador. During his time in this part of South America, he made many scientific observations. His most famous discovery turned out to be the element platinum. He had seen native men mining for ores, and he collected samples of the whitish rocks they were digging out. He studied these ores when he got back to Spain, and published a paper giving the results of his research, and the interesting metal they contained, but then he stopped research and went on to do other things, including being the first Spanish governor of the what became the US state of Louisiana. Other scientists picked up the research on platinum. For years there was a lot of confusion surrounding the identity of platinum because the ores also contained small amounts of the other members of the platinum group.

Today, pure platinum is produced as a by-product of nickel and copper mining. South Africa has the best natural deposits of platinum-rich ores and produces 80 percent of the world's supply. 45% of all the pure platinum produced is used to make catalytic converters for cars. The converters take toxins out of the exhaust fumes. Over 30% of the platinum goes into jewelry. About 10% is used by the petroleum industry for refining crude oil and turning it into useful products such as waxes, oils, fuels, and natural gas. (Refineries use metal plates as catalysts to encourage long carbon chains to break up into smaller ones.) 3% of the platinum goes to the computer manufacturing industry to make parts like hard drives. The remaining percentage goes to a variety of uses including medicines for treating cancers, glass-making equipment, corrosion-resistant tools (such as surgical tools), dental implants, oxygen sensors, and spark plugs. It is also used as a catalyst for the process of hydrogenating oils. In hydrogenation, hydrogen atoms are added to the chains of carbon atoms that form the oil molecules.

In the world of high-tech, platinum is used to make thermogravimetric machines that measure an object's mass over time as temperature changes. Platinum and cobalt are alloyed to make medium-strength magnets. Hexachloropatinic acid, Cl_6H_2Pt, has uses in photography, mirrors, porcelain coatings, and, most famously, as "voting ink" applied to one finger as the voter exits the polling place, thus preventing them from voting twice.

Platinum is relatively rare and is sometimes more expensive than gold. Many people buy platinum as a "safe" investment, because precious metals (gold, silver, platinum) will always hold their value; they are actually usable, unlike paper or digital money. During times of economic security, the price of platinum can be twice the price of gold. During times of economic uncertainty, the price of platinum falls below that of gold.

$PtCl_4$ Platinum (IV) chloride

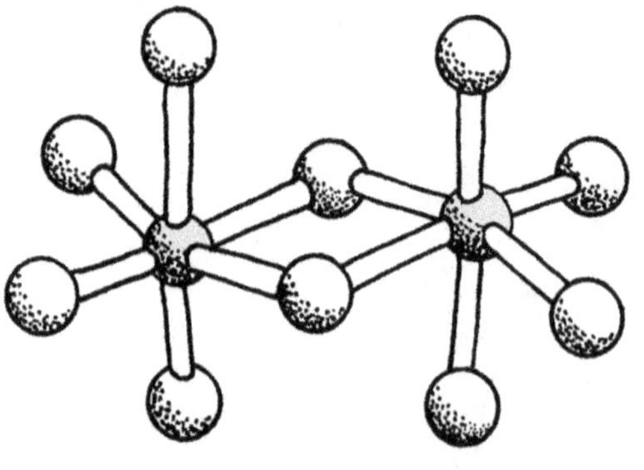

Gray balls are Pt.. White balls are Cl.

PtF_6 Platinum hexafluoride

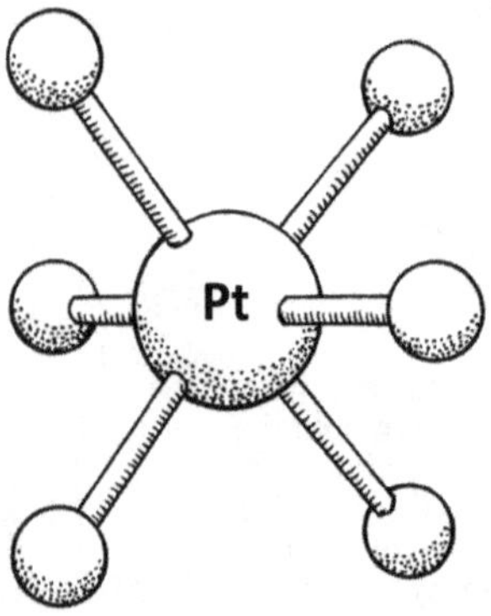

The unmarked balls are F, fluorine.

$Pt(NH_3)_2Cl_2$ Cisplatin

This is a medicine used to kill cancer cells.

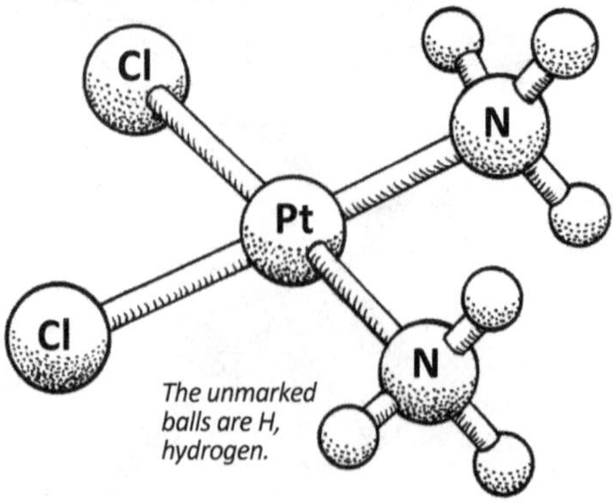

The unmarked balls are H, hydrogen.

78 Pt

44 Ru Ruthenium	45 Rh Rhodium	46 Pd Palladium
76 Os Osmium	77 Ir Iridium	78 Pt Platinum

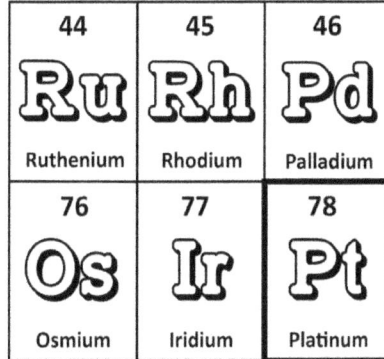

Platinum group

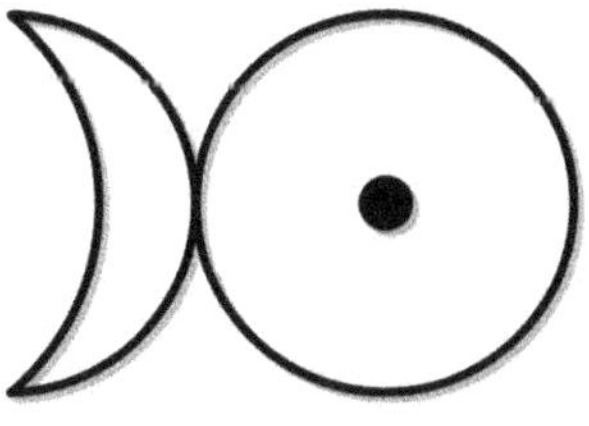

This symbol was used in the 1700s to represent platinum. It combines the symbols for moon (silver) and sun (gold).

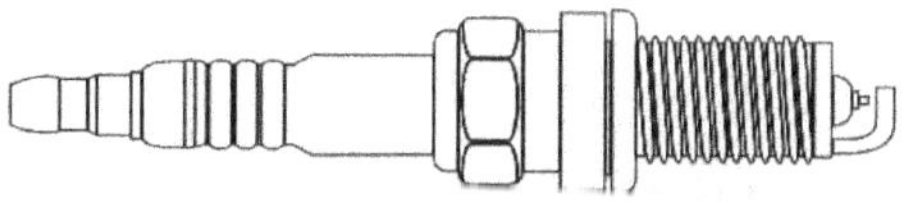

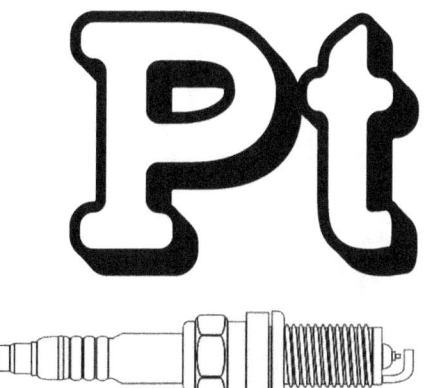

Platinum alloys are used in spark plugs.

Platinum

Made in 1937, the crown has 2,800 diamonds.

The frame of the crown of the Queen Mother (Elizabeth II's mother) is made of platinum. The fabric is purple velvet.

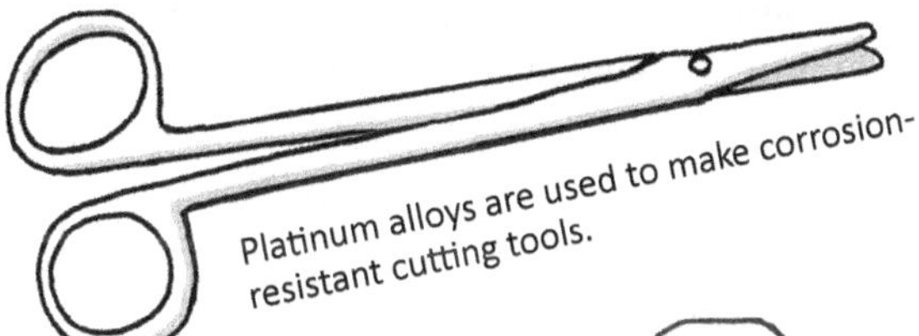

Platinum alloys are used to make corrosion-resistant cutting tools.

Platinum gives watches a shiny, durable surface.

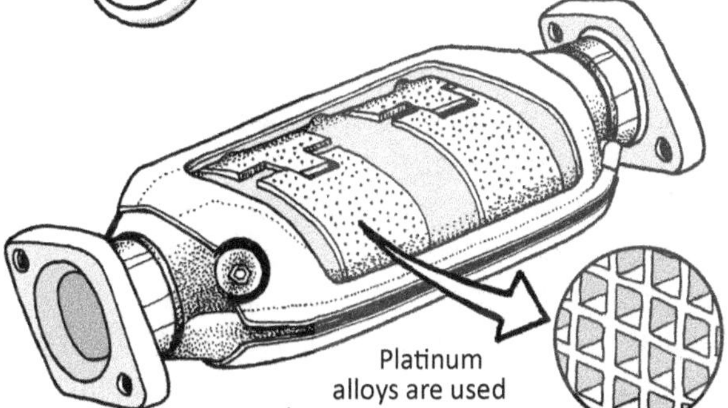

Platinum alloys are used in catalytic converters in cars. The converters remove toxins from the exhaust.

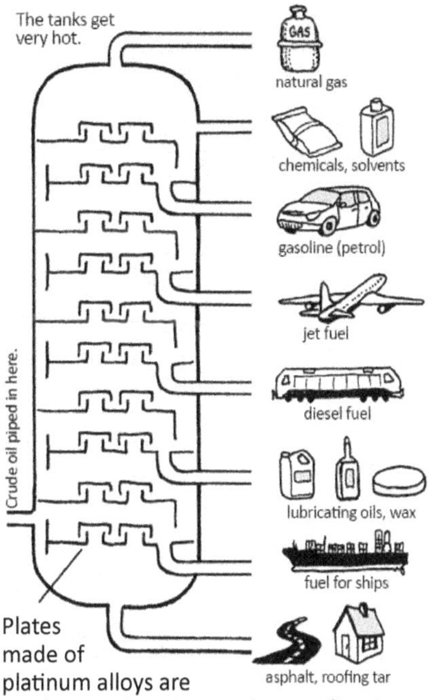

Plates made of platinum alloys are used as catalysts in petroleum refineries.

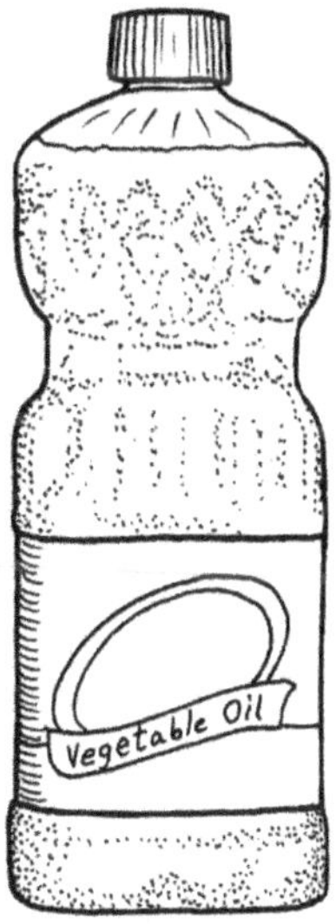

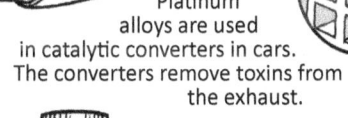

Platinum alloy plates are also used in process of hydrogenating oils (adding H atoms to strings of carbons).

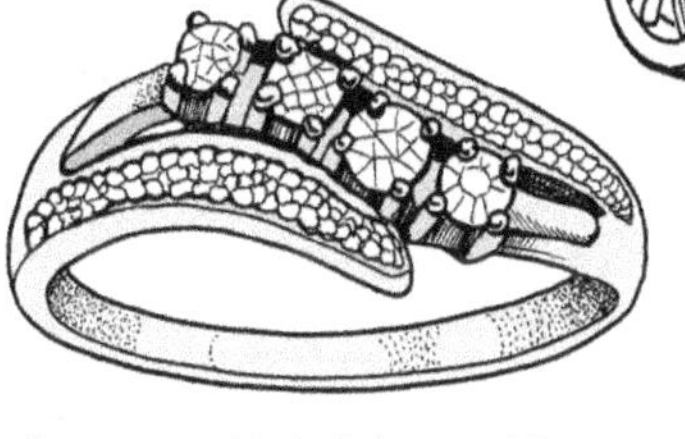

About one third of the world's supply of platinum is used to make jewelry.

Dental implants use platinum alloys.

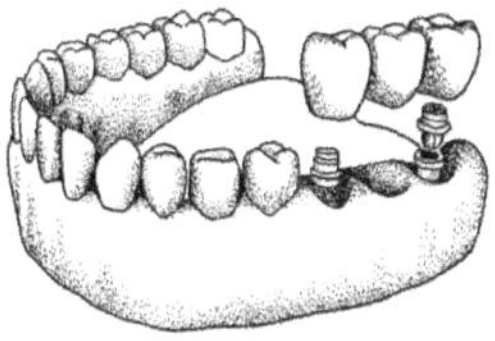

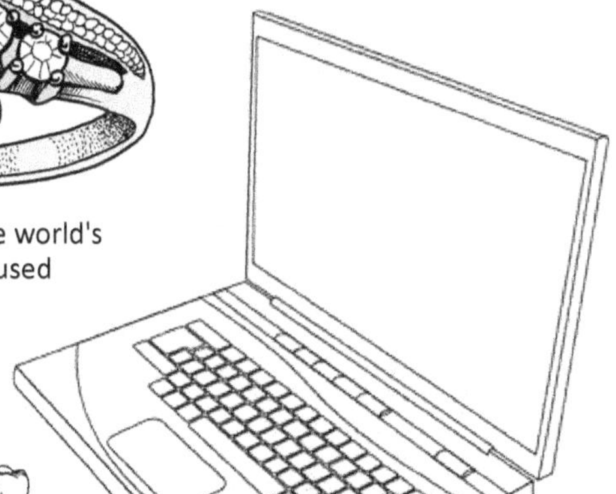

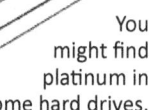

You might find platinum in some hard drives.

Au 79

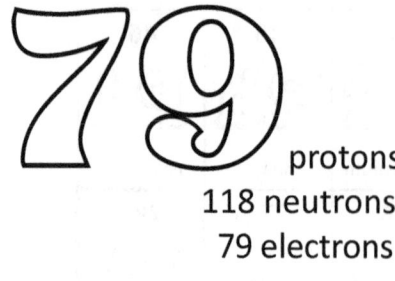

protons
118 neutrons
79 electrons

Gold

Atomic mass: 196.9

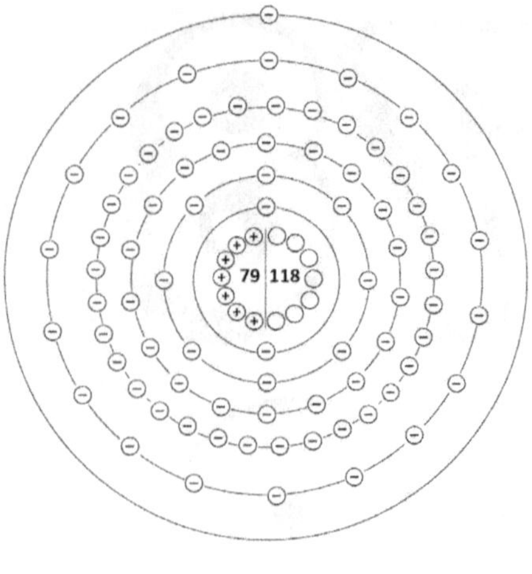

Gold is from an ancient word shared by many languages.
The symbol, Au, is from the Latin word for gold, "aurum"

Gold is one of the elements that ancient peoples discovered. It is one of the few elements that can be found in its pure ("native") form in the crust of the earth, often mixed into quartz or silver deposits. The particles are usually small, sometimes even smaller than sand grains. The largest nugget on record was found in Australia in 1869 and weighed 78 kg (173 lbs). People today still value gold as much as ancient peoples did. It has a beautiful shine and it is easy to work with. Gold can be rolled into sheets that are thinner than paper and it can be pulled and drawn into wires that are only molecules wide. Because it will make thin wires that can carry electricity, gold is very useful to the electronics industry and has been used to make circuit boards and microchips.

Because gold is so soft, items made of gold can be dented and scratched. This poses a problem for its use in jewelry and coins. Silver, copper, nickel or zinc can be added to gold to make it harder. A blend of silver and gold is called "electrum." If the other metals change the color of the gold, jewelers just advertise it as colored gold, such as "white gold," or "red gold." Sometimes gold is used as a thin coating over harder metals.

Since gold will reflect ultraviolet light, a thin layer of gold was sprayed on the inside of the helmets that the Apollo astronauts wore. In many of the pictures from the Apollo missions, the helmets look shiny and gold. Gold's ability to reflect heat (infrared rays) was tested in a sports car named the McLaren F1. Certain parts of the engine were wrapped in thin layers of gold. This did not add much to the price of the car, as only a few grams of gold were used.

People throughout time have purchased gold as a financial investment. Even today, gold is considered to be one of the safest investments because it never loses value. It is useful, beautiful, and doesn't rust or decay.

The oceans of the world contain 15,000 tons of gold if you could collect all of the atoms into one place. Gold washes down through rivers and out in to the sea. Riverbeds in some parts of the world are excellent sites to "pan" for gold, straining out small gold flakes from the sand and dirt. Gold deposits are often called "veins."

Gold is not toxic. You can buy extremely thin sheets of gold, called "gold leaf," to decorate desserts. Gold leaf was also used in the Middle Ages to decorate fancy handwritten manuscripts.

Cu_3Au Auricupride

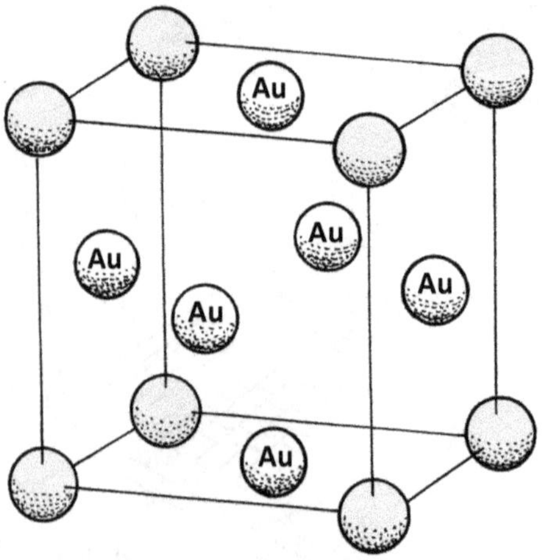

The gray balls are Cu, copper.

Au_2Br_6 Gold bromide

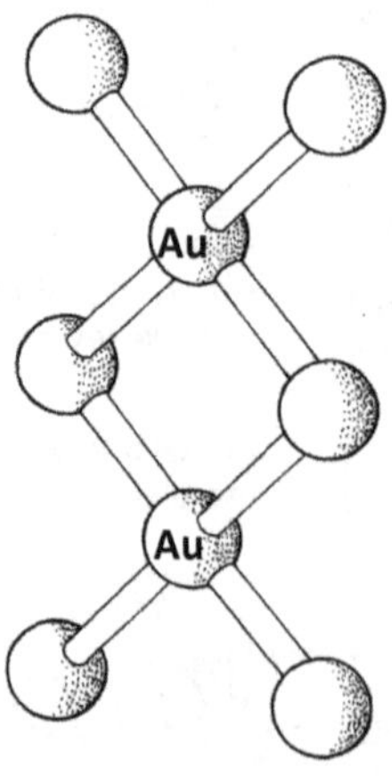

The unmarked balls are Br, bromine.

AuCl Gold chloride

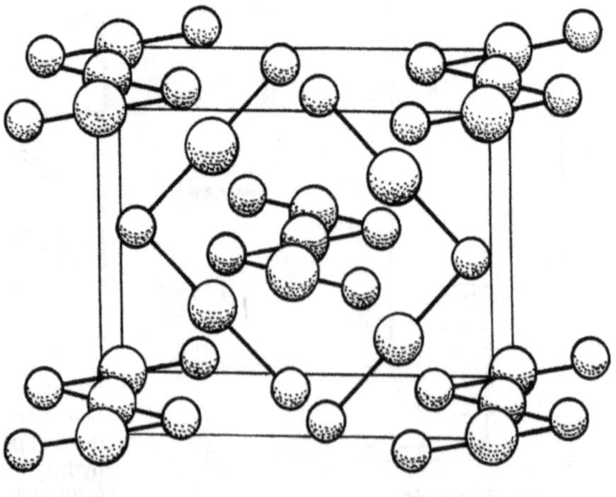

The large balls are Au, gold. The small balls are Cl, chlorine.

79 Au Gold

Gold is molded into ingots (bars) for storage.

The mask has blue and gold stripes.

Many artifacts from King Tut's tomb are decorated with gold.

Athena *Athena's owl.*

Greek gold coin from about 400 BC.

The "Gold Eagle" is an investment coin that first appeared in 1986.

This drawing is from a manuscript written in the 1500s, showing Spanish explorers (led by Cortez) during their quest for gold in South America.

The McLaren F1 has engine parts covered in gold, to dissipate the heat made by the V12 engine.

The inside of this astronaut's helmet was sprayed with a very thin layer of gold. Gold reflects the harmful UV light rays.

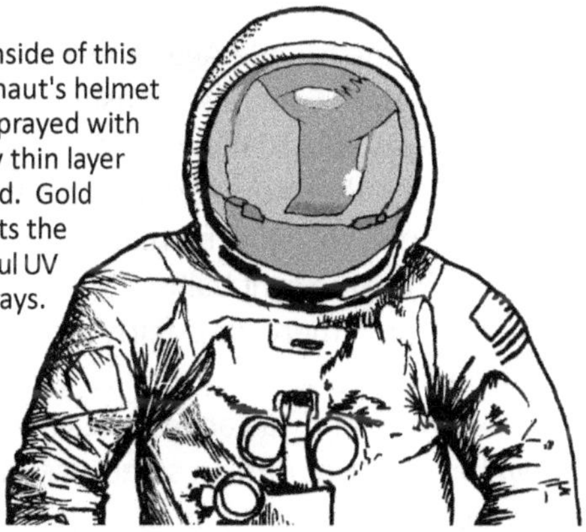

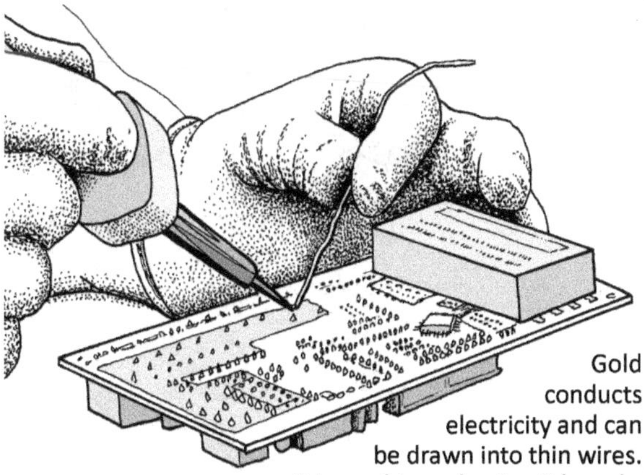

Gold conducts electricity and can be drawn into thin wires. It is used to make circuit boards.

Gold is non-toxic. In fact, you can buy extremely thin sheets of "gold leaf" and use them to decorate your desserts!

In the Middle Ages, scribes copied manuscripts by hand. They would often decorate the first letter of each paragraph and gold leaf was used along with paints.

Hg 80

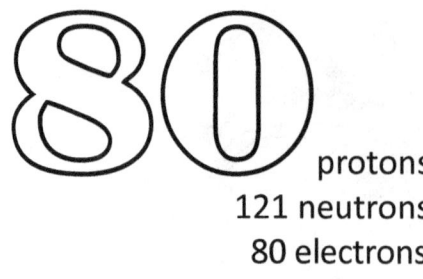

80 protons
121 neutrons
80 electrons

Atomic mass: 200.6

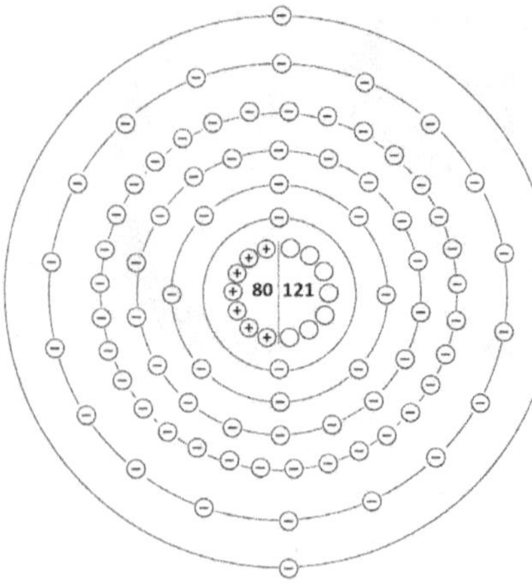

Mercury

Named after the Roman messenger god
"Hg" is from Latin "hydragyrum" meaning "water-silver"

This element has been known since ancient times. It is usually found in a mineral called cinnabar (HgS). If cinnabar is put into a fire, liquid mercury will come oozing out of the rock. This must have seemed almost magical to ancient peoples-- liquid coming out of a rock! The liquid was a beautiful silver color and it rolled around in a most unusual way, almost as if it was alive, giving rise to its nickname: quicksilver. ("Quick" used to mean "living"). Because mercury was so strange and interesting, it wasn't long until superstitious beliefs about it began to arise. Mercury was said to have amazing powers to cure many diseases and maybe even to extend life. Sadly, this was far from the truth. Mercury is highly toxic. It is amazing is how many people survived taking mercury "medicines." Much less harmful has been the use of mercury "amalgams" (alloys with Ag, Cu, Sn) by dentists for filling cavities.

The first Chinese emperor, Qin Shi Huangdi (259-210 BC),was obsessed with mercury. He thought it would make him live forever. His doctors agreed and prepared mercury pills for him to take every day. Before he died (of mercury poisoning, of course) he ordered that a secret underground tomb be built, with a large scale model of his city, featuring rivers of liquid mercury. The tomb has never been opened, but scientists do detect high levels of mercury coming out of the ground above it.

Cinnabar's first practical use was as a pigment. Vermilion red, made of powdered cinnabar, was used by painters in the Middle Ages and Renaissance. It was later replaced with cadmium red.

In the 1600s, a scientist named Torricelli discovered that a tall (80 cm) tube of mercury standing in a dish of mercury was able to measure air pressure. These mercury "barometers" were used until the digital age. A similar device is the thermometer, which also uses the predictable expansion of liquid mercury for measurement.

Dishes of liquid mercury were also used in lighthouses to float the heavy fresnel *(freh-nel)* lenses that increased the light from the light bulbs. The liquid mercury made an almost friction-free surface on which the lens could turn.

Liquid mercury is fairly easily vaporized. Two uses for mercury vapor are in compact fluorescent bulbs (still a current use) and as a propellant for ion propulsion engines for satellites (now replaced by noble gases, like Xe).

Mercury was one of the chemicals used in the 1800s by hat makers. Unfortunately, many hat makers developed brain problems because of exposure to mercury fumes. This unfortunate phenomenon is what inspired author Lewis Carroll to create a character called "the Mad Hatter" in his book *Alice in Wonderland.*

$Hg(NO_3)_2$ Mercuric nitrate

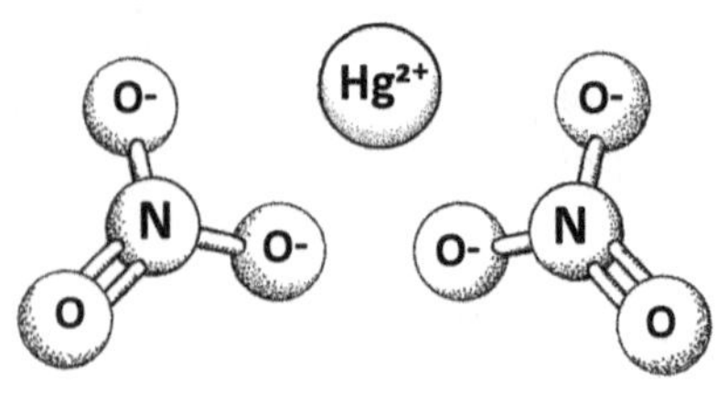

Hg_2Cl_2 Mercury (I) chloride

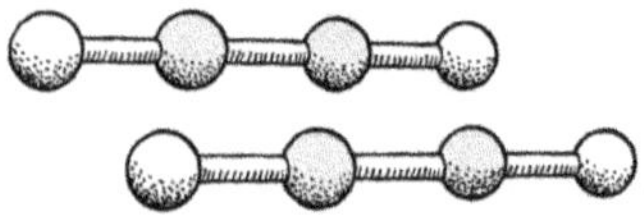

Gray balls are Hg. White balls are Cl.

HgO Mercuric oxide

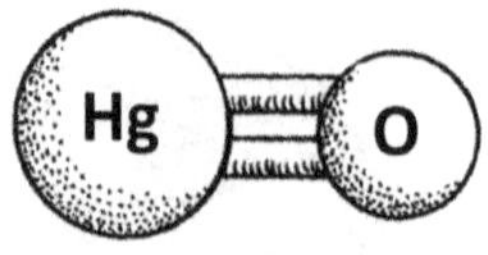

HgO is a red solid. This compound played an important role in chemical history. In 1774, Joseph Priestly used mercuric oxide to discover the element oxygen, though the name "oxygen" was given later, by Lavoisier.

HgTe Mercury telluride

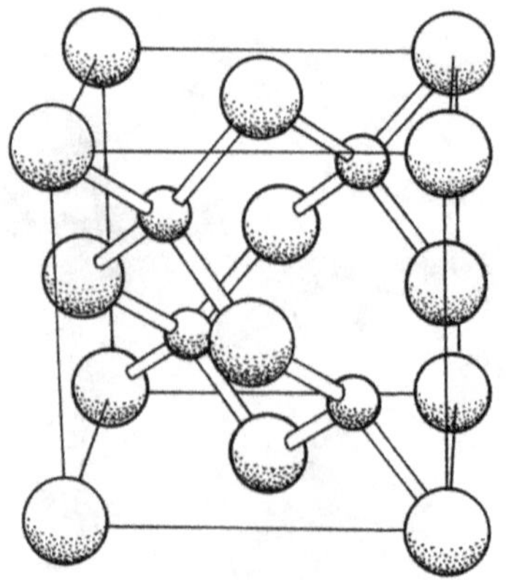

Small balls are Hg. Large balls are Te.

80 Hg

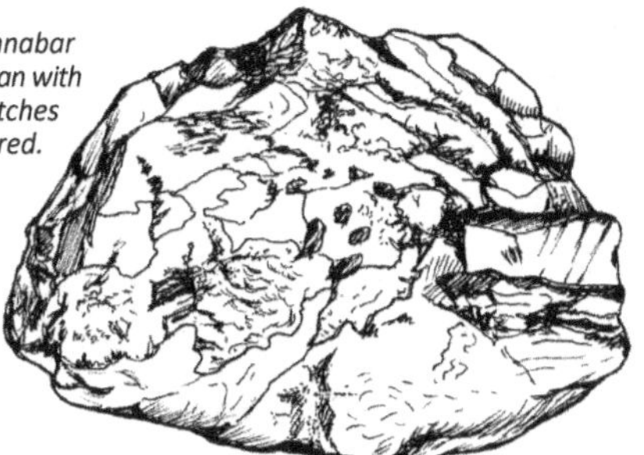

Cinnabar is tan with patches of red.

Cinnabar, HgS, is the primary ore of mercury.

"Hydragyrum"

Mercury

Fluorescent bulbs often contain mercury vapor.

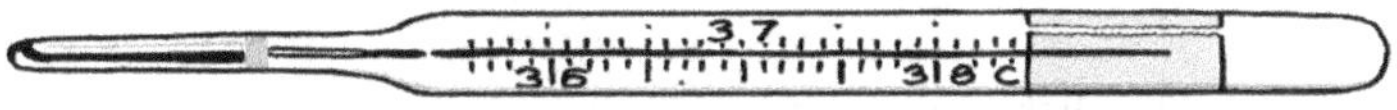

Mercury was used in thermometers until the late 1900s.

In 1643, Evangelista Torricelli demonstrated that an 80 cm tall tube of mercury could be used to make a barometer for measuring air pressure.

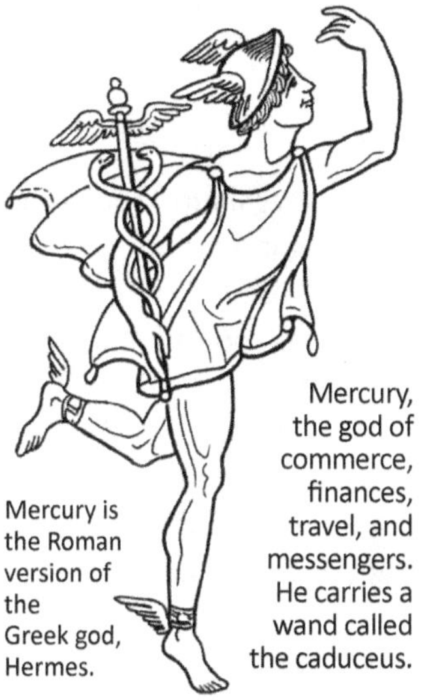

Mercury is the Roman version of the Greek god, Hermes.

Mercury, the god of commerce, finances, travel, and messengers. He carries a wand called the caduceus.

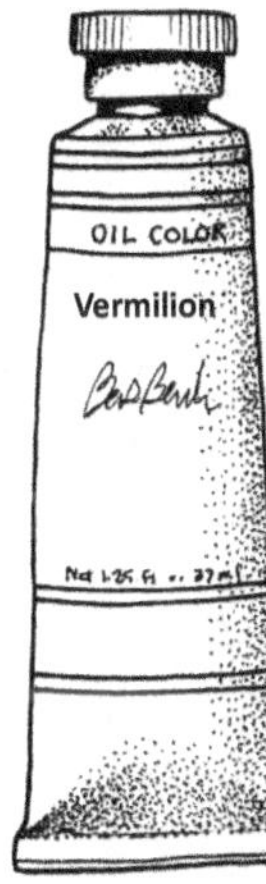

Vermilion red is made with powdered cinnabar. It is toxic so it must be used carefully.

The "Mad Hatter" from *Alice in Wonderland* was based on cases of mercury poisoning that occurred in the hat making industry.

Mercury amalgams were used to fill teeth.

The powerful fresnel lenses in the lights of lighthouses used to float on a bowl of friction-free liquid Hg.

Chinese Emperor Qin Shi Huangdi (259-210 BC) ate mercury pills every day, and built a secret underground tomb with rivers of mercury.

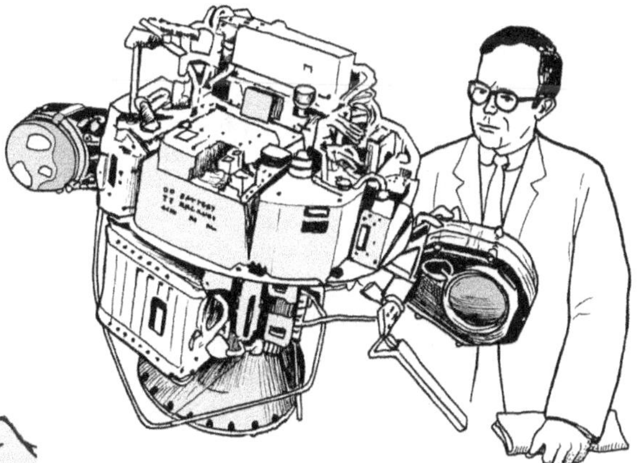

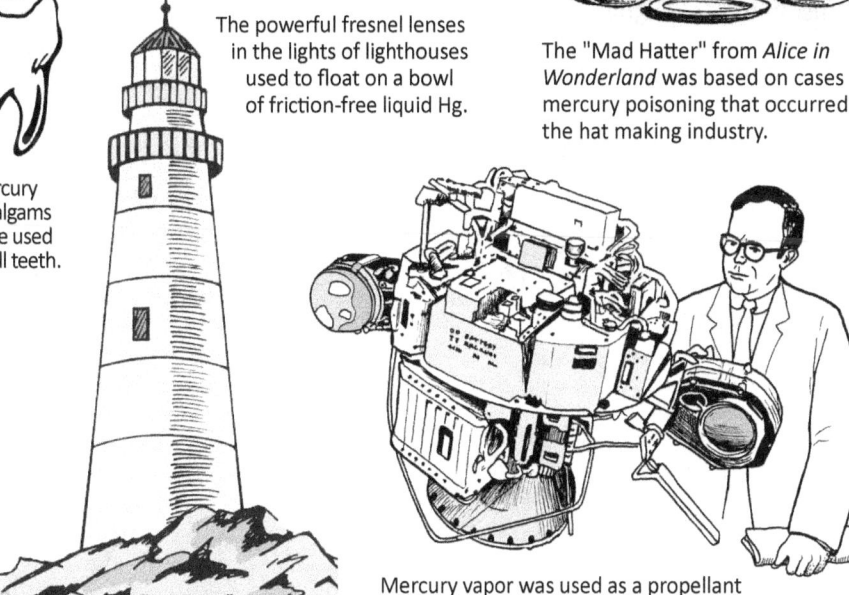

Mercury vapor was used as a propellant in early ion propulsion units in satellites in the 1960s, such as SERT-1. (We now use harmless noble gases.)

Tl

81

protons
123 neutrons
81 electrons

Atomic mass: 204.4

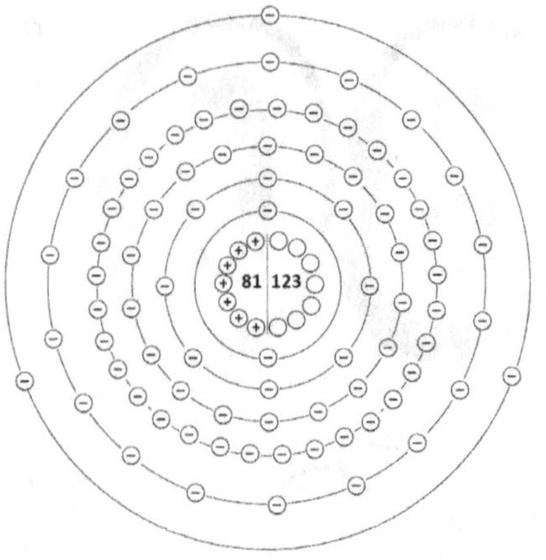

Thallium

Named using the Greek word for a green twig, "thallos"

William Crookes and Claude Lamy both discovered this element in 1861, using spectroscopy. They were working independently, not together, so this led to an ongoing controversy as to who should be recognized as the official discoverer. Crookes managed to publish his findings first and chose to name the new element thallium, using the Greek word for "green twig" because of the bright green line in thallium's emission spectrum pattern. Lamy received several awards for his work, so in the end, both men ended up receiving recognition.

A number of mineral ores contain thallium, all of which are considered relatively rare. Lorándite, $TlAsS_2$, is found in Macedonia (north of Greece) and in Tajikistan. Other thallium ores are found in copper and lead mines.

Thallium is one of the most toxic elements. Thallium sulfate was used as rat poison until it was banned in the 1970s. Thallium compounds were also used to kill germs (oral and topical antibiotics) until the mid-1900s.

If a human ingests too much thallium (usually as a result of an accident in a factory that uses thallium in its manufacturing processes) the cure is to eat a blue pigment powder called "Prussian blue." The pigment molecule, which is made of iron, carbon, and nitrogen, has the ability to grab the thallium atoms and hold them long enough that the entire molecule is flushed out of the body.

Thallium is used in the electronics industry to make photoresistors, which can sense intensity of light and adjust circuits accordingly. Photoresistors allow outdoor lights to turn on automatically after sunset. Thallium can be "doped" into crystals such as sodium iodide to make gamma ray detectors, useful in equipment inspection. More recently, some thallium compounds have been found to be superconductors, which may useful in the future.

Gold-plating is a process where a thin layer of gold can be put onto the surface of metal objects. If thallium salts are added to the electroplating solution, the process will go faster and produce a smoother surface.

The glass industry uses thallium oxide to make glass that is more dense and therefore better for certain applications, such as specialty lenses for cameras. Thallium bromide and iodide are used to make lenses that will transmit infrared light. Metal-halide lamps that produce intense light sometimes add thallium to their metal mixes.

Thallium-201, a radioactive isotope, has been used for decades in nuclear cardiography, as Tl-201 will reveal tissues that are low in oxygen. Technetium has slowly been replacing thallium for this purpose, but thallium is still used.

TlCl Thallium chloride

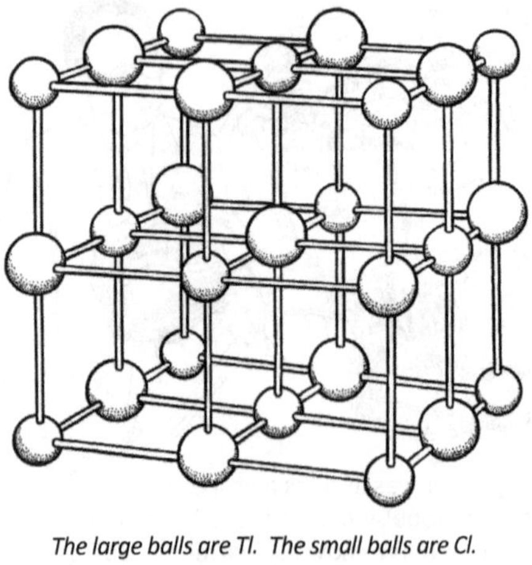

The large balls are Tl. The small balls are Cl.

Tl_2SO_4 Thallium sulfate

Thallium sulfate was used to poison rats.

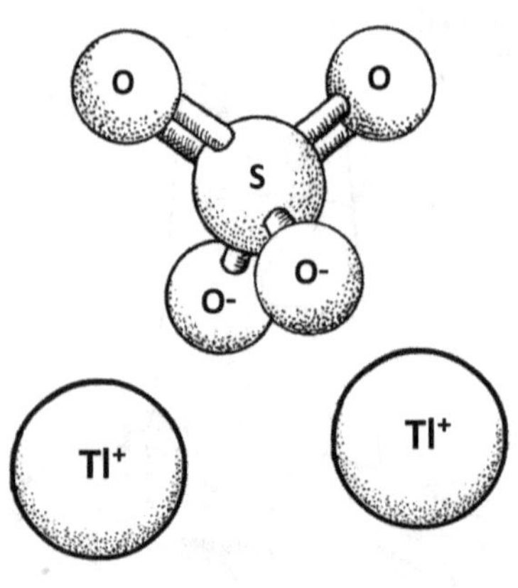

TlI Thallium iodide

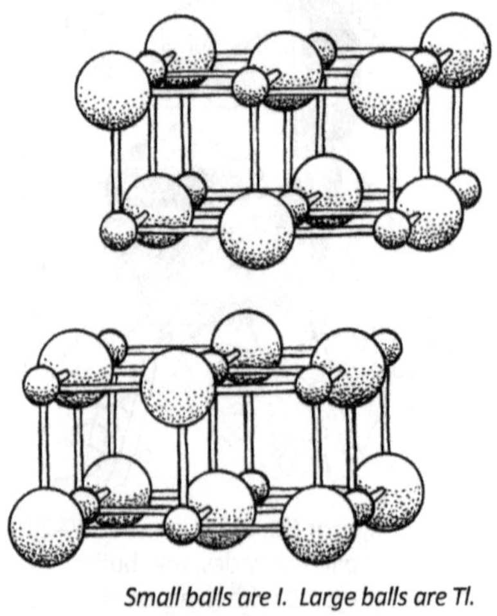

Small balls are I. Large balls are Tl.

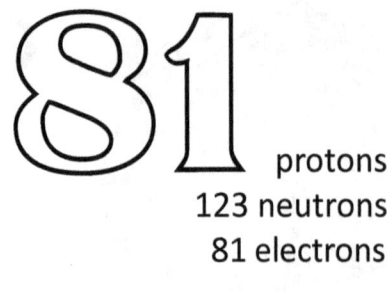

81 Tl

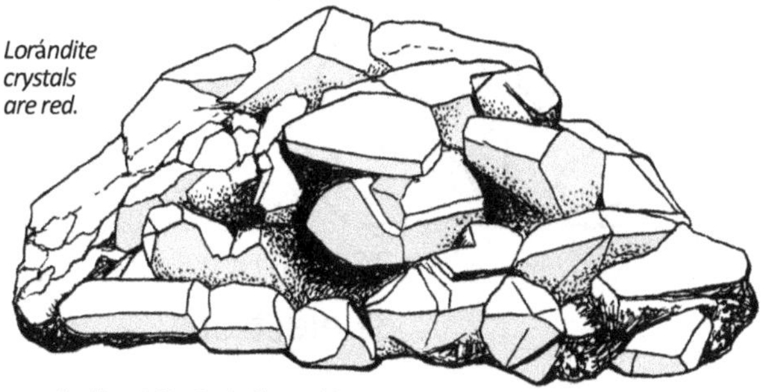

Lorándite, $TlAsS_2$, is found in Macedonia and Tajikistan.

Thallium

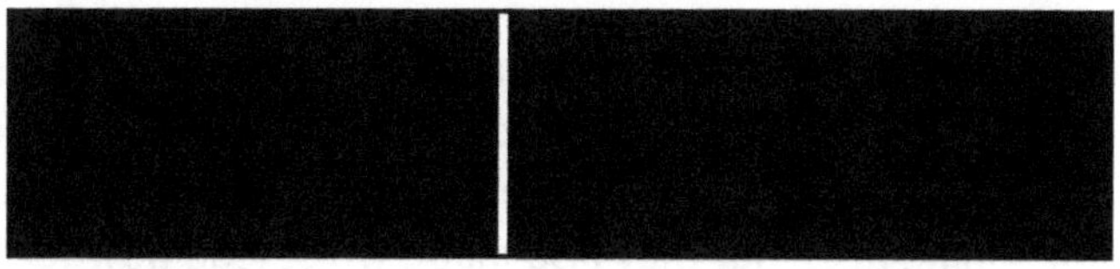

Thallium's name came from its unusual spectrum: a bright green line.

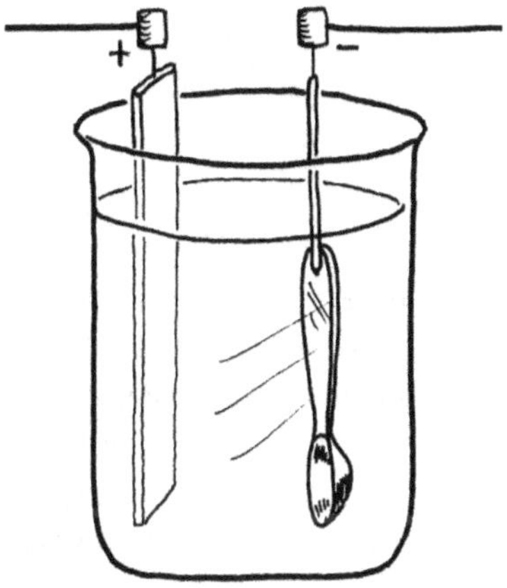

This spoon is being electroplated with gold (from the gold bar on the left). Thallium in the solution helps to speed up the process.

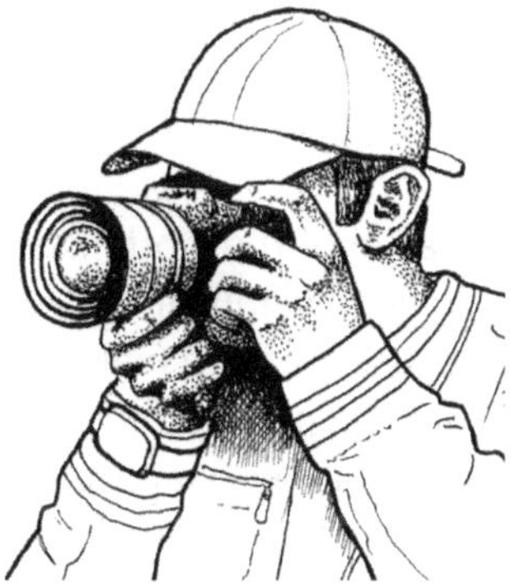

Thallium is used to make special lenses for cameras.

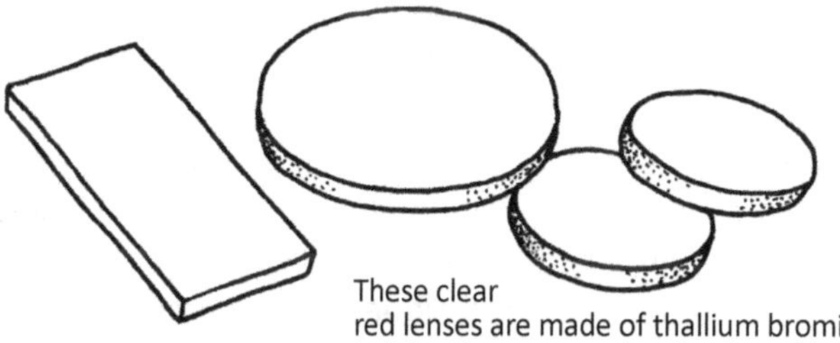

These clear red lenses are made of thallium bromide iodide. They are used with infrared light.

B=black P=purple O=orange

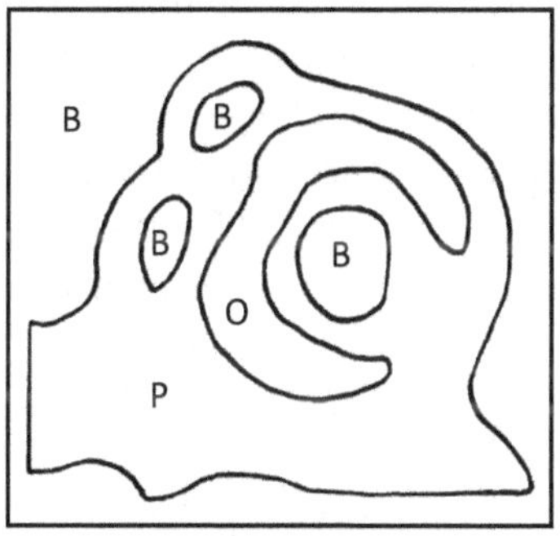

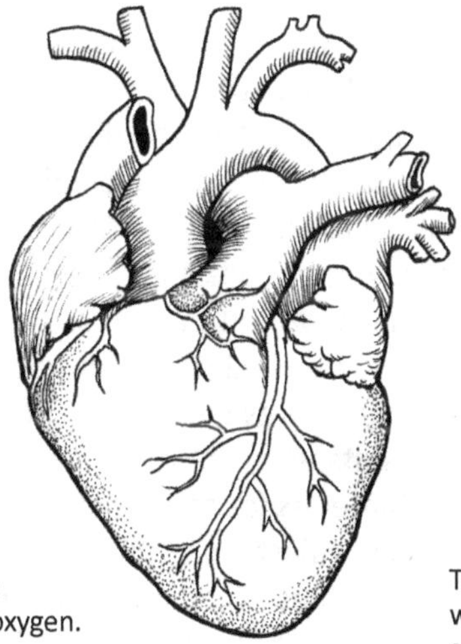

Thallium-201 is used in nuclear cardiography (heart imaging). Thallium highlights tissues not getting enough oxygen.

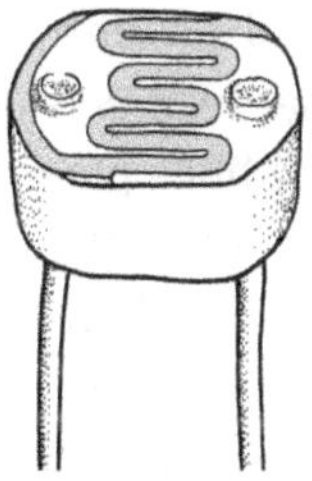

photoresistor

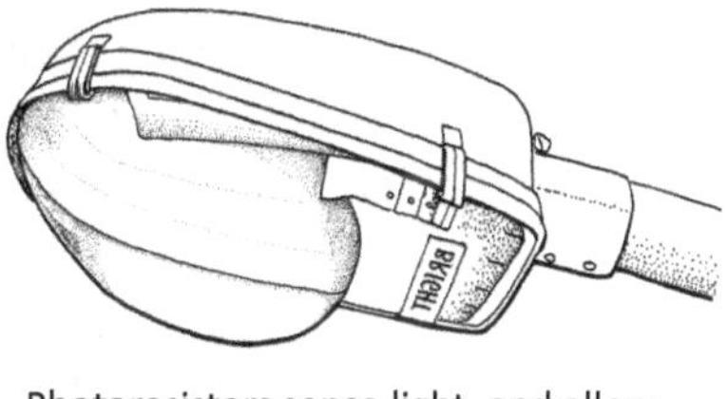

Photoresistors sense light, and allow streetlights to turn on automatically.

Thallium sulfate was used as rat and ant poison. It was banned in the 1970s because of its toxicity.

Gamma ray detectors can use thallium-doped sodium iodide crystals.

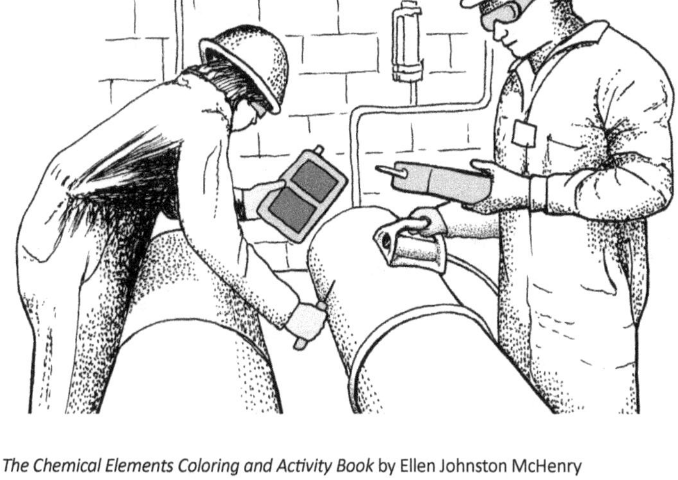

When humans accidentally consume a toxic dose of thallium, the cure is to eat "Prussian blue," a pigment powder used by artists.

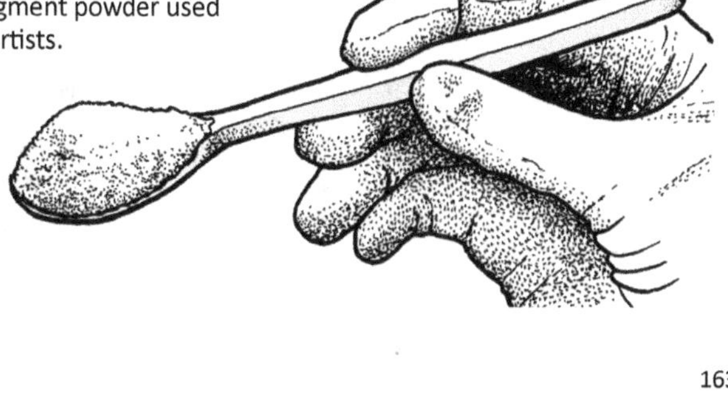

Pb 82

82 protons
126 neutrons
82 electrons

Lead

Atomic mass: 207.2

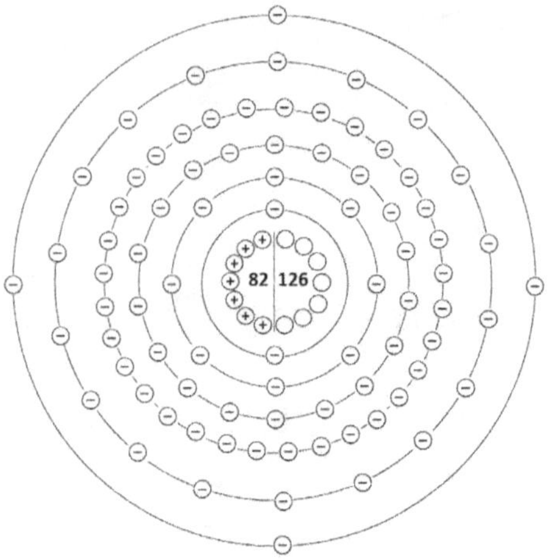

From the old English word "lēad"
"Pb" is from the Latin word for lead, "plumbum"

The element lead has a long history. Ancient cultures all over the world used it because it was easy to find and was not hard to extract from its ores. The primary ore is galena, a mix of lead and sulfur, PbS, which was often found in areas where they were mining for silver. The ancients used lead to make beads and coins, and for weights for fishing nets, but they also made white lead powders for painting pots and for making cosmetics and medicines.

By the time of the Roman empire, lead had many uses: roof tiles, water pipes, coins, writing tablets, weights for fishing nets, weights used as ammunition (in slings), and even, unfortunately, containers for holding food. It was not until the 20th century that the toxicity of lead was completely understood.

During the Middle Ages, lead was used to hold the pieces of glass in the giant stained glass windows in the cathedrals. Also during this time, lead saw widespread use in roof tiles and water pipes. When the printing press was invented, lead alloys were used to create the small blocks on which the letters were carved.

During the Renaissance, lead compounds were used to make white cosmetic powder. It was considered high fashion, and a symbol of modesty, for women to whiten their faces. Lead was also used to make white paint. Lead paints continued to be used well into the 20th century because they had the desirable quality of being very opaque.

Beginning in the 1700s, tin mixed with lead was used to make toys, including "tin soldiers." By the end of the 1800s, it was becoming clear that lead was dangerous, so toy manufacturers stopped using lead. Other uses for lead during this time included making huge pipes for pipe organs, as ballast in the bottom of sailboats, and for making spherical bullets ("shot") for muskets and flintlock guns. (Galena was often used as a source of lead for musket balls.)

In the 1920s, a new use was found for lead. A lead compound called tetraethyl lead was found to reduce the knocking sound make by gasoline engines. This "leaded gas" was banned by the US in the 1970s, and by the year 2000 most other countries had also passed laws against its use.

The largest use of lead today is in making lead-acid batteries for cars, boats, and back-up power supplies. Another major use is for making shielding devices that protect against x-rays and gamma rays. Lead aprons provide protection for patients and technicians using medical x-ray machines. Lead is a minor ingredient in various compounds used in acoustics and fiber optics. For example, lead-zirconium-titanate is used in ultrasound equipment.

$(CH_3CH_2)_4Pb$ Tetraethyl lead

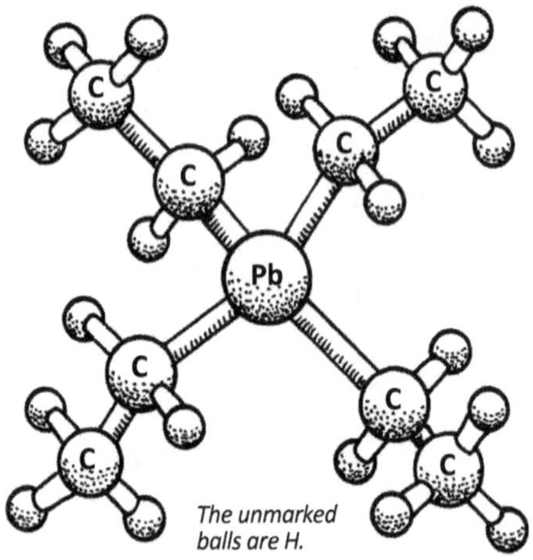

The unmarked balls are H.

$PbSO_4$ Lead sulfate

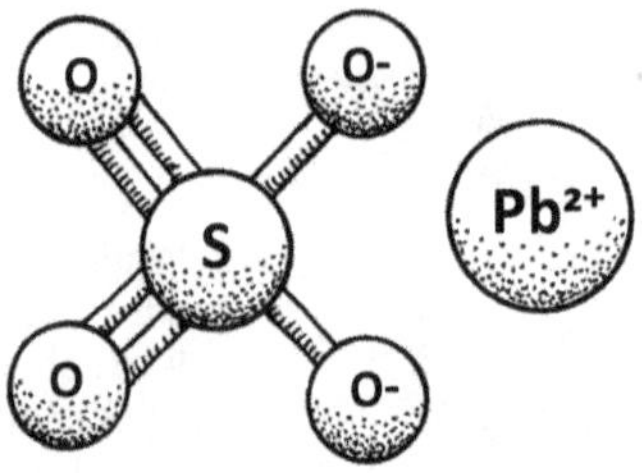

PbO Lead (II) oxide

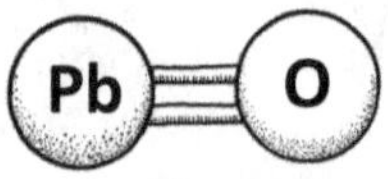

This molecule will attach to others like it and create a large crystal shape.

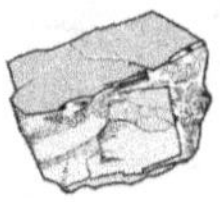

Galena

PbS Lead (II) sulfide (Galena)

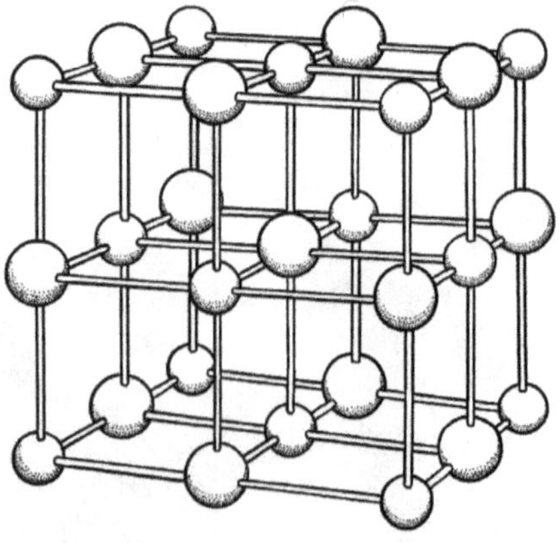

The large balls are Pb. The small balls are S.

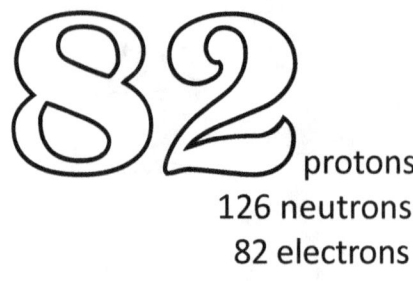

82 Pb

Lead

Organ pipes are made of an alloy of lead and tin.

In the Middle Ages, lead was used to make stained glass windows.

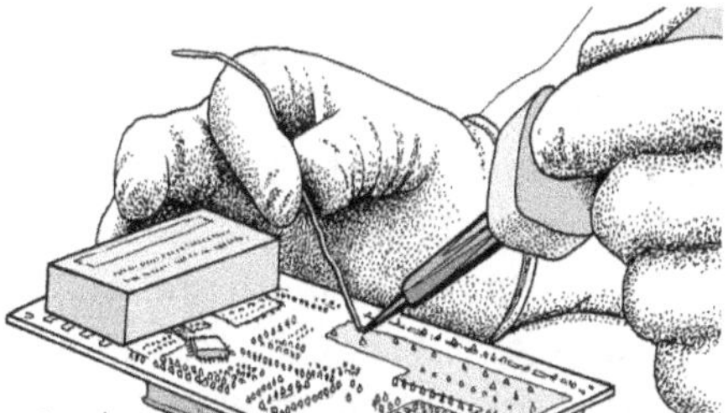

Lead alloys are used for soldering.

Deep sea divers would wear heavy lead belts and shoes to stop them from floating.

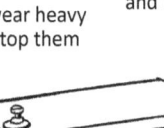

Toy tin soldiers were made of a mixture of tin and lead.

600 tons of lead were used to stabilize the Leaning Tower of Pisa.

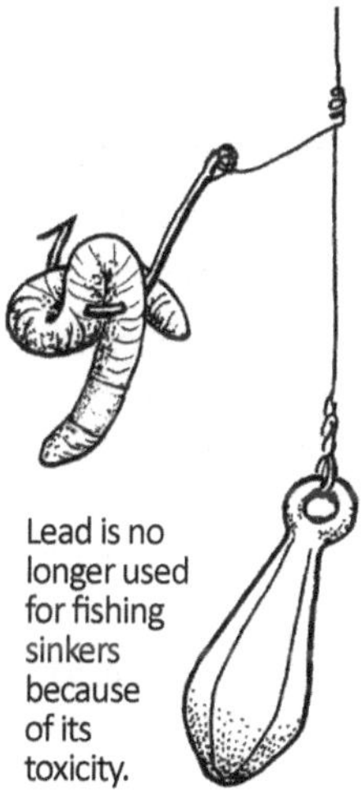

Lead is no longer used for fishing sinkers because of its toxicity.

Queen Elizabeth I had her face painted white, according to the fashion of the day. White cosmetics contained lead and probably made many people sick.

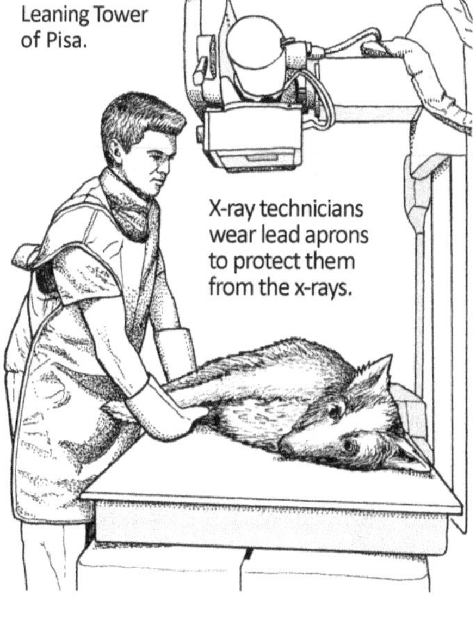

X-ray technicians wear lead aprons to protect them from the x-rays.

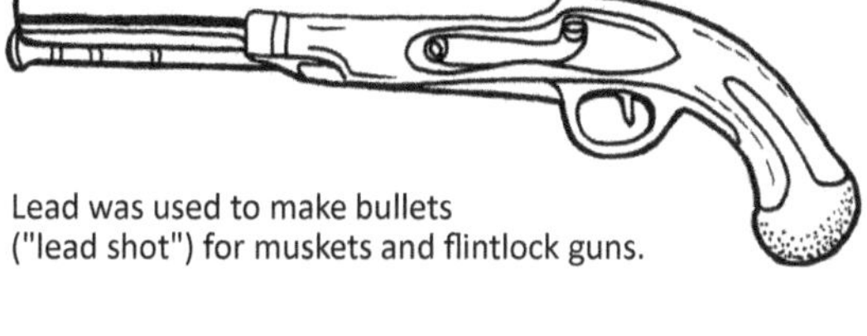

Lead was used to make bullets ("lead shot") for muskets and flintlock guns.

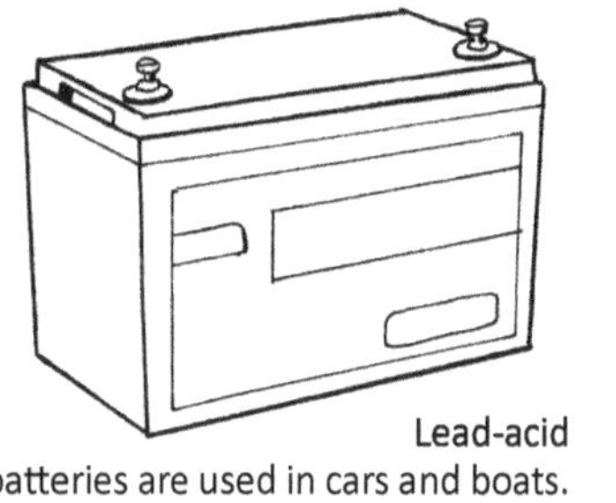

Lead-acid batteries are used in cars and boats.

Bi 83

Bismuth

83 protons
126 neutrons
83 electrons

Atomic mass: 208.9

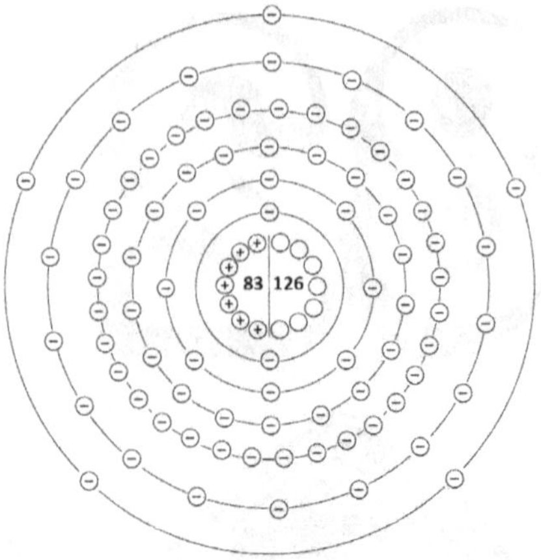

From the German word "bismuth" (or "wismut")

Bismuth has been known since ancient times and was used in alloys that included copper, tin or lead. Even the ancient Incas (in Peru) used bismuth alloys to make tools and weapons. Because bismuth looked a lot like silver, miners during the Middle Ages called bismuth "tectum argenti" meaning "silver being made." The miners thought silver deposits were the result of a long geological process and bismuth was one early step in this process.

Natural bismuth ores don't look spectacular, but modern chemists have found a way to make pure bismuth crystals (called "hopper crystals") that have an almost unbelievable square geometry and an amazing iridescent surface that shimmers with all the colors of the rainbow. The iridescence comes from oxygen reacting with the surface. Iridescent bismuth compounds, such as bismuth oxychloride, are used to make iridescent paints and nail polishes. Bismuth oxychloride gives a pearly white that is used to simulate "mother of pearl."

The use of bismuth for medical purposes began hundreds of years ago. Bismuth compounds were used to treat digestive problems and skin diseases. We still use some of these medicines today. The "bis" in Pepto-Bismol® represents the active ingredient, bismuth subsalicylate. Another bismuth compound is used for eye infections.

Bismuth vanadate is used to make yellow pigments, especially "lemon yellow." Bismuth yellows can be used instead of cadmium yellows, as bismuth is less toxic than cadmium.

Because bismuth is almost as dense as lead it has been used as a replacement for lead in things like bullets. fishing sinkers, and soldering wire. It was discovered that duck hunters would shoot dozens of lead "shot" into ducks which would fall into bodies of water and not be recovered. The lead would eventually dissolve into the water and begin to contaminate the ecosystem. Lead bullets have been banned in most countries. Soldering wire used to have a high lead content, but other metals, including bismuth, are now often used to make lead-free solder. Bismuth's similarity to lead also makes it able to protect against x-rays, so it can be used to make protective shields.

Bismuth germanate glows when exposed to radiation, so it is useful in medical imaging such as CAT scans.

Bismuth alloys are found in the sensors in sprinkler systems because they will melt when fire touches them.

Bi_2S_3 Bismuth sulfide

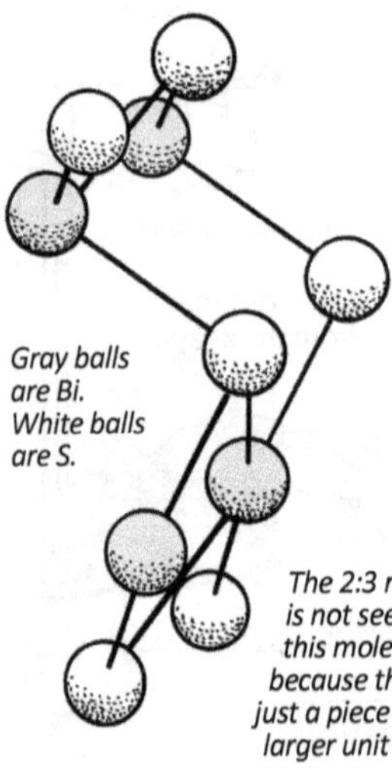

Gray balls are Bi. White balls are S.

The 2:3 ratio is not seen in this molecule because this is just a piece of a larger unit cell.

$BiCl_3$ Bismuth chloride

Unmarked balls are Cl.

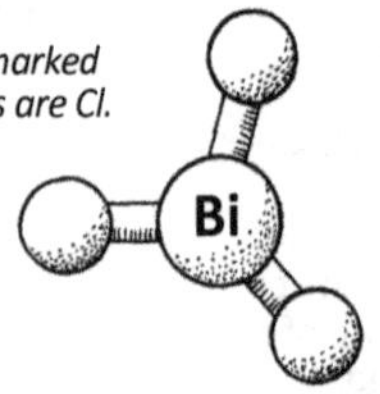

BiF_5 Bismuth fluoride

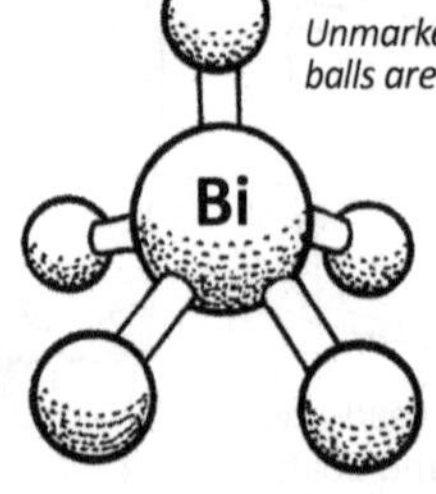

Unmarked balls are F.

BiClO Bismuth oxychloride

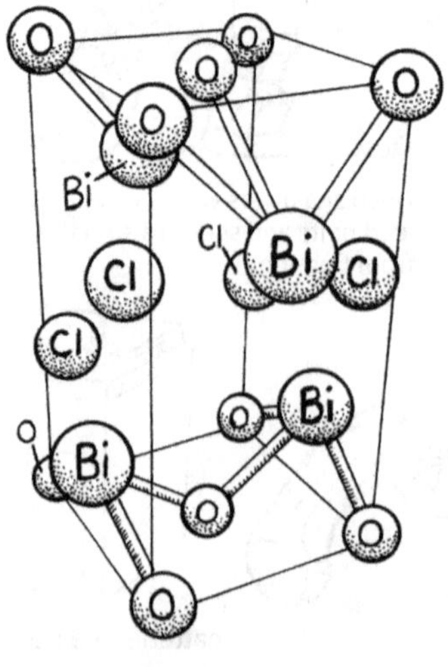

$C_7H_5BiO_4$ Bismuth subsalicylate

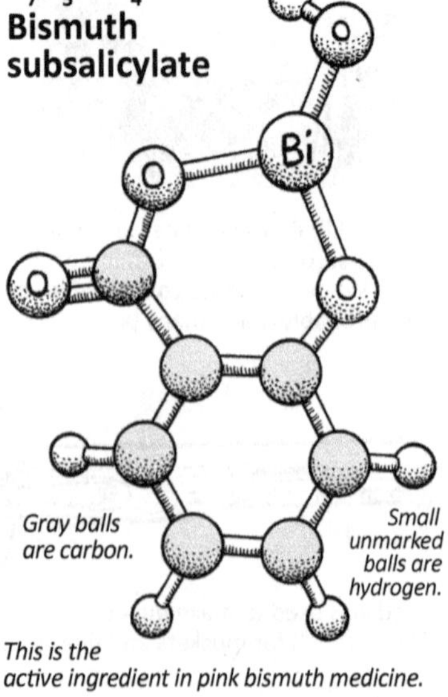

Gray balls are carbon.

Small unmarked balls are hydrogen.

This is the active ingredient in pink bismuth medicine.

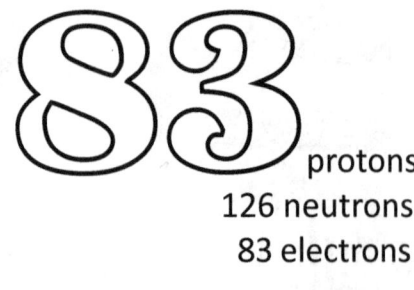

83 Bi

Bismuth

Bismuth is the "bis" in Pepto-Bismol®, an over-the-counter medicine that soothes upset stomach.

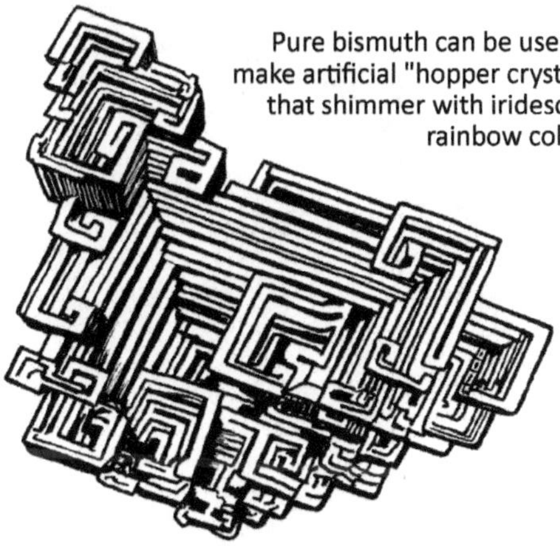

Pure bismuth can be used to make artificial "hopper crystals" that shimmer with iridescent rainbow colors.

Bismuth vanadate is used to make several shades of yellow paint.

Iridescent nail polish can be made with bismuth.

Bismuth is used in alloys that replace lead solder.

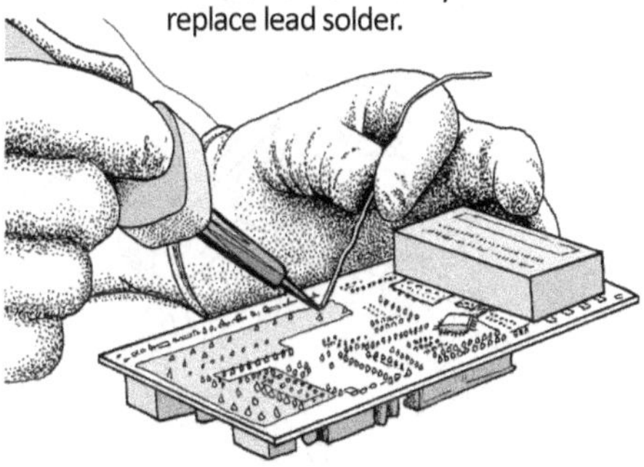

This engraving shows the processing of bismuth in the 1700s.

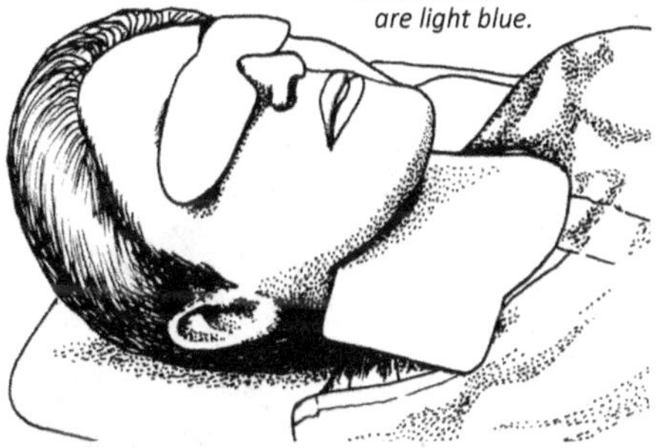

Bismuth is similar enough to lead that it can be used to make radiation shields.

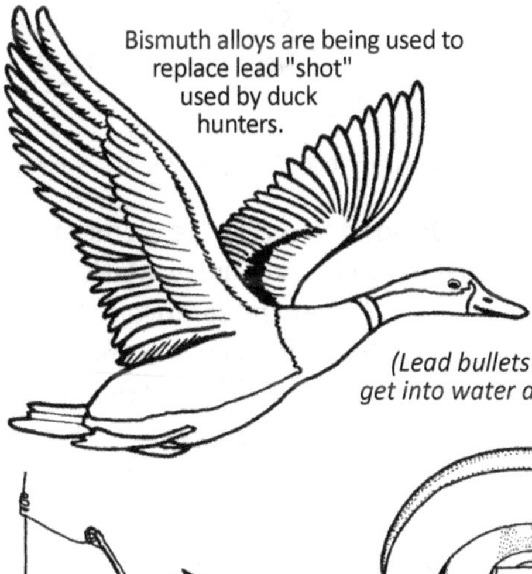

Bismuth alloys are being used to replace lead "shot" used by duck hunters.

(Lead bullets are one way that lead can get into water and affect ecosystems.)

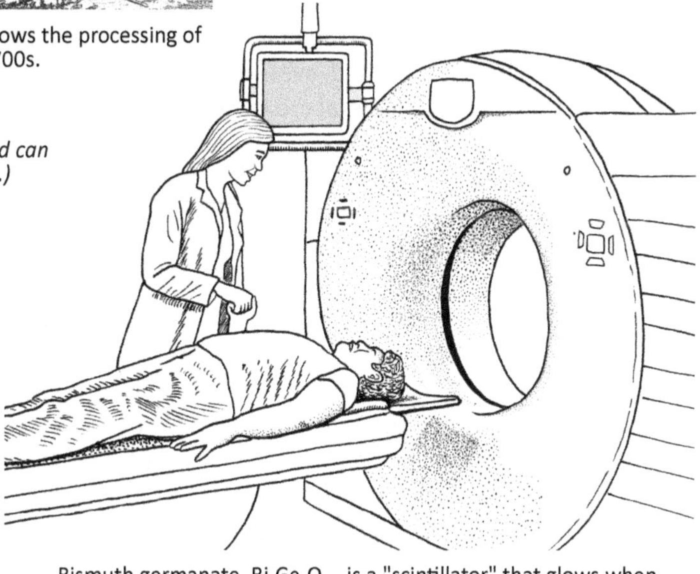

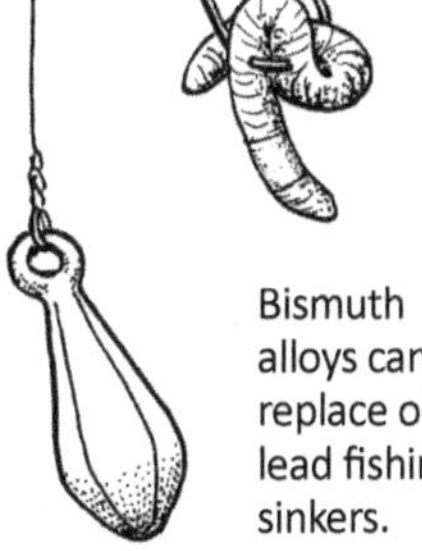

Bismuth alloys can replace old lead fishing sinkers.

Bismuth alloys are used to make fire sensors for sprinkler systems.

Bismuth germanate, $Bi_4Ge_3O_{12}$, is a "scintillator" that glows when exposed to x-rays, making it useful in CAT scan machines.

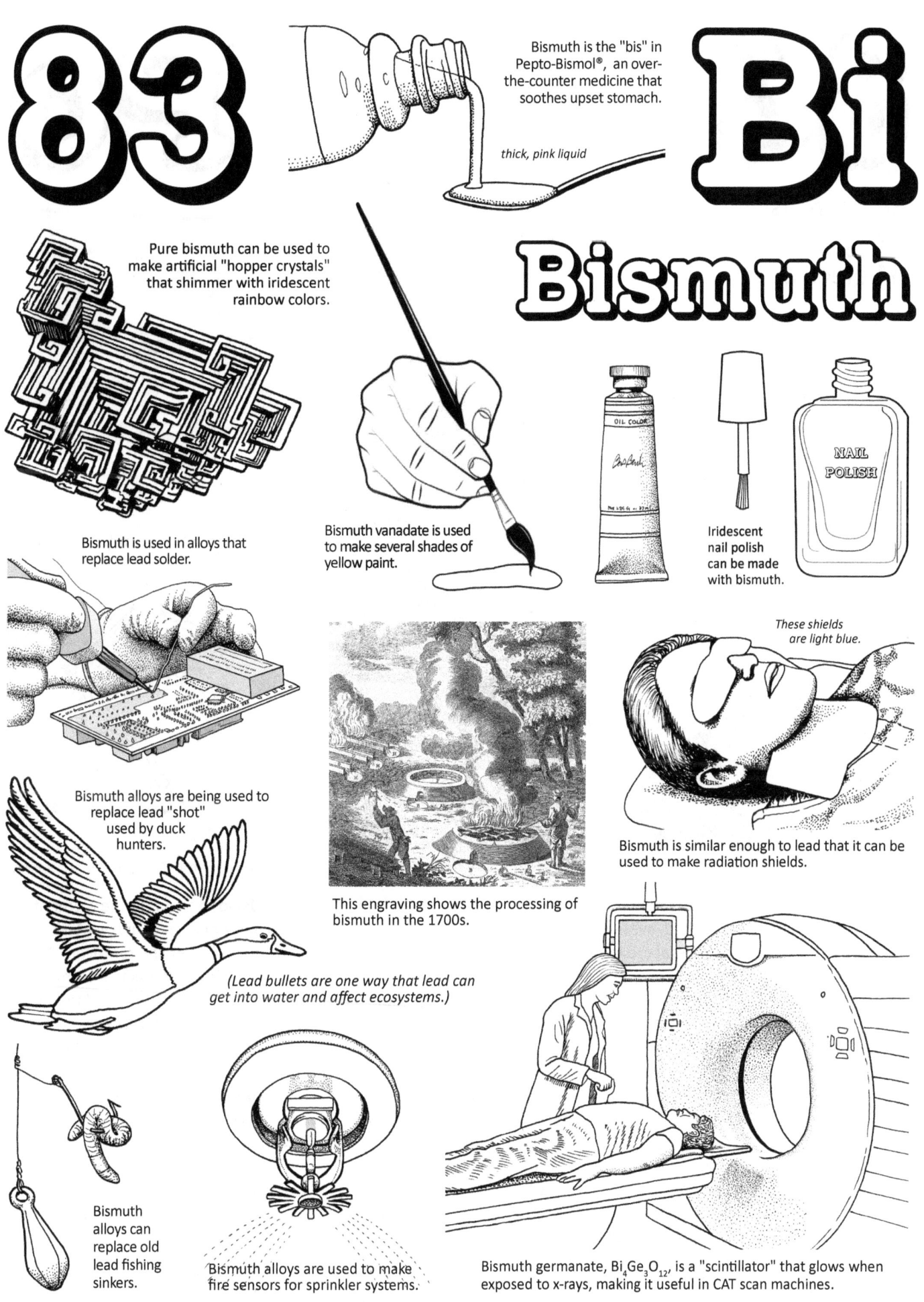

84 Polonium Po

FLAG OF POLAND

Top stripe is white, bottom stripe is red. Area around eagle is red.

84 protons
125 neutrons
84 electrons

Atomic mass: 209

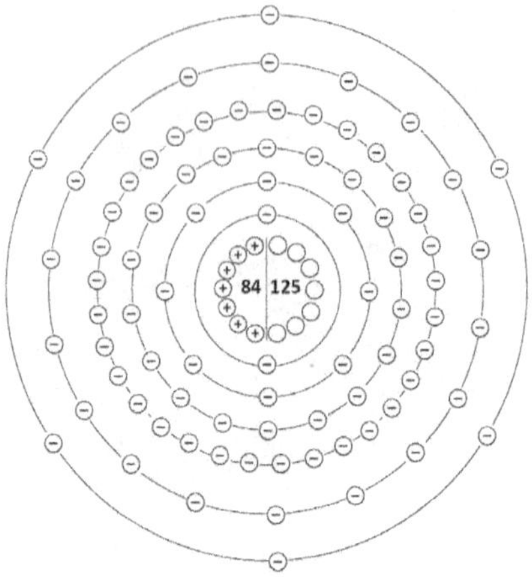

This symbol means "radio-active." It is often black and yellow.

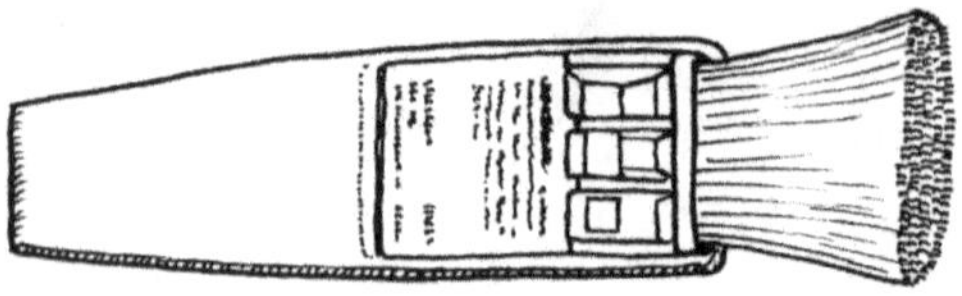

This is an anti-static brush, used to remove electrical charges from surfaces such as photographic plates and sheet plastics.

Marie Curie was born in Poland and moved to France.

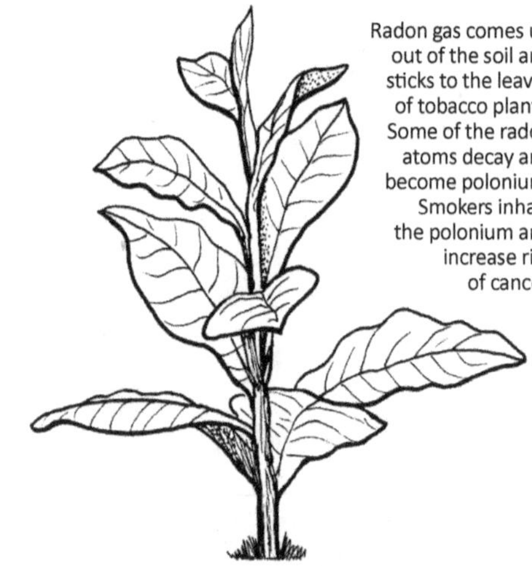

Radon gas comes up out of the soil and sticks to the leaves of tobacco plants. Some of the radon atoms decay and become polonium. Smokers inhale the polonium and increase risk of cancer.

PoO_2 Polonium (IV) oxide

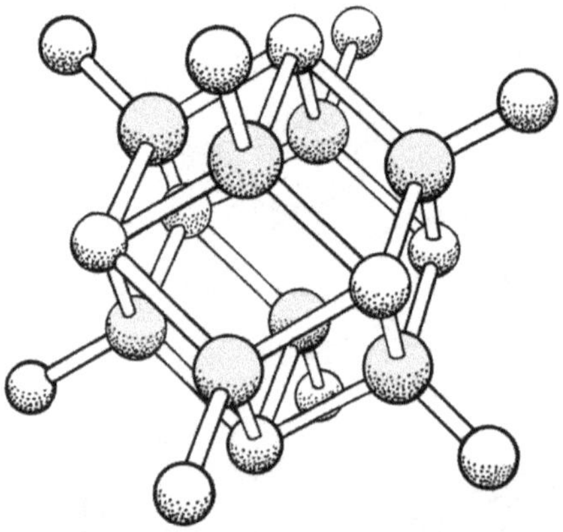

The light gray balls are O, oxygen. The others are Po.

Polonium was discovered In France by Marie and Pierre Curie in 1898. They decided to name the element after Marie's homeland, Poland, even though at that time Poland did not exist as an independent country. Marie had been investigating a mineral ore called pitchblende, which contained a lot of uranium. After she removed all the uranium and thorium from the pitchblende, it was still radioactive, so there had to be another radioactive element still in the ore. After extracting all the polonium, the ore was still radioactive! Five months later they found radium.

Polonium is extremely rare because all but one of its isotopes exist for only a few minutes. Essentially, the element polonium is nothing more than a temporary step in the decay process of larger radioactive elements. Only Po-210 exists long enough to be useful. Today, Po-210 is produced in nuclear labs, usually by bombarding bismuth (Bi-209) with protons. If a proton sticks to a bismuth atom, it immediately becomes polonium, since it is the number of protons that determines the identity of an element. Po-210 was produced in a research facility in Dayton, Ohio, as part of the Manhattan Project during World War II, and was combined with BeO, beryllium oxide, to make the detonators (starting the fission process) for the atomic bombs.

Today, Po-210 can be used safely in labs for educational purposes. Po-210 was used in the 1960s and 70s in Russia's designs for lunar rovers and some satellites, using the heat created by the polonium to keep the equipment from freezing. Also, Po-210 can be used safely in anti-static devices that remove electrical charges from surfaces such as photographic plates and plastic sheets. As polonium decays, it emits alpha particles (made of 2 protons and 2 neutrons) and these alpha particles cause the removal of the charged ions from the surfaces.

85 Astatine At

from Greek "astatos" meaning "unstable"

85 protons
125 neutrons
85 electrons

Atomic mass: 210

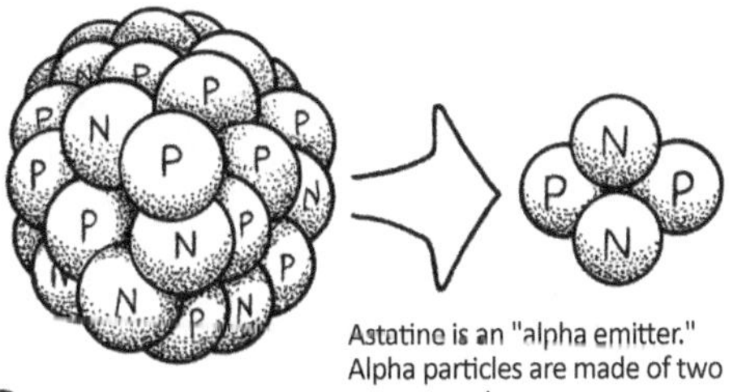

Astatine is an "alpha emitter." Alpha particles are made of two protons and two neutrons.

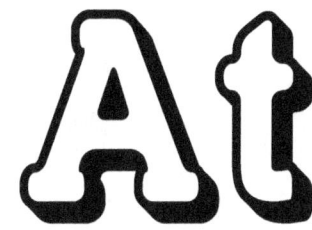

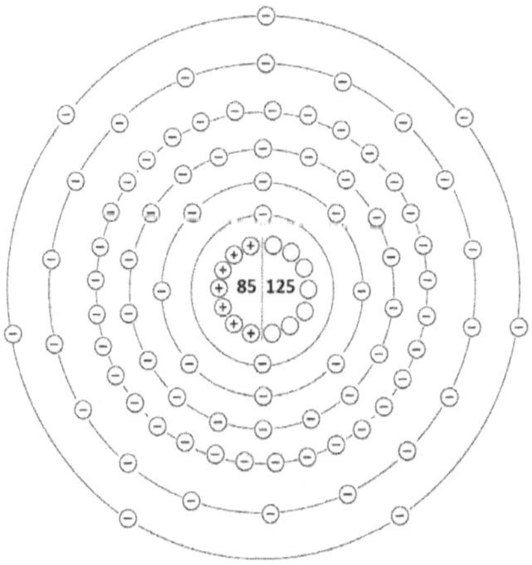

This symbol means "radioactive." It is often black and yellow.

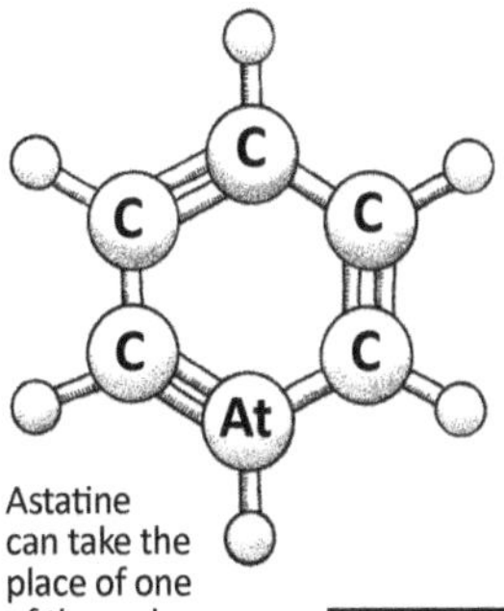

Astatine can take the place of one of the carbons in the benzene molecule.

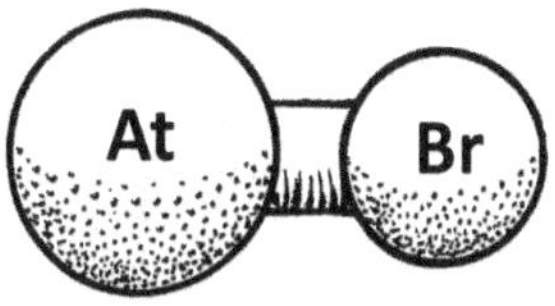

Halogens (F, Cl, Br, I) don't usually connect to form molecules, but astatine will bond with other halogens.

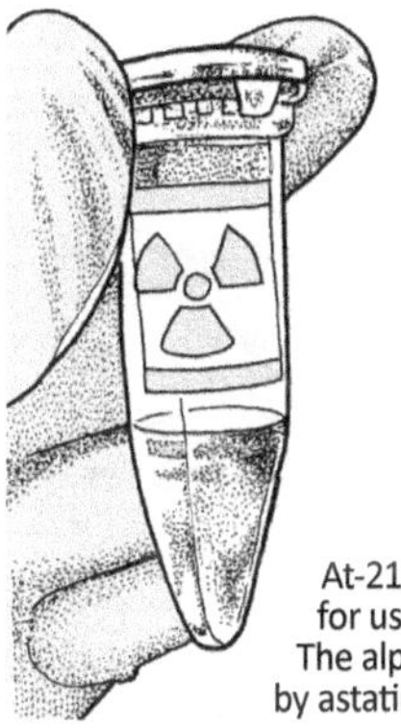

At-211 is being investigated for use in nuclear medicine. The alpha particles produced by astatine can kill tumor cells.

This picture shows the nuclear lab at the University of California at Berkeley in 1940. Their cyclotron machine was used to create the first manmade atoms of astatine.

Astatine is the most rare naturally-occurring element on earth. If you searched for astatine in the combined land masses of North and South America, to a depth of 10 miles (16 km), you would only be able to collect about a trillion atoms, which would be a microscopic speck. Any astatine you found would have been recently produced because astatine is very unstable; even the longest-lived isotope has a half-life of only about 8 hours. As larger radioactive atoms such as uranium and thorium begin to fall apart they emit alpha particles which contain two protons. Since the number of protons defines the identity of an element, the loss of protons causes the atom to change its identity. This can happen many times, in a process called a decay chain. Astatine is one "stop" in the decay chain. The decay process stops when the number of protons reaches a stable number (usually 82, lead, or 83, bismuth).

Many scientists in the early 1900s claimed to have discovered element 85, but the discovery could never be solidly verified. Then, in 1940, three scientists working at the University of California in Berkeley did an experiment with the new cyclotron machine. This machine could "throw" alpha particles at atoms, and thereby increase the number of protons in a nucleus. When they bombarded bismuth (83) with alpha particles, atoms with 85 protons were created. This was the first time in history that a new element had been discovered by creating it in a lab. Of course, their newly discovered element soon decayed and turned back into bismuth again. No one has ever been able to produce enough astatine so that its properties can be studied. We don't know what a large sample of astatine would look like. However, we can take a good guess because it is the bottom element in the halogen column, and we know the properties of the other halogens. Their color gets darker as you go down the column, so astatine would probably be very dark, perhaps even black.

86 Radon Rn

86 protons
136 neutrons
86 electrons

Atomic mass: 222

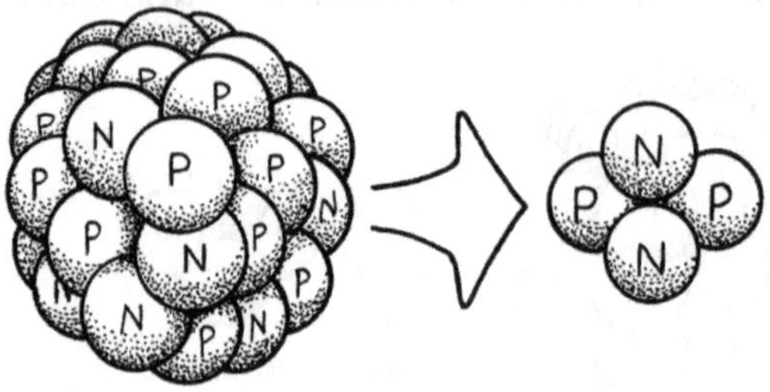

Rn-222 emits an alpha particle and turns into Po-218.

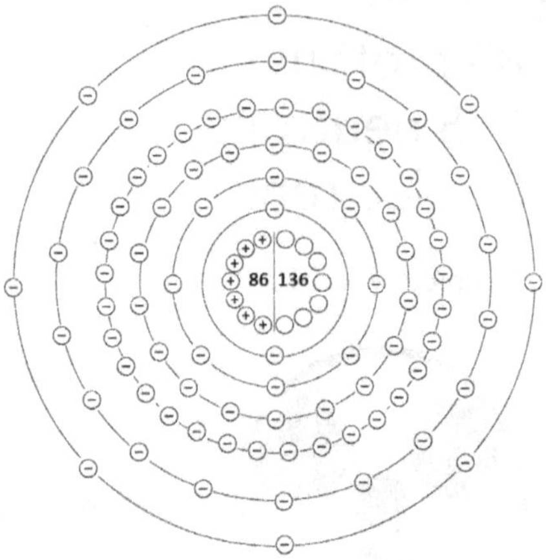

This symbol means "radioactive." It is often black and yellow.

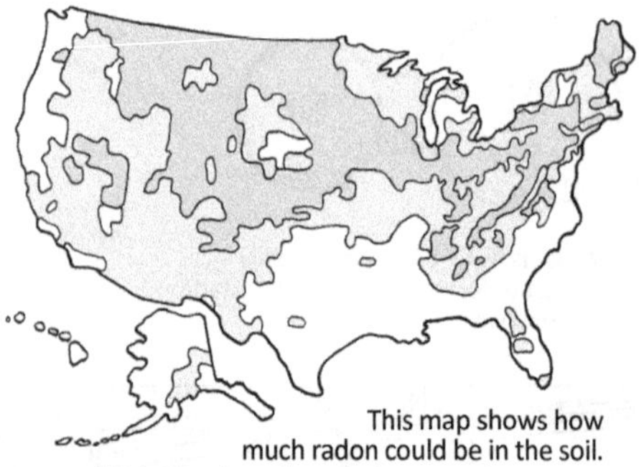

This map shows how much radon could be in the soil. Light areas have less, darker areas have more

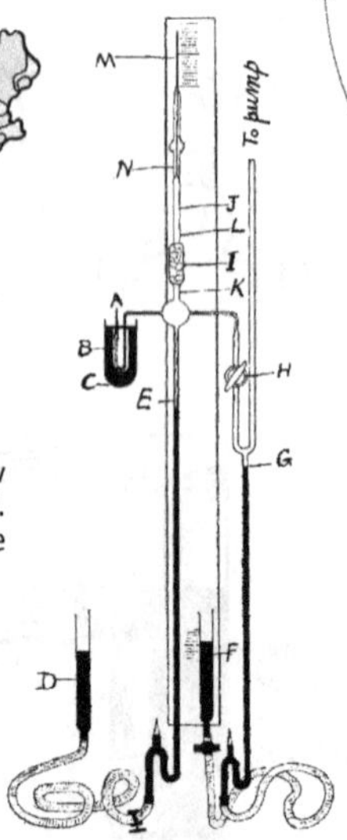

The discoverers of radon made this drawing of the equipment they used. The tubes are glass. The dark areas are mercury. Radon is in the "M" tube.

RnF_2 Radon difluoride

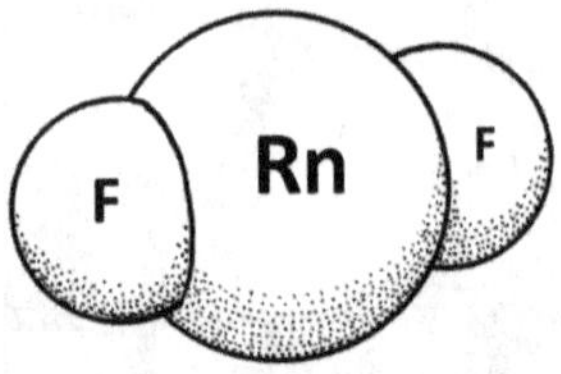

This is a "space filling" model that doesn't have sticks.

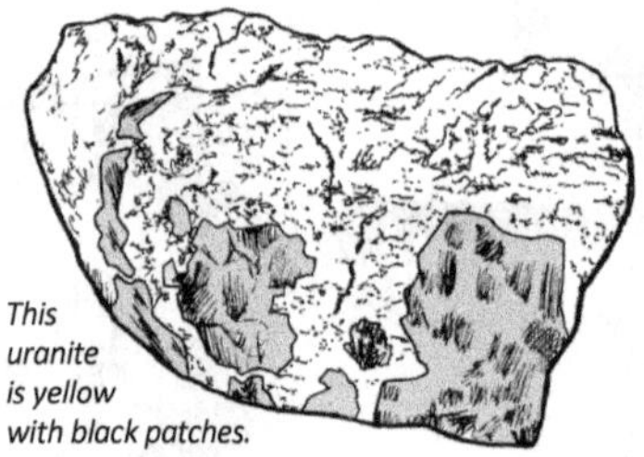

This uranite is yellow with black patches.

Uranite, UO_2, is one of the minerals that produces radon gas as the uranium decays.

Radon will seep into groundwater and come up into bodies of water. Scientists can measure radon in water to learn about the rocks underneath.

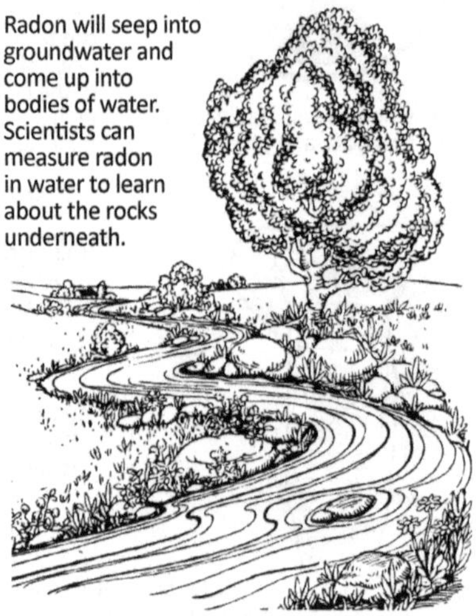

Radon is the largest and heaviest of the noble gases. Radon atoms are very unstable and will begin to fall apart in about 4 days, ejecting an alpha particle (two protons and two neutrons) to become polonium-218. If you had a balloon filled with radon gas, four days later half the radon in your balloon would have turned into polonium. In another four days, you'd have only a quarter of the radon you started with. Your radon would have originally started out as uranium, radium or thorium. These solid elements decay to produce radon gas.

In 1899, Marie and Pierre Curie noticed that their radium samples seemed to be producing a radioactive gas. They did not know that this was actually a new element. In the decade following, many scientists observed similar radioactive gases coming from other radioactive elements. Then, in 1909, William Ramsay and Robert Whytlaw-Gray isolated enough radon to be able to determine its atomic number, mass, density and melting point.

Like the other noble gases, radon is not reactive and does not like to bond to other atoms. However, in 1962, researchers in the U.S. were able to force radon to bond to fluorine, making radon difluoride.

Radon-222 comes mainly from the decay of uranium atoms. Large amounts of uranium are found in an ore called uranite, UO_2, also known as pitchblende. Smaller amounts of uranium are found in granites and in some shales. Areas that have a lot of uranium underground will have radon gas coming up out of the rocks and soil and entering the groundwater. If the radon gas goes off into the atmosphere it causes little harm. However, if radon gas leaks into mines or buildings, humans can be exposed to unsafe levels of radioactivity. Radon-222, a gas, decays by losing an alpha particle, and becomes polonium-218, a solid which will stick to lung tissue. Po-218 will continue to decay into smaller radioactive elements and with each decay, dangerous particles, or gamma rays, will shoot out from the atom, damaging the nearby cells. Longterm exposure to radon has been linked to higher risk of lung cancer.

87 Francium Fr

87 protons
136 neutrons
87 electrons

Atomic mass: 223

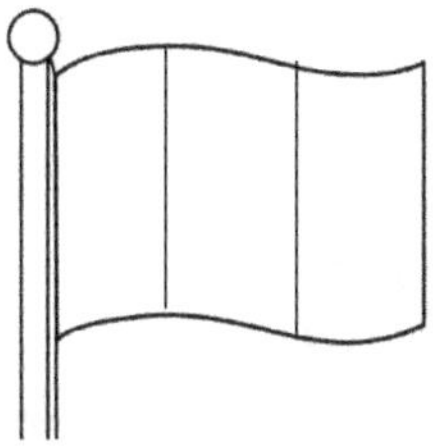

Marguerite Perey was a student of Marie Curie.

This uranite is yellow and black.

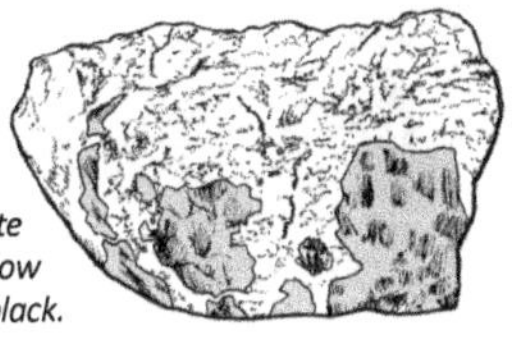

This sample of uranite contains about 100,000 atoms of francium.

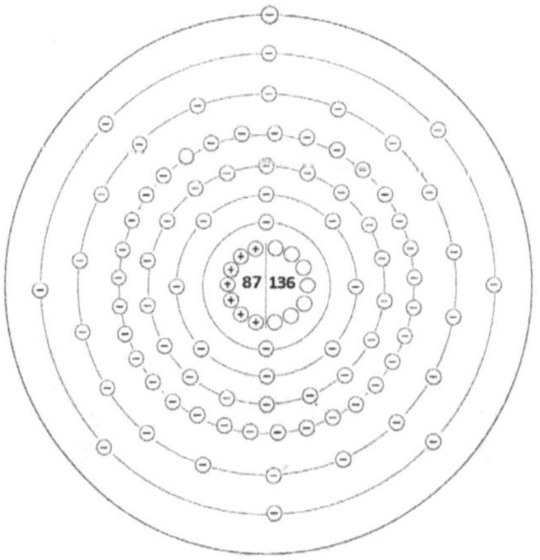

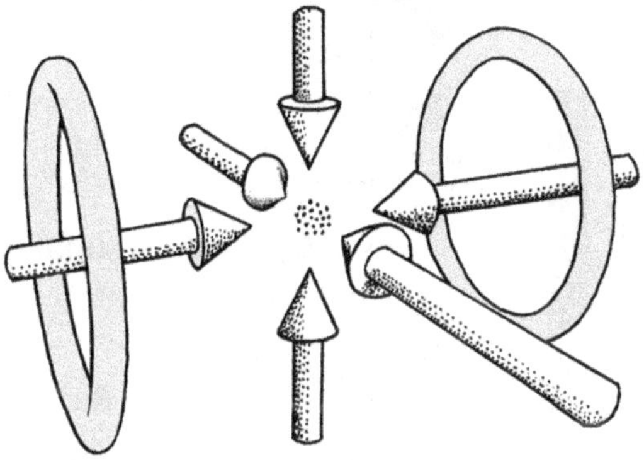

Researchers have been able to catch and hold up to 300,000 francium atoms in a device called a magneto-optical trap (MOT).

Francium can be made in particle accelerators by smashing oxygen atoms into gold atoms.

The existence of francium was predicted by Dmitri Mendeleyev, who invented the Periodic Table in 1869. He thought there should be an element underneath cesium in the first column. He knew that the element would have one electron in its outer electron shell and be very reactive. What he could not have imagined is that element 87 would be radioactive, because radioactivity had not been discovered yet.

In the early 1900s, many scientists claimed to have discovered element 87, but each time their claim was eventually proven to be false. The names they proposed for this element included russium, alkalinium, virginium, moldavium, and catium. In 1939, Marguerite Perey, working under Marie Curie's daughter, Irène, at the Curie Institute in Paris, identified a decay product that came from the actinium she was working with. (As you might guess, this was a very dangerous job, and almost everyone in the lab eventually died of an illness caused by the radioactivity.) It was confirmed that what she had found was indeed a few atoms of element 87, and she decided to name it francium.

Francium is the second most rare element found in nature. There is only one francium atom for every 1,000,000,000,000,000,000 (quintillion) uranium atoms. Francium is constantly appearing and disappearing as it is created and destroyed as part of the decay process of uranium. The longest-lived isotope has a half-life of only 22 minutes. Francium can lose an alpha particle to become astatine, or it can emit a beta particle and turn into radium.

Francium can now be made by using a particle accelerator that bombards gold atoms with oxygen atoms. (Gold has 79 protons, oxygen has 8; combine them and you get 87 protons.) The largest sample ever produced was only about 300,000 atoms. This was enough to do some experiments (quickly, before it decayed!) but not enough to be able to determine what francium would look like if we had a lump large enough to see.

88 Radium Ra

Ra-226 emits an alpha particle and turns into Rn-222.

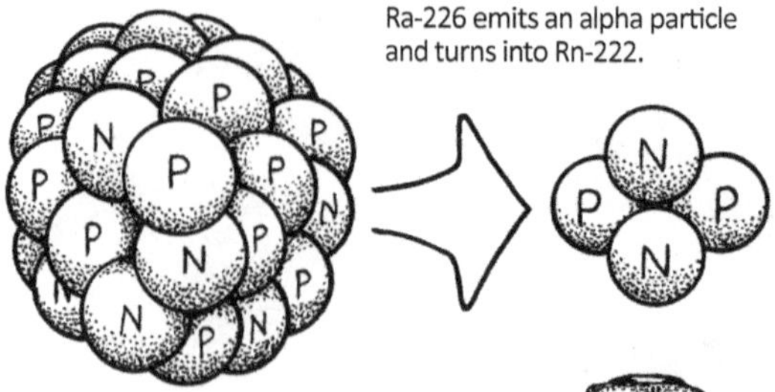

88 protons
138 neutrons
88 electrons

Atomic mass: 226

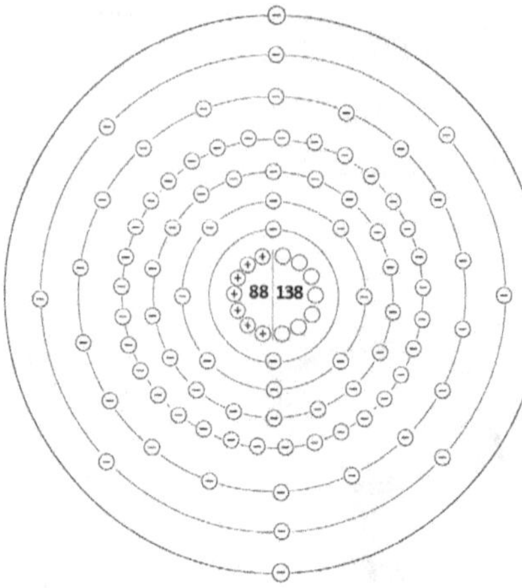

This symbol means "radioactive." It is often black and yellow.

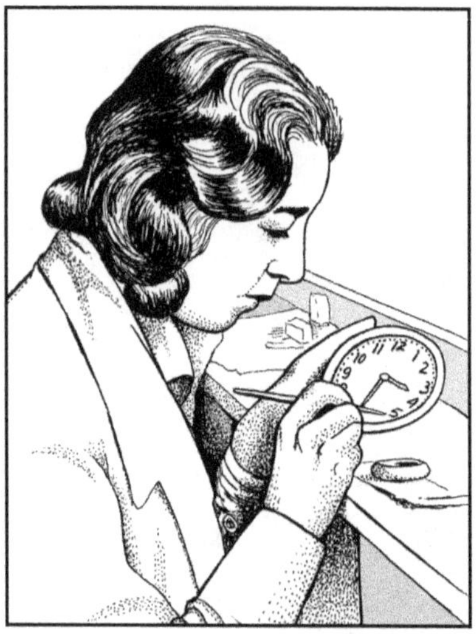

From 1900-1920 young women were hired to paint clock faces with radium paint. They all died of diseases caused by the radiation.

Radium was once sold as medicine.

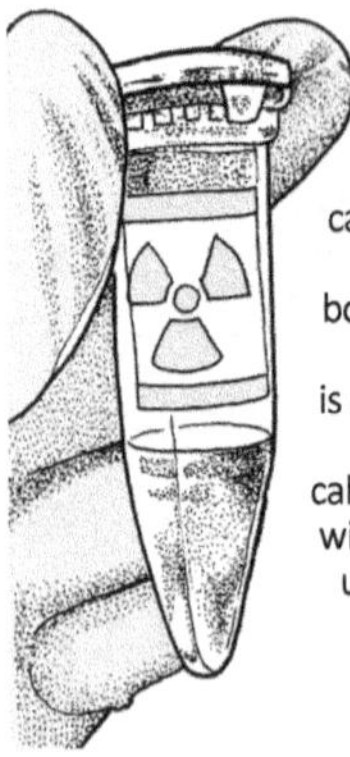

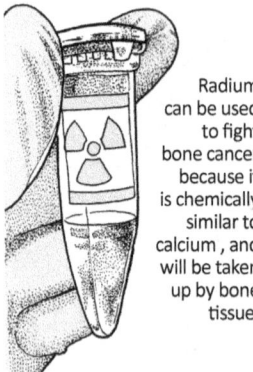

Radium can be used to fight bone cancer because it is chemically similar to calcium , and will be taken up by bone tissue.

In the early 1900s, you could buy radium toothpaste.

Radium was put into food products before radioactivity was found to be harmful.

black crystals

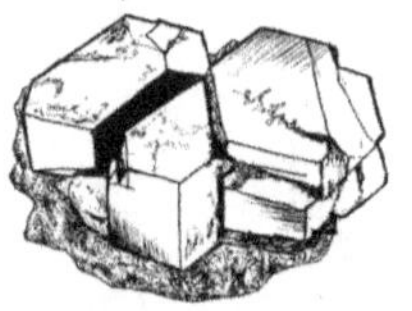

Radium was discovered in uranite (pitchblende).

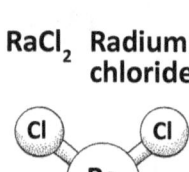

$RaCl_2$ Radium chloride

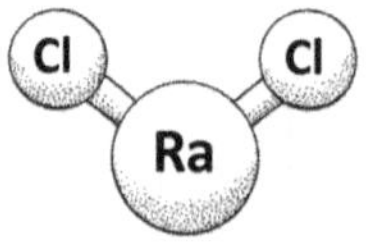

$RaSO_4$ Radium sulfate

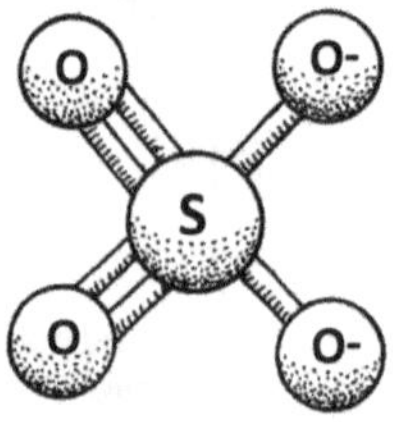

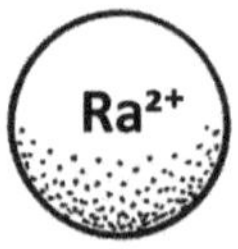

Marie Curie discovered radium in December of 1898, just five months after she discovered polonium. She had been working with samples of pitchblende (uranite) that contained large amounts of uranium. To extract the radium, Marie would crush and boil the rocks, and add various chemicals, such as sulfuric acid and sodium hydroxide. After all the uranium was gone, the ore still registered as being radioactive. She continued the series of chemical extraction procedures and managed to isolate a sample of polonium. But with the polonium gone, the ore was still radioactive! She continued the extraction procedures until she isolated a tiny bit of dust that glowed in the dark. She was sure she had found element 88 because this new element's chemistry was very similar to that of barium, the element right above 88 on the Periodic Table. All the elements in a column will have similar properties. However, radium could do something that none of the other elements in its column could do: it glowed in the dark.

There are 33 known isotopes of radium, with half-lives ranging from a few seconds to 1,600 years for Ra-226. When radium-226 decays, it emits an alpha particle and turns into radon-222. (Rn-222 then decays into Po-218.)

In the early 1900s, the terrible dangers of radioactivity were not yet known. Companies in the U.S. and Europe began to produce quantities of radium and they found dozens of fun ways to sell it. They painted watch dials, made glowing cosmetics, and put it into toothpaste and candy. Some people claimed radium could cure many diseases, so health spas offered radium baths. Companies that made watches and clocks hired young girls (known as the "radium girls") to paint glow-in-the-dark numbers on the clock faces. They were told to lick the tips of their brushes to keep the tips pointed. About ten years later, all the girls had died of radiation poisoning. The companies continued to use radium paint, but established safety procedures that prevented contact with the radium. In the 1960s, substitutes for radium were found: promethium and tritium (radioactive "heavy" hydrogen).

89 Actinium Ac

89 protons
136 neutrons
89 electrons

Atomic mass: 225

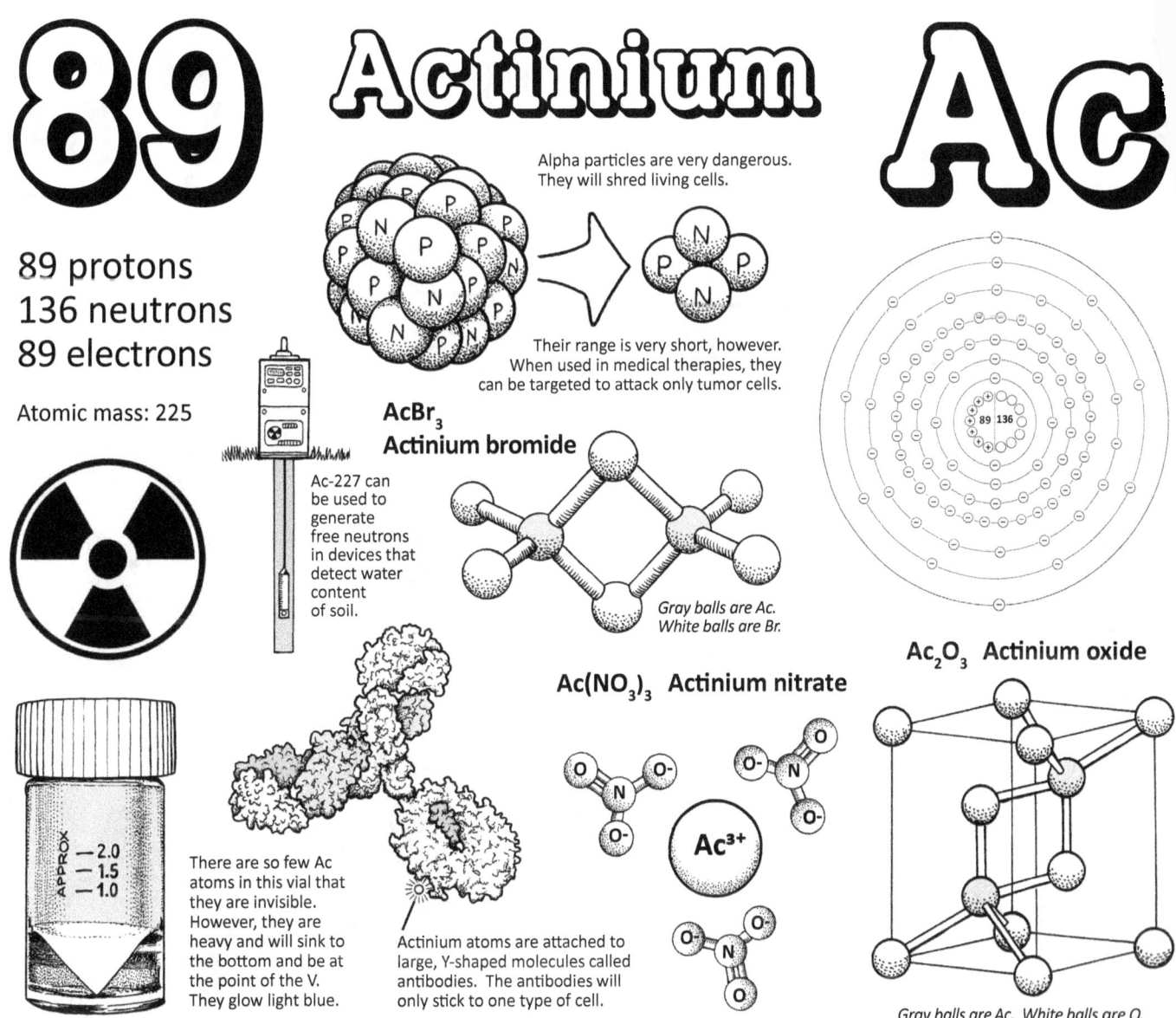

Alpha particles are very dangerous. They will shred living cells.

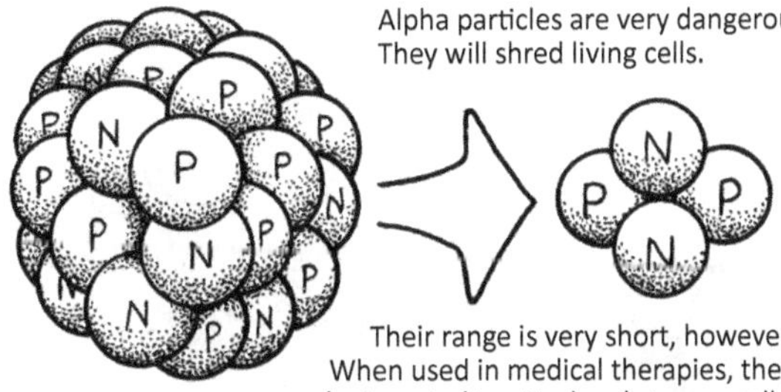

Their range is very short, however. When used in medical therapies, they can be targeted to attack only tumor cells.

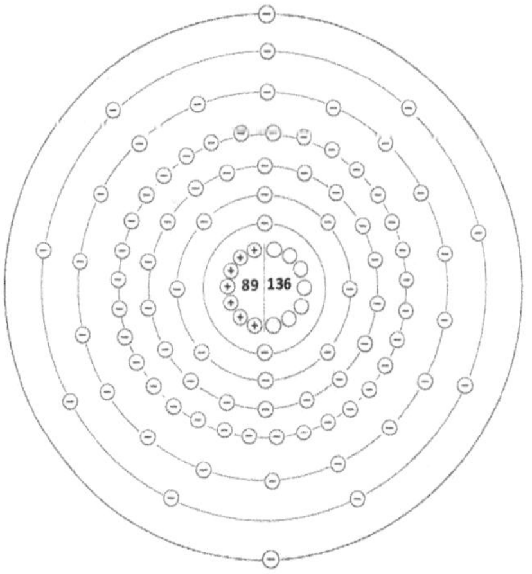

Ac-227 can be used to generate free neutrons in devices that detect water content of soil.

$AcBr_3$
Actinium bromide

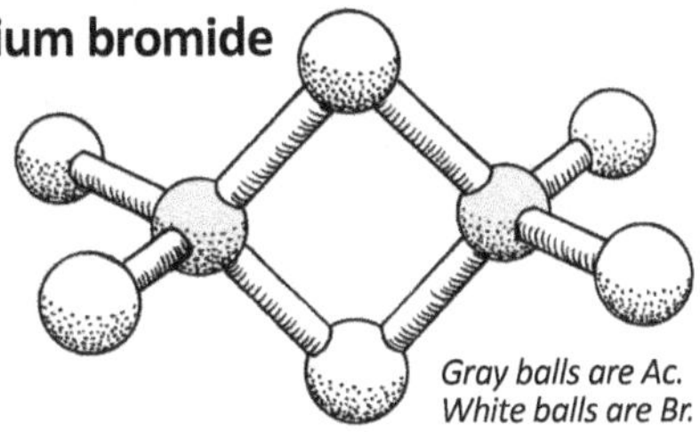

Gray balls are Ac. White balls are Br.

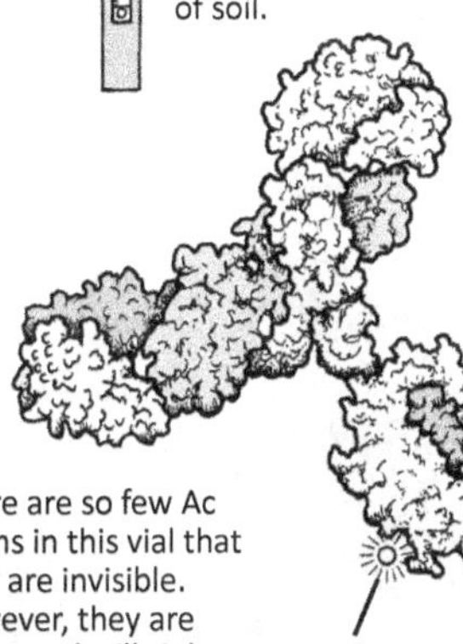

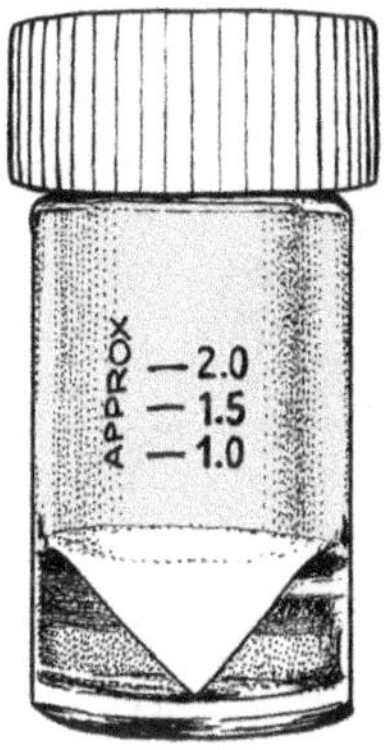

There are so few Ac atoms in this vial that they are invisible. However, they are heavy and will sink to the bottom and be at the point of the V. They glow light blue.

Actinium atoms are attached to large, Y-shaped molecules called antibodies. The antibodies will only stick to one type of cell.

$Ac(NO_3)_3$ Actinium nitrate

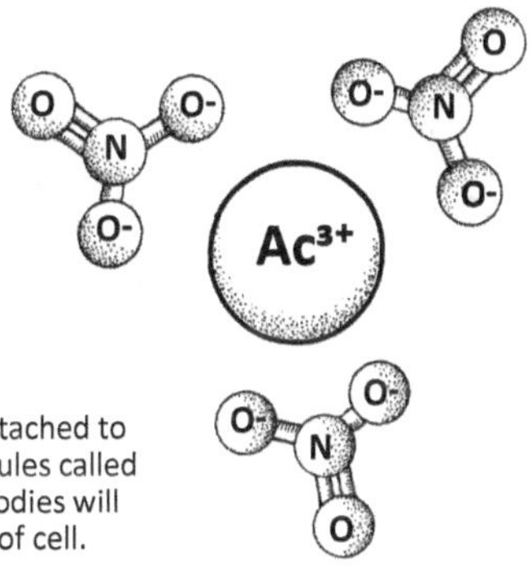

Ac_2O_3 Actinium oxide

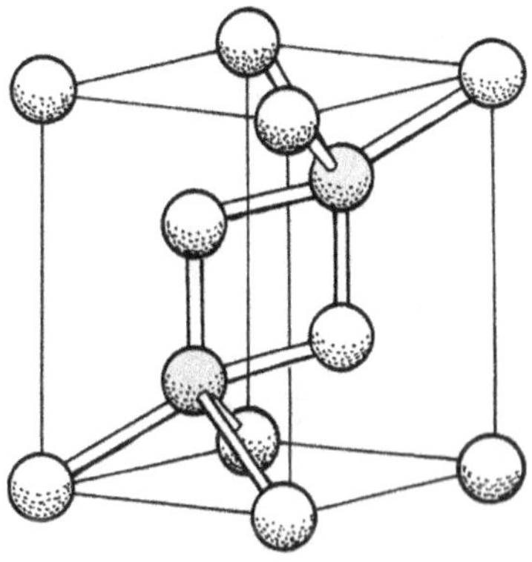

Gray balls are Ac. White balls are O.

Actinium sits under lanthanum in the Periodic Table, and thus shares similar chemical properties. It can form molecules like the ones lanthanum makes. Actinium, however, is radioactive, which means that any molecule it forms will also be unstable. Ac-225, one of the most studied isotopes, has a half-life of ten days, and will emit an alpha particle to become francium-221. Fr-221 will decay in a split second and become bismuth-213. Ordinary bismuth is very stable, but Bi-213 is not. In less than an hour, it will decay into either polonium-213 or thallium-209. Within minutes, both of these will turn into lead-209, which will quickly modify itself into bismuth-209. Bismuth-209 is stable, and the decay will then stop. Oak Ridge National Lab makes Ac-225 using that they call a "thorium cow." They have a large tube of thorium solution that continually produces Ac-225 (by decay to radium then to actinium).

Actinium was probably discovered by André-Louis Debierne, who had been working with Marie and Pierre Curie and watched them discover polonium and radium. It is said that he took the final waste residues and was able to prove there was yet another radioactive element that had been overlooked. Research by Friedrich Giesel in 1902 made it uncertain whether Debierne had found actinium or protactinium, and there was no way to go back and check because the samples had, of course, long since decayed. Thus, Giesel and Debierne are often considered co-discoverers.

Recently, medical researchers have realized that Ac-225 is the ideal radioactive isotope for fighting cancers. Researchers jokingly call Ac-225 the "Goldilocks" isotope because it is so perfect for this use. Its half-life of ten days is long enough to allow for manufacturing and shipping, but short enough not to cause long term damage to patients. Ac-225 can be attached to a biological molecule (antibody) that will only stick to cancer cells, not to normal cells, and this avoids many unwanted side effects. The problem is generating enough of it. Right now only about a thousand patients a year can be treated. World leaders in manufacturing Ac-225 are TRIUMF, in Canada, and Oak Ridge National Lab, in Tennessee. They bombard thorium with protons, and it decays into Ac-225.

0 Thorium Th

protons
2 neutrons
electrons

omic mass: 232

Thorium was named after Thor, the old Norse god of thunder.

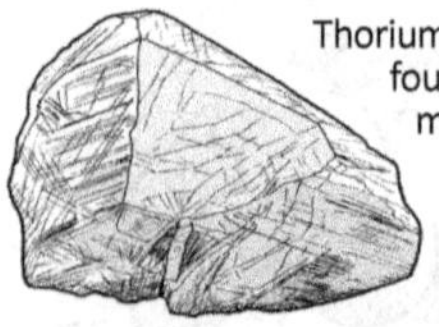

Thorium is found in monazite sands.

Thorianite is black and cubic. It has both ThO_2 and UO_2.

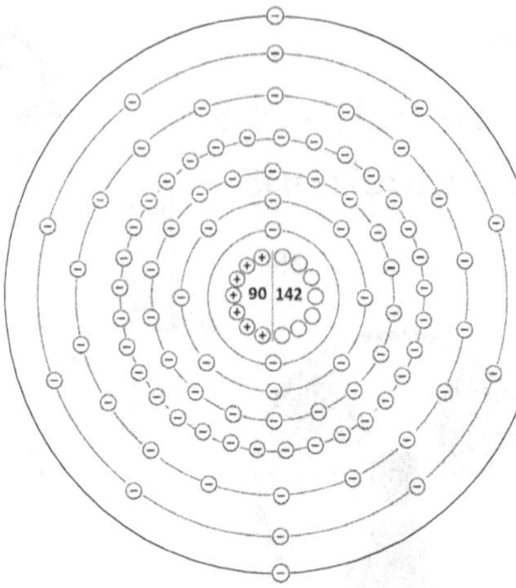

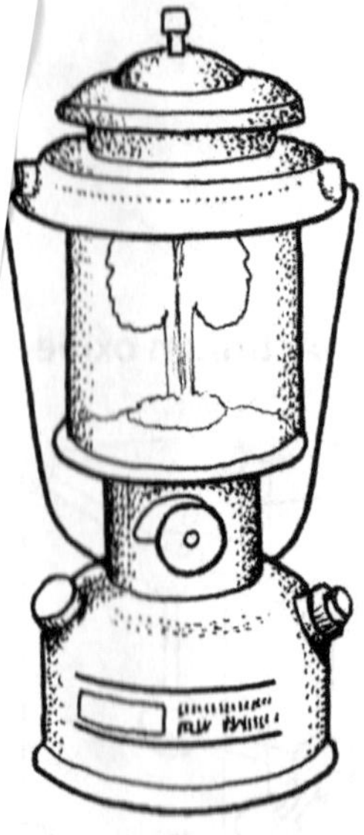

Thorium oxide was used in the mantles of gas lanterns because it made the light much brighter.

ThO_2 Thorium oxide

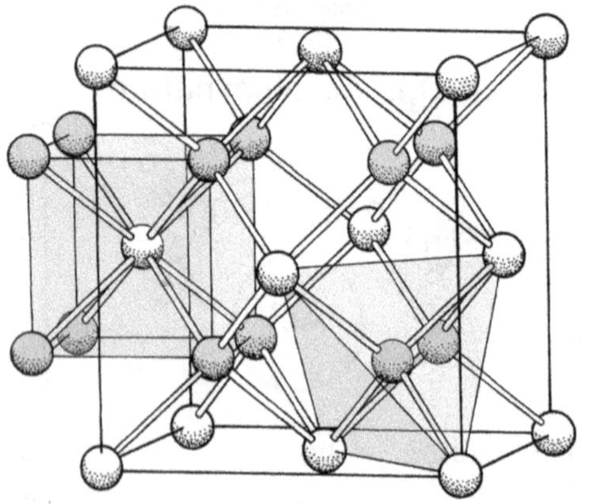

White balls are thorium. Gray balls are oxygen.

Thorium was used to make electrodes for gas-metal welding.

During the mid-1900s, thorium was added to glass to make specialty camera lenses.

Thorium was discovered in 1828 by Morten Esmark, a Norwegian priest who liked to collect rocks. He found an unusual, cube-shaped black rock and sent it to his geologist father, who then sent it to the Swedish chemist Jöns Jacob Berzelius (discoverer of cerium and selenium). Berzelius confirmed that this rock did, indeed, contain a new element. Since the rock had come from Norway, it seemed fitting to choose a Norwegian name, so Berzelius named the element thorium after the old Norse god of thunder, Thor.

Thorium is very similar to the elements in the lanthanide series. It is often found in the same monazite sands as yttrium, cerium, praseodymium, neodymium, europium and samarium. Companies that mine and refine the lanthanides can extract thorium, as well. The most common isotope of thorium, Th-232, is very stable, with a half-life of over 14 billion years; so for all practical purposes, it is not radioactive. However, thorium's nucleus is more easily split than the nuclei of the lanthanides, giving thorium a place in the nuclear energy industry. It can be used to create safer isotopes of uranium that will not decay into plutonium (which is used to make bombs). Other isotopes of thorium have half-lives ranging from days to years, and are found in the decay chains of larger elements such as uranium and neptunium. Any atom with 90 protons is, by definition, thorium, so if uranium loses two protons, it becomes thorium, though not the stable 232 isotope, but an unstable one that will decay into radium, then radon.

Thorium found in rocks is mostly stable Th-232, so thorium was not immediately identified as a radioactive element. It was used in many of the same applications as the lanthanide elements. Thorium oxide produced a bright, white light when used in the mantles of gas lanterns. It was also used to make metal electrodes for gas-metal arc welding, as an additive in metal alloys (notably for jet engines), and in glass manufacturing, especially for camera lenses. Thorium did not pose a major health risk to consumers, but it has gradually been replaced by other elements, such as yttrium, in order to protect workers in factories that were in daily contact with thorium.

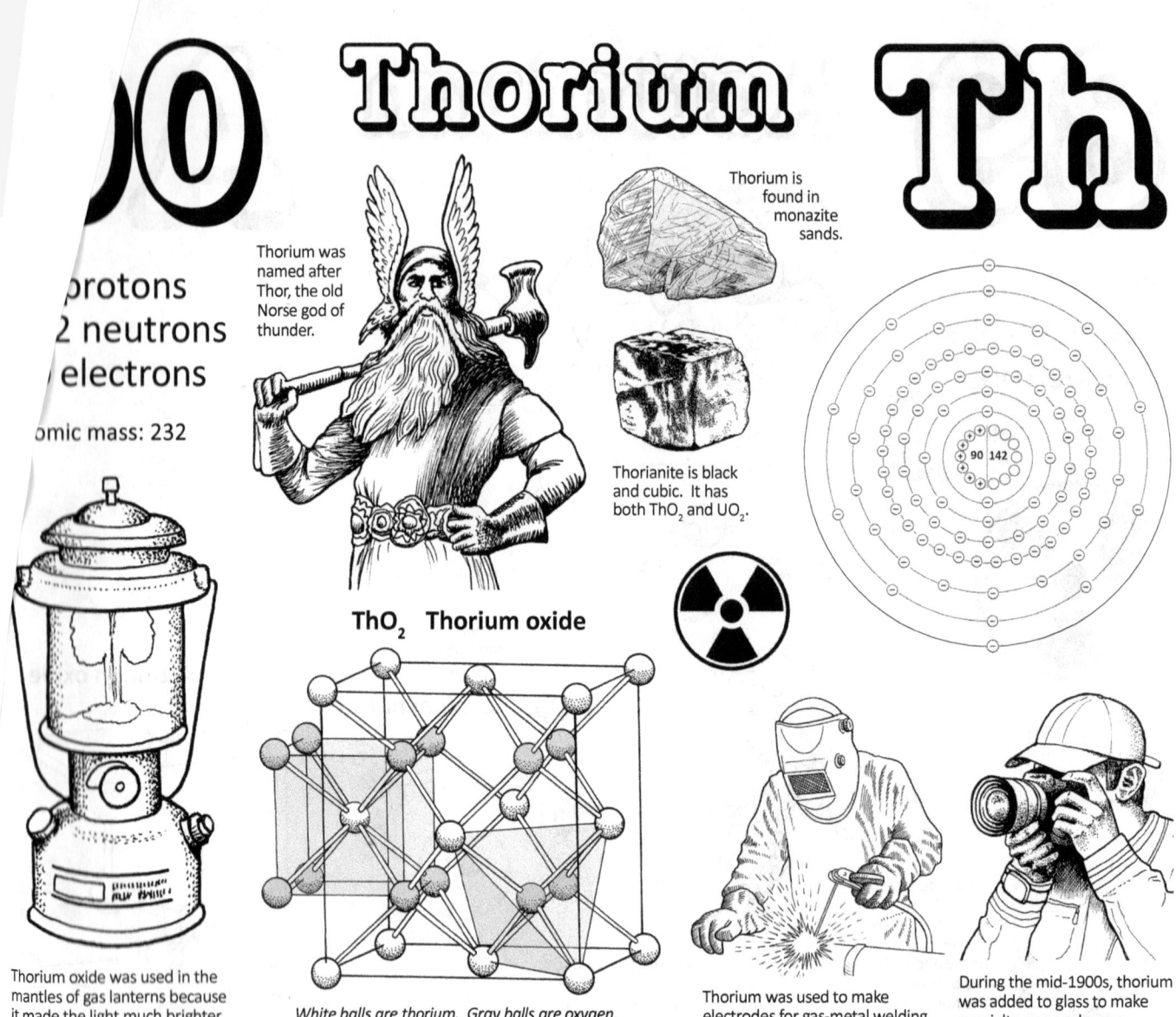

91 Protactinium Pa

91 protons
140 neutrons
91 electrons

Atomic mass: 231.03

This device collects falling particles in ocean water.

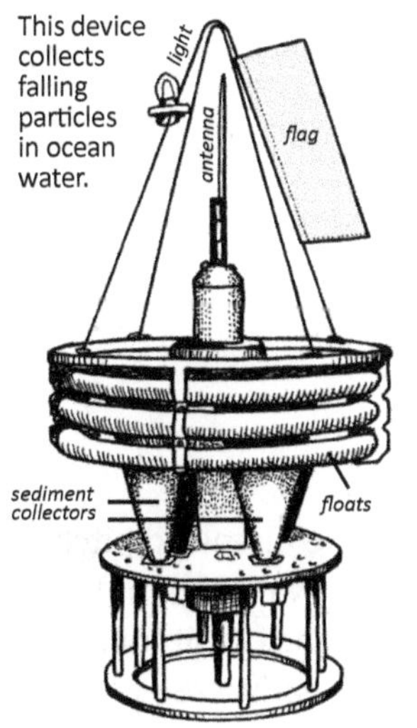

The sediments might be tested for thorium and protactinium in order to determine their age.

Torbernite is green.

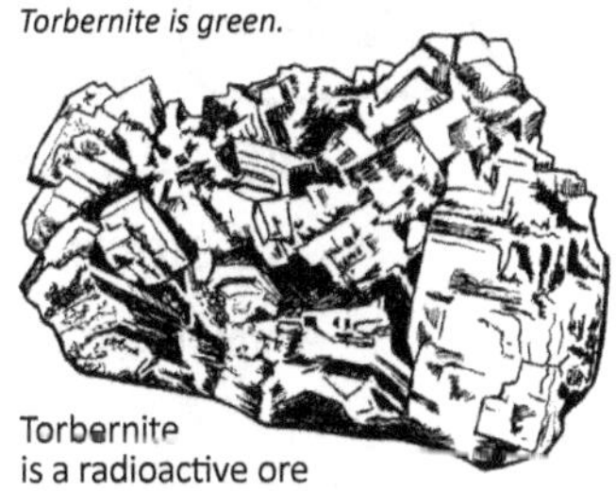

Torbernite is a radioactive ore that contains protactinium from the decay of uranium.

Lise Meitner and Otto Hahn are two of the discoverers of protactinium.

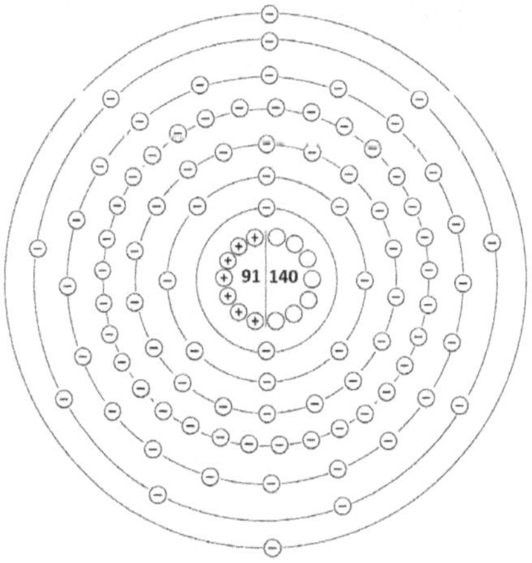

Protactinium is found in the waste produced by fission of uranium in nuclear power plants.

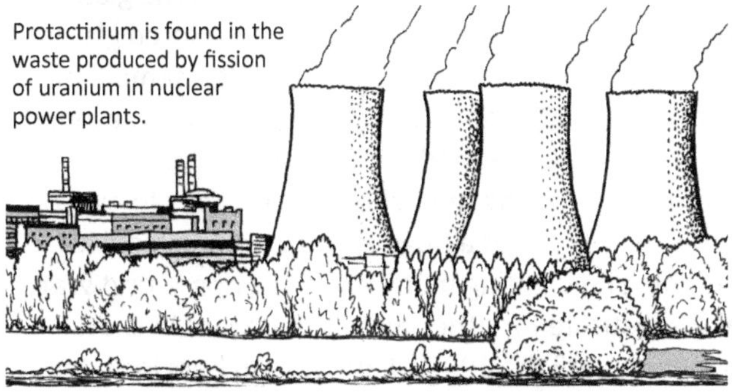

Pa_2O_5 Protactinium oxide

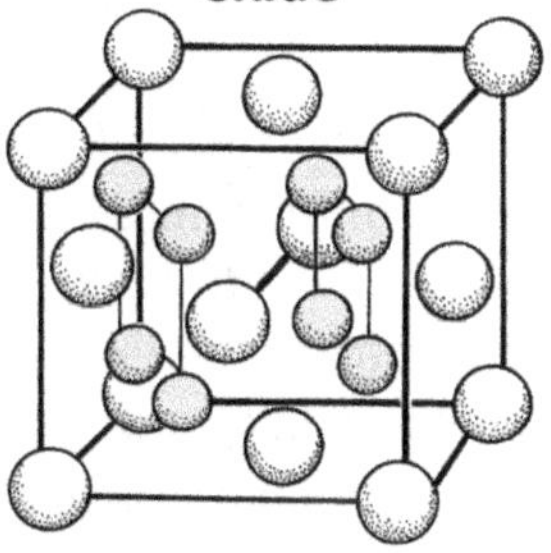

White balls are Pa. Gray balls are O.

$PaBr_5$ Protactinium bromide

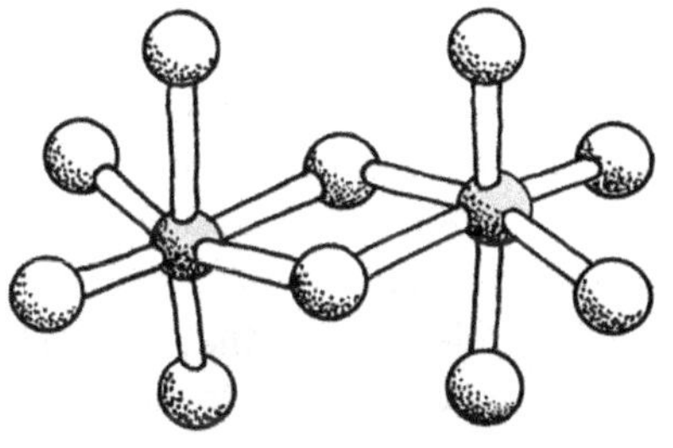

The gray balls are Pa. The white balls are Br.

In 1871, Dmitri Mendeleyev predicted that there should be an element between thorium and uranium. This seems obvious to us now, as we can look at the complete table, but Mendeleev's prediction in 1871 was quite visionary. There wasn't any "actinide series" yet, and radioactivity had not been formally discovered. Element 91 was first identified by Kasimir Fajans and Oswald Helmuth Göhring in 1913, when they found the isotope Pa-234 while they were researching the decay of uranium. It had a very short half-life, so they named it "brevium" meaning "brief." In 1918, two teams of scientists, working independently, both discovered the isotope Pa-231, which has a half-life of 33,000 years. One of these teams, Lise Meitner and Otto Hahn (who would later become famous for their roles in the invention of the atomic bomb) decided to use the name "proto-actinium" because it came before actinium in the decay chain. Eventually, in 1949, the International Union of Pure and Applied Chemistry (IUPAC) decided to change the name to "protactinium" because it was easier to pronounce.

Protactinium has very few practical uses. Pa-234 has a half-life of just over a minute and is relatively safe to use, so college and high school chemistry labs often use it for an experiment where students have to watch a Geiger counter tally the number of decays every ten seconds, then they do the math themselves to calculate the half-life. The other practical use is in trying to track ocean sediments. Collected sediments will contain very small amounts of Pa-231 and Th-230, which are part of the decay chain in one isotope of uranium. Uranium salts get dissolved into the ocean, where they decay into all the daughter elements. By measuring the ratio of Pa-231 to Th-230, oceanographers can guess at which ocean sediments are older than others. One of the most studied areas is the North Atlantic, where glaciers may have played a large role in the past.

Nuclear power produces Pa-231 and Pa-233 as by products of the splitting of uranium atoms. Both isotopes are undesirable. Pa-233 absorbs neutrons and slows down the fission process. Pa-231 has a very long half-life and creates waste storage problems. There is a chemical process involving molten bismuth and lithium that attempts to take these isotopes out of the nuclear waste.

U 92

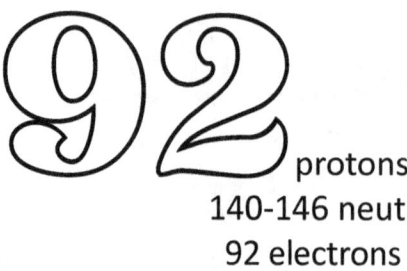

protons
140-146 neut.
92 electrons
Atomic mass: 238.02

Uranium

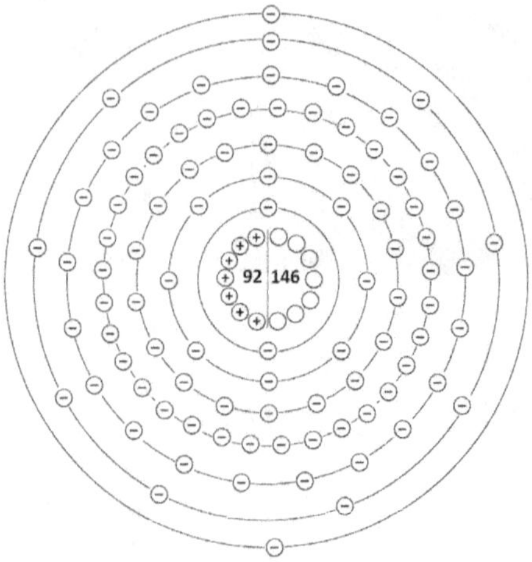

Named after the planet Uranus

Uranium was officially discovered by Martin Klaproth in 1789. He dissolved pitchblende in nitric acid then added some sodium hydroxide. The result was a compound of uranium ($Na_2U_2O_7$), not pure uranium, but since he correctly determined that there was a new element present, he gets the credit for its discovery. He named the new element after the planet Uranus, which had been discovered in the early 1780s. Before its official discovery, people had been using uranium without knowing what it was. Pitchblende, the most common uranium ore, had been mined in Europe (Austria and Czech Republic) since the Middle Ages, and was used to make a yellow pigment for coloring glass. This use of uranium by glass makers continued into the 1900s. Since the most common isotope found in pitchblende is U-238, with a half-life of over 4 billion years, the amount of radiation coming from the glass was minimal. The ceramic tile industry in the early 1900s also used uranium to make a wide variety of colored glazes.

Uranium's radioactivity was not discovered until 1896 when Henri Becquerel accidentally left a sample of uranium salts sitting on top of a photographic plate, and noticed that the plate had become "fogged." He reasoned that the uranium sample must have been emitting invisible rays of some kind that had affected the plate.

Marie Curie's discovery of radium had led to a large increase in the mining of pitchblende in order to extract the radium to make glow-in-the-dark paint. The extraction process produced a huge stockpile of inexpensive pitchblende "waste" that was still usable for making uranium pigments and creating metal alloys. During World War I, the pitchblende waste was used to make uranium-iron alloys as a substitute for other metals that were in short supply. Uranium's potential for making atomic weapons was not known until the 1930s. Early researchers included Enrico Fermi in Italy, Irène Curie in France, and Lise Meitner and Otto Hahn in Germany. In the 1940s, during WW II, both sides raced to figure out how to build an atomic bomb, using uranium as the explosive fuel. Uranium-238, the isotope found in pitchblende, doesn't naturally explode; the uranium had to be "enriched" by bombarding it with neutrons to create plutonium. In nuclear power plants, U-235 is used, but with much care, so that the fission process doesn't get out of control. Control rods made of other elements can slow the reaction whenever necessary.

After WW II, there was again a surplus of pitchblende waste "depleted uranium." It was assumed that most of the dangerous uranium was gone. The military began making bullets, shells, and plating for tanks out of alloys containing depleted uranium. Years later, the countries where these weapons were used began to complain about health problems in the civilian population, and blamed it on the depleted uranium which they say was still radioactive.

U_3O_8 Triuranium octoxide

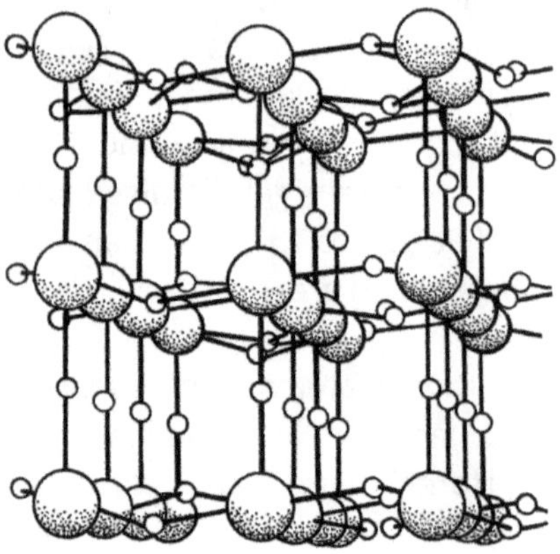

U_3O_8 is often referred to as "yellowcake." It is very stable and can be found in ores.

UO_2 is also fairly stable, but has some qualities that make it better for the fission process used in nuclear power plants.

UO_2 Uranium oxide

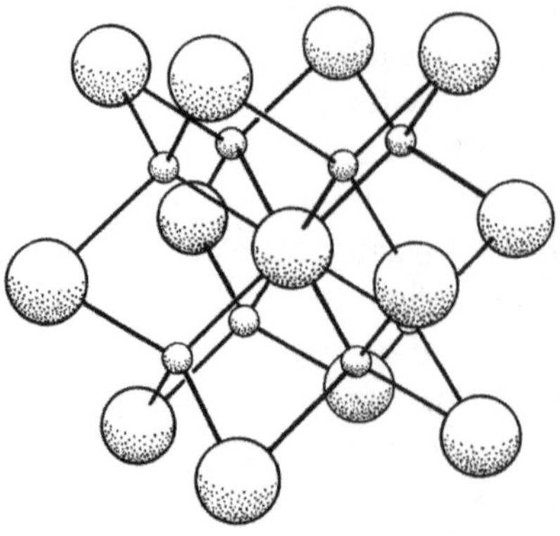

Large balls are U. Small balls are O.

UF_6 Uranium hexafluoride

This is the least stable uranium molecule and is part of the process of "enriching" uranium to make bombs.

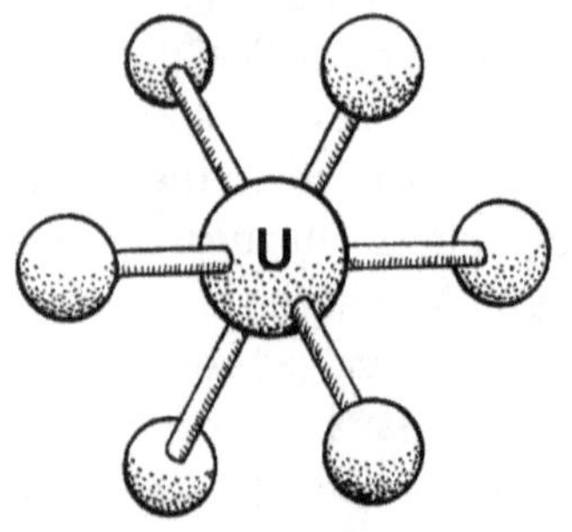

Unmarked balls are F.

All the unmarked balls are O, oxygen.

92 Uranium U

Uranium emits an alpha particle and turns into thorium.

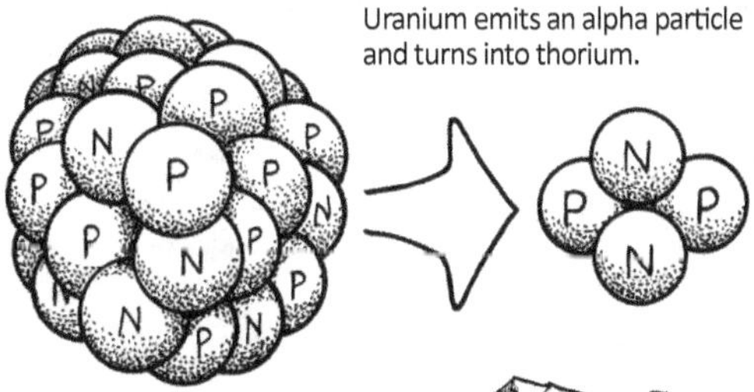

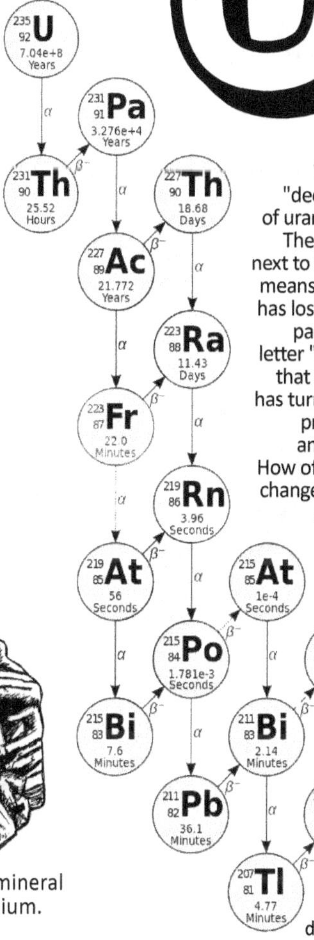

This is the "decay chain" of uranium-235. The letter "α" next to the arrow means the atom has lost an alpha particle. The letter "β" means that a neutron has turned into a proton plus an electron. How often these changes occur is called the half-life.

Pb is stable, so decay stops.

This symbol means "radioactive." It is often black and yellow.

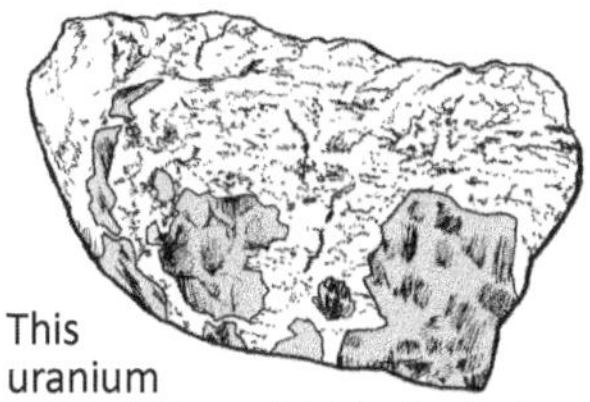

This uranium ore is yellow with black patches.

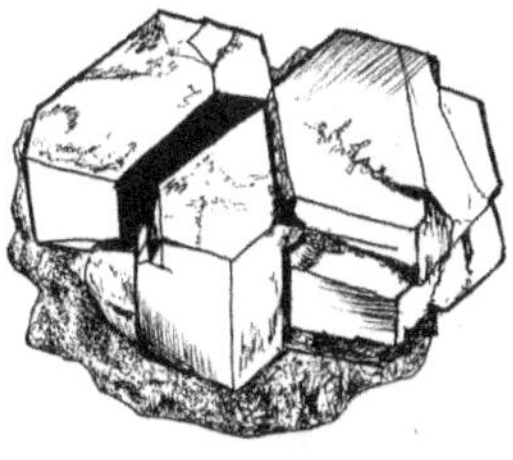

This uranite crystal is black.

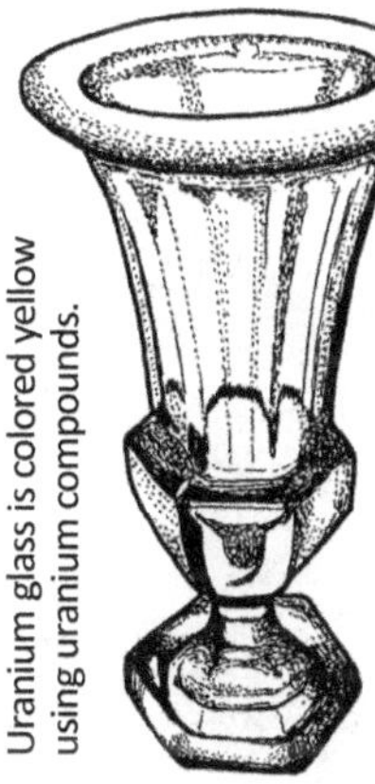

Uranium glass is colored yellow using uranium compounds.

Colors include yellow, orange, red, green, tan, and light blue.

Uranium from pitchblende ore has been used to make pigments for ceramic glazes.

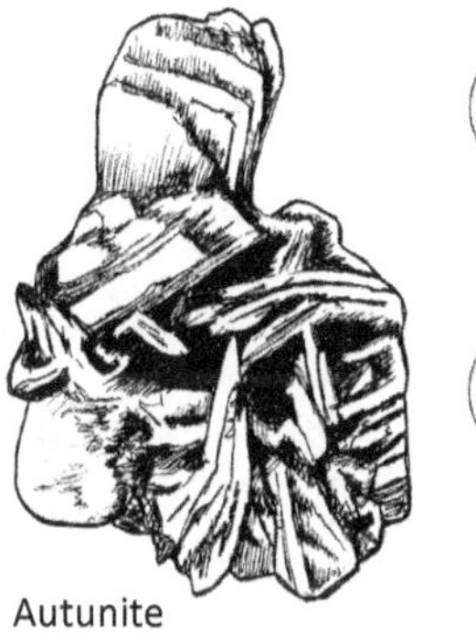

Autunite is a yellow-green mineral that contains uranium.

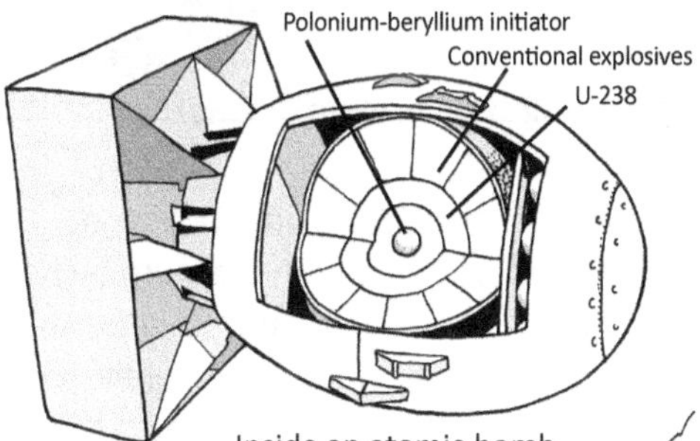

Inside an atomic bomb

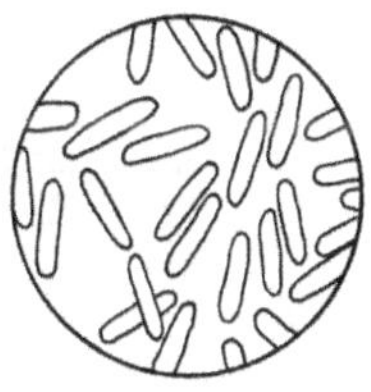

Citrobacter is a species of bacteria that absorbs and uses uranium. It absorbs pink stain.

Depleted uranium is a very dense metal and was used to make various military items.

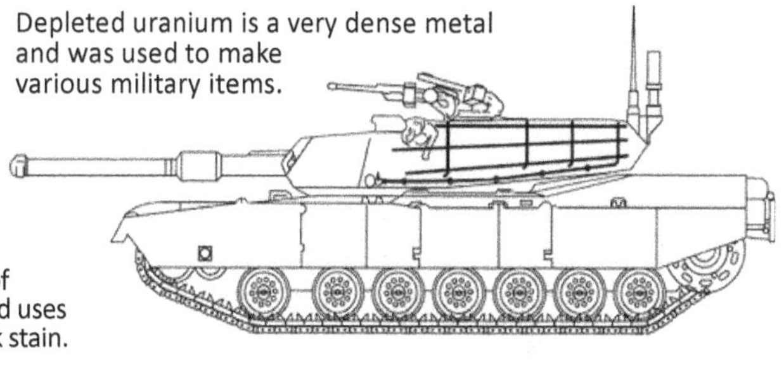

Nuclear power plants use uranium as fuel. The uranium is bombarded with particles that make it split, releasing energy.

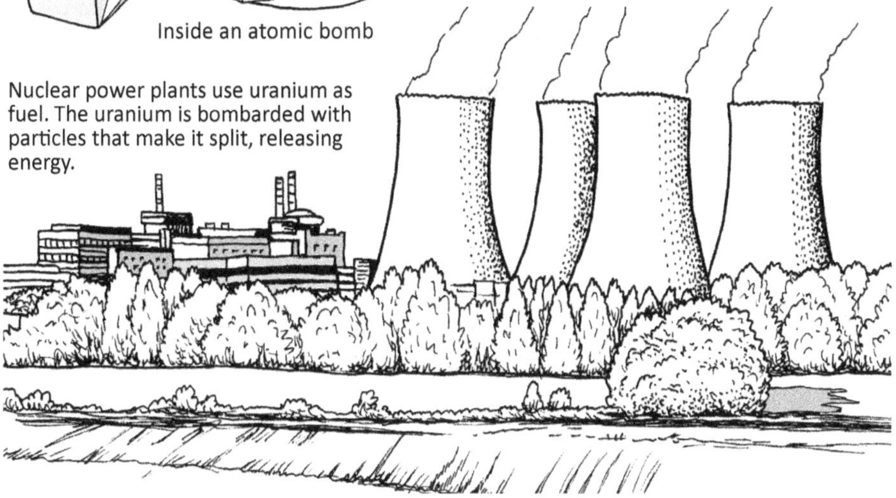

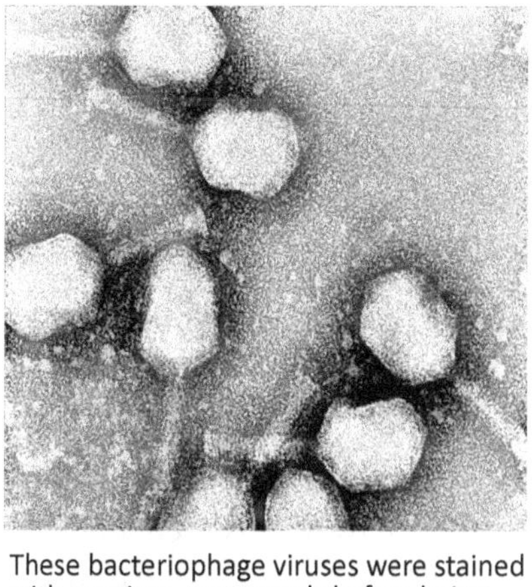

These bacteriophage viruses were stained with uranium compounds before being viewed under an electron microscope.

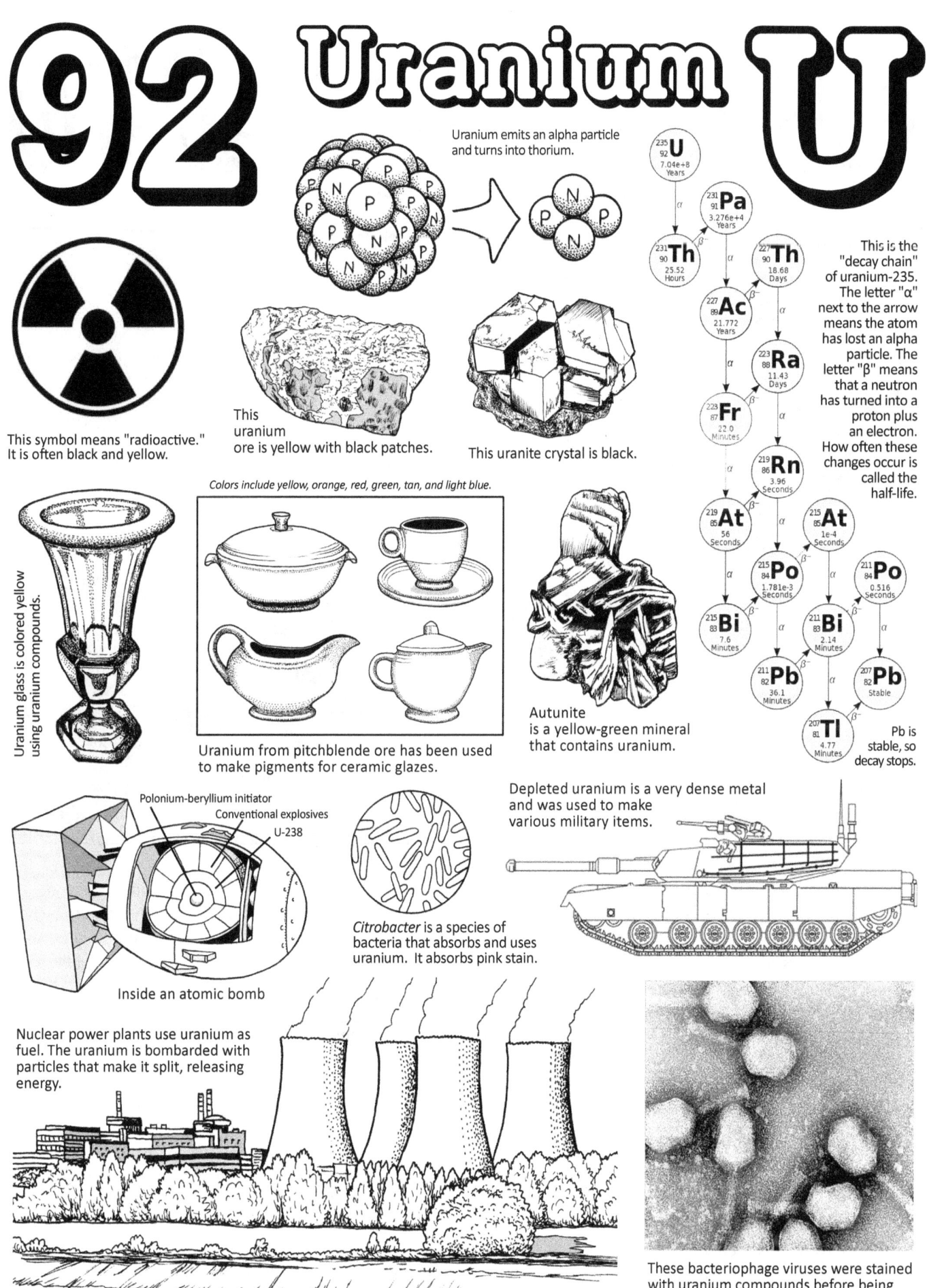

93 Neptunium Np

Neptunium-237 emits an alpha particle and turns into protactinium-233.

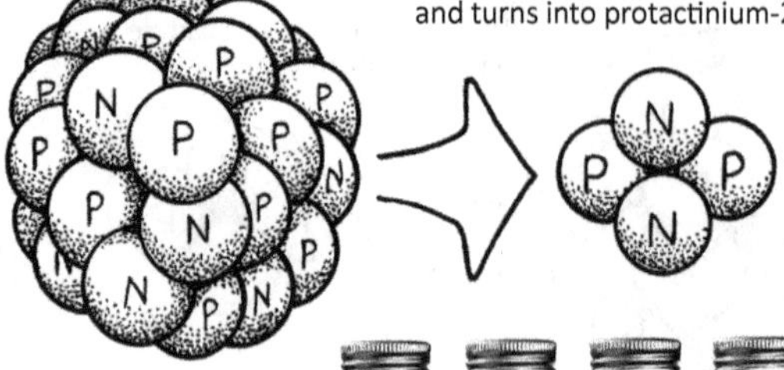

93 protons
142-146 neutrons
93 electrons

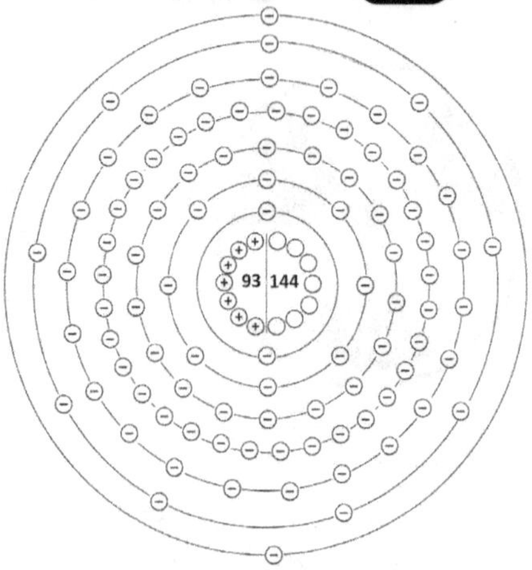

Atomic mass: 237

NpF_6 Neptunium hexafluoride

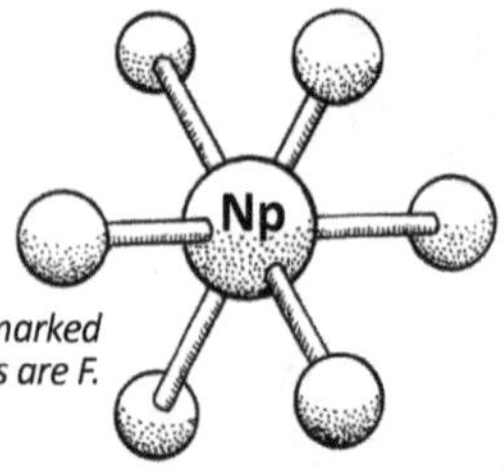

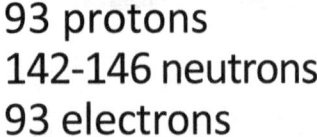

Unmarked balls are F.

Np will dissolve into solutions and color shades of yellow, green or blue.

NpO_2 Neptunium dioxide

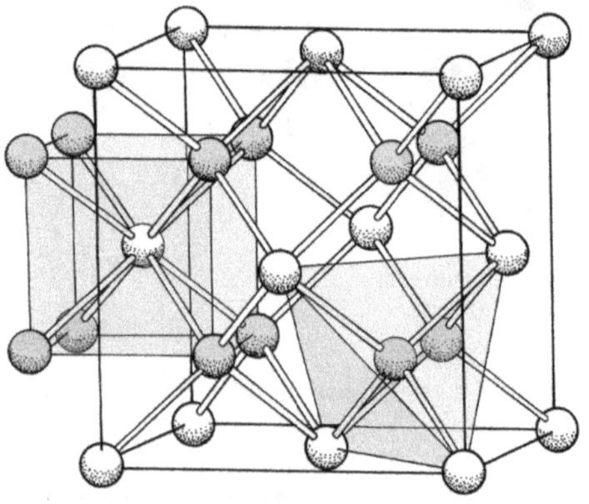

White balls are Np. Gray balls are O.

The Berkeley Radiation Lab as it looked in 1940.

Neptunium was discovered in 1940 by Edwin McMillan and Philip H. Abelson in the cyclotron machine at Berkeley Radiation Lab at the University of California in Berkeley. They were doing experiments with uranium by hitting it with fast-moving neutrons. They knew this would cause some of the uranium atoms to split, and they wanted to look at the atoms and particles that this process produced. They observed an unknown atom that had a half-life of 2.3 days. This half-life measurement did not match any isotope of any element known at that time. They did more experiments and finally came to the conclusion that it must be an isotope of element 93. Since uranium had been named after Uranus, it seemed logical to name the next element "neptunium," after the planet Neptune.

Neptunium was the first element to be discovered by manufacturing it artificially. Neptunium does exist in nature in small quantities, as the result of the decay of uranium, but this was not yet known when it was observed in the lab at Berkeley. The most stable isotope of neptunium is Np-237, with a half-life of over 2 million years. This allows it to be collected and studied, but also makes it a long-term hazard. Neptunium is a silvery metal, and will make various compounds, some of them displaying a variety of colors when dissolved in acidic or alkaline solutions.

Neptunium is always the result of the decay of other radioactive elements. Smoke detectors often use the radioactive isotope americium-241 as part of the detection device. After about 20 years, 3% of the Am-241 will decay into neptunium. Also, neptunium isotopes are found in the "spent" (used) fuel rods from nuclear reactors.

If Np-237 is irradiated with neutrons, it will produce plutonium-238, which is used as an energy source in radioisotope thermal generators that power satellites such as Voyager, Cassini, and Galileo, and the Curiosity Mars Rover. Because it can be turned into plutonium, neptunium plays an important role in space exploration.

94 Plutonium Pu

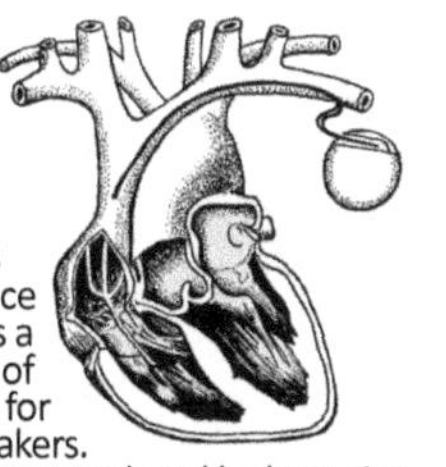

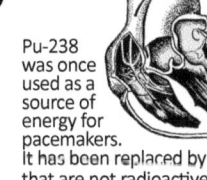

Pu-238 was once used as a source of energy for pacemakers. It has been replaced by batteries that are not radioactive.

Plutonium-238 emits an alpha particle and turns into uranium-234.

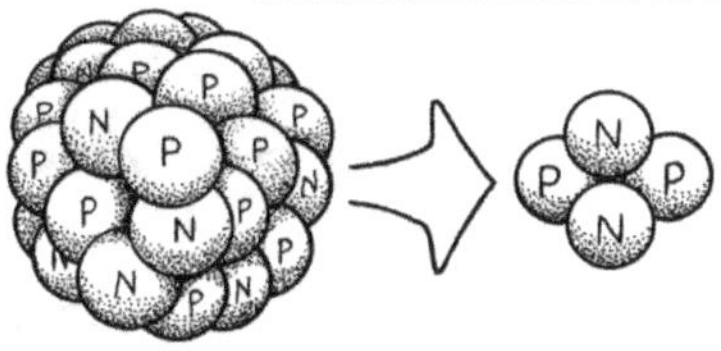

94 protons
144-150 neut.
94 electrons

Atomic mass: 232

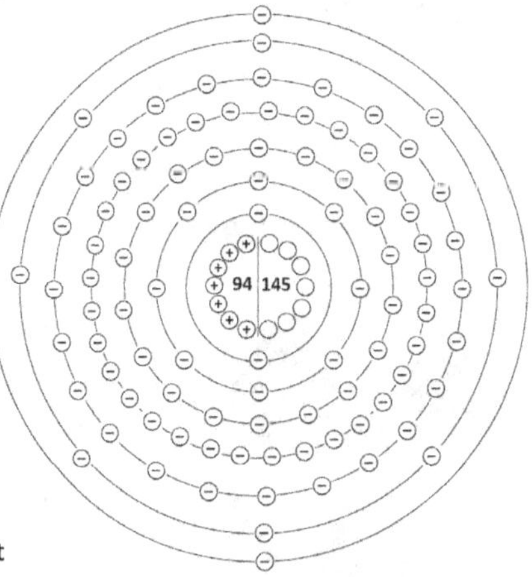

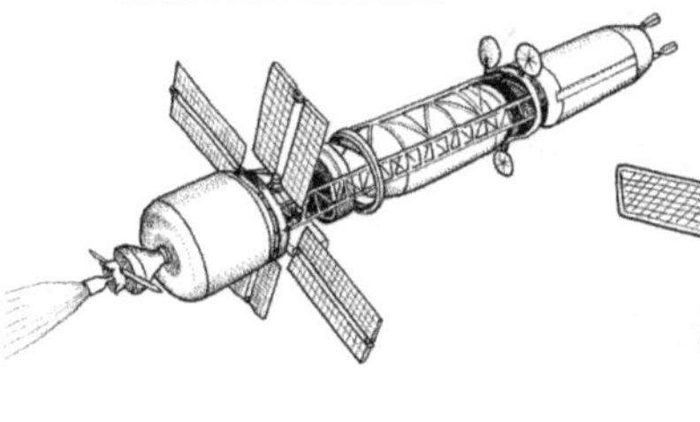

Plutonium-238 is used as a source of energy for satellites that travel far out into space, beyond the range of solar power.

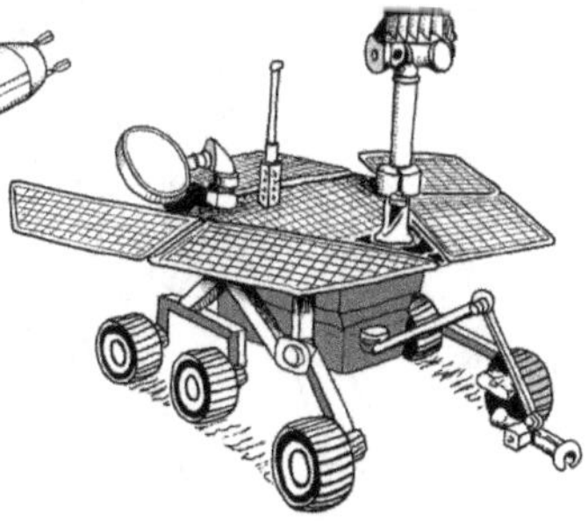

Pu-238 is used as a power source in rovers that explore the surface of the moon or Mars.

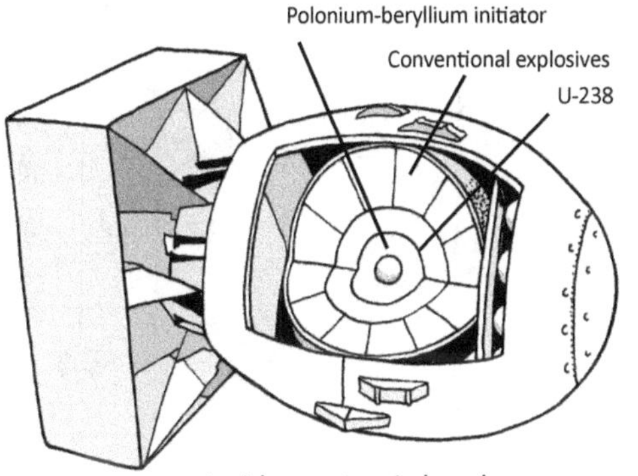

Inside an atomic bomb

PuO_2 Plutonium dioxide

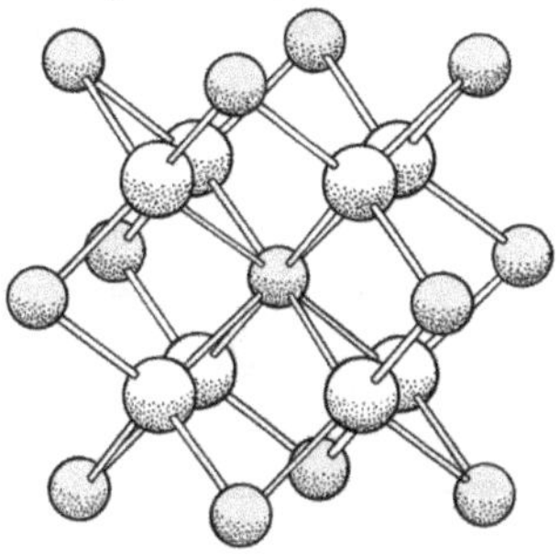

Gray balls are Pu. White balls are O.

Plutonium was discovered at the Berkeley Radiation Lab, not long after they found neptunium.

Plutonium was manufactured and discovered at the Berkeley Radiation Lab in February of 1941. A number of physicists were working on this project, including Edwin McMillan who initiated the discovery of neptunium. The leader of the team was Glenn Seaborg. Another famous physicist, Emilio Segrè, was also on the team. They used the cyclotron machine to hit uranium with "deuterons" (a neutron stuck to a proton). it didn't take long for the physicists to realize that plutonium had the potential to be a powerful explosive, and the key to being able to make atomic weapons. During World War II, the secret "Manhattan Project" in the U.S. used plutonium to make an atomic bomb that would eventually be dropped on Japan in 1945. Plutonium was manufactured at reactors in Oak Ridge, Tennessee, and at Hanford, Washington, and was shipped to Los Alamos, New Mexico, where the bombs were assembled.

Plutonium is a heavy, silvery metal and can be alloyed with other elements. To make the atomic bomb, both beryllium and gallium were added to the plutonium. Beryllium provided an extra source of free neutrons to speed up fission, and gallium made the normally hard and brittle plutonium easier to shape. The plutonium had to be shaped into a perfect sphere in order for the bomb to work correctly.

Pu-238, the most common isotope of plutonium, has a half-life of about 87 years. It emits alpha particles and can generate quite a lot of heat. A lump of plutonium would be very warm if you could touch it. Pu-238 has been used as an energy source for satellites that travel very far from the sun, too far to rely on solar energy. The Mars rovers also use Pu-238. In the late 1900s, Pu-238 was used in heart pacemakers. This was not as unsafe as it sounds. The alpha particles emitted by plutonium were easily stopped by the case around the pacemaker.

Today, Pu-238 is made from Np-237 in facilities that keep workers safe by having robots and robotic arms do the processing inside radiation-proof work areas.

95 Americium Am

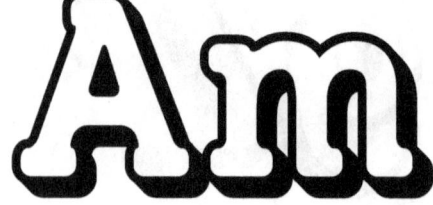

Am-241 emits an alpha particle and turns into Np-237.

95 protons
146-148 neutrons
95 electrons

Atomic mass: 208.9

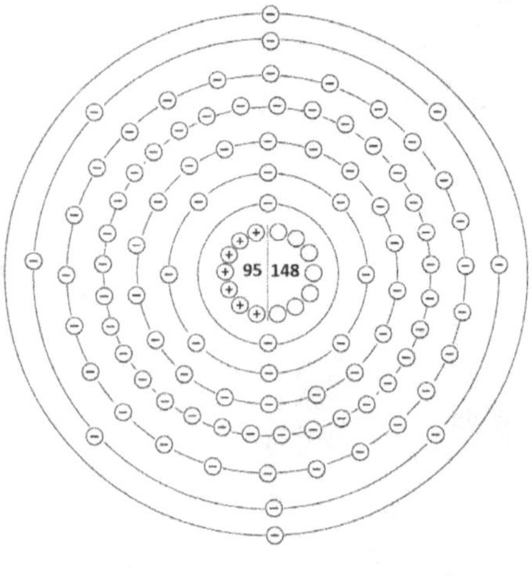

The symbol for radioactivity

These men are using a density gauge to evaluate a road surface. The device uses neutrons produced by Am-241 and beryllium.

AmO_2 Americium dioxide

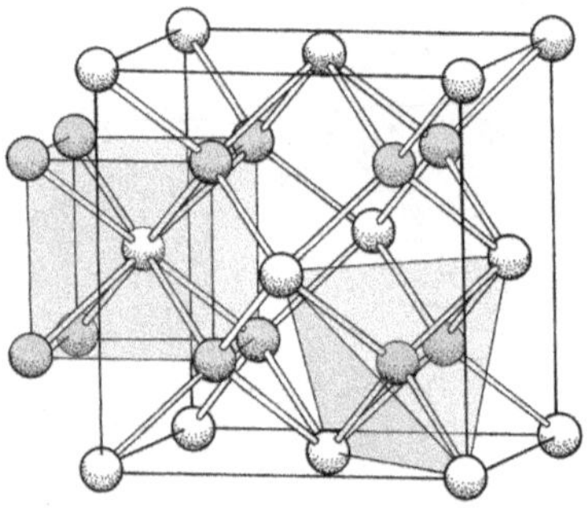

White balls are Am. Gray balls are O.

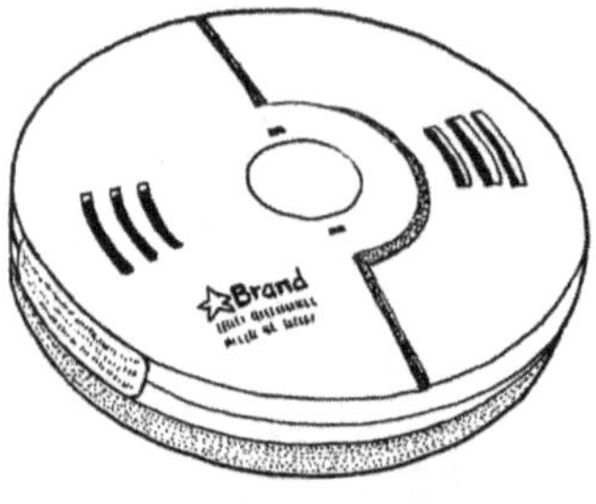

Am-241 is used in smoke detectors.

Am is one of the elements that can be recovered from nuclear waste.

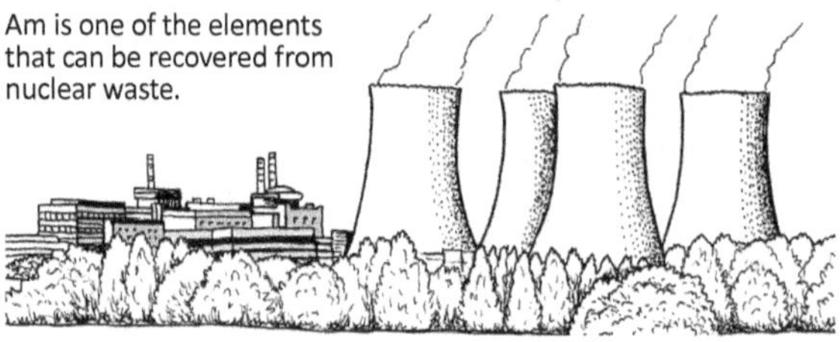

Americium *(am-er-ISS-ee-um)* is named after America, where it was discovered in 1944 during the secret "Manhattan Project" that made the first atomic bomb. Glenn Seaborg was the leader of a team that was working to identify the heavy elements that were being created as a result of this atomic research. After uranium was bombarded with neutrons in a cyclotron machine, a wide variety of radioactive atoms and isotopes were produced. They had just identified neptunium, plutonium and curium, so americium was the fourth "transuranic" (beyond uranium) element to be discovered. After the discovery of these four new elements, Glenn Seaborg decided that the Periodic Table needed to be rearranged. He put the actinide series right under the lanthanide series, as separate rows beneath the main table. With this new arrangement, element 95 would be right under europium. Seaborg thought it was fitting to have the continents, Europe and America, together as a top/bottom pair.

The most common and useful isotope is Am-241. It decays by emitting an alpha particle and turning into Np-237. Its half-life is 432 years, which is long enough to make it useful. Most of us have a small amount of Am-241 in our homes because it is used in smoke detectors. The tiny amount of radiation coming from the Am-241 is continually completing an electronic circuit that prevents the alarm from going off. If smoke enters the detector, it blocks the rays and prevents them from completing the electrical circuit. When the circuit is disrupted, the alarm goes off.

Am-241 can be combined with beryllium to create a source of free neutrons. The alpha particles from the americium cause the beryllium atoms to lose one of their neutrons. Beryllium is still stable with one less neutron, so the Am/Be combination makes a safe source of neutrons for use in devices like density gauges and groundwater sensors. Hydrogen atoms slow down neutrons. Water and oil contain a lot of hydrogen, so the sensor will be able to detect water or oil as deep as several meters underground. These gauges are also used to test the density of asphalt.

In nuclear physics research labs, americium is a popular choice for creating super heavy elements. If you hit americium with neon atoms, you can make element 105, dubnium. (95+10=105) If you hit americium with calcium atoms, you can make element 115, moscovium. (95+20=115)

96 Curium Cm

Curium emits alpha particles and turns into plutonium.

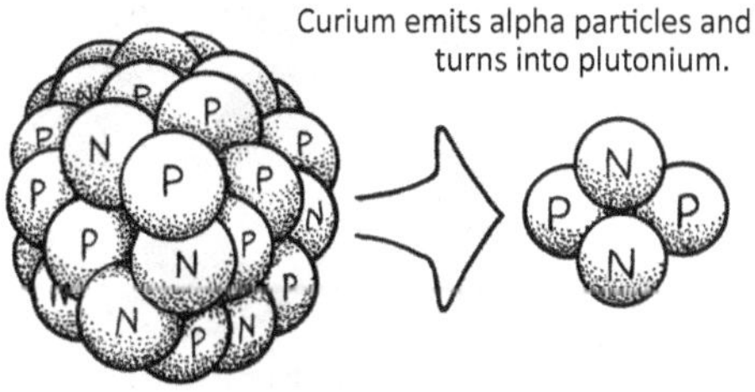

96 protons
146-154 neutrons
96 electrons

Atomic mass: 232

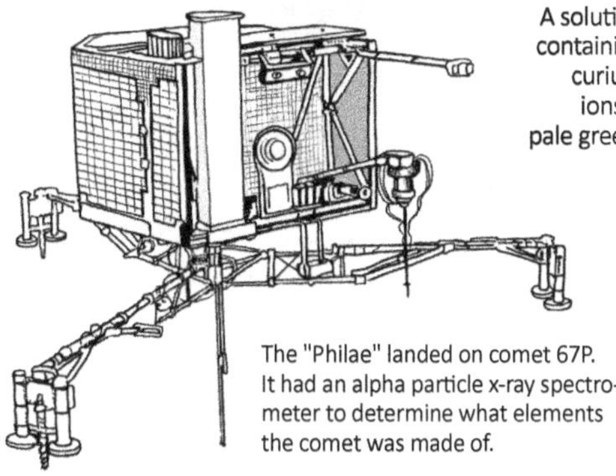

The "Philae" landed on comet 67P. It had an alpha particle x-ray spectrometer to determine what elements the comet was made of.

A solution containing curium ions is pale green.

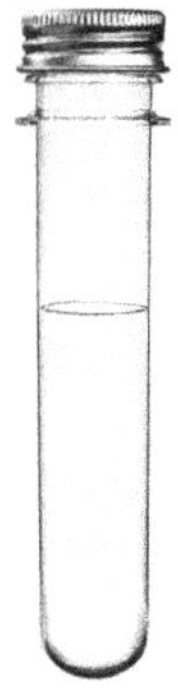

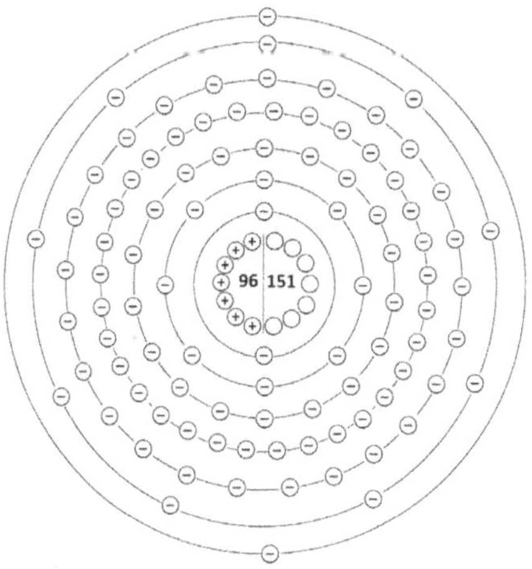

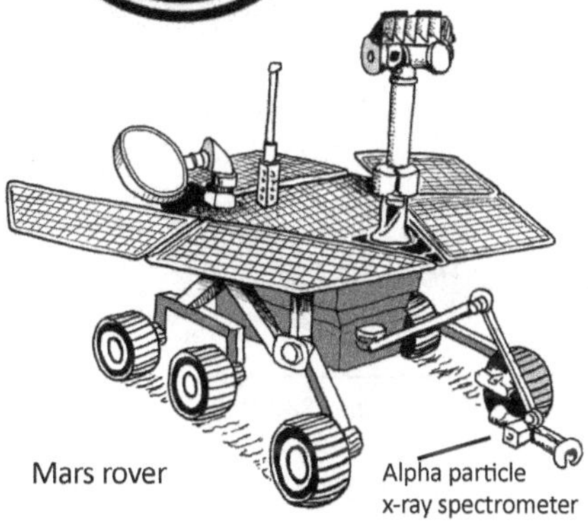

Mars rover

Alpha particle x-ray spectrometer

Cm_2O_3 **Curium (III) oxide**

White is O. Gray is Cm.

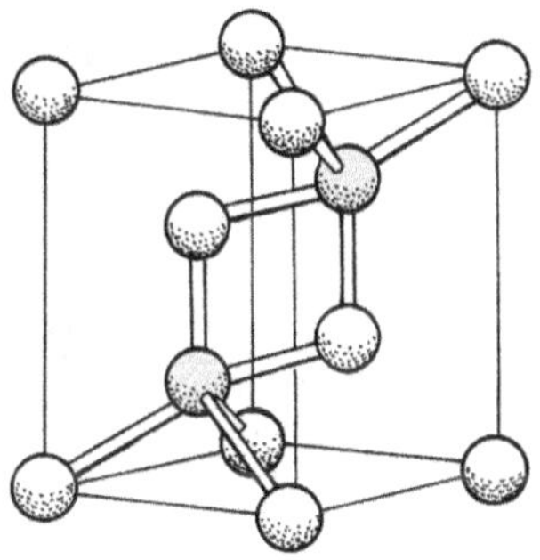

Curium was named to honor Marie and Pierre Curie. This antique stamp also commemorates their work.

Curium is named in honor of Marie and Pierre Curie, for all their pioneering research on radioactivity. This element was first observed in 1944 by Glenn Seaborg and his team of researchers at the University of Berkeley. They used their cyclotron machine to bombard plutonium atoms with alpha particles. When an alpha particle sticks to plutonium, the number of protons goes from 94 to 96. (Element 96 was actually identified before element 95.) When any radioactive element is hit with neutrons or alpha particles, a variety of isotopes of many elements are formed. Separating just one of them to create a pure sample is very tricky. It involves a lot of complicated chemistry, though the chemical principles are basically the same ones they use for non-radioactive elements. A pure sample of curium looks like a silvery metal, though in the dark it can be seen glowing purple. Experiments seem to indicate that it may have similar magnetic properties to the element right above it, gadolinium.

Curium is one of the heavy elements found in "spent" (used) fuel rods in nuclear reactors. Various isotopes of curium will be produced. Some will become additional fuel and will undergo fission. Other isotopes are more stable and have very long half-lives (millions of years) and will become problematic, hazardous nuclear waste. In the past, the plan was always to permanently bury hazardous waste, but recently new methods have been invented for dealing with particularly dangerous atoms. They are irradiated with particles that change them into safer isotopes.

Curium is a strong alpha particle emitter, which means it can be used in alpha particle x-ray spectrometers. These devices are an important tool that is always included on any satellite or spacecraft that is sent out to gather information about a planet or comet. For example, Mars rovers use these devices to identify the elements present in Martian rocks and soil.

Curium was studied as a possible source of energy for radioisotope thermoelectric generators that provide heat and motion for satellites in deep space, but price of making the right isotope (Cm-242) was too high, and other isotopes produced daughter elements that emitted a lot of beta and gamma radiation, instead of alpha particles.

97 Berkelium Bk

97 protons
148-152 neutrons
97 electrons

Atomic mass: 247

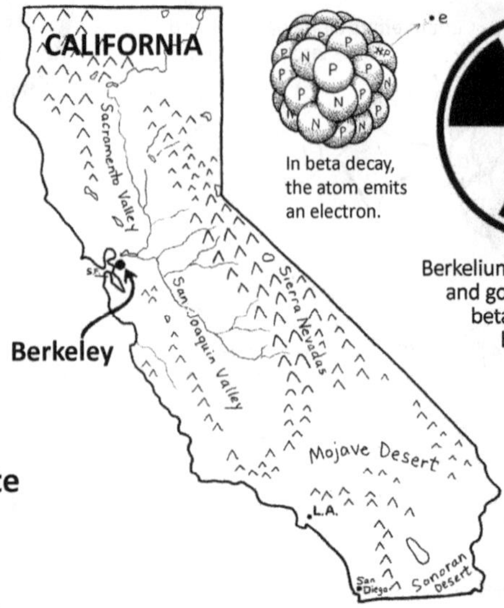

In beta decay, the atom emits an electron.

Berkelium is radioactive and goes through beta decay to become Cf-249.

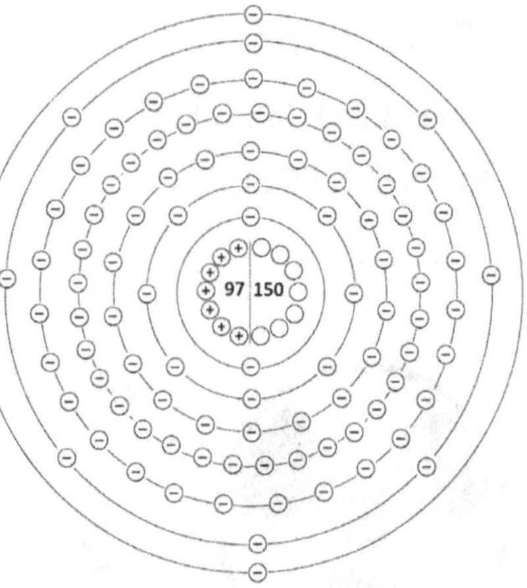

Berkelium (III) borate

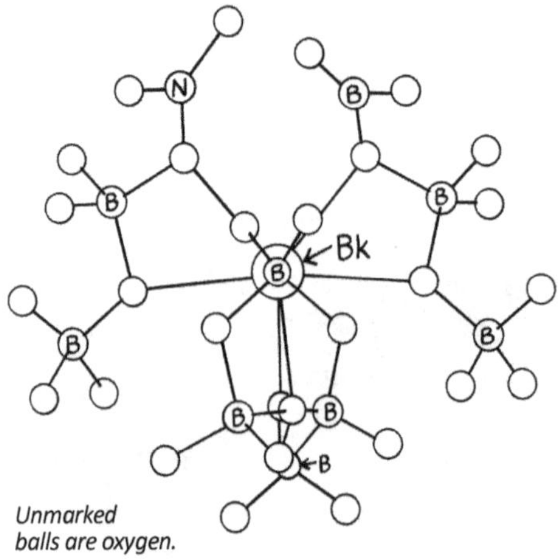

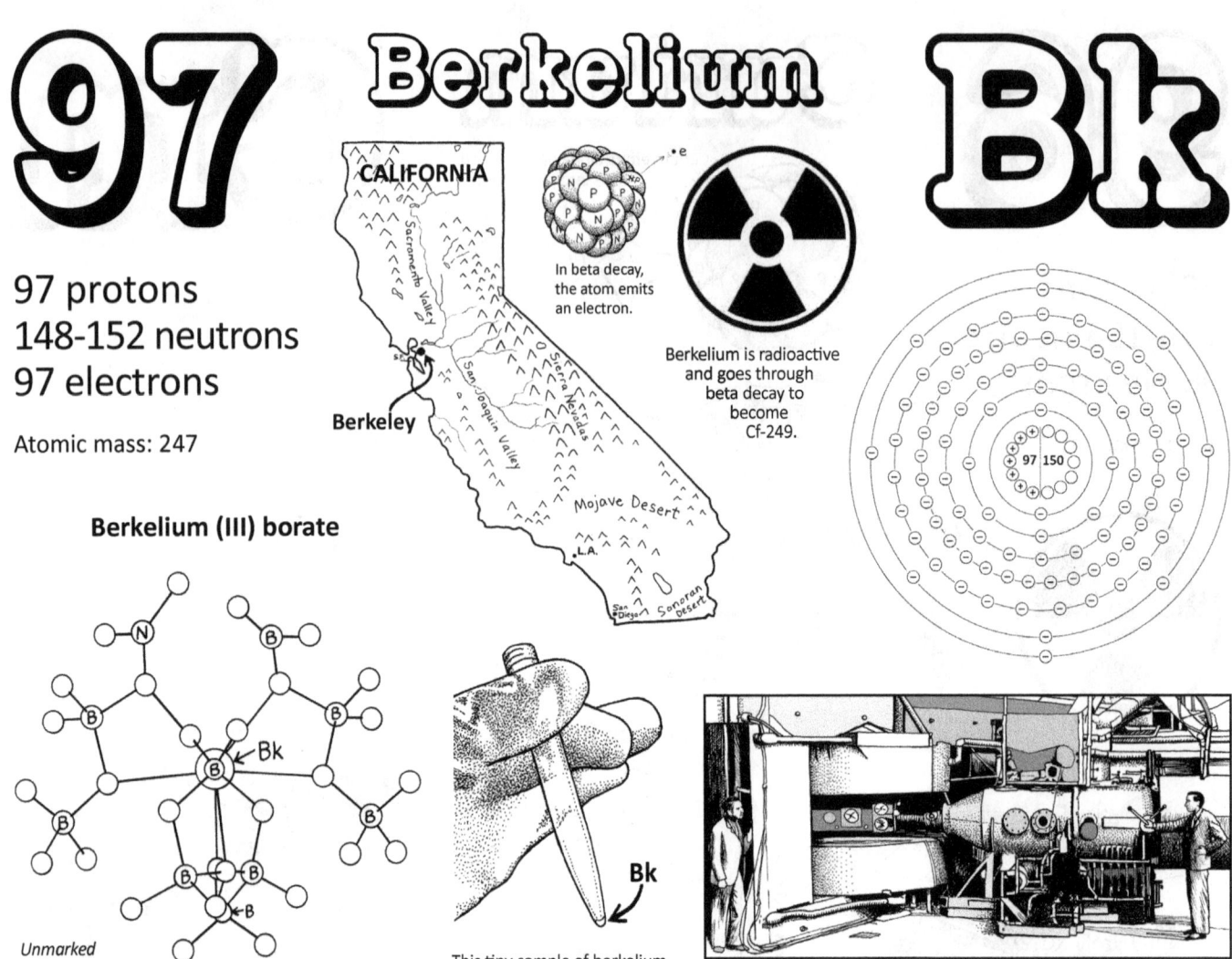

Unmarked balls are oxygen.

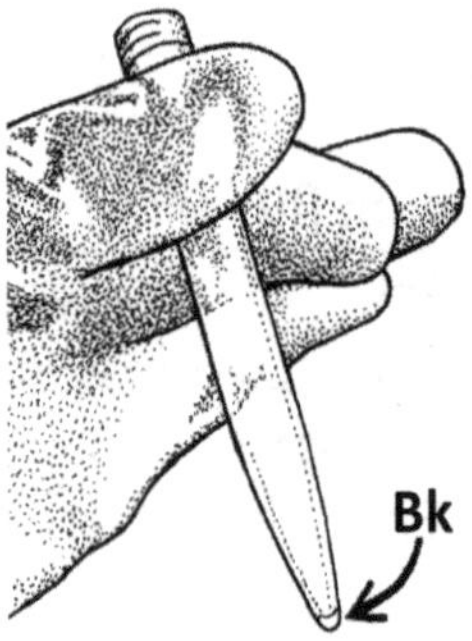

This tiny sample of berkelium looks like a light blue liquid.

Berkeleum was made at the Berkeley Radiation Lab in 1949.

Berkelium *(BERK-lee-um)* was created in the cyclotron machine at Berkeley, California, in 1949. Glenn Seaborg was the leader of the research team that year, but right beside him was Albert Ghiorso, who would be put in charge of the new particle accelerator at Berkeley in the 1950s. Ghiorso would help to discover 12 new elements.

Seaborg and his team manufactured berkelium in order to discover it. They used a "target" made of Am_2O_3. (This target had been made using a solution of americium-241 nitrate coated onto a platinum foil. After the solution had evaporated and created the Am_2O_3, the target was put into the cyclotron and bombarded with alpha particles for 6 hours. The target was removed, and then put through a series of chemical processes that involved chemicals such as nitric acid, ammonia, ammonium sulfate, hydrofluoric acid, potassium hydroxide, and perchloric acid. The intended result of all this chemistry was that the berkelium atoms would separated out from all other types of atoms, and the pure sample (however small) could be analyzed using a spectrometer. Every element has a unique spectral pattern, almost like its "fingerprint," so if they saw a new pattern, they knew they had discovered a new element. When the discovery was confirmed, they looked at where it would be located on the table, right under terbium, and thought that since terbium had been named after a city, they would name element 97 after a city. Since this new element was discovered in the city of Berkeley, the name "berkelium" was the obvious choice.

Berkelium doesn't have any practical applications. It isn't suitable for medical use, nor for any application in space technology or in nuclear energy. This is mostly due to the half-life and the decay products of its isotopes, but also because it is so hard to manufacture. Only 1 gram of berkelium has been produced since 1967. Berkelium is mainly used by researchers to create even larger super-heavy elements, such as tennessine. To create tennessine, a tiny batch of only 22 milligrams of berkelium was made (in 2009) by the Oak Ridge National Lab, then sent to the Joint Institute for Nuclear Research in Dubna, Russia, where it was exposed to fast-moving calcium atoms (Ca-48) for 150 days. Some of the calcium atoms stuck to the berkelium atoms to make an atom with 117 protons.

Currently, Oak Ridge Lab makes berkelium by putting plutonium-239 into their High Flux Isotope Reactor to create americium which decays to curium, which then decays into berkelium. (Berkelium decays into californium.)

98 Californium Cf

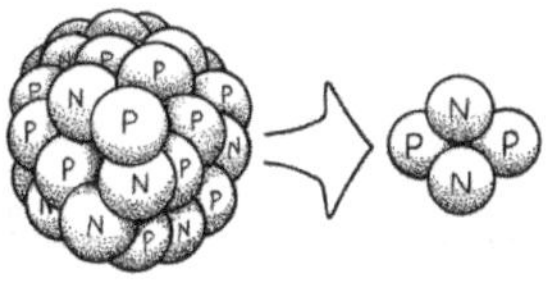

Californium atoms emit alpha particles and free neutrons.

98 protons
150-156 neutrons
98 electrons

Atomic mass: 251

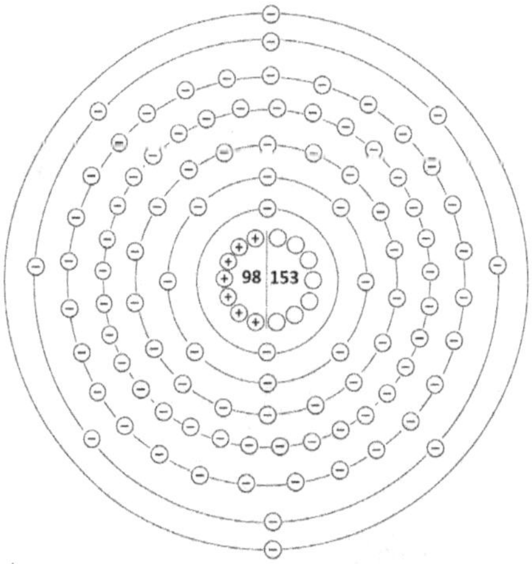

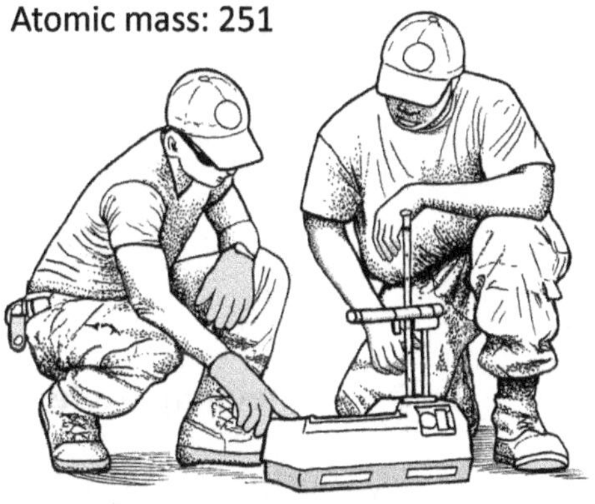

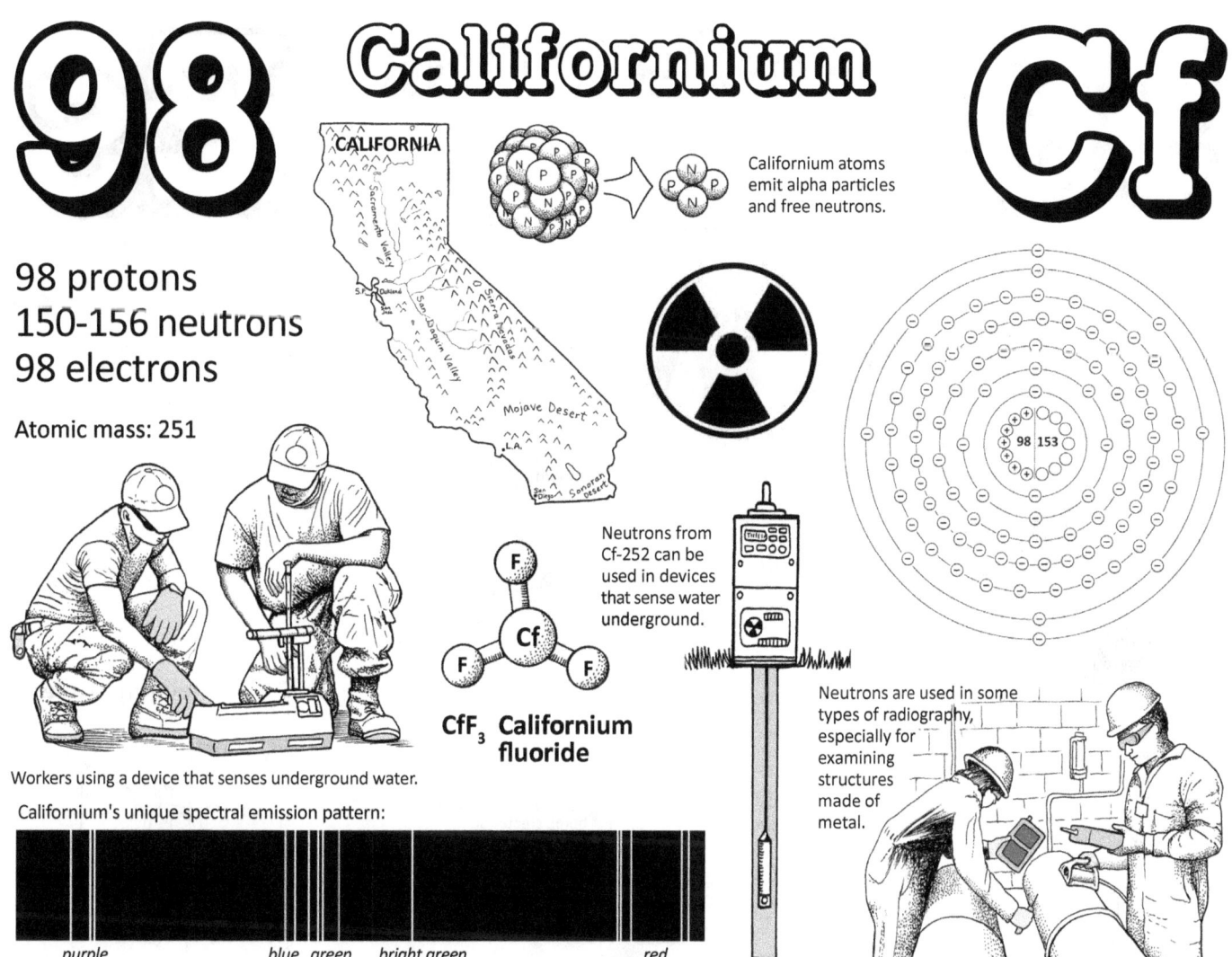

Workers using a device that senses underground water.

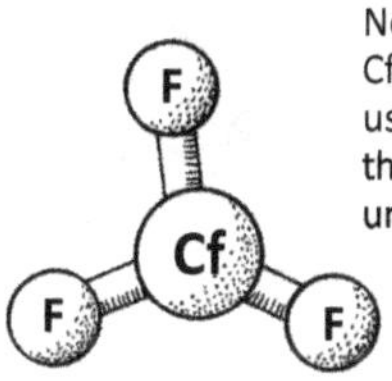

CfF_3 **Californium fluoride**

Neutrons from Cf-252 can be used in devices that sense water underground.

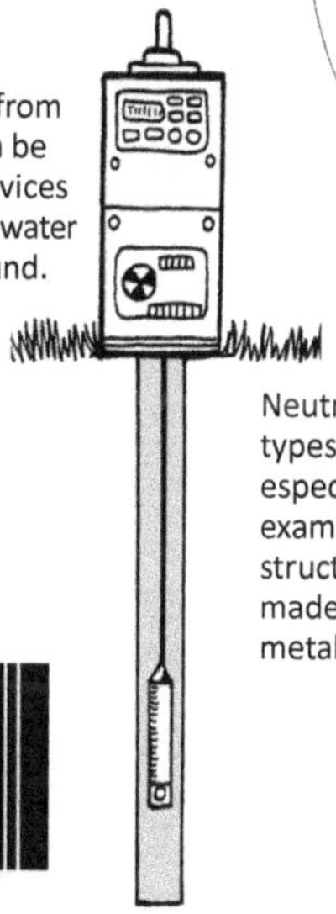

Neutrons are used in some types of radiography, especially for examining structures made of metal.

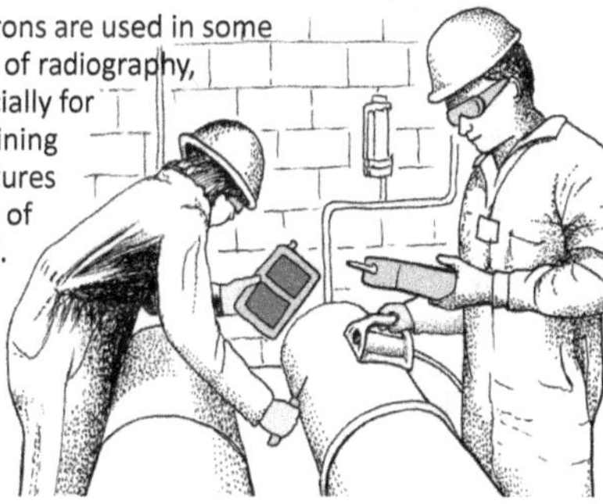

Californium's unique spectral emission pattern:

Only weeks after the discovery of berkelium, Glenn Seaborg and his team at the Berkeley Radiation Lab (now called the Berkeley National Lab) decided to keep going along the actinide series and see if they could make (and thus discover) element 98. They put a sample of curium-242 into their cyclotron machine and bombarded it with alpha particles. Curium has 96 protons, so when an alpha particle containing two protons stuck to a curium atom, the number of protons jumped to 98. This first experiment produced only about 5,000 atoms of element 98, but that was enough to make the necessary observations and be able to announce the discovery of a new element. In the naming of the three previous actinide elements, an effort had been made to coordinate with the element in the lanthanide series right above it. In this case, the lanthanide above element 98 is dysprosium, whose name means "hard to get at." Seaborg's team decided it wasn't possible to choose a name that was in any way similar to dysprosium, so they named it after the state where it was discovered, California. After the name was chosen, someone pointed out that when people first tried to travel to California in the early 1800s, it was, indeed, very hard to get to, so the name (sort of) coordinated with dysprosium after all.

Californium (Cf-252 in particular) is one of the few super-heavy actinides that has practical applications since it is a good source of neutrons. To make Cf -252, Bk-249 is bombarded with neutrons to make Bk-250 which quickly decays into Cf-250. (This is beta decay, where a neutron turns into a proton and an electron.) Cf-250 can continue to be saturated with neutrons until Cf-251 and Cf-252 are produced. Therefore, Cf-252 has several extra neutrons it can lose. A microgram of Cf-252 can produce as much as 139 million neutrons per minute.

Neutrons from Cf-252 can be used as starting fuel in nuclear reactors, as a fuel rod scanner (for analysis of remaining elements in nuclear fuel rod), and as a portable source of neutrons for detection devices that look for cracks, bad welds and corrosion in large metal structures, including airplanes. It is also used in moisture gauges that look for water reservoirs and oil deposits underground. The neutrons interact with the H atoms in the water or oil.

Like curium and berkelium, californium can be used as a starting point to manufacture even larger elements. Three atoms of oganesson were made in 2006 by exposing Cf-249 to fast-moving calcium atoms (Ca-48).

99 Einsteinium Es

99 protons
153-156 neut.
99 electrons

Atomic mass: 208.9

Many nations have produced stamps that honor Einstein.

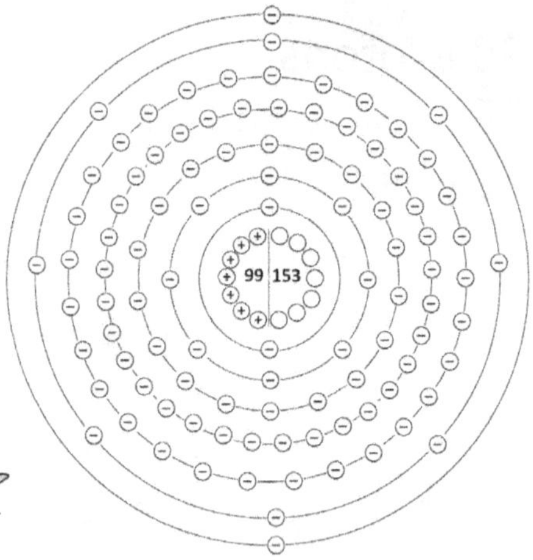

B-17 planes were flown around the edges of the mushroom cloud created by the test of the first H bomb. They collected radioactive atoms.

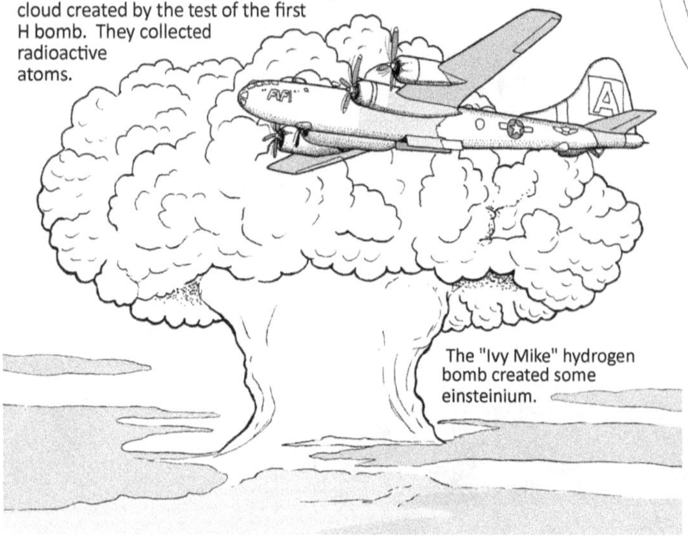

The "Ivy Mike" hydrogen bomb created some einsteinium.

EsF_3 Einsteinium trifluoride

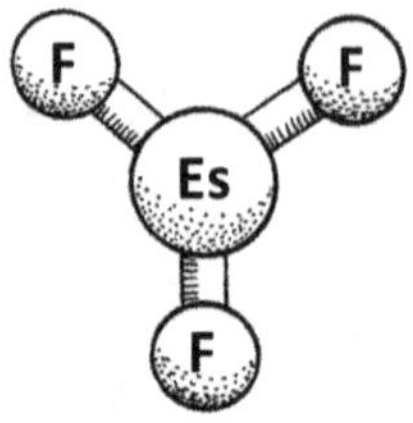

Es_2O_3 Einsteinium oxide

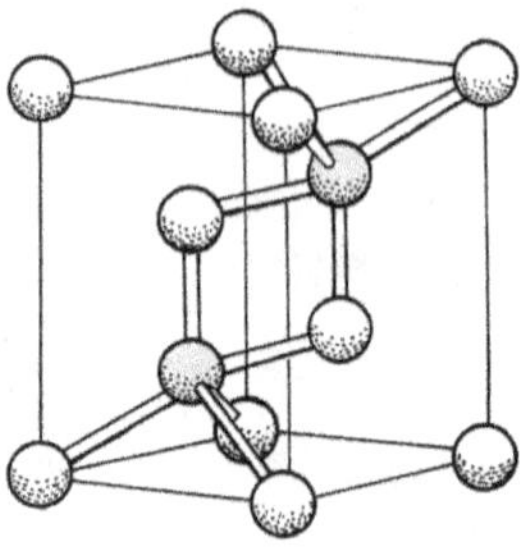

Gray is Es. White is O.

Element 99 was first discovered in radioactive waste that was collected during the explosion of the first hydrogen bomb in 1952. Los Alamos National Lab (in New Mexico) had developed a new type of atomic bomb that used heavy isotopes of hydrogen called deuterium (one neutron) and tritium (two neutrons). The bomb was nicknamed "Ivy Mike" and the test site was the Enewetak Atoll in the Pacific Ocean. This bomb was found to be 500 times more powerful than the bombs that were dropped on Japan at the end of World War II. The scientists who had designed the bomb wanted information about what types of super heavy elements were produced in the explosion, so airplanes fitted with filter paper sampling devices were sent to fly through the edges of the giant mushroom cloud. The filter papers were then sent to Berkeley National Lab where Albert Ghiorso and his team analyzed them and observed many heavy elements, including some atoms with 99 protons. This information was kept secret for several years, however, because knowing what isotopes the explosion produced gave valuable information about the procession of nuclear fission, and the ability to make even better bombs. The U.S. did not want this information to be leaked to Russian scientists who were also working on making atomic bombs during this "Cold War" era.

Many suggestions were put forward for the naming of element 99, most of them honoring the Los Alamos or Argonne National Lab: losalium, losalamium, losalamosium, lasium, alamosium, laslucium, uclasium, argonnium, phoenicium, arconium, u calium, anlium, athenium. In the end, they decided to honor Albert Einstein, whose famous equation $e=mc^2$ played a key role in understanding the relationship between atoms and energy.

Today, einsteinium is produced in very small quantities (a few millionths of a gram) by the Oak Ridge National Lab in Tennessee and a similar facility in Dimitrovgrad, Russia. There are many isotopes of einsteinium (variation in the number of neutrons) and the most stable one has a half-life of about 471 days. Einsteinium is used by researchers to gain more knowledge about how electrons behave in the atoms of the lanthanide and actinide groups, and also to produce even heavier elements. In 1955, the element mendelevium was made at the Berkeley National Lab by irradiating an Es target with alpha particles. The only practical application ever found for einsteinium was the use of Es-254 to calibrate the spectrometer used by the unmanned Surveyor 5 Lunar Probe in 1967.

100 Fermium Fm

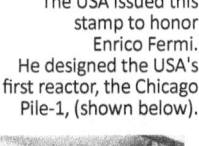

The USA issued this stamp to honor Enrico Fermi. He designed the USA's first reactor, the Chicago Pile-1, (shown below).

100 protons
152-157 neut.
100 electrons

Atomic mass: 208.9

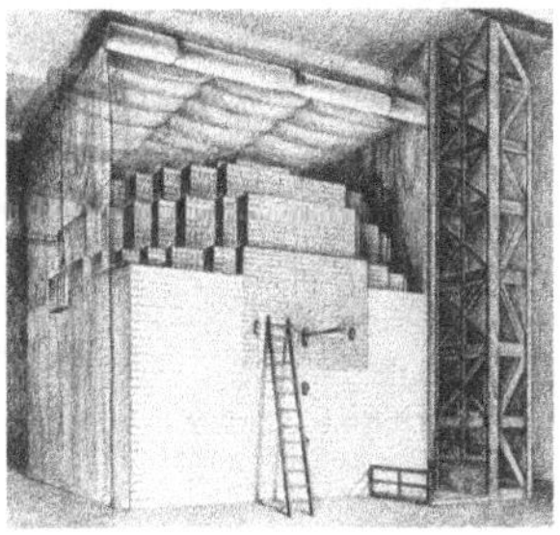

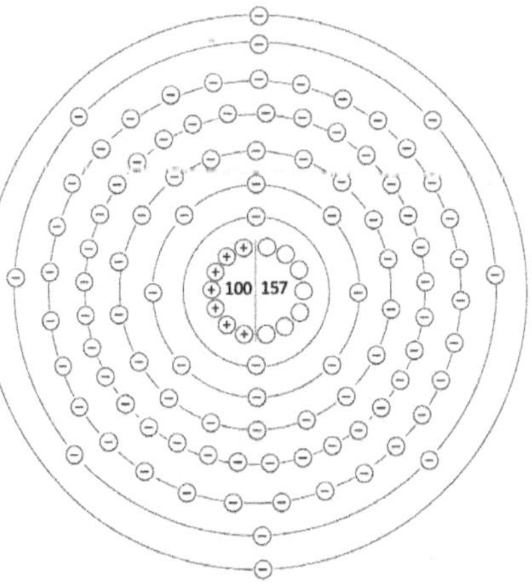

B-17 planes collected samples from the H bomb mushroom cloud in 1952.

$FmCl_2$
Fermium chloride

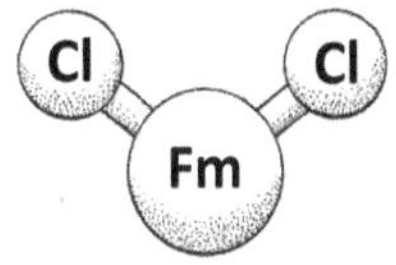

Very few compounds of fermium are known.

This stamp from 1955 honored a speech by President Eisenhower encouraging nuclear science to be used for peaceful purposes.

Element 100 was discovered at the same time as element 99. Both elements were found on the filter papers that had been designed to collect samples after the explosion of the Ivy Mike H bomb in 1952 at Enewetak Atoll in the Pacific Ocean. The collection papers had been mounted on the exterior of B-17 planes that flew in and around the edges of the mushroom cloud. The filter papers were analyzed at the Berkeley National Lab by Albert Ghiorso and his team and they quickly identified two previously unknown elements that had 99 and 100 protons. They named the new elements after famous scientists, with element 100 honoring Enrico Fermi, who had pioneered the use of neutrons to induce radioactivity, and then designed America's first nuclear reactor.

Fermi came to the U.S. in 1938 because of political unrest in Italy. He and his family flew to Sweden where he accepted the Nobel Prize in Physics, then instead of returning to Italy, they flew to New York and applied for political asylum. America happily received this outstanding physicist and he was immediately hired by Columbia University in New York City, where he continued his research on radioactivity and the splitting of uranium atoms. Hearing that German scientists were studying nuclear fission using uranium, Fermi was transfered to the University of Chicago where his task was to design and build (with help) a reactor. The result was the "Chicago Pile-1," a very simple reactor compared to the ones we use today. It was basically some rods of uranium surrounded by many blocks of graphite. The graphite would absorb the radioactive particles produced by the fission (splitting) of the uranium atoms. With the success of this project, Fermi was transfered to the "Manhattan Project," based in Los Alamos, New Mexico, where he and other scientists were tasked with building a bomb using radioactive atoms that would release an immense amount of energy. Although he participated in the science, Fermi did not like war and was opposed to the use of atomic power to make weapons of mass destruction.

Fermium atoms vary in their number of neutrons. The most stable of these isotopes, Fm-257, has a half-life of about 100 days. When fermium is produced by researchers (often at Oak Ridge National Lab), they make Fm-255 which has a half-life of only 20 hours, but easier to separate out from all the other isotopes produced. Fermium does not have any practical applications and is only used to do research on radioactivity.

101 Mendelevium Md

101 protons
143-152 neutrons
101 electrons

Russia issued this stamp in 2009 to honor the 175th birthday of Dmitri Mendeleev, inventor of the Periodic Table. Sadly, during his lifetime Mendeleev was persecuted by the Russian government because he dared to criticize government policies.

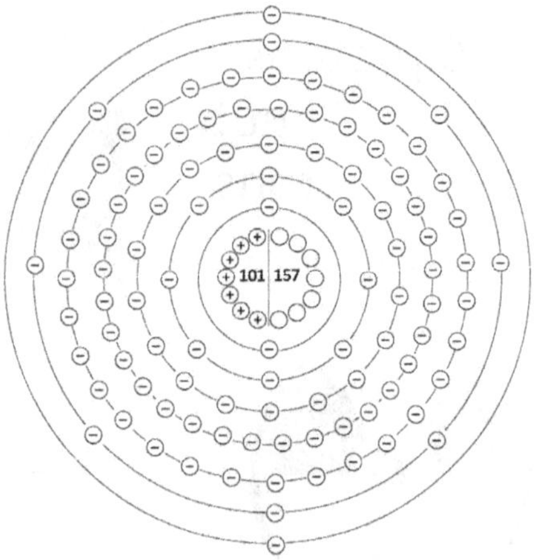

Alpha particles were used to make mendelevium atoms.

Mendelevium was created in the cyclotron machine at the Berkeley Lab in 1955 by Albert Ghiorso, Glenn Seaborg, Stanley Thompson and their team.

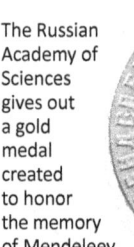

The Russian Academy of Sciences gives out a gold medal created to honor the memory of Mendeleev.

By the time element 101 was discovered, scientists began to realize that the Periodic Table did not yet have an element named to honor the man who created it. So in February of 1955, when the team at Berkeley University confirmed the existence of an atom with 101 protons, they decided they should take this opportunity to honor Dmitri Mendeleev. They had used a target made of about a billion atoms of einsteinium-253 and bombarded it with alpha particles using their cyclotron machine. If the alpha particles stuck to the nuclei of the einsteinium atoms, the proton count would go from 99 to 101. The number of neutrons did not matter, since it is the protons that determine the identity of the atom. The isotope of mendelevium produced was Md-256 and it had a half-life of only 77 minutes. The scientists had to literally run from one room to another (and even drive to another lab) to quickly do all the experiments necessary to analyze the new samples. It was a very complicated process to separate all the atoms and try to isolate just the mendelevium. However, they successfully documented their discovery and published their results.

Mendelevium can be made in various ways. Each process will create a different set of isotopes (number of neutrons). To get lighter isotopes (244 to 247) you can start with a bismuth target and hit it with argon atoms. To get medium-sized isotopes (248 to 253), you start with plutonium and americium targets and hit them with carbon and nitrogen atoms. To get the larger and more stable isotopes (254 to 258) it is best to do what the Berkeley team did and hit an einsteinium target with alpha particles.

Dmitri Mendeleev (sometimes spelled "Mendeleyev" so that we pronounce it correctly) was born in 1834 in the Siberian north of Russia. His family was poor and he was the youngest of at least 14 siblings. At age 15, after the death of his father, he and his mother moved to St. Petersburg so that Dmitri could attend a good school. He did well, and decided to focus on chemistry as his life's work. He also like to write, so he began to write textbooks about chemistry. As he wrote, he thought a lot about each known element (56 at that time) and began to realize that there were small groups of elements that had similar characteristics. He wondered how to use the similarities to arrange the elements into a table. One day he spent many hours trying to line up the little element cards he had made. That evening he fell asleep while he was still thinking, and in his dream he saw all the cards line up in a way he had not previously thought of. This was the beginning of the table we know today.

102 Nobelium No

A Korean stamp commemorating Alfred Nobel.

102 protons
148-160 neutrons
102 electron

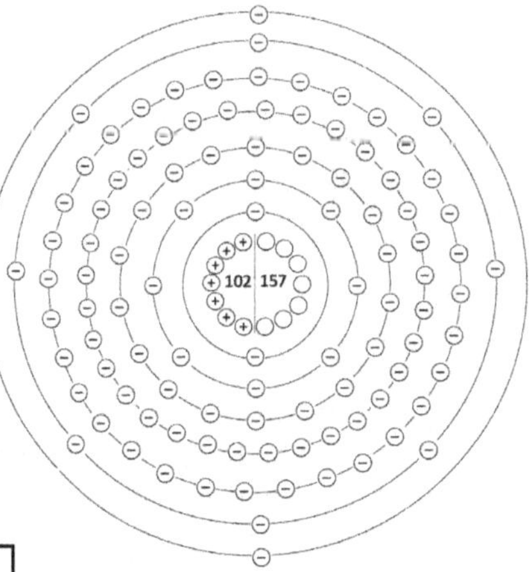

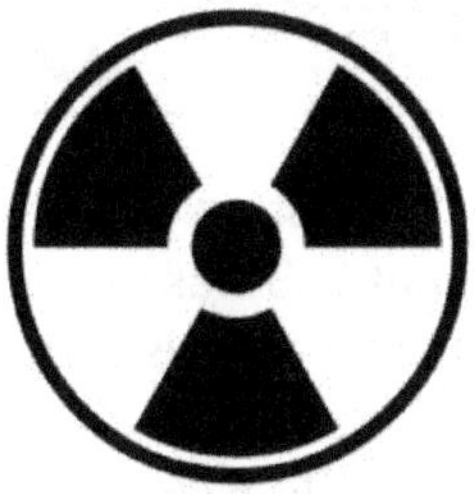

Alfred Nobel invented dynamite, a much safer explosive than most used in the 1800s.

$NoCl_2$ Nobelium (II) chloride

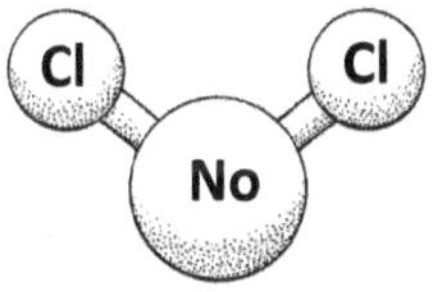

Nobelium does form molecules with several elements, but they have to be studied quickly before the No decays.

This is the front of the Joint Institute of Nuclear Research. The walls are yellow, the columns white, and the flags are mostly red, white, and blue.

$NoCl_3$ Nobelium (III) chloride

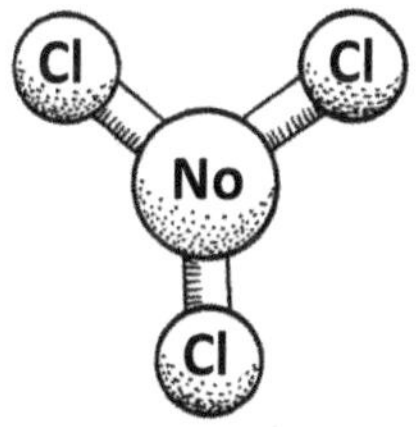

The discovery of element 102 is a complicated story. Scientists from three countries all claimed to be the first to manufacture, and thus discover, this element. The first announcement came from the scientists at the Nobel Institute in Sweden in 1957. They had bombarded a curium target with carbon atoms. Naturally, they named the new element after Alfred Nobel, the Swedish scientist after whom their institute was named. Upon hearing this, the team at Berkeley Lab tried to repeat this experiment to see if they could achieve the same results. They were unable to produce elements with 102 protons, so they challenged the claim of the Swedish scientists. Over the next several years, scientists at the Joint Institute of Nuclear Research in Dubna, Russia, tried different methods to produce element 102. In one experiment they used a Pu-239 target and hit it with oxygen atoms. The JINR then claimed to be the first to make element 102 and proposed that it be named joliotium, to honor Marie Curie's daughter, Irène Joliot-Curie, who was an excellent chemist like her mother. The JINR continued their work on element 102 for several more years, using different elements as targets and "bullets," and in 1966 presented their most convincing evidence for the creation of element 102. However, the Americans and the Swedes still thought their claims were valid.

It was not until 1992 that the mess was sorted out. The organization that holds the ultimate authority for naming elements, the International Union of Pure and Applied Chemistry (IUPAC), gave the JINR credit for the discovery, but determined that the original name, nobelium, would be kept, since it had already been in use for many years.

Alfred Nobel was born in Sweden in 1833. He followed in his father's footsteps and became an expert in the chemistry of explosives. In the 1800s, explosives were much more dangerous than they are today; the chemicals were unstable and would explode at the wrong time. Alfred's brother was killed in an accidental explosion, so Alfred was determined to make explosives safer. He found a way to combine nitroglycerin with very stable minerals, and received a patent for "dynamite." The sale of dynamite led to Alfred gaining much wealth. When one of his brothers died, a newspaper in France made a huge mistake and printed an obituary for Alfred, not the brother. The paper called Alfred "the merchant of death" because some people were using dynamite for evil purposes. Alfred was horrified that anyone would remember him in this way, so he used all his wealth to set up a foundation that would give yearly prizes (the Nobel Prizes) to people who had made outstanding contributions to science, economics, literature and world peace.

103 Lawrencium Lr

103 protons
151-163 neut.
103 electrons

Ernest O. Lawrence (1901-1958)

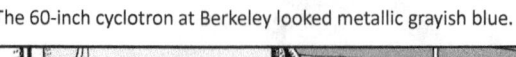

The 60-inch cyclotron at Berkeley looked metallic grayish blue.

Lawrencium was named in honor of Ernest Lawrence, inventor of the cyclotron. The 60-inch cyclotron at the Berkeley Lab was used to discover many super heavy elements. The claim of first discovery of element 103 was made by both the Berkeley lab and by the Joint Institute for Nuclear Research in Dubna, Russia. Albert Ghiorso and his team at Berkeley announced they had made element 103 in February, 1961, using a californium target and bombarding it with boron atoms. The scientists in Dubna looked at their research results and raised some questions as to whether these isotopes with a half-life of only 8 seconds were, in fact, element 103, but the research was convincing enough for the International Union of Pure and Applied Chemistry (IUPAC) to accept the discovery and approve the name. Later, as the years went on and more experiments were done by both the U.S. and Russian labs, the measurements and facts about this element were refined. In 1992, the IUPAC looked at all the research done by these labs and decided that they should share the credit and be listed as co-discoverers.

104 Rutherfordium Rf

104 protons
157-163 neut.
104 electrons

This stamp is light purple.

Rutherford is the "father of nuclear physics."

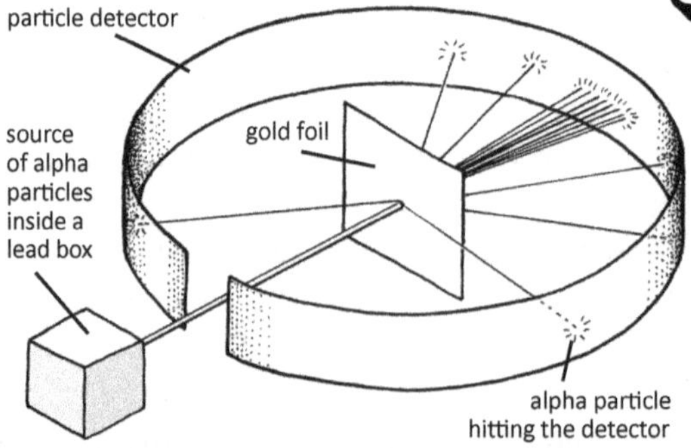

This shows the concept behind Rutherford's famous gold foil experiment. Most of the alpha particles went right through the gold foil, demonstrating that atoms were mostly empty space.

Even with their short half-lives of less than an hour, researchers have still been able to make Rf atoms form some molecules.

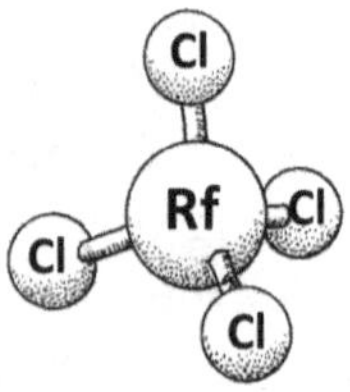

Rutherfordium was named to honor Ernest Rutherford who worked with J. J. Thomson to discover the electron. Rutherford went on to investigate radioactivity and gave us the names of the alpha and beta particles. He also worked with Hans Geiger (for whom the "Geiger counter" is named) to develop new ways of sensing and counting alpha and beta particles. His most famous experiment came in 1908 when he and Geiger and Ernest Marsden shot alpha particles through gold foil. Most of the particles went right through, with only a few bouncing back. This led to the conclusion that atoms were mostly empty space. Rutherford suggested that atoms had a small "nucleus" made of positive protons and uncharged neutrons. Rutherford received many awards during his life, and spent 5 years as president of the Royal Society.

The fact that rutherfordium will form molecules with atoms like chlorine and bromine is due to the fact that it is right under titanium, zirconium and hafnium. Chemists expect that elements in the same column will have similar chemical properties. The half-lives of isotopes of Rf range from about an hour to a few millionths of a second.

Rutherfordium was first reported by the JINR lab in Dubna, Russia, in 1964, and then by Berkeley in 1969.

105 Dubnium Db

105 protons
157-165 neutrons
105 electrons

Joint Institute of Nuclear Research.

Walls: yellow
Columns: white
Flags: mostly red, white, and blue.

The flag of Dubna, Russia

The atom and triangle: yellow
Tree top: green
Water: blue

The existence of element 105 was first reported in 1968 by the Joint Institute for Nuclear Research (JINR) in Dubna, Russia (just a little north of Moscow). The researchers bombarded a target of americium atoms (95 protons) with neon atoms (10 protons). In 1970, the team at Berkeley lab used californium and nitrogen to create atoms with 105 protons. Both teams heavily relied on the observation of alpha decay producing element 103. This implied that something had started with 105 protons, and lost 2 to become element 103.

The Russian team proposed the name "nielsbohrium," after Danish physicist Niels Bohr, and the American team proposed the name hahnium, after Otto Hahn, who had worked with Lise Meitner to discover the fission process. After two decades of arguments between Berkeley lab, JINR, and the IUPAC, and suggestions of renaming many of these new elements, it was finally decided that both labs should share credit for discovery, but the name would be dubnium.

106 Seaborgium Sg

106 protons
152-165 neut.
106 electrons

Glenn Seaborg standing in front of the ion exchange equipment that isolated actinides.

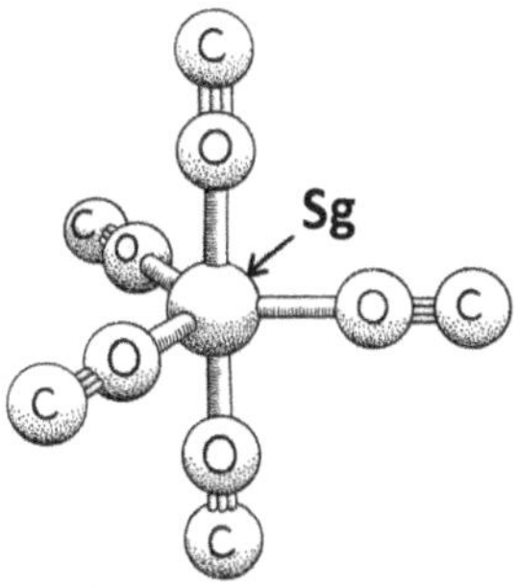

Seaborgium hexacarbonyl was created in a lab in 2014. The Sg soon decayed, however.

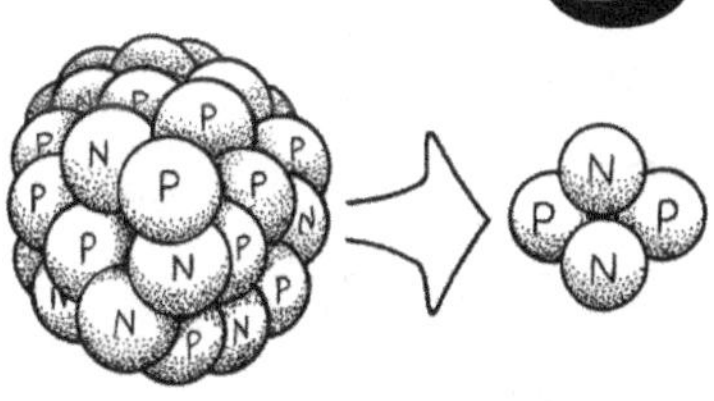

Observing alpha decays was the primary method used to determine whether element 106 had been made. They were able to accurately predict how each isotope of 106 would decay, so if they observed a certain decay pattern, they assumed it had come from element 106.

In 1974, a few atoms of element 106 were observed at both Berkeley lab and the JINR in Dubna, Russia. The Russians used lead and chromium (82+24) in their experiments, and the Americans used californium and oxygen (98+8). The naming of this element was disputed for years, until the issue was resolved in 1997 when the IUPAC decided that this element could be named after Glenn T. Seaborg who was the leader of the Berkeley lab in the late 1950s and helped to discover americium, berkelium, californium and lawrencium. Seaborg was still alive when the name was proposed, and this became controversial because no previous element had been named after a living person. In 1997, the IUPAC evaluated many naming controversies and renamed several elements, but kept seaborgium for 106.

Research on seaborgium is difficult because it must be produced one atom at a time, and the longest lived isotope has a half-life of only 14 minutes. The least stable isotope only exists for a few millionths of a second. However, in 2014 scientists were able to make a molecule called seaborgium hexacarbonyl, $Sg(CO)_6$. They expected Sg to be able to form this compound because the elements right above it on the Periodic Table are able to do so.

107 Bohrium Bh

107 protons
153-171 neutrons
107 electrons

Denmark made several commemorative stamps honoring one of the premier scientists, Niels Bohr.

25	Mn
43	Tc
75	Re
107	Bh

Bohrium is the last element in the group 7 columns so it should have chemical similarities with Mn, Tc and Re.

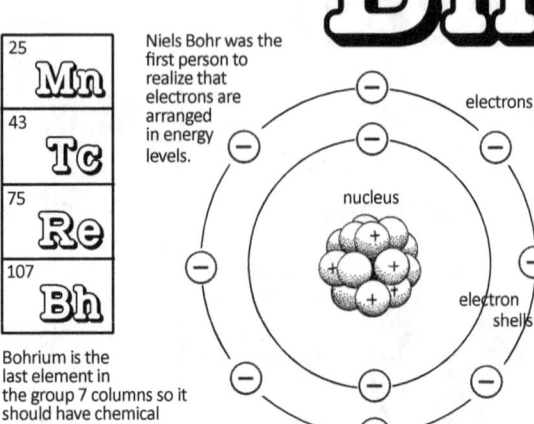

Niels Bohr was the first person to realize that electrons are arranged in energy levels.

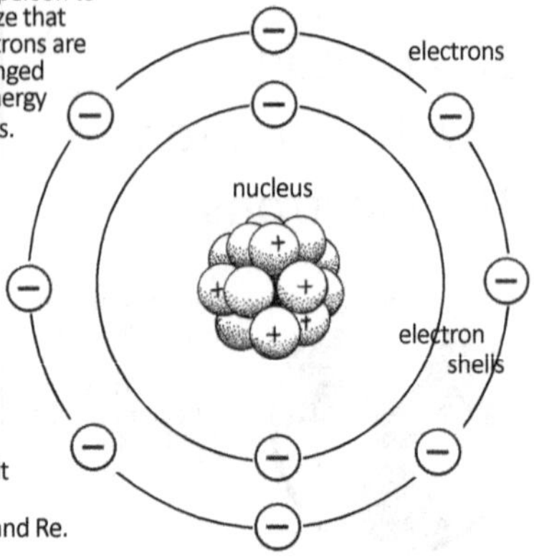

Element 107 was first reported by a team of Russian scientists in 1976. They used targets of lead and bismuth and hit them with chromium and manganese, respectively. In 1981, a German team at a nuclear lab in Darmstadt said that they had also manufactured 5 atoms of element 107 using bismuth and chromium. When the IUPAC investigated the claims of both groups, they found the research of the German team to be more convincing, so they gave the credit of discovery to Darmstadt. The German team proposed the name "nielsbohrium" but IUPAC said that no previous element had included someone's first name, so they insisted on naming it just "bohrium." Niels Bohr was a Danish physicists who ended up visiting and living in many other countries due to the upheavals of World War II. He is best known for figuring out that electrons are found in specific energy levels, or shells. He created the atomic "solar system model."

The longest half-life of any bohrium isotope is Bh-274, with a half-life of 40 seconds. Other isotopes vanish in a matter of thousandths of a second. So far no one has been able to confirm molecules made with bohrium.

---

108 Hassium Hs

108 protons
155-169 neut.
108 electrons

GERMANY

The German researchers used a linear particle accelerator to fuse lead atoms with iron atoms. (82+26= 108)

HsO_4
Hassium tetroxide

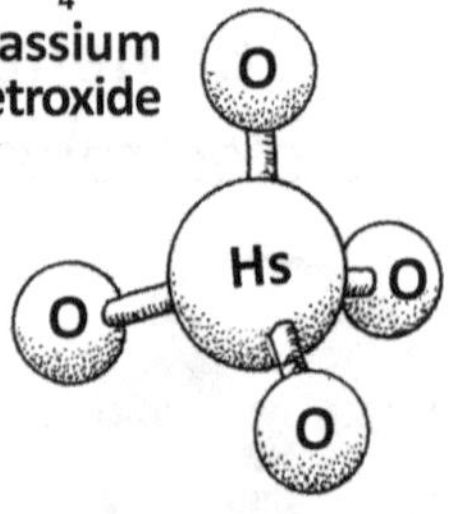

This is one of the very few Hs molecules that has been made.

Hassium is named after Hesse, the German state in which the GSI Helmholtz Institute for Heavy Ion Research is located. Germany wanted to catch up with the U.S. in the naming game. The researchers at Berkeley had used the city, state, and country of their research facility. The Germans already had an element named after their country (germanium) so they wanted to get a city and state as well. Hassium would be their state element, and darmstadtium would eventually become their city element. Researchers at JINR and Berkeley had been trying to synthesize element 108 for several years, but could not produce adequate documentation for the IUPAC. Although the German research was done in 1984, it was not until 1997 that the IUPAC officially announced hassium as element 108.

The longest-lived isotope of hassium has a half-life of only 110 seconds. Most isotopes decay much faster than this. However, despite this very small window of opportunity, scientists have still been able to briefly observe hassium bonding to oxygens to create hassium tetroxide. This is not surprising since hassium belongs to the platinum group.

109 Meitnerium Mt

109 protons
157-173 neut.
109 electrons

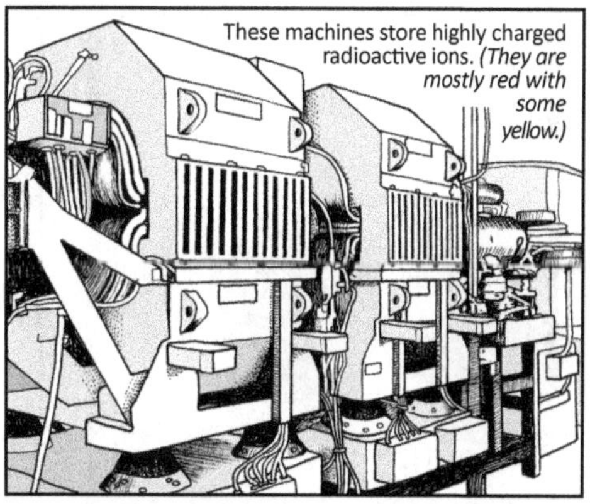

These machines store highly charged radioactive ions. *(They are mostly red with some yellow.)*

These are some of the machines at the Darmstadt facility.

Alpha particles are often ejected by super heavy elements. Each isotope of each heavy element has a unique decay pattern of precisely timed half-lives.

Meitnerium is the first element on the Periodic Table whose chemical properties have not been investigated. This is primarily because the half-lives of its isotopes are so short; most are thousandths of a second. This element was first synthesized in 1982 by the German team at the GSI Institute for Heavy Ion Research in Darmstadt. They used a bismuth target and hit it with iron atoms; the result was the detection of a single atom of element 109. The element was officially named in 1997, and everyone agreed that Austrian physicist Lise Meitner should receive the honor of having this element named after her. Meitner worked with Otto Hahn to discover the element protactinium and the process of nuclear fission. The only other element named after a woman is curium.

Since meitnerium is in the same column as cobalt, iridium and rhodium, it would most likely be a solid if we could gather enough atoms of it. Meitnerium is in the decay chain of larger elements with odd numbers, such as tennessine and nihonium. These atoms release alpha particles, dropping their atomic number by 2 each time.

---

110 Darmstadtium Ds

110 protons
157-171 neut.
110 electrons

The flag of the city of Darmstadt. *Top: blue, bottom: white. Lion is red on a yellow background. Crown is red and gold. The fleur-de-lis is white on a blue background.*

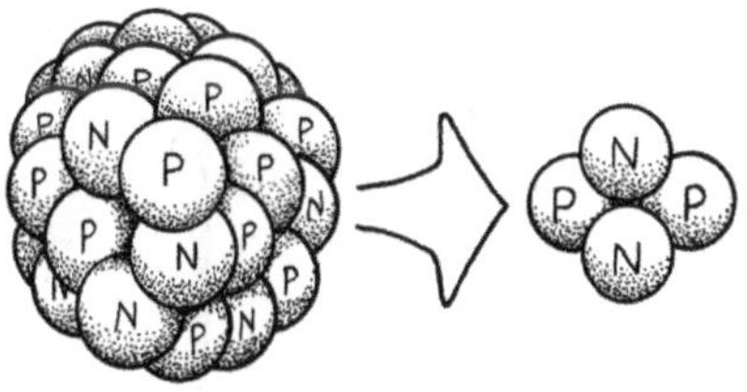

Researchers don't actually "see" the atoms they are making. The proof that they have created a new element is based on analyzing the nuclear decay patterns they see, and trying to figure out where all the alpha particles are coming from. Each isotope of every element has a unique half-life which gives them additional clues.

The credit for discovery of element 110 goes to the research team at the GSI Institute for Heavy Ion Research in Darmstadt, Germany. In 1994, they used lead (82) and nickel (28) to make a single atom that had 110 protons. It is important in the creation of super heavy elements to choose the right isotopes for your targets and your ion streams. For example, the German scientists found that switching from nickel-62 to nickel-64 (adjusting the number of neutrons) resulted in ability to make more atoms of element 101.

The German team, which included Sigurd Hofmann, Peter Armbruster, and Gottfried Münzenberg, were ready with their name suggestion. Many German scientists somewhat resented the fact that the scientists at Berkeley lab were able to name three elements after the location of their lab: berkelium, californium, and americium (city, state, and country). Now Germany had the chance to even the score. After choosing the name "darmstadtium" they now had their set of three elements: darmstadtium, hassium and germanium.

111 Roentgenium Rg

111 protons
161-175 neut.
111 electrons

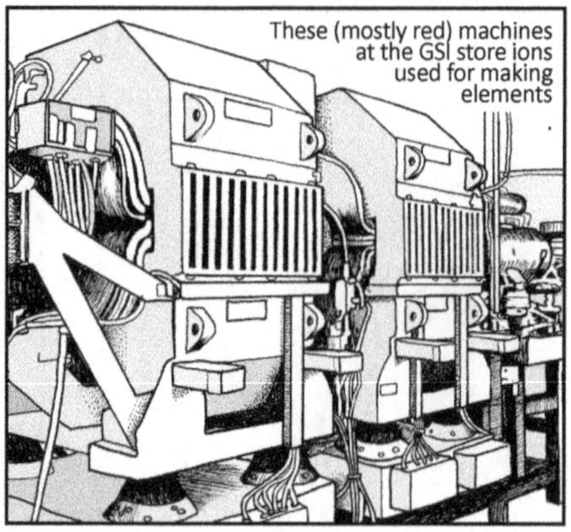

These (mostly red) machines at the GSI store ions used for making elements

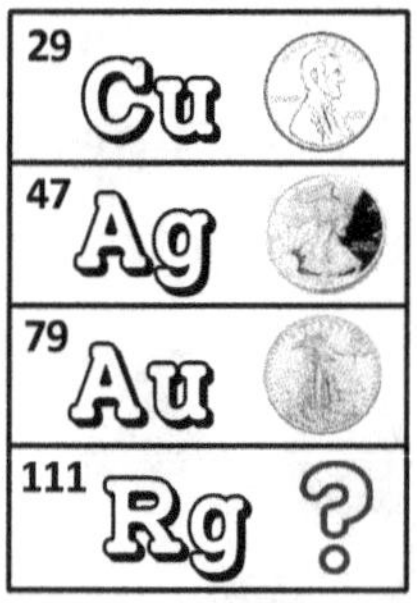

If we had enough Rg, could we make a radioactive coin?

The first lab to synthesize a few atoms of element 111 was the GSI Helmholtz Institute for Heavy Ion Research in Darmstadt, Germany, under the direction of Sigurd Hofmann in 1994. They used a bismuth target and bombarded it with fast-moving nickel atoms. (83+28=111) The most stable isotope seems to be Rg-282 with a half-life of 100 seconds.

The German team decided to name this element after the scientist who discovered x-rays, Wilhelm Roentgen (or Röntgen). As with many discoveries, his initial observation came quite unexpectedly. Roentgen was experimenting with a "Crookes" tube (the pear-shaped object shown on the stamp above) and realized that the tube was producing some kind of previously unknown rays, which he called x-rays, with "x" standing for "unknown." It wasn't long until he was demonstrating how these rays could take pictures of bones. One of his famous images is shown on the stamp.

Roentgenium *(ront-GEN-ee-um)* is in the same column as copper, silver and gold, so we can guess that it would share similar chemical properties, even though it is radioactive. Should we classify it as a radioactive precious metal?

---

112 Copernicium Cn

112 protons
165-174 neut.
112 electrons

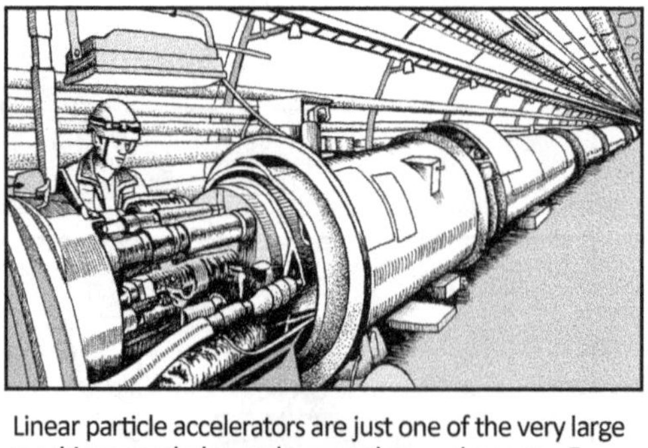

Linear particle accelerators are just one of the very large machines needed to make super heavy elements. For element 112, GSI Institute used atoms of zinc and lead. The target (zinc) was inside a large metal sphere.

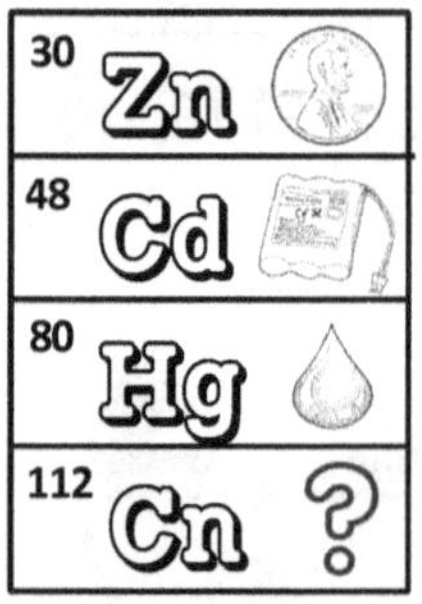

If we had enough Cn, would it be solid, or a liquid like Hg?

Element 112 was first created in 1996 at the GSI Helmholtz Institute for Heavy Ion Research in Darmstadt, by Sigurd Hofmann and his team. They fired zinc atoms at a target made of lead using a group of large machines collectively known as a particle accelerator. (The target area is small, but the supporting machinery can fill large buildings.) Most of the zinc atoms did not stick to the lead atoms, but one did, and that one new atom with 112 protons was detected. This experiment was repeated in the year 2000, and a few more atoms of element 112 were detected. The German team decided to name this element after a famous scientist from Poland, Nicolas Copernicus, who figured out that the earth goes around the sun. Copernicium *(ko-per-NISS-ee-um)* is the only element named after a scientist who was not a chemist or physicist and therefore had no connection to the Periodic Table.

Of all the isotopes of copernicium that have been produced, Cn-285 is the most stable, with a half-life of 29 seconds. If a large enough sample could be collected, copernicium would probably be a dense liquid, like mercury.

113 Nihonium Nh

113 protons
165-177 neut.
113 electrons

日本
The Japanese call their country "Nihon." This is how you spell Nihon with Japanese characters.
JAPAN

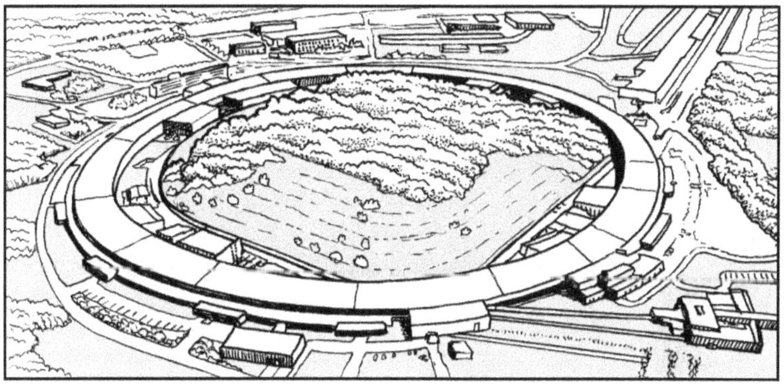

The particle accelerator at the RIKEN facility in Wako, Japan.

The manufacturing and discovery of element 113 was first claimed by an American-Russian team working at the JINR in Dubna. A year later, a Japanese team at the RIKEN facility in Japan also filed a claim for first discovery. In the next few years, teams in Sweden, Germany, China, and America also claimed to have synthesized this element. The IUPAC (International Union for Pure and Applied Chemistry) had to look at everyone's research and determine who had been first. In 2015, IUPAC announced that the Japanese team would receive official credit for discovery.

The RIKEN team used bismuth and zinc to make an atom with 113 protons. The JINR team had tried this previously and been unsuccessful, although they said they detected element 113 as a decay product of element 115. The RIKEN team used a technique called "cold fusion" and were able to produce one atom of 113. Further experiments have shown that the most stable isotope of nihonium is Nh-286, with a half-life of about ten seconds.

114 Flerovium Fl

114 protons
170-176 neut.
114 electrons

Georgy Flyorov, namesake of the Flerov lab, which is part of JINR.

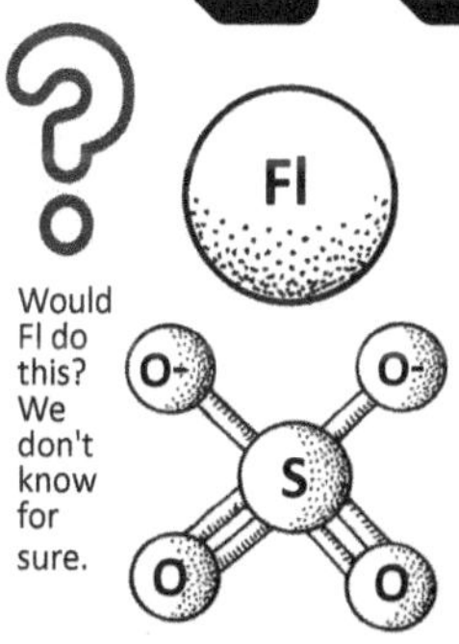
?
Fl
Would Fl do this? We don't know for sure.
O-
O-
S
O
O

Chemists suspect that Fl might act like the elements above it (C, Si, Sn) and form molecules with SO_4.

Flerovium was first synthesized in 1998 by researchers at the Joint Institute for Nuclear Research (JINR) in Dubna, Russia, under the direction of Yuri Oganessian. They used plutonium as the target and hit it with fast-moving calcium atoms (94+20) to make a single atom of 114. Since then, only 90 atoms of this element have ever been made. 58 have been manufactured directly, and the others were simply observed as part of the decay of even larger elements. The most stable isotope appears to be Fl-289 with a half-life of about 2 seconds. Flerovium is at the bottom of the column that contains carbon, silicon, germanium, tin and lead, so we can guess that it might have some similar chemical properties if it didn't decay so fast, although some experiments have indicated that is much less reactive than any of these elements. Scientists keep making these super heavy elements, hoping to find an isotope (number of neutrons) that is stable enough to make the element useful. Flerovium was named after the Flerov Lab, which is a part of the JINR facility. The lab was named after one of the researchers, Georgy Flyorov.

115 Moscovium Mc

115 protons
171-174 neut.
115 electrons

The entrance to the JINR in Dubna, Russia.

Technically, moscovium was named after the state (oblast) of Moscow, in which the city is located.

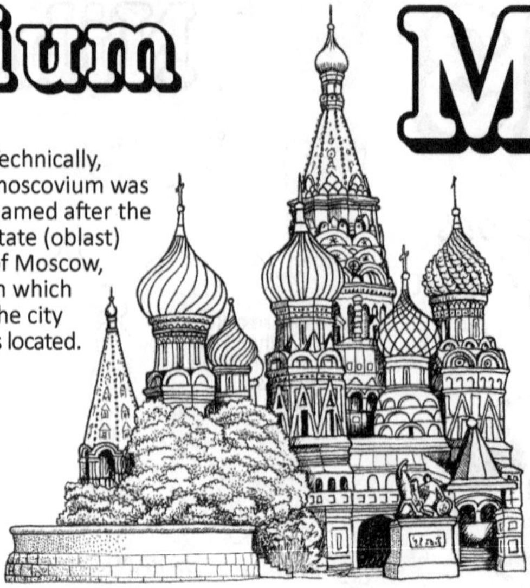

St. Basil's Cathedral is Moscow's most famous building. It is very colorful. You might want to find a color photo.

Element 115 was first made at the Joint Institute of Nuclear Research in Dubna, Russia, in 2004. American scientists from Lawrence Livermore National Lab were there, too, as this was a joint project between the two labs. They used a target made from americium and bombarded it with fast-moving atoms of calcium. The result was 4 atoms of element 115. However, when these results were evaluated by the IUPAC, they were not convincing enough. The Dubna lab had to do more experiments over the next 6 years. It was not until 2015 that the IUPAC finally gave the JINR credit for the discovery. Since elements 115 and 116 were both the result of a collaboration between the Dubna and Livermore labs, the IUPAC decided one element should have a Russian name and the other an American name. In March of 2017 there was an official naming ceremony in Moscow for three elements: moscovium, tennessine, and oganesson.

The most stable isotope of Mc has a half-life of about half a second. Mc is in the same column as N, P, As and Sb, but so far no molecules containing Mc have been made.

---

116 Livermorium Lv

116 protons
174-178 neut.
116 electrons

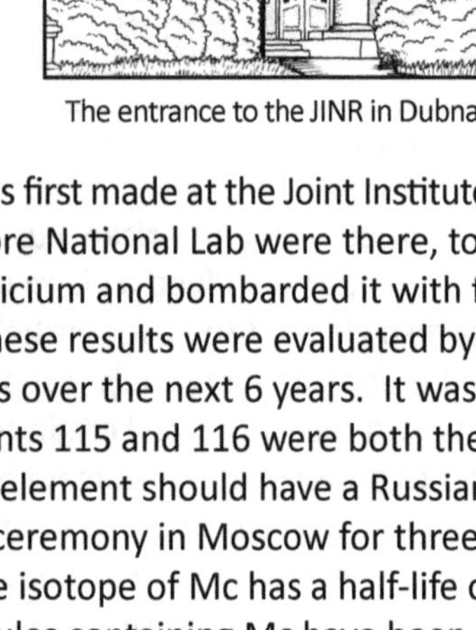

Robt. Livermore was a rancher.

Robert Livermore

Ernest Lawrence

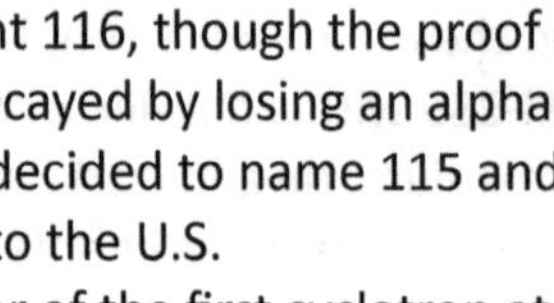

Lv decays into Fl.

Lv was made in Dubna, not Livermore.

The manufacturing and discovery of element 116 was a collaborative effort between the Lawrence Livermore National Lab in California, and the Joint Institute of Nuclear Research in Dubna, Russia. The experiment was carried out at the JINR facility in the year 2000. They used a target made of curium and hit it with fast-moving atoms of calcium. The result was the detection of a single atom of element 116, though the proof actually was based on the detection of an atom of element 114 (flerovium), as element 116 decayed by losing an alpha particle. The JINR originally proposed that element 116 be called moscovium, but the IUPAC decided to name 115 and 116 at the same time, giving one name to the U.S. and one to Russia, and assigning 116 to the U.S.

The lab is named after Ernest Lawrence, inventor of the first cyclotron at Berkeley and namesake of the element lawrencium. The lab is in the town of Livermore, which was named after its founder, Robert Livermore, an Englishman who came to California in the early 1800s to be one of the first cattle ranchers. CA was part of Mexico at that time.

117 Tennessine Ts

117 protons
176/177 neut.
117 electrons

The shaded state is Tennessee. The dot is the location of Oak Ridge National Lab.

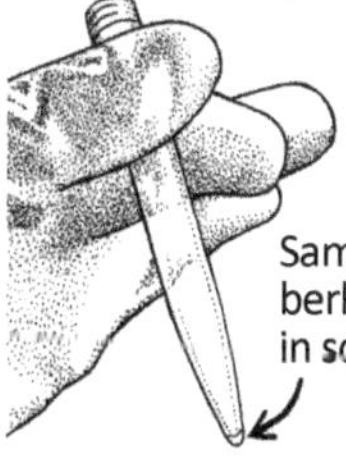

Sample of berkelium in solution

The Bk was flown to Russia.

Yuri Oganessian was in charge of making the Ts at the JINR lab.

The discovery of element 117 took about two years, and was a cooperation between Oak Ridge National Lab in Tennessee, JINR in Russia, and some consultants from Lawrence Livermore National Lab in California. The leaders of these labs met at a conference and talked about a plan to make element 117. They would use berkelium as the target and calcium atoms as the "bullets." Oak Ridge was the only lab making berkelium at that time (2008) and it took months to make all the preparations. When 22 mg of Bk had been made, it was put into a lead-shielded container and flown to Russia. Unfortunately, the customs office in Russia was not satisfied with the paperwork, and the berkelium had to cross the Atlantic several times before it finally was delivered to the JINR lab in Dubna. The calcium ions were made at a lab in another part of Russia and transported to the JINR where they were used in their particle accelerator. After six months of trying to fuse berkelium with calcium, they were finally able to announce their success in January of 2010. The element was named after the state in which the Oak Ridge lab is located. The ending "ine" was chosen because this element is in the same column as fluorine, chlorine, bromine, iodine and astatine.

---

118 Oganesson Og

118 protons
176 neutrons
118 electrons

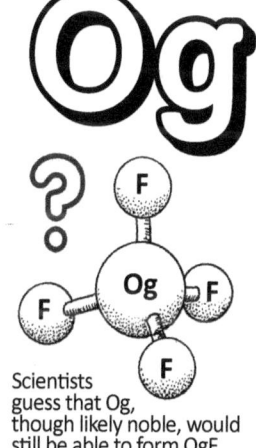

Yuri Oganessian, physicist at JINR

Joint Institute of Nuclear Research in Dubna, Russia

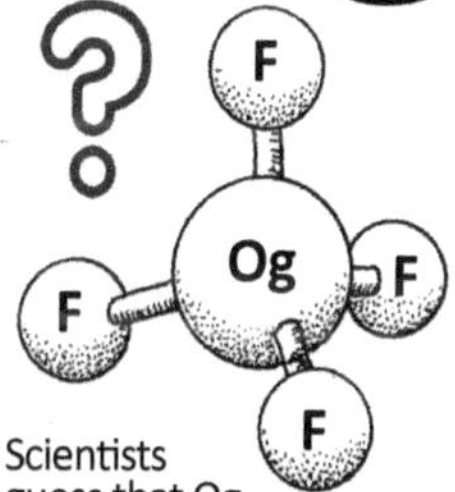

Scientists guess that Og, though likely noble, would still be able to form OgF_4.

Walls: yellow Columns: white
Flags: mostly red, white, and blue.

Element 118 had been predicted as far back as the late 1800s, after all the noble gases had been put into the last column of the Periodic Table. Scientists wondered if there could be an element below radon. In 2002, the JINR lab in Russia began a collaboration project with scientists from Livermore National lab. The Livermore scientists traveled to Dubna to participate in the attempt to make element 118. They thought they had succeeded several times, but it is the IUPAC that makes the final determination. It was not until 2015 that the IUPAC saw enough evidence to confirm the existence of element 118 and give credit to the international team that had worked under Yuri Oganessian.

The element is obviously named after Yuri Oganessian, but the ending "on" was chosen in order to match the noble gases above it on the table: neon, argon, krypton, xenon, radon. Oganessian had worked at the lab for decades and had helped to pioneer techniques for the synthesis of elements beyond 106. Very little is known about oganesson right now. Scientists continue to research these super heavy elements.

WORD PUZZLES

The answer key is at the back of this book.

SYMBOL PRACTICE

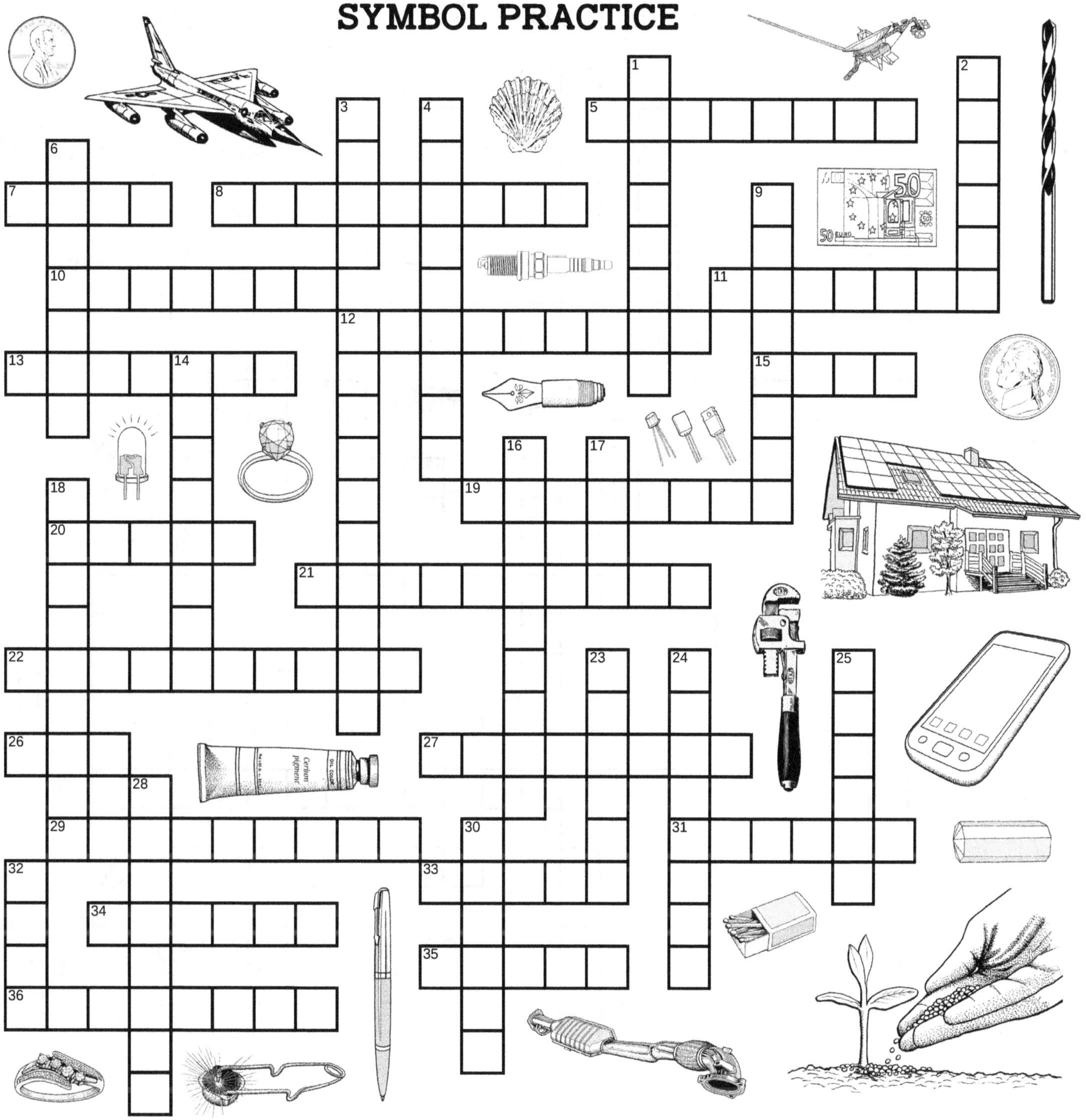

ACROSS

5) Tl
7) Pb
8) Te
10) Ca
11) Cd
12) K
13) As
15) Fe
19) Se
20) Ar
21) Mo
22) Tc
26) Sn
27) F
29) Mn
31) Na
33) B
34) Ni
35) Rn
36) Cl

DOWN

1) Cr
2) Ra
3) Au
4) Sr
6) Hg
9) Pt
12) P
14) N
16) Nd
17) Ne
18) Mg
23) C
24) W
25) S
28) Sb
30) Co
32) Zn

1) Which two letters do not appear in any symbol? ____, ____

2) Which element's name has only three letters? ___ ___ ___

3) Which 4 letters appear only once as the first (or only) letter of a symbol? ____, ____, ____, ____

4) Which letter appears most frequently as the first letter of a symbol? ____

ELEMENTS FROM HYDROGEN TO KRYPTON

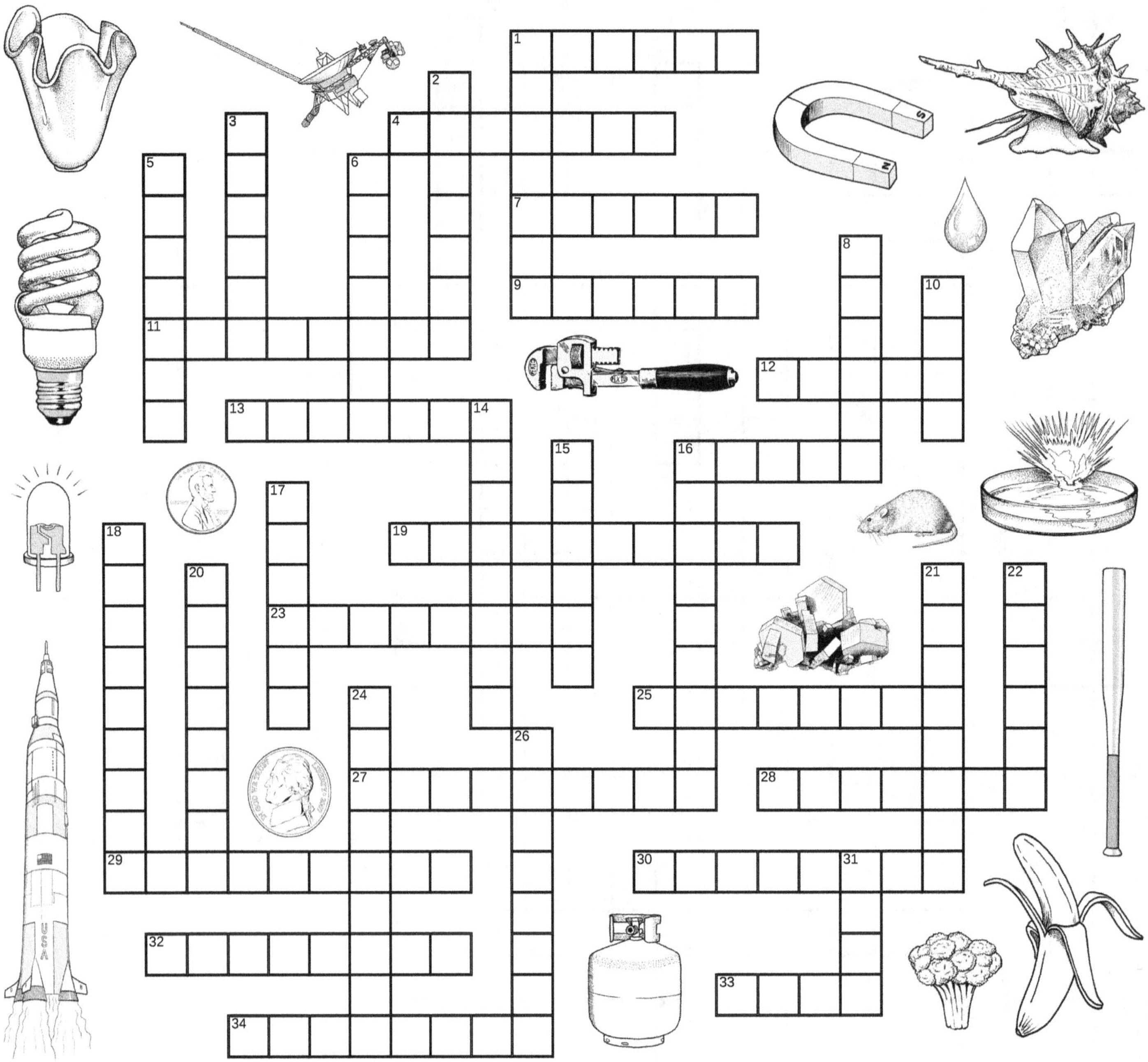

ACROSS

1) Alkali element found in bleach, baking soda and salt.
4) A lump of this element will melt in your hand.
7) Diamonds, coal, and plastics are all made of this element.
9) The center of the earth is likely made of this element and iron.
11) Found in drill bits, rockets, boat and plane parts, artificial hips.
12) This inert gas is used in light bulbs and in lasers for eye surgery.
13) Famous for use as poison, but used to make LEDs and lasers.
16) Found in fiberglass insulation, washing powder, and "slime"
19) Needed for strong bones and teeth; glows when heated.
23) Used to make non-stick pans, Teflon® tape and toothpaste.
25) Found in gun powder, window cleaner and air bags in cars.
27) Used to make transistors and diodes, named after a country.
28) This light metal is in lubricants, medicines and batteries.
29) Found in cave paintings, and nodules on ocean floor.
30) Named after the moon, is it sensitive to light.
32) This is the "C" in PVC plastic; also found in salt and bleach.
33) This gas is combined with helium to make red lasers.
34) Best known for foil, it is also used in magnets and deodorants.

DOWN

1) Combines with oxygen to make quartz.
2) Needed to make sea shells, chalk, bones, and milk.
3) Used to make magnets, and to color glass and ceramics blue.
5) Used for lasers, bright flash bulbs, and as satellite propellant
6) Was used to make purple dyes, old photographs, medicines.
8) Needed for processes of rusting and combustion.
10) Used to galvanize (protect) other metals from weathering.
14) Used to make stainless steel and yellow paint.
15) When this element tarnishes it turns green (Statue of Liberty).
16) Used to make strong springs, tools, and non-sparking metals.
17) Used to vulcanize rubber; found in many smelly things.
18) Found in gun powder, fertilizer, and molecular pumps in cells.
20) Stars use this element as fuel.
21) This light metal is used for baseball bats and lacrosse sticks.
22) This element is used in lasers, scuba tanks, rockets, balloons.
24) This element is the central atom in the chlorophyll molecule.
26) The first car to use this element was the Ford model T.
31) Primary ingredient of steel, and used in magnets.

ALKALIS, METALS, METALLOIDS, RARE EARTH METALS

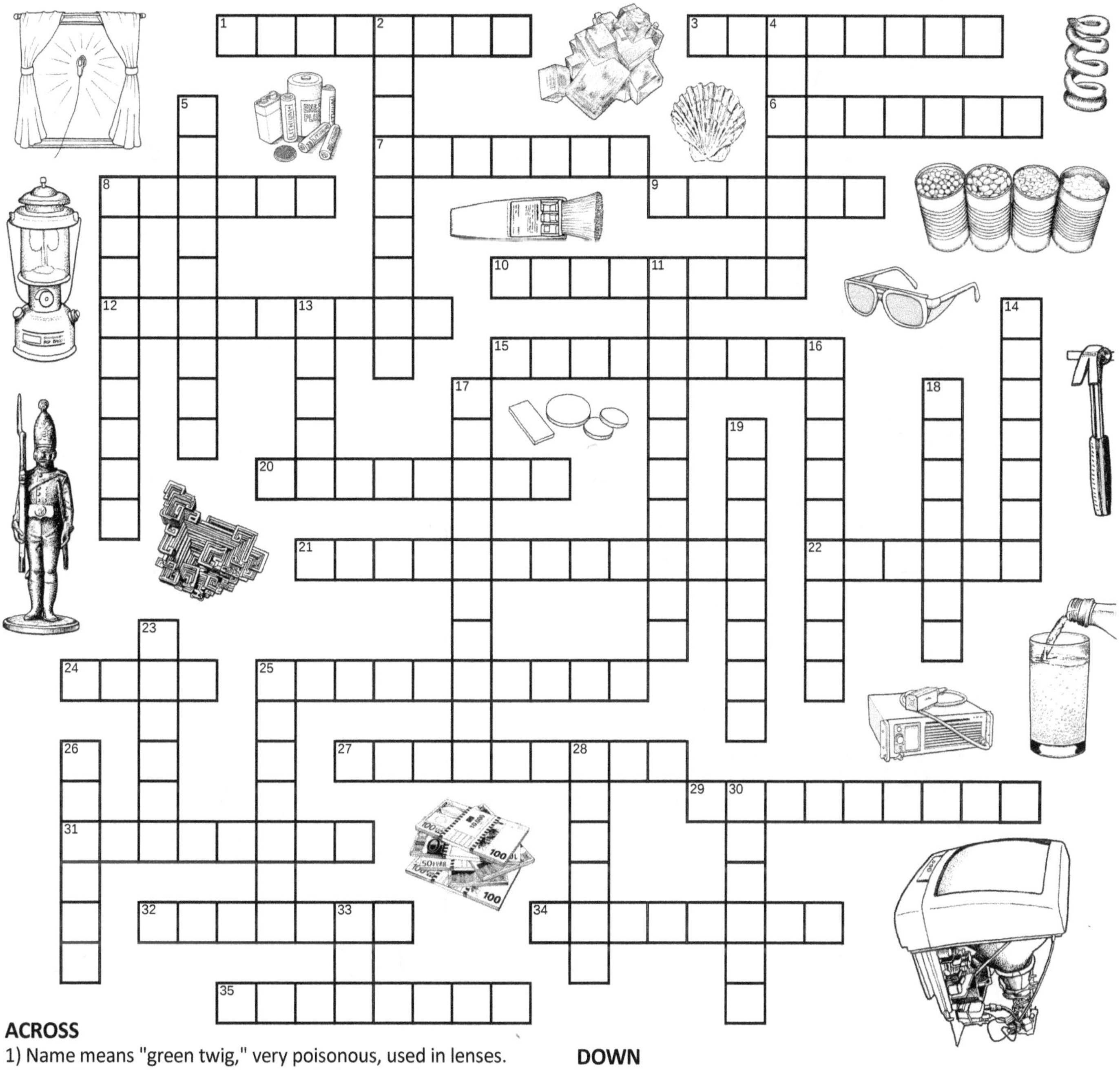

ACROSS

1) Name means "green twig," very poisonous, used in lenses.
3) First element discovered by Marie Curie.
6) The "Ter" in Terfenol-D® which turns a surface into a speaker.
7) Name comes from a very old name for Scandinavia.
8) Best known for its use in x-rays of the digestive system.
9) Named after an asteroid; found in self-cleaning ovens.
10) Name means "deep red," burns in water, used in atomic clocks.
12) Doped into YAG crystals for precision lasers; used in decoy flares.
15) One of two elements needed in a molecular pump in our cells.
20) Used by ancients for make up; mineral ore is stibnite.
21) Name means "green twin," used in purple glasses for welders.
22) Has one eletron in outer shell, has 2 yellow lines in spectrum.
24) Used in organ pipes, stained glass windows, bullets, and solder.
25) Makes green light; radioactive isotopes used in medical scanners.
27) Used to make diodes for crystal radios, and in infrared sensors.
29) Used in strong magnets in headphones and guitar pickups.
31) Named after a person; used in magnets found in solar planes.
32) Used as a lead substitute; used to treat upset stomach.
34) Used in fluorescent stripes in the euro, and for red in CRT TVs.
36) Discovered in France by woman who was student of Marie Curie.

DOWN

2) Name means "hidden," hybrid car batteries have 30 lbs of it.
4) Used in bright red fireworks and in long-life batteries.
5) Used in red fireworks and flares, and in CRT television screens.
8) Green emeralds are made of a mineral containing this element.
11) Is the "D" in Terfenol-D®, also used in magnets for wind turbines.
13) Can be used as ant poison, also used to make fiberglass.
14) Doped into YIG crystals for medical lasers, named after a city.
16) The sulfate compound of this element is called Epsom salt.
17) The only radioactive rare earth element.
18) Found in bones and teeth; used by mollusks to make shells.
19) Very light; used for bicycle frames and beverage cans.
23) Used to make glowing paint before its radioactivity was known.
25) Made famous by its use for making "disappearing spoons."
26) Satellites use small atomic clocks that use this alkali element.
28) This is the "I" in ITO, a coating for touch-screens and windshields.
30) Used to color glass pink; used in relays in fiberoptic cables.
33) Combined with copper to make bronze.

TRANSITION METALS

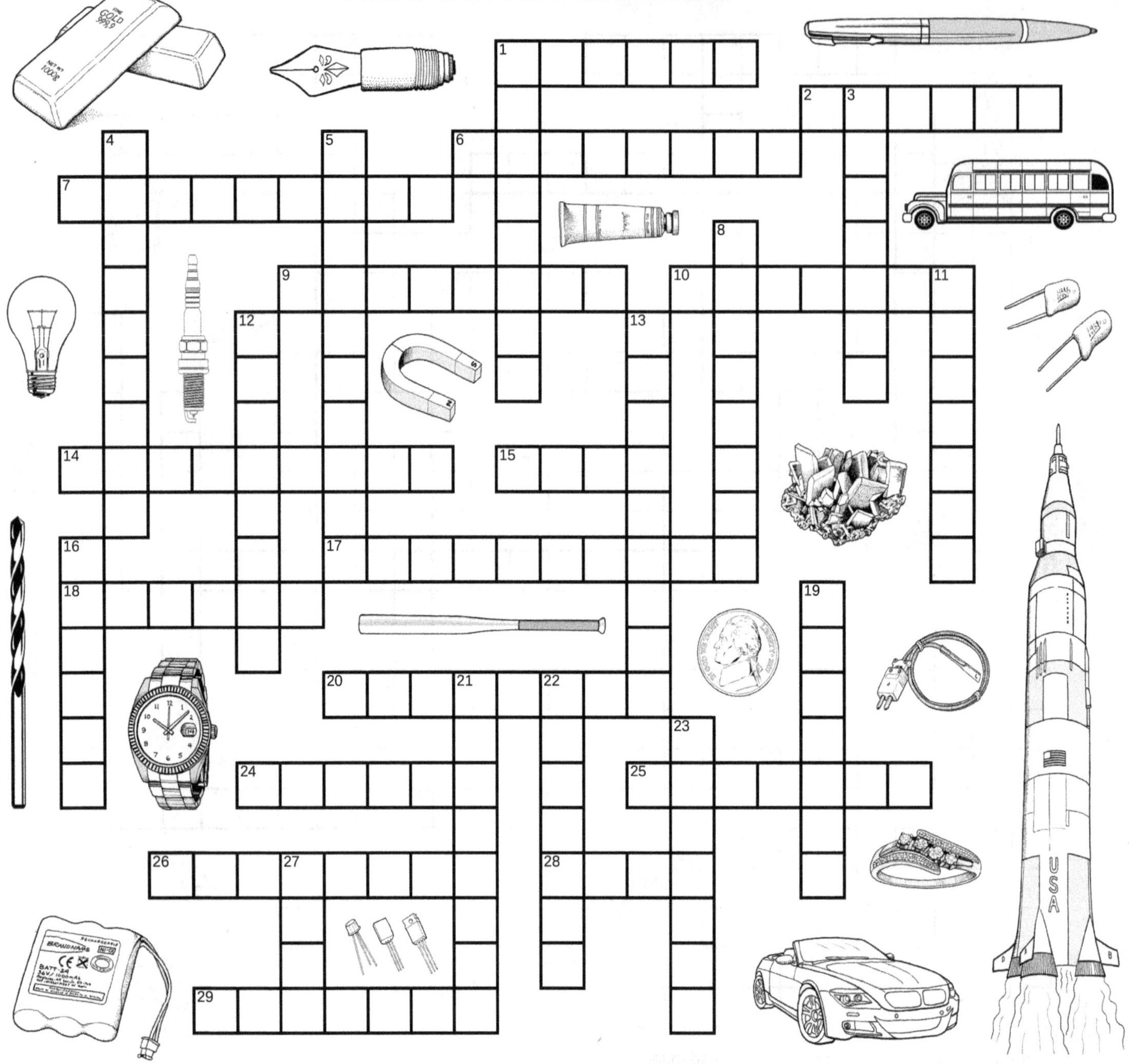

ACROSS

1) Used with bromine in early photography; used to kill germs.
2) Used for guitar strings and to coat the keys of wind instruments.
6) Under niobium on the Table; used for capacitors in electronic devices.
7) Required by many enzymes; used to make YInMn blue.
9) Makes yellow and green pigments. Makes steel "stainless."
10) Named after Copenhagen, Denmark. Used in plasma cutters.
14) Used in pen points, for lifting fingerprints, and in resistor chips.
15) Alloyed with copper to make brass; was used in voltaic piles.
17) Nitrogen-fixing bacteria need this to make an important enzyme.
18) In mid-1900s, was used for pen points and phonograph needles.
20) Used for jewelry, durable tools, and in catalytic converters in cars.
24) Alloyed with aluminum and nickel to make magnets.
25) Used to make jewelry shiny; also found in catalytic converters.
26) Primary ore is wolframite; used to make steel durable.
28) Found in hemoglobin molecules in our blood cells.
29) Alloyed with nickel to make rechargeable batteries.

DOWN

1) Found in high intensity bulbs, sports equipment, and airplanes.
3) Named after goddess of rainbow; used in spark plugs, crucibles.
4) Named after an asteroid; used to make flutes and jewelry.
5) Does not exist in nature. Made in labs for medical purposes.
8) Found in red mushroom caps and blue tunicate sea creatures.
11) Cinnabar is a common ore; used for the first barometer in 1643.
12) Named after Paris; used with aluminum garnet in lasers.
13) Found in quartz. Oxide compound is used as diamond substitute.
16) Alloyed with zinc to make brass.
19) Replaced thorium for mantles in gas lanterns; in is YInMn blue.
21) Rutile is the primary mineral ore; used to make airplane parts.
22) Used to make iridescent jewelry; alloys used for rocket nozzles.
23) Named after the Rhine River; is used as catalyst in oil refineries.
27) Used for tiny wires on circuit boards; expensive desserts are decorated with it.

URANIUM AND BEYOND

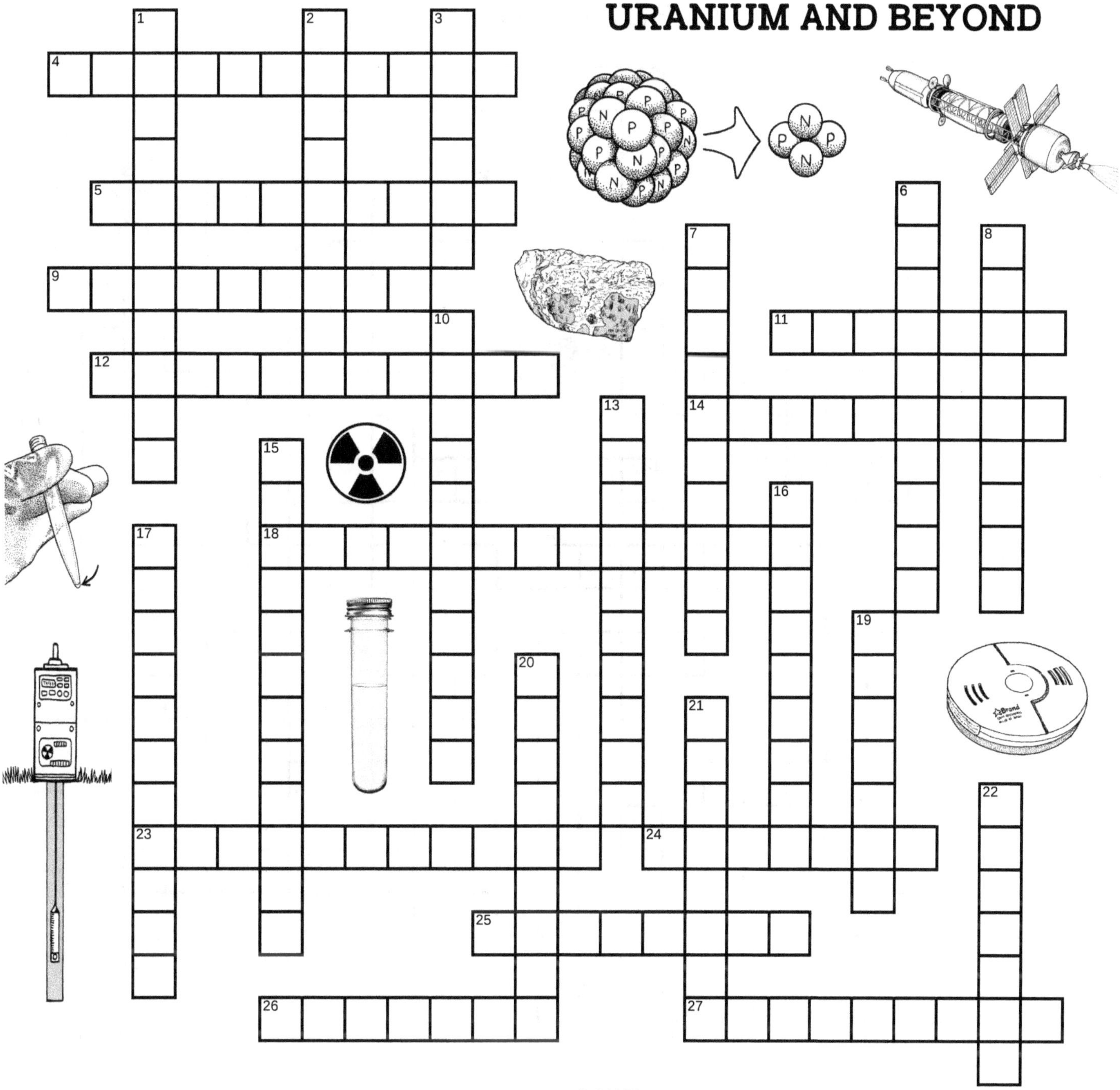

ACROSS

4) Named after the discoverer of x-rays.
5) Named for the woman who helped to discover uranium fission.
9) This element was made by hitting Pu atoms with Ca nuclei.
11) Named for the city where the Russia's JINR lab is located.
12) Named for a city that was named after a cattle rancher.
14) If we could make enough of this, it might be a noble gas.
18) Named after the scientist who did the gold foil experiment that showed that an atom is mostly empty space.
23) The U.S. state where Berkeley National Lab is located.
24) Along with Es, this element was discovered in the mushroom cloud of the Ivy Mike atomic bomb, detonated in 1952.
25) Named after Japan, the country were it was manufactured.
26) The principle ore of this element is pitchblende.
27) Named after the Russian state in which the JINR is located.

DOWN

1) Named for the inventor of the Periodic Table.
2) Named after the lab where U, Np and Pu were discovered.
3) Named after the two people who discovered Po and Ra.
6) The next-to-last element on the Periodic Table.
7) First element to be named after a living person.
8) Used as a power source in satellites, rovers, and pacemakers.
10) Named after the scientist who said "$E=mc^2$."
13) Named after the inventor of the cyclotron.
15) German scientists made it by fusing nuclei of lead and nickel.
16) This element can be found in most smoke detectors.
17) The only element named after an astronomer.
19) Named after the discoverer of electron shells.
20) This element always as 93 protons.
21) Named after the inventor of dynamite.
22) This element always has 108 protons.

LEAST ABUNDANT ELEMENTS

The least common, naturally-occuring elements (No super-heavy manufactured elements)

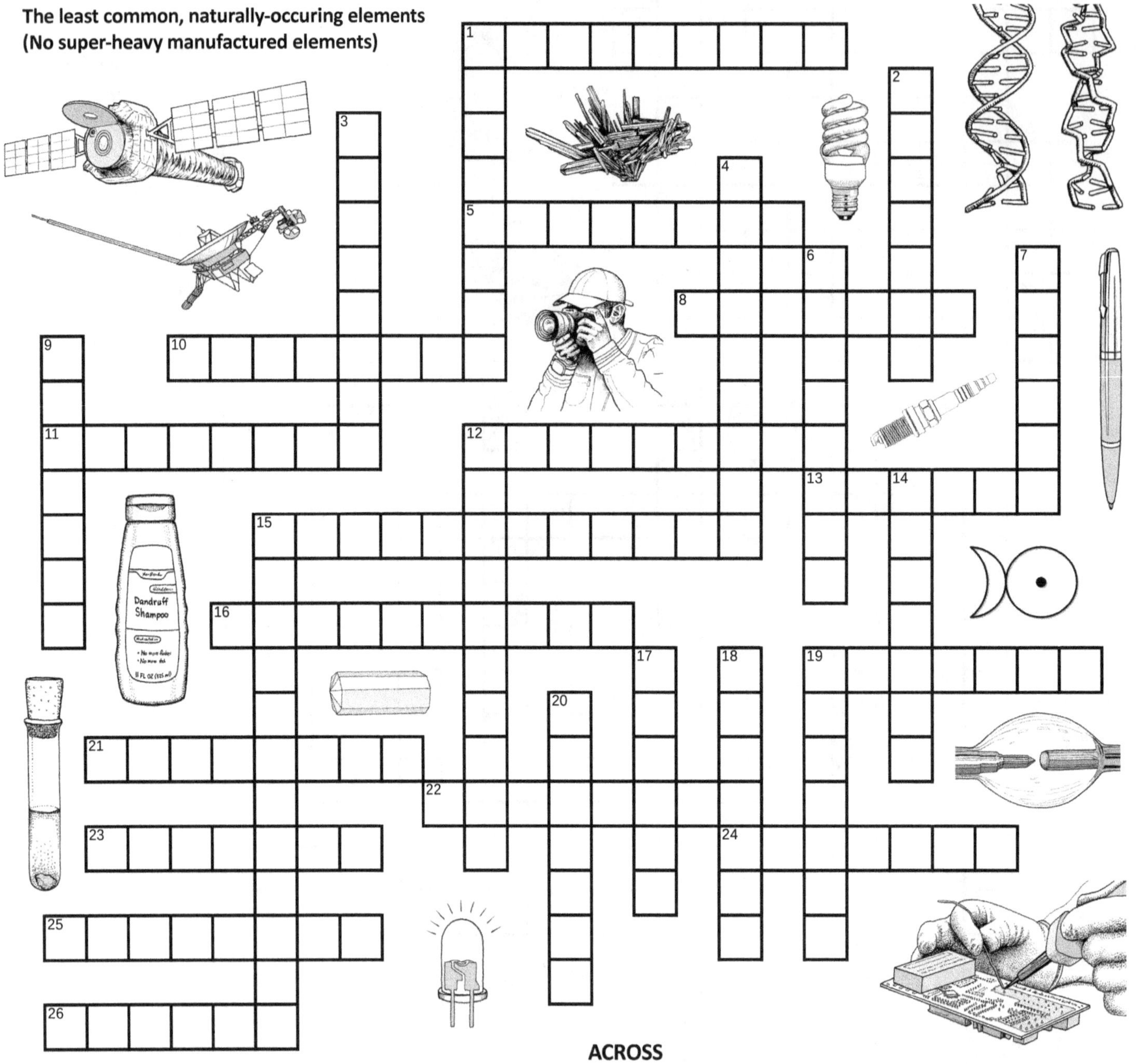

DOWN

1) This poisonous element has one bright green line in its spectrum.
2) Volvo was first to use this element in catalytic converters in its cars.
3) When radon decays and loses an alpha particle, it becomes this.
4) "Ruth and Rhoda" are neighbors on the table. This is "Ruth."
6) This element gets its symbol from its mineral ore, stibnite.
7) Combined with tin oxide to make ITO, an electrically conductive transparent surface for touch screens.
9) Used as less-toxic substitute for lead; used in stomach medicine.
12) Used to make buttons glow on lunar rover for Apollo 15-17.
14) First Chinese emperor died from using this as daily health tonic.
15) When this element does alpha decay it turns into actinium.
17) Named for the "sky blue" lines in its emission spectrum.
18) Used to dope crystals, making Ho-YIG lasers for medical use.
19) Makes bacterial endospores glow green. Used in Terfenol-D®.
20) Only non-metallic element that is liquid at room temperature.

ACROSS

1) Used to grow diphtheria bacteria, and to make Blu-ray discs.
5) Doped into crystals to make LuAG lasers.
8) Used in thermocouples and in oil refining. Chile is leading producer.
10) Name means "little silver."
11) Alexander Graham Bell used this to make his "photophone."
12) Sits above Pt on the Table; used in CO detectors, pens, spark plugs.
13) Member of Pt group; was a key to discovering right-handed DNA.
15) This element always has 59 protons.
16) Used as dry lubricant, and to make "Chromoly" steel alloy.
19) Used in arc bulbs for green light; in euro notes for blue stripes.
21) Start with Np. Do three alpha decays to get this element.
22) Compounds of this rare earth can make a wide variety of colors.
23) Used in pigments; combined with Te to make solar panels.
24) Used to make the mirrors inside the Chandra X-ray Observatory.
25) The name of this element means "unstable."
26) When this goes through alpha decay, it turns into radon.

WHO AM I?

1) I am a member of the transition metal group. I will happily form alloys with a wide range of metals, but the most popular alloy throughout history has been with copper. I can also combine with oxygen. One of my oxides is good at absorbing ultraviolet light. My ability to combine with oxygen allows me to be used as a protective coating for other metals. When combined with sulfur, I am able to absorb light then release it again slowly, giving a "glow in the dark" effect if you turn off the lights. Who am I? _______________

2) In my pure form I am a poisonous gas, yet I am necessary in your diet because I participate in important functions in your body. I am found in some water purification products, and when combined with an alkali element I can make a compound that has been used for centuries to preserve meat. Since the invention of plastic, I have been used to make corrosion-resistant plastic pipes that can be used in plumbing. Who am I? _________________

3) When in my pure form, I am very toxic. In the past, one way my toxicity was put to good use was to preserve wood from insects and fungi. When combined with gallium, I can be very useful in the electronics industry: I can generate light in LEDs or in tiny lasers, or I can help to absorb the sun's energy in solar panels. When combined with copper I can make a beautiful green pigment, although this pigment is no longer used due to toxicity. Who am I? ________________

4) I was officially discovered in the early 1800s, although one of my compounds had been used for centuries to treat minor skin problems. Another of my compounds is used as an antacid for the stomach. Though I am not a transition metal, I can be alloyed with some of them to make the metals lighter and stronger. Some of my compounds can produce white sparks, making me useful in fireworks and fire starting devices. The first scientist to produce a pure sample of me did so by using electricity to pull my atoms out of a solution. Who am I? ____________________

5) In my pure form, I am dark gray but have a glassy luster, and feel very light compared to many other elements. Because my outer shell of electrons is exactly half full, I am useful to the electronics industry. In the mid-1900s I was used to create an electronic part that changed the world: the transistor. Diodes containing me were used to make the first radios. Today, silicon has replaced me in many electronic devices, but you will still find me in satellites being used in gamma ray spectrometers and solar panels. Who am I? ____________________

6) I am often found in clear quartz crystals that are made of silicon and oxygen. However, when combined with just oxygen (no silicon) I can make clear crystals that look almost as nice as diamonds. These crystals can be useful as well as decorative and are often ground to small pieces and used to make sandpaper. When combined with lead and titanium I can make a ceramic compound that generates low frequency vibrations used in sonar and ultrasound. I was used to make pipes for nuclear reactors until the Fukushima disaster showed that I burst into flames when I come into contact with burning hydrogen. Who am I? ___________________

7) I am a soft metal, and because I melt at relatively low temperatures I am used (with other metals) as the "fuse" in sprinkler systems. I can be combined with helium to make lasers that emit blue light. Some of my compounds have long been used to make yellow and red pigments for paint. I am used in certain types of batteries. My name comes from an obscure (not well known) Greek mythological hero. Who am I? ___________________

8) Because of the arrangement of electrons in my outer shells, I am strongly magnetic. The strength of my magnetism means that very tiny magnets made of me can be as strong as larger conventional magnets. Magnets made of me have been used in a wide variety of electronic devices. I have other talents, as well. I can be used to color glass, and the glass will appear green, blue, pink or purple depending on the light it is in. I am doped into YAG laser crystals that are used for optical tweezers. In nature I am almost always found in the company of elements similar to myself and sometimes end up along with them in products designed to produce sparks. Who am I? ________________

9) I am a heavy metal but not nearly as toxic as some of my heavy metal "cousins." I am dense enough to be able to protect you against x-rays. I have a low melting point and therefore can be found in the fire sensors in sprinkler systems. In my pure form I can make artificial crystals that have numerous square shapes and shimmer with iridescence. Some of my compounds can be used to make yellow paint. I am found in over-the-counter medicine and in nail polish. You might find one of my alloys lying in a field that was used for duck hunting. Who am I? ________________

10) I am found in ocean water and in seaweed, and help to give the ocean its "fishy" smell. When combined with potassium, I make a light-sensitive compound that can also kill germs. When combined with silver, I can make a compound used for "seeding" clouds so that they will rain. I am essential to your body, especially in one of your glands. I can be used as a test for starch. Who am I? _______________

SYMBOL PRACTICE WITH SILLY RIDDLES

RIDDLE #1

_____ _____ _____ did _____e _____ _____ _____ _____a _____
tungsten hydrogen yttrium thorium molybdenum uranium selenium sulfur yttrium

"_____ _____e_____, _____ _____e_____," _____ _____ _____ the
carbon helium phosphorus carbon helium phosphorus tungsten helium nitrogen

_____ _____d'_____ _____ _____e _____ll a_____art? _____ _____ _____
boron iridium sulfur carbon silver iron phosphorus Helium tungsten arsenic

_____ _____l _____ _____g _____ _____ _____r the _____ _____d
fluorine iodine lithium nitrogen indium fluorine oxygen boron iridium

_____ _____ _____ _____ad _____e da_____ _____ _____ _____.
tungsten hydrogen oxygen hydrogen thorium yttrium oxygen fluorine fluorine

RIDDLE #2

_____ _____ _____ do _____ _____ _____ _____t _____ _____ _____
tungsten hydrogen astatine yttrium oxygen uranium germanium tungsten helium nitrogen

_____ _____ _____ _____r _____ _____ a _____ _____ _____ _____e
yttrium oxygen uranium carbon osmium sulfur vanadium americium phosphorus iridium

_____i_____ a _____ _____ _____? A _____rr _____ _____ _____ed _____ _____!
tungsten thorium molybdenum uranium selenium tellurium iodine fluorine iodine carbon astatine

RIDDLE #3

_____ _____ _____ _____ _____e _____ _____a _____ a_____ R_____ _____ _____'_____
tungsten hydrogen astatine tungsten erbium Barium thulium nitrogen neodymium oxygen bismuth nitrogen sulfur

_____ _____ _____ _____ _____ a_____ _____r t_____ _____ _____ _____e
neon tungsten nitrogen americium einsteinium fluorine tellurium helium yttrium tungsten erbium

_____ _____ _____ _____ _____ _____ _____ a t_____ _____?
ruthenium nitrogen oxygen vanadium erbium boron yttrium radium indium

_____ _____ _____a_____ a_____ R_____ _____ _____ _____ _____!
fluorine lanthanum thulium nitrogen neodymium iodine boron boron oxygen nitrogen

---

Which one of these elements can't be spelled using actual symbols? (Ex: tin: Ti N [titanium, nitrogen])

Arsenic Astatine Bismuth Carbon Copper Phosphorus Niobium Silicon Silver

---

Write the name of the element on the line.

1) Which element describes what dogs do with bones? ______________________
2) Which element is Superman's least favorite? __________________
3) Which element plays an equestrian sport? ___________________
4) Which element went to a clown convention? __________________
5) Which element is the least exciting? ___________________
6) Which element is multilingual? ____________________
7) Which element do you need when your clothes are wrinkled? _____________
8) Which element goes up and down? _______________
9) Which element describes the dinosaurs? ____________
10) Which element cheers for the Los Angeles Dodgers? ______________

Possible answers:

Bh	Eu
Si	Cf
Ar	Kr
Ba	Fe
Hg	Po

CREATE A QUOTE

Fill in the answers in ***alphabetical order***, then transfer the numbered letters to make the quote, below.

The elements whose names have only four letters:

39 76 1 56 73 3 63 53 6 5

The elements that are named after an object in the solar system:

33 12 62 25 10 66 54

7 15 8 68 14 2 67 38

43 46 9 70 11

The elements that have a symbol made of only one letter:

34 13 16 17 71 49 44 18

42 75 31 22 19 45 77 36

23 64 61 24 26 72 20

21 65 69 35 74 32

The elements named after a country:

51 28 59 37 47 48 57 30 55

29 50 60 27 58 40 41

The only element that has a W:

4

The only element that starts with a K:

52 78

A QUOTE BY DMITRI MENDELEEV, WHO MADE THE FIRST PERIODIC TABLE:

1 2 3 4 5 6 7 8 9 10 11 12 13 14 15 16 17 18

4 19 20 21 22 23 24 25 26 27 28 29 30 31 32 33 34 35 36 37 38 39 40

41 42 43 44 45 46 47 48 49. 50 4 51 52 53 54 55 56 57, 58

59 60 61 62 63 64 65 66 67 68 69 4 70 71 72 73 74 75 76 77 4 78.

ELEMENTS THAT ARE EASILY CONFUSED

1) Back in the early 1800s, famous chemist Humphry Davy, the first to isolate a pure sample of one of these elements, warned that the name choice for the other element would cause confusion because the names were too similar. They didn't listen to him and now we have to try not to confuse these elements!

___ ___ _G_ ___ ___ ___ ___ ___ ___ and ___ ___ _N_ ___ ___ ___ ___ ___ ___

2) The names of these elements sound very much the same. One is in the lanthanide series (a rare earth element) but the other is a heavy metal similar to lead and bismuth.

___ ___ ___ ___ ___ ___ ___ and ___ ___ ___ ___ ___ ___ ___ ___

3) Both of these elements are radioactive, but one is a solid and the other is a gas. Their names start with the same letter and they sit fairly close together on the Periodic Table.

___ ___ ___ ___ ___ and ___ ___ ___ ___ ___ ___

4) These transition metals have similar chemical characteristics and can be used for some of the same purposes. Their names start with the same letter. One name comes from a Spanish word, the other does not.

___ ___ ___ ___ ___ ___ ___ ___ and ___ ___ ___ ___ ___ ___ ___ ___ ___

5) These two elements are both rare metals that are difficult to extract from ore rocks. One is in the lanthanide series but the other is not. Both are named after capital cities of Scandinavian countries.

___ ___ ___ ___ ___ ___ ___ and ___ ___ ___ ___ ___ ___

6) Both of these transition metals start with the same two letters. They are not close on the Periodic Table.

___ ___ ___ _N_ ___ ___ ___ and ___ ___ ___ _D_ ___ ___ ___

7) Both of these elements start with the same letter, but one is a transition metal and the other is a "true" metal. One of the names comes from a color but the other does not.

___ ___ _D_ ___ ___ ___ and ___ ___ ___ _D_ ___ ___ ___

8) Both of these elements are transition metals but they are not close on the Periodic Table. One is found in copper and lead ores, the other in quartz. Their names start with the same letter.

___ ___ ___ ___ and ___ ___ ___ ___ ___ ___ ___ ___ ___

9) Though these elements start with the same letter they are very far apart on the Periodic Table. One of them is not natural but is made artificially in labs.

___ ___ ___ ___ ___ and ___ ___ ___ ___ ___ ___ ___

10) All three of these are in the lanthanide series (rare earth elements) and they are all named after the same small Swedish town.

___ ___ ___ ___ ___ ___, ___ ___ ___ ___ ___ ___ ___ and ___ ___ ___ ___ ___ ___ ___ ___ ___

ACTIVITY IDEAS

Some of these activities use the cards and/or the questions provided in the next section.

Feel free to adapt the formats and questions to make them more appropriate for the ages, abilities, and interest levels of your students.

ASSEMBLE A PERIODIC TABLE POSTER

The goal of this activity is to provide a hands-on experience with the Periodic Table. Handling the cards and arranging them in the right order will encourage a sense of familiarity with the individual elements and the table as a whole.

<u>You will need</u>:
- copies of the element picture card pages (copied onto card stock if you want the final product to be more sturdy)

NOTE: You can print the cards on your home printer using this digital file: www.ellenjmchenry.com/periodictableactivitycards.pdf
- colored pencils or markers
- scissors and tape
- a large, flat surface to work on
- a roll of paper if you want to make a permanent poster (wrapping paper, brown paper, etc.)
- a glue stick if you will be making a permanent poster

<u>What to do</u>:
1) Cut apart the cards on the thin lines, leaving a white "frame" around each card.
2) Separate the cards into groups according to their families. (See lists on next page.)
3) Color the outer edges (the "frames") of each element, making all the cards in each family group the same color.
4) Decide whether you want to put the lanthanides and actinides below the main table or inside the table.

Option #1: Put those rows below the main table. The final product will be about 115 cm (3 feet 8 inches) wide and 60 cm (24 inches) tall.

Option #2: Put these rows where they actually belong, between the first two columns (the "s" block) and the transition metals (the "d" block). This will make the table very wide. The total width of this format will be 200 cm (6 feet 6 inches). The height will be 45 cm (about 18 inches).

5) Lay out the cards so that they form the Periodic Table. If you have trouble keeping them straight, draw a few guide lines with a ruler.
6) Turn over a few cards at a time, taping the seams on the back side of the cards. Or, if you are putting them onto a large piece of paper, you can just glue them onto the paper.

OPTION #1: Putting the lanthanides and actinides below the table:

1 H																	2 He
3 Li	4 Be											5 B	6 C	7 N	8 O	9 F	10 Ne
11 Na	12 Mg											13 Al	14 Si	15 P	16 S	17 Cl	18 Ar
19 K	20 Ca	21 Sc	22 Ti	23 V	24 Cr	25 Mn	26 Fe	27 Co	28 Ni	29 Cu	30 Zn	31 Ga	32 Ge	33 As	34 Se	35 Br	36 Kr
37 Rb	38 Sr	39 Y	40 Zr	41 Nb	42 Mo	43 Tc	44 Ru	45 Rh	46 Pd	47 Ag	48 Cd	49 In	50 Sn	51 Sb	52 Te	53 I	54 Xe
55 Cs	56 Ba	71 Lu	72 Hf	73 Ta	74 W	75 Re	76 Os	77 Ir	78 Pt	79 Au	80 Hg	81 Tl	82 Pb	83 Bi	84 Po	85 At	86 Rn
87 Fr	88 Ra	103 Lr	104 Rf	105 Db	106 Sg	107 Bh	108 Hs	109 Mt	110 Ds	111 Rg	112 Cn	113 Nh	114 Fl	115 Mc	116 Lv	117 Ts	118 Og
				57 La	58 Ce	59 Pr	60 Nd	61 Pm	62 Sm	63 Eu	64 Gd	65 Tb	66 Dy	67 Ho	68 Er	69 Tm	70 Yb
				89 Ac	90 Th	91 Pa	92 U	93 Np	94 Pu	95 Am	96 Cm	97 Bk	98 Cf	99 Es	100 Fm	101 Md	102 No

OPTION #2: Putting the lanthanides and actinides into the main part of the table where they actually belong:

1 H																															2 He
3 Li	4 Be																									5 B	6 C	7 N	8 O	9 F	10 Ne
11 Na	12 Mg																									13 Al	14 Si	15 P	16 S	17 Cl	18 Ar
19 K	20 Ca															21 Sc	22 Ti	23 V	24 Cr	25 Mn	26 Fe	27 Co	28 Ni	29 Cu	30 Zn	31 Ga	32 Ge	33 As	34 Se	35 Br	36 Kr
37 Rb	38 Sr															39 Y	40 Zr	41 Nb	42 Mo	43 Tc	44 Ru	45 Rh	46 Pd	47 Ag	48 Cd	49 In	50 Sn	51 Sb	52 Te	53 I	54 Xe
55 Cs	56 Ba	57 La	58 Ce	59 Pr	60 Nd	61 Pm	62 Sm	63 Eu	64 Gd	65 Tb	66 Dy	67 Ho	68 Er	69 Tm	70 Yb	71 Lu	72 Hf	73 Ta	74 W	75 Re	76 Os	77 Ir	78 Pt	79 Au	80 Hg	81 Tl	82 Pb	83 Bi	84 Po	85 At	86 Rn
87 Fr	88 Ra	89 Ac	90 Th	91 Pa	92 U	93 Np	94 Pu	95 Am	96 Cm	97 Bk	98 Cf	99 Es	100 Fm	101 Md	102 No	103 Lr	104 Rf	105 Db	106 Sg	107 Bh	108 Hs	109 Mt	110 Ds	111 Rg	112 Cn	113 Nh	114 Fl	115 Mc	116 Lv	117 Ts	118 Og

NOTE: There is disagreement about some of these categories, especially the true and semi-metals. Use an Internet search engine to find several tables and compare. If you find one you like better than this list, go ahead and use it.

Alkali metals: Group 1 elements (first column on left)
Alkali earth metals: Group 2 elements (second column on left)
Transition metals: Elements 21 to 30, 39 to 48 and 71 to 80 (also called the "d" block)
"True" metals: Al, Ga, In, Tl, Sn, Pb, Bi
Semi-metals, or metalloids: B, Si, Ge, As, Sb, Te, Po (some lists include At)
Non-metals: C, N, O, P, S, Se
Halogens: F, Cl, Br, I (some lists include At)
Noble gases: He, Ne, Ar, Kr, Xe, Rn
Lanthanides (often referred to as the rare earths): 57 to 70
Actinides: 89 to 102
Super-heavy elements: 102 to 118

NOTE: Sometimes the super-heavy elements are put into the categories of the elements right above them.

"THE MISSING ELEMENT"

The goal of this activity is to become so familiar with the Periodic Table that you can determine which element is missing if one element card is removed.

NOTE: An alternative way to do this activity is to use a large Periodic Table that can have a piece of paper temporarily taped to it (blocking out an element rather than removing it). For example, you can buy a shower curtain with the table printed on it, or get a laminated table that is covered with a protective layer of plastic. In this case, all you need to do is make a black paper rectangle the same size and shape as one of the element squares. The black paper rectangle can be (loosely) taped on top of an element to block it out. Students take turns moving the black paper around.

<u>You will need</u>:
- one set of cards (either with or without pictures, your choice)
  (You can print the cards on your printer by using this digital files: www.ellenjmchenry.com/periodictableactivitycards.pdf)
- a flat surface large enough to assemble the cards into a complete Periodic Table
  OR a large space on a wall or whiteboard to which the cards can be taped
- tape loops or sticky dots if you want to put the cards on a wall surface

<u>How to prepare</u>:
1) Choose whether you want to lay out the cards on a flat surface or tape the cards to a wall or whiteboard. If you want to put them on a wall, have tape loops prepared that can be put onto the backs of the cards. You could also use the mildly adhesive "sticky dots" that are designed to stick to papers for a short time.
2) Choose whether you want to put the lanthanide and actinide series as separate rows at the bottom, or include them in the main part of the table. If they will appear in the main table, you will need a much wider space.
3) Have your students work as a team to make the cards form the entire table. Tell them where the lanthanide and actinide series will go. You may want to start by setting a central card in place, such as technetium. Starting with a central card in place will prevent the table from ending too far to the left or right.

<u>What to do</u>:
1) Choose one player to go first. All other players must turn around so their backs are to the table. No peeking! The player removes one element card and holds it secretly.
2) When this has been accomplished allow the other players to turn around and face the table. They will try to determine which element has been removed.
3) You may choose whether to let players call out their guesses, or whether to require them to raise their hands first.
4) Once the correct guess has been made, the student puts the element card back into its place.
5) A new player is chosen, and this process is repeated.

NOTE: You might also want to start out by letting the players have 30 seconds to "cram" before an element is taken away. Give them a chance to try to quickly store visual information. If they do this each time, before an element is removed, they will become familiar with the table in a surprisingly short time.

"COUNTDOWN"

The goal of this activity is to review and reinforce facts learned from the coloring pages. This activity can fill any amount of time. You can do just one or two elements, or many of them.

<u>You will need</u>:
- a piece of paper for each player
- pencils or pens
- the lists of clues from the QUESTION BANK at the end of this book

<u>How to prepare</u>:
1) Give a sheet of paper and a pencil or pen to each participant.
2) Choose which elements you want to use in the game and have those clues marked and ready to read.

<u>What to do</u>:
1) Tell the players to write the numbers 1 to 10 as if they were making a list, but write them in reverse order, from 10 down to 1.
2) For your first element, read the last and hardest clue first. The players will make a guess as to which element it is. They can leave the space blank if they don't know, but they will not be penalized for a wrong guess. Warn them that once we move on to the next clue they are strictly forbidden to go back and write anything in the previous space.
3) Read the next-to-last clue, which will also be a hard clue. The players fill in their guess next to the number 9 on their paper. It can be the same as their guess for number 10 or they can change their mind and write in another element.
4) Read the third-to-last clue, which will be the remaining hard clue. Again, the players fill in a guess. This clue might confirm a previous guess, or it might change their mind. They write their current guess next to the number 8.
5) Play continues like this, up the list of clues. Keep reminding the players they may NOT go back and change any previous answers. (In a classroom, students usually do a good job of policing each other. Most students will keep a sharp eye out for cheaters!)
6) Once the last and easiest clue has been read, the players make their final guess.
7) The answer is then revealed. The players look to see which number was the first time they made the correct guess. This will be their score. If they guessed it on the first and hardest clue, they score a 10. If the only guessed it on the last and easiest clue, they score a 1.
8) Go on to the next round. Players write the numbers 1 to 10 again (in reverse order, of course) and the next round of clues begins.
9) A cumulative score can be kept, and the player with the highest score wins.

"SYMBOL CHALLENGE"

The goal of this activity is to learn the symbols for the elements.
It should be played in small groups of 2-6 participants.

<u>You will need</u>:
- one copy of the set of symbol cards per small group (2 to 6 players)
- pencils
- paper clips
- tiny slips of paper

<u>How to prepare</u>:
1) Make a copy of the card set for each small group. Make each set a different color, or put an identifying mark on the cards of each set so that if the cards get mixed up they will be easy to sort out.
2) Choose which cards you want to include in the game. If your players are beginners, choose several dozen well-known elements. Follow this up with another game using slightly harder cards. This game can be played many times, and each time you can choose which elements to review and which new elements to add in.
3) Write the names of the elements on the backs of the cards. You can do this ahead of time or have the players do it before you start the game.

<u>What to do</u>:
1) Hand out a small slip of paper to each player and have them write their name on it. Give each player a paper clip.
2) Lay the chosen cards on the table so that all cards can be seen. Symbols should be facing up, names facing down.
3) The first player must "call" the card he wants to play by saying the letter symbol. For example, the player might say, "C." Then the player has a choice: he can either "guess" or "peek."

--If the player chooses "guess" he must say the name of the element out loud. He checks the answer by turning over the card, but making sure no one else sees the back of the card. If his guess was correct, he keeps the card. If not, he must put the card back on the table.

--If the player feels like he doesn't know enough to make a reasonable guess, he can choose the "peek" option. After he says the name of the symbol, and calls "peek," he may turn the card over to see the name on the back, but making sure no other players see the correct answer. He then returns the card to the table (name hidden!) and uses the paper clip to attach his name paper to it. This has the effect of reserving that card for the next turn.

No other player may "call" a card with someone else's name on it. When this player's turn comes around again, he "calls" the card that he has reserved and then uses the "guess" option. Hopefully, he will remember the name of the element and get it right this time and be able to keep the card. If not, he simply puts the card back as he would on any failed "guess" turn.
4) The game is over when there are no cards left on the table. The player with the most cards wins.

Group → / ↓ Period	1	2	3	4	5	6	7	8	9	10	11	12	13	14	15	16	17	18
1	1 H																	2 He
2	3 Li	4 Be											5 B	6 C	7 N	8 O	9 F	10 Ne
3	11 Na	12 Mg											13 Al	14 Si	15 P	16 S	17 Cl	18 Ar
4	19 K	20 Ca	21 Sc	22 Ti	23 V	24 Cr	25 Mn	26 Fe	27 Co	28 Ni	29 Cu	30 Zn	31 Ga	32 Ge	33 As	34 Se	35 Br	36 Kr
5	37 Rb	38 Sr	39 Y	40 Zr	41 Nb	42 Mo	43 Tc	44 Ru	45 Rh	46 Pd	47 Ag	48 Cd	49 In	50 Sn	51 Sb	52 Te	53 I	54 Xe
6	55 Cs	56 Ba		72 Hf	73 Ta	74 W	75 Re	76 Os	77 Ir	78 Pt	79 Au	80 Hg	81 Tl	82 Pb	83 Bi	84 Po	85 At	86 Rn
7	87 Fr	88 Ra		104 Rf	105 Db	106 Sg	107 Bh	108 Hs	109 Mt	110 Ds	111 Rg	112 Cn	113 Nh	114 Fl	115 Mc	116 Lv	117 Ts	118 Og

Lanthanides	57 La	58 Ce	59 Pr	60 Nd	61 Pm	62 Sm	63 Eu	64 Gd	65 Tb	66 Dy	67 Ho	68 Er	69 Tm	70 Yb	71 Lu
Actinides	89 Ac	90 Th	91 Pa	92 U	93 Np	94 Pu	95 Am	96 Cm	97 Bk	98 Cf	99 Es	100 Fm	101 Md	102 No	103 Lr

Group → / ↓ Period	1	2	3	4	5	6	7	8	9	10	11	12	13	14	15	16	17	18
1	1 H																	2 He
2	3 Li	4 Be											5 B	6 C	7 N	8 O	9 F	10 Ne
3	11 Na	12 Mg											13 Al	14 Si	15 P	16 S	17 Cl	18 Ar
4	19 K	20 Ca	21 Sc	22 Ti	23 V	24 Cr	25 Mn	26 Fe	27 Co	28 Ni	29 Cu	30 Zn	31 Ga	32 Ge	33 As	34 Se	35 Br	36 Kr
5	37 Rb	38 Sr	39 Y	40 Zr	41 Nb	42 Mo	43 Tc	44 Ru	45 Rh	46 Pd	47 Ag	48 Cd	49 In	50 Sn	51 Sb	52 Te	53 I	54 Xe
6	55 Cs	56 Ba		72 Hf	73 Ta	74 W	75 Re	76 Os	77 Ir	78 Pt	79 Au	80 Hg	81 Tl	82 Pb	83 Bi	84 Po	85 At	86 Rn
7	87 Fr	88 Ra		104 Rf	105 Db	106 Sg	107 Bh	108 Hs	109 Mt	110 Ds	111 Rg	112 Cn	113 Nh	114 Fl	115 Mc	116 Lv	117 Ts	118 Og

Lanthanides	57 La	58 Ce	59 Pr	60 Nd	61 Pm	62 Sm	63 Eu	64 Gd	65 Tb	66 Dy	67 Ho	68 Er	69 Tm	70 Yb	71 Lu
Actinides	89 Ac	90 Th	91 Pa	92 U	93 Np	94 Pu	95 Am	96 Cm	97 Bk	98 Cf	99 Es	100 Fm	101 Md	102 No	103 Lr

TRADITIONAL "20 QUESTIONS"

The goal of this activity is for the players to use their knowledge base to guess a mystery element.
It is best played after the students have been studying the elements for a while.
This activity works with any number of participants.

You will need:
- either a large Periodic Table for everyone to see, or a small copy for each player

How to prepare:
This game can be played with no preparation ahead of time.

What to do:
1) If there are more than two participants doing this activity, you will need to decide a fair way to take turns asking questions. You can "go around the circle" allowing each player one question, or use another method that will work well in your situation.
2) Choose one player to be the chooser. They will secretly choose one element. This player might also be asked to keep track of how many questions have been asked.
3) The other players will take turns asking questions. They should start out with broad questions such as "Is this element a transition metal?" or "Is this element a gas at room temperature?" or "Is this element used by the human body?" If they can't remember the names of the "families" (noble gas, halogen, alkali, alkali earth, transition, non-metal, semi-metal, lanthanide, actinide) they can still ask if it is in a certain row or column. The columns are called "groups" and the rows are called "periods."
You can copy the tables shown here to the left of this page, which have the groups and periods labeled.
4) After they either run out of guesses, or guess correctly, select a new player to be the element chooser.

NOTE: You might find that 20 questions is too many. If your players are very good at posing questions, you may need to limit the number of questions to 15 or even 12.

TRADITIONAL BINGO

The goal of this activity is to review and reinforce the learning they have been doing about individual elements, and to use critical thinking skills to make good guesses in cases where they are not sure.

NOTE: For this game, you will use only a few dozen of the element picture cards, not the entire set. Choose which elements you want to include in the game, and use the clues for those elements. Also, if you are working with advanced players and don't want them to have any pictures as clues, you can use the symbol (letter) cards instead.

Before you start preparing, decide how large you want to make the Bingo boards. You can use 16 cards to make a 4 by 4 square, or you can use 25 cards to make a 5 by 5 square. In the 5 by 5 arrangement, if you want to make the center square a "free" square, you can simply leave it open and not put a card there, or you can use the FREE squares shown after element 118. (Note that each page of picture cards has 12 elements. If you want to limit the number of pages that need to be copied, you can choose just 2 pages to print, giving you 24 elements, and simply leave the central spot open as a "free" square.)

<u>You will need</u>:
- a set of cards for each player (You may want to put an identifying mark on each set so you can sort them out easily if they get mixed up.)
- tokens for each player to set on top of their cards (ex: pennies, paper squares, candies, nuts, uncooked macaroni, dry beans...)
- appropriate clues

<u>How to prepare</u>:
1) Make a card set for each player.
2) Make sure each player has access to a supply of tokens (enough to almost cover their board)
3) Decide which clues you want to use and how you want to use them. You can cut the page into strips and have one clue per strip, then put the clue strips into a bag or box so you can pull them out randomly. You might want to consider having more than one bag of clues, so that after one or two rounds you can switch to a set of new clues. If you will be having the players take turns reading clues, you will want to have clue strips. However, if you will have an adult reading all the clues, you can leave the pages intact and have the adult reading through the clues (on the spot) and choose them randomly, keeping track of which ones they have used by writing numbers next to the clues as they are used.

<u>How to play</u>:
1) The players should set their element cards out to form a square, either 4x4 or 5x5 (as you have decided ahead of time). Each player's arrangement will be unique, reducing the possibility of a tie. If there are more elements than squares (which will happen if you use a 4x4 square) the players will choose which elements to use and which to set aside. Between rounds, they can change their arrangements and use some of the cards they had set aside.
2) Read out clues one at a time. Players place tokens on squares that fit the clues.
3) When someone gets 4 in row (or 5 in a row) they call out "Bingo!" then read their answers, which are checked by the clue giver. If this player is successful, you can either keep this game going until a few more players get a Bingo, or you can have the players clear their boards and everyone starts a new round. (The advantage to keeping the game going and allowing more than one winner is that you can keep going on your current list of clues, upping the chance that most of the clues will be used.)

"ELEMENT CONNECTIONS"

The goal of this activity is to review facts about various elements. It can be played with any number of participants. Instructors should choose questions from the question bank (or create their own) that target the knowledge base of the players. The question bank has easy, medium and hard clues, and all of the clues are based on the information given in the coloring section of this book.

<u>You will need</u>:
- one copy of the following pattern page for each player
- pencils or markers so the players can write a symbol on each circle
- tokens for each player (pennies work well)
- appropriate clues

<u>How to prepare</u>:
1) Make a copy of the pattern page for each player.
2) Choose at least 30 elements to use in this game. You can select more than 30, giving the players some choice about which ones they would like to use on their board. Bear in mind that the larger the number of possible elements, the more turns each player will have to "pass" because they don't have that element on their board. However, it is okay if they have to pass once in a while. If you have been going through the elements sequentially, consider starting with the first 36 (up to krypton). You could also choose a theme such as "transition metals." If you want to cover additional elements, just make more copies of the pattern page and have the players make a second board with new elements.
3) Get clues ready for each of the elements you have chosen. Select several clues for each element so you can play more than one round. If you decide to use the QUESTION BANK provided, select the clues that are best suited to the knowledge base of your players. (If you have students with very different levels of knowledge, you might want to create two groups and use separate clues for each.) You can make copies of the question pages and cut them into strips with one question per strip, or you can keep the pages intact and simply check off the clues you use as you go along.

<u>What to do</u>:
1) Tell the players which elements will be included in this game. Give them pencils or markers and have them fill in the circles on their game board with element symbols. Use each element only once. If the players work on their own, each board should be unique. This will reduce the chances of a tie.
2) Make sure each player has at least two dozen tokens.
3) Tell the players that the goal is to connect the top bar and the bottom bar with a continuous line of tokens. Notice how the circles are connected by small black lines. Imagine that the top bar is carrying electricity and you want to run a wire to the bottom bar. The wire would be a path of filled circles all connected by small lines. This means that your wire does not have to be straight. As long as there is a line between the circles, you can make a path. You can't do diagonals because there are no diagonal lines. The path can go left, right, up, down, as long as it is continuous from the top bar to the bottom bar.
4) Begin reading clues. If a player thinks the clue matches one of the elements on their board, they put a token on it.
5) When a player has completed a continuous path of tokens, they yell "Connection!" They will then read off the elements in that pathway and the person calling clues will check the answers. If they do not get them all correct, the game continues until someone succeeds.
6) When someone wins, players take the tokens off their board and then a new round starts. (Or, if you are working with younger students, you may want to continue the same game and let a few more players get their connection made.)

Here are some samples of "connections." You want a continuous pathway. It can be straight, but it doesn't have to be.

Of course, actual game boards will have more tokens on them than the paths shown here. As in Bingo, there will be tokens on the board that won't end up being part of the winning pathway.

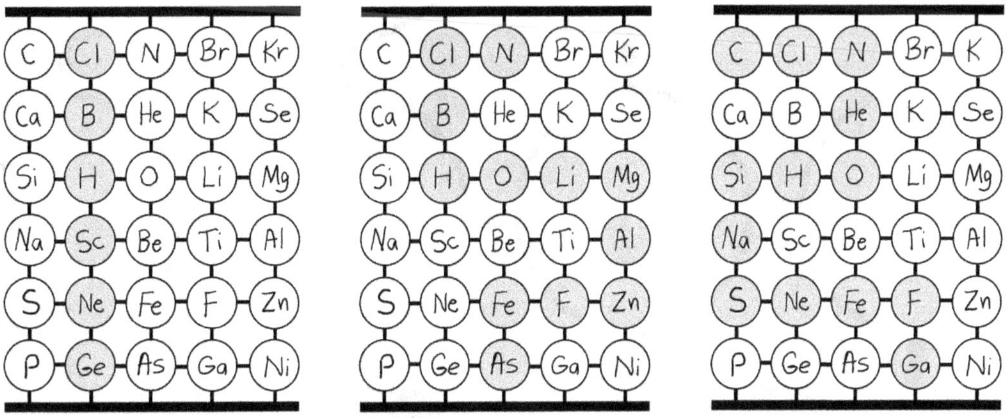

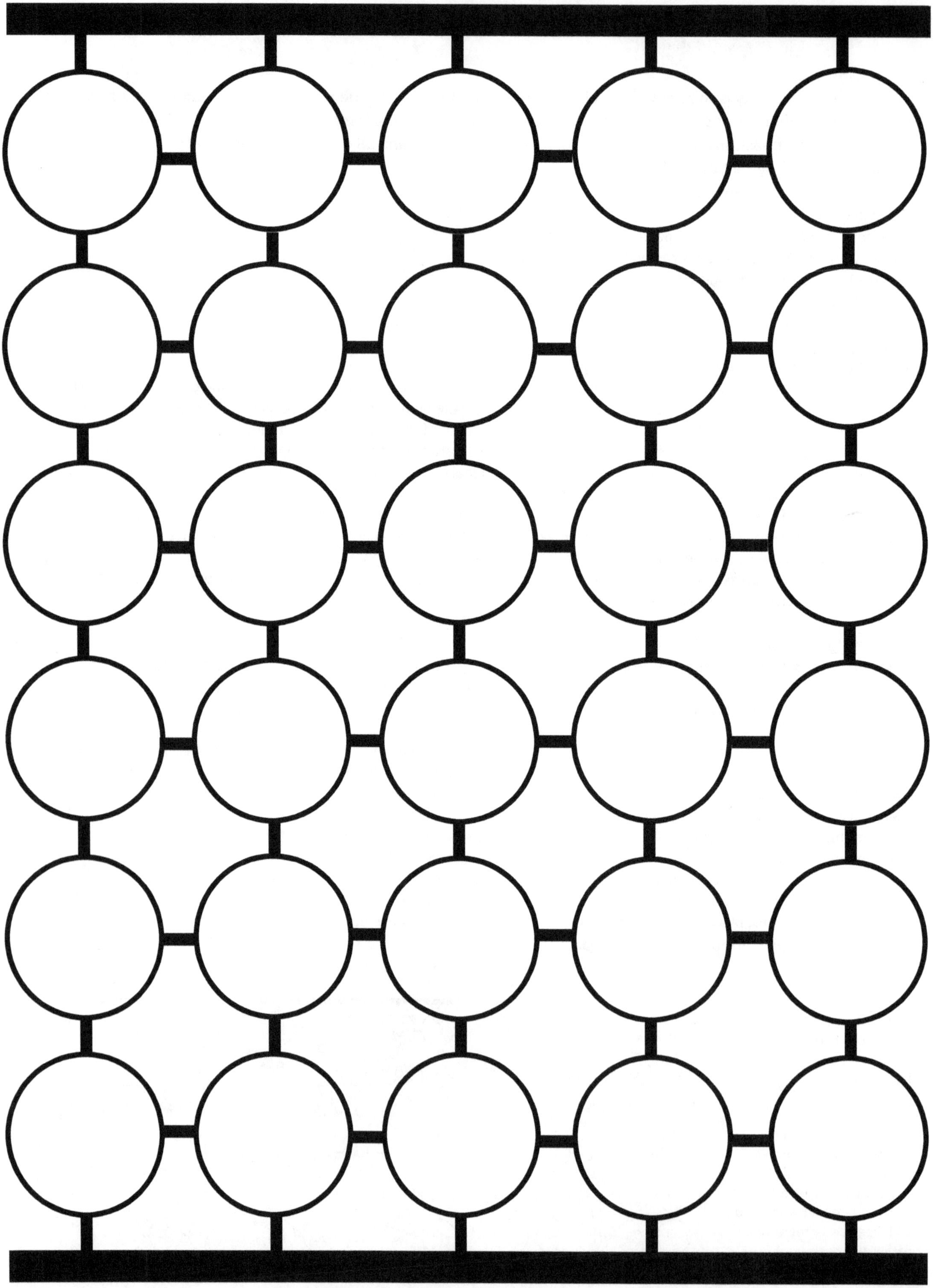

"THREE NEIGHBORS"

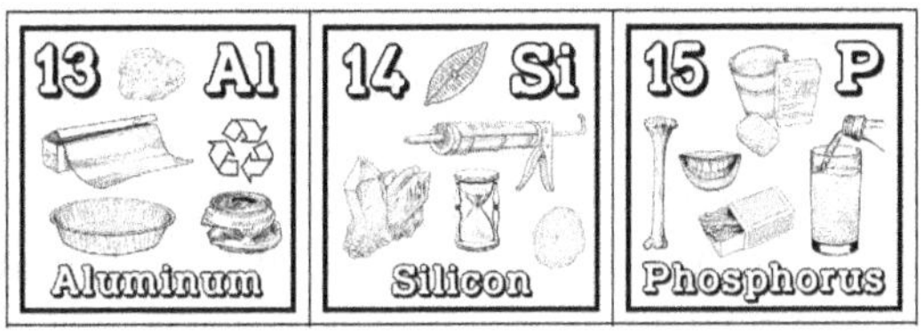

The goal of this activity is to make it fun to study the arrangement of the elements on on the table. It should be played in small groups of 2-6 participants.

The format of the game restricts us to using just the main section of the table, from H to Rn. We will also need to ignore the lanthanide and actinide series and the super heavy elements. However, the elements we'll be using are the most abundant, well-known, and essential elements in the world of chemistry.

Why is the arrangement so important? Elements in the same columns share similar chemical properties, and in the transition block, side-by-side neighbors are also often similar. Similar chemical properties were the clues Mendeleyev used to figure out how to set up the very first Periodic Table.

<u>You will need</u>:
- several dice per small group (ideally, a die for each player, but dice can be shared)
- very small tokens that can be placed on the squares of the small Periodic Tables to help keep track of the location of your cards (Avoid round things that will roll, or light things (like paper) that might be easily blown off. Suggestions: mini M&Ms, very small pebbles [often used in fish tanks], pine nuts, peanuts, dry raisins, mini Altoid® mints, Smarties® candies, Cheerios®) The number of tokens needed per game will vary, but each player should have access to at least 20 tokens.
- one copy of the Periodic Table/number rules for each player
- one copy of the first seven pages of picture cards for each small group
- one copy of the set of clue cards (3 pages total) per small group

NOTE: Use card stock, if possible, for the card sets. Most home printers can handle light/medium weight card stock (up to 65 lb)

<u>How to prepare</u>:
1) Make all the copies, and cut apart all the cards.
2) Gather the picture cards for the elements shown on the Periodic Table charts used in this game. Shuffle the picture cards and put them into a pile face down. TIP: Make sure they are shuffled really well before you begin. You don't want sequential cards stacked together.
3) Shuffle the clue cards and put them into a pile, face down.
4) Give each player one of the half-pages showing the Periodic Table and the number rules.
5) Make sure each player has access to at least 20 small tokens.

<u>How to play</u>:
1) The goal of the game is to get three cards that are all next to each other on the table, either horizontally or vertically.

2) You will have two piles of cards in the center of your playing area, the clue cards and the picture cards. Both of these are draw piles and the cards should be face down. You will also have two discard piles that will be face-up piles. Turn over three picture cards and lay them face up; this will be the start of the discard pile. You don't need to do this for the clue cards.

3) Deal five picture cards to each player. Keep these cards face up. It's okay for other players to see your cards. Look at your cards and put tokens on the corresponding squares on your copy of the table. Seeing the location of your elements on the table will be an immense help as you try to get three in a row, either side by side or up and down.

4) Players take turns turning over and reading the clue cards. However, the clues will be for everyone. All players listen to the clue and look to see if any of their cards qualify. If you have more than one card that fits a clue, you make the choice of which one to play. After a clue card is used, it is put into a face-up discard pile.

5) The pictures are there to help you remember facts about that element, but if you know other facts that are not represented in the picture, you may use your knowledge if it is to your advantage. If you know extra information but it would put you at a disadvantage, you don't have to say anything. However, you can't ignore the picture clues. If another player points out a picture clue on your card that you are not using, you must then use that information.

For example, one clue card says, "If you have an element that is used in any type of clock or watch..." The silicon card does not show a picture of a watch, but silicon is found in quartz, which is used to make quartz watches. If you know about quartz watches, you can play this card. (Actually, the silicon card has a picture of an hourglass-- you could consider an hourglass a type of clock!) Another example: If the clue is "an element that is used for something that burns..." and you know that aluminum is used to make thermite (which burns so hot that it is used to weld train tracks), you may make use this knowledge even though thermite is not shown on that card. However, if knowing about thermite will force you to make a move that you think will be to your disadvantage, you are not required to reveal that fact about aluminum.

A few cards will require some knowledge not shown in a picture. For example, "an element that is found in the air around us." You can't draw a picture of air, so you will need to remember the elements found in air: nitrogen, oxygen, carbon (in carbon dioxide), hydrogen (in water molecule), and the noble gases.

Silly answers should be discouraged. For example, the element hydrogen could be used as an answer for many of the questions since it is found in so many molecules. Yes, hydrogen atoms are probably lurking somewhere in sports equipment, but it's obvious that the point of the question is to think of metals that are used particularly for this purpose.

6) If you receive a card from another player as the result of them rolling a 2 or 3, you can't use that card on that turn. You have to wait until your next turn to play that card.

7) One of the clue cards asks about elements whose names start with a vowel. You may use yttrium and ytterbium as correct answers since, in this instance, the "y" is being used as a vowel, to make the "ih" sound.

8) Note that some of the clues may result in a group discussion of its interpretation. This is actually a wonderful side benefit of the game and may result in knowledge being shared between players. However, you will need to keep these discussion short, and make sure you have someone to be the "final word" if the group cannot come to a consensus.

9) Remember to move your tokens around on your table when necessary. If you lose a card, its token will need to be removed, and as you gain cards, new tokens will need to added. The tokens let you see at a glance which cards you would like to get and which one are the best candidates for discarding.

10) The game is over when someone gets three cards that are adjacent to each other on the Periodic Table, either up and down or side to side.

Group → / ↓ Period	1	2	3	4	5	6	7	8	9	10	11	12	13	14	15	16	17	18
1	1 H																	2 He
2	3 Li	4 Be											5 B	6 C	7 N	8 O	9 F	10 Ne
3	11 Na	12 Mg											13 Al	14 Si	15 P	16 S	17 Cl	18 Ar
4	19 K	20 Ca	21 Sc	22 Ti	23 V	24 Cr	25 Mn	26 Fe	27 Co	28 Ni	29 Cu	30 Zn	31 Ga	32 Ge	33 As	34 Se	35 Br	36 Kr
5	37 Rb	38 Sr	39 Y	40 Zr	41 Nb	42 Mo	43 Tc	44 Ru	45 Rh	46 Pd	47 Ag	48 Cd	49 In	50 Sn	51 Sb	52 Te	53 I	54 Xe
6	55 Cs	56 Ba	57 La	72 Hf	73 Ta	74 W	75 Re	76 Os	77 Ir	78 Pt	79 Au	80 Hg	81 Tl	82 Pb	83 Bi	84 Po	85 At	86 Rn

DICE RULES:

1 ...keep this card, AND take the top card from <u>either</u> the draw pile or the discard pile

2 ...give this card to the player on your left AND take a card from the draw pile.

3 ...give this card to the player on your right AND take a card from the draw pile.

4 ...put this card on the discard pile AND take the top card from the draw pile.

5 ...put this card face up on the discard pile.

6 ...keep this card AND take the top card on the discard pile.

Group → / ↓ Period	1	2	3	4	5	6	7	8	9	10	11	12	13	14	15	16	17	18
1	1 H																	2 He
2	3 Li	4 Be											5 B	6 C	7 N	8 O	9 F	10 Ne
3	11 Na	12 Mg											13 Al	14 Si	15 P	16 S	17 Cl	18 Ar
4	19 K	20 Ca	21 Sc	22 Ti	23 V	24 Cr	25 Mn	26 Fe	27 Co	28 Ni	29 Cu	30 Zn	31 Ga	32 Ge	33 As	34 Se	35 Br	36 Kr
5	37 Rb	38 Sr	39 Y	40 Zr	41 Nb	42 Mo	43 Tc	44 Ru	45 Rh	46 Pd	47 Ag	48 Cd	49 In	50 Sn	51 Sb	52 Te	53 I	54 Xe
6	55 Cs	56 Ba	57 La	72 Hf	73 Ta	74 W	75 Re	76 Os	77 Ir	78 Pt	79 Au	80 Hg	81 Tl	82 Pb	83 Bi	84 Po	85 At	86 Rn

DICE RULES:

1 ...keep this card, AND take the top card from <u>either</u> the draw pile or the discard pile

2 ...give this card to the player on your left AND take a card from the draw pile.

3 ...give this card to the player on your right AND take a card from the draw pile.

4 ...put this card on the discard pile AND take the top card from the draw pile.

5 ...put this card face up on the discard pile.

6 ...keep this card AND take the top card on the discard pile.

If you have an element that is named after a country or a geographical region... (not a town or city)	If you have an element that is named after a town or a city...	If you have an element whose symbol does not begin with the same letter as its name...
If you have an element whose symbol is made of just one letter...	If you have an element whose atomic number begins with a 3...	If you have an element whose name has four letters...
If you have an element whose name has two syllables...	If you have an element that has something to do with sparks...	If you have an element whose atomic number is divisible by 5...
If you have an element that is used in batteries...	If you have an element that is found in some kind of gemstone...	If you have an element that is used in magnets of any kind...
If you have an element that is used in lasers...	If you have an element whose name starts with a vowel...	If you have an element that is used to make jewelry... (meaning the jewelry itself)
If you have an element that is used in medicines...	If you have an element whose atomic number is a prime number...	If you have an element that is used to make something designed to burn or explode...

CLUE CARDS FOR "THREE NEIGHBORS"

If you have an element that is used in some kind of light bulb...	If you have an element that is found in the air around us...	If you have an element whose name ends with the letters O-N...
If you have an element that is (or was) used to make coins...	If you have an element whose name has a double letter (ex: "tt" or "ll")...	If you have an element whose name comes from a color, or a word that refers to color...
If you have an element that has something to do with teeth...	If you have an element that has something to do with glass... (not glass in light bulbs)	If you have an element whose name contains one of these letters: X, Y, Z...
If you have an element that was named after a mythological god or goddess...	If you have an element that is named after something in the solar system...	If you have an element whose isotopes are all radioactive...
If you have an element that is used to make tools...	If you have an element whose atomic number has only one digit...	If you have an element that is on the edge of the table and not surrounded by other elements...
If you have an element that is used to make pigments for paint...	If you have an element that is used in rockets or satellites...	If you have an element that is used to make a cleaning product...

CLUE CARDS FOR "THREE NEIGHBORS"

If you have an element that has a connection to teeth or bones...	If you have an element that is used to make musical instruments...	If you have an element that is used to make containers or pans for food product...
If you have an element that is used to make a hygiene product... (soap, toothpaste, etc)	If you have an element that is known for being smelly...	If you have an element that is used to make a medical device or a tool used by doctors...
If you have an element that is used to repair bones or joints...	If you have an element that is used to make small electronic parts...	If you have an element that is used to make sports equipment...
If you have an element that is used to make parts for planes, boats or subs...	If you have an element that is used in welding or soldering...	If you have an element that has been known since ancient times...
If you have an element that has a reputation for being poisonous...	If you have an element that is essential to making some type of glasses, lenses, or goggles...	If you have an element that is, or was, used in making pottery or glassware...
If you have an element that is used in any type of clock or watch...	If you have an element known for its use in a military device... (not a vehicle)	If you have an element whose atomic number ends with a 7...

CLUE CARDS FOR "THREE NEIGHBORS"

PLAYING CARDS

These cards can be used for many games and activities. You can use the ideas listed in this book, or you can make your own.

NOTE: If you are making multiple copies of these cards so that each player can have their own set, consider making each set a different color (use colored paper or card stock) or put a number, letter or colored dot on the back or side of each card in the set. If the players will be keeping their cards, have them put their name or initials on the back of each card. It is surprisingly easy for cards to get mixed up, and it can be a time-consuming task to sort them out if they are not labeled.

There is a digital file of these cards available so you can easily print them out on your home printer. Most home printers can print on 110 lb card stock, which is the most widely used type of card stock and therefore likely to be found in your local stores.
Go to: **www.ellenjmchenry.com/periodictableactivitycards**

1
H
Hydrogen

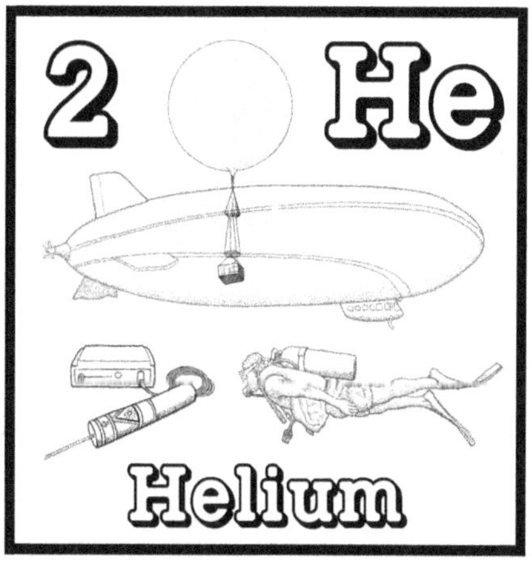
2
He
Helium

3
Li
WHITE LITHIUM
GREASE
Lithium

4
Be
W153
BeCu
Beryllium

5
B
BORAX
DETERGENT BOOSTER
Boron

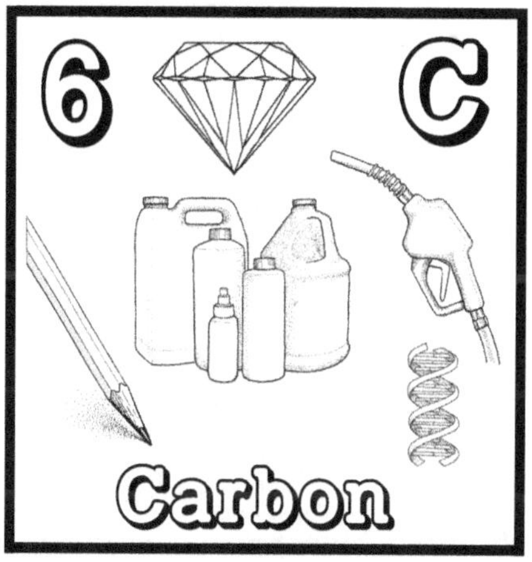
6
C
Carbon

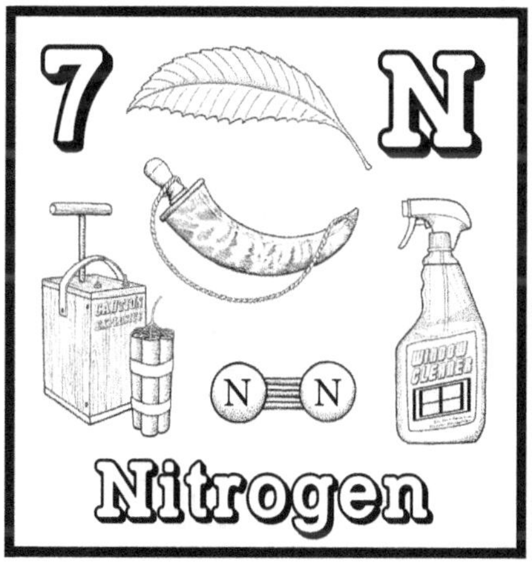
7
N
N
N
Nitrogen

8
O
Oxygen

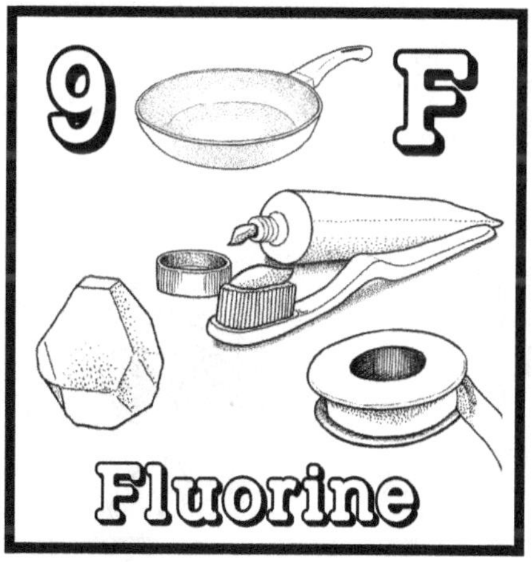
9
F
Fluorine

10
Ne
Café
Neon

11
Na
BLEACH
Sodium

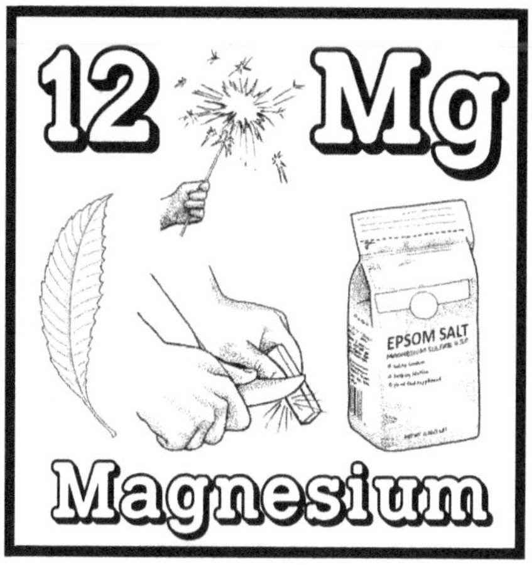
12
Mg
EPSOM SALT
Magnesium

13
Al
Aluminum

14
Si
CAULK
Silicon

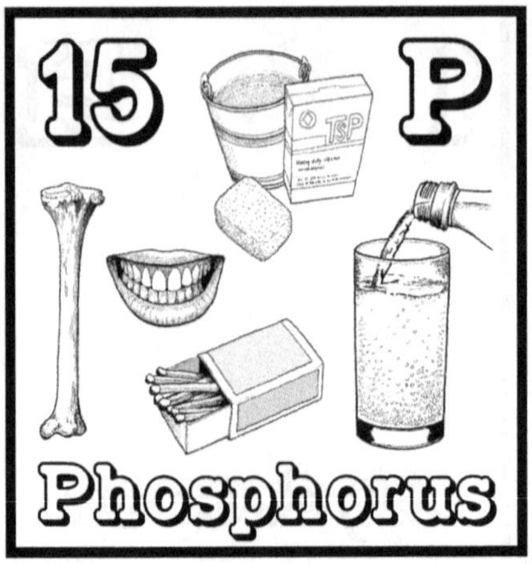
15
P
Phosphorus

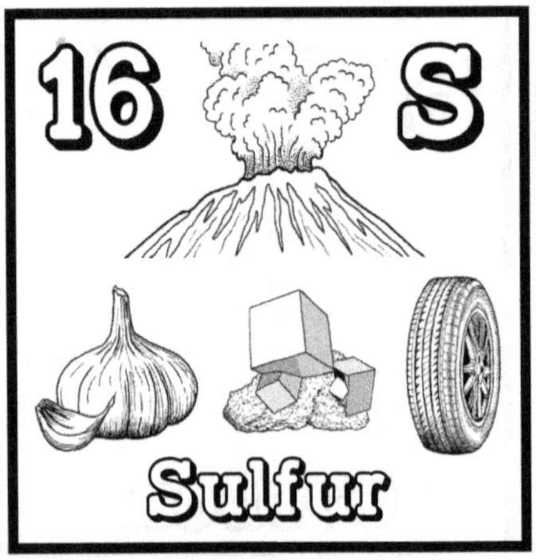
16
S
Sulfur

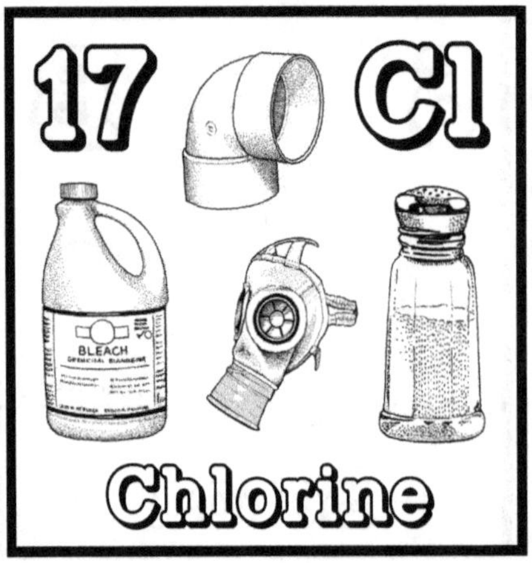
17
Cl
BLEACH
Chlorine

18
Ar
Argon

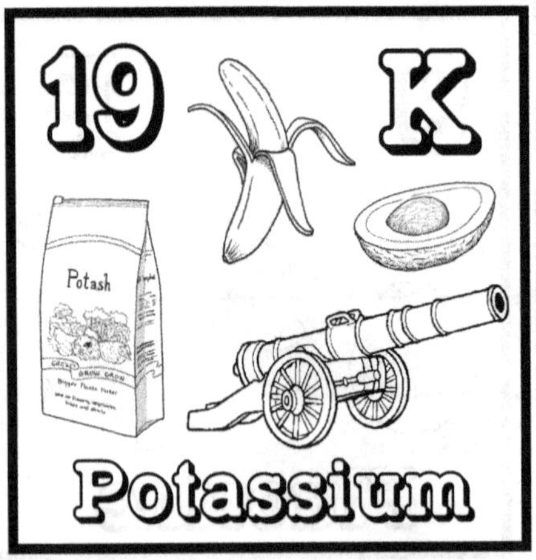
19
K
Potash
Potassium

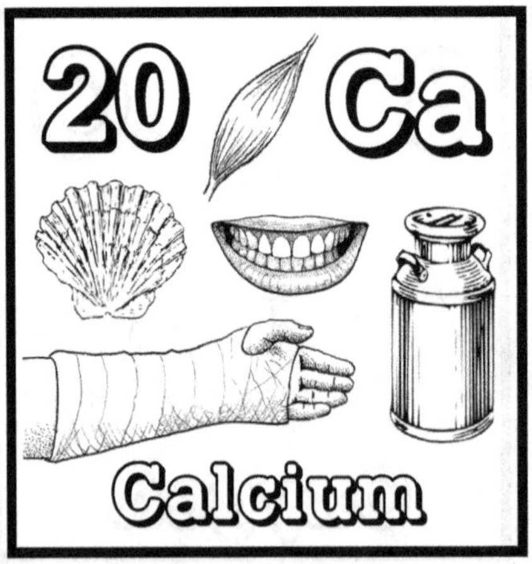
20
Ca
Calcium

21
Sc
Scandium

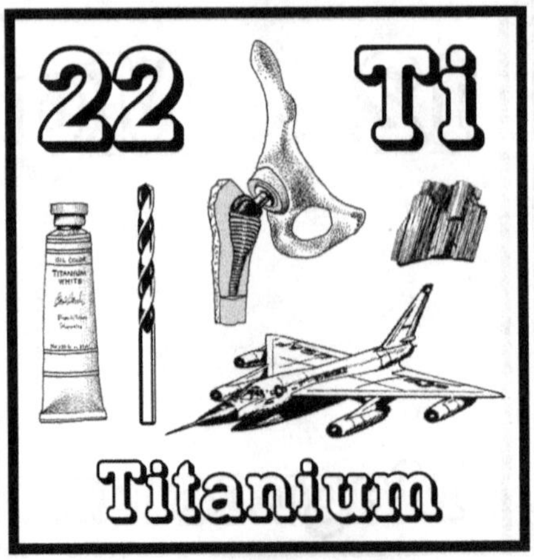
22
Ti
Titanium

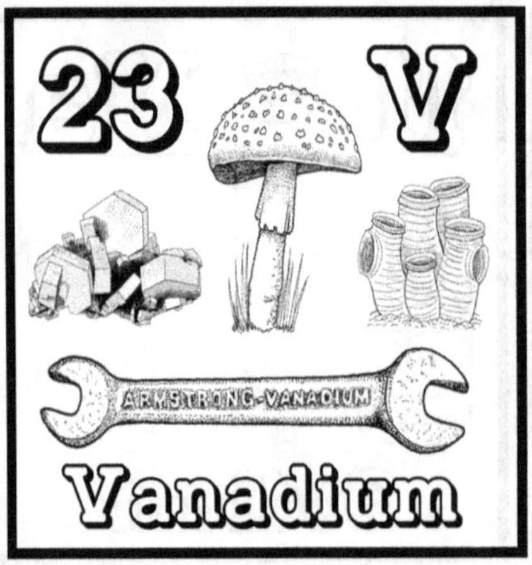
23
V
ARMSTRONG-VANADIUM
Vanadium

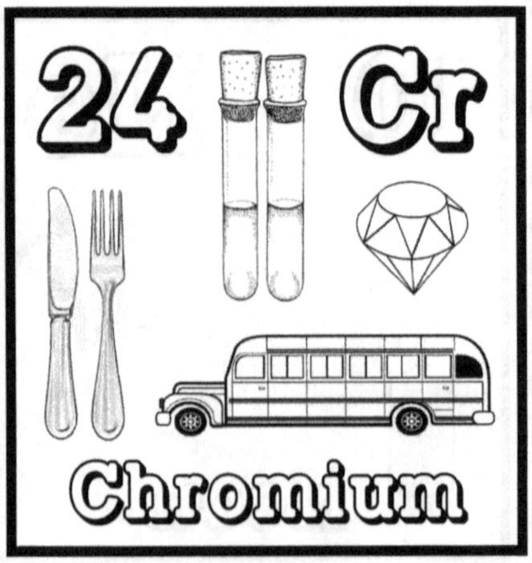
24
Cr
Chromium

25 Mn
Manganese
26 Fe
Iron
27 Co
Cobalt
28 Ni
Nickel
29 Cu
Copper
30 Zn
Zinc
31 Ga
Gallium
32 Ge
Germanium
33 As
Arsenic
34 Se
Dandruff Shampoo
Selenium
35 Br
BROMO SELTZER
Acid Indigestion and Headache
Bromine
36 Kr
Café
Kr-Ar
Krypton

37
Rb
Rubidium

38
Sr
Strontium

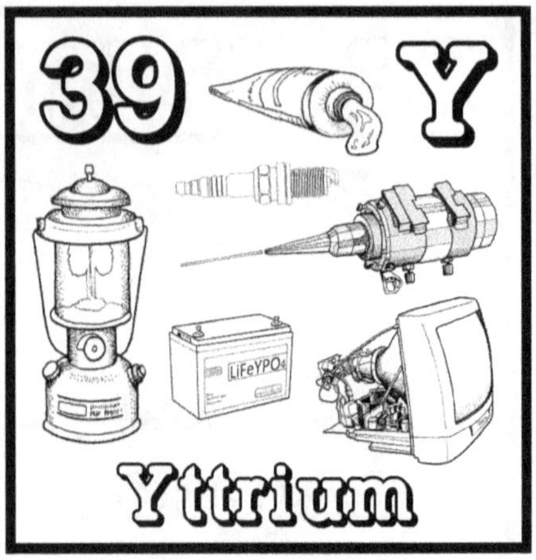
39
Y
LiFeYPO4
Yttrium

40
Zr
Special
ANTIPERSPIRANT
Zirconium

41
Nb
Niobium

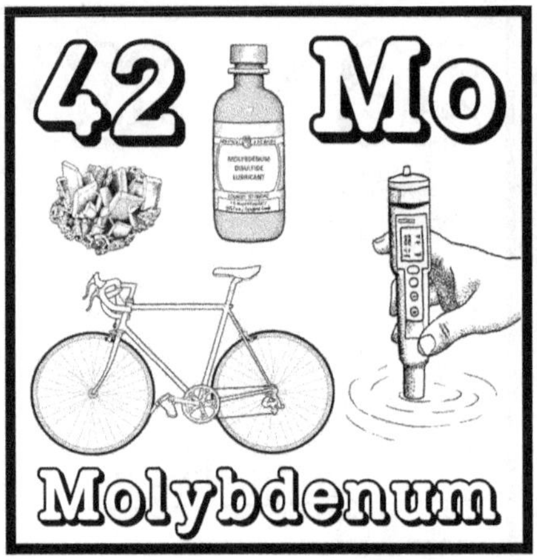
42
Mo
Molybdenum

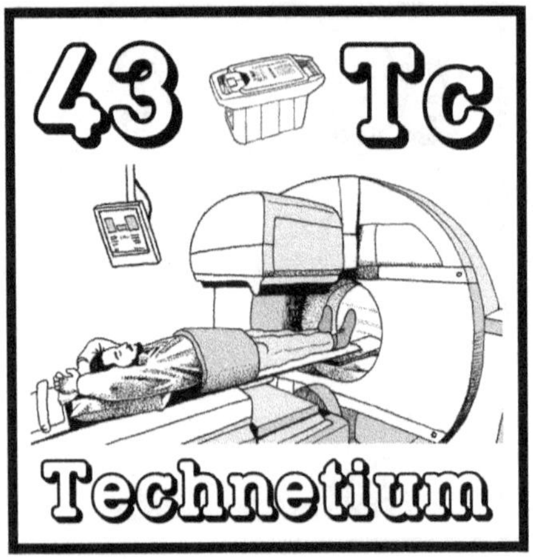
43
Tc
Technetium

44
Ru
Ruthenium

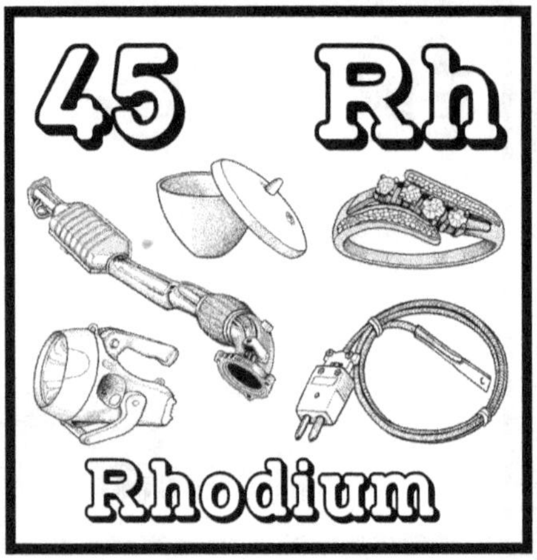
45
Rh
Rhodium

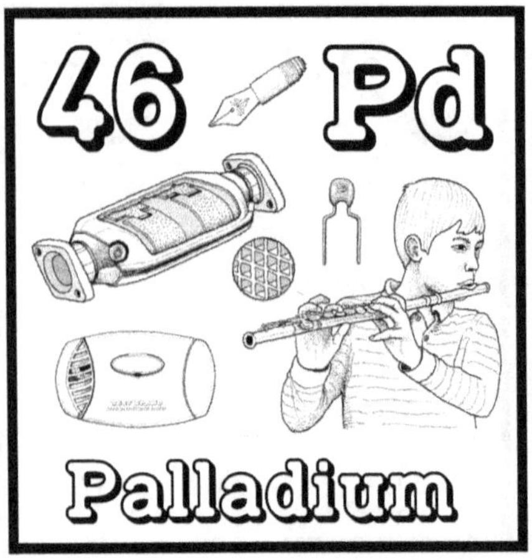
46
Pd
Palladium

47
Ag
Silver

48
Cd
Cadmium

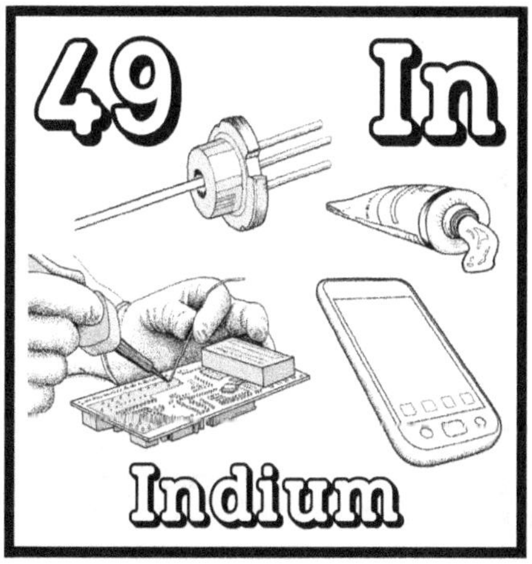
49
In
Indium

50
Sn
Tin

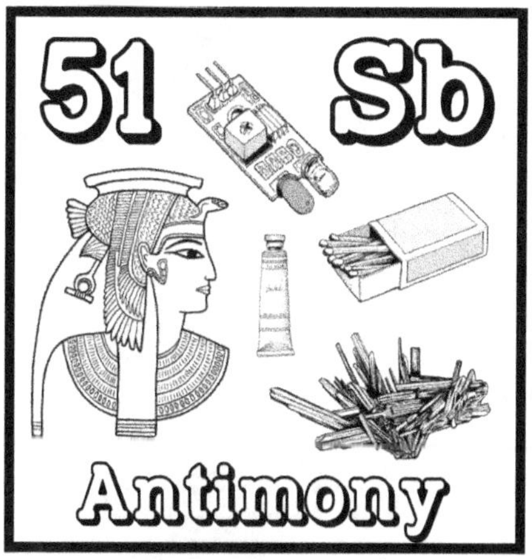
51
Sb
Antimony

52
Te
Tellurium

53
I
Iodine

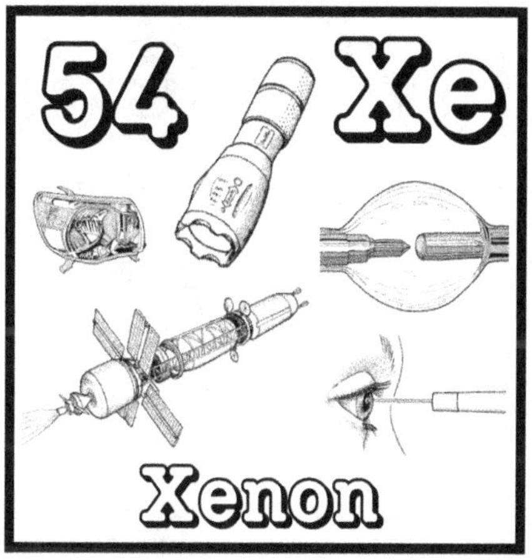
54
Xe
Xenon

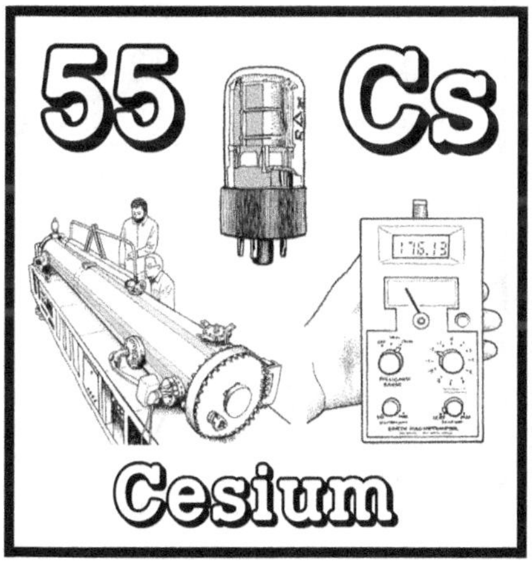
55
Cs
176.13
Cesium

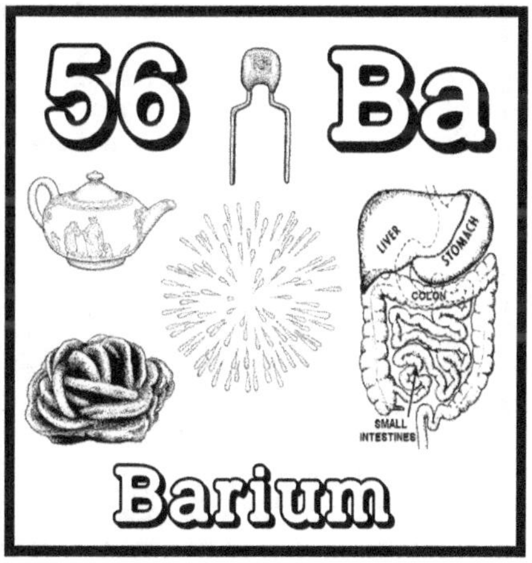
56
Ba
LIVER
STOMACH
COLON
SMALL INTESTINES
Barium

57
La
Lanthanum

58
Ce
Cerium

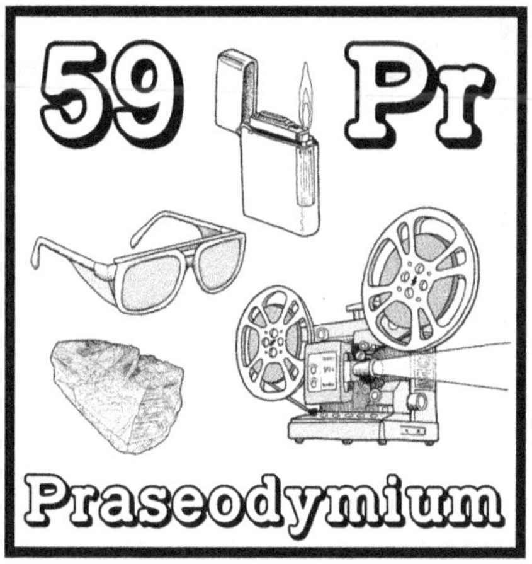
59
Pr
Praseodymium

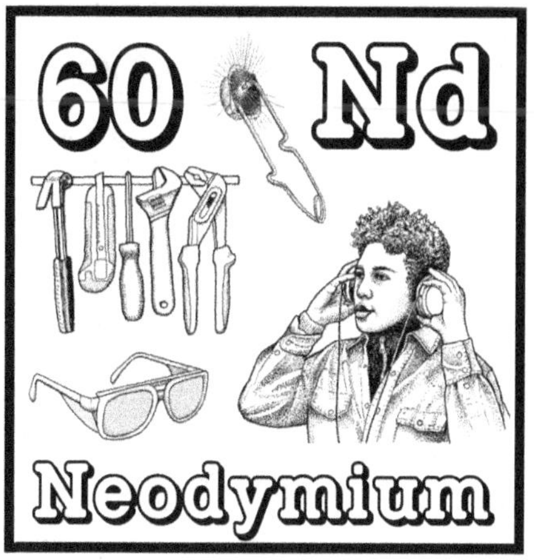
60
Nd
Neodymium

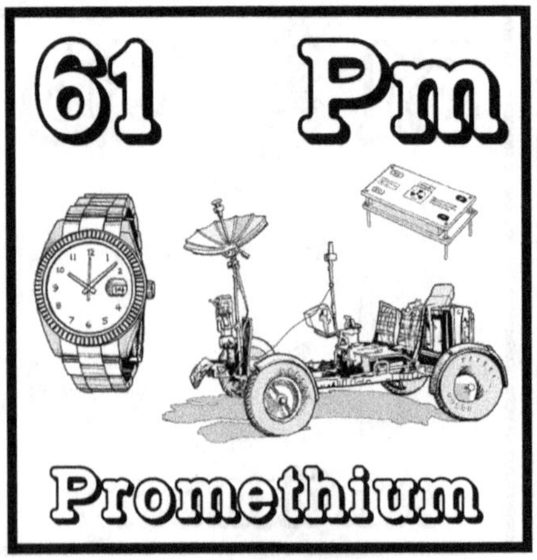
61 Pm
Promethium

62 Sm
Samarium

63 Eu
Europium

64 Gd
Gadolinium
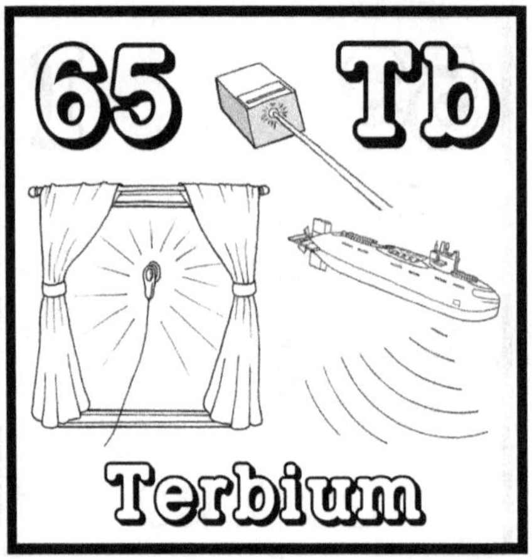
65 Tb
Terbium
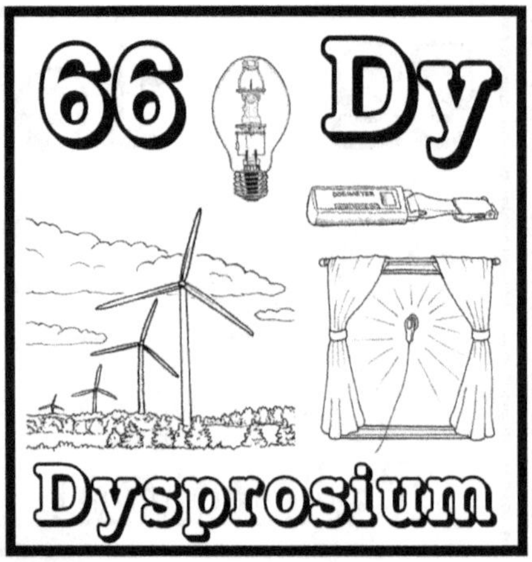
66 Dy
Dysprosium
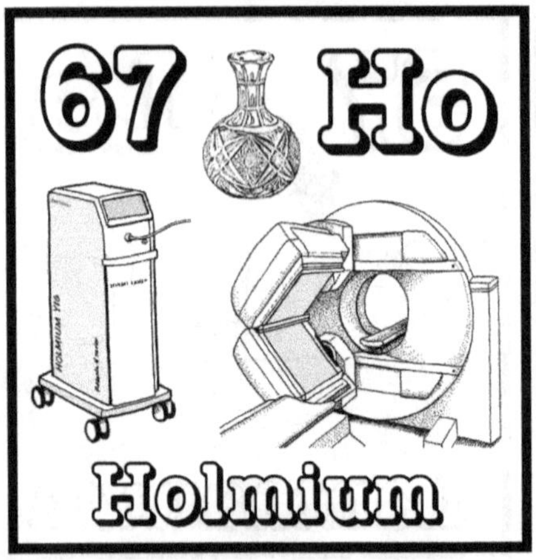
67 Ho
Holmium
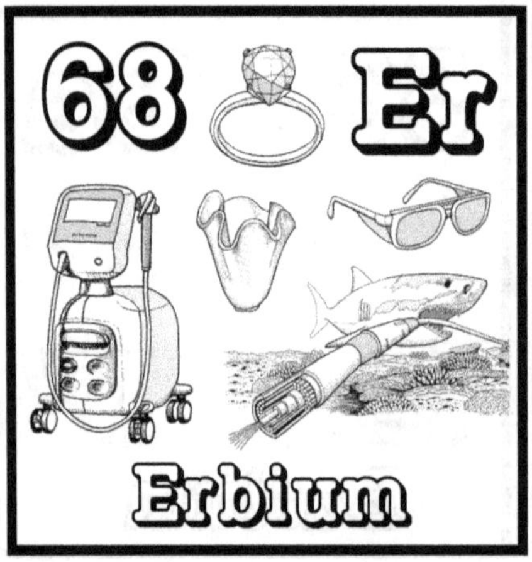
68 Er
Erbium

69 Tm
Thulium
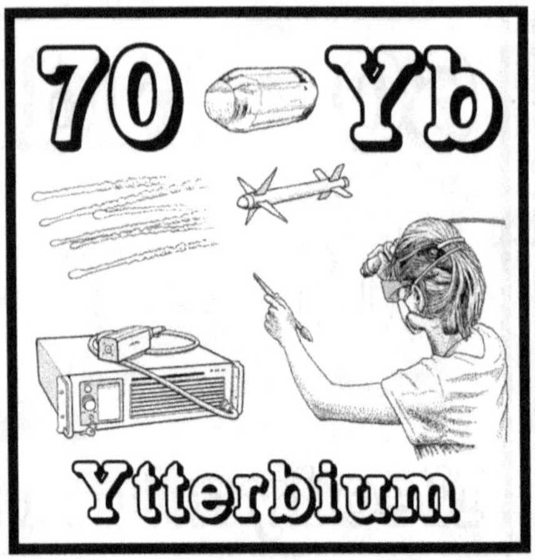
70 Yb
Ytterbium

71 Lu
Lutetium

72 Hf
Hafnium

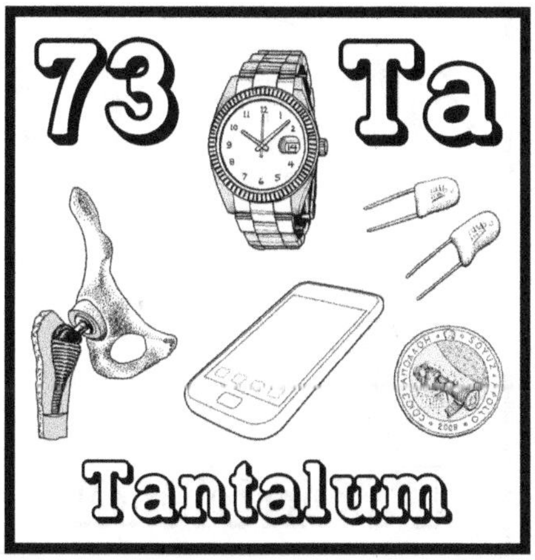
73
Ta
Tantalum

74
W
Tungsten

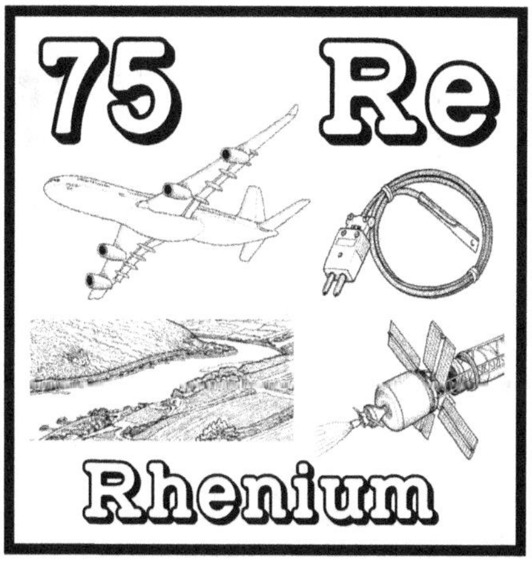
75
Re
Rhenium

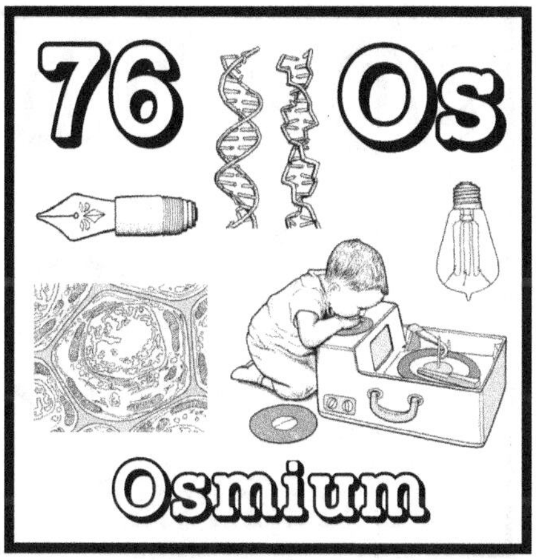
76
Os
Osmium

77
Ir
SEM images
Iridium

78
Pt
Platinum

79
Au
Gold

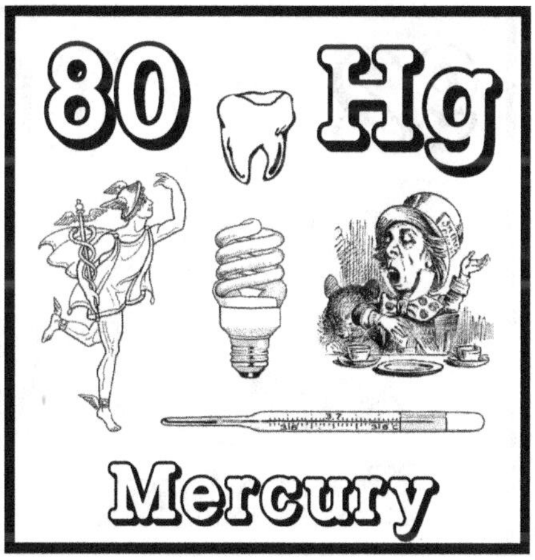
80
Hg
Mercury

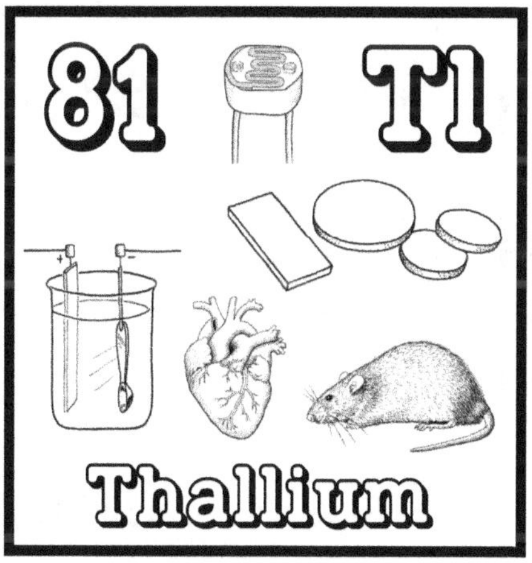
81
Tl
Thallium

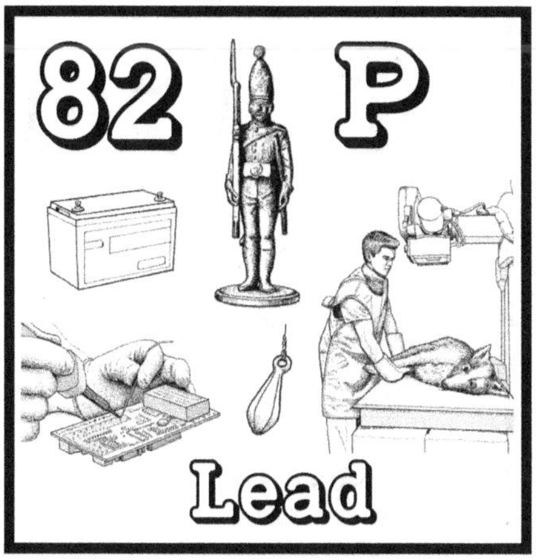
82
P
Lead

83
Bi
Bismuth

84
Po
Polonium

85 At

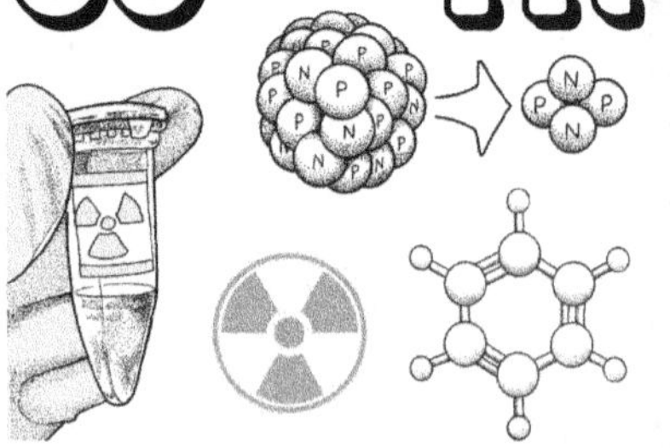

Astatine

86 Rn

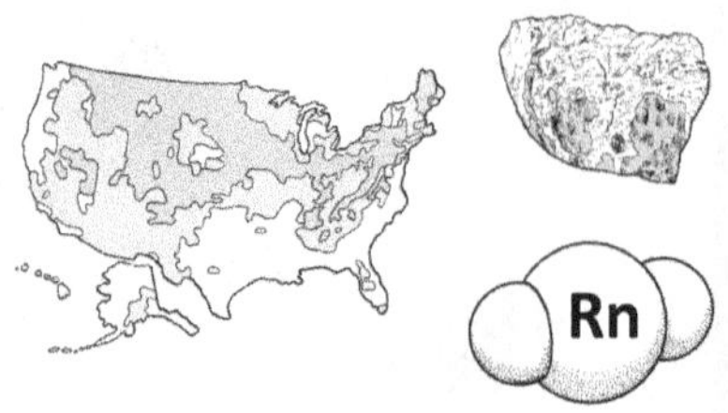

Radon

87 Fr

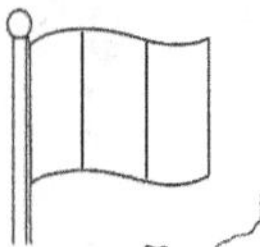

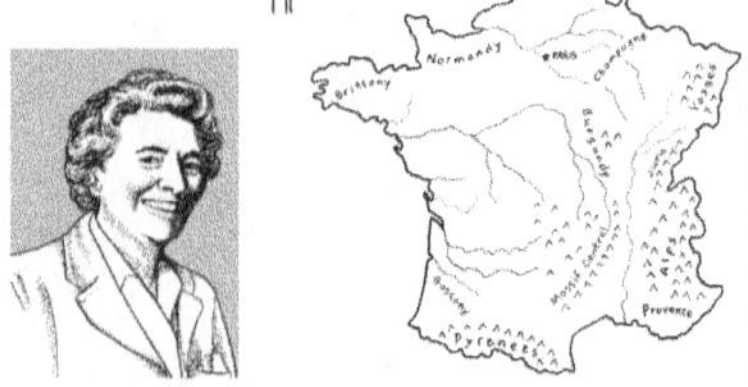

Francium

88 Ra

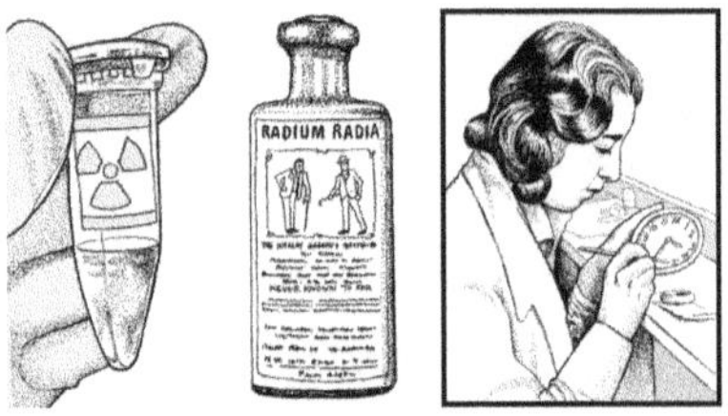

Radium

89 Ac

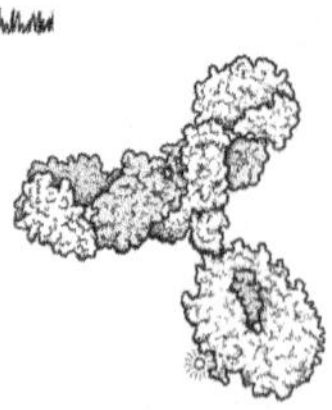

Actinium

90 Th

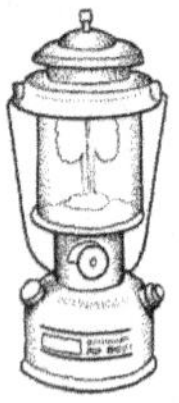

Thorium

91 Pa

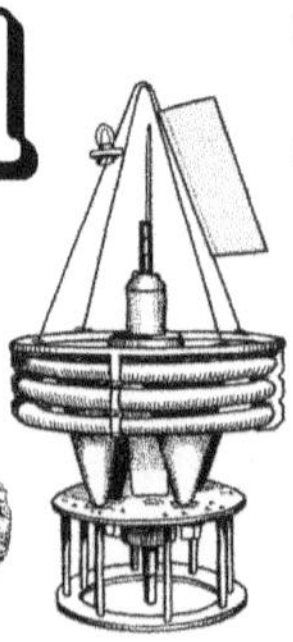

Protactinium

92 U

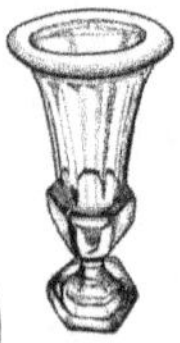

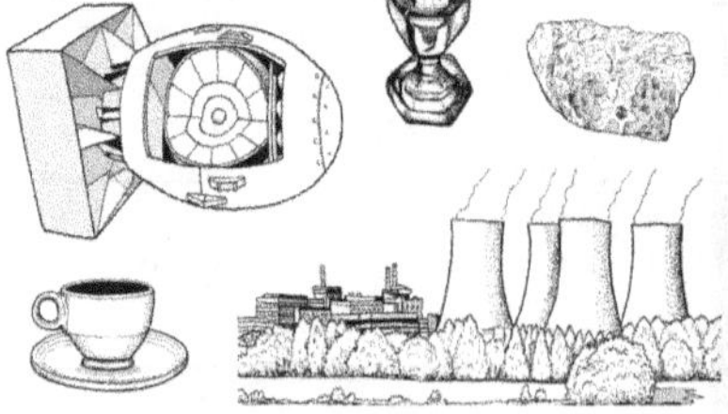

Uranium

93 Np

Neptunium

94 Pu

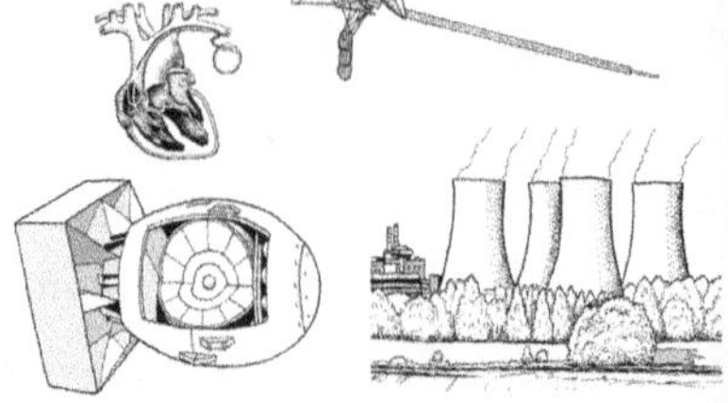

Plutonium

95 Am

Americium

96 Cm

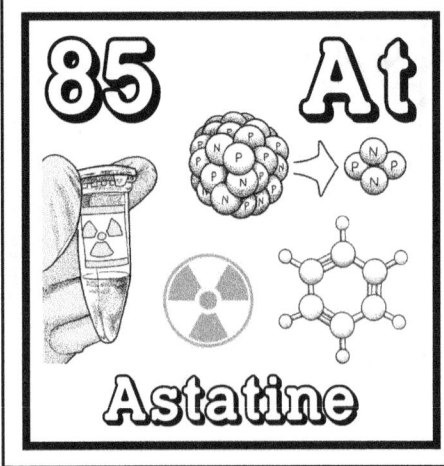

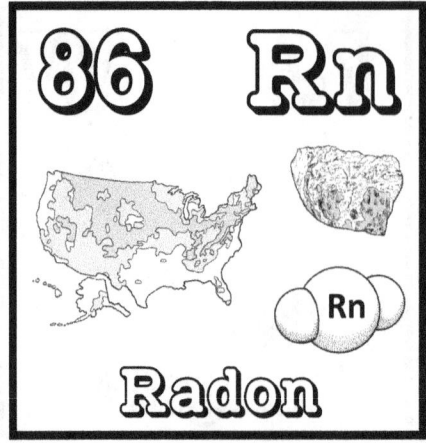

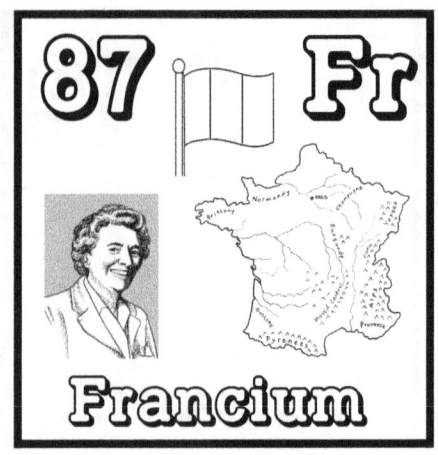

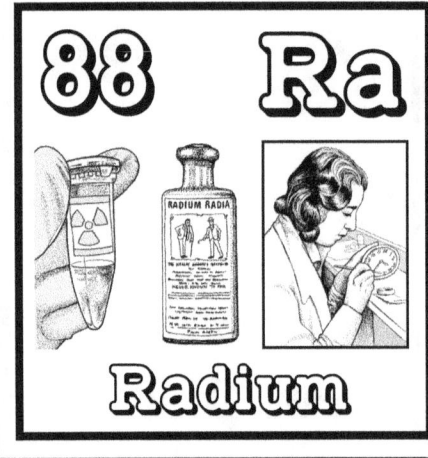

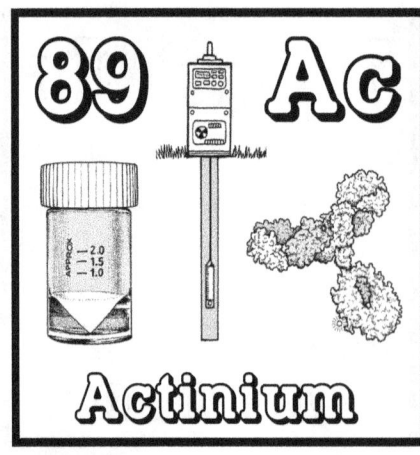

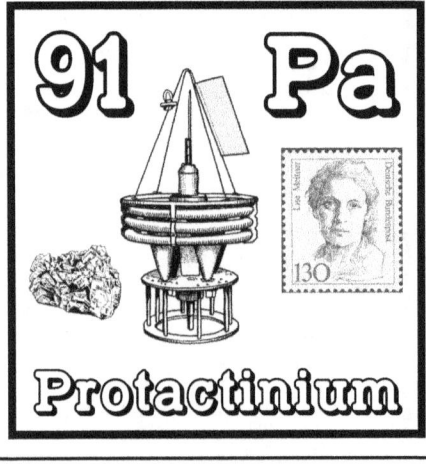

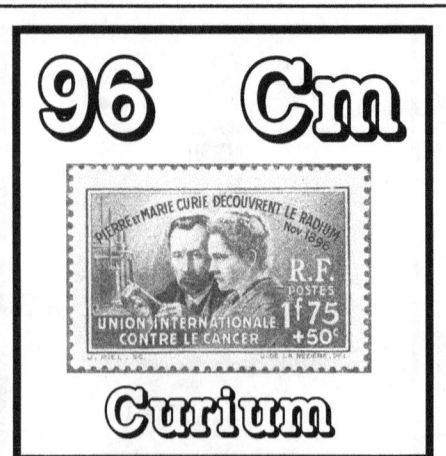

Curium

97 Bk

Berkelium

98 Cf

Californium

99 Es

Einsteinium

100 Fm

Fermium

101 Md

Mendelevium

102 No

Nobelium

103 Lr

Lawrencium

104 Rf

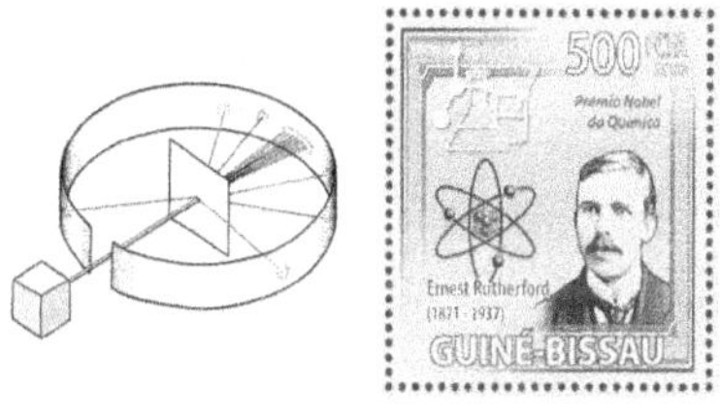

Rutherfordium

105 Db

Dubnium

106 Sg

Seaborgium

107 Bh

Bohrium

108 Hs

Hassium

109 Mt

Meitnerium

110 Ds

Darmstadtium

111 Rg

Roentgenium

112 Cn

Copernicium

113 Nh

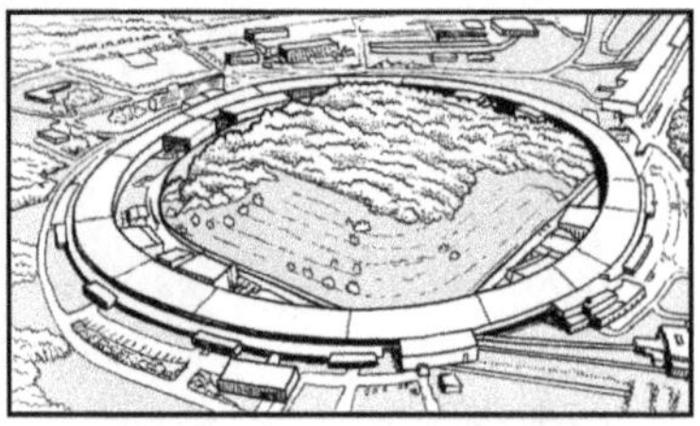

Nihonium

114 Fl

Flerovium

115 Mc

Moscovium

116 Lv

Lawrence

Livermore

Livermorium

117 Ts

Tennessine

118 Og

Oganesson

FREE

FREE

H	He	Li	Be
B	C	N	O
F	Ne	Na	Mg
Al	Si	P	S
Cl	Ar	K	Ca

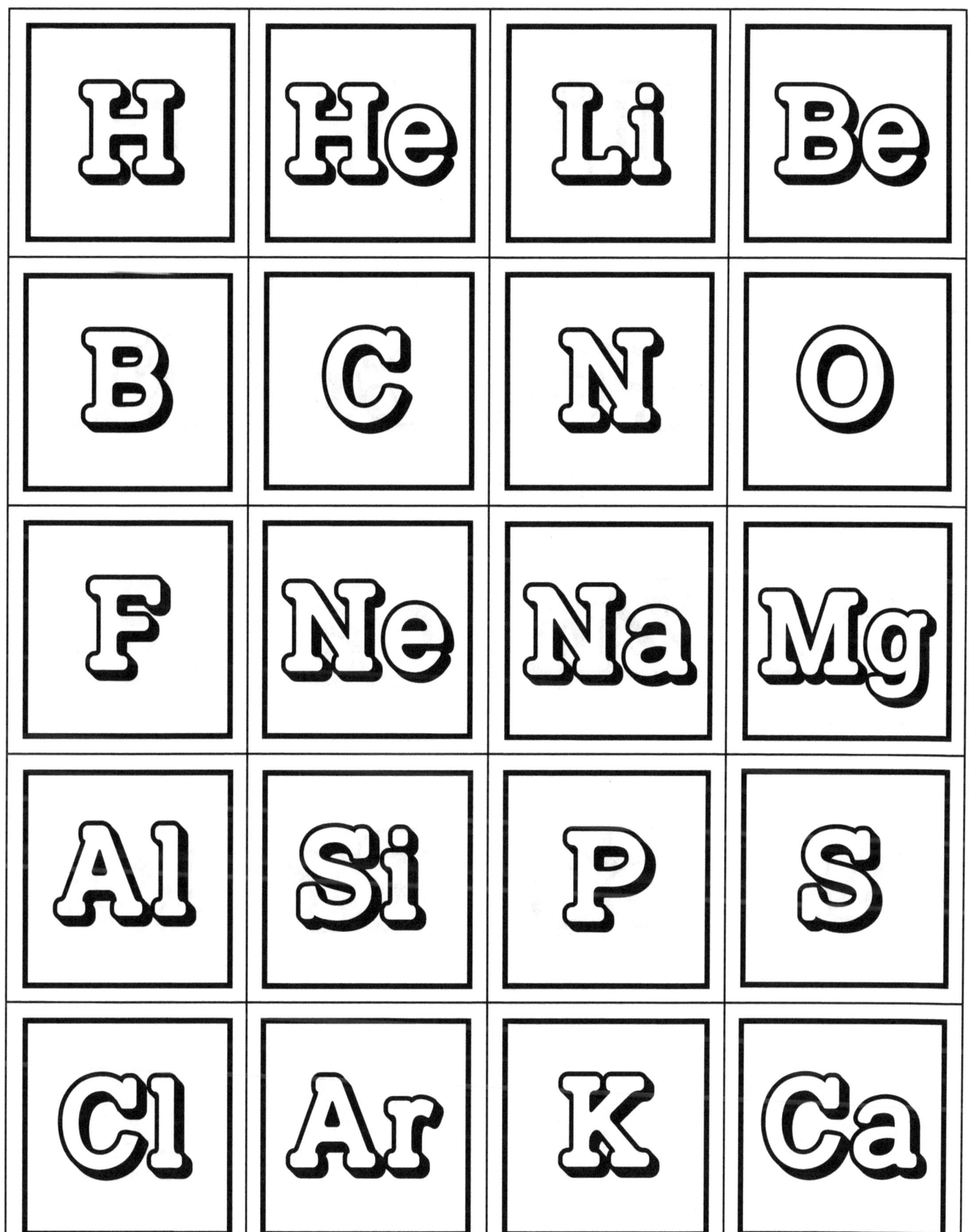

Sc	Ti	V	Cr
Mn	Fe	Co	Ni
Cu	Zn	Ga	Ge
As	Se	Br	Kr
Rb	Sr	Y	Zr

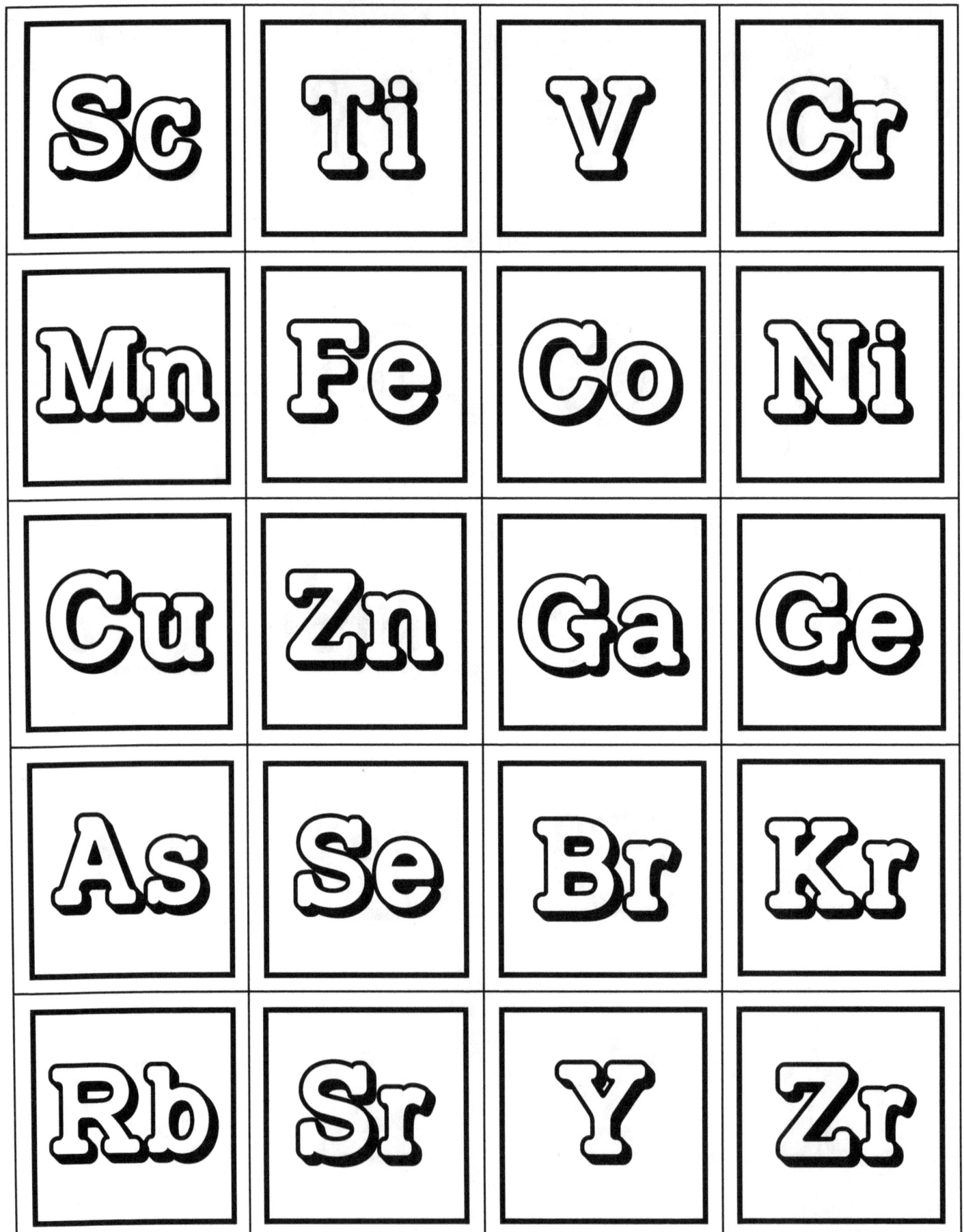

Nb	Mo	Tc	Ru
Rh	Pd	Ag	Cd
In	Sn	Sb	Te
I	Xe	Cs	Ba
La	Ce	Pr	Nd

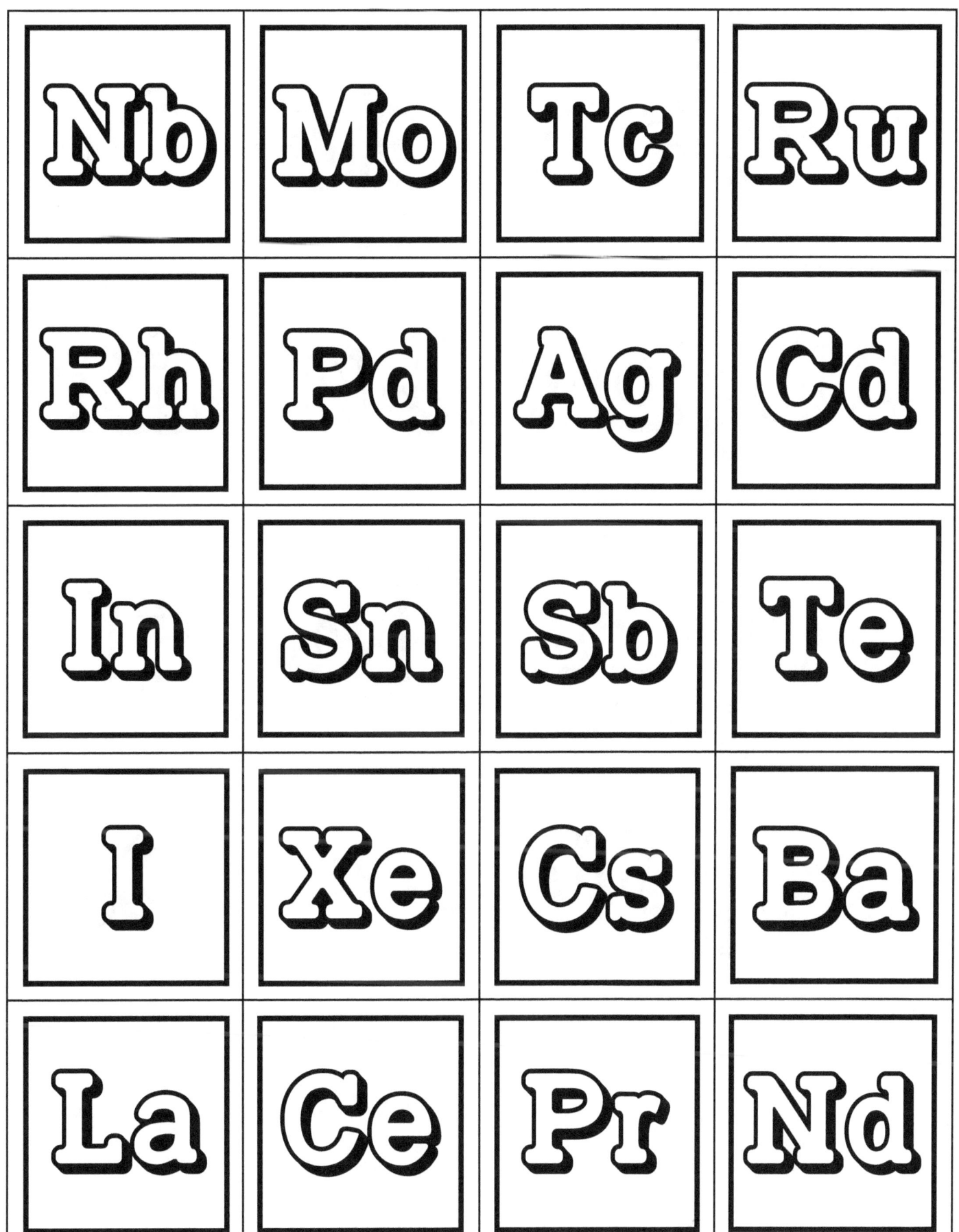

Pm	Sm	Eu	Gd
Tb	Dy	Ho	Er
Tm	Yb	Lu	Hf
Ta	W	Re	Os
Ir	Pt	Au	Hg

Tl	Pb	Bi	Po
At	Rn	Fr	Ra
Ac	Th	Pa	U
Np	Pu	Am	Cm
Bk	Cf	Es	Fm

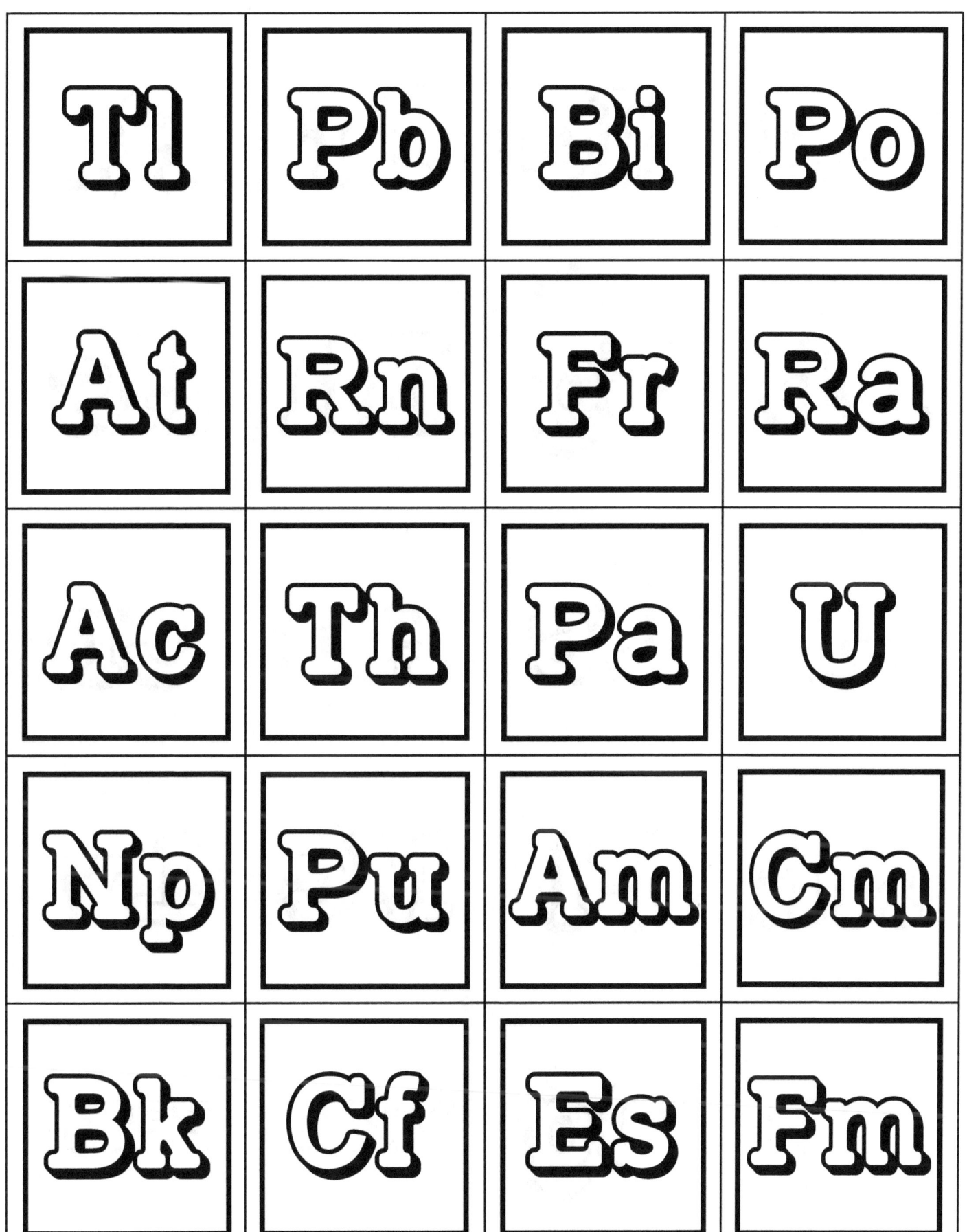

Md	No	Lr	Rf
Db	Sg	Bh	Hs
Mt	Ds	Rg	Cn
Nh	Fl	Mc	Lv
Ts	Og	FREE	FREE

QUESTION BANK

The following lists of questions can be used for many games.

In front of each question is a letter, E, M, or H, indicating the approximate level of difficulty: easy, medium or hard. Choose which questions you feel are appropriate for your players.

The intent is for you to be able to choose which elements will be included in your game, and make copies of those pages.
For example, you might want to do a Bingo game on just the first two rows on the Periodic Table, or just the transition elements.

You can keep the clues intact and simply use a pencil to check off which questions have been read, or you can use scissors to cut the page into strips, with one question per strip. This will let you put the questions into a box or bag and pull out questions randomly.

E	This element is very abundant in our sun and is part of the fusion reaction that makes heat.	hydrogen 1
E	Two atoms of this element are found in every water molecule.	hydrogen 1
E	This element is often used as rocket fuel. It combines with liquid oxygen to create combustion.	hydrogen 1
E	This is the smallest and lightest of all elements.	hydrogen 1
M	Three atoms of this element are connected to a nitrogen atom to make an ammonia molecule.	hydrogen 1
M	Four atoms of this element are connected to a carbon atom to make a molecule of methane.	hydrogen 1
M	Hydrogenated oils are made of strings of carbons with atoms of this element attached.	hydrogen 1
H	When combined with chlorine atoms, this element makes a strong acid (found in our stomachs).	hydrogen 1
H	Two atoms of this element combined with two atoms of oxygen make a common first aid product.	hydrogen 1
H	Atomic welding uses this gaseous element between two metal electrodes.	hydrogen 1

E	This element was first discovered in the sun, using a spectrometer.	helium 2
E	This element is used to fill balloons and blimps because it is light and non-flammable.	helium 2
E	This element has 2 protons and 2 neutrons.	helium 2
E	This element is used as a shielding gas in arc welding.	helilum 2
M	This element is mixed with oxygen and put into scuba tanks (to replace nitrogen).	helium 2
M	This elemental gas is used to pressurize liquid hydrogen in rocket fuel tanks.	helium 2
M	This elemental gas is produced as the result of the decay of uranium atoms.	helium 2
H	In 1895, William Ramsay discovered this gaseous element in rocks.	helium 2
H	This element is mixed with neon and used in lasers.	helium 2
H	This gas can be liquefied and used to cool magnets in MRI machines and particle accelerators.	helium 2

E	This element is best known for its use in long-life batteries.	lithium 3
E	This very light element makes bright red spark in fireworks and flares.	lithium 3
E	This element has 3 protons and usually 4 neutrons, though sometimes it can have 3 neutrons.	lithium 3
E	If a pure sample of this element is put into water, it will burn while floating on the surface.	lithium 3
M	Compounds containing this element are used to make medicines, mostly for neurological problems.	lithium 3
M	This lightest of all metals is used to make industrial lubricants.	lithium 3
M	Because this element is so light, it is added to metal alloys that will be used to build airplanes.	lithium 3
H	When combined with fluorine, it makes a clear crystal that is used in optical lenses.	lithium 3
H	When combined with carbonate, CO_3, this element can be used in ceramic glazes and tile adhesives.	lithium 3
H	When combined with hydroxide (OH) it makes a compound that can remove CO_2 from the air inside airplanes.	lithium 3

E	When this element is added to copper it makes very strong bronze used for tools and springs.	beryllium, 4
E	This element was discovered in 1828 when it was extracted from a green mineral called beryl.	beryllium, 4
E	This is one of the elements found in emerald gemstones.	beryllium, 4
E	This element was used to discover neutrons in 1932.	beryllium, 4
M	In its pure form, this element is used to make small windows that allow x-rays to pass through.	beryllium, 4
M	Bronze that contains this element is different from most metals because it can't create a spark.	beryllium, 4
M	This very tough, spark-proof element is used to make contacts for spot welders.	beryllium, 4
H	When combined with oxygen, this element makes a compound that is used for telescope mirrors.	beryllium, 4
H	Bronze tools that contain 2% of this element are used in coal mines, oil rigs, and around MRI machines.	beryllium, 4
H	Because this element is a good source of free neutrons, it is used inside atomic bombs.	beryllium, 4

E	This element can be extracted from the mineral borax.	boron, 5
E	This element is used to make a laundry washing powder.	boron, 5
E	When white glue is added to a compound of this element, you get "slime." ("goop")	boron, 5
E	Most atoms of this element have 6 neutrons, but some have 5.	boron, 5
M	This element forms a compound silicon to make a heat resistant glass used in kitchens and labs.	boron, 5
M	Glass beads made from silicon and this element are spun into fiberglass or used in reflective paints.	boron, 5
M	This element has been used as ant poison, though it is relatively safe for us to use in crafts and labs.	boron, 5
H	This element is found in a mineral called ulexite, which displays fiber-optic properties.	boron, 5
H	Atoms of this element that have only 5 neutrons are used as neutron absorbers in nuclear power plants.	boron, 5
H	This element can be dissolved into a solution and then used to grow decorative crystals.	boron, 5

E	This element's name comes from the Latin word for charcoal.	carbon, 6
E	In its pure form, this element can take the form of graphite, coal, or a diamond.	carbon, 6
E	This element combines with hydrogen to make fuels such as gasoline, methane, and cooking oils.	carbon, 6
E	This element can form long molecules called polymers, that are used to make plastics.	carbon, 6
M	When we exhale, a gas containing this element leaves our lungs.	carbon, 6
M	Sixty atoms of this element can make a molecule that looks like a soccer ball.	carbon, 6
M	Proteins, fats, sugars, and DNA molecules are all structured around atoms of this element.	carbon, 6
H	Octane, which is a liquid fuel, as 8 atoms of this element.	carbon, 6
H	Limestone rocks contain calcium, oxygen, and this element.	carbon, 6
H	Our houses should have detectors for the dangerous gas made of this element and oxygen.	carbon, 6

E	Most of the air around us is made of this element.	nitrogen, 7
E	This gaseous element is used in air bags in cars.	nitrogen, 7
E	If plants don't get enough of this element, the leaves will lose their green color.	nitrogen, 7
E	This element is a primary ingredient in cleaning products that contain ammonia (ex: window cleaners).	nitrogen, 7
M	This element is found in compounds used to preserve meats.	nitrogen, 7
M	This element is found in gunpowder and dynamite (TNT).	nitrogen, 7
M	When one atom of this element combines with 3 atoms of hydrogen an ammonia molecule is formed.	nitrogen, 7
H	When 2 atoms of this element combine with an atom of oxygen, a molecule of "laughing gas" is formed.	nitrogen, 7
H	This gaseous element is used to protect fresh fruit in long-term storage (to keep oxygen out).	nitrogen, 7
H	When one atom of this element combines with 2 atoms of oxygen, the result is air pollution (not CO_2).	nitrogen, 7

E	This element makes 20% of our atmosphere.	oxygen, 8
E	This element will make wet metal rust.	oxygen, 8
E	This is the third most abundant element in the universe, after hydrogen and helium.	oxygen, 8
E	This element is necessary for combustion (burning).	oxygen, 8
M	Plants produce this element as a by-product (waste) during the process of photosynthesis.	oxygen, 8
M	When this element combines with silicon, it makes the mineral quartz.	oxygen, 8
M	When liquefied, this gaseous element is used in rockets to burn the fuel.	oxygen, 8
H	When three atoms of this element join together, we call the molecule "ozone."	oxygen, 8
H	The name of this element comes from a Greek word meaning "sour."	oxygen, 8
H	When this gaseous element attaches to NaCl (salt) it forms a type of bleach.	oxygen, 8

E	This element is most well-known for its use in toothpaste, to help prevent cavities.	fluorine, 9
E	When this element is combined with carbon, you can make a non-stick surface (brand name Teflon).	fluorine, 9
E	When combined with calcium, this element makes a clear green or light purple crystal.	fluorine, 9
E	This element's name means "to flow" because it is added to hot metals to make them flow more easily.	fluorine, 9
M	This element is famous for being the most "electronegative" element on the Periodic Table.	fluorine, 9
M	This element is the lightest and smallest of the "halogen" elements (second column from right).	fluorine, 9
M	When this element is combined with hydrogen (1 to 1 ratio) you get an extremely dangerous acid.	fluorine, 9
H	The mineral formed by this element and calcium can be used to make high magnification camera lenses.	fluorine, 9
H	When combined with carbon, this element can make a slippery fabric used for rain gear. (ex: Gore-Tex®)	fluorine, 9
H	When 6 atoms of this element are combined with one atom of sulfur, it makes a non-toxic gas.	fluorine, 9

E	This element's name comes from the Greek word for "new."	neon, 10
E	A very bright type of fluorescent light is named after this element, though not all bulbs contain it.	neon, 10
E	This element is combined with helium to make red lasers.	neon, 10
E	This element has 10 electrons, 8 of which completely fill its outer shell.	neon, 10
M	This element is the second-lightest of the noble gas family, the last column on the right side of the table.	neon, 10
M	Fluorescent lights containing this element glow bright reddish-orange.	neon, 10
M	This element was discovered by Sir William Ramsay in 1898, along with krypton and xenon.	neon, 10
H	The spectral emission pattern produced by this gaseous element has mostly red and orange lines.	neon, 10
H	This element is used to make tiny glowing numbers in "nixie" tubes (used as novelty digital clocks).	neon, 10
H	This very light noble gas is used inside some structures that are designed to absorb lightning strikes.	neon, 10

E	When this element combines with chlorine, it produces molecules of table salt.	sodium, 11
E	This element is used in high-pressure vapor lamps, which are often used as street lights.	sodium, 11
E	When this element is combined with HCO_3, you get baking soda.	sodium, 11
E	The spectral emission pattern of this element features two very bright yellow lines.	sodium, 11
M	When this element combines with chlorine, it make salty crystals that have a cubic shape.	sodium, 11
M	When this element combines with chlorine and oxygen, it can make a type of bleach.	sodium, 11
M	This element has 11 protons.	sodium, 11
H	This element works with potassium in molecular pumps that carry signals through our nerve cells.	sodium, 11
H	In its pure form, this element is a light metal that will burn yellow while floating on water.	sodium, 11
H	When this element is combined with hydroxyl (OH), it can be used in scuba "re-breathers" to remove CO_2.	sodium, 11

E	This element is used in sparklers that make white sparks.	magnesium, 12
E	John Epsom accidentally discovered this element when the water from his new well tasted bitter.	magnesium, 12
E	This element has 12 protons.	magnesium, 12
E	When combined with SO_4, this element makes a compound we know as "Epsom salt."	magnesium, 12
M	This element is the central molecule in the chlorophyll molecule (found in plants).	magnesium, 12
M	Humphry Davy officially discovered this element in 1808, but named after a certain area in Greece.	magnesium, 12
M	This light metal is commonly found in metal alloys used for things that need to be strong but light.	magnesium, 12
H	This element can make sparks so it is used in manual fire starters.	magnesium, 12
H	This element was used in camera flashes during the early 1900s.	magnesium, 12
H	Tracer bullets use this light metal element to make a bright red flash.	magnesium, 12

E	This lightweight metallic element is used to make bicycle frames and parts for airplanes.	aluminum, 13
E	This very light metal can be rolled into thin sheets and used as foil.	aluminum, 13
E	This element is named after a compound called "alum."	aluminum, 13
E	This metal is used to make cans for beverages. These cans are easy to recycle.	aluminum, 13
M	This element is found in antiperspirants. It discourages the sweat glands from producing sweat.	aluminum, 13
M	When combined with nickel and cobalt, this element is used to make magnets.	aluminum, 13
M	The primary mineral ore of this element is bauxite.	aluminum, 13
H	When combined with oxygen, this element makes the mineral "corundum." (Rubies are made of corundum.)	aluminum, 13
H	When combined with hydroxide (OH) this element forms a compound found in some antacid medicines.	aluminum, 13
H	This light metal is pronounced as a 4-syllable word in the U.S., but as a 5-syllable word in the U.K.	aluminum, 13

E	When this element combines with oxygen, it can create the mineral quartz.	silicon, 14
E	This element is used to make "caulk" which is used to fill cracks around windows and sinks.	silicon, 14
E	This element is used by the semi-conductor industry to build microchips for computers and phones.	silicon, 14
E	This element can be used to make flexible baking pans.	silicon, 14
M	Glass is made primarily of this element combined with oxygen.	silicon, 14
M	This element is used by microscopic organisms known as diatoms to build their shells.	silicon, 14
M	The valley in California that is known for making high-tech devices is named after this element.	silicon, 14
H	This element is similar to carbon because it can make 4 bonds, due to the 4 electrons in its outer shell.	silicon, 14
H	This element is found in quartz, and gives it piezo-electric properties useful in making clocks and sonars.	silicon, 14
H	This element is a primary ingredient in the polymer molecules found in Silly Putty®.	silicon, 14

E	Match tips contain this element in its red form.	phosphorus, 15
E	Along with calcium, this element is an important element in the structure of bones and teeth.	phosphorus, 15
E	The name of this element means "light bearer."	phosphorus, 15
E	A pure white sample of this element will glow yellowish-green when heated.	phosphorus, 15
M	This element was discovered by an alchemist who was trying to extract gold from urine.	phosphorus, 15
M	This element is a primary ingredient in the mineral "apatite."	phosphorus, 15
M	Carbonated beverages contain an acid based on this element.	phosphorus, 15
H	This element is a key ingredient in TSP, a cleaning solution that dissolves grease.	phosphorus, 15
H	This element was extracted from bird droppings to make fertilizers for plants.	phosphorus, 15
H	When combined with calcium, this element forms a compound used to make china dishes.	phosphorus, 15

E	The ancient world called this element "brimstone" which means "burning stone."	sulfur, 16
E	Compounds that contain this element are very smelly. (examples: garlic and skunks)	sulfur, 16
E	A pure sample of this element looks like a yellow rock, and smells like a burning match.	sulfur, 16
E	When this element combines with iron, Fe, it can make pyrite, also known as "fool's gold."	sulfur, 16
M	When this element is combined with barium and oxygen, it can form the mineral barite.	sulfur, 16
M	This element is the key to "vulcanizing" rubber-- making it stable in both hot and cold temperatures.	sulfur, 16
M	When this element combines with oxygen and magnesium, it can make Epsom salt crystals.	sulfur, 16
H	This element was officially recognized as an element in 1777 by famous chemist Antoine Lavoisier.	sulfur, 16
H	Car batteries contain a well-known acid based on this element.	sulfur, 16
H	Jupiter's moon, Io, is yellowish-orange due to the many volcanoes that release this element.	sulfur, 16

E	When this element combines with sodium it can make crystals of table salt.	chlorine, 17
E	In its pure form this element is a toxic green gas.	chlorine, 17
E	This element can be combined with oxygen and sodium to make bleach.	chlorine, 17
E	The name of this element comes from the Greek word for "light green."	chlorine, 17
M	PVC plastic pipes contain this element. (It is the "C" in PVC.)	chlorine, 17
M	The gaseous form of this element was used as a weapon during World War 1.	chlorine, 17
M	This element is one of the "C's" in CFCs, which were once used in refrigerators. (The other C is carbon.)	chlorine, 17
H	When this element combines with hydrogen, it makes an acid found in our stomachs.	chlorine, 17
H	Humphry Davy was the first chemist to realize that this greenish gas was an element, not a compound.	chlorine, 17
H	When 4 atoms of this element are attached to a carbon atom, you get a fluid used in dry cleaning.	chlorine, 17

E	This gaseous element is put into light bulbs because it is inert and won't react with other molecules.	argon, 18
E	This element has 18 protons.	argon, 18
E	This element is the third noble gas. The first two are helium and neon.	argon, 18
E	The name of this element comes from a Greek word for "lazy."	argon, 18
M	This gaseous element is put into paint cans to push out reactive gases like oxygen and nitrogen.	argon, 18
M	This gaseous element is used in lasers that make bright blue light and are used for eye surgery.	argon, 18
M	This element is used as a shield gas in arc welding, since it won't react with any sparks.	argon, 18
H	This element is used to "butcher" chickens. The birds suffocate quickly because of lack of oxygen.	argon, 18
H	This element is put into "glove boxes" so chemists can work with compounds that can't touch oxygen.	argon, 18
H	Because this gaseous element is inert, it is put into graphite furnaces to prevent unwanted combustion.	argon, 18

E	This element is right below sodium on the Periodic Table, so it has similar chemical properties.	potassium, 19
E	When this element is combined with NO_3, you get saltpeter, an ingredient in gunpowder.	potassium, 19
E	This element used to be extracted from potash (wood ashes).	potassium, 19
E	Your body needs this element. Good dietary sources are bananas, potatoes, and avocados.	potassium, 19
M	This element works with sodium in one of the molecular pumps found in your nerve cells.	potassium, 19
M	The symbol for this element is from the Arabic word for potash: "kali."	potassium, 19
M	This element can be combined with chlorine to make a substitute for table salt (NaCl).	potassium, 19
H	The mineral orthoclase feldspar, or "K-feldspar," is known for its high content of this element.	potassium, 19
H	Compounds of this element are used in food preparation, such a bromide compound used in bread.	potassium, 19
H	When combined with chlorine, this element makes a compound used for de-icing sidewalks.	potassium, 19

E	Along with phosphorus, this element is found in teeth and bones.	calcium, 20
E	When this element combines with CO_3, it makes limestone.	calcium, 20
E	Our bodies need this element. Two good dietary sources of this element are milk and broccoli.	calcium, 20
E	Chalk is made of this element combined with oxygen and carbon.	calcium, 20
M	Snails and clams use a carbonate compound of this element to make their shells.	calcium, 20
M	When this element is combined with fluorine it makes a clear green or light purple crystal.	calcium, 20
M	When this element combines with SO_4, it makes the mineral "gypsum."	calcium, 20
H	Our muscle cells use this element to signal the start of a contraction.	calcium, 20
H	Though many of this element's compounds are white, its pure form is a soft, gray metal.	calcium, 20
H	This light metal looks and acts much like magnesium because of its location on the Periodic Table.	calcium, 20

E	This element is named after Scandinavia.	scandium, 21
E	This element has 21 protons and 21 electrons.	scandium, 21
E	This element is used inside high-intensity light bulbs found in large sports stadiums.	scandium, 21
E	Frames of expensive bicycles sometimes contain small amounts of this element, along with aluminum.	scandium, 21
M	Aluminum baseball bats often contain a small amount of this element.	scandium, 21
M	Lacrosse sticks are often made of alloys containing aluminum and this element.	scandium, 21
M	This is the first element in the block that we call the transition metal elements.	scandium, 21
H	The hull of the Russian MIG-29 military plane was made of aluminum alloyed with this element.	scandium, 21
H	Atoms of this element that have 25 neutrons instead of 24 are radioactive and are used as "tracers."	scandium, 21
H	Mendeleyev predicted the existence of this element. He called it eka-boron because of its location.	scandium, 21

E	This element was named after the Titans, characters from Greek mythology.	titanium, 22
E	An oxide compound of this element is widely used by artists in white paint.	titanium, 22
E	This element is non-toxic and is compatible with body tissues. It is used to repair bones and joints.	titanium, 22
E	This element makes metal alloys very strong; it is used in making drill bits, tools, and rocket engines.	titanium, 22
M	Metal tools that contain this element are stronger and last longer, but are more expensive.	titanium, 22
M	This element is very resistant to corrosion so it is added to metals used for boat propellers and engines.	titanium, 22
M	This element makes metal very strong. It is used in expensive golf clubs, horseshoes, and tools.	titanium, 22
H	This metallic element is not magnetic. Bones repaired with this element can safely go into MRI scanners.	titanium, 22
H	Rutile, a shiny, black mineral ore, is made of this element and oxygen.	titanium, 22
H	The pure form of this metallic element has a unique feature: it will burn in the presence of nitrogen.	titanium, 22

E	This element was named after the Norse goddess of beauty, Vanadis.	vanadium, 23
E	The poisonous, bright red "fly agaric" mushroom stores high levels of this element.	vanadium, 23
E	The name of this element comes from the fact that it can produce a wide range of beautiful colors.	vanadium, 23
E	The Ford Model-T was the first car to use metal alloys that contained this element.	vanadium, 23
M	This element is essential to a blue, tube-like sea creature called the tunicate.	vanadium, 23
M	You can find wrenches with the name of this element stamped on them because it is in the steel.	vanadium, 23
M	When combined with lead, oxygen and chlorine, this element forms bright red, hexagonal crystals.	vanadium, 23
H	The source of this element used to be mines in Peru, until it was discovered as an impurity in uranium ores.	vanadium, 23
H	This element was first discovered in Mexico in 1801 in a rock called "brown lead."	vanadium, 23
H	When this element is in an acetylacetonate compound, it can imitate our body's insulin molecule.	vanadium, 23

E	The name of this element comes from the Greek word for color.	chromium, 24
E	"School bus yellow" paint was originally made with this element.	chromium, 24
E	When tiny amounts of this element are present in the mineral corundum, it makes a red ruby.	chromium, 24
E	This element is added to steel to make it "stainless," meaning it won't rust.	chromium, 24
M	Similar to vanadium, this element can make a wide range of colors when put into various solutions.	chromium, 24
M	The red mineral "crocoite" contains lead and this element.	chromium, 24
M	This element makes steel harder. This type of steel is durable enough to be used as train tracks.	chromium, 24
H	This transition metal element is sometimes used in the leather tanning process.	chromium, 24
H	Adding this element to paint can make the painted object (ex: tanks) invisible to infrared sensors.	chromium, 24
H	The only elements that are harder than (the pure form of) this element are boron and carbon (diamond).	chromium, 24

E	People confuse this element with the element "magnesium."	manganese, 25
E	Parts of the ocean floor are littered with mineral balls, or "nodules," that contain a lot of this element.	manganese, 25
E	Some ancient peoples drew pictures on cave walls using brown pigments made from this element.	manganese, 25
E	The blue pigment known as "YInMn" contains yttrium, indium, and this element.	manganese, 25
M	When combined with K and O_4, this element makes a purple compound used to sterilize water.	manganese, 25
M	Compounds of this element create amazing patterns on rocks that look similar to fossilized ferns.	manganese, 25
M	Our bodies need a little bit of this element. Good sources are mussels, lima beans, and sweet potatoes.	manganese, 25
H	Some World War 1 combat helmets were made of a steel alloy containing this element.	manganese, 25
H	When mixed with carbonate, CO_3, this elements makes a dark pink mineral called "rhodochrosite."	manganese, 25
H	A protein recycling enzyme called "arginase" has atoms of this element at its center.	manganese, 25

E	The symbol for this element comes from the Latin word "ferrum."	iron, 26
E	Red blood cells are red because of the presence of this element.	iron, 26
E	When carbon is added to this element, you can make steel.	iron, 26
E	This element will rust when it comes into contact with oxygen.	iron, 26
M	This metallic element is used to make very heavy cooking pans.	iron, 26
M	Lodestone is a natural mineral that is magnetic because it contains this element.	iron, 26
M	The ancient world first encountered this element (in its pure form) in meteorites.	iron, 26
H	An atom of this element is at the center of the hemoglobin molecule.	iron, 26
H	When two atoms of this element join to three atoms of oxygen, you get a red mineral called hematite.	iron, 26
H	Our bodies need this element. Good dietary sources are meat, fish, eggs, beans, and leafy vegetables.	iron, 26

E	The name of this element comes from an old German word for "goblin."	cobalt, 27
E	This element, along with iron, aluminum, and nickel, is found in AlNiCo magnets.	cobalt, 27
E	This element can color glass bright blue.	cobalt, 27
E	This element has 27 protons.	cobalt, 27
M	Like titanium, this element is non-toxic and is used to make metal alloys for artificial joints.	cobalt, 27
M	Ancient peoples in China used this element to paint blue designs on their white porcelain vases.	cobalt, 27
M	A radioactive form of this element is used to sterilize medical equipment and to treat cancers.	cobalt, 27
H	An atom of this element is found at the center of the vitamin B-12 molecule (cobalamin).	cobalt, 27
H	The Impressionist painters loved the bright blue pigment made from this element.	cobalt, 27
H	This element is often added to lithium carbonate to make long-life batteries for electronic devices.	cobalt, 27

E	The U.S. has a coin whose name is the same as this element.	nickel, 28
E	The core of the earth is believed to be made of iron and this element.	nickel, 28
E	The name of this element comes from a German word for "devil" because of miners' superstitious beliefs.	nickel, 28
E	AlNiCo magnets are made of iron, aluminum, cobalt and this element.	nickel, 28
M	Guitar strings often made of wire of this element wrapped around a steel wire.	nickel, 28
M	Cadmium and this element are combined to make rechargeable batteries.	nickel, 28
M	The keys of musical instruments are coated with this element.	nickel, 28
H	The snaps on clothing are often coated with this element to keep them shiny.	nickel, 28
H	Like vanadium and chromium, solutions containing this element can be a wide variety of colors.	nickel, 28
H	Platinum and this element are alloyed to make metal plates used to hydrogenate vegetable oils.	nickel, 28

E	The pennies of many countries are coated with this element.	copper, 29
E	This element has been used to make kettles, coins and pipes.	copper, 29
E	The name of this element means "from Cyprus."	copper, 29
E	The Statue of Liberty is made of this element, and shows how this element can turn green over time.	copper, 29
M	If you add tin to this element, you get bronze.	copper, 29
M	The blood of snails and slugs looks slightly blue because it uses this element to carry oxygen.	copper, 29
M	Wire made of this element is wound around steel rods to make electromagnets.	copper, 29
H	The elements right below this element on the Periodic Table are silver and gold.	copper, 29
H	The mineral "malachite" gets its green color from this element.	copper, 29
H	This element is toxic to fungi and bacteria. It is often used in fungicides that we spray on plants.	copper, 29

E	This element is used to "galvanize" nails and screws to protect them from weathering.	zinc, 30
E	Brass is an alloy of copper and this element.	zinc, 30
E	Pennies in the U.S. are made of this element and coated with a thin layer of copper.	zinc, 30
E	The first "voltaic pile" battery was made of a stack of alternating discs of copper and this element.	zinc, 30
M	This element is often found mixed into ores of copper and lead.	zinc, 30
M	Like titanium and lead, this element is used to make white paint.	zinc, 30
M	When combined with sulfur, this element can make a compound that glows after being exposed to light.	zinc, 30
H	When combined with oxygen, this element forms a compound that can absorb UV light.	zinc, 30
H	Powder of this element is mixed with sulfur powder to create a launching explosive for model rockets.	zinc, 30
H	This element is found in a complex molecule called a "finger," designed to grab and hold DNA.	zinc, 30

E	This element is named after France, using its old Latin name.	gallium, 31
E	This element's melting point is so low that it will melt in your hand.	gallium, 31
E	This element has been combined with tin and indium to make a mercury substitute for thermometers.	gallium, 31
E	In the 1800s, people used this element to make "disappearing spoons" that would melt in your tea.	gallium, 31
M	When this element is combined with arsenic, it makes a compound used in solar panels on Mars rovers.	gallium, 31
M	Mendeleyev predicted the discovery of this soft metal element. It would be below Al on the table.	gallium, 31
M	When this element is combined with nitrogen to make a nitride, it is used in blue lasers for Blu-ray players.	gallium, 31
H	This element was discovered in 1875 by Paul Emile Lecoq de Boisbaudran, using a spectrometer.	gallium, 31
H	This soft, metallic element is often found with aluminum in bauxite ores.	gallium, 31
H	When this element is combined with nitrogen to make a nitride, it can be used in radio frequency devices.	gallium, 31

E	This element is named after Germany.	germanium, 32
E	This element has 32 protons and 32 electrons.	germanium, 32
E	This element is used to make diodes for crystal radios and other electronic devices.	germanium, 32
E	This element has 4 electrons in its outer shell, just like the elements above it: carbon and silicon.	germanium, 32
M	In 1869, Mendeleyev predicted the discovery of this element. It would be below silicon on the table.	germanium, 32
M	This element was used for electronics until it was mostly replaced by silicon in the late 1900s.	germanium, 32
M	Compounds of this element are used for wide angle camera lenses as well as radio diodes.	germanium, 32
H	Compounds of this element are used for many parts found on satellites, including infrared sensors.	germanium, 32
H	The semi-conducting properties of this element were discovered in 1948.	germanium, 32
H	When Clemens Winkler discovered this element in 1886, he wanted to name it neptunium.	germanium, 32

E	This element is famous for its use as a poison (mainly for rats, but sadly for people, too).	arsenic, 33
E	When this element is combined with gallium it makes a compound used to make solar panels.	arsenic, 33
E	This element has 32 protons and 42 neutrons.	arsenic, 33
E	This element was used by ancient peoples to make yellow paint and white glaze for pottery.	arsenic, 33
M	The name of this element comes from an ancient Syrian word meaning "gold colored."	arsenic, 33
M	Orpiment is a yellow mineral ore made of this element and sulfur.	arsenic, 33
M	Realgar is a red mineral ore made of this element and sulfur.	arsenic, 33
H	The largest use of this element today is in alloys with lead. Some car batteries use these alloys.	arsenic, 33
H	When combined with gallium, this element makes a compound that generates light in laser diodes.	arsenic, 33
H	This element was used to make treated (rot-resistant) lumber until it was determined to be too toxic.	arsenic, 33

E	This element's name comes from the Greek word for the moon.	selenium, 34
E	This element has a property no other element has. It will conduct electricity when light shines on it.	selenium, 34
E	A compound of this element is the active ingredient in dandruff shampoo.	selenium, 34
E	Alexander Graham Bell used this element to make a "photophone" before he invented the telephone.	selenium, 34
M	This element was used to make photocopiers during the late 1900s.	selenium, 34
M	This element is named after the moon because it is right above tellurium (earth) on the Periodic Table.	selenium, 34
M	In the 1900s, this element made possible the invention of light meters, solar cells and photocopiers.	selenium, 34
H	This element is combined with copper, indium and gallium to make solar panels.	selenium, 34
H	This element is needed to make an enzyme called glutathione, which neutralizes "free radicals."	selenium, 34
H	The largest use of this element is in glass manufacturing but it is also used in flat-panel digital x-rays.	selenium, 34

E	The name of this element comes from the Greek word for "stinky smell."	bromine, 35
E	This is the only non-metallic element that is a liquid at room temperature.	bromine, 35
E	An ocean mollusk called the murex uses this element to make its dark purple ink.	bromine, 35
E	This element won't burn so it is put into fabrics to make them fire resistant.	bromine, 35
M	This element is usually found in or near salty water, where you also find sodium, chlorine and iodine.	bromine, 35
M	During the mid 1900s, this element was added to gasoline (petrol) to catch and hold atoms of lead.	bromine, 35
M	This element was used to make a "seltzer" medicine that was sold in blue bottles during the 1900s.	bromine, 35
H	This element, along with iodine, contributes to the fishy smell along seashores.	bromine, 35
H	Until 1991, a methane compound of this element was used to fumigate soil, killing insects.	bromine, 35
H	In the early days of photography, metal plates were exposed to a gaseous form of this element.	bromine, 35

E	This gaseous element has 32 electrons, 8 of which completely fill its outer shell.	krypton, 36
E	The name of this element comes from a Greek word meaning "hidden."	krypton, 36
E	This noble gas was discovered by Sir William Ramsay, who also discovered neon and xenon.	krypton, 36
E	This gaseous element is used in photographic flash bulbs used in high-speed photography.	krypton, 36
M	This noble gas is used as a propellant in some satellites, such as the SpaceX Starlink.	krypton, 36
M	This gaseous element is used in light bulbs that need to be very bright. (It can be mixed with Ar and Xe.)	krypton, 36
M	This element is used to make very bright lasers for laser shows. (He, Ne and Ar are not bright enough.)	krypton, 36
H	From 1960 to 1983, the wavelength of an isotope of this element was used as the official length of a meter.	krypton, 36
H	When combined with fluorine, this noble gas is used in lasers that produce ultraviolet light.	krypton, 36
H	A radioactive isotope of this element can be inhaled by patients getting MRI scans of their lungs.	krypton, 36

E	This element's name comes from the Latin word for deep red.	rubidium, 37
E	This element is below potassium on the Periodic Table.	rubidium, 37
E	This alkali element is a very reactive soft metal that will burn with red flames while floating on water.	rubidium, 37
E	Our bodies will treat this element as if it is potassium, which make it useful in PET scans of the brain.	rubidium, 37
M	This element got its name from the bright red line that Bunsen and Kirchhoff saw in its emission spectrum.	rubidium, 37
M	Like cesium (just below it on the table), this element is used to grab gas molecules in vacuum tubes.	rubidium, 37
M	This first-column (alkali) element is used to make lenses for night vision goggles.	rubidium, 37
H	Like cesium (which is directly below it on the table), this element can be used in small atomic clocks.	rubidium, 37
H	Compounds of this element can make red fireworks, though its neighbor, strontium, is used more often.	rubidium, 37
H	This element can be used to make magnetometers that sense tiny amounts of magnetism.	rubidium, 37

E	This element has 38 protons.	strontium, 38
E	This element provides the bright red flames in emergency signal flares.	strontium, 38
E	Bones will absorb this element as if it were calcium. Archaeologists scan for this element in old bones.	strontium, 38
E	This element is below calcium on the Periodic Table.	strontium, 38
M	This element was named after a small Scottish town.	strontium, 38
M	An aluminate compound of this element will glow in the dark and is safe to use in toys.	strontium, 38
M	In the 1900s, this element was put into television screen glass to absorb x-rays from the cathode tubes.	strontium, 38
H	Russia used a radioactive isotope of this element to power lighthouses during the mid-1900s.	strontium, 38
H	This element used to be a key ingredient in a process that extracted sugar from sugar beets.	strontium, 38
H	A radioactive isotope of this element is used to fight bone cancer because bones absorb it like calcium.	strontium, 38

E	This element was named after the Swedish town of Ytterby, as were erbium, terbium and ytterbium.	yttrium, 39
E	This element, along with indium and manganese, is found in the blue pigment called YInMn.	yttrium, 39
E	This element is the first one in the second row of the central block of transition metals.	yttrium, 39
E	This element is the "Y" in YAG lasers. A is for aluminum and G is for garnet.	yttrium, 39
M	This element was discovered in a heavy, black rock that also contained erbium and terbium.	yttrium, 39
M	This element has largely replaced thorium as the material for making mantles for camping lanterns.	yttrium, 39
M	This element is sometimes classified as a "rare earth" though it is not in the row of lanthanides.	yttrium, 39
H	This element was combined with europium and terbium to make red and green in CRT color televisions.	yttrium, 39
H	This element is added to large $LiFeYPO_4$ lithium batteries to make them more energy-efficient.	yttrium, 39
H	When combined with barium, copper and oxygen, this element makes a superconducting material.	yttrium, 39

E	This element is found in silicon-based crystals called zircons.	zirconium, 40
E	This element is used as a diamond substitute when it combines with O_2 to form a "cubic" crystal.	zirconium, 40
E	This element is used to make sandpaper.	zirconium, 40
E	This element has 40 electrons and is used in steel alloys.	zirconium, 40
M	The name of this element comes from the Persian word "zargun" meaning "golden."	zirconium, 40
M	This element forms a clear crystal used as an abrasive when it combines with either SiO_4 or O_2.	zirconium, 40
M	The compound PZT contains lead, titanium and this element, and is used to make sonar and ultrasound.	zirconium, 40
H	This element is often a key ingredient in antiperspirants. Aluminum is also used in the same way.	zirconium, 40
H	An isotope of this element (atomic weight 89) is used in molecules that tag Y-shaped antibodies.	zirconium, 40
H	Pipes made with this element failed in the Fukushima power plant because of contact with hydrogen.	zirconium, 40

E	This element is named after the daughter of the Greek god Tantalus.	niobium, 41
E	The very first use of this element was for filaments in light bulbs, but titanium quickly replaced it.	niobium, 41
E	Jewelry made out of this element shimmers with iridescent rainbow colors.	niobium, 41
E	This element was used to make the nozzle for the Apollo 15 rocket.	niobium, 41
M	The metal cases of heart pacemakers have been made out of this element because it is non-reactive.	niobium, 41
M	Rocket nozzles have been made of an alloy of 89% this element combined with halfnium and titanium.	niobium, 41
M	For almost 100 years, this element was known as "columbium," from "Columbia," meaning "America."	niobium, 41
H	Brazil is the largest producer of this shimmering transition metal element.	niobium, 41
H	Superconducting wire made from this element is often used to make magnets for huge particle accelerators.	niobium, 41
H	The mineral sample from which this element was discovered came from Connecticut, USA.	niobium, 41

E	"Chromoly" is a type of steel that contains chromium and this element.	molybdenum, 42
E	This element is used by a bacteria that live in nodules on the roots of plants.	molybdenum, 42
E	When combined with sulfur, this element makes a compound used as a lubricant for machine parts.	molybdenum, 42
E	"Big Bertha" was a German howitzer that used this element in its steel alloys.	molybdenum, 42
M	The name of this element comes from a Greek word meaning "lead," though it is not lead.	molybdenum, 42
M	When this element is combined with lead and oxygen, it can make a mineral called "wulfenite."	molybdenum, 42
M	This element is at the center of an enzyme found in bacteria that pull nitrogen out of the air.	molybdenum, 42
H	This element is essential to making gauges that are used to measure water and air pollution.	molybdenum, 42
H	This element was discovered by Carl Scheele in 1778. For the next 100 years, it had no practical use.	molybdenum, 42
H	This element is used by most forms of life to make enzymes essential to oxygen use by cells.	molybdenum, 42

E	All forms of this element are radioactive and will eventually turn into molybdenum or ruthenium.	technetium, 43
E	Of all radioactive elements, this one has the smallest atomic number.	technetium, 43
E	The name of this element comes from a Greek word meaning "artificial."	technetium, 43
E	This element is the radioactive atom used to make "Cardiolite," a tracer used to image the heart.	technetium, 43
M	If an extra proton sticks to a molybdenum nucleus, the atom immediately turns into this element.	technetium, 43
M	A radioactive isotope of this element with atomic weight of 99 is used in many medical tests.	technetium, 43
M	This lightweight radioactive metal can be pulled out of spent nuclear fuel rods, or made in a cyclotron.	technetium, 43
H	This radioactive element is widely used in medical tests because it makes gamma rays for a short time.	technetium, 43
H	The only place you'll find this radioactive transition metal outside a lab is in rocks that contain U or Th.	technetium, 43
H	This element was discovered in Italy in a piece of molybdenum metal that had become radioactive.	technetium, 43

E	This is the lightest member of the platinum group.	ruthenium, 44
E	The discoverer named this element after a region of eastern Europe from which his ancestors had come.	ruthenium, 44
E	This element is most famous for its use as a coating over the gold nib in the Parker 51 fountain pen.	ruthenium, 44
E	This element is used to make a red stain for staining biological samples, including seeds.	ruthenium, 44
M	A radioactive isotope of this element is used to treat cancers inside the eye.	ruthenium, 44
M	An oxide compound of this element is used to reveal fingerprints on objects at crime scenes.	ruthenium, 44
M	This element has an atomic weight of 101, and it has 57 neutrons.	ruthenium, 44
H	This element is used to make photomasks for use in extreme ultraviolet lithography on silicon wafers.	ruthenium, 44
H	This element is used to make thick film resistor chips for electronic sensors in cars.	ruthenium, 44
H	The tetroxide of this element changes when exposed to oils, so it can pick up oily fingerprints.	ruthenium, 44

E	This element was named using the Greek word for "rose."	rhodium, 45
E	This element is found in platinum ores that also contain ruthenium, osmium, iridium and palladium.	rhodium, 45
E	Jewelry stores can electroplate their merchandise with this element to make them extra shiny.	rhodium, 45
E	This element is more expensive than gold or platinum. Some prestigious awards are coated with it.	rhodium, 45
M	This element can be used to coat the inside of headlights to make them more reflective.	rhodium, 45
M	This element was discovered in 1803 by William Wollaston, a few weeks after he discovered palladium.	rhodium, 45
M	Like Pt and Pd, this element has the ability to turn a car's exhaust fumes into less harmful gases.	rhodium, 45
H	Canada makes a one ounce silver coin that is coated (electroplated) with this element.	rhodium, 45
H	In 1976, Volvo was the first car company to use a catalytic converter based on this element.	rhodium, 45
H	This heat-resistant element can be found in thermocouples, spark plugs, crucibles, and pacemaker wires.	rhodium, 45

E	This platinum group element is named after an asteroid.	palladium, 46
E	This element is used to make professional quality flutes.	palladium, 46
E	Like Rh and Pt, this element is used in catalyitc converters to turn car exhaust into less harmful gases.	palladium, 46
E	This element has 46 protons.	palladium, 46
M	This platinum group element is used to make catalytic converters for cars and spark plugs for	palladium, 46
M	This element was discovered in 1803 by Willam Wollaston, a few weeks before he discovered rhodium.	palladium, 46
M	This durable member of the platinum group is used in dental implants, surgical tools and spark plugs.	palladium, 46
H	This element can be found in watch springs, pen points, and inside carbon monoxide detectors.	palladium, 46
H	Nickel, silver, or this element can be mixed with gold to make "white gold."	palladium, 46
H	This platinum group metal is mixed with silver to make ceramic capacitors for electronic devices.	palladium, 46

E	The ancient Greek navy was financed by coins made from this metal, mined near Athens.	silver, 47
E	This element is part of a trio, below copper and above gold.	silver, 47
E	The name of this element is used to refer to eating utensils.	silver, 47
E	Coins made of this element used to be put into barrels of water to kill germs.	silver, 47
M	The symbol for this element comes from the Latin word "argentum."	silver, 47
M	This element is toxic to bacteria and viruses so it is used for first aid, as a salve and in bandages.	silver, 47
M	When in a fulminate compound, this element is used to make harmless explosives called "bang snaps."	silver, 47
H	In a nitrate compound, this element has been used to prevent eye infections, even in newborn babies.	silver, 47
H	When combined with bromine, this metal makes a light-sensitive compound used in early photography.	silver, 47
H	This element conducts electricity better than any other and can be drawn into wires only a few atoms wide.	silver, 47

E	This element is combined with nickel to make rechargeable batteries.	cadmium, 48
E	This element has 48 protons.	cadmium, 48
E	This toxic element has been used as a pigment to make yellow and red paint, and to color glass.	cadmium, 48
E	This element forms a trio with zinc and mercury, and is usually found in ores that contain zinc.	cadmium, 48
M	Sprinkler systems have valves made of a soft metal alloy of bismuth and either indium or this element.	cadmium, 48
M	Batteries that contain this element are often bound together in a yellow plastic wrapper.	cadmium, 48
M	This element got its name from a mineral named after the Greek hero who founded the city of Thebes.	cadmium, 48
H	This toxic transition element is used with helium to make lasers that produce ultra-violet light.	cadmium, 48
H	This element can absorb neutrons and can be used to slow the fission process in nuclear power plants.	cadmium, 48
H	A telluride compound of this element is used to make solar panels for houses.	cadmium, 48

E	This element was named for the deep blue line seen in its emission spectrum.	indium, 49
E	This element, along with yttrium and manganese, is used to make the bright blue YInMn pigment.	indium, 49
E	This element is used to replace lead in solder because it is less toxic than lead.	indium, 49
E	This element and tin are used to make ITO, an electrically conductive clear coating for touch screens.	indium, 49
M	This element is combined with copper, gallium and selenium to make thin-film flexible solar cells.	indium, 49
M	A gallium arsenide compound of this element is widely used in electronics such as LEDs and transistors.	indium, 49
M	This element and tin are used to make ITO, which helps to keep frost off airplane windshields.	indium, 49
H	This element and tin are used to make ITO, which can be used as a antireflective coating for lenses.	indium, 49
H	This is one of the elements used to slow down nuclear fission reactions (along with Cd, B, and Ag).	indium, 49
H	This non-toxic, very soft metal is used to seal cracks around lids in air-tight vacuum chambers.	indium, 49

E	This is the only element whose name has only three letters.	tin, 50
E	When this element is combined with copper, you get bronze.	tin, 50
E	Before plastic was invented, this element, either alone or in alloys, was molded into small toys such as soldiers.	tin, 50
E	This element is most famous for its use in making cans for the preservation of food products.	tin, 50
M	This element's symbol comes from the Latin word "stannum."	tin, 50
M	When this element is mixed with antimony and copper, you get pewter.	tin, 50
M	This element, along with lead, was used to make huge pipes for pipe organs.	tin, 50
H	This element and indium are used to make ITO, a clear, electrically conductive coating for touch screens.	tin, 50
H	When combined with fluorine, this element makes a compound used by dentists to fluoridate teeth.	tin, 50
H	When heated, this element makes a liquid surface onto which you can pour glass, to make "float glass."	tin, 50

E	This element was used by ancient Egyptian men and women to draw dark lines around their eyes.	antimony, 51
E	The mineral "stibnite" is made of this element combined with oxygen.	antimony, 51
E	No one knows what the name of this element means. Some say "monk-killer." Others say, "Not alone."	antimony, 51
E	The symbol for this element comes from the Latin word "stibium."	antimony, 51
M	Naples yellow is a paint pigment that comes from this element.	antimony, 51
M	An oxide compound of this element is used in flame retardant fabrics for clothing and car seats.	antimony, 51
M	Some infrared detectors contain this element, and indium.	antimony, 51
H	Compounds of this element are used to treat some skin diseases in cattle.	antimony, 51
H	This element is sometimes used along with tin and lead to make solder.	antimony, 51
H	This element can be found in the tips of some matches, and is also used as a catalyst in making PETE plastic.	antimony, 51

E	This element's name is the Latin word for "earth."	tellurium, 52
E	This element is chemically similar to sulfur and can be used to vulcanize rubber.	tellurium, 52
E	An oxide compound of this element is used to make DVD and Blu-ray discs.	tellurium, 52
E	This element can replace selenium in various processes, because it lies right below it on the table.	tellurium, 52
M	A compound made of Cd, Hg, and this element is used in night vision goggles to detect infrared.	tellurium, 52
M	Only one type of bacteria (diphtheria) can grow on a gel made with this element.	tellurium, 52
M	This element was discovered in a gold mine in Transylvania, Romania; it was called "metallum problematicum."	tellurium, 52
H	Cadmium mixed with this element makes a compound used in highly efficient solar panels.	tellurium, 52
H	A compound made of Cd, Hg, and this element is used to make infrared detectors for satellites.	tellurium, 52
H	This element is added to BaO_2 to make "delay powder" for blasting caps used in dynamite detonators.	tellurium, 52

E	This element's name comes from the Latin word for purple.	iodine, 53
E	This element is needed by the thyroid to make thyroxine molecules.	iodine. 53
E	A compound of this element and potassium is used to sterilize water.	iodine, 53
E	A compound of silver and this element is used to "seed" clouds to make them rain.	iodine, 53
M	This element is often added to table salt to help prevent deficiency disease.	iodine, 53
M	Lugol's solution contains this element and is used in lab experiments that test for starch.	iodine, 53
M	This element penetrates into bacteria and kills them by destroying the amino acids they are made of.	iodine, 53
H	A radioactive isotope of this element is used to treat some thyroid diseases.	iodine, 53
H	In 1908 this element began to be used in operating rooms to sterilize equipment surgical sites.	iodine, 53
H	This element was discovered in 1811, using seaweed ash and sulfuric acid.	iodine, 53

E	This element is the largest noble gas that is not radioactive.	xenon, 54
E	The name of this element comes from the Greek word for "strange."	xenon, 54
E	This element was discovered in 1898, during the series of experiments that discovered neon and krypton.	xenon, 54
E	This gaseous element is used in very bright arc lamp bulbs and in military tactical flashlights.	xenon, 54
M	This element has 54 electrons, all of which are part of full shells that don't want to lose or gain electrons.	xenon, 54
M	This inert, gaseous element has been used in scuba tanks, as an anesthetic, and for lung imaging.	xenon, 54
M	Along with krypton, this element has been used as a propellant in the propulsion units of satellites.	xenon, 54
H	This is the most rare gaseous element in the earth's atmosphere.	xenon, 54
H	This gaseous element is sometimes breathed by athletes to stimulate production of red blood cells.	xenon, 54
H	This rare, gaseous element is used in eximer lasers that are used for eye surgery.	xenon, 54

E	This element's name means "sky blue" in Latin.	cesium, 55
E	This element is used to make extremely accurate clocks.	cesium, 55
E	One of the first uses for this element was as a "getter" in vacuum tubes, to catch unwanted atoms.	cesium, 55
E	This element was named after the light blue lines in its emission spectrum.	cesium, 55
M	This was the first element discovered by Bunsen and Kirchhoff, who invented the spectrometer.	cesium, 55
M	GPS satellites keep time using very small but highly accurate clocks based on this element.	cesium, 55
M	This element is very reactive (more than K or Rb) and will "burn" while floating on the surface of water.	cesium, 55
H	A second is defined as "the time it takes for an atom of this element to oscillate 9,192,631,770 times."	cesium, 55
H	The largest use of this element used to be in vacuum tubes, but now it is in the oil drilling industry.	cesium, 55
H	Radioactive isotopes of this element can be used to track contamination from nuclear disaster sites.	cesium, 55

E	A nitrate compound of this element burns bright green in fireworks.	barium, 56
E	A mineral called "the desert rose" is made of this element combined with SO_4.	barium, 56
E	A sulfate compound of this element is swallowed by patients who need x-rays of their intestines.	barium, 56
E	The name of this element comes from the Greek word for "heavy."	barium, 56
M	A type of rock containing this element was discovered in Italy. This rock can glow for months or years.	barium, 56
M	This element shares similar chemical properties with beryllium, calcium and magnesium.	barium, 56
M	This element is very useful in electronics, especially capacitors, when combined with TiO_3.	barium, 56
H	Though this element is generally non-toxic, it has been combined with CO_3 to make rat poison.	barium, 56
H	Clay that is high in a sulfate compound of this element is used to make "Jasperware."	barium, 56
H	YBCO, made of yttrium, copper, oxygen and this element, is being studied for its superconducting ability.	barium, 56

E	Only actinium and this element have a series named after them.	lanthanum, 57
E	The name of this element comes from the Greek word for "hidden."	lanthanum, 57
E	The first commercial use of this element was in 1886 when its oxide was used to make lantern mantles.	lanthanum, 57
E	This element, along with cerium, is one of the key ingredients in the "strikers" used by welders.	lanthanum, 57
M	Because this element can prevent algae from using phosphorus, it is found in algaecides for pools.	lanthanum, 57
M	This rare earth element is used by bacteria that live around volcanic fumaroles.	lanthanum, 57
M	Hybrid cars contain as much as 30 pounds of this rare earth element in their rechargeable batteries.	lanthanum, 57
H	Clay that contains a lot of this element can be put into lakes and streams to control growth of algae.	lanthanum, 57
H	This rare earth is alloyed with tungsten to make welding electrodes. Cerium is also used this way.	lanthanum, 57
H	This rare earth element is added to glass to make telescopic lenses for cameras.	lanthanum, 57

E	This element was named after an asteroid (but not asteroid Pallas).	cerium, 58
E	This is the second element in the rare earth series, but was the first one discovered.	cerium, 58
E	This element, along with lanthanum, is one of the primary ingredients in “strikers” used by welders.	cerium, 58
E	This element is used in making the liners inside self-cleaning ovens.	cerium, 58
M	An oxide compound of this element is used to polish glass.	cerium, 58
M	This rare earth is similar to the platinum group elements in that it can be used in catalytic converters.	cerium, 58
M	In the Periodic Table, this element sits on top of thorium, which is in the actinide series.	cerium, 58
H	This second element in the rare earth series is used for electrodes in bright “carbon arc” light bulbs.	cerium, 58
H	An oxide of this element, along with thorium oxide, was used to make lantern light brighter and whiter.	cerium, 58
H	This element was added to glass used in old-fashioned TV screens, to prevent the glass from darkening.	cerium, 58

E	This element has 59 protons.	praseodymium, 59
E	The name of this element comes from the Greek words for “green” and “twin.”	praseodymium, 59
E	This element is used to make purple glasses for welders. The glasses absorb bright yellow light.	praseodymium, 59
E	This third element in the lanthanide series is used in alloys for making airplane engines.	praseodymium, 59
M	This element was discovered at the same time as neodymium, by Carl von Welsbach.	praseodymium, 59
M	This element is strongly magnetic, though its neighbor, neodymium, is even more magnetic.	praseodymium, 59
M	Like its neighbor, cerium, this element can be extracted from monazite sand.	praseodymium, 59
H	Despite the fact that this metal will turn a solution of HCl green, it will color glass yellow.	praseodymium, 59
H	This element, along with Ce, Nd, and La, is often used in the “flints” found in lighters and strikers	praseodymium, 59
H	Like many other rare earths, this element can be used in arc bulbs used in movie projectors.	praseodymium, 59

E	This element is used to make very small magnets for headphones and microphones.	neodymium, 60
E	The name of this element means "new twin."	neodymium, 60
E	This element was discovered at the same time as praseodymium.	neodymium, 60
E	Magnets made from this element are used in workshops to hold heavy tools on the wall.	neodymium, 60
M	This element is added to glass to color it blue, green, pink or purple.	neodymium, 60
M	Magnets of this element are used to make the pickups on electric guitars.	neodymium, 60
M	This can be used instead of praseodymium to make "flints" for lighters and fire starters.	neodymium, 60
H	Glass tinted with this element will show different colors under different types of light.	neodymium, 60
H	This element in YAG lasers is used to make "optical tweezers" to hold microscopic things.	neodymium, 60
H	You can find this element in purple glasses that protect welders from bright yellow light.	neodymium, 60

E	This element was named after the Greek god who is said to have taught humans about fire.	promethium, 61
E	The only radioactive element with a lower atomic number than this element is technetium.	promethium, 61
E	Paint made with this radioactive element was used to mark the buttons on the Apollo lunar rovers.	promethium, 61
E	This radioactive element replaced radium to make glowing numbers on clocks and watches.	promethium, 61
M	The most stable isotope of this radioactive rare earth element has a half-life of 17 years.	promethium, 61
M	This radioactive member of the lanthanide series sits next to neodymium.	promethium, 61
M	An isotope of this element with atomic weight 147 is used to make atomic batteries that last 5 years.	promethium, 61
H	This radioactive lanthanide element emits beta radiation (electrons) that can make ZnS paint glow.	promethium, 61
H	The first name proposed for this element was "clintonium," but a researcher's wife had a better idea.	promethium, 61
H	This element was discovered as the result of research on uranium in what is now Oak Ridge Lab.	promethium, 61

E	This element was found in a mineral ore called Samarskite.	samarium, 62
E	An alloy of cobalt and this element has been used for small magnets in the motors of solar airplanes.	samarium, 62
E	This element was named after the chairman of the Russian Corps of Mining Engineers.	samarium, 62
E	This element has 62 protons and 88 neutrons.	smaarium, 62
M	This was the first element to be named after a person. It was discovered in the Ural Mountains in 1879.	samarium, 62
M	Like Nd, this rare earth can be used (often with Co) to make noiseless pickups for electric guitars.	samarium, 62
M	This rare earth element is used as a catalyst in chemical reactions, notably for breaking down PCBs.	samarium, 62
H	This rare earth was discovered by the same French chemist who discovered gallium and dysprosium.	samarium, 62
H	Along with Ho, Dy, this rare earth can be used in control rods to slow down neutrons in nuclear reactors.	samarium, 62
H	Magnets made of cobalt and this rare earth make motors for machines that generate a lot of heat.	samarium, 62

E	This rare earth element was named after a continent.	europium, 63
E	This element is used to make designs on Euro bank notes that only show up under UV light.	europium, 63
E	When this element is added to other compounds it can make them fluoresce red, green or blue.	europium, 63
E	The bright light of compact (coiled) fluorescent bulbs is caused by powders of terbium and this element.	europium, 63
M	This rare earth can be added to ZnS or $ASrAl_2O_4$ to make powders that glow in various colors in UV light.	europium, 63
M	Old CRT TVs used yttrium orthovanadate crystals doped with this element to create bright red light.	europium, 63
M	One ore of this element is Bastnäsite, a reddish crystal that also contains a lot of cerium.	europium, 63
H	Boisbaudran was looking at samarium with a spectrometer when he saw this element's spectral lines.	europium, 63
H	Crystals of calcium fluoride that contain trace amounts of this element will glow blue under UV light.	europium, 63
H	This rare earth is almost as reactive as lithium. It turns water yellow and creates hydrogen bubbles.	europium, 63

E	This element has 64 protons. It usually has 93 neutrons, though not always.	gadolinium, 64
E	This element was named after a mineral ore that was named after a Finnish chemist.	gadolinium, 64
E	This element is combined with gallium to make imitation diamonds called GG garnets (or GGGs).	gadolinium, 64
E	The radioactive "153" isotope of this element is used in portable x-ray units used by vets and zoos.	gadolinium, 64
M	Submarines powered by nuclear reactors can use this rare earth element to slow the reaction.	gadolinium, 64
M	Bone density scanners can use gamma rays made by the radioactive "153" isotope of this element.	gadolinium, 64
M	This rare earth element was used in old CRT televisions to make green light.	gadolinium, 64
H	Abnormal tissue will absorb this element in DOTA, making the tissues show up on an MRI scan.	gadolinium, 64
H	The magnetic properties of this element change at 20° C (68° F). Above this, it loses regular magnetism.	gadolinium, 64
H	This neighbor of terbium is the green component of white light produced by fluorescent bulbs.	gadolinium, 64

E	This element was named after the Swedish town of Ytterby and does not start with Y or E.	terbium, 65
E	This element is the "Ter" in Terfenol-D®, a compound that will respond to magnetic fields around it.	terbium, 65
E	Fe, Dy, and this element are used to make the SoundBug® device that turns a window into a loudspeaker.	terbium, 65
E	When this element weighs 159, it has 94 neutrons.	terbium, 65
M	This element was discovered at almost the same time as erbium and their names got mixed up.	terbium, 65
M	Like Gd, this element glows green and is combined in light bulbs with europium which makes red and blue.	terbium, 65
M	This neighbor of gadolinium can be doped into laser crystals to make them glow green.	terbium, 65
H	This element can be used by biologists to make bacterial endospores glow green.	terbium, 65
H	Some air pollutants will cause this rare earth to glow green, so it can be used in air pollution gauges.	terbium, 65
H	This element is combined with iron and dysprosium to make a compound that reacts to magnetism.	terbium, 65

E	This element is the "D" in Terfenol-D®, a metal that changes shape in response to magnetic fields.	dysprosium, 66
E	This is the only element whose symbol contains the letter Y but does not start with Y.	dysprosium, 66
E	Like its neighbor, holmium, this element can be used in control rods in nuclear reactors.	dysprosium, 66
E	Fe, Tb, and this element are used to make the SoundBug® device that turns a window into a loudspeaker.	dysprosium, 66
M	This element is used to make strong magnets for huge wind turbine generators.	dysprosium, 66
M	The name of this element means "hard to get" because it took 30 tries for Boisbaudran to discover it.	dysprosium, 66
M	This element sits on top of californium in the Periodic Table.	dysprosium, 66
H	This rare earth element can be found in electric cars, computers, dosimeters and wind turbines.	dysprosium, 66
H	Fe, Tb and this element are used to make actuators that are found in sonars in submarines.	dysprosium, 66
H	An iodide compound of this rare earth element can be used in high intensity light bulbs.	dysprosium, 66

E	This element is named after the birthplace of its discoverer, Stockholm, Sweden.	holmium, 67
E	This element sits on top of einsteinium in the Periodic Table.	holmium, 67
E	This neighbor to dysprosium can color glass either yellow or red.	holmium, 67
E	This element has 67 protons and often has 98 neutrons, although the number can vary.	holmium, 67
M	Like neodymium, samarium, and dysprosium, this rare earth is also used in small, strong magnets.	holmium, 67
M	This element is doped into yttrium garnet crystals used in medical lasers used to remove kidney stones.	holmium, 67
M	Like its neighbor, erbium, this element can add a pink color to cubic zirconia crystals.	holmium, 67
H	An oxide compound of this element looks yellow in daylight and pink under fluorescent light.	holmium, 67
H	Like samarium and dysprosium, this rare earth element is used in control rods in nuclear reactors.	holmium, 67
H	Glass with this element in it is used to calibrate spectrophotometers that use light to analyze samples.	holmium, 67

E	This symbol for this element, named after the Swedish town of Ytterby, starts with a vowel.	erbium, 68
E	Like einsteinium and europium, this element's symbol starts with the letter E.	erbium, 68
E	This element is used in amplifiers in fiber optic cables that lie on the bottom of the ocean.	erbium, 68
E	This element is widely used to color glass pink. Inexpensive glass jewels often contain this element.	erbium, 68
M	Like its neighbor, holmium, this element can add a pink color to cubic zirconia crystals.	erbium, 68
M	This element was discovered at almost the same time as terbium, and their names got mixed up.	erbium, 68
M	Like Sm, Ho, Dy, this rare earth element can be used in control rods in nuclear reactors.	erbium, 68
H	This element is put into lenses that are used in glasses that protect patients' eyes during laser surgery.	erbium, 68
H	Scientists are experimenting with carbon bucky balls that have this element (and a nitrogen) inside it.	erbium, 68
H	This element is doped into YAG crystals used in lasers used for dental surgery and to remove scars.	erbium, 68

E	The name of this element comes from an ancient word for Scandinavia.	thulium, 69
E	This element has 69 protons. When it has 100 neutrons it is stable. With 101 neutrons, it is radioactive.	thulium, 69
E	This neighbor to erbium and ytterbium is used in high intensity arc light bulbs.	thulium, 69
E	Along with Sm and Pm, this one of 3 rare earths whose symbols are the first and last letters of their name.	thulium, 69
M	This element is used in Euro bank notes (counterfeit prevention) because it glows blue under UV light.	thulium, 69
M	This element sits on top of mendelevium in the Periodic Table.	thulium, 69
M	The "170" radioactive isotope of this rare earth element is used in portable x-ray machines.	thulium, 69
H	This element was discovered in 1879 by Per Theodor Cleve, at the same time he discovered holmium.	thulium, 69
H	Crystals doped with Ho, Cr, and this element are found in lasers used for medical procedures on skin.	thulium, 69
H	Like dysprosium, this rare earth element is doped into crystals used in dosimeters that measure radiation.	thulium, 69

E	This is one of two elements whose symbols begin with the letter Y. The other element is yttrium.	ytterbium, 70
E	This element has 70 protons and 103 neutrons in its nucleus, and is stable (not radioactive).	ytterbium, 70
E	This element sits on top of nobelium in the Periodic Table.	ytterbium, 70
E	The name of this element has four syllables and comes from the name of a small Swedish town.	ytterbium, 70
M	YAG lasers doped with this element are used to clean old paintings and artifacts because of their precision.	ytterbium, 70
M	This neighbor of thulium is completely stable, but if neutrons are added it will become radioactive.	ytterbium, 70
M	This rare earth element can be used in decoy flares that military jets use to throw off heat-seeking missiles.	ytterbium, 70
H	This element is doped into YAG crystals used for industrial lasers (welding, cutting, engraving, cleaning).	ytterbium, 70
H	A new type of atomic clock uses atoms of this element suspended in a laser beam.	ytterbium, 70
H	This rare earth metal can be combined with fluoride to make fillings for teeth.	ytterbium, 70

E	The name of this element comes from an ancient word for the city of Paris.	lutetium, 71
E	$LuTaO_4$, a host material for phosphors that glow under x-rays, is made of tantalum, oxygen and this element.	lutetium, 71
E	LuAG crystals are made of aluminum, oxygen and this element. They are used in lasers and sensors.	lutetium, 71
E	This element sits on top of lawrencium in the Periodic Table.	lutetium, 71
M	This element has 71 protons and 71 electrons.	lutetium, 71
M	This element, named after Paris, is used in alloys that function as catalysts for "cracking" petroleum.	lutetium, 71
M	The discovery of this element is closely linked to the discovery of its neighbor, ytterbium.	lutetium, 71
H	This neighbor of hafnium was almost named cassiopeium, after the constellation Cassiopeia.	lutetium, 71
H	This element is often the last in the long line of lanthanides at the bottom of Periodic Tables.	lutetium, 71
H	If you alphabetized the symbols starting with the letter L, this element would come just before livermorium.	lutetium, 71

E	This element's name comes from an old word for Copenhagen, Denmark.	hafnium, 72
E	This element has similar chemical properties to Zr and Ti since it is underneath them on the table.	hafnium, 72
E	This element's atomic number is divisible by 2, 3, 4, 6, 8, 9, 12, 18, and 24.	hafnium, 72
E	This element is usually found inside zircon crystals because it can take zirconium's place in the crystal.	hafnium, 72
M	Both this element and the one below it on the table have symbols ending with the letter "f."	hafnium, 72
M	This was the last non-radioactive element to be discovered. The year was 1923, the site was Copenhagen.	hafnium, 72
M	This element, along with Nb, Ti, and Ta, was used to make the nozzles for the Apollo lunar module.	hafnium, 72
H	In 2007, this element was used to reduce the size of microprocessor gates from 90 nm to 45 nm.	hafnium, 72
H	This element is found in "heavy metal sands" with W, Ti, Th, and Fe, in Brazil, Malawi and Australia.	hafnium, 72
H	This element can take a lot of heat and is used to make electrodes (tips) for plasma cutters.	hafnium, 72

E	This element was named after a character in a Greek myth who was tantalized by things he couldn't reach.	tantalum, 73
E	This element sits below niobium in the Periodic Table, so it shares similar chemical properties.	tantalum, 73
E	The atomic # of this element is a prime number greater than 40 and its two digits add up to 10.	tantalum, 73
E	This element was used as filaments in light bulbs until its neighbor, tungsten, was found to be better.	tantalum, 73
M	This element is heat-resistant and is often found in alloys with titanium and with its neighbor, tungsten.	tantalum, 73
M	The largest use of this rather obscure (less-known) metal is in small capacitors used in electronics.	tantalum, 73
M	The Congo, in Africa, has been the site of wars over the rights to extraction of this lesser-known element.	tantalum, 73
H	This element is used to make pipes for harsh chemicals and EFPs (explosively formed penetrators).	tantalum, 73
H	Like Rh, Pt and Nb, this element can be electroplated onto metal items to make them extra shiny.	tantalum, 73
H	Kazakhstan minted a coin of silver and this element, showing the Apollo-Soyuz mission of 1975.	tantalum, 73

E	The name of this element means "heavy stone" in Swedish.	tungsten, 74
E	This element was extracted from a mineral called wolframite, the word from which its symbol is derived.	tungsten, 74
E	This element is famous for its use as the bright filament inside incandescent light bulbs.	tungsten, 74
E	This element is so heavy that it can often replace lead in many applications.	tungsten, 74
M	This element is used in a carbide compound used for tough applications like drill bits, tools, rockets.	tungsten, 74
M	This element has a weight similar to gold, so counterfeiters have filled gold bars with this cheaper element.	tungsten, 74
M	This element has a very high melting point and is alloyed with Ti, Mo, Nb, and Ta to make rocket nozzles.	tungsten, 74
H	This element is chemically similar to molybdenum; both are used to remove sulfur from petroleum.	tungsten, 74
H	Like molybdenum, above it on the table, a sulfide of this element can be used as a dry lubricant.	tungsten, 74
H	This element is well known for its use in gas arc welding. It is used to make the electrodes.	tungsten, 74

E	This element is named after the Rhine Valley region of Europe.	rhenium, 75
E	This element has 75 protons.	rhenium, 75
E	This element is often combined with its neighbor, tungsten, to make an extremely useful metal alloy.	rhenium, 75
E	Though the element directly above it on the table is radioactive, this element is not.	rhenium, 75
M	On the Periodic Table, this element sits on top of the element named after physicist Niels Bohr.	rhenium, 75
M	An alloy of this element and Pt is used for "cracking" petroleum to make gas, oil, wax, tar, etc.	rhenium, 75
M	Photographic flash units might contain wires made of a pure form of this element, a neighbor of Os.	rhenium, 75
H	The alloy of this metal and tungsten can be used to make wires that will generate x-rays.	rhenium, 75
H	Chile and Peru are the leading producers of this element, though it is named after a region in Germany.	rhenium, 75
H	This is the least abundant solid element that is not radioactive. It is not a rare earth.	rhenium, 75

E	The name of this element comes from the Greek word for "smell."	osmium, 76
E	This element is a member of the platinum group which also includes Ru, Rh, Pd, Ir and Pt.	osmium, 76
E	This element was used to make pen points for fountain pens during the 1900s. (It is not Ru or Ir.)	osmium, 76
E	This element was used to make needles for phonographs during the 1940s and 1950s.	osmium, 76
M	The tetroxide compound of this element is used to stain cells to highlight areas high in lipids (fats).	osmium, 76
M	This element is often found in ores with its neighbor, iridium, and are hard to separate.	osmium, 76
M	Like the element directly above it, ruthenium, this element can be used to expose oily fingerprints.	osmium, 76
H	This element played a key role in the discovery of right-handed DNA. (Most DNA is left-handed.)	osmium, 76
H	This element holds the record for being most dense, but it is brittle and hard to melt, so not as useful as Pb.	osmium, 76
H	Carl von Welsbach tried to use this element for filaments in light bulbs, but soon switched to tungsten.	osmium, 76

E	This element is named after the Greek goddess of the rainbow.	iridium, 77
E	This member of the platinum group comes right before platinum on the Periodic Table.	iridium, 77
E	The Chandra X-ray Observatory satellite has mirrors coated with this plantinum group element.	iridium, 77
E	This neighbor of osmium is used to make crucibles that can withstand more heat than most elements.	iridium, 77
M	This element and einsteinium are the only elements whose names have three "i"s.	iridium, 77
M	This member of the Pt group can be found in airplane spark plugs, satellites, and fountain pens.	iridium, 77
M	This element is famous for being more abundant in certain geological layers than in others.	iridium, 77
H	Containers made of this element hold the plutonium that provides power to the Voyager 2 satellite.	iridium, 77
H	This Pt group element can be used as a thin coat over specimens that will be scanned using SEM.	iridium, 77
H	Usually, this element has 116 neutrons. An isotope with 115 neutrons is used to make gamma rays.	iridium, 77

E	The name of this element comes from the Spanish word for "little silver."	platinum, 78
E	In the platinum group of elements, this element has the highest atomic number.	platinum, 78
E	About 1/3 of the world's supply of this precious metal goes into making jewelry. It's not Ag or Au.	platinum, 78
E	The frame of the crown made for Queen Elizabeth's mother is made of this precious heavy metal.	platinum, 78
M	Almost half the world's supply of this member of the platinum group is used in catalytic converters.	platinum, 78
M	This element can be used for many of the same purposes as the element right above it, palladium.	platinum, 78
M	This precious metal is harder than silver or gold and is put into alloys used for tools and cutting devices.	platinum, 78
H	This element is a key ingredient in Cisplatin, a medicine for treating cancer.	platinum, 78
H	This element was discovered in ores from Peru by a Spaniard who went on to be governor of Louisiana.	platinum, 78
H	The symbol for this element used in the 1700s was the moon symbol connected to the sun symbol.	platinum, 78

E	The symbol for this element comes from the Latin word "aurum."	gold, 79
E	Spanish soldiers came to Central and South America in the 1500s searching for this element.	gold, 79
E	This element is made into bars called ingots and often stored in warehouses guarded by soldiers.	gold, 79
E	This is one of the elements found in its pure form in nature, often as veins in quartz or silver deposits.	gold, 79
M	This element can be drawn into microscopically thin wires useful for making tiny circuit boards.	gold, 79
M	The largest nugget of this element ever found was in Australia and weighed 78 kg (173 lbs).	gold, 79
M	A thin layer of this element was sprayed on the inside of the glass of the helmets of the Apollo astronauts.	gold, 79
H	This element is not toxic and can be rolled into extremely thin sheets used to decorate fancy desserts.	gold, 79
H	In past centuries, this element was rubbed (not painted) onto fancy manuscripts.	gold, 79
H	When this element is alloyed with silver, the product is called "electrum."	gold, 79

E	This element was used in thermometers from the 1700s until the late 1900s.	mercury, 80
E	The name of this element comes from the Roman messenger god who had wings on his feet.	mercury, 80
E	In the 1800s, this toxic element was used to make felt hats, causing brain injuries to the hat makers.	mercury, 80
E	The symbol of this element comes from the Latin word "hydragyrum," meaning "water silver."	mercury, 80
M	The most well-known ore of this element is cinnabar, a compound of sulfur and this element.	mercury, 80
M	The first Chinese emperor mistakenly thought this toxic element was good medicine and took it daily.	mercury, 80
M	In past decades, this toxic element was put into "amalgams" used to fill cavities in teeth.	mercury, 80
H	Pools of this liquid element were once used in some lighthouses to float the heavy Fresnel lenses.	mercury, 80
H	In 1643 it was demonstrated that a tall tube filled with this element could be used as a barometer.	mercury, 80
H	An oxide compound of this element was used by Joseph Priestly to discover the element oxygen.	mercury, 80

E	The name of this element comes from a Greek word that means "green twig."	thallium, 81
E	This highly toxic element has 81 protons.	thallium, 81
E	This element sits underneath boron, aluminum, gallium and indium on the Periodic Table.	thallium, 81
E	This element's spectral emission pattern is unusual: it is mainly just one bright green line.	thallium, 81
M	A radioactive isotope of this element, weighing 201, is used to make images of the heart.	thallium, 81
M	A sulfate compound of this element was used as rat and ant poison until it was banned in the 1970s.	thallium, 81
M	This element is used to make photoresistors that allow outdoor lights to turn on automatically.	thallium, 81
H	The cure for poisoning from this element is to eat a pigment called Prussian blue.	thallium, 81
H	This toxic element, in a bromide/iodide compound, is used to make lenses used with infrared light.	thallium, 81
H	This element is used in solutions to help speed up the gold electroplating process.	thallium, 81

E	The symbol for this element comes from the Latin word "plumbum."	lead, 82
E	This element is so dense that a small amount is heavy enough to make fish hooks sink.	lead, 82
E	This element was often combined with tin to make toy tin soldiers.	lead, 82
E	For centuries, this element was used to hold the pieces of colored glass in stained glass windows.	lead, 82
M	A common ore for this element is galena, a mixture of this element with sulfur. Galena is shiny and heavy.	lead, 82
M	This element is combined with tin to make very tall bass pipes (low notes) for pipe organs.	lead, 82
M	Centuries ago, women used compounds of this element to paint their faces white.	lead, 82
H	This element is used to make protective gear for x-ray technicians, to protect them from the x-rays.	lead, 82
H	This toxic element makes excellent soldering wire. Less toxic replacements don't work quite as well.	lead, 82
H	This toxic heavy metal is used in large "acid" batteries, the kind used in cars and boats.	lead, 82

E	This element is the "bis" in Pepto-Bismol® stomach medicine.	bismuth, 83
E	This element is dense (heavy) enough to replace lead in fishing sinkers.	bismuth, 83
E	The atomic number of this element is the next prime number after gold's (prime) atomic number.	bismuth, 83
E	This element is less toxic than lead, so it has been used to make "shot" (bullets) for duck hunters.	bismuth, 83
M	This element is sometimes put into fingernail polish to give it a pearly, iridescent look.	bismuth, 83
M	This element is similar enough to lead that it can replace lead in protective x-ray shields.	bismuth, 83
M	This neighbor of lead is used in alloys found in the fuses of sprinkler systems.	bismuth, 83
H	Years ago, miners in Germany thought this element was silver that was still in the process of being made.	bismuth, 83
H	A pure form of this element is the substance that very square "hopper crystals" are made of.	bismuth, 83
H	A vanadate compound of this non-toxic heavy metal is used to make yellow pigments for paint.	bismuth, 83

E	This element was named after the country where Marie Curie was born.	polonium, 84
E	This element was discovered by Marie and Pierre Curie five months before they discovered radium.	polonium, 84
E	This radioactive element is used in anti-static brushes that remove electrical charges from surfaces.	polonium, 84
E	When radon gas decays (gives off an alpha particle) it turns into this element.	polonium, 84
M	This element is sometimes made in labs, by making an extra proton stick to a bismuth nucleus.	polonium, 84
M	On the Periodic Table, this element is underneath oxygen, sulfur, selenium and tellurium.	polonium, 84
M	When this radioactive element loses an alpha particle it turns into lead, which is not radioactive.	polonium, 84
H	This element was the first to be discovered by boiling down pitchblende, a mineral ore of uranium.	polonium, 84
H	This radioactive element can be found on tobacco leaves, the result of the decay of radon atoms.	polonium, 84
H	This element was combined with BeO to make the detonator for the plutonium in nuclear bombs.	polonium, 84

E	The name of this element comes from the Greek word "astatos" meaning "unstable."	astatine, 85
E	This element always has 85 protons.	astatine, 85
E	When this radioactive element loses an alpha particle, it turns into bismuth.	astatine, 85
M	On the Periodic Table, this element is in the halogen column, under chlorine, bromine, and iodine.	astatine, 85
M	This element was created in a cyclotron by adding alpha particles to bismuth atoms.	astatine, 85
H	If you bombarded this element with alpha particles and one stuck to its nucleus it would turn into francium.	astatine, 85
H	This is the most rare naturally occurring element on earth, a microscopic speck of it per continent.	astatine, 85

E	This element is the largest noble gas and sits under xenon and krypton in the Periodic Table.	radon, 86
E	This element is unstable; in just a few days it will eject an alpha particle and turn into polonium.	radon, 86
E	This radioactive gas is a decay product of uranium. It comes up out of the ground into soil and water.	radon, 86
M	This radioactive gas causes problems when it gets trapped in mines and in basements of houses.	radon, 86
M	This radioactive gas gets into groundwater and can give scientists clues about the rocks underneath.	radon, 86
H	This element's name comes from the fact that it was observed in and around samples of radium.	radon, 86
H	This radioactive gas is found more abundantly in the Appalachians, the Rockies and the northern midwest.	radon, 86

E	This element was discovered by Marguerite Perey and named after her native country.	francium, 87
E	This radioactive element always has 87 protons.	francium, 87
E	This radioactive element can lose an alpha particle and turn into astatine.	francium, 87
M	This element can be made in a particle accelerator by smashing oxygen atoms into gold atoms.	francium, 87
M	This discoverer of this element was a student of Marie Curie and also worked under her daughter, Irène.	francium, 87
H	This element was predicted by Mendeleyev. He said it would be underneath cesium in his table.	francium, 87
H	This radioactive element can have one of its neutrons turn into a proton, turning it into radium.	francium, 87

E	Before this element was known to be dangerous, it was used to paint numbers on watches and clocks.	radium, 88
E	This element seemed magical as it glowed in the dark and it was used everywhere in the early 1900s.	radium, 88
E	This element was discovered by Marie Curie, five months after she discovered polonium.	radium, 88
M	This radioactive element decays by emitting an alpha particle, which turns it into radon.	radium, 88
M	This dangerous radioactive element was sold in food and health products in the early 1900s.	radium, 88
H	This element will decay into radon, then into polonium, and then finally into lead, which is stable.	radium, 88
H	This element was found to be too toxic to be used in paint, so promethium and tritium were used instead.	radium, 88

E	This element has an entire series named after it. It is not a rare earth element.	actinium, 89
E	When this element loses an alpha particle, it turns into francium. The francium will then decay.	actinium, 89
E	One of its isotopes is called the "Goldilocks" isotope because its half-life and decay pattern are useful.	actinium, 89
M	Atoms of this radioactive element can be stuck to antibody molecules that attach to specific cells.	actinium, 89
M	This element is shipped to medical labs in V-shaped vials. The atoms will be at the bottom of the V.	actinium, 89
H	One isotope of this element generates free neutrons that are used to detect underground water sources.	actinium, 89
H	This element was discovered in the waste residues left by the Curies after discovering Po and Ra.	actinium, 89

E	This element was named after the Norse god of thunder because its mineral ore had come from Norway.	thorium, 90
E	This element was used to make mantles for gas lanterns but has been replaced by non-radioactive elements.	thorium, 90
E	This element always has 90 protons, but the number of neutrons can vary.	thorium, 90
M	If uranium loses an alpha particle it becomes one of the less stable isotopes of this element.	thorium, 90
M	This element can be used to make isotopes of U that will not decay into Pu, making safer nuclear energy.	thorium, 90
H	This mildly radioactive element has been used to make electrodes for welding machines.	thorium, 90
H	Although this element is not a rare earth, it acts much like them and is often found with them in ores.	thorium, 90

E	Mendeleyev predicted the discovery of this element and said it would be between thorium and uranium.	protactinium, 91
E	Lise Meitner and Otto Hahn named this element using a word meaning "before actinium."	protactinium, 91
E	This element always has 91 protons, but the number of neutrons can vary.	protactinium, 91
M	Thorium and this element are collected by devices floating in the ocean, to trap and test sediments.	protactinium, 91
M	This element, a decay product of actinium, is found in Torbernite mineral ores.	protactinium, 91
H	The 234 isotope of this element is safe enough to use in a classroom and is used to demo half-life.	protactinium, 91
H	The 231 and 233 isotopes of this element are produced by uranium fission and difficult to deal with.	protactinium, 91

E	Pitchblende is an oxide compound of this radioactive element.	uranium, 92
E	This was the first element to be named after a planet (other than the earth).	uranium, 92
E	This radioactive element is used to make yellow pigments for glass and ceramics.	uranium, 92
M	This element splits apart easily (fissions) to produce heat that can be used in turbines that make electricity.	uranium, 92
M	This ability of this element to split apart was a key part of the making of the first atomic bombs.	uranium, 92
H	This abundant and well-known radioactive element can be used to stain samples for electron microscopy.	uranium, 92
H	Citrobacter is a species of bacteria that absorbs and uses this well-known radioactive element.	uranium, 92

E	This element is in the middle of the series that was named after the outer planets.	neptunium, 93
E	This element always has 93 protons.	neptunium, 93
E	This radioactive element emits an alpha particle and turns into protactinium.	neptunium, 93
M	This element was discovered in 1940 in the cyclotron in the Berkeley lab, when a proton stuck to a U atom.	neptunium, 93
M	This element plays a role in space exploration only because it decays into Pu-238, a power source.	neptunium, 93
H	Americium will decay by losing an alpha particle and turn into this element.	neptunium, 93
H	When this radioactive element is dissolved into solutions it turns them yellow, green or blue.	neptunium, 93

E	This element is the third in the series that was named after the outer planets.	plutonium, 94
E	When this radioactive element decays by losing an alpha particle it turns into uranium.	plutonium, 94
E	The 238 isotope of this element is very useful as a power source in satellites and Mars rovers.	plutonium, 94
M	This element was combined with beryllium as a detonator inside uranium-based atomic bombs.	plutonium, 94
M	This element was once used as a power source in pacemakers for the heart.	plutonium, 94
H	This element can be made by adding an alpha particle to the nucleus of a uranium atom.	plutonium, 94
H	The discoverers of this element were worried that the symbol letters would sound too much like a silly word.	plutonium, 94

E	This element was named after the United States of America.	americium, 95
E	This element is used in smoke detectors. Smoke interferes with the detection of its alpha particles.	americium, 95
E	When this element decays and emits an alpha particle, it turns into neptunium.	americium, 95
M	On the Periodic Table, this element sits right below europium.	americium, 95
M	This element is combined with beryllium to generate free neutrons used in devices that detect water.	americium, 95
H	Labs hit this element with neon atoms to make element 105, dubnium.	americium, 95
H	If you can make an alpha particle stick to this element's nucleus, you turn it into berkelium.	americium, 95

E	This element was named in honor of Marie and Pierre Curie for their research on radioactivity.	curium, 96
E	This element always has 96 protons, though the number of neutrons can vary from 146 to 154.	curium, 96
E	This element is below gadolinium on the Periodic Table, suggesting it might have magnetic properties.	curium, 96
M	This element was first made in 1944 by hitting plutonium with alpha particles.	curium, 96
M	If an atom of californium decays and emits an alpha particle, it will turn into this element.	curium, 96
H	The Philae satellite landed on comet 67P and used alpha particles from this element to analyze its chemistry.	curium, 96
H	This transuranic element is used in alpha particle x-ray spectrometers on satellites and Mars rovers.	curium, 96

E	This element was named after the lab in California where elements 93 to 101 were discovered.	berkelium, 97
E	This element was named after a town because it is below terbium, which is also named after a town.	berkelium, 97
E	This element always has 97 protons. If the number increases or decreases it is no longer this element.	berkelium, 97
M	The atomic number of this element is the highest prime number under 100.	berkelium, 97
M	This element goes through beta decay. A neutron becomes a proton and the atom turns into Cf.	berkelium, 97
H	The starting point for making this element was americium oxide, which was hit with alpha particles.	berkelium, 97
H	Researchers used this element to make tennessine. They hit this element with calcium nuclei.	berkelium, 97

E	This element always has 98 protons, but the nucleus can be enriched with extra neutrons.	californium, 98
E	This element was named after the US state in which it was discovered by Glenn Seaborg in 1949.	californium, 98
E	This element is produced by the beta decay of berkelium. (A neutron in Bk turns into a proton.)	californium, 98
M	This element is below dysprosium. Seaborg invented a way to make the names match (sort of).	californium, 98
M	This element is used as a source of neutrons in devices that detect sources of underground water.	californium, 98
H	The neutrons released by this element can be used in sensors that inspect metal structures for cracks.	californium, 98
H	A microgram of this actinide element can produce 139 million free neutrons per minute.	californium, 98

E	This element was named after the physicist famous for the theory of relativity.	einsteinium, 99
E	This element, along with fermium, was collected in the mushroom cloud from an atomic bomb in 1952.	einsteinium, 99
E	The atomic number of this element is the highest two-digit number.	einsteinium, 99
M	If you hit the nucleus of this element with alpha particles you might get some mendelevium.	einsteinium, 99
M	If this element loses an alpha particle, it turns into berkelium.	einsteinium, 99
H	This element is below holmium on the Periodic Table.	einsteinium, 99
H	This element was discovered by the Berkeley lab a few weeks before fermium was discovered.	einsteinium, 99

E	This element was named after an Italian physicist who left Italy just before World War 2.	fermium, 100
E	This element, along with einsteinium, was collected in the mushroom cloud from an atomic bomb in 1952	fermium 100
E	This element always has 100 protons, but the number of neutrons can vary.	fermium, 100
M	If this element loses an alpha particle, it turns into californium.	fermium, 100
M	This element sits below erbium on the Periodic Table.	fermium, 100
H	This element is named after the physicist who made the very first nuclear reactor, the "Chicago pile."	fermium, 100
H	This element was discovered by the Berkeley lab a few weeks after einsteinium was discovered.	fermium, 100

E	This element is named after the scientist who made the first Periodic Table.	mendelevium, 101
E	The atomic number of this element is the smallest prime number greater than 100.	mendelevium, 101
M	Lighter isotopes of this element are made by shooting argon nuclei at a bismuth target.	mendelevium, 101
H	Heavier isotopes of this element are made by shooting alpha particles at an einsteinium target.	mendelevium, 101

E	This element was named after the scientist who invented dynamite.	nobelium, 102
E	The first lab to make a claim about the discovery of this element was the Nobel Institute in Sweden.	nobelium, 102
M	If this element were to emit an alpha particle, it would turn into fermium.	nobelium, 102
H	The JINR claimed to have made this element by hitting a plutonium target with oxygen nuclei.	nobelium, 102

E	This element is named after the inventor of the world's first cyclotron, at Berkeley Lab in California.	lawrencium, 103
E	This element always has 103 protons, although its atoms only exist for a few seconds.	lawrencium, 103
M	This element was made by hitting a californium target with boron nuclei.	lawrencium, 103
H	On many Periodic Tables, this element is the last one in the actinide line, under the rare earth lutetium.	lawrencium, 103

E	This element is named after the scientist who did the famous gold foil experiment.	rutherfordium, 104
E	This element sits under titanium, zirconium and hafnium, so it is chemically similar.	rutherfordium, 104
M	If this element loses an alpha particle, it turns into nobelium.	rutherfordium, 104
H	The namesake of this element showed that atoms are mostly empty space, with a tiny nucleus.	rutherfordium, 104

E	This element is named after the town in which Russia's Joint Institute for Nuclear Research is located.	dubnium, 105
E	This element was first made at the JINR by hitting an americium target with neon nuclei.	dubnium, 105
M	This element sits under Nb and Ta, so if we could collect enough of it, it might also be iridescent.	dubnium, 105
H	If this element were to lose an alpha particle, it would turn into lawrencium.	dubnium, 105

E	This element is named after the director of the Berkeley lab when Am, Bk, Cf and Lr were discovered.	seaborgium, 106
E	This element was the very first to be named after a living person.	seaborgium, 106
M	This element decays by losing an alpha particle and turns into rutherfordium.	seaborgium, 106
H	A hexacarbonyl molecule of this element was made in 2014, a rare example of a super heavy molecule.	seaborgium, 106

E	This element was named after the Danish scientists who figured out the electrons are arranged in shells.	bohrium, 107
E	This element always has 107 protons, though its atoms only exist for a few seconds.	bohrium, 107
M	The namesake of this element gave us the "solar system" model of the atom.	bohrium, 107
H	This atom is at the bottom of the column that contains the radioactive element with the lowest atomic #.	bohrium, 107

E	The element is named after the German state in which the GSI Helmholtz Institute is located.	hassium, 108
E	Germany wanted elements named after its country, one of its states, and a city. This is the state.	hassium, 108
M	This element is at the bottom of the column in which iron is at the top.	hassium, 108
H	If this element loses an alpha particle, it will turn into Sg. Gaining one will turn it into Ds.	hassium, 108

E	This element was named after the woman who worked with Otto Hahn to investigate uranium fission.	meitnerium, 109
E	If this element decays and ejects an alpha particle, it turns into bohrium.	meitnerium, 109
M	This element was created by hitting a bismuth target with fast-moving iron atoms.	meitnerium, 109
H	This is the only element to be named after a woman. Curium was named after both Marie and Pierre.	meitnerium, 109

E	This element is named after the German city where the GSI Helmholtz research center is located.	darmstadtium, 110
E	This element always has 110 protons and anywhere from 151 to 171 neutrons.	darmstadtium, 110
M	This element can be created by hitting a lead target with fast-moving nickel atoms.	darmstadtium, 110
H	The naming of this element settled a score with Berkeley lab which had 3 elements named after it.	darmstadtium, 110

E	This element was named after the scientist who discovered x-rays.	roentgenium, 111
E	The atomic number of this element is three digits and they are all the same number.	roentgenium, 111
M	This element is underneath copper, silver and gold, so is it a radioactive precious metal?	roentgenium, 111
H	This element can be made by hitting a bismuth target with fast-moving nickel atoms.	roentgenium, 111

E	This is the only element named after a scientist who was not a chemist or physicist.	copernicium, 112
E	12 elements start with the same letter as this element. The second letter of its symbol is "n."	copernicium, 112
M	This element is right under Hg on the table. If we could make enough atoms, would it also be a liquid?	copernicium, 112
H	If you combine zinc and lead atoms, you get atoms of this element.	copernicium, 112

E	This element is named after Japan.	nihonium, 113
E	This element has the highest atomic number that is a prime number.	nihonium, 113
M	This element was created by smashing bismuth and zinc atoms together.	nihonium, 113
H	This element was created in the particle accelerator at the RIKEN facility in Japan.	nihonium, 113

E	This element begins with the letter F, but it is not fluorine, iron or fermium.	flerovium, 114
E	This element always has 114 protons, though its atoms only exist for about 2 seconds.	flerovium, 114
M	This element was named after a lab at the JINR. The lab was named after one of its head scientists.	flerovium, 114
H	This element is at the bottom of the column that includes carbon, silicon, germanium, tin and lead.	flerovium, 114

E	This element always has 115 protons, but only exists for half a second.	moscovium, 114
E	This element is at the bottom of the column that contains N, P, As, Sb and Bi.	moscovium, 114
M	This element is named after the city and district in Russia in which the JINR is located.	moscovium, 114
H	If this element ejects an alpha particle, it turns into nihonium, which further decays into roentgenium.	moscovium, 114

E	This element is named after the city in which the Lawrence Livermore Nat'l Lab is located.	livermorium, 116
E	This is the only element whose symbol has "v" as its second letter.	livermorium, 116
M	Oddly enough, the name of this element traces back to an English-Mexican cattle rancher in the 1800s.	livermorium, 116
H	This element was made by hitting a curium target with fast-moving calcium atoms.	livermorium, 116

E	Californium and this element are the only ones named after U.S. states.	tennessine, 117
E	This element was named after the state in which the Oak Ridge National Lab is located.	tennessine, 117
M	This is the penultimate (next-to-last) element on the Periodic Table.	tennessine, 117
H	The berkelium target for this element was made at Oak Ridge lab. JINR hit the target with calcium nuclei.	tennessine, 117

E	This element was named after the Russian scientist who led the international team that discovered it.	oganesson, 118
E	This element begins with the letter O, but it is not oxygen or osmium.	oganesson, 118
M	This element is at the very bottom of the column that contains all the noble gases (He, Ne Ar, etc).	oganesson, 118
H	So far, no element heavier than this one has ever been made.	oganesson, 118

ANSWER KEYS TO CROSSWORD PUZZLES

SYMBOL PRACTICE

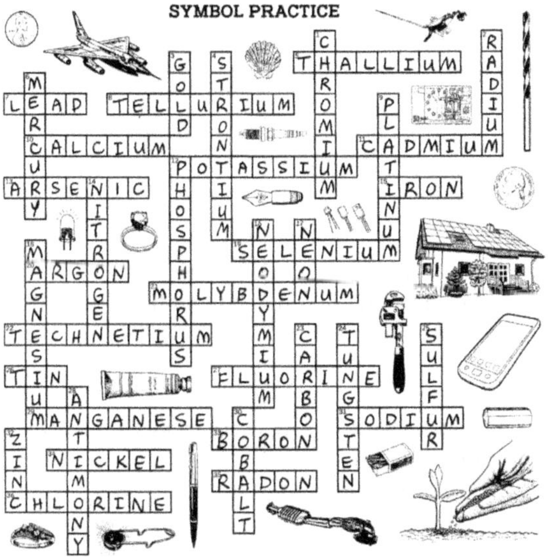

1) J, Q 2) tin 3) K, X, U, V 4) c

ELEMENTS FROM HYDROGEN TO KRYPTON

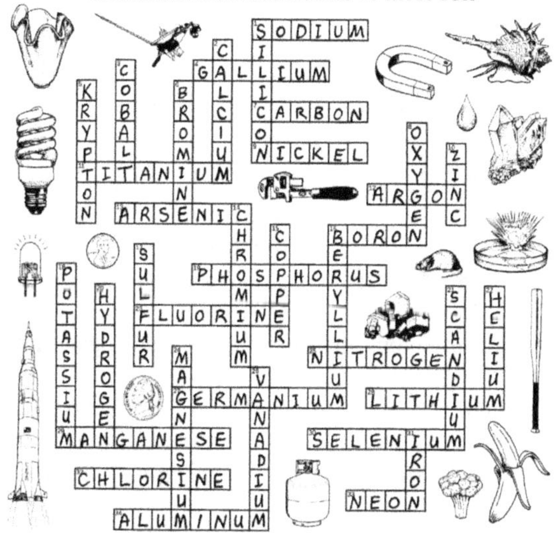

ALKALIS, METALS, METALLOIDS, RARE EARTH METALS

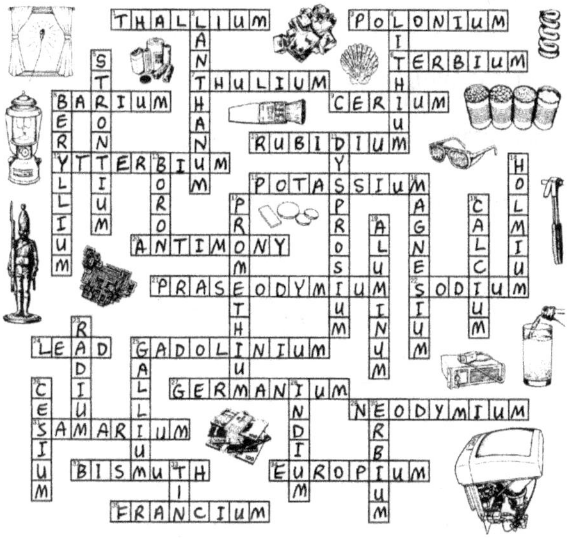

TRANSITION METALS

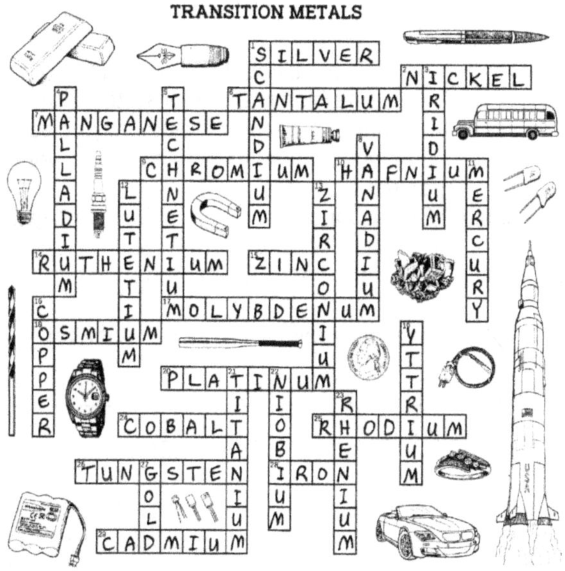

URANIUM AND BEYOND

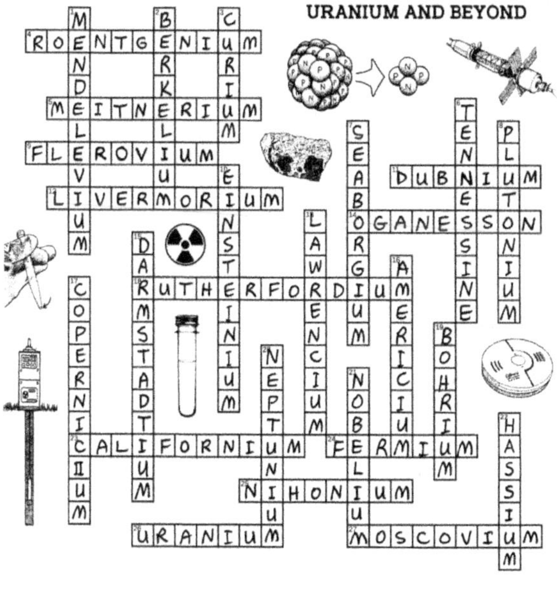

LEAST ABUNDANT ELEMENTS

The least common, naturally-occuring elements
(No super-heavy manufactured elements)

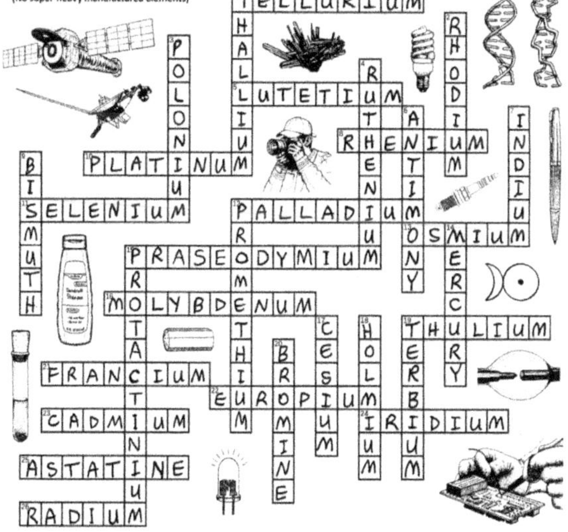

WHO AM I?

1) zinc 2) chlorine 3) arsenic 4) magnesium 5) germanium
6) zirconium 7) cadmium 8) neodymium 9) bismuth 10) iodine

SYMBOL PRACTICE WITH SILLY RIDDLES

RIDDLE #1: Why did the mouse say, "Cheap, cheap!" when the bird's cage fell apart?
He was filling in for the bird who had the day off.
RIDDLE #2: What do you get when you cross a vampire with a mouse? A terrified cat.
RIDDLE #3: What were Batman and Robin's new names after they were run over by a train? Flatman and Ribbon!

Which one of these elements can't be spelled using actual symbols? Niobium

1) barium (bury 'em) 2) krypton 3) polonium (polo is a sport played on horseback) 4) silicon (silly con)
5) bohrium (bore-ee 'em) 6) europium (many languages are spoken in Europe) 7) iron
8) mercury (used in thermometers) 9) argon (are gone) 10) californium (LA is in California)

CREATE A QUOTE

Elements with four letters: gold, iron, lead, neon, zinc
Elements named after something in the solar system: cerium (Ceres), helium (the sun), neptunium (Neptune), palladium (Pallas), plutonium (Pluto), selenium (the moon), tellurium (the earth), uranium (Uranus)
Elements that have a symbol made of only one letter: boron, carbon, fluorine, iodine, nitrogen, oxygen, phosphorus, potassium, sulfur, tungsten, uranium, yttrium
Elements named after a country: americium, francium, gallium (France), germanium, nihonium (Japan), polonium
The only element that has a W: lawrencium
The only element that starts with a K: krypton

Quote: "I saw in a dream a table where all the elements fell into place. Awakening, I immediately wrote it down."

ELEMENTS THAT ARE EASILY CONFUSED

1) MAGNESIUM and MANGANESE
2) THULIUM and THALLIUM
3) RADON and RADIUM
4) PLATINUM and PALLADIUM
5) HOLMIUM and HAFNIUM
6) RHENIUM and RHODIUM
7) INDIUM and IRIDIUM
8) ZINC and ZIRCONIUM
9) BORON and BOHRIUM
10) ERBIUM, TERBIUM and YTTERBIUM

www.ingramcontent.com/pod-product-compliance
Lightning Source LLC
Chambersburg PA
CBHW081935170426
43202CB00018B/2929